Frequency Response and Time-Dependent Circuit Behavior

This chapter introduces the concepts of frequency- and time-dependent response in electronic circuits. In contrast to previous chapters, where all elements of a circuit were assumed to respond instantly to applied signals, we now include the frequency limitations and time delays inherent to electronic devices. These effects are caused by the inevitable presence of energy-storage elements (capacitors and inductors) within the circuit. Capacitors are sometimes purposely added by the designer to shape a circuit's frequency response or to isolate bias voltages and currents from signal paths. Capacitance is also naturally present in the basic physical structure of transistors and diodes, and often affects circuit behavior. Lastly, the *stray* capacitance and inductance of wires or printed-circuit-board paths may become important in high-speed or high-frequency circuits. In all cases, predicting the effects of capacitance and inductance on circuit behavior requires a special set of design and analysis tools, which are the subject of this chapter.

We begin by reviewing the sources of internal device and interconnection capacitances. We next review Bode plots, an important frequency-response design and analysis tool. We then proceed to analyze the role of capacitance in shaping the frequency response of transistor circuits. The chapter concludes by discussing the effects of capacitance on transient response—an important concern in switched or digital circuits. Throughout the chapter, the reader is assumed to be familiar with phasors and with the fundamentals of resistor–capacitor circuits on the level covered in Chapter 1.

One final note about the topic of frequency response. The discussion of the chapter may seem to focus primarily on analysis, rather than on design. This approach, however, is actually a good path to successful design. By analyzing several key, representative circuits, as is done in this chapter, the student of electronics can gain valuable insight into the effect of a given capacitor or inductor on circuit behavior. This knowledge can then be used to make intelligent design choices when choosing a circuit topology or selecting element values.

9.1 SOURCES OF CAPACITANCE AND INDUCTANCE IN ELECTRONIC CIRCUITS

As stated in the introduction, capacitance plays a dominant role in shaping the time and frequency response of modern electronic circuits. Inductance is also found in circuits in distributed and parasitic form, but its effects are usually important only well above the breakpoint frequencies of the major circuit capacitances. Analysis of the frequency or time response of a given circuit invariably focuses solely on the role of its various capacitances. One exception to this rule is found in the design of power circuits using transformers, where inductance becomes extremely important. Other exceptions include the design and analysis of certain types of radio-frequency and oscillator circuits and high-speed digital circuits. We defer these topics to a later time.

The most common origins of capacitance include the discrete capacitors used in single-element design, the stray capacitances contributed by interconnections such as wires or printed-circuit-board paths, and the internal capacitances that originate within electronic devices. Each of these sources of capacitance will be examined for its effect on circuit behavior.

9.1.1 Stray Lead Capacitance

All discrete and integrated-circuit elements exhibit stray capacitance between the external device connections. These capacitances are sometimes called *package* or interconnect capacitances. Many take the form of shunt capacitances to ground contributed by the foil pathways of printed-circuit boards or the interconnection lines of an integrated circuit. Others are contributed by the wires leading to the device terminals. A given stray capacitance is customarily labeled by subscripts that indicate the two terminals across which it is found. This labeling scheme is depicted for an arbitrary three-terminal device in Fig. 9.1(a). As depicted in Fig. 9.1(b), it is also possible for stray capacitance to appear between a device lead and ground, or between leads of adjacent devices. The value of a given stray capacitance must be estimated from device layout geometry or measured in an actual circuit. As a general rule, stray lead capacitances are small—usually on the order of picofarads for discrete devices, to hundreds or tens of femtofarads for integrated-circuit devices. Despite these small values, a complete circuit analysis must consider stray capacitance, because the overall circuit frequency response is likely to be affected.

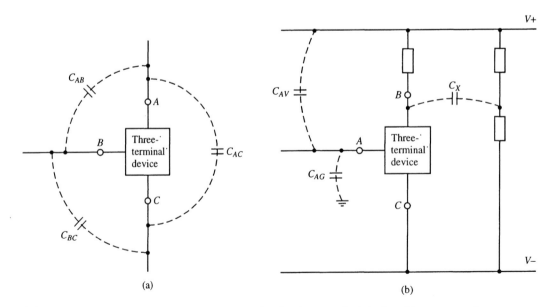

(a)   (b)

Figure 9.1 (a) Stray capacitance found between the leads of a three-terminal device; (b) stray capacitance between device leads and ground or other circuit elements.

9.1.2 Stray Lead Inductance

When devices are interconnected by wires or conduction paths, inductance is introduced into the circuit. These stray, or parasitic, inductances arise because a closed conduction path behaves much like a single-turn coil, as illustrated in Fig. 9.2(a). The inductance of such a circular current path is proportional to its enclosed area.

The effects of stray circuit inductance can be modeled by the addition of discrete "lead" inductors in series with each of the relevant paths, as illustrated in Fig. 9.2(b). The value of a given stray inductance can be estimated by considering the geometrical layout of its contributing connection paths.

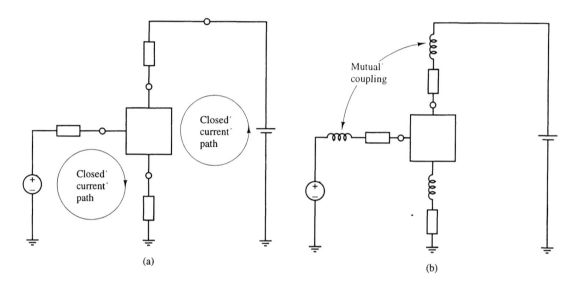

Figure 9.2 Stray inductance caused by the interconnections between devices: (a) current paths function as inductances; (b) equivalent circuit representation.

In modern analog design, the effects of stray inductance become important only at very high frequencies or in situations where the step response of a circuit is important. In both cases, currents with large time derivatives are experienced. Such currents cause significant voltages to develop across stray lead inductances. The magnetic field from one stray inductance may also couple to another stray inductance, as illustrated in Fig. 9.2(b). Such coupling leads to a mutual inductance that behaves much like a transformer and can cause unwanted signal transfer across different parts of the circuit. The effects of stray inductances are also important in digital circuits, where the switching of devices between conducting and nonconducting states creates currents with large time derivatives.

On a single integrated circuit, or "chip," interconnection paths have very small area, and stray inductance can often be neglected. When chips are interconnected, however, the connections *between* integrated circuits, especially connections to power-supply buses, can contribute nonnegligible inductance that must be considered in analyzing overall circuit behavior.

9.1.3 Internal Capacitance of the *pn* Junction

The *pn* junction forms the basis of the two-terminal diode and is also found in the structure of the BJT, MOSFET, and JFET. The *pn* junction exhibits nonnegligible capacitance under both forward-biased and reverse-biased conditions.

The origin of the internal reverse-bias junction capacitance can be understood by examining the physics of the *pn* junction and its associated depletion region. Under reverse-biased conditions, an electric field extends from the fixed positive ion cores on the *n*-type side of the depletion region to the fixed negative ion cores on the *p*-type side, as shown in Fig. 9.3(a). This field resembles that of the simple two-plate capacitor of Fig. 9.3(b), in which an electric field originates on the positive charge of the right-hand plate and terminates on the negative charge of the left-hand plate.

Figure 9.3
(a) Electric field of a reverse-biased *pn* junction. The field diminishes with distance on either side of $x = 0$.
(b) Electric field of a two-plate capacitor.

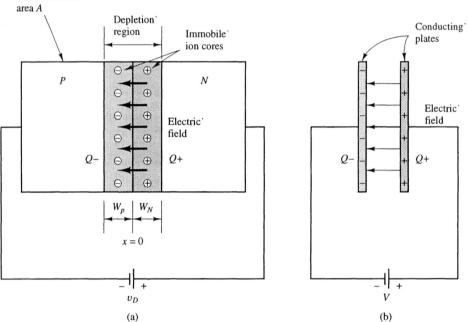

(a)

(b)

For the discrete capacitor of Fig. 9.3(b), the charges on the plates are equal to $Q_+ = CV$ and $Q_- = -CV$, where C is the total plate capacitance, and where $Q_+ = -Q_- = Q$. The electric field between the plates, well removed from the edges, is proportional to Q.

For the reverse-biased *pn* junction of Fig. 9.3(a), the total charge on the right-hand side of the depletion region (i.e., the charge upon which the electric field originates) can be conceptually represented by the quantity

$$Q_+ = qN_D W_N A \qquad (9.1)$$

where N_D is the donor concentration, A is the cross-sectional area of the junction, W_N the width of the right-hand half of the depletion region, and q the unit charge. Similarly, the total charge on the left-hand side of the depletion region (i.e., the charge upon which the electric field terminates) can be represented by

$$Q_- = -qN_A W_P A \qquad (9.2)$$

where N_A is the acceptor concentration, and W_P the width of the left-hand half of the depletion region.

The quantities described by Eqs. (9.1) and (9.2) represent the total positive charge and total negative charge, respectively, on opposite sides of the depletion region. The charges Q_+ and Q_- must again have equal magnitudes Q because the electric field must originate and terminate on equal amounts of charge. As in the capacitor structure, the electric field at $x = 0$ in the center of the *pn* junction will be proportional to Q, hence the *pn* junction will have a capacitance analogous to that of the discrete capacitor structure. Note that the electric field in the depletion region of

Fig. 9.3(a) is not constant, but decreases with distance away from $x = 0$. This effect occurs because charges are distributed throughout the depletion region, rather than being confined to the surfaces of capacitor plates, as in Fig. 9.3(b).

For both the discrete capacitor and the *pn* junction, a change in applied voltage causes a change in Q_+ and Q_-. When V is incremented by Δv, charges $\Delta Q = \pm C \Delta v$ flow to the capacitor plates, and the electric field between them increases in direct proportion. When the voltage applied to the reverse-biased *pn* junction is increased by Δv_D, the increase in Q_+ and Q_- is accommodated by the exposure of more charged donor and acceptor ion cores, which requires a widening of the depletion region and an increase in the depletion-region electric field at $x = 0$.

It is possible to show[1] that for an abruptly doped *pn* junction, the change ΔQ in Q_+ and Q_- per unit change Δv_D—a quantity equivalent to the *incremental* or small-signal capacitance of the reverse-biased *pn* junction —is given by

$$\frac{\Delta Q}{\Delta v_D} \equiv C_j = A \left(\frac{q\epsilon}{2} \frac{N_A N_D}{N_A + N_D} \right)^{1/2} (\Psi_o - v_D)^{-1/2} \tag{9.3}$$

In this equation, A is the cross-sectional area of the junction, v_D is the applied reverse-bias voltage, and ϵ is the permittivity of the semiconductor. The *built-in* voltage Ψ_o is a parameter of the semiconductor material, as discussed in Appendix A. The small-signal reverse-biased junction capacitance expressed by Eq. (9.3) is called the *depletion capacitance*. Its value varies as the inverse square root of v_D. The relatively slow variation of C_j with v_D allows the reverse-biased junction capacitance to be approximated by a constant value for small voltage excursions near the bias point of the *pn* junction. For junctions with doping gradients that are not abrupt, the depletion capacitance is more generally given by

$$C_j = k_j (\Psi_o - v_D)^{-n} \tag{9.4}$$

where n is a number ranging from 1/3 to 4, and k_j is a constant that has a form similar to the first term in parentheses in Eq. (9.3). For a discrete *pn* junction, the value of C_j can be as large as tens of picofarads. On an integrated circuit, the value of C_j is more likely to lie in the single-digit picofarad to femtofarad range. The voltage-dependent capacitance of a reverse-biased diode is actually put to good use in certain types of high-frequency *oscillator* circuits in which C_j is used to produce a sinusoid whose frequency can be changed by a dc bias voltage. As previously mentioned in Section 3.3.7, a diode used in this way is called a *varactor diode*.

When a *pn* junction is forward-biased by an applied voltage v_D, depletion capacitance is still present, but another internal capacitance, called the *charge-storage*, or *diffusion*, capacitance, becomes even more significant. The origin of the diffusion capacitance can be understood by examining the current flow mechanism through the junction. Current flow occurs when holes are injected by v_D from the *p*-side to the *n*-side and electrons are injected from the *n*-side to the *p*-side. These injected charges build up carrier-concentration gradients that decay away from the depletion region and cause diffusion current to flow. An increase in the applied external diode voltage and the subsequent increase in current flow through the junction must be accompanied by an increase in the slopes of these concentration gradients. These increased slopes require the accumulation of *additional* carriers in the concentration gradients. If the applied diode voltage increases from v_D to $v_D + \Delta v_D$, an increment of charge ΔQ must flow from the external device terminals. This additional flow of charge can be modeled by an equivalent diffusion capacitance C_d defined by

$$C_d = \frac{\Delta Q}{\Delta v_D} \tag{9.5}$$

[1] See, for example, B. G. Streetman, *Solid State Electronic Devices*. Englewood Cliffs, N.J.: Prentice Hall, 1990, p. 178.

It is possible to show that C_d in a forward-biased pn junction is given approximately by

$$C_d = k_d I_s e^{-v_D/\eta V_T} \approx k_d i_D \tag{9.6}$$

where k_d is a constant that depends on device geometry and doping concentrations, and i_D is the diode current. For a large pn junction, C_d can be as high as hundreds of picofarads.

9.1.4 Capacitance in the Bipolar Junction Transistor

The physical processes that contribute to incremental capacitance in the pn junction also contribute to incremental capacitance in the bipolar junction transistor. The BJT's collector–base junction, which is reverse-biased under active-region operation, exhibits the same depletion capacitance C_j as a reverse-biased pn junction, given by Eq. (9.3) or (9.4). Under forward-biased conditions, the base–emitter junction of a BJT has a diffusion capacitance C_d that differs from that of the simple pn junction. A well-designed BJT has a very narrow base region, which causes the injected carrier distribution in its base region to be approximately linear. This condition causes the incremental diffusion capacitance between the base and emitter of a BJT in the active region to be given approximately by

$$C_{d\text{-}be} = \frac{W^2}{2D} \frac{I_C}{\eta V_T} \tag{9.7}$$

where W is the width of the base region, I_C the transistor bias current, and D the minority-carrier diffusion constant in the base region. For an npn device, electrons are the minority carrier in the base. The forward-biased base–emitter junction of a BJT, like a forward-biased pn junction, also exhibits a depletion capacitance, but its value is generally small compared to the diffusion capacitance.

Figure 9.4
Revised small-signal BJT models include internal equivalent junction capacitances and transverse base resistance r_x:
(a) $g_m v_\pi$ representation;
(b) $\beta_o i_b$ representation.

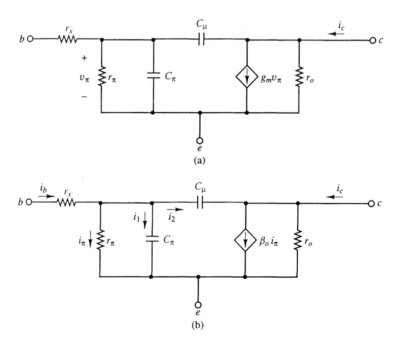

Both the depletion and diffusion capacitances affect BJT behavior at high frequencies and must be included in a more complete small-signal device model. Two such models, one for the $\beta_o i_b$ representation and one for $g_m v_\pi$, are shown in Fig. 9.4. The combined effects of the depletion and diffusion capacitances of the forward-biased base–emitter junction are represented in each model by a single capacitance C_π; the depletion capacitance of the reverse-biased base–collector junction is represented by C_μ. The symbols C_π and C_μ come from a BJT model called the *hybrid-pi*, of which the model presented in this book is a subset. These symbols are sometimes replaced by the more general symbols C_{be} and C_{bc}, respectively. On the data sheets of discrete BJTs, C_μ is also referred to as C_{ob}. This labeling signifies that C_{ob} is the output capacitance in the *common-base* configuration.

The lateral base resistance r_x, which interacts with C_π and C_μ at high frequencies, is also included in the device models of Fig. 9.4. At lower frequencies, where C_π and C_μ are unimportant, r_x has little effect on circuit behavior. Its omission from the BJT model up to now has been justified, but it must be included when considering the effects of C_π and C_μ.

Care must be taken when C_π is included in the β_o representation of the BJT, because i_b, defined as the *total* base current, includes the signal currents through C_π and C_μ. Specifically, as illustrated in Fig. 9.4(b),

$$i_b = i_\pi + i_1 + i_2 \tag{9.8}$$

The variable that drives the dependent source in the model consists only of the current i_π, which flows through r_π and gives rise to the voltage v_π. This relationship is summarized by the equation

$$g_m v_\pi = g_m(r_\pi i_\pi) = \beta_o i_\pi \neq \beta_o i_b \tag{9.9}$$

With C_π and C_μ included in the model, the dependent current source can still be considered a function of the total base current i_b if β_o is represented as a function of frequency $\beta(\omega)$, and i_b and i_π are represented as phasors, as shown in Fig. 9.5. In the discussion that follows, an appropriate expression for $\beta(\omega)$ is derived that accounts for the inequality between $\mathbf{I}_b$ and $\mathbf{I}_\pi$. Provided that certain conditions are met, the quantity $\beta(\omega)\mathbf{I}_b$ becomes equivalent to the quantity $g_m \mathbf{V}_\pi$.

Figure 9.5
Small-signal BJT model with frequency-dependent $\beta(\omega)$.

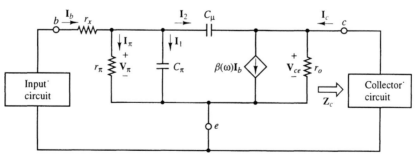

In the model of Fig. 9.5, the base current $\mathbf{I}_b$ consists of three components:

$$\mathbf{I}_b = \mathbf{I}_\pi + \mathbf{I}_1 + \mathbf{I}_2 \tag{9.10}$$

where

$$\mathbf{I}_\pi = \frac{\mathbf{V}_\pi}{r_\pi} \tag{9.11}$$

$$\mathbf{I}_1 = \frac{\mathbf{V}_\pi}{1/j\omega C_\pi} = \mathbf{V}_\pi(j\omega C_\pi) \tag{9.12}$$

and

$$\mathbf{I}_2 = \frac{\mathbf{V}_\pi - \mathbf{V}_{ce}}{1/j\omega C_\mu} = (\mathbf{V}_\pi - \mathbf{V}_{ce})(j\omega C_\mu) \tag{9.13}$$

In Eq. (9.13), $\mathbf{V}_{ce}$ is the small-signal collector-to-emitter voltage, as determined in part by the circuit connected to the collector. By substituting Eqs. (9.11) through (9.13) into (9.10), $\mathbf{I}_b$ can be expressed by the equation

$$\mathbf{I}_b = \frac{\mathbf{V}_\pi}{r_\pi} + \mathbf{V}_\pi (j\omega C_\pi) + \mathbf{V}_\pi (j\omega C_\mu) - \mathbf{V}_{ce}(j\omega C_\mu) \tag{9.14}$$

Equation (9.14) can be solved for $\mathbf{V}_\pi$, yielding

$$\mathbf{V}_\pi = \frac{\mathbf{I}_b r_\pi + (j\omega r_\pi C_\mu)\mathbf{V}_{ce}}{1 + j\omega r_\pi (C_\pi + C_\mu)} \tag{9.15}$$

In the limit of low frequency ($\omega \to 0$), Eq. (9.15) reduces to $\mathbf{V}_\pi = \mathbf{I}_b r_\pi$, so that the $g_m v_\pi$ and $\beta_o i_b$ representations become equivalent. At higher frequencies, the value of $\mathbf{V}_\pi$ given by (9.15) is dependent on $\mathbf{V}_{ce}$. If the inequality

$$\left| j\omega C_\mu \mathbf{V}_{ce} \right| \ll |\mathbf{I}_b| \tag{9.16}$$

is met, Eq. (9.15) becomes

$$\mathbf{V}_\pi \approx \frac{\mathbf{I}_b r_\pi}{1 + j\omega r_\pi (C_\pi + C_\mu)} \tag{9.17}$$

If $\mathbf{I}_2$ through C_μ is small, $\mathbf{V}_{ce}$ will be on the order of $\beta_o \mathbf{I}_\pi \mathbf{Z}_c \| r_o$, where $\mathbf{Z}_c$ is the small-signal impedance of the collector circuit. Under such conditions, the inequality (9.16) can be written as

$$\left| j\omega C_\mu \beta_o \mathbf{I}_\pi \mathbf{Z}_c \| r_o \right| \ll \mathbf{I}_b \tag{9.18}$$

Because $\mathbf{I}_b$ will always be larger than $\mathbf{I}_\pi$, Eq. (9.18) is equivalent to the condition

$$|\beta_o \mathbf{Z}_c \| r_o| \ll \frac{1}{\omega C_\mu} \tag{9.19}$$

This latter inequality states that if Eq. (9.17) is to be a valid approximation, the impedance of C_μ must be much larger than β_o times the impedance of the collector circuit. This condition also causes a small $\mathbf{I}_2$, so the formulation is self-consistent. Note that a small C_μ results in a large impedance $1/j\omega C_\mu$.

If Eq. (9.17) is used to represent $\mathbf{V}_\pi$, the value of the dependent current source in the small-signal model of Fig. 9.5 becomes

$$g_m \mathbf{V}_\pi \approx \frac{(g_m r_\pi)\mathbf{I}_b}{1 + j\omega r_\pi (C_\pi + C_\mu)} \equiv \beta(\omega)\mathbf{I}_b \tag{9.20}$$

where $\qquad \beta(\omega) = \dfrac{g_m r_\pi}{1 + j\omega r_\pi (C_\pi + C_\mu)} = \dfrac{\beta_o}{1 + j\omega r_\pi (C_\pi + C_\mu)} \tag{9.21}$

Equations (9.20) and (9.21) become exact when $\mathbf{V}_{ce} = 0$, that is, when the collector is shorted to the emitter. Hence $\beta(\omega)$ is called the *short-circuit* small-signal current gain. At frequencies below the single pole of $\beta(\omega)$, located at

$$\omega_H = \frac{1}{r_\pi (C_\pi + C_\mu)} \tag{9.22}$$

$\beta(\omega)$ is equivalent to β_o.

One figure of merit for a BJT under short-circuit current-gain conditions is the frequency at which the magnitude of $\beta(\omega)$ reaches unity. This so-called *unity-gain* frequency ω_T, which lies well above the pole frequency ω_H given by (9.22), can be found by determining the frequency at which $|\beta(\omega)|$ reaches unity:

$$|\beta(\omega)|_{\omega=\omega_T} = \left| \frac{\beta_o}{1 + j\omega_T r_\pi (C_\pi + C_\mu)} \right| = 1 \tag{9.23}$$

Near ω_T, the imaginary term in the denominator of Eq. (9.23) dominates, so that the magnitude of the denominator obeys the approximation

$$[1 + \omega_T^2 r_\pi^2 (C_\pi + C_\mu)^2]^{1/2} \approx \omega_T r_\pi (C_\pi + C_\mu) \tag{9.24}$$

Substituting this denominator into Eq. (9.23) and solving for ω_T results in

$$\omega_T = \frac{\beta_o}{r_\pi (C_\pi + C_\mu)} \equiv \frac{g_m}{C_\pi + C_\mu} \tag{9.25}$$

From Eq. (9.25), it follows that

$$C_\pi = \frac{g_m}{\omega_T} - C_\mu \tag{9.26}$$

which yields the value of C_π in terms of C_μ and the unity-gain frequency ω_T. This expression is useful, because C_μ and ω_T can be measured directly in the laboratory, whereas C_π is much more difficult to measure.

A plot of the short-circuit current gain $\beta(\omega)$, showing the relationship between β_o, ω_H, and ω_T, is shown in Fig. 9.6. On manufacturers' data sheets, the frequency-dependent current gain $\beta(\omega)$ is sometimes labeled with the symbol h_{fe} with the understanding that h_{fe} is shorthand notation for $h_{fe}(\omega)$. The value of $\beta(\omega)$ at low frequencies, which is equivalent to β_o, is also sometimes labeled with the symbol h_{fe}. The intended meaning of the symbol h_{fe} must be determined from the context in which it is used. As previously noted, the symbol h_{FE} refers to β_F. This *h-parameter* notation is a by-product of traditional two-port network theory.

Figure 9.6
Plot of the short-circuit current gain $\beta(\omega)$ showing the relationship between β_o, ω_H, and ω_T.

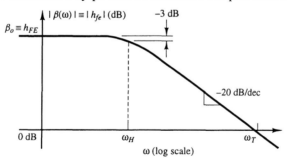

9.1.5 Capacitance in the MOS Field-Effect Transistor

In this section, the origins of the internal capacitances of the MOSFET are examined. The structure of the gate and substrate in the MOSFET of Fig. 9.7(a) closely resembles that of the simple capacitor in Fig. 9.7(b). In the latter case, two conducting plates of area A are separated by an insulating oxide of thickness t_{ox}. If fringing effects are neglected at the edges of the capacitor of Fig. 9.7(b), its capacitance can be expressed by

$$C = \frac{\epsilon_{ox} A}{t_{ox}} \tag{9.27}$$

In this expression, ϵ_{ox} is the dielectric permittivity of the oxide material.[2]

---

[2] For silicon dioxide, $\epsilon_{ox} = 3.9\,\epsilon_o$, where $\epsilon_o = 8.85 \times 10^{-12}$ F/m, is the permittivity of free space.

Figure 9.7
(a) MOSFET
structure with oxide
thickness t_{ox} and
gate area A;
(b) simple capacitor
plates with
separation t_{ox} and
area A.

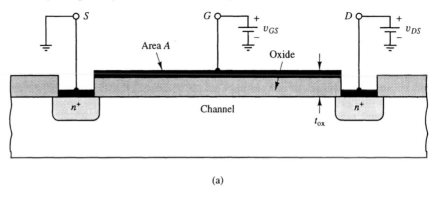

(a)

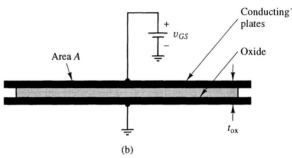

(b)

In the n-channel MOSFET of Fig. 9.7(a), if v_{DS} is small compared to v_{GS}, the entire channel will be at essentially the same potential as the source. Under such conditions, the electric field beneath the gate will be identical to that of the capacitor of Fig. 9.7(b). The gate-to-substrate capacitance (i.e., gate-to-source capacitance) for $v_{DS} \ll v_{GS}$ will be given by

$$C_{gs} = \frac{\epsilon_{ox} A}{t_{ox}} \qquad \text{for} \qquad v_{DS} \ll v_{GS} \qquad (9.28)$$

where $A = WL$, the area of the gate, is equal to the gate width times its length, t_{ox} is the oxide-layer thickness, and ϵ_{ox} is the permittivity of the oxide layer.

When v_{DS} is comparable to, or greater than, v_{GS}, the MOSFET channel can no longer be approximated as an equipotential having the same voltage as the source. Rather, the distribution of v_{DS} down the length of the channel must be taken into account when computing the total gate-to-substrate capacitance. Because the surface potential of the substrate increases toward the drain end of the channel, the net gate-to-substrate surface voltage drop, and the related gate-to-substrate field, become smaller near the drain end. The weaker field at the drain end initiates and terminates on a smaller density of charge than does the stronger field near the source end of the channel, as depicted in Fig. 9.8. For the same v_{GS}, the total charges $\pm Q$ associated with the oxide-layer field decrease as v_{DS} is increased. Since the gate-oxide capacitance is proportional to the total charge Q, the equivalent value of C_{gs} is reduced as v_{DS} is increased. It can be shown that the resulting equivalent gate-to-source capacitance C_{gs} in a MOSFET is given approximately by

$$C_{gs} = \frac{2}{3} \frac{\epsilon_{ox} A}{t_{ox}} \qquad \text{for} \qquad v_{DS} \sim (v_{GS} - V_{TR}) \qquad (9.29)$$

This capacitance is slightly smaller than the value given by Eq. (9.28). In either case, the internal capacitance C_{gs} given by Eq. (9.28) or (9.29) appears in parallel with the stray lead capacitances of the MOSFET.

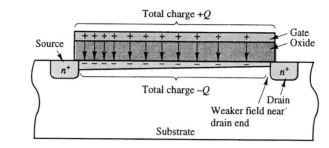

igure 9.8
lectric field in the
xide layer is
eaker near the
rain end of the
annel, resulting
a smaller charge
ensity on the gate
ad substrate. As a
onsequence, the
tal charges ±Q
e reduced, and the
fective value of
e capacitance
$_{gs} \equiv Q/v_{GS}$ is
naller.

In contrast to the BJT, which exhibits a relatively large capacitance across the collector-to-base depletion region, the MOSFET exhibits a small capacitance across the drain-to-channel depletion region. The planar nature of the MOSFET also minimizes the capacitance between the drain and source. Thus, only stray lead capacitances are generally of importance between the drain and the other two terminals.

A revised small-signal model for the MOSFET, including both the gate-to-source capacitance C_{gs} and the stray lead capacitances C_{ds} and C_{gd}, is shown in Fig. 9.9. For typical discrete MOSFET devices, these capacitances are in the picofarad range. In integrated-circuit devices, their values are much smaller, approaching the femtofarad range for the smallest available devices. The value of r_x in a MOSFET is usually small and is often omitted from the small-signal device model.

igure 9.9
evised
nall-signal
OSFET model
cluding its
rious
pacitances.

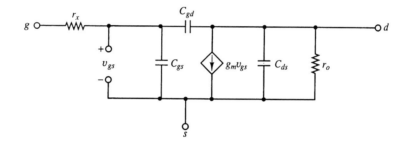

.1.6 Capacitance in the Junction Field-Effect Transistor

Under normal operating conditions, the reverse-biased junction between the gate and substrate in a JFET contributes a depletion-region capacitance that resembles the capacitance of a reverse-biased diode. For small v_{DS}, the channel is essentially an equipotential and the gate-to-substrate (i.e., gate-to-source) depletion capacitance is given for abrupt junction-doping gradients by Eq. (9.3) with v_{GS} substituted for v_D:

$$C_{gs} = A \left(\frac{q\epsilon}{2} \frac{N_A N_D}{N_A + N_D} \right)^{1/2} (\Psi_o - v_{GS})^{-1/2} \tag{9.30}$$

For other doping gradients, the gate-to-source depletion capacitance is given by Eq. (9.4):

$$C_{gs} = k_f (\Psi_o - v_{GS})^{-n} \tag{9.31}$$

For larger v_{DS}, the channel will no longer be an equipotential. Although a picture similar to Fig. 9.8 can be a valid model, the situation is further complicated in the JFET because the capacitance between gate and channel is a nonlinear function of gate voltage, as given by Eq. (9.3) or (9.4). Analysis of this more complex problem, subject to certain assumptions, yields values of C_{gs} in the range 1 to 3 pF for most discrete devices designed for signal applications.

Like the MOSFET, the JFET exhibits negligible internal capacitance between its drain and source. Only the stray lead capacitances are of importance between these device terminals. A revised small-signal model for the JFET, including its various capacitances, is identical to the MOSFET model shown in Fig. 9.9.

9.2   SINUSOIDAL STEADY-STATE AMPLIFIER RESPONSE

The various capacitances described in Section 9.1, as well as any discrete capacitors specifically added by the designer, all influence the response of an electronic circuit. Indeed, the body of this chapter deals with methods for dealing with and predicting the effect of capacitance on circuit response. Before embarking on a study of these effects, however, we first review several key concepts and definitions that pertain to the *frequency domain*. In the frequency domain, a circuit is assumed to have been excited for some time by a sinusoidal input, such that all natural, transient responses have decayed to zero. Under such *sinusoidal steady-state* excitation, every voltage and current signal in the circuit acquires the frequency of the input and can be represented by a phasor. More importantly, each capacitor in the circuit can be represented by a frequency-dependent impedance of value $1/j\omega C$. This feature transforms the differential equations that normally govern capacitive circuits into simple algebraic equations. Any arbitrary input signal can always be represented as a Fourier-series superposition of sinusoids of different frequencies and amplitudes. Knowledge of the circuit's response to the individual sinusoidal Fourier components of the input allows the designer to predict the circuit's response to a complex periodic signal. The next three sections are devoted to a review of concepts that are important to the frequency domain. The study of actual circuits that contain capacitance begins in Section 9.3.

9.2.1   Bode Plot Representation in the Frequency Domain

The input–output response of a circuit in the frequency domain under sinusoidal steady-state conditions is called the circuit's *system function*, or sometimes the *transfer function*.[3] The system function contains a wealth of information about the circuit's steady-state behavior under sinusoidal excitation. This information is neatly expressed in the compact, graphical form of a *Bode plot* (pronounced "Bo-dee"). When a linear circuit has a frequency-dependent system function, both the magnitude and phase angle of the response are variables of great interest. It is often useful to know their values over very large ranges in frequency spanning several orders of magnitude. Similarly, it is often desirable to assign equal importance to the lower and higher ends of the frequency spectrum. The Bode plot consists of a set of straight lines placed on a graph with the frequency on the horizontal axis and either the output amplitude or phase angle on the vertical axis. The straight lines serve as asymptotes that closely represent the actual circuit response, but are much easier to manipulate and analyze. We shall first develop the Bode plot for the simple circuits of Figs. 9.10 and 9.11. These simple circuits highlight the key role of capacitors in many electronic circuits. We then extend the concept to encompass more complicated circuits having system functions of arbitrary complexity.

---

[3] More accurately, the term *transfer function* is used to describe the frequency-domain relationship between input and output signals appearing in different parts of the circuit. The more general term *system function* includes transfer functions, but can also be used to describe the impedance or admittance of a single port.

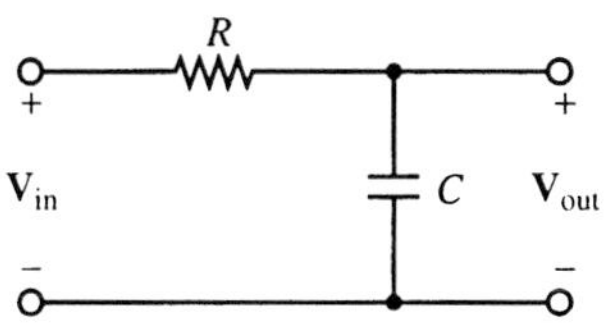

Figure 9.10
Simple RC circuit
with the capacitor
as a shunt element.

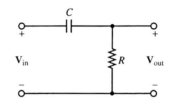

In general, the use of Bode plots is limited to linear circuits. Many nonlinear circuits, however, including the amplifier circuits of this chapter, can be represented by frequency-dependent piecewise-linear or small-signal circuit models. The Bode-plot formulation is useful for describing the small-signal frequency response of these circuits as well.

Figure 9.11
Simple RC circuit
with the capacitor
as a series element.

A complete Bode plot consists of two separate parts. The first shows the magnitude of the output variable relative to the input variable as a function of frequency. The second part shows the phase angle of the output variable relative to the input variable as a function of frequency. The angle of the input variable is arbitrarily (and for convenience) taken as the zero-angle reference.

As an example, consider the Bode plot for the simple circuit of Fig. 9.10, which consists of a series resistor and a *shunt*, or parallel, capacitor. The system function of this circuit becomes, via voltage division

$$\frac{\mathbf{V}_{out}}{\mathbf{V}_{in}} = \frac{1/j\omega C}{R + 1/j\omega C} = \frac{1}{1 + j\omega RC} \tag{9.32}$$

where the capacitor is treated as an element having impedance $1/j\omega C$. As an aid in drawing the Bode plot, we note the behavior of the system function at three extremes of frequency. In the low-frequency limit $\omega \ll 1/RC$, the imaginary part of the denominator becomes negligible, and the system function (9.32) reduces to $\mathbf{V}_{out}/\mathbf{V}_{in} = 1$ so that

$$\left|\frac{\mathbf{V}_{out}}{\mathbf{V}_{in}}\right| = 1 \tag{9.33}$$

and
$$\angle\, \mathbf{V}_{out} = 0 \tag{9.34}$$

where the angle of $\mathbf{V}_{in}$ is taken as the zero-angle reference.

In the high-frequency limit $\omega \gg 1/RC$, the imaginary term in the denominator of Eq. (9.32) becomes larger than the real term, so that the system function reduces to

$$\frac{\mathbf{V}_{out}}{\mathbf{V}_{in}} \rightarrow \frac{1}{j\omega RC} \tag{9.35}$$

with
$$\left|\frac{\mathbf{V}_{out}}{\mathbf{V}_{in}}\right| = \frac{1}{\omega RC} \quad \text{and} \quad \angle\, \mathbf{V}_{out} = -90° \tag{9.36}$$

In this limit of large ω, the magnitude $|\mathbf{V}_{out}/\mathbf{V}_{in}|$ *decreases* by a factor of 10 for every factor-of-10 *increase* in ω.

At the boundary between high- and low-frequency extremes, which occurs at the point $\omega = 1/RC$, the magnitude of the real and imaginary terms of the denominator of Eq. (9.32) become equal to each other, so that the magnitude and angle of the system function become

$$\left|\frac{\mathbf{V}_{out}}{\mathbf{V}_{in}}\right| = \left|\frac{1}{1+j}\right| = \frac{1}{\sqrt{2}} = 0.707 \tag{9.37}$$

and
$$\measuredangle \, \mathbf{V}_{out} = - \measuredangle \, (1+j) = -45° \tag{9.38}$$

In Fig. 9.12, the magnitude and phase angle of the circuit of Fig. 9.10 are plotted as functions of frequency on logarithmic scales. The plots include the three limiting region cases described above. Both magnitude and frequency are plotted logarithmically, so that the high and low ends of the axes are given equal graphical weighting. Note that a logarithmic scale has no zero point and a logarithmic graph has no origin; hence the point at which the vertical and horizontal axes cross on the magnitude plot is arbitrary.

Figure 9.12
Plot of the frequency response of the circuit of Fig. 9.10:
(a) magnitude plot;
(b) phase-angle plot.

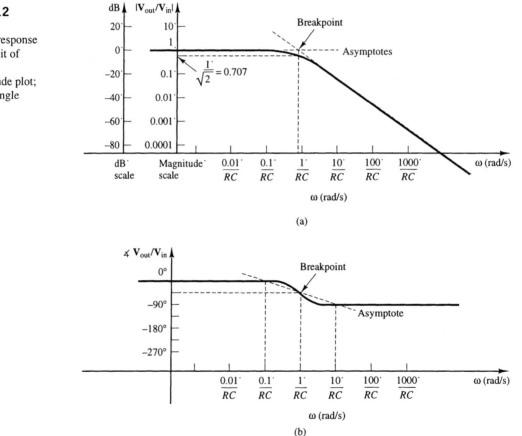

(a)

(b)

For $\omega \ll 1/RC$, the magnitude plot approaches the horizontal asymptote $|\mathbf{V}_{out}/\mathbf{V}_{in}| = 1$ given by Eq. (9.33). For $\omega \gg 1/RC$, the plot approaches the asymptote given by Eq. (9.36). These asymptotes together constitute the circuit's magnitude Bode plot. Above their point of intersection at $\omega = 1/RC$, the right-hand asymptote slopes downward by a factor of 10 for every factor-of-10 increase in ω. It can be shown that at the breakpoint $\omega = 1/RC$, the actual magnitude curve falls by $1/\sqrt{2}$ from the value at the point of intersection. The phase-angle plot has two horizontal asymptotes—one for $\omega \ll 1/RC$ and one for $\omega \gg 1/RC$—located at $0°$ and $-90°$, respectively. The phase-angle plot passes through the $-45°$ point at the breakpoint $\omega = 1/RC$.

It is often convenient to express the logarithmic magnitude scale of the Bode plot with a unit called the magnitude *decibel*, defined by

$$\text{dB} = 20 \log_{10} \left| \frac{\mathbf{V}_{out}}{\mathbf{V}_{in}} \right| \tag{9.39}$$

The decibel is a logarithmic unit; hence a dB scale used in a logarithmic plot appears linear, as in Fig. 9.12(a).

We next consider the circuit of Fig. 9.11, which consists of a series capacitor and a shunt resistor. The system function of this circuit is given, again using voltage division, by

$$\frac{\mathbf{V}_{out}}{\mathbf{V}_{in}} = \frac{R}{R + 1/j\omega C} = \frac{j\omega RC}{1 + j\omega RC} \tag{9.40}$$

Figure 9.13
Plot of the frequency response of the circuit of Fig. 9.11:
(a) magnitude plot:
(b) phase-angle plot.

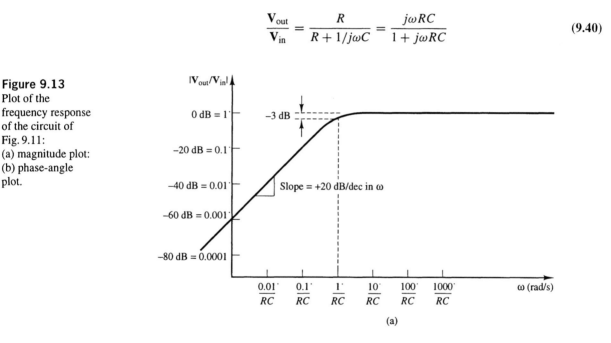

(a)

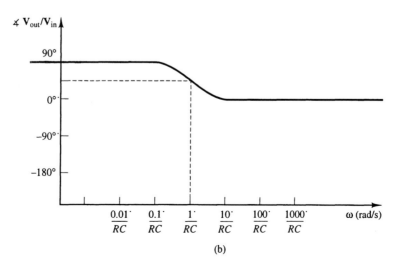

(b)

The Bode-plot asymptotes can be found from the three limits at $\omega \ll 1/RC$, $\omega \gg 1/RC$, and $\omega = 1/RC$:

$$\text{At} \quad \omega \ll 1/RC : \left| \frac{\mathbf{V}_{\text{out}}}{\mathbf{V}_{\text{in}}} \right| \rightarrow \left| \frac{j\omega RC}{1} \right| = \omega RC \qquad \text{and} \; \measuredangle \, \mathbf{V}_{\text{out}} \rightarrow +90° \tag{9.41}$$

$$\text{At} \quad \omega \gg 1/RC : \left| \frac{\mathbf{V}_{\text{out}}}{\mathbf{V}_{\text{in}}} \right| \rightarrow \left| \frac{j\omega RC}{j\omega RC} \right| = 1 \qquad \text{and} \; \measuredangle \, \mathbf{V}_{\text{out}} \rightarrow 0° \tag{9.42}$$

$$\text{At} \quad \omega = 1/RC : \left| \frac{\mathbf{V}_{\text{out}}}{\mathbf{V}_{\text{in}}} \right| = \left| \frac{j}{1+j} \right| = \frac{1}{\sqrt{2}} \qquad \text{and} \; \measuredangle \, \mathbf{V}_{\text{out}} = 90° - 45° = 45° \tag{9.43}$$

The low-frequency limit (9.41) has a factor of ω in the numerator. For $\omega \ll 1/RC$, the magnitude plot thus approaches an asymptote with an upward slope of 20 dB per factor-of-10 change in ω, as shown in Fig. 9.13. Similarly, for $\omega \gg 1/RC$, the magnitude plot asymptotically approaches a constant value of unity. It can be shown for this system function that the low- and high-frequency asymptotes cross at the breakpoint $\omega = 1/RC$, where the actual magnitude plot passes $1/\sqrt{2}$ below the breakpoint crossing. The factor of $1/\sqrt{2} = 0.707$ can also be expressed in decibels as

$$\text{dB} = 20 \, \log_{10} \, 0.707 \approx -3 \, \text{dB} \tag{9.44}$$

At low frequencies, the Bode plot of Fig. 9.13 has an upward slope of $+20$ dB per decade in ω. This slope results because the low-frequency limit (9.41) has a factor of ω in the numerator. Suppose, for example, that at some low frequency $\omega_1 \ll 1/RC$, the magnitude has a decibel value of

$$\text{dB}_1 = 20 \, \log_{10} \, |\mathbf{V}_{\text{out}}/\mathbf{V}_{\text{in}}| = 20 \, \log_{10} \, \omega_1 RC \tag{9.45}$$

where $|\mathbf{V}_{\text{out}}/\mathbf{V}_{\text{in}}|$ is expressed using the limiting case (9.41). At some higher frequency $\omega_2 = 10\omega_1$ that still satisfies the limit $\omega_2 \ll 1/RC$, the decibel value becomes

$$\text{dB}_2 = 20 \, \log_{10} \, 10\omega_1 RC = 20 \, \log_{10} \, \omega_1 RC + 20 \, \log_{10} \, 10 = \text{dB}_1 + 20 \tag{9.46}$$

This value is 20 decibels more than the decibel value at ω_1.

EXERCISE 9.1 Draw the magnitude and angle Bode plots for the circuits of Figs. 9.10 and 9.11 if the capacitor is replaced by an *inductor* of value L.

9.2 Show that the slopes of the nonhorizontal portions of the magnitude plots of Figs. 9.12 and 9.13 have values equal to ± 6 dB per *octave*, where an octave is a factor-of-2 change in frequency.

9.2.2 Bode-Plot Representation of System Functions of Arbitrary Complexity

In later sections of this chapter, we will examine circuits with system functions that are far more complex than those of Eqs. (9.32) and (9.40). Fortunately, the task of constructing the Bode plot of any circuit, no matter how complex, is greatly simplified if its system function can be expressed in the general form

$$H(j\omega) = A \frac{j\omega(1 + j\omega/\omega_2)(1 + j\omega/\omega_4) \cdots}{(1 + j\omega/\omega_1)(1 + j\omega/\omega_3)(1 + j\omega/\omega_5) \cdots} \tag{9.47}$$

The numbered frequencies $\omega_1 \cdots \omega_n$ are the *breakpoints* of the system function, and A is a constant. The solitary factor of $j\omega$ in the numerator is not present for all circuits. If the binomial containing a given breakpoint frequency ω_n appears in the numerator, then ω_n is called a *zero* of the system function. If the binomial appears in the denominator, then ω_n is called a *pole*.

Regardless of its type, a binomial term containing ω_n will affect the circuit's magnitude and phase response as the driving frequency approaches and passes through the value ω_n.

Suppose that the frequency ω of the input signal driving the circuit initially lies well below ω_n. In such a case, the binomial term containing ω_n will alter neither the magnitude nor the phase of the system function, but will simply multiply the system function by unity. This statement can be verified by observing the characteristics of a single binomial term for frequencies well below ω_n:

$$\left| 1 + \frac{j\omega}{\omega_n} \right| \approx 1 \qquad \text{for} \qquad \omega \ll \omega_n \tag{9.48}$$

and
$$\measuredangle \left(1 + \frac{j\omega}{\omega_n} \right) \approx 0° \tag{9.49}$$

Conversely, if the driving frequency ω lies well *above* a given breakpoint frequency ω_n, the binomial term associated with ω_n will contribute a factor of ω/ω_n to the magnitude of the system function and an angle factor of $90°$. This statement can be verified by noting that

$$\left| 1 + \frac{j\omega}{\omega_n} \right| \approx \frac{\omega}{\omega_n} \qquad \text{for} \qquad \omega \gg \omega_n \tag{9.50}$$

and
$$\measuredangle \left(1 + \frac{j\omega}{\omega_n} \right) \approx \measuredangle \frac{j\omega}{\omega_n} = 90° \tag{9.51}$$

If the binomial appears in the numerator as a zero, the factor of ω/ω_n will appear in the numerator, and the angle contribution of $90°$ will be *added* to the overall angle. If the binomial appears in the denominator as a pole, the contributed factor of ω/ω_n will appear in the denominator, and the angle contribution of $90°$ will be *subtracted* from the overall angle.

The transition between the extremes $\omega \ll \omega_n$ and $\omega \gg \omega_n$ occurs at $\omega = \omega_n$. At this frequency, the binomial of ω_n contributes a factor of $\sqrt{2}$ to the magnitude of the system function and an angle of $45°$. The validity of this statement can be shown by noting that at $\omega = \omega_n$,

$$\left| 1 + \frac{j\omega}{\omega_n} \right| = |1 + j| = \sqrt{1^2 + 1^2} = \sqrt{2} \tag{9.52}$$

and
$$\measuredangle \left(1 + \frac{j\omega}{\omega_n} \right) = \measuredangle (1 + j) = 45° \tag{9.53}$$

When the numerator of the system function contains a single non-binomial factor of $j\omega$, a factor of ω will be contributed to the magnitude and a constant factor of $90°$ will be contributed to the phase angle at all values of the driving frequency ω.

Given these guidelines, the Bode-plot asymptotes that describe the magnitude and phase response of a system function of the form (9.47) are easily constructed. We briefly review the procedure here. The process begins by considering frequencies well below the lowest breakpoint of the system function. At such frequencies, the response will be flat (zero slope) with magnitude A and phase angle zero. (If the numerator contains a solitary factor of $j\omega$, the response at low frequencies will instead have a magnitude of $A\omega$, a phase angle of $90°$, and an initial slope of $+20$ dB/decade.) The system function is next evaluated as the frequency is increased. As the frequency passes through a given breakpoint ω_n, its binomial term will begin to contribute a factor of ω/ω_n to the magnitude of the system function. If the binomial appears in the numerator, the slope of the asymptote describing the magnitude response will increase by $+20$ dB/decade. If the binomial term appears in the denominator, the slope of the asymptote will decrease by -20 dB/decade.

The angle portion of the Bode plot can be constructed in a similar fashion. When the binomial term appears in the numerator, the angle of the system function will undergo a phase shift of $+90°$ as the frequency passes through the breakpoint. If the binomial term appears in the denominator, the phase shift will be $-90°$. The phase shift contributed *at* the breakpoint will be equal to $+45°$ or $-45°$, respectively. If a solitary factor of $j\omega$ appears in the numerator of the system function, the Bode plot will *begin* with an upward slope of $+20$ dB/decade and a phase angle of $+90°$ at low frequencies.

In the following examples, the techniques for constructing a Bode plot are illustrated for two cases. The first involves a system function whose response is flat at low frequencies. The second involves a system function with a factor of $j\omega$ in the numerator.

EXAMPLE 9.1

Draw the magnitude and angle Bode plots of a circuit that has a frequency-domain system function given by

$$H(j\omega) = \frac{\mathbf{V}_{out}}{\mathbf{V}_{in}} = \frac{100}{(1 + j\omega/10^2)(1 + j\omega/10^6)} \tag{9.54}$$

Solution

The system function (9.54) has one pole at $\omega = 10^2$ rad/s and one at 10^6 rad/s. At frequencies well below the lowest pole at $\omega = 10^2$ rad/s, the magnitude of the system function is flat and approaches the limit $|H| = 100 \equiv +40$ dB, as shown in Fig. 9.14. Above the pole at $\omega = 10^2$ rad/s, the asymptote describing the magnitude response acquires a slope of -20 dB/decade. The actual magnitude curve lies -3 dB below the asymptote intersection at point A. Above the second pole at $\omega = 10^6$ rad/s, the asymptote acquires an additional slope of -20 dB/decade, for a total slope of -40 dB/decade. With no other poles or zeros in the system function, this new slope continues indefinitely for all higher frequencies. The actual magnitude curve again lies -3 dB below the asymptote intersection at point B.

Figure 9.14
Magnitude plot of the system function of Eq. (9.54).

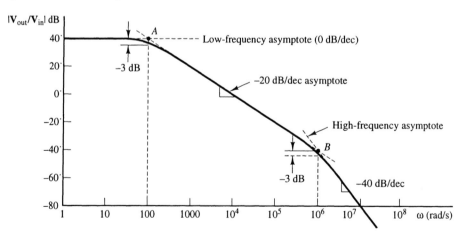

The angle plot of Eq. (9.54) is shown in Fig. 9.15. Well below $\omega = 10^2$ rad/s, the angle of the system function approaches zero. As the first pole at $\omega = 10^2$ rad/s is passed, the angle undergoes a net phase shift of $-90°$, with its value precisely at $\omega = 10^2$ rad/s equal to $-45°$. As the second pole at $\omega = 10^6$ rad/s is passed, it contributes an additional phase shift of $-90°$, for a total phase shift of $-180°$ well above $\omega = 10^6$ rad/s. The total phase shift precisely at $\omega = 10^6$ rad/s is $-135°$, with $-90°$ contributed from the pole at $\omega = 10^2$ rad/s and $-45°$ contributed by the pole at $\omega = 10^6$ rad/s

Figure 9.15
Angle plot of the system function of Eq. (9.54).

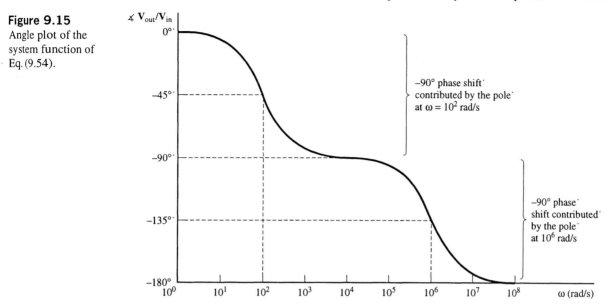

EXAMPLE 9.2

Construct the Bode plot of a circuit whose input–output system function is given by

$$H(j\omega) = \frac{\mathbf{V}_{out}}{\mathbf{V}_{in}} = 50\frac{j\omega(1 + j\omega/10)}{(1 + j\omega/10^4)(1 + j\omega/10^7)} \qquad (9.55)$$

This system function has a solitary factor of $j\omega$ in the numerator.

Figure 9.16
Magnitude plot of the system function of Eq. (9.55).

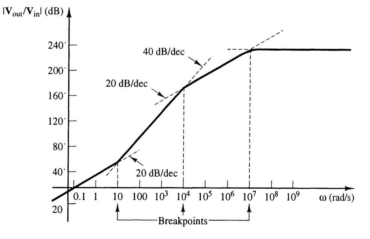

Solution

The magnitude Bode plot of $|\mathbf{V}_{out}/\mathbf{V}_{in}|$ for the system function (9.55) is shown in Fig. 9.16. The point of intersection of the two axes is arbitrary. The system function contains a solitary factor of $j\omega$ in the numerator, hence the plot begins with a positive slope of $+20$ dB/decade for frequencies below the lowest breakpoint $\omega = 10$ rad/s. At the frequency $\omega = 10$, the zero in the numerator takes effect and the slope of the Bode-plot asymptote acquires an additional factor of $+20$ dB/decade to become $+40$ dB/decade. At the frequency $\omega = 10^4$ rad/s, the first pole in the denominator is encountered, and the asymptote slope is reduced by -20 dB/decade to again become $+20$ dB/decade. Finally, at the second pole frequency $\omega = 10^7$ rad/s, the asymptote acquires another factor of -20 dB/decade and becomes horizontal for all frequencies greater than

$\omega = 10^7$ rad/s, which is the highest breakpoint of the system function. At each of the breakpoints in the system function (9.55), the actual frequency-response curve falls $+3$ dB or -3 dB above or below the intersection points of the asymptotes.

Well above the highest breakpoint frequency $\omega = 10^7$, the magnitude of the system function can be approximated by

$$\frac{|\mathbf{V}_{out}|}{|\mathbf{V}_{in}|} \approx 50\frac{\omega(\omega/10)}{(\omega/10^4)(\omega/10^7)} = \frac{50(10^4)(10^7)}{10} = 5(10^{11}) \equiv 234\,\text{dB} \qquad (9.56)$$

Note that the factors of ω^2 cancel out in the numerator and denominator in Eq. (9.56), leaving a term that is constant with frequency.

The angle portion of the Bode plot of Eq. (9.55) is shown in Fig. 9.17. In this case, the solitary factor of $j\omega$ in the numerator contributes an initial angle of $+90°$ to the plot. Above the zero at $\omega = 10$ rad/s, an additional angle of $+90°$ is contributed, making the total angle $+180°$. Above the next breakpoint at $\omega = 10^4$, which is a pole, the angle is reduced by $-90°$ to $+90°$. Above the highest pole at $\omega = 10^7$, the total system function angle is again reduced by $-90°$ to zero, which is a result consistent with the horizontal slope of the magnitude plot at high frequencies. Note that precisely at the location of each of the breakpoints, the system function angle is shifted by half the overall $90°$ angle shift contributed by the breakpoint.

Figure 9.17
Angle plot of the
system function of
Eq. (9.55).

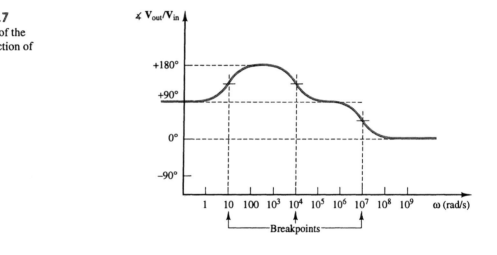

EXERCISE 9.3   Draw the magnitude and angle Bode plots of the circuits of Figs. 9.10 and 9.11 if $R = 5\,\text{k}\Omega$ and $C = 10\,\mu\text{F}$.

9.4   A circuit has a system function with poles at $\omega = 500$ rad/s and 3×10^5 rad/s. At $\omega = 0$, the system function has a value of 50. Draw its magnitude and angle Bode plots.

9.5   Draw the magnitude and angle Bode plots of the system function

$$H(j\omega) = 9.5\frac{j\omega(1 + j\omega/50)}{[1 + j\omega/(3 \times 10^2)][1 + j\omega/(2 \times 10^3)][1 + j\omega/10^6]}$$

9.2.3 High-, Low-, and Midband-Frequency Limits

Many signal-processing applications require a circuit or system to have a flat, constant response over a specified range of frequencies called the *midband*. If the frequency components of the input signal are confined to this range, the output will replicate the form of the input and have the same spectral content. For such circuits, the locations of the specific poles and zeros of the system function are of less interest than the frequency range over which the response may be considered flat.

The flat-response region is usually the portion of the Bode plot with maximum magnitude. Its limits are therefore defined as those frequencies ω_L and ω_H at which the magnitude of the system function first falls by a factor of $1/\sqrt{2}$, or $-3\,dB$, from the horizontal. A magnitude reduction of $1/\sqrt{2}$ corresponds to halving of the power delivered to a resistive load.

The limits of the midband region may not always coincide with individual poles. Multiple poles may contribute simultaneously to the degradation of the circuit's output amplitude. This concept is illustrated in Fig. 9.18, which depicts the magnitude Bode plot of the system function:

$$H(j\omega) = 1000 \frac{j\omega/10}{(1 + j\omega/10)(1 + j\omega/10^4)[1 + j\omega/(2 \times 10^4)]} \tag{9.57}$$

Equation (9.57) has a low-frequency pole at $\omega_a = 10\,rad/s$ and two high-frequency poles—one at $\omega_1 = 10^4\,rad/s$ and one at $\omega_2 = 2 \times 10^4\,rad/s$.

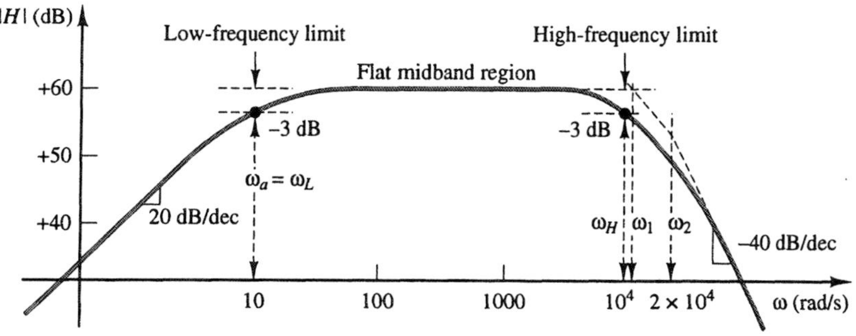

Figure 9.18
Magnitude plot of the system function (9.57) showing two closely spaced poles at $\omega_1 = 10^4\,rad/s$ and $\omega_2 = 2 \times 10^4\,rad/s$.

The low- and high-frequency limits ω_L and ω_H are used to designate the ends of the flat midband region, which has a magnitude of $|H| = 1000 \equiv +60\,dB$. One might assume from the discussion of Section 9.2.1 that $\omega_1 = 10^4\,rad/s$, the first pole to be encountered above the midband, represents ω_H. The system function (9.57) has another nearby pole at $\omega_2 = 2 \times 10^4\,rad/s$, however, which also contributes to the reduction of the Bode plot magnitude at ω_1.

The exact value of ω_H can be computed by solving for the frequency ω_H at which $|H|$ falls by $1/\sqrt{2}$ from its midband value of 1000:

$$|H|_{\omega=\omega_H} \equiv \frac{100\omega_H}{[1 + (\omega_H/10)^2]^{1/2}[1 + (\omega_H/10^4)^2]^{1/2}[1 + (\omega_H/2 \times 10^4)^2]^{1/2}} = \frac{1000}{\sqrt{2}} \tag{9.58}$$

This equation can be solved for ω_H to yield

$$\omega_H \approx 0.84 \times 10^4\,rad/s \tag{9.59}$$

This frequency is lower than the breakpoint $\omega_1 = 10^4$ because the nearby pole at $\omega_2 = 2 \times 10^4\,rad/s$ also degrades the system response at frequencies near ω_1.

9.2.4 Superposition of Poles

For a system function like Eq. (9.57), which exhibits a clearly defined midband region, the locations of ω_H and ω_L can always be found exactly by solving an equation of the form (9.58). Such calculations, however, become tedious for system functions with many closely spaced poles. In such cases, a simplifying technique called the *superposition-of-poles* approximation provides reasonable estimates of ω_L and ω_H while eliminating much of the tedious algebra.

The superposition-of-poles approximation can be applied to any system function that can be put in the form of a midband-gain multiplied by separate low-frequency and high-frequency system functions. Such a system function will have the overall form

$$
H(j\omega) = A_o \cdot H_L \cdot H_H
$$

$$
= A_o \underbrace{\left[\frac{(j\omega/\omega_a)}{(1+j\omega/\omega_a)} \frac{(j\omega/\omega_b)}{(1+j\omega/\omega_b)} \cdots \frac{(j\omega/\omega_m)}{(1+j\omega/\omega_m)} \right]}_{}
$$

$$
\times \underbrace{\left[\frac{1}{(1+j\omega/\omega_1)(1+j\omega/\omega_2)\cdots(1+j\omega/\omega_n)} \right]}_{} \quad \text{(9.60)}
$$

$$H_L \underline{\hspace{5cm}} \qquad\qquad H_H \underline{\hspace{3cm}}$$

Here A_o is a constant equal to the magnitude of the system function in the flat midband region, and H_L and H_H constitute the low- and high-frequency contributions, respectively, to $H(j\omega)$. The breakpoints $\omega_a \cdots \omega_m$ of H_L jointly define the low-frequency limit of the midband. The breakpoints $\omega_1 \cdots \omega_n$ of H_H define the high-frequency limit of the midband. Note that Eq. (9.57) is of the form (9.60), with $A_o = 1000$, $\omega_a = 10$, $\omega_1 = 10^4$, and $\omega_2 = 2 \times 10^4$.

High-Frequency Limit

At the high-frequency end of the midband, the poles of H_L have no effect on the response. At $\omega = \omega_H$, for example, each of the binomials in the denominator of H_L approaches the value $j\omega_H/\omega_m$, canceling the corresponding factor $j\omega_H/\omega_m$ in the numerator of H_L, so that $|H_L| \to 1$. At frequencies near ω_H, $H(j\omega)$ therefore can be approximately expressed by

$$
H(j\omega) \approx A_o H_H = \frac{A_o}{(1+j\omega/\omega_1)(1+j\omega/\omega_2)\cdots(1+j\omega/\omega_n)} \quad \text{(9.61)}
$$

The denominator of Eq. (9.61) consists of a product of binomials that can be multiplied out and put in the form

$$
1 + j\omega \left(\frac{1}{\omega_1} + \frac{1}{\omega_2} + \cdots + \frac{1}{\omega_n} \right) + (j\omega)^2 \left(\frac{1}{\omega_1\omega_2} + \frac{1}{\omega_1\omega_n} + \frac{1}{\omega_2\omega_n} + \cdots + \frac{1}{\omega_j\omega_n} \right)
$$

$$
+ (j\omega)^3 \left(\text{terms of the form} \quad \frac{1}{\omega_j\omega_k\omega_n} \right) + \cdots + \frac{(j\omega)^n}{\omega_1\omega_2\cdots\omega_n} \quad \text{(9.62)}
$$

The second term in Eq. (9.62) contains the factor $j\omega/\omega_n$ from each binomial; the third term contains all possible combinations of $\omega^2/\omega_j\omega_k$; the fourth term contains all possible combinations of order ω^3, and so on. The final term is equal to $(j\omega)^n/(\omega_1\omega_2\cdots\omega_n)$.

By definition, all of the poles ω_1 through ω_n of Eq. (9.61) are higher than the midband endpoint ω_H. Thus, at frequencies near ω_H, terms of order ω^2 or higher in Eq. (9.62) may be ignored, because these terms will be much smaller than terms of order ω. This approximation is weakest when two poles coincide exactly near ω_H, but can be shown to yield moderately good results even in such a case (see Problem 9.36). Neglecting terms of order ω^2 or higher in the denominator of equation (9.62) allows the approximate high-frequency system function (9.61) to be further approximated by

$$H(j\omega) \approx \frac{A_o}{1 + j\omega \left(\frac{1}{\omega_1} + \frac{1}{\omega_2} + \cdots + \frac{1}{\omega_n}\right)} \tag{9.63}$$

The denominator of Eq. (9.63) contains a single binomial term that causes $|H|$ to fall by $-3\,\text{dB}$ when the imaginary part of the denominator equals the real part. The high-frequency $-3\,\text{dB}$ point ω_H of the system function (9.60), which constitutes the upper limit of the midband region, will thus be given approximately by

$$\omega_H = \left(\frac{1}{\omega_1} + \frac{1}{\omega_2} + \cdots + \frac{1}{\omega_n}\right)^{-1} \tag{9.64}$$

As Eq. (9.64) suggests, ω_H can be expressed in "parallel combination" notation as $\omega_1 \| \omega_2 \cdots \| \omega_n$ and can be thought of as the "parallel" superposition of all the individual high-frequency poles ω_1 through ω_n. Equation (9.64) is known as the *superposition-of-poles* approximation at the high-frequency end of the midband.

According to (9.64), the high-frequency poles with the lowest frequency will make the most contribution to ω_H. If one pole is significantly lower in frequency, it will dominate ω_H. Similarly, poles located near each other will make nearly equal contributions to ω_H. Any poles located well above ω_H will make little contribution to the value of ω_H.

Low-Frequency Limit

A similar approach applies at the low-frequency end of the midband. Near the low-frequency end of $H(j\omega)$, the poles of the high-frequency function H_H have little effect on the response. At such frequencies, each of the binomial terms in H_H approaches unity. At frequencies near the low-frequency limit ω_L, the system function $H(j\omega)$ given by Eq. (9.60) thus can be approximately expressed by

$$H(j\omega) \approx A_o H_L = A_o \frac{(j\omega/\omega_a)}{(1 + j\omega/\omega_a)} \frac{(j\omega/\omega_b)}{(1 + j\omega/\omega_b)} \cdots \frac{(j\omega/\omega_m)}{(1 + j\omega/\omega_m)} \tag{9.65}$$

If each of the factors $j\omega/\omega_a$ through $j\omega/\omega_m$ is divided into numerator and denominator, Eq. (9.65) becomes

$$H(j\omega) \approx A_o H_L = A_o \frac{1}{(\omega_a/j\omega + 1)(\omega_b/j\omega + 1) \cdots (\omega_m/j\omega + 1)} \tag{9.66}$$

The denominator of Eq. (9.66) can be expressed in polynomial form as

$$1 + \frac{1}{j\omega}(\omega_a + \omega_b + \cdots + \omega_m) + \frac{1}{(j\omega)^2}(\omega_a\omega_b + \omega_a\omega_m + \omega_b\omega_m + \cdots + \omega_j\omega_m) +$$
$$\cdots + \left[\text{terms of the form} \frac{1}{(j\omega)^3}(\omega_j\omega_k\omega_m)\right] + \cdots + \frac{1}{(j\omega_L)^m}(\omega_a\omega_b \cdots \omega_m) \tag{9.67}$$

By definition, all the poles $\omega_a \cdots \omega_m$ are lower in frequency than the actual low-frequency midband endpoint ω_L. Hence, at frequencies near $\omega = \omega_L$, the terms of order $1/\omega^2$ or higher may be ignored. These terms are presumed to be much smaller than terms of order $1/\omega$. This approximation is weakest when two poles coincide exactly near ω_L. It can be shown, however, that the approximation yields good results even in such a case (see Problem 9.36).

Neglecting terms of order $(1/\omega)^2$ or higher allows the approximate low-frequency system function (9.65) to be further approximated by

$$H(j\omega) \approx A_o \frac{1}{1 + (1/j\omega)(\omega_a + \omega_b + \cdots + \omega_m)}$$

Multiplying numerator and denominator by $j\omega$ and dividing both by $(\omega_a + \omega_b + \cdots + \omega_m)$ results in

$$H(j\omega) = \frac{j\omega/(\omega_a + \omega_b + \cdots + \omega_m)}{1 + [j\omega/(\omega_a + \omega_b + \cdots + \omega_m)]} \tag{9.68}$$

The denominator of this expression contains a single complex binomial that describes the lower -3-dB endpoint of the system function (9.60). According to Eq. (9.68), the value of this low-frequency limit will be given approximately by

$$\omega_L \approx \omega_a + \omega_b + \cdots + \omega_m \tag{9.69}$$

As this expression suggests, the low-frequency -3-dB limit of the midband region may be expressed as an additive "series" superposition of all the low-frequency poles $(\omega_a \cdots \omega_m)$. As indicated by Eq. (9.69), the low-frequency poles with the highest value will make the most contribution to ω_L. If one pole is significantly higher in frequency, it will dominate. Poles located near each other will contribute in nearly equal amounts to ω_L. Similarly, any poles located well below ω_L will make little contribution to the value of ω_L.

Summary of Method

In summary, when a system function has a clearly defined midband region, the superposition-of-poles approximation may be applied by classifying all poles as either high- or low-frequency types. The upper -3-dB point ω_H of the flat midband region can be estimated by a parallel superposition of poles:

$$\omega_H \approx \frac{1}{1/\omega_1 + 1/\omega_2 + \cdots + 1/\omega_n} \equiv \omega_1 \| \omega_2 \cdots \| \omega_n \tag{9.70}$$

The lower -3-dB point ω_L of the flat midband region can be estimated by a series superposition of poles:

$$\omega_L \approx \omega_a + \omega_b + \cdots + \omega_m \tag{9.71}$$

If multiple poles exist at either end of the midband, the superposition-of-poles approximation will always slightly *underestimate* the actual value of ω_H and slightly *overestimate* the actual value of ω_L.

EXAMPLE 9.3　Use the superposition-of-poles approximation to estimate the upper -3-dB endpoint ω_H of the system function:

$$H(j\omega) = 1000 \frac{j\omega/10}{(1 + j\omega/10)(1 + j\omega/10^4)[1 + j\omega/(2 \times 10^4)]} \tag{9.72}$$

Compare the result to the true value (9.59) obtained from Eq. (9.58). (This system function contains only one low-frequency pole, hence the superposition-of-poles approximation is not needed to find ω_L.)

Solution

$H(j\omega)$ begins at low frequencies with a solitary factor of $j\omega$ and an initial slope of $+20$ dB/decade. The flat midband region thus begins at the lowest-frequency pole $\omega = 10$ rad/s. The upper -3-dB limit of the midband region can be estimated by superimposing the two remaining high-frequency poles:

$$\omega_H = \omega_1 \| \omega_2 = \frac{1}{1/10^4 + 1/(2 \times 10^4)} \approx 0.67 \times 10^4 \, \text{rad/s} \qquad (9.73)$$

This estimated value for ω_H is slightly lower than the true value $\omega_H = 0.84 \times 10^4$ rad/s obtained from Eq. (9.58).

EXERCISE 9.6   Find ω_L, ω_H, and the midband gain of the system function

$$(j\omega) = 5 \frac{j\omega/2}{(1 + j\omega/2)(1 + j\omega/10^5)(1 + j\omega/10^6)}$$

Answer: $\omega_L = 2$ rad/s; $\omega_H \approx 9.1 \times 10^4$ rad/s; $A_o = 5 \equiv 14$ dB

9.7   Find ω_L, ω_H, and the midband gain of the system function of Exercise 9.5.
Answer: $\omega_L = 2.3 \times 10^4$ rad/s; $\omega_H = 10^6$ rad/s; $A_o = 1.1 \times 10^5 \equiv 101$ dB

9.3 FREQUENCY RESPONSE OF CIRCUITS CONTAINING CAPACITORS

The concepts presented in Section 9.2 provide powerful tools for working in the frequency domain. With these tools mastered we can now understand the effects of capacitance (and inductance, where important) on circuit behavior. In the sections that follow, we shall use these tools to analyze and design real electronic circuits. To facilitate the connection between the abstract concepts of Section 9.2 and the real circuits of the rest of the chapter, we first provide several key definitions that help categorize the role of each capacitor in shaping circuit response.

9.3.1 High- and Low-Frequency Capacitors

The influence of a given capacitance often occurs at a frequency that lies either above or below a circuit's midband region. Conversely, the midband represents the frequency range over which circuit behavior is unaffected by circuit capacitance. From a frequency-domain point of view, it is often useful to categorize a given capacitor as either a *high-frequency* or *low-frequency* type, depending on whether its effects are felt above or below the midband range. In an amplifier, a high-frequency capacitor is defined as one that degrades the gain above the midband range. Similarly, a low-frequency capacitor is defined as one that degrades the gain below the midband range. Because capacitive impedance is inversely proportional to frequency, it follows that a low-frequency capacitor must behave as a short circuit in the midband, while a high-frequency capacitor must behave as an open circuit in the midband.

As a general rule, a given capacitor will function as a low-frequency type if it appears in series with a circuit's input or output terminal. Conversely, a capacitor will function as a high-frequency type if it shunts an input or output node to small-signal ground. According to this categorization, all internal device and stray lead capacitances are high-frequency types, while all external series capacitors specifically added to the circuit by the designer act as low-frequency capacitors. Common exceptions to this series/shunt rule include external bypass capacitors that connect the common terminal of a transistor to ground. As will be shown in Section 9.3.6, a bypass capacitor appears as a shunt to ground but is a low-frequency type because it degrades amplifier gain *below* the midband range.

To help illustrate the concept of capacitance type, consider the BJT and FET amplifier circuits of Fig. 9.19. The coupling capacitors C_S and C_C and the bypass capacitor C_E are all low-frequency types. The load capacitor C_L and the internal capacitances C_π and C_μ or C_{gs} and C_{gd} of the transistors are high-frequency types. In general, low-frequency capacitors tend to be large—in the microfarad range for discrete designs and in the nanofarad range for integrated-circuit designs. High-frequency capacitors tend to be small—in the picofarad to femtofarad range for discrete and integrated designs, respectively.

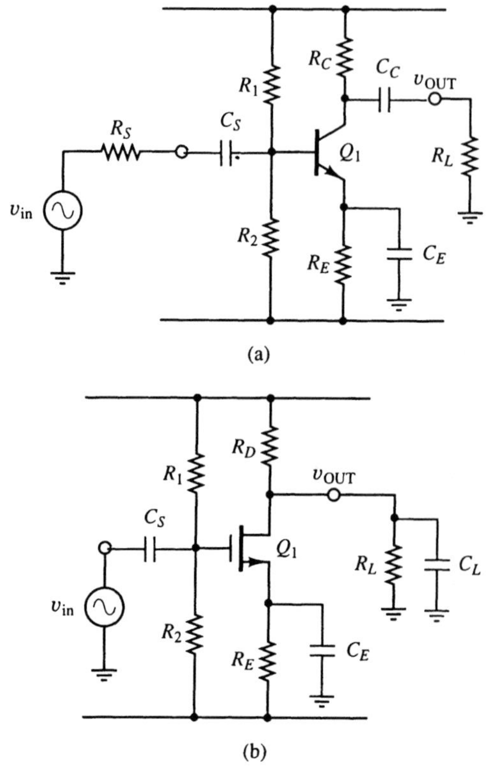

Figure 9.19
Amplifier circuits containing high- and low-frequency capacitors: (a) BJT inverter circuit; (b) MOSFET inverter circuit.

(a)

(b)

The categorization of capacitances as either high-frequency or low-frequency types greatly simplifies the design and analysis of linear amplifiers. Each capacitor type is important to only one end of the frequency spectrum. Specifically, at frequencies well below the high-frequency end of the midband, all high-frequency capacitors may be treated as open circuits and may be ignored in the circuit analysis. Similarly, at frequencies well above the low-frequency end of the midband region, all low-frequency capacitors may be treated as short circuits and may be ignored in the circuit analysis. For such a simplification to be possible, the breakpoints of high- and low-frequency capacitors must be well removed from each other. A flat midband region where the gain is constant must clearly exist. The small-signal model of a circuit in which high-frequency capacitors are represented by open circuits and low-frequency capacitors by short circuits is called the *midband model*.

EXAMPLE 9.4

The circuit of Fig. 9.20 functions as a single-stage BJT amplifier fed from a signal source v_{in} and series resistance $R_S = 1\,k\Omega$. The amplifier drives load elements $R_L = 1\,M\Omega$ and $C_L = 14\,pF$, which represent the shunt resistance and capacitance, respectively, of a typical oscilloscope. The circuit contains two external capacitors C_S and C_C that behave as open circuits at dc and allow Q_1 to be properly biased without influence from v_{in} or R_L. Were these capacitors not in place, the v_{in} source, which acts as a dc short circuit, and the load resistor R_L, would both affect the biasing of Q_1.

Figure 9.20
BJT circuit containing high- and low-frequency capacitors.

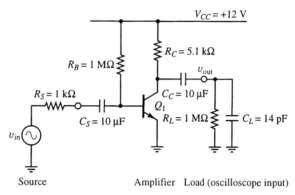

Source   Amplifier   Load (oscilloscope input)

Identify the high- and low-frequency capacitors in the circuit, compute the midband-gain, and separately examine the high- and low-frequency ends of the Bode plot. For the purpose of illustration, assume a BJT at room temperature ($V_T = 25\,mV$) with parameters $C_\mu = 0$, $r_x = 10\,\Omega$, $f_T = 450\,MHz$, $\beta_F = \beta_o = 100$, $V_f = 0.7\,V$, and $r_o = \infty$. (The analysis of the circuit with $C_\mu \neq 0$, which is a bit complicated, will be covered in a later section. In practice, most all BJTs have a nonzero C_μ.)

Solution

• Identify the high- and low-frequency capacitors

Capacitors C_S and C_C appear in series with the transistor's input and output signal paths. As the frequency is reduced, the impedances of these capacitors increase, thereby reducing the amplifier gain. Capacitors C_S and C_C thus function as low-frequency types that behave as short circuits in the midband. Conversely, the load capacitor C_L and internal BJT capacitance C_π both shunt the signal path to ground, thereby reducing amplifier gain at higher frequencies. These capacitances function as high-frequency types that behave as open circuits in the midband.

• Compute the bias current through Q_1 so that g_m, r_π, and C_π can be evaluated

Analysis of the circuit yields the bias current,

$$I_C \approx \frac{\beta_F(V_{CC} - V_f)}{R_B} = \frac{100(12\,V - 0.7\,V)}{1\,M\Omega} \approx 1.13\,mA \tag{9.74}$$

For this value of I_C, the small-signal parameters of the BJT become

$$g_m = \frac{I_C}{\eta V_T} = \frac{1.13\,mA}{(1)(0.025\,V)} \approx 45\,mA/V \tag{9.75}$$

$$r_\pi = \frac{\beta_o}{g_m} = \frac{100}{45\,mA/V} \approx 2.2\,k\Omega \tag{9.76}$$

where the value $\eta = 1$ has been assumed for $I_C \sim 1\,mA$, and

$$C_\pi = \frac{g_m}{2\pi f_T} - C_\mu = \frac{45\,mA/V}{2\pi(450\,MHz)} - 0 \approx 16\,pF \tag{9.77}$$

Figure 9.21
Small-signal model
of the circuit of
Fig. 9.20 valid at
low frequencies.

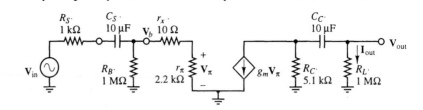

• Evaluate the low-frequency model of the circuit

In the low-frequency small-signal model of the amplifier, shown in Fig. 9.21, capacitors C_π and C_L are set to open circuits because they have little effect near the low-frequency end of the midband. With these capacitors absent, voltage division can be used to find the voltage $\mathbf{V}_b$:

$$\mathbf{V}_b = \mathbf{V}_{in} \frac{R_B \| (r_x + r_\pi)}{R_B \| (r_x + r_\pi) + 1/j\omega C_S + R_S} \tag{9.78}$$

In this expression, C_S is treated as a complex impedance of value $1/j\omega C_S$, and R_B, r_x, and r_π are lumped together into a single resistance of value $R_B \| (r_x + r_\pi)$.

For the given resistor values, the approximations $(r_\pi + r_x) \approx r_\pi$ and $R_B \| (r_x + r_\pi) \approx r_\pi$ apply. Because r_x is much smaller than r_π, the voltage $\mathbf{V}_\pi$ is almost equal to $\mathbf{V}_b$. Consequently, Eq. (9.78) can be approximated by

$$\mathbf{V}_\pi \approx \mathbf{V}_b \approx \mathbf{V}_{in} \frac{r_\pi}{r_\pi + R_S + 1/j\omega C_S} \equiv \mathbf{V}_{in} \frac{j\omega C_S r_\pi}{1 + j\omega C_S (r_\pi + R_S)}$$

$$= \mathbf{V}_{in} \frac{r_\pi}{r_\pi + R_S} \frac{j\omega C_S (r_\pi + R_S)}{1 + j\omega C_S (r_\pi + R_S)} \tag{9.79}$$

The output voltage $\mathbf{V}_{out}$ can be found using current division to find the current through R_L. Specifically, $\mathbf{I}_{out}$ will be equal to the portion of dependent source current $g_m \mathbf{V}_\pi$ that flows through the series combination of R_L and $1/j\omega C_C$, rather than through R_C:

$$\mathbf{V}_{out} = \mathbf{I}_{out} R_L = -g_m \mathbf{V}_\pi \underbrace{\frac{R_C}{R_C + R_L + 1/j\omega C_C}}_{\text{current-divider term}} R_L$$

$$= -g_m R_C \mathbf{V}_\pi \frac{j\omega C_C R_L}{1 + j\omega C_C (R_C + R_L)} \tag{9.80}$$

$$= -g_m \frac{R_C R_L}{R_L + R_C} \frac{j\omega C_C (R_L + R_C)}{1 + j\omega C_C (R_L + R_C)} \mathbf{V}_\pi$$

Using Eq. (9.79) for $\mathbf{V}_\pi$, Eq. (9.80) becomes

$$\frac{\mathbf{V}_{out}}{\mathbf{V}_{in}} = -g_m R_C \frac{j\omega C_S r_\pi}{1 + j\omega C_S (r_\pi + R_S)} \frac{j\omega C_C R_L}{1 + j\omega C_C (R_C + R_L)}$$

$$= -g_m R_C \frac{r_\pi}{r_\pi + R_S} \frac{R_L}{R_L + R_C} \frac{j\omega C_S (r_\pi + R_S)}{1 + j\omega C_S (r_\pi + R_S)} \frac{j\omega C_C (R_L + R_C)}{1 + j\omega C_C (R_L + R_C)} \tag{9.81}$$

This expression has the form $A_o H_L(j\omega)$, where A_o is a constant, and $H_L(j\omega)$ is the low-frequency portion of the system function. The latter is in the standard product-of-binomials Bode-plot form, with poles at

$$f_1 = \frac{\omega_1}{2\pi} = \frac{1}{2\pi C_S(r_\pi + R_S)} = \frac{1}{2\pi(10\,\mu F)(2.2\,k\Omega + 1\,k\Omega)} \approx \frac{31.3\,\text{rad/s}}{2\pi} \qquad (9.82)$$
$$= 5\,\text{Hz}$$

$$\text{and } f_2 = \frac{\omega_2}{2\pi} = \frac{1}{2\pi C_C(R_C + R_L)} = \frac{1}{2\pi(10\,\mu F)(5.1\,k\Omega + 1\,M\Omega)} \approx \frac{0.1\,\text{rad/s}}{2\pi} \qquad (9.83)$$
$$= 0.016\,\text{Hz}$$

These computed frequencies represent the two low-frequency poles of the amplifier, and will be used at the end of the example to construct the circuit's Bode plot.

Figure 9.22
Small-signal model of the circuit of Fig. 9.20 valid at high frequencies.

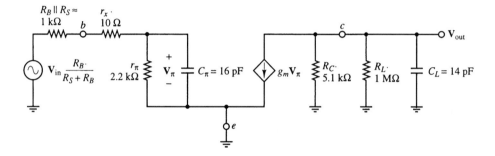

• Evaluate the high-frequency model of the circuit

The high-frequency small-signal model for this amplifier is shown in Fig. 9.22. In this case, C_S and C_C have been replaced by short circuits, and R_B, R_S, and V_{in} have been represented by an appropriate Thévenin equivalent. The circuit can be further simplified by taking a second Thévenin equivalent that also incorporates r_x and r_π, as shown in Fig. 9.23. In this latter circuit, resistors R_C and R_L have also been combined into a single parallel resistance $R_C \| R_L$.

Figure 9.23
Circuit of Fig. 9.22 with the resistive elements of the input circuit replaced by a second Thévenin equivalent.

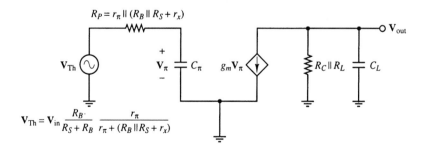

The effect of C_π on the voltage V_π in Fig. 9.23 can be assessed using voltage division. Specifically,

$$V_\pi = \frac{1/j\omega C_\pi}{1/j\omega C_\pi + R_P} V_{th} \equiv \frac{1}{1 + j\omega R_P C_\pi} V_{th} \approx \frac{1}{1 + j\omega R_P C_\pi} \frac{r_\pi}{r_\pi + R_S} V_{in} \qquad (9.84)$$

where $R_P = r_\pi \| (R_B \| R_S + r_x)$. In formulating the factor multiplying V_{in} in Eq. (9.84), the approximations $R_B \gg R_S$ and $r_x \ll R_S$ have been used, so that $V_{th} \approx V_{in} r_\pi / (r_\pi + R_S)$.

The effect of C_L on the circuit can be assessed by combining the three output-load elements in parallel:

$$\mathbf{V}_{out} = -g_m \mathbf{V}_\pi [R_C \| R_L \| (1/j\omega C_L)]$$

$$= -g_m \mathbf{V}_\pi \frac{(R_C \| R_L)/j\omega C_L}{R_C \| R_L + 1/j\omega C_L} = -g_m \mathbf{V}_\pi \frac{R_C \| R_L}{1 + j\omega (R_C \| R_L) C_L} \quad (9.85)$$

Combining Eqs. (9.84) and (9.85) results in

$$\frac{\mathbf{V}_{out}}{\mathbf{V}_{in}} = -g_m (R_C \| R_L) \frac{r_\pi}{r_\pi + R_S} \frac{1}{1 + j\omega R_P C_\pi} \frac{1}{1 + j\omega (R_C \| R_L) C_L} \quad (9.86)$$

The last two factors in the expression describe $H_H(j\omega)$, the high-frequency portion of the system function. $H_H(j\omega)$ is again in the standard product-of-binomials Bode-plot form, with poles at

$$f_3 = \frac{\omega_3}{2\pi} = \frac{1}{2\pi (R_C \| R_L) C_L} = \frac{1}{2\pi (5.1\,\text{k}\Omega \| 1\,\text{M}\Omega)(14\,\text{pF})} = 2.24\,\text{MHz} \quad (9.87)$$

and $\quad f_4 = \frac{\omega_4}{2\pi} = \frac{1}{2\pi R_P C_\pi} = \frac{1}{2\pi (2.2\,\text{k}\Omega \| 1\,\text{M}\Omega \| 1\,\text{k}\Omega)(16\,\text{pF})} = 14.5\,\text{MHz} \quad (9.88)$

These computed frequencies represent the two high-frequency poles of the amplifier and will be used at the end of the example to construct the circuit's Bode plot.

• **Evaluate the midband-gain of the amplifier**

The midband-gain of the circuit can be evaluated from either Eq. (9.81), with ω assumed much larger than the two low-frequency poles ω_1 and ω_2, or from Eq. (9.86), with ω assumed much smaller than the two high-frequency poles ω_3 and ω_4. In either case, the midband-gain becomes

$$a_v = -g_m R_C \frac{r_\pi}{r_\pi + R_S} \frac{R_L}{R_C + R_L} = -(45\,\text{mA/V})(5.1\,\text{k}\Omega)(0.69)(0.995) \approx -157 \equiv 44\,\text{dB}$$

$$\quad (9.89)$$

probe loading factor
input loading factor
basic midband gain

Figure 9.24

Magnitude plot of the response of the circuit of Fig. 9.20. The actual response curve is shown as a solid line beneath the dashed Bode-plot asymptotes.

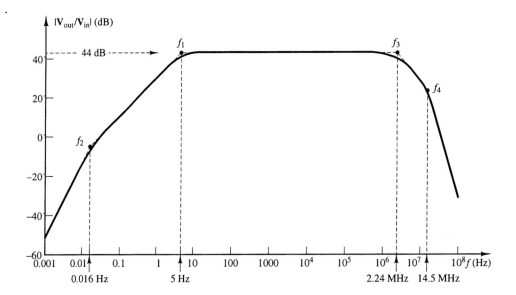

• **Draw the magnitude Bode plot of the amplifier response**

The magnitude Bode plot of the amplifier response is shown in Fig. 9.24. This plot is governed by the four poles f_1 through f_4 computed previously and shows a clearly defined midband region where the gain is equal to 44 dB, as expressed by Eq. (9.89). The midband region extends from $f_1 = 5\,\text{Hz}$, which is the highest of the low-frequency poles, to $f_3 = 2.24\,\text{MHz}$, which is the lowest of the high-frequency poles.

EXERCISE 9.8   Draw the angle portion of the Bode plot of the response of the circuit of Example 9.4.

9.9   Compute the new low-frequency breakpoints in the circuit of Fig. 9.20 is R_B is changed from 1 MΩ to 500 kΩ. Note that the values of r_π will also change due to the change in bias current I_C.
Answer: 7.6 Hz; 0.016 Hz

9.10   Compute the new high-frequency breakpoints in the circuit of Fig. 9.20 if $f_T = 200\,\text{MHz}$.
Answer: 2.24 MHz; 6.4 MHz

9.3.2 The Dominant-Pole Concept

In many signal-processing applications, an amplifier must process an input signal whose spectrum is confined to the amplifier's midband range. In such situations, the nature of the gain outside the midband region is not of interest. The endpoints of the midband region are formally defined as the two frequencies, one high and one low, at which the gain first falls by −3 dB from its midband value, as illustrated on the Bode plot of Fig. 9.25. Each endpoint marks a *reduction* in the gain from the maximum value of the midband, hence both low- and high-frequency endpoints must be associated with a pole of the amplifier's system function, not a zero. The choice of the value −3 dB to define the midband region endpoints is motivated by amplifier output-power considerations. If the circuit feeds a resistive load, a reduction in gain by $1/\sqrt{2} \approx -3\,\text{dB}$ corresponds to an output-power reduction of one-half.

Figure 9.25
Typical Bode plot showing the midband region and its high- and low-frequency −3-dB endpoints. The locations of other poles and zeros outside the midband region, which are seldom of interest, are also shown.

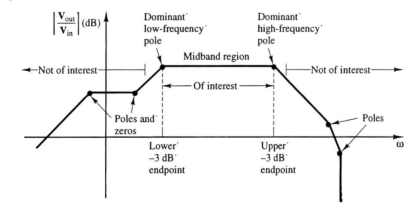

It should be evident from the preceding section that the high- and low-frequency −3 dB points are determined by the behavior of the high- and low-frequency capacitors, respectively. If no two capacitors have poles at the same frequency, the values of the low- and high-frequency endpoints will each be determined by one low-frequency and one high-frequency capacitor, respectively. These capacitors are called the *dominant capacitors* of the circuit because they contribute *dominant poles* to the circuit response. If duplicate pole frequencies exist, the midband endpoints may be determined by superposition of poles, as discussed in Section 9.2.3.

To illustrate the dominant-pole idea, imagine an amplifier driven at an input frequency somewhere in the midband range. In the midband, all low-frequency capacitors behave as incremental short circuits to signals, and all high-frequency capacitors as open circuits. If the input frequency is reduced, a point will be reached where the low-frequency capacitor with the highest pole frequency—the dominant capacitor—will cease to behave as an incremental short circuit. At the pole frequency of this capacitor, the amplifier response will begin to be degraded and the low-frequency −3-dB endpoint of the midband-region response will be initiated. All other low-frequency capacitors will still behave as incremental short circuits, however, because their poles lie at still lower frequencies. The dominant capacitor and its pole frequency can be identified in the following way: A pole frequency value is computed for each low-frequency capacitor while considering all other low-frequency capacitors to be incremental short circuits and all high-frequency capacitors to be open circuits. The low-frequency capacitor with the highest breakpoint frequency computed in this way will actually *be* the dominant low-frequency capacitor, because all other low-frequency capacitors will truly behave as short circuits at its breakpoint. The nondominant breakpoint frequencies of other low-frequency capacitors obtained by this method will be correct provided that their computed values are not affected by the "opening up" of the dominant-pole capacitor or other capacitors with higher breakpoint frequencies.

The dominant-pole principle also applies to the high-frequency capacitors, which behave as *open* circuits as the high-frequency −3-dB midband endpoint is approached from within the midband. The high-frequency capacitor and its pole frequency can be identified by examining, in turn, each high-frequency capacitor while considering all other high-frequency capacitors to be *open* circuits, and all low-frequency capacitors to be short circuits. The capacitor with the *lowest* pole frequency so computed will *be* the dominant high-frequency capacitor, because all other high-frequency capacitors will truly behave as open circuits at its breakpoint. The computed breakpoint frequencies of other nondominant high-frequency capacitors will be correct provided that their computed values are not affected by the "shorting out" of the dominant or other high-frequency capacitors with lower breakpoint frequencies.

Determining the Dominant-Capacitor Pole Frequency by the Thévenin-Resistance Method

The identities of a circuit's low- and high-frequency dominant capacitors, if they exist, can be found using a simple technique called the *Thévenin-resistance method*. In this method, the Thévenin equivalent seen by each capacitor is evaluated with all low-frequency capacitors set to short circuits and all high-frequency capacitors set to open circuits, except for the capacitor under evaluation. With all other capacitors set to either open or short circuits, the Thévenin impedance seen by the capacitor under evaluation becomes purely resistive and equal to the small-signal Thévenin-resistance r_{th} seen at the capacitor's terminals in the midband model of the circuit. This scenario is depicted in Fig. 9.26. As the frequency of the circuit's driving signal is increased, the frequency of the Thévenin voltage $\mathbf{V}_{th}$ seen by the capacitor will increase as well. The capacitor will attain an impedance comparable to r_{th} at a frequency given by

$$\omega = \frac{1}{r_{th}C} \tag{9.90}$$

The capacitor being evaluated will begin to short together its own terminals above this pole frequency. At the pole frequency of the *actual* dominant capacitor, all other low- and high-frequency capacitors will *in fact* behave as short and open circuits, respectively. Hence the dominant-pole capacitor at both high- and low-frequency ends of the midband can be found by simply evaluating the Thévenin resistance seen at the terminals of each capacitor using the midband circuit model, and computing a tentative value for each capacitor pole using the resulting value of r_{th}. The lowest high-frequency pole and highest low-frequency pole so computed will represent the actual high- and low-frequency dominant poles, respectively.

Figure 9.26
The pole of a given capacitor is found by evaluating the midband small-signal Thévenin resistance seen at its own terminals. All other capacitors are set to either short circuits (if a low-frequency type) or open circuits (if a high-frequency type).

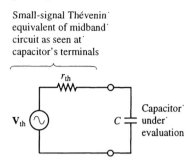

Small-signal Thévenin equivalent of midband circuit as seen at capacitor's terminals

r_{th}

V_{th}

C — Capacitor under evaluation

Note that the poles of nondominant capacitors computed in this way may not be accurately computed if the Thévenin-resistance seen by the capacitor depends on the open or closed status of other capacitors in the circuit. Additionally, if two capacitors have closely spaced poles, they will share dominance, and the −3-dB endpoint associated with them must be computed using superposition of poles. We shall explore these situations later in the chapter. When coupled with the superposition-of-poles principle, the Thévenin-resistance method is sometimes called the method of *open-circuit time constants* at high frequencies and the method of *short-circuit time constants* at low frequencies.

EXAMPLE 9.5

Find the low- and high-frequency −3-dB endpoints of the circuit of Fig. 9.27 using the dominant-pole principle. The MOSFET has internal capacitance $C_{gs} = 8$ pF, with negligible C_{gd} and C_{ds} and an output resistance of $r_o = 100$ kΩ. The device is biased in the constant-current region with $K = 1$ mA/V^2 and $V_{TR} = 4$ V, so that $I_D = 1$ mA and $g_m = 2\sqrt{KI_D} = 2$ mA/V.

Figure 9.27
MOSFET circuit with high- and low-frequency capacitors.

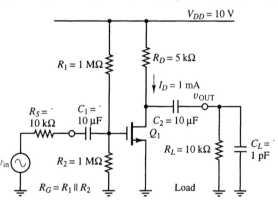

$V_{DD} = 10$ V

$R_1 = 1$ MΩ

$R_D = 5$ kΩ

$I_D = 1$ mA

v_{OUT}

$R_S = 10$ kΩ

$C_1 = 10$ μF

$C_2 = 10$ μF

Q_1

$R_L = 10$ kΩ

$C_L = 1$ pF

v_{in}

$R_2 = 1$ MΩ

$R_G = R_1 \| R_2$

Load

Solution

• Find the midband-gain and identify the high- and low-frequency capacitors

Capacitors C_1 and C_2 appear in series with the signal path and therefore behave as low-frequency capacitors that reduce the amplifier gain as the frequency is decreased. Conversely, shunt capacitors C_L and C_{gs} act to reduce the gain as the frequency is increased, and therefore behave as high-frequency capacitors. A straightforward small-signal analysis, in which C_1 and C_2 are treated as short circuits and C_{gs} and C_L as open circuits, yields the midband gain of the amplifier, given by

$$\frac{v_{out}}{v_{in}} = -g_m(R_D \| r_o \| R_L)\frac{R_G}{R_G + R_S} \tag{9.91}$$

where $R_G = R_1 \| R_2$. Because $r_o \gg (R_D \| R_L)$ and $R_G \gg R_S$, Eq. (9.91) reduces to

$$\frac{v_{out}}{v_{in}} \approx -g_m(R_D \| R_L) = -(2\,\text{mA/V})(3.33\,\text{k}\Omega) \approx -6.7 \tag{9.92}$$

• Use the dominant-pole technique to find the low-frequency endpoint of the midband region

The dominant low-frequency pole is found by constructing the small-signal model with C_L and C_{gs} set to open circuits. The model is then evaluated, first with C_2 set to a short, then with C_1 set to a short. The small-signal model with C_2 set to a short is shown in Fig. 9.28(a). The Thévenin resistance seen at the terminals of C_1, with $\mathbf{V}_{in}$ set to zero, is equal to $R_S + R_G$; hence the computed, tentative pole frequency of C_1 is given by

$$f_1 = \frac{\omega_1}{2\pi} = \frac{1}{2\pi} \frac{1}{C_1(R_S + R_G)} = \frac{1}{2\pi} \frac{1}{(10\,\mu\text{F})(510\,\text{k}\Omega)} \approx 0.03\,\text{Hz} \qquad (9.93)$$

A similar consideration yields the tentative pole frequency for C_2, which is computed with C_1 taken as a short, as in Fig. 9.28(b). With $\mathbf{V}_{in}$ set to zero, $\mathbf{V}_{gs}$ becomes zero as well, so that the $g_m v_{gs}$ source becomes an open circuit, and the Thévenin resistance seen at the terminals of C_2 becomes $(R_D \| r_o) + R_L$. The computed pole of C_2 is thus given by

$$f_2 = \frac{\omega_2}{2\pi} = \frac{1}{2\pi} \frac{1}{C_2(R_D \| r_o + R_L)} = \frac{1}{2\pi} \frac{1}{(10\,\mu\text{F})(14.8\,\text{k}\Omega)} \approx 1.1\,\text{Hz} \qquad (9.94)$$

Because $f_2 > f_1$, the low-frequency -3-dB endpoint of this circuit's midband occurs at 1.1 Hz, and C_2 is the dominant low-frequency capacitor.

Figure 9.28

Small-signal model of the circuit of Fig. 9.27 with (a) C_2 set to a short; (b) C_1 set to a short.

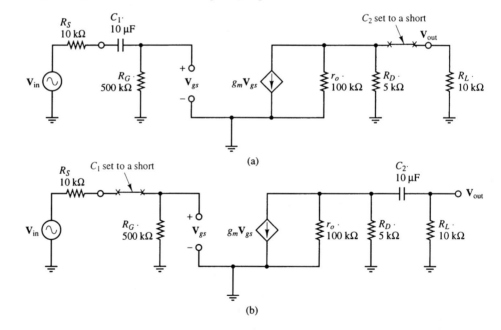

• Use the dominant-pole technique to find the high-frequency endpoint of the midband region

The high-frequency small-signal model of this amplifier, found by setting all low-frequency capacitors to short circuits, is shown in Fig. 9.29. The possibly dominant pole of C_{gs} can be found by setting C_L to an open circuit and finding the Thévenin-resistance seen at the C_{gs} terminals:

$$f_A = \frac{\omega_A}{2\pi} = \frac{1}{2\pi (R_S \| R_G)C_{gs}} = \frac{1}{2\pi (9.8\,\text{k}\Omega)(8\,\text{pF})} \approx 2.0\,\text{MHz} \qquad (9.95)$$

Similarly, the possibly dominant pole of C_L, found with C_{gs} set to an open circuit, is given by

$$f_B = \frac{\omega_B}{2\pi} = \frac{1}{2\pi} \frac{1}{(R_D\|r_o\|R_L)C_L} = \frac{1}{2\pi(3.23\,\text{k}\Omega)(1\,\text{pF})} \approx 49\,\text{MHz} \qquad (9.96)$$

The -3-dB endpoint of the circuit's midband occurs at $2\,\text{MHz}$, which is the lower of these two values, and C_{gs} is the dominant high-frequency capacitor.

Figure 9.29
Small-signal model of the circuit of Fig. 9.27 valid at high frequencies.

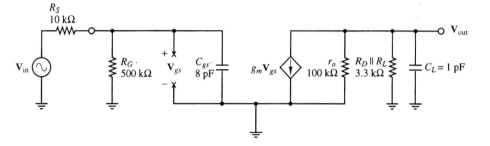

Given the pole values (9.93) to (9.96) computed using the Thévenin-resistance method, the midband region is seen to extend from the highest low-frequency pole at $1.1\,\text{Hz}$ to the lowest high-frequency pole at $2.0\,\text{MHz}$. The computed pole values found in this example are widely separated, such that one pole is clearly dominant at each end of the midband. This clear ordering of poles does not always occur, however. When two or more computed pole values are close in value, superposition of poles must be used to find the -3-dB endpoints of the midband region. This situation is illustrated in the next example.

EXAMPLE 9.6

Find the -3-dB midband endpoints of the BJT amplifier of Fig. 9.30 using the dominant-pole method and the superposition-of-poles technique. The load to the amplifier consists of a meter with input parameters $R_M = 1\,\text{M}\Omega$ and $C_M = 4\,\text{pF}$. As in Example 9.4, assume the BJT to be at room temperature with parameters $C_\mu = 0$, $r_x = 10\,\Omega$, $f_T = 450\,\text{MHz}$, $\beta_F = \beta_o = 100$, $V_f = 0.7\,\text{V}$, $\eta = 1$, and $r_o = \infty$.[4]

Figure 9.30

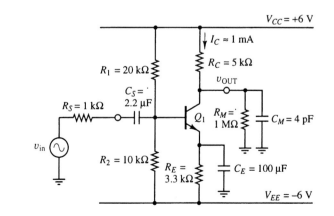

<hr />

[4] As previously stated, the assumption $C_\mu = 0$ is not realistic; it is made here for simplification purposes only. The effect of nonzero C_μ will be explored in the next section. The assumption $r_o = \infty$ is justified here because r_o appears in parallel with the much smaller R_C in the midband model.

Solution

An analysis of the circuit's bias network, using the methods outlined in Chapter 7, yields an I_C of about 1 mA. This current value provides a means to evaluate the small-signal parameters g_m, r_π, and C_π:

$$g_m = I_C/\eta V_T = (1\,\text{mA})/(1)(0.025\,\text{V}) = 40\,\text{mA/V} \tag{9.97}$$

$$r_\pi = \beta_o/g_m = 100/(40\,\text{mA/V}) = 2.5\,\text{k}\Omega \tag{9.98}$$

and $\quad C_\pi = g_m/2\pi f_T - C_\mu = (40\,\text{mA/V})/(2\pi \times 450\,\text{MHz}) - 0 \approx 14\,\text{pF} \tag{9.99}$

With these values calculated, the small-signal model can be analyzed to determine the low- and high-frequency -3-dB midband endpoints.

• Compute values for the low-frequency poles using the Thévenin-resistance method. Find the lower -3-dB endpoint f_L

A low-frequency small-signal model for the circuit, in which the high-frequency capacitors C_π and C_M have been set to open circuits, is shown in Fig. 9.31. As required by the dominant-pole technique, a pole value is computed for C_S, taken with C_E set to a short circuit, and r_x is assumed much smaller than r_π:

$$f_S = \frac{\omega_S}{2\pi} = \frac{1}{2\pi[R_S + (R_B\|r_\pi)]C_S} = \frac{1}{2\pi(1\,\text{k}\Omega + 1.8\,\text{k}\Omega)(2.2\,\mu\text{F})} \approx 26\,\text{Hz} \tag{9.100}$$

Note that R_E does not appear in this equation because it is shorted out by C_E. Similarly, the computed pole of C_E, taken with C_S set to a short circuit, becomes

$$f_E = \frac{\omega_E}{2\pi} = \frac{1}{2\pi\left[R_E\left\|\dfrac{r_\pi + (R_S\|R_B)}{\beta_o + 1}\right.\right]C_E}$$

$$= \frac{1}{2\pi\left[3.3\,\text{k}\Omega\left\|\dfrac{2.5\,\text{k}\Omega + 0.87\,\text{k}\Omega}{101}\right.\right](100\,\mu\text{F})} \approx 48\,\text{Hz} \tag{9.101}$$

In the latter equation, the Thévenin-resistance seen by C_E is computed by shorting C_S, setting $\mathbf{V}_{\text{in}}$ to zero, and applying a test source to the emitter node of the BJT, as in Chapter 7.

Figure 9.31
Small-signal model of the circuit of Fig. 9.30 valid at low frequencies. Capacitors C_π and C_L have been set to open circuits.

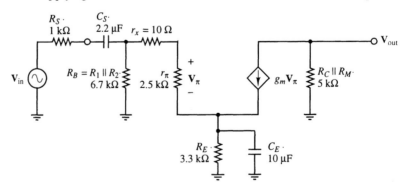

The two computed pole values (9.100) and (9.101) are closely spaced, with neither clearly dominant. In order to compute the low-frequency -3-dB endpoint of the midband region, super-position of poles must be used. Invoking Eq. (9.69) (expressed in hertz, rather than radians/second) results in

$$f_L = f_S + f_E = 26\,\text{Hz} + 48\,\text{Hz} = 74\,\text{Hz} \tag{9.102}$$

• **Compute values for the high-frequency poles using the Thévenin-resistance method. Find the upper −3-dB endpoint f_H**

The high-frequency poles of the amplifier can be found using the small-signal model of Fig. 9.32, in which C_S and C_E have been set to short circuits, and a Thévenin equivalent has been made of $\mathbf{V}_{in}$, R_S, and R_B. Note that setting C_E to a short circuit grounds the emitter of Q_1, thereby bypassing emitter resistor R_E and greatly increasing the amplifier gain.

Figure 9.32
Small-signal model of the circuit of Fig. 9.30 valid near the high-frequency end of the midband. Capacitors C_S and C_E have been set to short circuits, thereby bypassing the emitter of Q_1 to ground and increasing the amplifier gain.

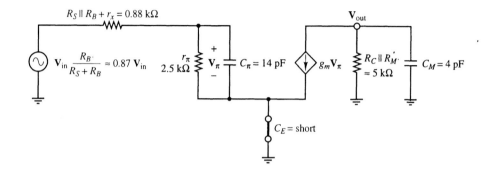

The net Thévenin resistance seen by C_π will be equal to the parallel combination $[(R_S \| R_B) + r_x] \| r_\pi \approx R_S \| R_B \| r_\pi$, hence the pole associated with C_π becomes

$$f_\pi = \frac{1}{2\pi(R_S \| R_B \| r_\pi)C_\pi} = \frac{1}{2\pi[(1\,\text{k}\Omega) \| (6.7\,\text{k}\Omega) \| (2.5\,\text{k}\Omega)](14\,\text{pF})} \approx 17.6\,\text{MHz} \quad \textbf{(9.103)}$$

Similarly, the Thévenin resistance seen by C_M will be equal to $R_C \| R_M$, hence the pole of C_M becomes

$$f_M = \frac{1}{2\pi(R_C \| R_M)C_M} = \frac{1}{2\pi[(5\,\text{k}\Omega) \| (1\,\text{M}\Omega)](4\,\text{pF})} \approx 8.0\,\text{MHz} \quad \textbf{(9.104)}$$

These two computed pole values are again closely spaced, with neither one clearly dominant. The high-frequency −3-dB endpoint of the midband region must therefore be found by superposition of poles, this time using Eq. (9.70) (again expressed in hertz, rather than in radians/second):

$$f_H = f_\pi \| f_M = \frac{(17.6\,\text{MHz})(8.0\,\text{MHz})}{17.6\,\text{MHz} + 8.0\,\text{MHz}} = 5.5\,\text{MHz} \quad \textbf{(9.105)}$$

Discussion. The use of the dominant-pole method provides information about the −3-dB endpoints of the midband region, but it does not always yield correct pole values for the nondominant capacitors. Suppose, for example, that the element values in Fig. 9.30 are changed such that C_E is the clearly dominant capacitor. It then will be the first to cease behaving as a short circuit when the frequency is reduced below the midband. In the dominant-pole method, the pole frequency of C_S is computed for *possible dominance* by assuming C_E to be a short, yielding a Thévenin resistance (with r_x neglected) of $R_S + R_B \| r_\pi$. At the *actual*, nondominant pole frequency of C_S, however, which lies well below the pole of C_E, the latter will behave as an *open* circuit. The true Thévenin resistance seen by C_S near its own pole frequency thus will be equal to $R_S + R_B \| [r_\pi + (\beta_o + 1)R_E]$, which is much larger in value. This larger Thévenin resistance will yield a pole frequency for C_S that is smaller than the tentative value computed with C_E set to a short. If the complete response of the circuit is of interest, including those frequency regions lying outside the midband, the true location of the nondominant pole of C_S becomes important. This issue is explored in more detail in Problem 9.45. ∎

EXERCISE 9.11	Revise the estimates of f_L and f_H in Example 9.5 by invoking superposition of poles. **Answer:** $f_L \approx 1.1\,\text{Hz}$; $f_H \approx 1.95\,\text{MHz}$
9.12	Show that the bias current in the BJT of Fig. 9.30 is equal to about 1 mA.
9.13	Draw the magnitude Bode plot of the response of the circuit of Fig. 9.30.
9.14	Compute the low-frequency -3-dB midband endpoint f_L for the circuit of Fig. 9.30 if C_S is changed to $5\,\mu\text{F}$ and C_E to $470\,\mu\text{F}$. **Answer:** $f_L \approx 22\,\text{Hz}$
9.15	Compute the high-frequency midband endpoint f_H the circuit of Fig. 9.30 if C_M is changed to $10\,\text{pF}$ and f_T of the transistor is equal to $350\,\text{MHz}$. **Answer:** $f_H \approx 2.6\,\text{MHz}$
9.16	For the circuit of Example 9.6, find the low- and high-frequency poles of the circuit by direct computation of the frequency-dependent system function. Compare with the values found by the dominant-pole technique.
9.17	For the MOSFET circuit of Fig. 9.27, let $R_1 = R_2 = 100\,\Omega$, $R_S = 50\,\Omega$, $R_D = 500\,\Omega$, and $R_L = 1\,\text{k}\Omega$. Estimate the high- and low-frequency -3-dB endpoints of the midband region using the dominant-pole technique. For this exercise, assume the MOSFET to have parameters $K = 1.0\,\text{mA/V}^2$, $C_{gs} = 1\,\text{pF}$, $C_{gd} = 0$, $C_{ds} = 1\,\text{pF}$, and $I_D = 4\,\text{mA}$.

9.3.3 Effect of Transverse Capacitance on Amplifier Response

In the examples of the preceding section, the transverse capacitance spanning the transistor's input and output ports (C_μ in the BJT or C_{gd} in the MOSFET) was assumed to be negligible. In a real transistor, this internal transverse capacitance is almost always present. We have delayed an examination of its effects until this time because the analysis is involved and somewhat complex. It can also be approached several different ways. One possible approach is illustrated in Example 9.7, where the Thévenin-resistance method is used to find the contribution of C_μ to the dominant high-frequency pole of a BJT circuit. Another approach, illustrated in Example 9.8, makes use of a network principle known as *Miller's Theorem* to estimate the effect of the transverse capacitor. Finally, the results of these approximate methods are compared with an exact solution of the circuit's system function in Example 9.9.

As a matter of convenience, the examples and discussions in the next few sections focus on BJT amplifiers. The methods presented, however, apply equally well to FET circuits. One need only set r_π to infinity and substitute C_{gs} and C_{gd} for C_π and C_μ to make the various equations apply to an FET circuit having the same connection topology.

EXAMPLE 9.7	Find the upper -3-dB endpoint of the midband region of the BJT amplifier of Fig. 9.30 using the Thévenin-resistance method and superposition of poles. In this case, assume the BJT to have parameters $C_\mu \equiv C_{ob} = 2\,\text{pF}$, $f_T = 450\,\text{MHz}$, $r_x = 10\,\Omega$, $r_o = \infty$, $\eta = 1$, and $\beta_o = 100$. Additionally, to help simplify the analysis and focus on the role of C_μ, omit load elements C_M and R_M. The frequency response of this circuit with $C_\mu = 0$ was analyzed previously in Example 9.6.

Solution

The analysis with nonzero C_μ begins with the small-signal model of Fig. 9.33, in which the low-frequency capacitors C_S and C_E have been set to short circuits, and the high-frequency capacitor C_μ is included. Capacitor C_μ is recognized as a high-frequency type because it shunts current away from r_π in the inverting amplifier topology, thereby reducing the value of v_π at higher frequencies. This shunting action and the resulting gain reduction become more pronounced as the frequency is increased and the impedance of C_μ becomes smaller.

Figure 9.33
Small-signal model of the circuit of Fig. 9.30 with C_μ present and R_M and C_M omitted. This model is valid near the high-frequency end of the midband.

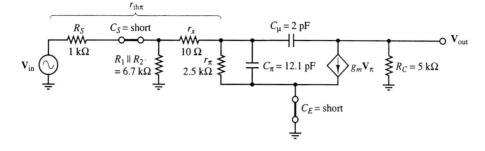

• Find the bias current and small-signal parameters of the BJT

A high-frequency analysis of the circuit begins with a determination of the small-signal parameters of the BJT. An analysis of the circuit's bias network, using the methods of Chapter 7, yields a collector current of $I_C = 1\,\text{mA}$, with resulting small-signal parameters $g_m = 40\,\text{mA/V}$ and $r_\pi = 2.5\,\text{k}\Omega$. With C_μ equal to 2 pF, the value of C_π becomes, from Eq. (9.26),

$$C_\pi = \frac{g_m}{\omega_T} - C_\mu = \frac{40\,\text{mA/V}}{2\pi(450\,\text{MHz})} - 2\,\text{pF} \approx 12.1\,\text{pF} \tag{9.106}$$

The next step in the process involves computing pole frequencies for each of the high-frequency capacitors using the Thévenin-resistance method.

• Find the small-signal Thévenin resistance seen by C_π

With $\mathbf{V}_{\text{in}}$ set to zero and C_μ set to an open circuit, the small-signal resistance seen by C_π becomes

$$\begin{aligned} r_{\text{th}\pi} &= r_\pi \| [r_x + (R_S \| R_1 \| R_2)] \\ &= 2.5\,\text{k}\Omega \| [10\,\Omega + (1\,\text{k}\Omega) \| (6.7\,\text{k}\Omega)] \approx 650\,\Omega \end{aligned} \tag{9.107}$$

• Compute the Thévenin-resistance pole frequency of C_π

Based on the computed value of $r_{\text{th}\pi}$, the computed Thévenin-resistance pole frequency of C_π becomes

$$f_\pi = \frac{\omega_\pi}{2\pi} = \frac{1}{2\pi r_{\text{th}\pi} C_\pi} = \frac{1}{2\pi(650\,\Omega)(12.1\,\text{pF})} \approx 20.2\,\text{MHz} \tag{9.108}$$

• Compute the Thévenin resistance seen by C_μ

The presence of the $g_m v_\pi$ dependent source complicates the calculation of the Thévenin resistance seen by C_μ. Its value is most easily found by applying a test-voltage source across the terminals of C_μ with C_π set to an open circuit, as in Fig. 9.34. Our goal is to express v_{test} in terms of other voltages in the circuit that relate to i_{test} so that the ratio $v_{\text{test}}/i_{\text{test}}$ can be found.

Figure 9.34
Test-voltage source applied to the terminals of the transverse capacitance C_μ with all other high-frequency capacitors treated as open circuits.

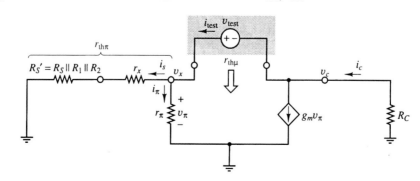

The value of v_π resulting from i_{test} can be found by combining parallel resistances at the v_x node:

$$v_\pi = i_{\text{test}}[r_\pi \| (r_x + R_S \| R_1 \| R_2)] = i_{\text{test}} r_{\text{th}\pi} \tag{9.109}$$

where $r_{\text{th}\pi} = r_\pi \| (r_x + R_S \| R_1 \| R_2)$. Note that $r_{\text{th}\pi}$, the net resistance appearing between the v_x node and ground, is the same $r_{\text{th}\pi}$ computed in Eq. (9.107).

The value of $v_c = -i_c R_C$ is next found by computing the current i_c. Application of KCL to the node labeled v_c results in

$$i_c = i_{\text{test}} + g_m v_\pi \tag{9.110}$$

In this circuit, v_{test} is equal to $v_x - v_c$, which can be expressed as

$$v_{\text{test}} = v_\pi - v_c = i_{\text{test}} r_{\text{th}\pi} - (-i_c R_C) \tag{9.111}$$

where Eq. (9.109) has been used for v_π. Substituting Eq. (9.110) for i_c into this last equation, together with Eq. (9.109) for v_π, results in

$$\begin{aligned} v_{\text{test}} &= i_{\text{test}} r_{\text{th}\pi} + (i_{\text{test}} + g_m v_\pi) R_C \\ &= i_{\text{test}} r_{\text{th}\pi} + i_{\text{test}} R_C + g_m R_C i_{\text{test}} r_{\text{th}\pi} \end{aligned} \tag{9.112}$$

Solving Eq. (9.112) for $v_{\text{test}} / i_{\text{test}}$ results in

$$r_{\text{th}\mu} = \frac{v_{\text{test}}}{i_{\text{test}}} = r_{\text{th}\pi} + R_C + g_m R_C r_{\text{th}\pi} = R_C + r_{\text{th}\pi}(1 + g_m R_C) \tag{9.113}$$

This expression shows $r_{\text{th}\mu}$ to be the sum of the two physical resistances to ground on either side of the v_{test} source plus an additional term $g_m R_C r_{\text{th}\pi}$ contributed by the dependent source.

• Compute the Thévenin-resistance pole frequency of C_μ

The pole frequency computed using $r_{\text{th}\mu}$, with $r_{\text{th}\pi} = 650\,\Omega$ from Eq. (9.107), becomes

$$\begin{aligned} f_\mu = \frac{\omega_\mu}{2\pi} &= \frac{1}{2\pi[R_C + r_{\text{th}\pi}(1 + g_m R_C)]C_\mu} \\ &= \frac{1}{2\pi[5\,\text{k}\Omega + (650\,\Omega)(1 + 40\,\text{mA/V} \cdot 5\,\text{k}\Omega)](2\,\text{pF})} \approx 587\,\text{kHz} \end{aligned} \tag{9.114}$$

The pole f_μ is more than an order of magnitude lower in frequency than $f_\pi = 20.2\,\text{MHz}$ and is clearly the dominant high-frequency pole, hence, to a first approximation, $f_H \approx 587\,\text{kHz}$.

A more accurate value for the -3-dB endpoint f_H can be obtained using superposition of poles. Applying Eq. (9.70) for high-frequency capacitors results in

$$f_H = f_\mu \| f_\pi = \left(\frac{1}{587\,\text{kHz}} + \frac{1}{20.2\,\text{MHz}} \right)^{-1} = 570\,\text{kHz} \tag{9.115}$$

The preceding example makes use of the Thévenin-resistance method to assess the effect of each high-frequency capacitor on the amplifier's frequency response. One surprising result of the analysis is that C_μ, though small, makes the most contribution, via f_μ, in the superposition-of-poles expression (9.115). As a general rule, transverse capacitance across a high-gain inverting stage has a profound effect on amplifier frequency response due to a phenomenon known as the *Miller effect*. In order to appreciate the phenomenon of the Miller effect, we digress in the next section to discuss Miller's theorem and its ramifications.

9.3.4 Miller's Theorem and Miller Multiplication

Miller's theorem is a useful equivalence principle that can be applied to any two-port linear circuit in which a transverse element connects the input port to the output port. Miller's theorem is valid for all resistive, capacitive, and inductive circuits as well as circuits containing general linear complex impedance elements. According to Miller's theorem, a transverse impedance element such as $\mathbf{Z}_\mu$ in the circuit of Fig. 9.35(a) can be modeled by the equivalent parallel impedance $\mathbf{Z}_A$, shown in Fig. 9.35(b). As seen from the terminals of port A, this second circuit models the behavior of the original circuit in every way if an appropriate value of $\mathbf{Z}_A$ is chosen.

Figure 9.35
(a) Circuit with transverse impedance $\mathbf{Z}_\mu$ connected between ports A and B; (b) equivalent circuit at port A with $\mathbf{Z}_A = \mathbf{Z}_\mu/(1 - \mathbf{V}_b/\mathbf{V}_a)$.

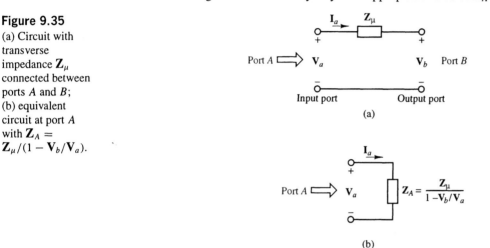

The correct value of $\mathbf{Z}_A$ can be found by noting the v–i relation of port A in the original circuit of Fig. 9.35(a). This equation can be written, using phasor notation and KVL, as

$$\mathbf{V}_a = \mathbf{I}_a \mathbf{Z}_\mu + \mathbf{V}_b \qquad (9.116)$$

where the expression for $\mathbf{V}_b$ must include the effects of any loading on the output port. Equation (9.116) can also be expressed in the form

$$\mathbf{I}_a = \frac{\mathbf{V}_a - \mathbf{V}_b}{\mathbf{Z}_\mu} \qquad (9.117)$$

The v–i relation of port A in the equivalent network of Fig. 9.35(b) is given by

$$\mathbf{I}_a = \frac{\mathbf{V}_a}{\mathbf{Z}_A} \qquad (9.118)$$

If the networks of Fig. 9.35(a) and 9.35(b) are to be equivalent at port A, their v–i equations (9.117) and (9.118) must be identical. The same $\mathbf{I}_a$ must flow for the same applied $\mathbf{V}_a$ in each case. The equations will indeed be identical if $\mathbf{Z}_A$ is chosen such that

$$\mathbf{Z}_A = \mathbf{Z}_\mu \frac{\mathbf{V}_a}{\mathbf{V}_a - \mathbf{V}_b} = \frac{\mathbf{Z}_\mu}{1 - \mathbf{V}_b/\mathbf{V}_a} \qquad (9.119)$$

For this value of $\mathbf{Z}_A$, the v–i equation at port A of the equivalent network becomes

$$\mathbf{I}_a = \frac{\mathbf{V}_a}{\mathbf{Z}_A} = \frac{\mathbf{V}_a}{\mathbf{Z}_\mu} \frac{\mathbf{V}_a - \mathbf{V}_b}{\mathbf{V}_a} = \frac{\mathbf{V}_a - \mathbf{V}_b}{\mathbf{Z}_\mu} \qquad (9.120)$$

This last equation is identical to the v–i relation of the actual network, as expressed by Eq. (9.117).

The general equation (9.119) may be applied to the case where the transverse element is a capacitor C_μ, and $\mathbf{V}_a$ is a sinusoid of known magnitude and frequency. Given that $\mathbf{Z}_\mu = 1/j\omega C_\mu$ in this case, Eq. (9.119) becomes

$$C_A = C_\mu(1 - \mathbf{V}_b/\mathbf{V}_a) \qquad (9.121)$$

As Eq. (9.121) suggests, the transverse capacitance C_μ, as observed from port A, can be modeled as an equivalent parallel capacitance whose value depends on the voltage ratio $\mathbf{V}_b/\mathbf{V}_a$. When Miller's theorem is applied to an amplifier in which port A is the input port, the ratio $\mathbf{V}_b/\mathbf{V}_a$ represents the amplifier gain and is generally quite large. If the gain is also negative, this characteristic causes the equivalent parallel capacitance C_A to become very large as well, inevitably making the pole associated with C_A the dominant pole. The equivalent capacitance C_A, in parallel with a shunt capacitance such as C_π or C_{gs}, makes up the total capacitive portion of the amplifier's input impedance.

It is important to note that C_A represents a model for the input port of the amplifier *as a whole*, including the effects of the load R_C. It is not part of an equivalent transistor model. Thus, although Miller's theorem can be used in principle to represent C_μ as an equivalent capacitance C_B observed from the output port, such an equivalence cannot be used to determine amplifier poles. Finding the equivalent Miller capacitance C_B requires knowledge of the ratio $\mathbf{V}_a/\mathbf{V}_b$, with the latter computed by exciting the voltage $\mathbf{V}_b$ from the amplifier's output port and monitoring the voltage $\mathbf{V}_a$ at its input port. This scenario has little to do with the way in which the real amplifier is excited and does not correctly model the effect of C_μ on the output port of an amplifier that is excited from its input port. The effect of C_μ on the $g_m v_\pi$ source will be examined in more detail later in this section.

Despite its limited use at the output port, Miller's theorem is extremely useful for assessing the effect of transverse capacitance at the input port. Fortunately, the phenomenon of *Miller multiplication* almost always causes C_A to produce the amplifier's dominant high-frequency pole. If the -3-dB limit to the midband region is all that is of interest, any higher-frequency poles related to the transverse capacitance need not be computed. This situation is illustrated in the next example.

EXAMPLE 9.8

The upper -3-dB endpoint frequency of the amplifier of Fig. 9.30 was found in Example 9.7 using the Thévenin-resistance method. In this example, find the approximate frequency of the endpoint using Miller's theorem. As in Example 9.7, omit the load elements C_M and R_M for simplicity, and assume the BJT to be biased at $I_C = 1$ mA with parameters $C_\mu = 2$ pF, $f_T = 450$ MHz, $r_o = \infty$, $r_x = 10\,\Omega$, $\beta_o = 100$, $g_m = 40$ mA/V, and $r_\pi = 2.5$ kΩ.

Solution

The analysis begins with the equivalent circuit seen by the $\mathbf{V}_{in}$ input source, shown in Fig. 9.36. As suggested by Eq. (9.121), the transverse capacitance C_μ of Fig. 9.33 is modeled by an equivalent parallel Miller capacitance C_A. We assume the large combined parallel capacitance $C_\pi \| C_A = (C_\pi + C_A)$ to be the circuit's dominant high-frequency capacitor, hence C_μ will behave as an open circuit with respect to the $g_m v_\pi$ source near the pole frequency of $(C_\pi + C_A)$. Under these conditions, $\mathbf{V}_{out}$ will be given by $-g_m \mathbf{V}_\pi R_C$. This assumption is tantamount to ignoring the actual current through C_μ compared to the dependent-source current $g_m \mathbf{V}_\pi$; the latter is assumed to flow solely through the load resistance R_C. With $\mathbf{V}_{out} = -g_m \mathbf{V}_\pi R_C$, the value of the Miller capacitance C_A becomes

$$\begin{aligned} C_A &= C_\mu(1 - \mathbf{V}_{out}/\mathbf{V}_\pi) = C_\mu(1 + g_m R_C) \\ &= (2\,\text{pF})[1 + (40\,\text{mA/V})(5\,\text{k}\Omega)] = 402\,\text{pF} \end{aligned} \qquad (9.122)$$

The value of $g_m R_C$ in this case results in a large C_A, hence $(C_\pi + C_A)$ will indeed be the circuit's dominant high-frequency capacitance.

Figure 9.36
Equivalent Miller capacitance seen by the input source.

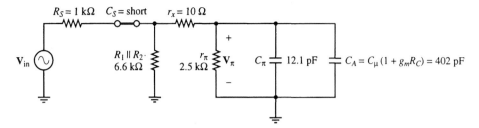

$R_S = 1\,\text{k}\Omega$ $\quad C_S = \text{short}$ $\quad r_x = 10\,\Omega$

V_{in}

$R_1 \| R_2 \cdot$ $6.6\,\text{k}\Omega$

r_π $2.5\,\text{k}\Omega$ $\quad V_\pi$

C_π $12.1\,\text{pF}$

$C_A = C_\mu (1 + g_m R_C) = 402\,\text{pF}$

At the given bias current of 1 mA, the value of C_π, as computed in Example 9.6, again becomes 12.1 pF. For the value $r_{th\pi} = 650\,\Omega$ previously found in Eq. (9.107), the computed pole of $(C_\pi + C_A)$ becomes

$$f_A = \frac{1}{2\pi r_{th\pi}(C_\pi + C_A)} = \frac{1}{2\pi(650\,\Omega)(12.1\,\text{pF} + 402\,\text{pF})} \approx 591\,\text{kHz} \qquad (9.123)$$

This frequency, found using Miller's theorem, lies close to the value $f_H = 570\,\text{kHz}$ computed in Example 9.7 via the Thévenin-resistance method. It is slightly higher due to the neglect of the current through C_μ in evaluating the ratio V_{out}/V_π. (This effect is examined in more detail in Problem 9.62.)

It is instructive to compare the complete system function of the amplifier of Fig. 9.33, derived using KVL and KCL, to the approximate results of Examples 9.7 and 9.8. This analysis is reasonably straightforward, as shown in the next example.

EXAMPLE 9.9

Using basic circuit equations, derive the complete high-frequency system function of the circuit of Examples 9.7 and 9.8. The starting point should be the simplified circuit of Fig. 9.37, in which V_{in}, R_S, $R_1 \| R_2$, r_x, and r_π have all been represented by the Thévenin elements V_{th} and $r_{th\pi}$. Note that $r_{th\pi}$ is the same Thévenin-resistance seen by the base terminal in Examples 9.7 and 9.8.

Figure 9.37
Circuit of Fig. 9.33 in which the input elements are represented by V_{th} and $r_{th\pi}$.

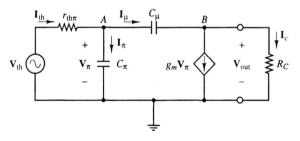

I_{th} $\quad r_{th\pi}$ $\quad A$ $\quad I_\mu$ $\quad C_\mu$ $\quad B$

V_{th} $\quad V_\pi$ $\quad C_\pi$ $\quad I_\pi$ $\quad g_m V_\pi$ $\quad V_{out}$ $\quad R_C$ $\quad I_c$

Solution
Applying KCL at node A, where $I_{th} = I_\pi + I_\mu$, results in

$$\frac{V_{th} - V_\pi}{r_{th\pi}} = V_\pi(j\omega C_\pi) + (V_\pi - V_{out})j\omega C_\mu \qquad (9.124)$$

Here each capacitor has been represented as an impedance of value $1/j\omega C$. Equation (9.124) may also be put in the form

$$V_\pi \left[j\omega(C_\pi + C_\mu) + \frac{1}{r_{th\pi}} \right] - V_{out}(j\omega C_\mu) = \frac{V_{th}}{r_{th\pi}} \qquad (9.125)$$

Similarly, applying KCL at node B, where $\mathbf{I}_\mu = g_m \mathbf{V}_\pi + \mathbf{I}_c$, results in

$$(\mathbf{V}_\pi - \mathbf{V}_{\text{out}})j\omega C_\mu = g_m \mathbf{V}_\pi + \frac{\mathbf{V}_{\text{out}}}{R_C} \tag{9.126}$$

This second equation can be solved for $\mathbf{V}_\pi$ in terms of $\mathbf{V}_{\text{out}}$, yielding

$$\mathbf{V}_\pi = -\frac{\mathbf{V}_{\text{out}}(j\omega C_\mu + 1/R_C)}{g_m - j\omega C_\mu} \tag{9.127}$$

Substituting Eq. (9.127) into Eq. (9.125) and solving for $\mathbf{V}_{\text{out}}$ results in, after some algebra,

$$\frac{\mathbf{V}_{\text{out}}}{\mathbf{V}_{\text{th}}} = \frac{g_m R_C(1 - j\omega C_\mu/g_m)}{1 + j\omega(C_\pi r_{\text{th}\pi} + C_\mu r_{\text{th}\pi} + R_C C_\mu + C_\mu g_m r_{\text{th}\pi} R_C) + (j\omega)^2 C_\pi C_\mu R_C r_{\text{th}\pi}} \tag{9.128}$$

The denominator of this expression is a second-order equation, indicating that the system function has two poles. This result is to be expected, because the circuit contains two capacitors. The system function also exhibits a zero at $\omega = g_m/C_\mu$, which corresponds to the frequency at which the current flow through C_μ becomes comparable in magnitude to that of the $g_m v_\pi$ source.

The poles ω_1 and ω_2 predicted by Eq. (9.128) can be found by expressing its denominator in the form of a product of binomials:

$$\left(1 + \frac{j\omega}{\omega_1}\right)\left(1 + \frac{j\omega}{\omega_2}\right) = 1 + j\omega\left(\frac{1}{\omega_1} + \frac{1}{\omega_2}\right) + \frac{(j\omega)^2}{\omega_1\omega_2} \tag{9.129}$$

By making the assumption that $\omega_1 \ll \omega_2$, so that $(1/\omega_1 + 1/\omega_2) \approx 1/\omega_1$, the pole ω_1 can be identified as the reciprocal of the factor multiplying $j\omega$ in Eq. (9.128). Specifically,

$$\omega_1 = \frac{1}{r_{\text{th}\pi} C_\pi + [r_{\text{th}\pi}(1 + g_m R_C) + R_C]C_\mu} \tag{9.130}$$

We recognize this frequency as the superposition of the two poles

$$\omega_\pi = 2\pi f_\pi = 1/r_{\text{th}\pi} C_\pi \tag{9.131}$$

and
$$\omega_\mu = 2\pi f_\mu = 1/[R_C + r_{\text{th}\pi}(1 + g_m R_C)]C_\mu \tag{9.132}$$

obtained using the Thévenin-resistance method in Example 9.7. The pole given by Eq. (9.130) is also nearly equivalent to the dominant pole $f_A = 1/2\pi r_{\text{th}\pi}(C_\pi + C_A)$ computed using Miller's theorem in Example 9.8. It differs only by the addition of the term $R_C C_\mu$ in the denominator of ω_1. As previously noted, this additional term accounts for the reduction by C_μ of the voltage ratio $\mathbf{V}_{\text{out}}/\mathbf{V}_\pi$ that was not considered when computing the value of C_A in Example 9.8. In either case, for the circuit parameters specified previously, Eq. (9.130) predicts a dominant pole occurring at a frequency of about 570 kHz, which is the same value found in Example 9.7.

Equation (9.129) also can be used to find the second pole ω_2 of the system function. If the previous assumption $\omega_1 \ll \omega_2$ is to be valid, this pole must occur at a much higher frequency than the pole ω_1. Equating the $(j\omega)^2$ term in Eq. (9.129) with the $(j\omega)^2$ term in the denominator of Eq. (9.128) results in

$$\begin{aligned}
\omega_2 &= \frac{1}{\omega_1} \frac{1}{R_C r_{\text{th}\pi} C_\pi C_\mu} \\
&= \frac{r_{\text{th}\pi} C_\pi + [r_{\text{th}\pi}(1 + g_m R_C) + R_C]C_\mu}{R_C r_{\text{th}\pi} C_\pi C_\mu} \\
&\equiv \frac{1}{R_C C_\mu} + \frac{1}{R_C C_\pi} + \frac{g_m}{C_\pi} + \frac{1}{r_{\text{th}\pi} C_\pi}
\end{aligned} \tag{9.133}$$

The third term in this expression is equal to g_m/C_π, which is, by inspection, larger than the parameter $\omega_T = g_m/(C_\pi + C_\mu)$ of the BJT. Indeed, substituting numbers into Eq. (9.133) leads to the value

$$f_2 = \frac{\omega_2}{2\pi} = \frac{1}{2\pi(5\,\text{k}\Omega)(2\,\text{pF})} + \frac{1}{2\pi(5\,\text{k}\Omega)(12.1\,\text{pF})} + \frac{40\,\text{mA/V}}{2\pi(12.1\,\text{pF})} + \frac{1}{2\pi(650\,\Omega)(12.1\,\text{pF})}$$

$$= 15.9\,\text{MHz} + 2.6\,\text{MHz} + 526\,\text{MHz} + 20.2\,\text{MHz} \approx 565\,\text{MHz}$$

$$\textbf{(9.134)}$$

This pole is larger in value than the parameter $f_T = 450\,\text{MHz}$ of the BJT and is very much larger than the dominant pole $f_1 = \omega_1/2\pi = 570\,\text{kHz}$. The second pole caused by a transverse capacitor usually lies at a frequency so high that it is seldom of interest in amplifier design. The one exception to this rule occurs in the design of operational amplifiers, where a transverse capacitor is intentionally connected between the output and input terminals of an internal op-amp stage to achieve desired frequency-response properties. This feature of op-amp design, called *compensation*, will be discussed in Chapter 12.

As illustrated by the examples of this section, Miller multiplication of a transistor's transverse capacitance can severely limit the high-frequency performance of an amplifier. This limitation can be overcome using a topology called the *cascode* configuration. The frequency response of the cascode configuration is discussed in detail in Section 9.4.3.

EXERCISE 9.18 Compute the dominant-pole frequency of the circuit of Example 9.7 if a BJT with an f_T of 350 MHz is used. **Answer:** 565 kHz

9.19 Compute the dominant-pole frequency of the circuit of Example 9.7 if the bias current is changed to 2 mA while all external resistor values are kept at the same value. **Answer:** 367 kHz

9.20 Show that Eq. (9.130) can be expressed as the superposition $f_\pi \| f_\mu$ of the two poles found by the Thévenin-resistance method in Example 9.7.

9.21 Show that Eq. (9.128) can be derived from Eqs. (9.127) and (9.125).

9.22 Find the midband gain of the amplifier described in Examples 9.7 and 9.8. **Answer:** -129

9.3.5 High-Frequency Poles with Feedback Resistor

When a device is connected in the follower configuration, or when a resistor is shared between the input and output loops of an inverter, as in the feedback-bias configuration, Miller's theorem must be used with care. The difficulty can be best illustrated by considering the small-signal circuit of Fig. 9.38, which can represent either type of amplifier. For simplicity, the output resistance r_o of the BJT has been assumed infinite. In principle, Miller's theorem could be used to represent C_μ as a parallel capacitance C_A. The two-port network to which the theorem must be applied, however, does not include the feedback resistor R_E. As a consequence, C_A appears only across the base–emitter terminals of the BJT, as in Fig. 9.39, and not between node A and ground. Similarly, the voltages $\mathbf{V}_a$ and $\mathbf{V}_b$ are those measured across the ports of the transistor alone, and not across the input and output ports of the amplifier as a whole. This complication leads to tedious algebra when computing the equivalent Miller capacitance C_A. Moreover, when the circuit of Fig. 9.38 represents a follower with the output taken across R_E, the resistance R_C is often absent, and the $\mathbf{V}_c$ node is connected directly to signal ground. The large amplifier gain that causes Miller multiplication, resulting in large C_A, is also absent, so that the capacitance $(C_\pi + C_A)$ may not

Figure 9.38
High-frequency
small-signal model
of a BJT follower
circuit.

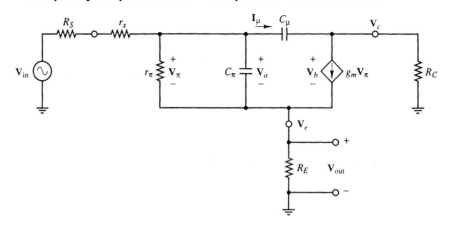

dominate the high-frequency pole. In such cases, the use of Miller's theorem provides little insight into circuit behavior. The high-frequency response of an amplifier with feedback resistor is best derived using the Thévenin-resistance method and superposition of poles, as illustrated in the following example.

Figure 9.39
Transverse
capacitance C_μ in
Fig. 9.46 appears as
input capacitance
C_A.

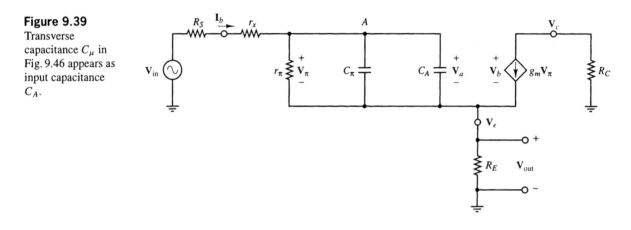

EXAMPLE 9.10

Find the upper -3-dB endpoint of the BJT circuit of Fig. 9.40, for which $C_\mu \equiv C_{ob} = 2\,\mathrm{pF}$, $f_T = 400\,\mathrm{MHz}$, and $r_x = 10\,\Omega$. For the purpose of illustration, suppose that $\beta_o = 100$ and $I_C = 2.5\,\mathrm{mA}$, so that $g_m = 100\,\mathrm{mA/V}$ and $r_\pi = 1\,\mathrm{k}\Omega$ at room temperature. Assume that C_S behaves as a short circuit at frequencies of interest. Note that the poles determined using the Thévenin-resistance method apply whether the amplifier is used as a follower, with the output taken from v_e, or as an inverter, with the output taken from v_c.

Solution

This amplifier can be represented by the small-signal model of Fig. 9.38, with $R_S = R_g \| R_1 \| R_2 = 845\,\Omega$, $v_{in} = v_g[(R_1 \| R_2)/(R_g + R_1 \| R_2)]$, and

$$C_\pi = \frac{g_m}{\omega_T} - C_\mu = \frac{100\,\mathrm{mA/V}}{2\pi(400\,\mathrm{MHz})} - 2\,\mathrm{pF} \approx 38\,\mathrm{pF} \qquad (9.135)$$

The high-frequency -3-dB point of the amplifier can be found by evaluating the small-signal Thévenin-resistances seen by C_π and C_μ.

Figure 9.40

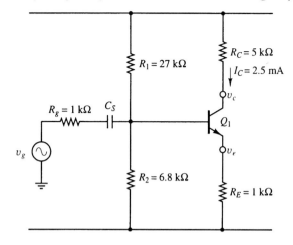

• Find the Thévenin-resistance seen by C_π

The Thévenin-resistance seen by C_π is readily evaluated by probing the terminals of r_π in the midband model with a v_{test} source, as in Fig. 9.41. Note that the voltage $(v_{test} + v_e)$ appears across $(R_S + r_x)$ when v_{test} is applied. Applying KCL to node x in Fig. 9.41 results in

$$i_{test} = \frac{v_{test}}{r_\pi} + \frac{v_{test} + v_e}{R_S + r_x} \qquad (9.136)$$

Application of KCL to node e yields an expression for v_e:

$$v_e = i_e R_E = (g_m v_\pi + i_\pi - i_{test})R_E = \left(g_m v_{test} + \frac{v_{test}}{r_\pi} - i_{test}\right)R_E \qquad (9.137)$$

Substituting this expression for v_e into Eq. (9.136) results in

$$i_{test} = \frac{v_{test}}{r_\pi} + \frac{v_{test} + R_E[g_m v_{test} + (v_{test}/r_\pi) - i_{test}]}{R_S + r_x} \qquad (9.138)$$

Equation (9.138) can be solved for i_{test} to yield

$$i_{test} = \frac{v_{test}\left[\dfrac{1}{r_\pi} + \dfrac{1}{R_S + r_x}\left(1 + R_E g_m + \dfrac{R_E}{r_\pi}\right)\right]}{1 + R_E/(R_S + r_x)} \qquad (9.139)$$

The small-signal Thévenin resistance seen by C_π is thus given by the expression

$$
\begin{aligned}
r_{th\pi} = \frac{v_{test}}{i_{test}} &= \frac{1 + R_E/(R_S + r_x)}{\dfrac{1}{r_\pi} + \dfrac{1}{R_S + r_x}\left[1 + R_E\left(g_m + \dfrac{1}{r_\pi}\right)\right]} \\[2ex]
&= \frac{(R_S + r_x + R_E)r_\pi}{R_S + r_x + r_\pi + R_E(g_m r_\pi + 1)} \qquad (9.140) \\[2ex]
&= \frac{(845\,\Omega + 10\,\Omega + 1\,k\Omega)(1\,k\Omega)}{845\,\Omega + 10\,\Omega + 1\,k\Omega + 1\,k\Omega[(100\,mA/V)(1\,k\Omega) + 1]} \approx 18\,\Omega
\end{aligned}
$$

Figure 9.41
Probing the location of C_π with a test-voltage source. The input source v_{in} has been set to zero.

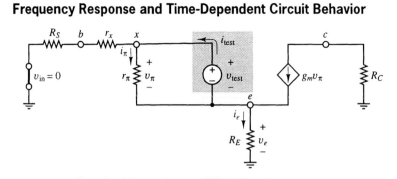

• Compute the contribution of C_π to the high-frequency -3-dB point

With $r_{th\pi}$ computed, the Thévenin-resistance method can be used to compute the contribution of C_π to the high-frequency -3-dB point. Specifically, using Eq. (9.140) for $r_{th\pi}$ results in

$$f_\pi = \frac{1}{2\pi r_{th\pi} C_\pi} = \frac{1}{2\pi (18\,\Omega)(38\,\text{pF})} = 233\,\text{MHz} \tag{9.141}$$

• Find the Thévenin resistance seen by C_μ

The Thévenin-resistance seen by C_μ can be found by applying a test voltage source across the terminal connections of C_μ in the midband model, as in Fig. 9.42. Applying KCL to the v_c node results in

$$i_c = i_{test} + g_m v_\pi \tag{9.142}$$

Similarly, applying KCL to the v_x node yields

$$i_s = i_{test} - i_\pi = i_{test} - \frac{v_\pi}{r_\pi} \tag{9.143}$$

Meanwhile, $v_{test} = v_x - v_c$, which can be expressed in the form

$$v_{test} = i_s R'_S + i_c R_C \tag{9.144}$$

where $R'_S = R_S + r_x$.
  Combining Eqs. (9.142) through (9.144) yields

$$\begin{aligned} v_{test} &= \left(i_{test} - \frac{v_\pi}{r_\pi} \right) R'_S + (i_{test} + g_m v_\pi) R_C \\ &= i_{test}(R'_S + R_C) + v_\pi \left(\frac{-R'_S}{r_\pi} + g_m R_C \right) \end{aligned} \tag{9.145}$$

An expression for v_π can be found by taking KVL around loop A:

$$i_s R'_S = v_\pi + \left(\frac{v_\pi}{r_\pi} + g_m v_\pi \right) R_E \tag{9.146}$$

This equation can be combined with Eq. (9.143) to yield

$$\left(i_{test} - \frac{v_\pi}{r_\pi} \right) R'_S = v_\pi \left[1 + \left(\frac{1}{r_\pi} + g_m \right) R_E \right] \tag{9.147}$$

or

$$v_\pi = \frac{i_{test} R'_S r_\pi}{r_\pi + (1 + g_m r_\pi) R_E + R'_S} \tag{9.148}$$

Substituting Eq. (9.148) for v_π into Eq. (9.145) results in

$$v_{test} = i_{test} \left[R'_S + R_C + \frac{R'_S(-R'_S + g_m r_\pi R_C)}{r_\pi + (1 + g_m r_\pi)R_E + R'_S} \right]$$

$$= i_{test} \frac{(R'_S + R_C)[r_\pi + (1 + g_m r_\pi)R_E + R'_S] + R'_S(-R'_S + g_m r_\pi R_C)}{r_\pi + (1 + g_m r_\pi)R_E + R'_S} \tag{9.149}$$

Simplification of this expression leads to

$$r_{th\mu} = \frac{v_{test}}{i_{test}} = R_C + R'_S \| [r_\pi + (1 + g_m r_\pi)R_E] + \frac{R'_S(g_m r_\pi)R_C}{r_\pi + R'_S + (1 + g_m r_\pi)R_E} \tag{9.150}$$

where $R'_S = R_s + r_x = 855\,\Omega$. The $r_{th\mu}$ given by Eq. (9.150) is equal to the sum of resistances to ground measured on either side of the v_{test} source, plus an additional term that reflects the coupling between the base and collector loops by R_E. Substitution of resistance values into Eq. (9.150) results in

$$r_{th\mu} = 5\,k\Omega + (855\,\Omega)\|(1\,k\Omega + 101\,k\Omega) + \frac{(855\,\Omega)(100)(5\,k\Omega)}{1\,k\Omega + 855\,\Omega + 101\,k\Omega} \approx 10\,k\Omega \tag{9.151}$$

Figure 9.42
Test-voltage source applied to the terminals of transverse capacitance C_μ.

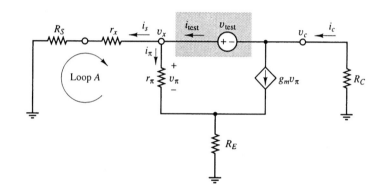

• **Compute the contribution of C_μ to the high-frequency −3-dB point**

With $r_{th\mu}$ computed, the contribution of C_μ to the high-frequency −3-dB point becomes

$$f_\mu = \frac{1}{2\pi r_{th\mu} C_\mu} = \frac{1}{2\pi(10\,k\Omega)(2\,pF)} \approx 7.96\,MHz \tag{9.152}$$

Because f_μ is much lower in value than $f_\pi = 233\,MHz$, the former dominates the high-frequency pole of the amplifier.

• **Use superposition of poles to find the upper −3-dB endpoint**

Superposition of f_π and f_μ results in

$$f_H = f_\pi \| f_\mu = \left(\frac{1}{233\,MHz} + \frac{1}{7.96\,MHz} \right)^{-1} \approx 7.7\,MHz \tag{9.153}$$

Discussion. One interesting by-product of the feedback resistor R_E in the circuit of Fig. 9.40 is illustrated by the results (9.140) and (9.150). As these equations show, the circuit designer can reduce the Thévenin-resistances seen by C_π and C_μ by increasing the value of R_E. Reducing $r_{th\pi}$ and $r_{th\mu}$ also reduces the time constants associated with C_π and C_μ, thus increasing the bandwidth of the amplifier. Such bandwidth improvement comes at the expense of midband gain, which decreases with increasing R_E. This issue is explored in Problem 9.64. ∎

EXERCISE 9.23   Compute the midband gain of the amplifier of Example 9.10.   **Answer:** -4.9

9.24   Suppose that the circuit of Fig. 9.40 is used as a follower with $R_C = 0$. Compute the resulting high-frequency -3-dB point using the equations of Example 9.10.   **Answer:** $\approx 67\,\text{MHz}$

9.25   Show that Eq. (9.150) results from Eq. (9.149).

9.26   For the circuit of Fig. 9.40, find expressions for $r_{\text{th}\pi}$ and $r_{\text{th}\mu}$ that would apply if a MOSFET were substituted for the BJT.   **Answer:** $r_{\text{th}\pi} \approx (R_S + R_E)/(1 + g_m R_E)$; $r_{\text{th}\mu} \approx R_C + R_S' + g_m R_S' R_C/(1 + g_m R_E)$

9.3.6   Frequency Response with Bypass Capacitor

The simultaneous goals of stable bias and large amplifier gain are often not compatible, as shown by the examples of Chapters 7 and 8. A large resistance shared by the amplifier's input and output loops, such as R_E in the feedback-bias circuit of Fig. 9.43, produces stable bias, but results in low amplifier gain. The gain can be improved if the feedback resistor is shunted to signal ground by a large capacitor that behaves as an incremental short circuit at frequencies of interest and acts to increase amplifier gain. This bypass method often involves the use of a large discrete capacitor, and is somewhat obsolete as a design technique. Nevertheless, it is worthy of study, because the issues involved apply universally to the study of capacitors and frequency response in electronic circuits. In this section, the capacitor bypass technique is examined to determine the exact frequencies at which effective bypass takes place. The analysis focuses on the BJT in the feedback-bias configuration, but the concepts are applicable to FET amplifiers as well.

Figure 9.43
BJT inverter in the feedback-bias configuration with bypass capacitor C_E.

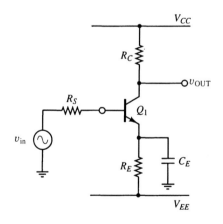

BJT Inverter with Bypass Capacitor

We begin by considering the BJT inverter of Fig. 9.43. The small-signal model of this circuit, with v_{in} and v_{out} represented by phasor voltages, is shown in Fig. 9.44. For simplicity, the resistances r_x and r_o have been omitted from the model, because we presume r_x to be much smaller than R_S, and r_o to be much larger than R_C. Similarly, because the focus is on the circuit's low-frequency response, the internal capacitances C_π and C_μ, which affect only the high-frequency response, are assumed to behave as open circuits.

Figure 9.44
Small-signal model
of the inverter of
Fig. 9.43.

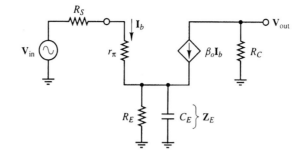

As a small-signal analysis shows [see, e.g., Eq. (7.132) in Chapter 7] the gain of this circuit *without* C_E present is given by

$$\frac{\mathbf{V}_{out}}{\mathbf{V}_{in}} = \frac{-\beta_o R_C}{R_S + r_\pi + (\beta_o + 1)R_E} \tag{9.154}$$

This equation is also valid at frequencies where C_E behaves as an open circuit. In the limit $(R_S + r_\pi) \ll (\beta_o + 1)R_E$, Eq. (9.154) reduces to

$$\frac{\mathbf{V}_{out}}{\mathbf{V}_{in}} \approx \frac{-R_C}{R_E} \tag{9.155}$$

If the bypass capacitor C_E *is* present in the circuit, and if the frequency spectrum of the input is sufficiently high such that C_E acts as an incremental short circuit to signals, the gain is instead given by

$$\frac{\mathbf{V}_{out}}{\mathbf{V}_{in}} = \frac{-\beta_o R_C}{R_S + r_\pi} \tag{9.156}$$

This equation is valid at frequencies where C_E behaves as a short circuit.

The values of R_E and $(R_S + r_\pi)$ are generally in the same range; hence the bypassed gain (9.156) is on the order of β_o times larger than the unbypassed gain (9.155). The frequency or frequencies at which C_E undergoes the transition from "open"- to "short"-circuit behavior are of particular interest. The problem can be addressed by treating the parallel combination of R_E and C_E in Fig. 9.44 as a complex impedance of value

$$\mathbf{Z}_E = R_E \| \frac{1}{j\omega C_E} = \frac{R_E/j\omega C_E}{R_E + 1/j\omega C_E} = \frac{R_E}{1 + j\omega R_E C_E} \tag{9.157}$$

Note that $\mathbf{Z}_E$ approaches the value R_E well below the pole at $\omega = 1/R_E C_E$ and approaches the value $1/j\omega C_E$ well above the pole.

By substituting $\mathbf{Z}_E$ for R_E in the gain equation (9.154), an exact expression for the small-signal gain of the amplifier with C_E present can be found:

$$\frac{\mathbf{V}_{out}}{\mathbf{V}_{in}} = \frac{-\beta_o R_C}{R_S + r_\pi + (\beta_o + 1)\mathbf{Z}_E} = \frac{-\beta_o R_C}{R_S + r_\pi + \frac{(\beta_o+1)R_E}{1+j\omega R_E C_E}} \tag{9.158}$$

Multiplying the numerator and denominator of Eq. (9.158) by $(1 + j\omega R_E C_E)$ results in

$$\frac{\mathbf{V}_{out}}{\mathbf{V}_{in}} = \frac{-\beta_o R_C(1 + j\omega R_E C_E)}{[R_S + r_\pi + (\beta_o + 1)R_E] + j\omega R_E C_E(R_S + r_\pi)} \tag{9.159}$$

Dividing the numerator and denominator of Eq. (9.159) by $[R_S + r_\pi + (\beta_o + 1)R_E]$ reduces the expression to standard complex binomial form:

$$\frac{\mathbf{V}_{out}}{\mathbf{V}_{in}} = \frac{-\beta_o R_C}{R_S + r_\pi + (\beta_o + 1)R_E} \frac{1 + j\omega R_E C_E}{1 + \frac{j\omega R_E C_E(R_S+r_\pi)}{[R_S+r_\pi+(\beta_o+1)R_E]}} \tag{9.160}$$

The imaginary term in the denominator of Eq. (9.160) can be written as

$$\frac{j\omega C_E R_E (R_S + r_\pi)}{R_S + r_\pi + (\beta_o + 1)R_E} = \frac{j\omega C_E R_E [(R_S + r_\pi)/(\beta_o + 1)]}{[(R_S + r_\pi)/(\beta_o + 1)] + R_E} = j\omega C_E \left[R_E \left\| \frac{R_S + r_\pi}{\beta_o + 1} \right. \right] \quad \textbf{(9.161)}$$

Equation (9.160), which describes the gain of the circuit, thus can be expressed as

$$\frac{\mathbf{V}_{out}}{\mathbf{V}_{in}} = \frac{-\beta_o R_C}{R_S + r_\pi + (\beta_o + 1)R_E} \frac{1 + j\omega/\omega_1}{1 + j\omega/\omega_2} \quad \textbf{(9.162)}$$

where the pole ω_2 and zero ω_1 are given by

$$\omega_2 = \frac{1}{C_E \left[R_E \left\| \frac{R_S + r_\pi}{\beta_o + 1} \right. \right]} \quad \textbf{(9.163)}$$

and

$$\omega_1 = \frac{1}{R_E C_E} \quad \textbf{(9.164)}$$

The pole (9.163) involves R_E in parallel with another resistance, whereas the zero (9.164) involves just R_E. Both include the same capacitance C_E; hence it will always be true that $\omega_2 > \omega_1$.

At frequencies well above ω_2, both imaginary terms in Eq. (9.162) dominate, and the gain becomes

$$\frac{\mathbf{V}_{out}}{\mathbf{V}_{in}} \approx \frac{-\beta_o R_C}{R_S + r_\pi + (\beta_o + 1)R_E} \frac{j\omega/\omega_1}{j\omega/\omega_2} = \frac{-\beta_o R_C}{R_S + r_\pi + (\beta_o + 1)R_E} \frac{\omega_2}{\omega_1} \quad \textbf{(9.165)}$$

where the ratio ω_2/ω_1 is given by

$$\frac{\omega_2}{\omega_1} = \frac{\frac{1}{C_E [R_E \| (R_S + r_\pi)/(\beta_o + 1)]}}{\frac{1}{R_E C_E}} = \frac{R_E C_E}{C_E \left[R_E \left\| \frac{(R_S + r_\pi)}{(\beta_o + 1)} \right. \right]} = \frac{R_E C_E}{C_E \frac{R_E (R_S + r_\pi)/(\beta_o + 1)}{R_E + (R_S + r_\pi)/(\beta_o + 1)}}$$

$$= \frac{R_E C_E [(\beta_o + 1)R_E + R_S + r_\pi]}{R_E C_E (R_S + r_\pi)} = \frac{(\beta_o + 1)R_E + R_S + r_\pi}{R_S + r_\pi} \quad \textbf{(9.166)}$$

This factor is precisely the one required to turn the unbypassed gain term in Eq. (9.165), as given by Eq. (9.154), into the bypassed gain (9.156). As shown in the Bode plot of Fig. 9.45, the zero ω_1 represents the frequency at which C_E *begins* to bypass the emitter node to ground, thereby increasing the magnitude of the gain. The pole ω_2 represents the frequency at which complete bypassing is achieved.

Figure 9.45
Bode plot of the gain function given by Eq. (9.162). The pole frequency ω_2 will always be greater than ω_1, as Eqs. (9.163) and (9.164) show.

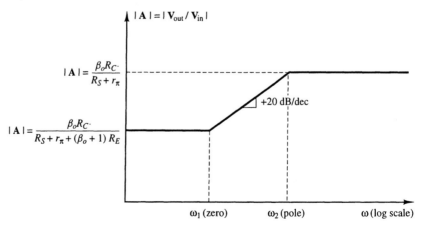

The preceding mathematical derivation yields precise values for the pole and zero frequencies of the gain expression (9.162). A closer examination of the circuit provides insight into their physical significance. As Eq. (9.158) indicates, the gain of the amplifier will increase if the impedance $\mathbf{Z}_E$ between the emitter and ground is reduced. Suppose the amplifier to be driven by a sinusoidal input at a frequency well below the zero ω_1, so that C_E appears as an open circuit and $\mathbf{Z}_E \approx R_E$.

As the input frequency is raised, a point will be reached where the impedance of C_E, which *decreases* with frequency, becomes comparable to R_E. Above this point, the decreasing magnitude of $\mathbf{Z}_E$ will cause the gain to increase in magnitude. (The gain will also experience a phase shift, which we shall ignore for now.) The transition will occur when the capacitor's impedance is *equal* in magnitude to R_E, that is, when

$$\left| \frac{1}{j\omega C_E} \right| = R_E \tag{9.167}$$

or

$$\omega = \frac{1}{R_E C_E} \tag{9.168}$$

This condition predicts the zero frequency ω_1 of the gain equation (9.162).

As the input frequency is increased still further, the improvement in gain will continue until the continually reduced impedance of C_E has no further effect on the circuit. At this point, the impedance of $\mathbf{Z}_E$ will be dominated by C_E and will be so small that the emitter node will be essentially held at ground potential. The frequency at which this second condition occurs can be found by examining the Thévenin equivalent of the circuit as seen by C_E, shown in Fig. 9.46. Of primary interest is the small-signal Thévenin-resistance $r_{\text{th}E}$ measured at the emitter terminal. Applying a small-signal test source between node e and ground yields the value

$$r_{\text{th}E} = R_E \left\| \frac{R_S + r_\pi}{\beta_o + 1} \right. \tag{9.169}$$

(We recognize $r_{\text{th}E}$ as the small-signal output resistance of the voltage-follower output of the amplifier.) When C_E is connected, some portion of $\mathbf{V}_{\text{th}}$ will appear across the capacitor as $\mathbf{V}_e$. The voltage $\mathbf{V}_e$ will be reduced toward zero as frequency is increased and the impedance of C_E reduced. Node e will be bypassed to ground when the impedance of C_E becomes much smaller than $r_{\text{th}E}$. Complete bypassing will occur at frequencies where the condition

$$\left| \frac{1}{j\omega C_E} \right| \ll r_{\text{th}E} \tag{9.170}$$

is met. The transition to the bypass condition begins at the frequency where $1/j\omega C_E$ is *equal* in magnitude to $r_{\text{th}E}$:

$$\left| \frac{1}{j\omega C_E} \right| = r_{\text{th}E} = R_E \left\| \frac{R_S + r_\pi}{\beta_o + 1} \right. \tag{9.171}$$

or

$$\omega = \frac{1}{C_E \left[R_E \left\| \frac{R_S + r_\pi}{\beta_o + 1} \right. \right]} \tag{9.172}$$

This latter condition describes the pole frequency ω_2 of the gain equation (9.162).

Figure 9.46
(a) Small-signal circuit seen by C_E at its own terminals; (b) Thévenin equivalent of the small-signal circuit seen by C_E.

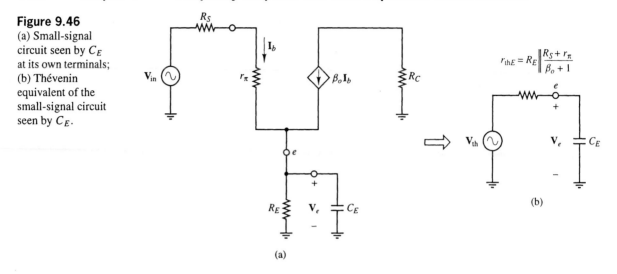

As a final note, we examine the frequency-dependent role of C_E from a bias point of view. When the bias circuit is first energized, C_E is charged to the dc voltage imposed on it by the bias circuit (i.e., to the dc voltage drop across R_E). Thereafter, the capacitor will undergo a voltage change only if charged or discharged through the small-signal resistance seen by its terminals. If the small-signal circuit is incapable of charging or discharging C_E on a time scale comparable to changes in the input signal, the charged capacitor will behave like a "voltage source". The emitter node will remain at the fixed voltage of the capacitor and will function as a small-signal ground. Such a condition will occur if $\omega \gg r_{thE}C_E$, or

$$\frac{1}{\omega C_E} \ll r_{thE} \tag{9.173}$$

where r_{thE} is the small-signal Thévenin-resistance seen by C_E. In this case, r_{thE} is given by Eq. (9.169); hence the onset of the condition (9.173) occurs at

$$\frac{1}{\omega C_E} = R_E \left\| \frac{R_S + r_\pi}{\beta_o + 1} \right. \tag{9.174}$$

which is identical to Eq. (9.172).

DESIGN

EXAMPLE 9.11   The amplifier of Fig. 9.47 must provide a midband gain magnitude of at least 120, beginning at a frequency no higher than 100 Hz. Resistors R_1, R_2, and R_E have been chosen to set the BJT bias current to about 1.4 mA. Choose appropriate values for R_C, C_E, and C_S such that the gain objectives are met. Capacitor values should be chosen such that the pole of C_E dominates the low-frequency response.

Solution

• Assess the goals of the problem

The value of R_C must be chosen first such that the magnitude of the midband gain is at least 120. Capacitors C_E and C_S are then chosen such that C_E produces a dominant-pole frequency of no more than 100 Hz.

Figure 9.47

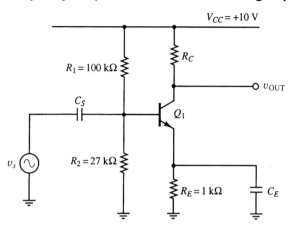

• Set the midband gain by choosing the value of R_C

The midband gain of the amplifier is obtained from the small-signal model of Fig. 9.48 (with r_o assumed much larger than R_C and omitted). Straightforward analysis yields a midband gain of $-g_m R_C$. At $I_C = 1.4\,\text{mA}$ and $\eta = 1$, g_m will equal $I_C/\eta V_T = 56\,\text{mA/V}$, hence setting the midband gain to -120 requires an R_C of $120/(56\,\text{mA/V}) = 2.14\,\text{k}\Omega$. Rounding this value up to the nearest standard resistor value of $2.2\,\text{k}\Omega$ yields a midband gain of $-(56\,\text{mA/V})(2.2\,\text{k}\Omega) = -123$, or about $42\,\text{dB}$.

Figure 9.48
Small-signal model
of the circuit of
Fig. 9.47 when C_E
and C_S behave as
short circuits.

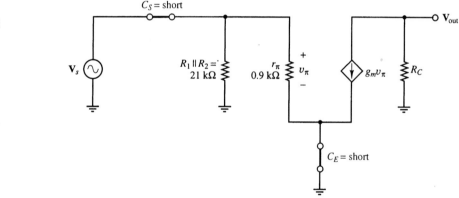

• Choose a value for C_S

In order to make C_E the dominant low-frequency capacitor, the pole frequency of C_S must be much lower than that of C_E. If C_E is truly dominant, it will behave as an open circuit near the pole of C_S, hence the Thévenin resistance seen by C_S near its own pole frequency will be equal to $R_1 \| R_2 \| [r_\pi + (\beta_o + 1) R_E]$. An estimate of the minimum value of C_S, computed for a low-end value of $\beta_o = 50$ and a frequency 1000 times lower than the dominant pole at $100\,\text{Hz}$, becomes

$$C_S \approx \frac{1}{2\pi(f_L/10^3)R_1 \| R_2 \| [r_\pi + (\beta_o + 1)R_E]}$$

$$= \frac{1}{2\pi(0.1\,\text{Hz})[21\,\text{k}\Omega \| (0.9\,\text{k}\Omega + 51\,\text{k}\Omega)]} = 106\,\mu\text{F} \approx 100\,\mu\text{F} \qquad (9.175)$$

where $r_\pi = \beta_o/g_m = 0.9\,\text{k}\Omega$. If the actual value of β_o is larger, the pole of C_S will occur at an even lower frequency than $0.1\,\text{Hz}$. The value (9.175) is conveniently equal to a standard capacitor value.

- **Choose a value for C_E**

The design requirements ask that C_E be chosen to be the dominant low-frequency capacitor with the highest pole frequency. The value of C_S was therefore selected by assuming C_E to behave as an open circuit at the pole frequency of C_S. Conversely, C_S will behave as a short at the pole frequency of C_E if the design requirements are met. With C_S a short, and with no series resistance between v_s and C_S, the small-signal resistance appearing between the base of the BJT and ground becomes zero. The pole of C_E therefore can be computed using Eq. (9.163) with $R_S = 0$:

$$\omega_E = \frac{1}{C_E[R_E \| r_\pi/(\beta_o + 1)]} \approx \frac{1}{C_E(R_E \| 1/g_m)} \tag{9.176}$$

where $r_\pi/\beta_o = 1/g_m$. According to the right-hand expression in (9.176), if the pole of C_E is to lie no higher than $100\,\text{Hz}$, the value of C_E must be at least

$$C_E = \frac{1}{(2\pi \cdot 100\,\text{Hz})[(1\,\text{k}\Omega) \| (56\,\text{mA/V})^{-1}]} = 90.7\,\mu\text{F} \tag{9.177}$$

The next *highest* standard value of $100\,\mu\text{F}$ should be chosen, resulting in an actual dominant-pole frequency of about $91\,\text{Hz}$.

- **Evaluate the design and revise if necessary**

The zero of C_E, given by Eq. (9.164), describes the frequency at which C_E first begins to bypass R_E to ground. This frequency lies at

$$f = \frac{1}{2\pi R_E C_E} = \frac{1}{2\pi(1\,\text{k}\Omega)(100\,\mu\text{F})} = 1.6\,\text{Hz} \tag{9.178}$$

which is higher in value than the computed pole frequency of C_S. Because both the pole and zero of C_E lie above the computed pole of C_S, the capacitor C_E indeed behaves as an open circuit near the pole of C_S, as previously assumed. Similarly, the pole of C_S is low enough in frequency that it does not appreciably affect the -3-dB endpoint frequency created by the pole of C_E. Hence no further revision of the design is necessary. The low-frequency gain magnitude Bode plot of the amplifier, reflecting its various pole and zero frequencies, is shown in Fig. 9.49.

Figure 9.49
Low-frequency-magnitude Bode plot of the circuit of Fig. 9.47.

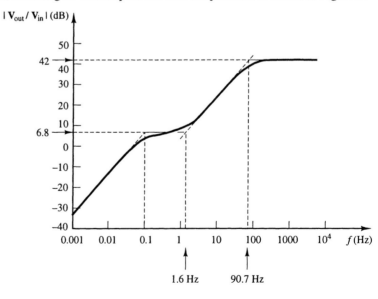

EXERCISE 9.27   For the case $R_S = 0$ in Fig. 9.43, derive an approximate expression for ω_2 in Eq. (9.162) in terms of the BJT bias current I_C. Take the limit of the expression for large g_m.
Answer: $\omega_2 = (1 + I_C R_E/\eta V_T)/R_E C_E \approx I_C/\eta V_T C_E$

9.28   Based on the results of Exercise 9.27, find the minimum C_E necessary if the emitter in Fig. 9.43 is to be bypassed to ground above 10 Hz. The minimum desirable I_C is 0.5 mA.
Answer: $\approx 320\,\mu F$

9.29   Using superposition of poles and the Thévenin-resistance method, find the low-frequency -3-dB endpoint of the circuit of Example 9.11 if $C_S = 1\,\mu F$.   **Answer:**   197 Hz

9.30   Repeat the design procedure of Example 9.11, but choose capacitor values such that C_S is the dominant low-frequency capacitor.

9.31   For the circuit of Example 9.11, show that the intermediate gain between the pole of C_S at 0.1 Hz and the zero of C_E at 1.6 Hz is equal to about $-2.2 \equiv 6.8$ dB.

9.32   Consider the MOSFET circuit of Fig. 9.50, in which a capacitor is used to bypass the source to signal ground. Show that the pole introduced by C_E occurs at $\omega_2 = 1/C_E(R_E \| 1/g_m)$.

Figure 9.50

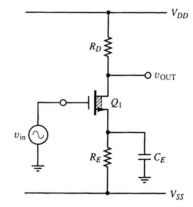

9.33   For the MOSFET amplifier of Fig. 9.50, derive an expression for the pole frequency ω_2 of C_E in terms of the MOSFET parameter K and the bias current I_D.   **Answer:** $\omega_2 = [1 + 2R_E(KI_D)^{1/2}]/R_E C_E$

9.34   If the largest capacitance available is $0.1\,\mu F$ and the smallest allowable MOSFET current is 0.1 mA, what is the minimum lower midband limit that can be achieved in the circuit of Fig. 9.50 with $R_E = 1\,k\Omega$? Assume that the MOSFET has a conductance parameter of $K = 1\,mA/V^2$.
Answer: 2.6 kHz

9.35   Using the feedback-bias configuration of Fig. 9.50, design a MOSFET amplifier that has a gain of at least 10 dB and a lower midband limit of at least 10 Hz.

9.4 FREQUENCY RESPONSE OF THE DIFFERENTIAL AMPLIFIER

The analysis techniques of this chapter are readily applied to the differential amplifier of Chapter 8. Most differential amplifiers are designed to operate at dc, so the "midband" region usually extends all the way to zero frequency without experiencing a low-frequency -3-dB endpoint. At the high-frequency end of the midband, however, the differential amplifier exhibits many of the same features of a single-transistor inverter. In this section, we examine the high-frequency −3-dB endpoint of a differential amplifier under differential- and common-mode excitation. The discussion again focuses primarily on BJT amplifiers, but the analyses apply equally well to differential amplifiers made from MOSFETs and JFETs.

9.4.1 Differential-Mode Frequency Response

The basic BJT differential amplifier of Fig. 9.51 was analyzed for its bias and gain properties in Chapter 8. For simplicity, the current mirror of the original circuit is represented here by a current source of value I_o in parallel with a Norton resistance r_n. Under balanced differential-mode excitation, the emitter node E functions as a small-signal ground to each half of the amplifier, causing both the left and right sides to resemble a single-transistor inverting amplifier with the emitter grounded. The differential-mode frequency response will thus be dominated by Miller multiplication of the transverse capacitance C_μ in each half of the amplifier.

Figure 9.51
Balanced BJT differential amplifier biased by current source I_o.

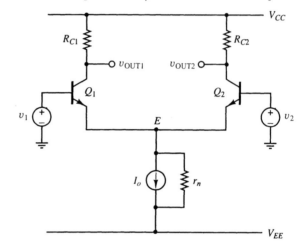

DESIGN

EXAMPLE 9.12

Using the basic differential amplifier configuration of Fig. 9.51, design a circuit with single-ended differential-mode gains of about ±100 and a high-frequency −3-dB point of at least 10 MHz. Assume the BJTs to be matched with parameters $\beta_o = 100$, $C_\mu = C_{ob} = 1$ pF, $f_T = 400$ MHz, $r_o = \infty$, and $r_x = 50\,\Omega$. Note that v_1 and v_2 are connected directly to Q_1 and Q_2 without source resistances R_S, hence the high-frequency response will depend strongly on the BJT base resistances r_x.

Solution

• Assess the goals of the problem

The parameters to be set in this problem are the values of the matched resistors R_C and bias current I_o. The value of r_n does not affect the differential-mode gain because node E acts as a small-signal ground under differential-mode excitation.

• Choose a design strategy

The single-ended gains of the amplifier, computed previously in Chapter 8, have magnitude $g_m R_C/2$. This quantity can also be expressed as $I_C R_C/2\eta V_T$, where $I_C R_C$ is the voltage drop across R_C. The gain thus can be set over a wide range of I_C values as long as the product $I_C R_C$ is appropriately chosen. Meanwhile, the capacitance C_π, which contributes to the dominant high-frequency pole, depends on the value of I_C via g_m, hence I_C first should be set to meet the bandwidth constraints of the amplifier.

• Find an expression for the dominant high-frequency pole

We assume at the outset that the Miller-multiplied capacitance C_μ, together with C_π, contribute the dominant high-frequency pole. As shown in the small-signal model of Fig. 9.52, the virtual ground that appears at node E under differential-mode excitation allows C_μ to be represented on either side of the amplifier by an equivalent parallel capacitance of value

$$C_A = C_\mu(1 + g_m R_C).$$

The circuit of Fig. 9.52 is balanced with respect to differential-mode signals, hence only one side of the amplifier need be analyzed. Application of the Thévenin resistance method with $r_{thA} = r_x \| r_\pi$ results in a dominant-pole frequency of

$$f_H = \frac{1}{2\pi(r_x \| r_\pi)(C_\pi + C_A)} \tag{9.179}$$

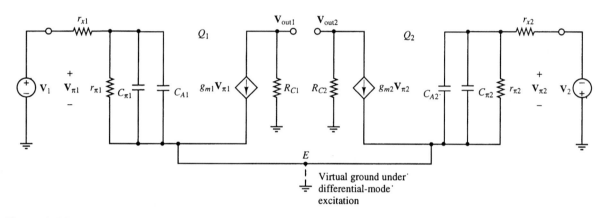

Figure 9.52 High-frequency small-signal model of the differential amplifier of Fig. 9.51 under differential-mode excitation. The transverse capacitances C_μ have been represented by equivalent Miller-multiplied capacitances of value $C_A = C_\mu(1 + g_m R_C)$. Node E acts as a small-signal ground under differential-mode excitation.

• Estimate the maximum allowed $(C_\pi + C_A)$ for a bandwidth of $10\,\text{MHz}$

The value of r_π will depend on I_C, which is the parameter to be determined. However, given the condition $r_x \| r_\pi \approx r_x$, the maximum allowed $(C_\pi + C_A)$ for a minimum bandwidth of $10\,\text{MHz}$ can be determined from Eq. (9.179):

$$(C_\pi + C_A)_{\text{max}} = \frac{1}{2\pi r_x f_H} = \frac{1}{2\pi(50\,\Omega)(10\,\text{MHz})} = 318\,\text{pF} \tag{9.180}$$

The sum $(C_\pi + C_A)$ can also be expressed as

$$C_\pi + C_A = (g_m/\omega_T - C_\mu) + C_\mu(1 + g_m R_C) = g_m/\omega_T + C_\mu g_m R_C \tag{9.181}$$

The last term in this expression consists of C_μ multiplied by the product $g_m R_C$, which we know will equal 200 from the gain specifications (gain $= g_m R_C/2$). Hence the maximum allowed value of g_m/ω_T becomes

$$(g_m/\omega_T)_{\text{max}} = (C_\pi + C_A)_{\text{max}} - C_\mu g_m R_C = 318\,\text{pF} - (1\,\text{pF})(200) = 118\,\text{pF} \tag{9.182}$$

- **Determine the value of the bias currents I_C**

Because $g_m = I_C/\eta V_T$, Eq. (9.182) can be expressed in terms of the maximum allowed I_C in each transistor if the bandwidth specifications are to be met:

$$I_{C\text{-max}} = \eta V_T g_m = \eta V_T \omega_T (g_m/\omega_T)_{\max} = (0.025\,\text{V})(2\pi \cdot 4 \times 10^8\,\text{Hz})(118\,\text{pF}) \approx 7.4\,\text{mA} \tag{9.183}$$

Since C_π increases with I_C, we choose (somewhat arbitrarily) the value $I_C = 5\,\text{mA}$, which provides an ample safety margin. For this value of I_C, the current-source current should be set to $I_o = 10\,\text{mA}$.

- **Choose the value of R_C**

Meeting the gain specification $g_m R_C/2 = I_C R_C/2\eta V_T = 100$ requires that the voltage drop across each collector resistor be $I_C R_C = 2\eta V_T(100) = 5\,\text{V}$. With $I_C = 5\,\text{mA}$, each R_C must be set to $1\,\text{k}\Omega$.

- **Evaluate the design and revise if necessary**

We now analyze the circuit by substituting the chosen values of I_C and R_C into the various circuit equations. Specifically, for $I_C = 5\,\text{mA}$ and $g_m = (5\,\text{mA})/(1)(0.025\,\text{V}) = 200\,\text{mA/V}$, the actual value of C_π becomes

$$C_\pi = g_m/\omega_T - C_\mu = (200\,\text{mA/V})/(2\pi \cdot 4 \times 10^8\,\text{Hz}) - 1\,\text{pF} \approx 78.6\,\text{pF} \tag{9.184}$$

For the chosen values of g_m and R_C, the Miller-multiplied capacitance becomes

$$C_A = C_\mu[1 + g_m R_C] = (1\,\text{pF})[1 + (200\,\text{mA/V})(1\,\text{k}\Omega)] = 201\,\text{pF} \tag{9.185}$$

Combining these capacitance values, with $r_\pi = \beta_o/g_m = 500\,\Omega$, yields an estimate of the actual dominant high-frequency pole:

$$f_H = \frac{1}{2\pi(r_x\|r_\pi)(C_\pi + C_A)} = \frac{1}{2\pi(50\,\Omega\|500\,\Omega)(78.6\,\text{pF} + 201\,\text{pF})} \approx 12.5\,\text{MHz} \tag{9.186}$$

This frequency actually exceeds the design specification of 10 MHz, hence no further design revision is necessary. ⚡

EXERCISE 9.36   Show via direct calculation that the midband gain of the amplifier of Example 9.12 meets design specifications.

9.37   Find f_H for differential-mode excitation in the circuit of Fig. 9.51 if the BJTs have an Early voltage $V_A = 100\,\text{V}$.   **Answer:**   13 MHz

9.38   Determine values for I_o and R_C in the circuit of Fig. 9.51 if the midband gain is to be 150 and the bandwidth equal to 10 MHz.

9.4.2 Common-Mode Frequency Response

In this section, we examine the frequency response of the circuit of Fig. 9.51 under common-mode excitation. When a common-mode signal is applied, node E will no longer function as a signal ground, and signal current will flow through r_n. Under these conditions, the effect of the capacitance C_n appearing in parallel with r_n must also be considered. Although no such capacitance is included in Fig. 9.51, it will almost always be present in a real circuit because the I_o source will be implemented using another transistor that will contribute its output capacitance to C_n. The circuit's frequency response under common-mode excitation can be analyzed using the model of Fig. 9.53, which represents either symmetrical half of the differential amplifier. The shared elements r_n and C_n have each been split into parallel halves of value $2r_n$ and $C_n/2$, respectively.

Figure 9.53
High-frequency
small-signal
half-circuit that
represents the
amplifier of
Fig. 9.51 under
common-mode
excitation. The
biasing current
source has been
represented by the
parallel
combination of r_n
and C_n, so that $2r_n$
and $C_n/2$ appear in
each half circuit.

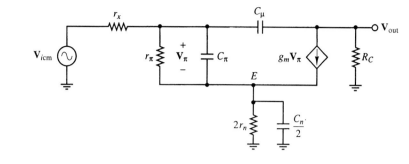

Exact analysis of this circuit is actually very complicated and best left to computer-aided design tools such as SPICE or MICROCAP. Considerable insight into the effect of C_n on the circuit, however, can be gained by a few fundamental observations. The capacitance $C_n/2$ behaves much like the emitter-bypass capacitance of the circuit of Fig. 9.44, which was analyzed in detail in Section 9.3.6. As the frequency is increased, $C_n/2$ reduces the impedance between node E and ground, thereby increasing the amplifier gain. This gain increase will begin at the zero of $C_n/2$, located at $\omega = 1/(2r_n)(C_n/2) = 1/r_nC_n$, and will end when the impedance of C_n is much smaller than the Thévenin resistance seen looking back into the emitter of the BJT. In a differential amplifier, such a gain increase is *not* desirable, because a small common-mode gain is one of the key features of the differential configuration. The onset of the gain increase initiated by the zero of $C_n/2$ therefore constitutes an important limit to the common-mode response.

The frequency at which $C_n/2$ completely bypasses the emitter node to ground usually occurs at an extremely high value and is seldom of relevance. The poles of C_π and C_μ, however, cause a decrease in the common-mode gain at high frequencies that reverses the gain increase initiated by the zero of $C_n/2$. This behavior is illustrated in SPICE Example 9.16 at the end of the chapter.

The location of the common-mode zero in the amplifier of Example 9.12 is easily calculated. For the purpose of analysis, we assume that $C_n = 1\,\text{pF}$ and $r_n = 1\,\text{M}\Omega$. These parameters are typical of a current source implemented using the Widlar configuration. For the various circuit parameters specified in Example 9.12, the zero of $C_n/2$ occurs at

$$f_Z = \frac{1}{2\pi r_n C_n} = \frac{1}{2\pi (1\,\text{M}\Omega)(1\,\text{pF})} \approx 160\,\text{kHz} \tag{9.187}$$

The common-mode rejection ratio, equal to the ratio of the differential- and common-mode gains, will begin to degrade above this frequency. The f_Z given by Eq. (9.187) is low in value compared to the differential-mode bandwidth of 12.5 MHz computed in Example 9.12.

9.4.3 Frequency Response of the Cascode Configuration

As illustrated by the analysis of Section 9.4.1, the basic differential amplifier suffers from the same high-frequency limitations as a single-transistor amplifier. In both cases, Miller multiplication of the transverse capacitance C_μ (or C_{gs} in FET circuits) at the input node severely limits the ability of the circuit to amplify differential high-frequency signals. In this section, we revisit the cascode configuration, originally introduced in Chapter 6. This circuit topology has the ability to extend high-frequency response by minimizing Miller multiplication at the amplifier input terminals. The cascode configuration is discussed here in the context of differential amplifiers but also can be used to extend the frequency response of single-ended amplifiers.

As discussed in Section 6.3.2, the cascode configuration is formed when an inverting device acts as the current-source input to a second device connected in the current-follower configuration. A basic cascoded BJT differential amplifier employing this concept is shown in Fig. 9.54. The transistors are matched, with near-equal bias currents and small-signal parameters. Input devices Q_1 and Q_2, driven by v_1 and v_2, provide current signals to the current-follower devices Q_3 and Q_4. These latter devices appear in the common-base connection with bases held at small-signal ground by zener D_1. The zener holds the bases of Q_3 and Q_4 at a positive voltage relative to ground, hence Q_1 and Q_2, whose collectors lie one V_f drop below the bases of Q_3 and Q_4, remain biased in the constant-current region with a positive V_{CE}.

Figure 9.54
BJT version of the cascoded differential-amplifier configuration.

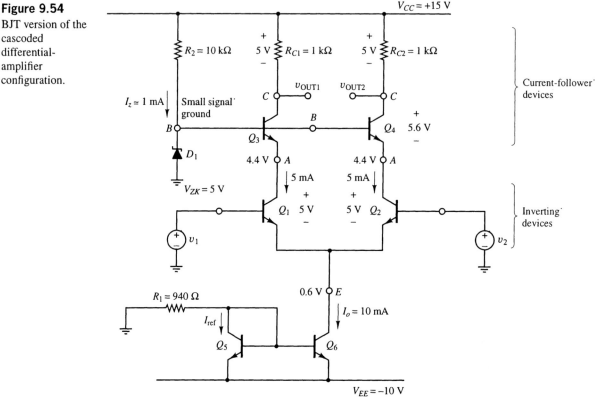

Differential-Mode Midband Gain

Under balanced differential-mode excitation, node E will remain at signal ground, hence the half-circuit small-signal model of Fig. 9.55 can be used to find the midband gain. For simplicity, the output resistances r_{o1} and r_{o3} of Q_1 and Q_3 are assumed large enough to be ignored. (This assumption is valid for r_{o1}, but only approximately valid for r_{o3}, as shown in Problem 9.90.) Similarly, the small-signal resistances r_x of each BJT and r_z of the zener are assumed small. If pure differential-mode inputs $v_1 = v_a$ and $v_2 = -v_a$, where $v_{idm} = 2v_a$, are applied to the amplifier, the small-signal collector current through Q_1 will be given by

$$i_{c1} = \beta_{o1}i_{b1} = \frac{\beta_{o1}}{r_{\pi 1}}v_a \qquad (9.188)$$

This collector current will equal the emitter current of Q_3:

$$i_{c1} = i_{e3} = (\beta_{o3} + 1)i_{b3} \qquad (9.189)$$

Equation (9.189) can be used to find an expression for the base current in the current-follower device Q_3 in terms of the input current i_{b1}:

$$i_{b3} = \frac{i_{c1}}{\beta_{o3} + 1} = \frac{\beta_{o1} i_{b1}}{\beta_{o3} + 1} \approx i_{b1} \tag{9.190}$$

where $\beta_{o1} = \beta_{o3} = \beta_o$. Multiplying Eq. (9.190) by β_{o3} provides an expression for the collector current through Q_3:

$$i_{c3} = \beta_{o3} i_{b3} = \frac{\beta_{o3} \beta_{o1} i_{b1}}{(\beta_{o3} + 1)} \approx \beta_{o1} i_{b1} \equiv i_{c1} \tag{9.191}$$

As this equation shows, the base and collector currents of Q_1 are replicated in the current-follower device Q_3, so that the single-ended output measured at the collector of Q_3 becomes

$$v_{\text{out}1} = -i_{c3} R_{C1} \approx -\beta_{o1} i_{b1} R_{C1} = \frac{-\beta_{o1} R_{C1}}{r_{\pi 1}} v_a = -g_m R_{C1} v_a \tag{9.192}$$

Given that $v_{idm} = 2v_a$, the single-ended differential-mode gain of the full amplifier becomes

$$A_{\text{dm-se1}} = \frac{v_{\text{out}1}}{2v_a} = -\frac{g_{m1} R_{C1}}{2} \tag{9.193}$$

A similar analysis yields

$$A_{\text{dm-se2}} = \frac{v_{\text{out}2}}{2v_a} = +\frac{g_{m1} R_{C1}}{2} \tag{9.194}$$

Analysis of the biasing current mirror in Fig. 9.54 shows that $I_o = 10\,\text{mA}$, so that $I_C = 5\,\text{mA}$, $g_m = 200\,\text{mA/V}$, $A_{\text{dm-se1}} = -100$, and $A_{\text{dm-se2}} = +100$. This amplifier has exactly the same midband response as the diff-amp of Example 9.12, in which Q_1 and Q_2 drive the pull-up loads R_{C1} and R_{C2} directly. A straightforward analysis shows these two amplifiers to have the same common-mode gain as well. The principal importance of the cascode configuration lies in its ability to extend the high-frequency range of the differential-mode response, as we now show.

Figure 9.55
Small-signal half-circuit for analyzing the midband behavior of the cascode amplifier of Fig. 9.54.

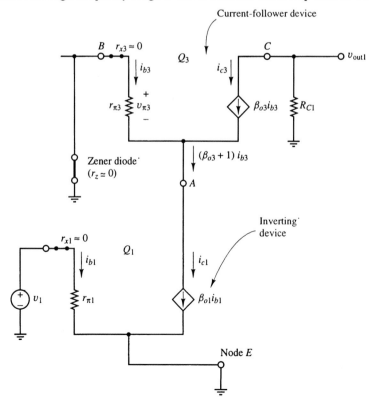

High-Frequency Performance

A high-frequency small-signal model valid for the half-circuit on either side of node E, which remains at signal ground under differential-mode excitation, is shown in Fig. 9.56. We shall find the high-frequency −3-dB point of this circuit using the dominant-pole technique. For the purposes of comparison, we assume the BJTs to have the same parameters as those of Example 9.12: $\beta_o = 100$, $C_\mu = 1\,\text{pF}$, $f_T = 400\,\text{MHz}$, $r_o = \infty$ and $r_x = 50\,\Omega$. With $I_C = I_o/2 = 5\,\text{mA}$, the value of C_π for each device is again equal to $g_m/2\pi f_T - C_\mu \approx 78.6\,\text{pF}$. Similarly, each device has an r_π of $\beta_o/g_m = 500\,\Omega$.

Figure 9.56
High-frequency
small-signal model
valid for either half
of the cascode
amplifier of
Fig. 9.54. The
transverse
capacitance $C_{\mu 1}$ is
replaced by an
equivalent parallel
capacitance C_A.
Node E functions
as a small-signal
ground under
differential-mode
excitation.

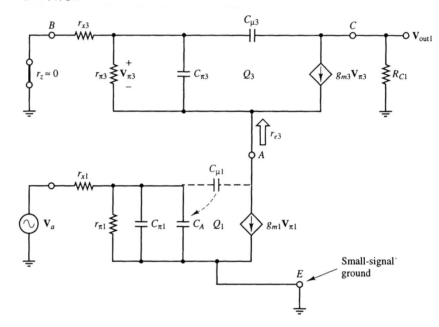

In Fig. 9.56, $C_{\mu 1}$ is replaced by an equivalent Miller capacitance C_A. The value of C_A can be computed using Miller's theorem. With $C_{\pi 3}$ and $C_{\mu 3}$ set to open circuits, the load resistance seen by the collector of Q_1 is equal to the small-signal resistance presented by the emitter of Q_3:

$$r_{e3} = \frac{r_{\pi 3} + r_{x3}}{\beta_{o3} + 1} \approx \frac{r_{\pi 3}}{\beta_{o3}} \equiv \frac{1}{g_{m3}} \tag{9.195}$$

Applying this resistance to the Miller approximation of Section 9.3.4 yields

$$C_A \approx C_{\mu 1}\left(1 + \frac{g_{m1}}{g_{m3}}\right) = 2C_{\mu 1} = 2\,\text{pF} \tag{9.196}$$

where $g_{m1} \approx g_{m3}$ because Q_1 and Q_3 have approximately the sa : I_C. This small value of C_A should be compared to the value of 201 pF obtained in Example 9.12 [see Eq. (9.185)]. The reduction in C_A is a consequence of the low small-signal resistance presented to the collector of Q_1 by the emitter of Q_3.

The contribution of $(C_{\pi 1} + C_A)$ to the high-frequency −3-dB point can be computed using this new value of C_A:

$$f_A = \frac{1}{2\pi (r_{\pi 1}\|r_{x1})(C_{\pi 1} + C_A)} = \frac{1}{2\pi (45\,\Omega)(78.6\,\text{pF} + 2\,\text{pF})} \approx 43\,\text{MHz} \tag{9.197}$$

where $(r_{\pi 1} \| r_{x1})$ is the Thévenin resistance driving C_A. The small value of C_A caused by the low resistance r_{e3}—the trademark of the cascode configuration—leads to a large value of f_A. Note that r_{x1}, which was justifiably omitted from the midband model, must be included now because it interacts directly with $(C_{\pi 1} + C_A)$. For completeness, we also include r_{x2} in the high-frequency model of Fig. 9.56.

The contribution of $C_{\mu 3}$ to the -3-dB point can be found by setting all other capacitances to open circuits and setting $\mathbf{V}_a$ to zero, as shown in Fig. 9.57. KVL taken around the base loop of Q_1 yields the result $\mathbf{I}_{\pi 1} = 0$. Similarly, $\mathbf{I}_{\pi 3} = 0$ as well. With $\mathbf{I}_{\pi 3} = 0$ and $\mathbf{I}_{\pi 1} = 0$, the dependent sources of Q_3 and Q_1 become open circuits, isolating the lower terminal of $r_{\pi 3}$. The Thévenin-resistance seen by $C_{\mu 3}$ therefore becomes simply $R_{C1} + r_{x3}$; hence the computed pole of $C_{\mu 3}$ is given by

$$f_\mu = \frac{1}{2\pi (R_{C1} + r_{x3})C_{\mu 3}}$$

$$= \frac{1}{2\pi (1.05\,\text{k}\Omega)(1\,\text{pF})} \approx 152\,\text{MHz} \qquad (9.198)$$

Similarly, the Thévenin resistance seen by $C_{\pi 3}$ in Fig. 9.56 can be found by applying a test voltage to the $C_{\pi 3}$ terminals with all other capacitances set to open circuits, as shown in Fig. 9.58. Note that i_x is equivalent to the current $g_{m3}v_{\pi 3}$ when $i_{c1} = 0$. Application of KCL leads to

$$i_{\text{test}} = i_{\pi 3} + i_x = \frac{v_{\text{test}}}{r_{\pi 3}} + g_{m3}v_{\text{test}} \qquad (9.199)$$

so that $$r_{\text{th}\pi} = \frac{v_{\text{test}}}{i_{\text{test}}} = r_{\pi 3} \left\| \frac{1}{g_{m3}} = 50\,\Omega \| 5\,\Omega \approx 4.6\,\Omega \right. \qquad (9.200)$$

Figure 9.57
Circuit seen by $C_{\mu 3}$ with $\mathbf{V}_a$ set to zero and all other capacitors set to open circuits.

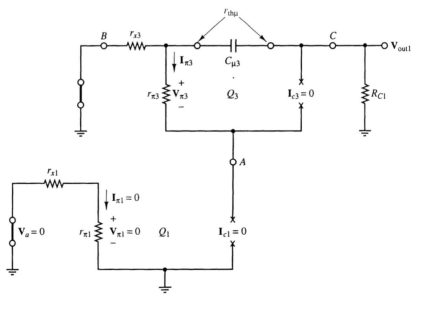

Given this value of $r_{\text{th}\pi}$, the computed pole of $C_{\pi 3}$ becomes

$$f_\pi = \frac{1}{2\pi r_{\text{th}\pi} C_{\pi 3}}$$

$$= \frac{1}{2\pi (4.6\,\Omega)(78.6\,\text{pF})} \approx 445\,\text{MHz} \qquad (9.201)$$

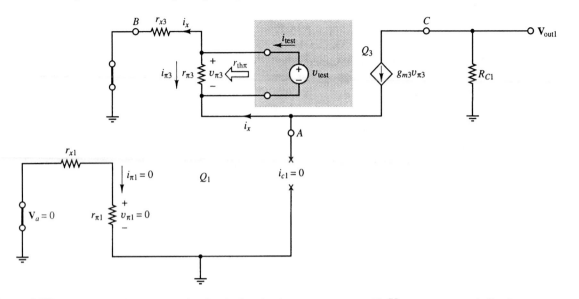

Figure 9.58 Thévenin resistance seen by $C_{\pi 3}$ is found using a v_{test} source with $\mathbf{V}_a$ set to zero and all other capacitors set to open circuits.

The three frequencies f_A, f_π, and f_μ, none of which is clearly dominant, may now be combined using superposition of poles to find the -3-dB point f_H:

$$f_H = f_A \| f_\pi \| f_\mu = \left(\frac{1}{43\,\text{MHz}} + \frac{1}{445\,\text{MHz}} + \frac{1}{152\,\text{MHz}} \right)^{-1} \approx 31\,\text{MHz} \tag{9.202}$$

This value should be compared to the -3-dB point of the amplifier without cascode, obtained in Example 9.12 as

$$\begin{aligned} f_H = \frac{\omega_H}{2\pi} &= \frac{1}{2\pi (r_x \| r_\pi)[C_\pi + C_\mu(1 + g_m R_C)]} \\ &= \frac{1}{2\pi (45\,\Omega)(78.6\,\text{pF} + 201\,\text{pF})} \approx 12.5\,\text{MHz} \end{aligned} \tag{9.203}$$

In the latter case, the Miller multiplication of C_μ by the factor $(1 + g_m R_C)$ substantially decreases the dominant upper -3-dB pole frequency by a factor of about 2.5 compared to the identical cascode configuration.

EXERCISE **9.39**   Draw the differential-mode Bode plots for the circuit of Fig. 9.54.

9.40   Analyze the current mirror of Fig. 9.54 and show that it produces a bias current of $I_o = 10\,\text{mA}$.

9.41   Show that the high-frequency poles of the cascode amplifier of Fig. 9.54 are not affected by r_z of the zener as long as $r_z \ll r_{\pi 3}$ and $r_z \ll R_C$.

9.42   If the BJTs in the circuit of Fig. 9.54 are matched, why are their small-signal parameters not all exactly the same?

9.43   Estimate the high-frequency -3-dB point of the amplifier of Fig. 9.54 if I_o is reduced to 1 mA.
Answer: 87 MHz

9.4.4 Integrated-Circuit Design Considerations

Many of the circuits described in this chapter make use of capacitors to couple signals, isolate bias networks from input and output terminals, and achieve large gain. In contrast, the differential amplifiers of this section have no series coupling or shunt bypass capacitors. Unlike amplifiers made from discrete components, differential amplifiers are often fabricated as part of large multistage circuits built entirely on an integrated-circuit (IC) "chip." Fabrication techniques that are well within the state of the art allow micron-sized transistors to be made, and entire circuits now routinely occupy chip surface areas formerly occupied by just a single transistor. In this realm, any capacitor larger than a few tens of nanofarads requires too much surface area to be fabricated on the IC itself. Large-valued capacitors must be located off-chip and connected to the IC by external wires. Because of this limitation, capacitors of the coupling or bypass type are usually avoided in the design of integrated-circuit amplifiers. The dc coupling and bypass techniques illustrated by the diff-amp examples of this section are instead used exclusively.

9.5 TIME RESPONSE OF ELECTRONIC CIRCUITS

The presence of stray, internal, or discrete circuit capacitance greatly affects the way in which an electronic circuit responds to transient, time-varying input signals. In previous chapters, the response of a circuit to step and other inputs was assumed to be independent of time. Before leaving this chapter, we reexamine circuit behavior from a time-domain point of view and consider the effects of capacitance on circuit transient response.

9.5.1 Internal Diode Capacitance and the Half-Wave Rectifier

The half-wave rectifier of Chapter 4 is often used in peak detector circuits and dc power supplies. In this section, the half-wave rectifier is analyzed more precisely by including the effects of diode depletion capacitance under reverse-biased conditions. The effects of the forward-bias diffusion and depletion capacitances are examined in later problems.

Suppose that the input v_{IN} to the half-wave rectifier of Fig. 9.59(a) consists of one cycle of a square wave of amplitude V_o. Analysis of this circuit follows readily from the circuit models of Figs. 9.59(b) and 9.59(c), in which appropriate piecewise linear models have been substituted for the diode under forward- and reverse-biased conditions, respectively. Note the presence of the depletion capacitance C_j in Fig. 9.59(c).

Figure 9.59
Half-wave rectifier revisited: (a) basic circuit;
(b) piecewise linear model when $v_{IN} = V_o$ (diode forward-biased);
(c) piecewise linear model when $v_{IN} = -V_o$ (diode reverse-biased).

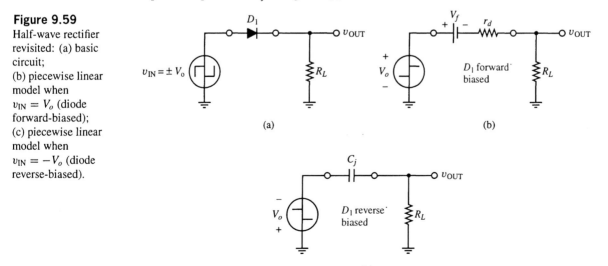

When v_{IN} is positive and the diode forward-biased, the response of the circuit, as found in Chapter 4, is given by

$$v_{OUT} = (V_o - V_f)\frac{R_L}{R_L + r_d} \tag{9.204}$$

Recall that if $r_d \ll R_L$, v_{OUT} becomes

$$v_{OUT} \approx V_o - V_f \tag{9.205}$$

As v_{IN} heads toward zero from its positive value of V_o, the model of Fig. 9.59(b) will apply until v_{IN} reaches V_f, at which point v_{OUT} will equal zero. As v_{IN} crosses through zero on its way to $-V_o$, the model of Fig. 9.59(c) takes over. As the diode becomes reverse-biased, the capacitance C_j begins to affect the circuit's behavior. The response of the new circuit with C_j in place can be found by assuming v_{OUT} to begin at the value $v_{OUT} = 0$ and by setting the input to a step function of magnitude $-V_o$. The resulting output becomes the transient step response found in Chapter 1 (see Section 1.9.1):

$$v_{OUT} = -V_o e^{-t/R_L C_j} \tag{9.206}$$

Although not strictly correct, we presume C_j to be a constant in Eq. (9.206). In reality, C_j is a function of the diode voltage, as given by Eq. (9.3) or (9.4). The actual differential equation describing the circuit is a nonlinear one, and difficult to solve in closed form. By assuming C_j to be a constant, the approximate result (9.206) allows us to gain insight into the operation of the circuit and the limitations imposed by the depletion capacitance.

The resulting v_{OUT} versus time, showing both positive and negative portions of the v_{IN} cycle, is shown in Fig. 9.60. For simplicity, we assume that $r_d \ll R_L$, so that the approximation (9.205) is valid under forward-bias conditions.

Figure 9.60
Output versus time for the circuit of Fig. 9.59. C_j is approximated as constant, and $r_d \ll R_L$. Dashed line is the input voltage.

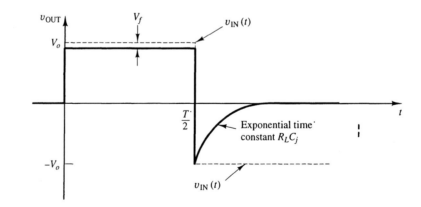

The validity of assuming C_j to be a constant may be tested by comparing Eq. (9.206) with the exact solution shown in Fig. 9.61 for the case $V_o = 10\,\text{V}$, $R_L = 1\,\text{M}\Omega$, and $r_d \approx 0$. In the exact solution, the voltage dependence of C_j was assumed to be given by $C_j = k_j(\Psi_o - v_D)^{1/2}$, where $k_j = 400\,\text{pF-V}^{1/2}$ and $\Psi_o = 0.9\,\text{V}$. The exact solution shown in Fig. 9.61 was obtained by numerical integration for the given set of parameters. The approximate solution (9.206) was plotted after evaluating C_j at $v_D = -V_o/e$, so that $C_j \approx 186\,\text{pF}$. As seen from the plots, Eq. (9.206) provides a reasonable estimate of circuit behavior.

Figure 9.61
Plots of v_{OUT}
versus time for the
circuit of Fig. 9.59
under reverse-bias
conditions.
$V_o = 10\,\text{V}$;
$R_L = 1\,\text{M}\Omega$;
$r_d \approx 0$. Solid line:
approximate
solution with
constant
$C_j = 186\,\text{pF}$;
dashed line: exact
solution with $C_j = k_j(\Psi_o - V_o)^{-1/2}$;
$k_j = 400\,\text{pF-V}^{1/2}$;
$\Psi_o = 0.9\,\text{V}$.

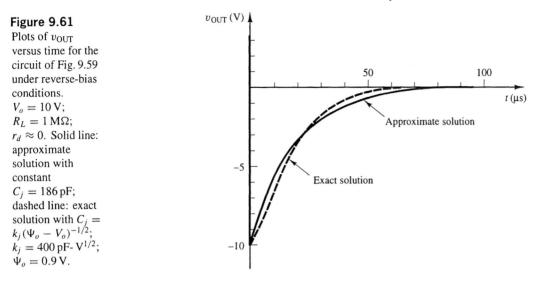

9.5.2 Incremental Step Response of a Transistor Amplifier

In many situations, transistor circuits must linearly amplify pulses, square waves, and other transient signals. The fidelity with which these tasks are performed is directly related to the presence of both internal and external circuit capacitances. Such circuits are best analyzed in the time domain, because transient signals often contain complicated frequency spectra that make analysis in the frequency domain difficult.

In the time domain, the high- and low-frequency capacitor definitions used in the frequency domain also apply. A low-frequency capacitor causes "droop," or decay, in the amplifier step response. In contrast, a high-frequency capacitor limits the rise time of the step response. If the time constant of a low-frequency capacitor is much longer than the time scale of interest, the capacitor will retain its dc bias voltage indefinitely after the transient and will behave as a small-signal short circuit. Similarly, if the time constant of a high-frequency capacitor is much shorter than the time scale of interest, the capacitor will behave as an open circuit that is charged instantaneously.

Figure 9.62
Simple BJT
inverter with input
coupling capacitor.

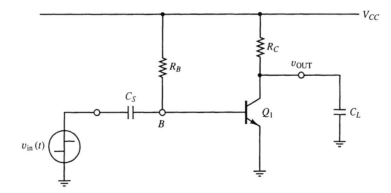

The simple BJT inverter of Fig. 9.62 provides a good vehicle for illustrating the principles of small-signal transient behavior in a transistor amplifier. In this case, the fixed bias network of R_B and V_{CC} is used to bias the device in its active, constant-current region. The series capacitor C_S isolates the bias network from the v_{in} voltage source, which would otherwise appear as a dc ground to node B. For the moment, we shall ignore the effect of the internal BJT capacitances C_π and C_μ and the external capacitor C_L and instead concentrate on the effect of C_S on circuit behavior. Only the signal components of the input and output voltages are of interest, hence we may represent the circuit by the small-signal model of Fig. 9.63, where $r_{in} = R_B \| r_\pi$. For simplicity, we omit the transverse base resistance r_x from the small-signal model, which is justified if $r_x \ll R_B \| r_\pi$. Likewise, the BJT output resistance r_o, which appears in parallel with R_C, can be omitted if $r_o \gg R_C$.

Figure 9.63
Small-signal model of the circuit of Fig. 9.62 with small-signal step voltage applied. The internal BJT capacitances C_π and C_μ and transverse base resistance r_x have been omitted.

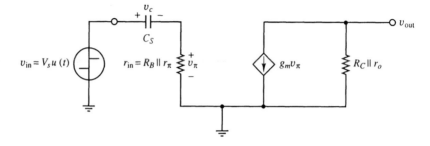

The small-signal output of this inverter was analyzed without the coupling capacitor C_S in Chapter 7 and was found to be

$$v_{out} = -g_m(R_C \| r_o)v_\pi \approx -g_m R_C v_\pi \tag{9.207}$$

In the present case, C_S and r_{in} form a simple series resistor–capacitor circuit with voltage taken across r_{in}. If $v_{in}(t)$ is a small-signal step function of amplitude V_s, then v_π will be given by

$$v_\pi = V_s e^{-t/r_{in}C_S} \qquad \text{for} \qquad t > 0 \tag{9.208}$$

where $r_{in} = R_B \| r_\pi$. This exponential expression follows from the analysis of resistor-capacitor circuits in Chapter 1.

The incremental output of the amplifier in response to the input step can be obtained from Eq. (9.207) and is given by

$$v_{out} = -g_m R_C v_\pi = -g_m R_C(V_s e^{-t/r_{in}C_S}) \qquad \text{for} \qquad t > 0 \tag{9.209}$$

This signal component decays to zero with exponential time constant $r_{in}C_S$, causing the total output voltage v_{OUT} to return to its dc bias value several time constants after the step function. This result is not surprising; C_S must eventually appear as an open circuit to the new dc value that v_{in} acquires after its initial step function transient.

The step response given by Eq. (9.209) is an important indicator of the amplifier's ability to amplify signals that are transient in nature. If the $r_{in}C_S$ time constant is short compared to time durations of interest, the amplifier will initially respond to v_{in} but the output will quickly relax back to its dc bias point. Conversely, in the limit $r_{in}C_S \to \infty$, the exponent in Eq. (9.209) approaches unity, and the signal component of the output approaches a constant value given by

$$\left. \frac{v_{out}}{V_s} \right|_{\lim r_{in}C_S \to \infty} = -g_m R_C \qquad \text{for} \qquad t > 0 \tag{9.210}$$

In this limit, the signal component of the output consists of an amplified version of the step function $V_s u(t)$. In practice, the condition $r_{in}C_S \to \infty$ must be achieved by making C_S large, because $r_{in} = R_B \| r_\pi$ is usually fixed to a finite value by the bias constraints of the circuit.

EXAMPLE 9.13

For the circuit of Fig. 9.62, suppose that $V_{CC} = 5\,\text{V}$, $C_S = 1\,\mu\text{F}$, $R_B = 1\,\text{M}\Omega$, and $R_C = 3.3\,\text{k}\Omega$, with $\beta_F = \beta_o = 210$, $V_f = 0.6\,\text{V}$, and $\eta = 1$. Find the total output voltage (signal plus bias component) if v_{in} is a periodic square wave of peak amplitude $V_p = 5\,\text{mV}$ and period $T = 100\,\text{ms}$, as depicted in Fig. 9.64. Repeat the analysis for $T = 0.1\,\text{ms}$.

Figure 9.64
Square-wave $v_{in}(t)$ of period T and peak amplitude V_p. At any value of time, t' denotes the time elapsed since the most recent square-wave transition.

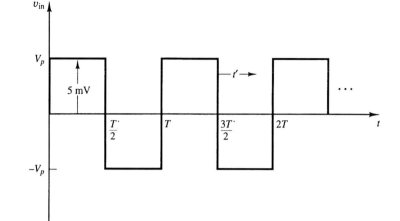

Solution

The component values of this circuit yield the following bias and small-signal parameters at room temperature:

$$I_B = \frac{V_{CC} - V_f}{R_B} = \frac{5\,\text{V} - 0.6\,\text{V}}{1\,\text{M}\Omega} = 4.4\,\mu\text{A} \qquad (9.211)$$

$$I_C = \beta_F I_B = 0.92\,\text{mA} \qquad (9.212)$$

$$\begin{aligned} V_{OUT} = V_{CE} &= V_{CC} - I_C R_C \\ &= 5\,\text{V} - (0.92\,\text{mA})(3.3\,\text{k}\Omega) = 1.95\,\text{V} \approx 2\,\text{V} \end{aligned} \qquad (9.213)$$

$$g_m = \frac{I_C}{\eta V_T} = \frac{0.92\,\text{mA}}{(1)(0.025\,\text{V})} \approx 37\,\text{mA/V} \qquad (9.214)$$

$$r_\pi = \frac{\beta_o}{g_m} = \frac{210}{37\,\text{mA/V}} \approx 5.7\,\text{k}\Omega \qquad (9.215)$$

$$r_{in} = R_B \| r_\pi = 1\,\text{M}\Omega \| 5.7\,\text{k}\Omega \approx 5.7\,\text{k}\Omega \qquad (9.216)$$

The circuit will treat every transition of the square wave as either a positive or negative step function of magnitude $V_s = 2V_p$, initiating a signal response of the form

$$v_{out} = \mp g_m R_C (2V_p e^{-t'/r_{in} C_S}) \qquad (9.217)$$

where t' expresses the time elapsed since the most recent square-wave transition. At any time t, the value of v_{out} will be a superposition of all responses of the form (9.217) initiated by previous square-wave transitions. Note that superposition works because the small-signal model of the BJT is a linear one.

When $T = 100\,\text{ms}$, the time constant $r_{in} C_S = (5.7\,\text{k}\Omega)(1\,\mu\text{F}) = 5.7\,\text{ms}$ will be much shorter than the period of the square wave. Each transient response of the form (9.217) will thus decay to zero well before the next square-wave transition. The resulting total $v_{OUT}(t)$, including both bias and signal components, is shown in Fig. 9.65. With $g_m R_C \approx 120$, the output makes a jump of magnitude $g_m R_C (2V_p) = (120)(10\,\text{mV}) = 1.2\,\text{V}$ after each square-wave transition.

A similar consideration yields the result for $T = 0.1\,\text{ms} \equiv 100\,\mu\text{s}$. In this case, the time constant $r_{in} C_S = 5.7\,\text{ms}$ is much longer than the period of the square wave. The individual responses of the form (9.217) decay only slightly between square-wave transitions. At any given time, the superimposed previous responses will alternately cancel one another. The net output will thus appear as in Fig. 9.66. The square wave appears as an amplified replica of the input

Figure 9.65

Total output of the amplifier of Fig. 9.62 for $T = 100$ ms. The effects of C_π, C_μ, and C_L have been ignored.

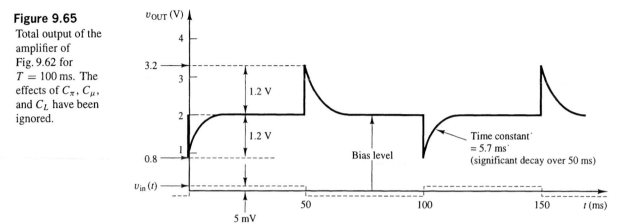

signal, because the capacitor behaves as a short circuit to the signal $v_{in}(t)$ over the time span T. Note that the output again makes a jump of magnitude 1.2 V at each square-wave transition, as in Fig. 9.65.

Figure 9.66

Total output of the amplifier of Fig. 9.62 for $T = 100\,\mu$s. The effects of C_π, C_μ, and C_L have been ignored.

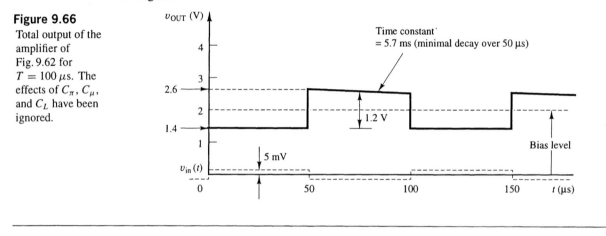

EXERCISE 9.44   Find v_{out} in Example 9.13 if $v_{in}(t)$ has a period of 100 ms and C_S is set to 2.5 μF.

9.45   Find v_{out} in Example 9.13 if $v_{in}(t)$ has a period of 100 ms and R_B is changed to 500 kΩ with $V_{CC} = 10$ V.

9.46   Find v_{out} in the circuit of Fig. 9.62 if $v_{in}(t)$ is a 5-mV peak square wave connected to the amplifier via a source resistance $R_S = 1$ kΩ in series with C_S.

A complete description of the step response of the amplifier of Fig. 9.62 must also include the effects of the internal capacitances C_π and C_μ as well as the load capacitance C_L. These capacitive elements do not affect the "droop," or decay, of the output, but do affect the time required for the step transition to fully occur. This latter interval is often called the *rise time* of the circuit. Because C_π, C_μ, and C_L are usually small compared to C_S, the time constant $r_{in}C_S$ will be long compared to the rise time. Capacitor C_S thus may be treated as an incremental short circuit (voltage across C_S constant) over the duration of the rise time. In the small-signal model of the circuit, shown in Fig. 9.67, the series base resistance r_x is included because it strongly affects the circuit's rise time. Miller's theorem can be applied in the time domain as well as in the frequency domain, so that the transverse capacitance C_μ can be represented here by an equivalent Miller-multiplied capacitance $C_A = C_\mu[1 + g_m(R_C \| r_o)]$, as was done in Section 9.3.4.

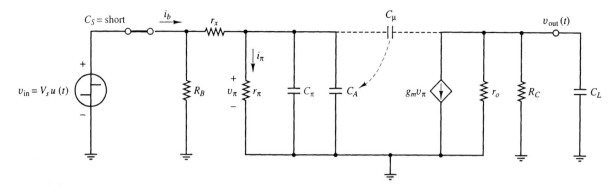

Figure 9.67 Small-signal model of the amplifier of Fig. 9.62, including the capacitances C_π, C_μ, and C_L and series base resistance r_x. The series capacitor C_S is treated as a short circuit over the time span of interest.

The circuit of Fig. 9.67 can be further simplified by taking the Thévenin equivalent of v_{in}, R_B, r_x, and r_π, as in Fig. 9.68. This simplification reduces the input loop to a simple RC circuit. Note that v_π still appears across C_π even though r_π has been absorbed into r_{th}. Also, with C_S taken as a short, R_B is connected directly in parallel with v_{in} and has no effect on the circuit's rise time.

Figure 9.68
Simplified version
of the small-signal
model of Fig. 9.67.

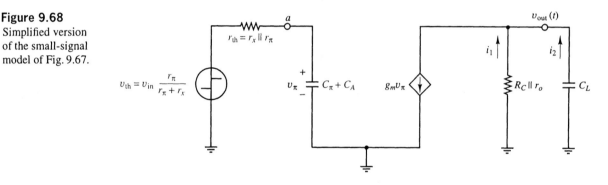

The input loop in Fig. 9.68 has the form of a simple RC circuit. If $v_{in}(t)$ is a step function of value $V_s u(t)$, where $u(t)$ is a unit step function, then v_π will be given by

$$v_\pi = v_{th}[1 - e^{-t/r_{th}(C_\pi + C_A)}] \qquad \text{for} \qquad t > 0 \qquad \text{(9.218)}$$

where $v_{th} = V_s[r_\pi/(r_\pi + r_x)]u(t)$ and $r_{th} = r_\pi \| r_x$. If $r_x \ll r_\pi$, such that $r_{th} \approx r_x$ and $v_{th} \approx v_{in}$, then Eq. (9.218) becomes

$$v_\pi \approx V_s[1 - e^{-t/r_x(C_\pi + C_A)}]u(t) \qquad \text{(9.219)}$$

Now suppose that the effect of $(C_\pi + C_A)$ dominates, so that the effect of C_L on the rise time can be neglected. In such a case, the output voltage of the circuit becomes

$$v_{out} = -g_m v_\pi (R_C \| r_o) \approx -g_m R_C[1 - e^{-t/r_x(C_\pi + C_A)}]V_s u(t) \qquad \text{(9.220)}$$

where $R_C \| r_o \approx R_C$. Conversely, if the time constant $r_x(C_\pi + C_A)$ is short compared to the time constant $(R_C \| r_o)C_L$, the effect of $(C_\pi + C_A)$ can be neglected, and the principal time delay in the circuit will occur in the output loop. Under such conditions, the circuit response can be found by solving for the output voltage v_{out} in terms of v_π. This calculation is complicated somewhat due to the presence of C_L. Applying KCL to the v_{out} node in Fig. 9.68 results in

$$i_2 + i_1 = g_m v_\pi \qquad (9.221)$$

where i_1 and i_2 flow up through $R_C \| r_o$ and C_L, respectively. The current i_2 pulled up through C_L is given by

$$i_2 = -C_L dv_{out}/dt \qquad (9.222)$$

Substituting Eq. (9.222) into (9.221) and substituting $-v_{out}/(R_C \| r_o)$ for i_1 leads to the differential equation:

$$\frac{dv_{out}}{dt} + \frac{v_{out}}{(R_C \| r_o)C_L} = \frac{-g_m v_\pi}{C_L} \frac{(R_C \| r_o)}{(R_C \| r_o)} \qquad (9.223)$$

where the last term on the right has been multiplied and divided by $(R_C \| r_o)$. Equation (9.223) has the solution

$$v_{out} = -g_m(R_C \| r_o)[1 - e^{-t/(R_C \| r_o)C_L}]V_s u(t) \qquad (9.224)$$

where $v_\pi \approx V_s u(t)$.

A complete expression for the output that accounts for the simultaneous effects of both $(C_\pi + C_A)$ and C_L can be found by convolution or by direct solution of the differential equation (9.223), with (9.218) substituted for v_π. Either method produces the result

$$\frac{v_{out}}{V_s} = -g_m(R_C \| r_o)\frac{r_\pi}{r_\pi + r_x}\left(1 - \frac{\tau_1}{\tau_1 - \tau_2}e^{-t/r_{th}(C_\pi + C_A)} - \frac{\tau_2}{\tau_2 - \tau_1}e^{-t/(R_C \| r_o)C_L}\right) \qquad (9.225)$$

where $\tau_1 = r_{th}(C_\pi + C_A)$ and $\tau_2 = (R_C \| r_o)C_L$. In the limit $\tau_2 \ll \tau_1$ [the case where $(C_\pi + C_A)$ dominates over C_L], Eq. (9.225) reduces to (9.220). In the limit $\tau_1 \ll \tau_2$ [the case where C_L dominates over $(C_\pi + C_A)$], the equation reduces to (9.224).

EXAMPLE 9.14

Find the output of the inverter of Fig. 9.62 with the effects of C_π, C_μ, and C_L included if the input signal is equal to a 5-mV voltage step. Assume the BJT to have the parameters given in Example 9.13, with $f_T = 400$ MHz, $\beta_o = 210$, $C_{ob} = C_\mu = 1$ pF, $r_x = 200\,\Omega$, and $r_o = \infty$. Let the load capacitance C_L be equal to 5 pF. Furthermore suppose that an external stray capacitance of value $C_{be} = 1$ nF exists between the base and the emitter.

Solution

From the bias values calculated in Example 9.13, we find that

$$C_\pi = \frac{g_m}{2\pi f_T} - C_\mu = \frac{37 \text{ mA/V}}{2\pi(400 \text{ MHz})} - 1\text{ pF} = 13.7\text{ pF} \qquad (9.226)$$

and
$$C_A = C_\mu(1 + g_m R_C) = (1\text{ pF})[1 + (37 \text{ mA/V})(3.3 \text{ k}\Omega)] = 123\text{ pF} \qquad (9.227)$$

Because $r_x \ll r_\pi$, the input-loop time constant can be approximated by

$$r_x(C_\pi + C_A + C_{be}) = (200\,\Omega)(13.7\text{ pF} + 123\text{ pF} + 1000\text{ pF}) \approx 227\text{ ns} \qquad (9.228)$$

Similarly, the output-loop time constant is given by

$$R_C C_L = (3.3 \text{ k}\Omega)(5\text{ pF}) = 16.5\text{ ns} \qquad (9.229)$$

The time constant of $(C_\pi + C_A + C_{be})$ clearly dominates, hence the RC time constant associated with the circuit's rise time is approximately 227 ns. Note, as a very general rule of thumb, that a high-frequency time constant will be negligible if it is at least five times shorter than the circuit's dominant high-frequency time constant. Over a time span equal to one dominant time constant, the rising exponential associated with the shorter time constant will rise to at least the value $(1 - e^{-5}) = 0.993$, and will thus add little to the overall rise-time delay created by the dominant time constant.

A plot of the signal component of v_{OUT} versus time, including the effects of C_π, C_A, C_{be}, and C_S, is shown in Fig. 9.69. The gain factor of the circuit is equal to $-g_m R_C \approx -120$. Given this gain value, the magnitude of the small-signal step change in v_{OUT} becomes

$$-g_m R_C \frac{r_\pi}{r_\pi + r_x} V_s = -(120) \frac{5.7\,\text{k}\Omega}{5.7\,\text{k}\Omega + 200\,\Omega} (5\,\text{mV}) \approx -0.58\,\text{V} \qquad (9.230)$$

Figure 9.69

Small-signal response of the circuit of Fig. 9.68 to a 5-mV step input. The effects of C_π, C_L, and C_S are evident.

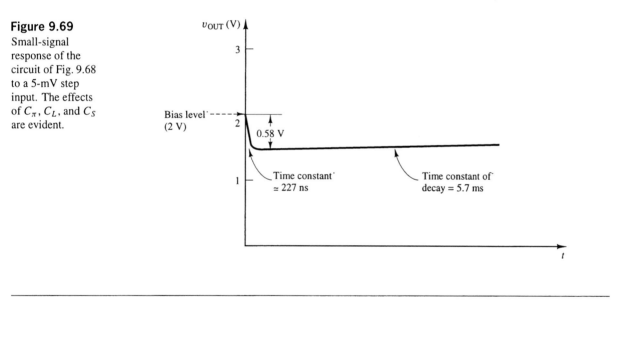

EXERCISE 9.47 Find the response of the circuit of Fig. 9.62 to an impulse of magnitude 10 mV.

9.48 How large can r_x become before the approximation in Eq. (9.219) ceases to be valid?

9.49 Suppose that the external capacitance C_{be} is absent in Example 9.14. Use Eq. (9.225) to determine the resulting output voltage versus time. In this case, neither $(C_\pi + C_A)$ nor C_L will clearly dominate in determining the rise time of the circuit.

9.50 Use Eq. (9.225) to plot v_{out} versus time in the circuit of Example 9.14 without neglecting the time constant contributed by C_L.

SUMMARY

- The appearance of capacitance (and, to some extent, inductance) in electronic circuits has a profound effect on circuit behavior.

- Capacitance can be contributed to a circuit from a variety of sources. Stray lead capacitance, internal device capacitance, and discrete capacitance must all be considered in evaluating circuit performance.

- Inductance is often contributed to a circuit from stray lead effects.

- In the sinusoidal steady-state, the frequency-response behavior of a circuit is determined using the impedance relationships $\mathbf{V} = j\omega L \mathbf{I}$ and $\mathbf{V} = \mathbf{I}/j\omega C$. Algebraic equations, obtained using these relationships and KVL and KCL, can be solved to find voltages and currents as a function of frequency.

- A Bode plot represents the sinusoidal steady-state behavior of a circuit as a function of frequency. A complete Bode plot consists of a magnitude plot and an angle plot.

- The Bode plot provides a quick and reasonably accurate method for approximating a circuit's system function.

- Superposition of poles can be used to approximate the effect of more than one capacitor or inductor at a given frequency.

- The time and frequency dependencies contributed by a nonlinear element can be modeled by adding capacitors to its piecewise linear or small-signal device model.

- A "low-frequency capacitor" affects amplifier response below the midband region. A "high-frequency capacitor" affects amplifier response above the midband region.

- Low-frequency capacitors are usually connected in series with other circuit elements. High-frequency capacitors are connected in parallel with other circuit elements.

- One low-frequency capacitor often dominates low-frequency circuit behavior and determines the low-frequency endpoint of the midband. One high-frequency capacitor often dominates high-frequency circuit behavior and determines the high-frequency endpoint of the midband.

- Transverse capacitance in an inverting amplifier often dominates the high-frequency response.

- Miller's theorem provides a convenient means of evaluating the effects of transverse capacitance.

- The high-frequency poles of the follower and feedback-bias configurations are more easily computed directly, without applying Miller's theorem.

- A differential amplifier has different frequency responses under differential-mode and common-mode excitation.

- The cascode configuration provides a way of extending amplifier bandwidth.

- Large capacitors are seldom found in integrated-circuit designs.

- In the time domain, the transient behavior of a circuit is determined using the v–i relations $i_C = C \, dv_C/dt$ for the capacitors and $v_L = L \, di_L/dt$ for the inductors.

◆ SPICE EXAMPLES

EXAMPLE 9.15 Use SPICE to find the frequency response of the BJT amplifier of Fig. 9.30. As in Example 9.6, assume the transistor to have parameters $C_\mu = 0$, $r_x = 10\,\Omega$, $f_T = 450\,\text{MHz}$, $\beta_F = \beta_o = 100$, $V_f = 0.7\,\text{V}$, and $r_o = \infty$. Assess the change in the low-frequency -3-dB point of the amplifier as C_E is varied between zero (no bypass) and $100\,\mu\text{F}$.

Solution

A diagram of the circuit with nodes suitably numbered for SPICE is shown in Fig. 9.70. An appropriate input file listing follows. For the stated BJT parameters, the BJT bias current again will be equal to about 1 mA, yielding a C_π of approximately 14 pF. This capacitance value is set in the .MODEL statement via the parameter CJE.

In order to vary C_E, its value is set via the .MODEL statement Ctest using the parameter CAP. The last entry in the element statement for C_E consists of a multiplier of value $100\,\mu\text{F}$, which indicates that the actual value of C_E will be 10^{-4} times the value of CAP set in the .MODEL statement. Meanwhile, the .MODEL statement is linked by the label Ctest to a .STEP statement that changes the value of CAP from 0 to 1 in steps of 0.25, thus changing the value of C_E from 0 to $100\,\mu\text{F}$ in steps of $25\,\mu\text{F}$.

Input file:

```
FREQUENCY RESPONSE of BJT AMPLIFIER of FIG. 9.30
*CHANGE CE TO SEVERAL VALUES

*Specify the power supplies and resistors:
        VCC 1 0 6V
        VEE 7 0 -6V
        RC  1 2 5k
        RE  4 7 3.3k
        R1  1 3 20k
        R2  3 7 10k
        RS  5 6 1k
        RM  2 0 1Meg

*Set the input to a 0.1-V ac source:
        Vin 5 0 ac 0.1V 0

*Specify the fixed-value capacitors:
        CS  6 3 2.2uF
        CM  2 0 4pF

*Specify a capacitor CE whose value is 10⁻⁴ (100u) times the value
*specified by CAP in the .MODEL statement labeled Ctest:
        CE 4 0 Ctest 100u
*Label a .MODEL statement Ctest and assign to it the value CAP
        .MODEL Ctest CAP
*Change the parameter CAP from 0 to 1 in steps of 0.25; The parameter
*Ctest(C) links the .STEP command to the .MODEL statement labeled Ctest.
*The value of CE will be changed from 0 to 100 µF in steps of 25 µF:
        .STEP CAP Ctest(C) 0 1.0 0.25

*Specify the parameters of the BJT:
        Q1 2 3 4 BJT
        .MODEL BJT NPN(BF=100 CJE=14p RB=50)

*Calculate and list the bias parameters of the circuit:
        .OP

*Plot the magnitude Bode plot (DEC frequency scale) from 0.01 Hz to
*100 MHz with five data points per decade:
        .AC DEC 5 0.01 100e6
        .PROBE v(2)
        .END
```

Figure 9.70
Circuit of Fig. 9.30
with nodes
numbered for
SPICE.

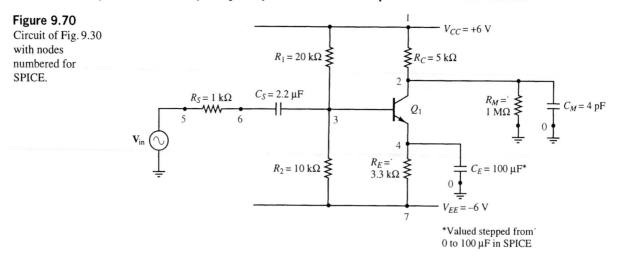

*Valued stepped from
0 to 100 µF in SPICE

Results. The output at node 2, as obtained from the .PROBE utility with the y-axis set to a logarithmic scale, is shown in Fig. 9.71. The high-frequency -3-dB point lies at about 5.7 MHz, which is approximately equal to the value 5.5 MHz obtained in Example 9.6. The SPICE value is somewhat higher because the superposition-of-poles technique actually underestimates the true pole frequency.

The low-frequency -3-dB point computed by SPICE lies at about 64 Hz, as shown in Fig. 9.71. This value lies somewhat below the value 74 Hz computed in Example 9.6. This discrepancy occurs, in part, because the r_π calculated by SPICE equals 2.7 kΩ, as can be seen by observing the output listing of the simulation. The value of $r_\pi = 2.5$ kΩ was used in Example 9.6. In addition, r_x was neglected in Example 9.6, whereas $r_x \equiv$ RB $= 50$ Ω in the SPICE simulation.

Figure 9.71
Magnitude
response plots for
the circuit of
Fig. 9.70 for
various values of
C_E from zero to
100 µF.

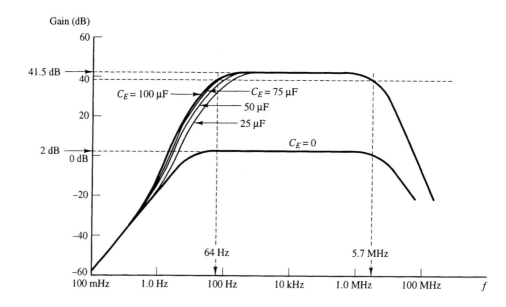

With C_E set to zero, the midband gain magnitude is equal to 2 dB, which is close to the approximate value of unbypassed gain predicted by the ratio $R_C/R_E = (5\,\text{k}\Omega)/(3.3\,\text{k}\Omega) \approx 1.5 \equiv 3.6$ dB. With C_E nonzero, the gain is increased to the bypassed value 41.5 dB in the midband. As these latter plots show, the desired low-frequency response is reached for C_E values larger than about 50 µF. ∎

EXAMPLE 9.16

In Section 9.4.2, the common-mode performance of the differential amplifier of Fig. 9.51 was briefly examined. For the component and bias values specified in Example 9.12, the common-mode gain was shown to increase with frequency above the transmission zero at $f_Z = 160\,\text{kHz}$. Use SPICE to determine the common-mode gain over a complete range of frequencies, including those at which the common-mode gain first levels off and then decreases due to the effects of C_π and C_μ.

Solution

In order to focus on the fundamental nature of the common-mode response, we shall obtain the solution by simulating the linear small-signal half-circuit model of Fig. 9.72. This approach assures us that the results obtained are caused solely by the interaction between C_π, C_μ, and the g_m source, rather than by anomalies associated with nonlinearities or hidden parameters of the transistors. The various elements in Fig. 9.72 are assigned values as specified or calculated in Example 9.12 and Section 9.4.2. The input v_{icm} is arbitrarily set to an ac source of magnitude 1 V and phase angle zero. Its frequency is varied logarithmically (DEC scale) from 1 kHz to 100 GHz with 50 points per decade using the .AC statement.

Figure 9.72
Common-mode small-signal half-circuit model with nodes numbered for SPICE.

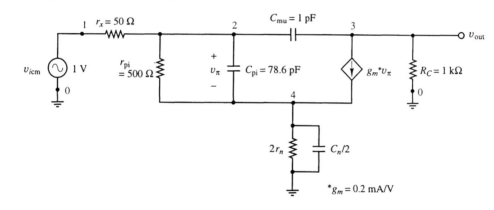

Input file:

```
FREQUENCY RESPONSE of COMMON-MODE HALF-CIRCUIT
*Specify the input source, resistors, and capacitors:
        Vicm  1 0 ac 1 0
        rx    1 2 50
        rpi   2 4 500
        RC    3 0 1k
        Cpi   2 4 78.5pF
        Cmu   2 3 1pF
*Specify the elements 2rn and Cn/2 in the common leg:
        r_(2rn) 4 0 2MEG
        Cn/2    4 0 0.5pF
*Specify a dependent source of value 200 mA/V:
        gm 3 4 2 4 0.2
*Sweep the frequency from 1 kHz to 100 GHz:
        .AC DEC 50 1k 100G
*Monitor the output voltage at node 3 using the .PROBE utility:
        .PROBE v(3)
        .END
```

Results. The magnitude Bode plot of Fig. 9.73, obtained from the .PROBE utility with logarithmic y-axis, shows the common-mode gain to begin at about $-66\,\text{dB}$ at low frequencies, and to indeed experience an increase in gain at $f_Z \approx 160\,\text{Hz}$. These values are consistent with those predicted in Section 9.4.2. The gain increase is not terminated by C_π and C_μ until the frequency reaches the gigahertz range, with the first pole located at about 316 MHz and the other at about 7.4 GHz. By the time these poles are reached, the common-mode gain increases from its dc value of -64 dB to almost 0 dB—a large value for common-mode gain in a differential amplifier. In most applications, the poles of the common-mode response occur beyond the practical frequency range of interest. In situations where the gigahertz range is of relevance, stray capacitances of subpicofarad magnitude that were formerly unimportant, and not included in Fig. 9.72, must also be considered in evaluating both the common- and differential-mode responses. ∎

Figure 9.73

Magnitude plot of the common-mode gain of the half-circuit of Fig. 9.72.

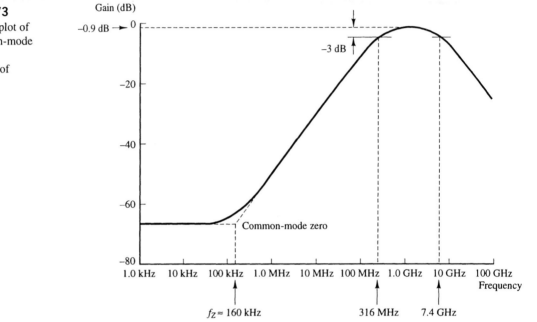

EXAMPLE 9.17

Use SPICE to examine the transient behavior and rise time of the circuit of Fig. 9.74. Excite the circuit with a square wave of magnitude ± 5 mV and period $T = 100$ ms, as in Example 9.13, and observe the output as a function of time. Assume the BJT parameters and elements values specified in Example 9.13.

Figure 9.74

Circuit of Fig. 9.62 with nodes numbered for SPICE and stray capacitance C_{be} included.

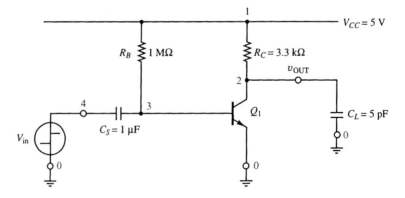

Solution

A version of the circuit with nodes suitably numbered for SPICE is shown in Fig. 9.74. An appropriate input file listing also follows. The ± 5-mV, 100-ms square wave is represented using a PULSE waveform. The time response is examined for 300 ms using the .TRAN statement, in which the "print" and sampling intervals are set to 100 μs.

Input file listing:

```
TRANSIENT RESPONSE OF BJT AMPLIFIER
VCC 1 0 5V
Vin 4 0 PULSE(5mV -5mV 0 1nS 1nS 50mS 100mS)
RB 1 3 1MEG
RC 1 2 3.3k
CS 4 3 1uF
CL 2 0 5pF
Q1 2 3 0 BJT
.MODEL BJT NPN(BF=210 CJC = 1pF CJE = 13.7pF RB=200)
.TRAN 100us   300ms   0 100us
.PROBE V(2)
.END
```

◆

Results. The output of the circuit, as observed at node 2 using the .PROBE utility, is shown in Fig. 9.75. Over this time scale, the rise time is too fast to be observed, but the droop caused by C_S is readily observed. The decay time constant of 5.7 ms, predicted in Example 9.13, appears to be confirmed, but the positive and negative transitions of the waveform, equal to 0.9 V and -1.34 V, respectively in the simulation, differ significantly from the balanced ± 1.2-V jumps predicted by the linear model of Example 9.13. This discrepancy occurs because V_{in} is applied directly to the base of the BJT, where is it permitted to forward bias the base–emitter junction without a series, current-limiting resistor. The base–emitter junction has an exponential v–i characteristic, hence positive excursions of V_{in} will cause more base current to flow than that predicted by the linearized small-signal model of Fig. 9.63. This phenomenon causes the true negative peaks of v_{OUT}, as evaluated by the SPICE simulation, to be larger than -1.2 V. Similarly, negative excursions of V_{in} will cause less base current to flow than that predicted by the linear model, hence the true positive peaks of v_{OUT} will be smaller than 1.2 V. This result illustrates that the small-signal model is only valid for limited excursions about the transistor bias point. The circuit configuration of Fig. 9.62, in which the BJT base is driven directly by V_{in}, causes the range of validity of the small-signal model to be limited to only extremely small changes in V_{in} that are much less than the thermal voltage V_T. ■

Figure 9.75
Transient response
of the circuit of
Fig. 9.74 with
$C_{be} = 1$ nF.

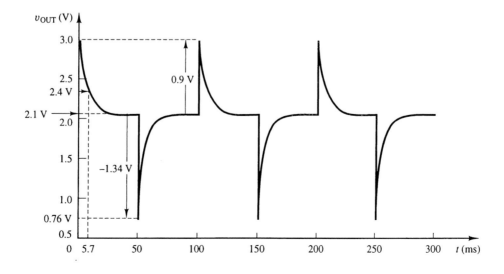

PROBLEMS

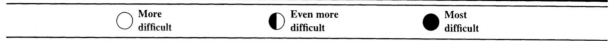

	More difficult		Even more difficult		Most difficult

9.1 Sources of Capacitance and Inductance in Electronic Circuits

9.1 Estimate the depletion capacitance in an abruptly doped reverse-biased silicon *pn* junction diode in which $\Psi_o \approx 0.6\,\text{V}$, $N_A = 10^{16}/cm^3$, $N_D = 10^{15}/cm^3$, $A = 10^{-8}\,\text{m}^2$, and $v_D = -5\,\text{V}$.

9.2 Suppose that an abruptly doped silicon *pn* junction diode is fabricated with $N_A = N_D = 10^{16}/cm^3$. If $\Psi_o = 0.67\,\text{V}$, what is the maximum area that the junction can have if its capacitance is to be no more than 100 pF at a reverse-bias voltage of $-10\,\text{V}$?

9.3 Estimate the diffusion capacitance in a silicon BJT with a base width of $W = 0.1\,\mu\text{m}$ and a bias current of $I_C = 1\,\text{mA}$.

9.4 A BJT with parameters $f_T = 350\,\text{MHz}$ and $C_\mu = 1.5\,\text{pF}$ is biased at 2 mA. What is the value of its small-signal base–emitter capacitance C_π?

9.5 A BJT with parameters $f_T = 420\,\text{MHz}$ and $C_\mu = 0.5\,\text{pF}$ is biased at 0.5 mA. At what frequency will the magnitude of $\beta(\omega)$ fall to unity?

9.6 A MOSFET with gate dimensions $W = 2\,\mu\text{m}$ and $L = 4\,\mu\text{m}$ has a gate oxide of thickness of 1000 Å. Estimate the value of C_{gs} at $v_{DS} = 0$ and $v_{DS} = 5\,\text{V}$.

9.7 Suppose that a MOSFET is fabricated with $t_{ox} = 500$ Å. What is the maximum allowed gate area if C_{gs} is to be no more than 1 pF at $v_{DS} = 0$?

9.8 Estimate the C_{gs} for a silicon JFET fabricated with $N_A = 2 \times 10^{15}/cm^3$, $N_D = 4 \times 10^{15}/cm^3$, and $A = 10^{-7}\,\text{m}^2$ if $v_{GS} = -2\,\text{V}$.

9.2 Sinusoidal Steady-State Amplifier Response

9.2.1 Bode-Plot Representation in the Frequency Domain

9.9 Draw the magnitude and angle Bode plots of v_{OUT} versus v_{IN} for each of the circuits of **Fig. P9.9**.

9.10 Draw the magnitude and angle Bode plots of v_{OUT} versus v_{IN} for each of the circuits of **Fig. P9.9** if each capacitor is replaced by an inductor having a value in microhenries equal to the capacitor's value in microfarads.

9.11 If the v_{IN} terminal in **Fig. P9.9(c)** is fed by a sinusoidal voltage source, derive an expression for the magnitude and angle Bode plots of v_{OUT} versus v_{IN}. Draw each plot for the following component values.

(a) $R_1 = R_2 = 10\,\text{k}\Omega$; $C_1 = 1\,\mu\text{F}$.

(b) $R_1 = R_2 = 2.2\,\text{k}\Omega$; $C_1 = 50\,\mu\text{F}$.

(c) $R_1 = 220\,\text{k}\Omega$, $R_2 = 470\,\text{k}\Omega$; $C_1 = 10\,\mu\text{F}$.

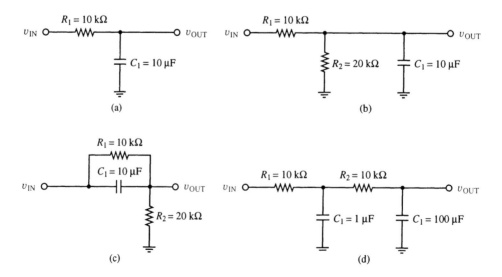

Figure P9.9

9.12 Draw the magnitude and angle Bode plots of v_{OUT} versus v_{IN} for the circuit of **Fig. P9.12**. For the purposes of this problem, assume that C_1 behaves as a short circuit at high frequencies, and C_2 behaves as an open circuit at low frequencies.

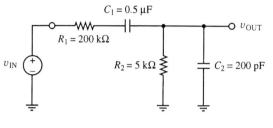

Fig. P9.12

9.13 In some applications, including those involving sound-reproduction equipment, the slopes of a magnitude Bode plot are often expressed in terms of decibels per *octave*, where an octave is defined as a factor-of-2 change in frequency. Determine the decibel/octave slope change in a magnitude Bode plot introduced by a single pole.

9.14 Suppose that two poles coincide at exactly the same frequency. Prove that the actual response curve will lie approximately −6 dB below the intersection of the two Bode-plot asymptotes lying on either side of the double pole. What is the exact angle of the response curve at the double pole frequency? What is the net change in angle realized well above the double pole frequency?

9.15 ⊇ Design a simple *RC* circuit that can be connected in series with an oscilloscope input to attenuate frequencies above 10 MHz with a rolloff of −40 dB/decade. Such circuits are used routinely in oscilloscopes to eliminate unwanted signals coupled into the probe from nearby radio and television stations.

9.16 ⊇ ○ A speech by a prominent politician is to be played on a local radio station news program. Due to a station error, the tape contains a 60-Hz noise signal of magnitude −10 dB relative to the voice signal. Although the noise does not mask out the voice, it is objectionable. The station manager wishes to use two tape recorders to record a second tape from the first, minus the 60-Hz noise. Design an *RC* circuit that can be placed between the tape recorders to reduce the 60-Hz noise to at most −40 dB while passing all signals above 200 Hz.

9.17 ○ A 10× probe increases the input resistance of an oscilloscope by a factor of 10. The increase in input resistance comes at the expense of a one-tenth reduction

in the overall oscilloscope gain. The 10× function is realized by adding a resistor R_X in series with the probe tip, as shown in **Fig. P9.17(a)**. When a voltage v_1 is created by the circuit under test, a voltage v_2 appears across the terminals of the oscilloscope's vertical input channel.

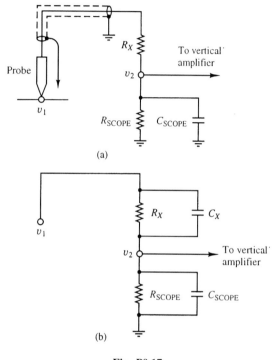

Fig. P9.17

(a) What load resistance will be presented to the circuit if $R_X = 0$?

(b) If R_{SCOPE} is equal to $1\,M\Omega$, choose R_X so that the 10× function is realized at dc. What load resistance will be presented to the circuit in this case?

(c) If $C_{SCOPE} = 14\,pF$, draw the magnitude Bode plot of v_2 versus v_1 for the circuit of **Fig. P9.17(a)**. The high-frequency response should fall off at −20 dB per decade above a breakpoint related to R_{SCOPE}, R_X, and C_{SCOPE}.

(d) The high-frequency rolloff observed in part (c) can be eliminated by the addition of a compensating capacitor C_X, as shown in **Fig. P9.17(b)**. Choose C_X such that the frequency variations of v_2/v_1 are eliminated. Draw the Bode plot of the modified circuit. In an actual scope probe, R_X and C_X are typically located in a small box at the scope end of the probe cable. The capacitor C_X is adjustable so that the probe can be precisely compensated.

(e) Using the guidelines outlined in Problem 9.17, design a $1000\times$ oscilloscope probe that can be used to make measurements on high-voltage equipment. Assume the oscilloscope input to have input specifications $R_{SCOPE} = 1\,M\Omega$ and $C_{SCOPE} = 14\,pF$. If the voltage across R_{SCOPE} must be limited to 100 V, how large a voltage can be measured using your high-voltage probe? (*Note:* Do not attempt to actually build such a probe in the lab. High voltage can be dangerous!)

9.2.2 Bode-Plot Representation of System Functions of Arbitrary Complexity

9.18 Draw magnitude and angle Bode plots for the system function $H(j\omega) = 900(j\omega)(1 + j\omega/90)/[(1 + j\omega/5000)(1 + j\omega/10^5)]$.

9.19 Draw magnitude and angle Bode plots for the system function $H(j\omega) = 10(j\omega)^2(1 + j\omega/50)/[(1 + j\omega/2000)^2(1 + j\omega/10^6)]$.

9.20 Draw magnitude and angle Bode plots for the system function $H(j\omega) = 10/[(1 + j\omega/25)(1 + j\omega/2500)(1 + j\omega/2 \times 10^5)]$.

9.21 Draw magnitude and angle Bode plots for the system function $H(j\omega) = 200(1 + j\omega/90)^2/(1 + j\omega/5000)^2$.

9.22 Draw magnitude and angle Bode plots for the system function $H(j\omega) = 1/(1 + j\omega/100)^3$. What are the actual values of $|H|$ and $\measuredangle H$ at $\omega = 100$?

9.23 A circuit is required with a maximum gain of 500, a $-20\,dB/decade$ rolloff above $20\,kHz$, and a $+20\,dB/decade$ rolloff below $10\,Hz$. Specify an appropriate system function for the circuit.

9.24 A circuit is required with a maximum gain of 20, a $-40\,dB/decade$ rolloff above $1\,MHz$, a $+20\,dB/decade$ rolloff from $100\,Hz$ to $0.1\,Hz$, and a flat response for frequencies below $0.1\,Hz$. Specify an appropriate system function for the circuit. What is the magnitude of such a system function at dc ($\omega = 0$)?

9.25 ○ A circuit is required with a flat response of value $0\,dB$ from dc to $500\,Hz$, a $+20\,dB/decade$ slope from $500\,Hz$ to $5\,kHz$, a flat response from $5\,kHz$ to $20\,kHz$, a $+20\,dB/decade$ slope from $20\,kHz$ to $1\,MHz$, a flat response from $1\,MHz$ to $100\,MHz$, and a $-60\,dB/decade$ rolloff above $100\,MHz$. Specify an appropriate system function for the circuit and draw its magnitude Bode plot. What is the magnitude of the system function over each flat region?

9.2.3 High-, Low-, and Midband-Frequency Limits

9.26 A circuit has the system function $H(j\omega) = 10(j\omega)/[(1+j\omega/50)(1+j\omega/2000)(1+j\omega/4000)]$. Using the direct calculation method, find the exact value of the -3-dB point caused by the closely spaced poles at $2\,krad/s$ and $4\,krad/s$.

9.27 A circuit has the system function $H(j\omega) = 10/[(1+j\omega/50)(1 + j\omega/200)(1 + j\omega/10^5)]$. Using the direct calculation method, find the exact value of the -3-dB point caused by the closely spaced poles at $50\,rad/s$ and $200\,rad/s$.

9.28 A system function has a constant gain of 5 at dc, a zero in its system function at $f = 8\,Hz$, and poles at $100\,Hz$, $20\,kHz$, and $1\,MHz$. Find the exact values of the upper and lower -3-dB endpoints of the midband region by direct calculation. What is the bandwidth of the response?

9.29 A system function has a flat 0-dB midband region that extends to zero frequency and four coincident poles at $f = 50\,kHz$.

(a) Draw the Bode plot of the response and find the high-frequency -3-dB point.

(b) Find the actual magnitude of the response at $50\,kHz$.

9.30 Prove that a voltage transfer function with a midband region and zeros only (no poles) cannot be realized by an actual physical circuit.

9.2.4 Superposition of Poles

9.31 Use the superposition-of-poles approximation to estimate the upper -3-dB endpoint ω_H of the system function $H(j\omega) = 500(j\omega)/[(1 + j\omega/75)(1 + j\omega/5000)(1 + j\omega/10^4)]$.

9.32 Use the superposition-of-poles approximation to estimate the upper -3-dB endpoint ω_H of the system function $H(j\omega) = 20/[(1+j\omega/4000)(1+j\omega/5000)(1+j\omega/6000)]$.

9.33 Use the superposition-of-poles approximation to estimate the lower -3-dB endpoint ω_L of the system function $H(j\omega) = 100(j\omega)/[(1+j\omega/20)(1+j\omega/50)(1+j\omega/10^4)]$.

9.34 Use the superposition-of-poles approximation to estimate the upper and lower -3-dB endpoints ω_L and ω_H of the system function $H(j\omega) = 100(j\omega)^2/[(1 + j\omega/20)(1+j\omega/50)(1+j\omega/75)(1+j\omega/10^4)(1+j\omega/2 \times 10^4)]$.

9.35 Use the superposition-of-poles approximation to estimate the upper and lower -3-dB endpoints ω_H and ω_L of the system function $H(j\omega) = 500(j\omega)(1 + j\omega/5)/[(1 + j\omega/75)(1 + j\omega/200)(1 + j\omega/5000)(1 + j\omega/10^4)]$.

9.36 ○ Show that in the worst case where two poles coincide exactly at the upper end of the midband, the superposition-of-poles approximation will yield a maximum error of 22% in the actual value of the -3-dB frequency ω_H. Similarly, show that when two poles coincide at the lower end of the midband, superposition of poles will yield a maximum error of 29% in the value of the low-frequency -3-dB frequency ω_L.

9.3 Frequency Response of Circuits Containing Capacitors

9.3.1 High- and Low-Frequency Capacitors

9.37 For each of the circuits of **Fig. P9.9**, determine which capacitors are low-frequency and which are high-frequency types. Repeat for the circuit of **Fig. P9.12**.

9.38 Consider the circuit of Fig. 9.20 with C_C replaced by a short circuit. Draw the magnitude Bode plot of the gain over the low- to midband-frequency range if $\beta_o = 100$.

9.39 Consider the circuit of Fig. 9.20 with $R_B = 1\,\mathrm{M}\Omega$, $R_C = 1\,\mathrm{k}\Omega$, and $C_S = 10\,\mu\mathrm{F}$. If $r_\pi = 5\,\mathrm{k}\Omega$, $C_\pi = 5\,\mathrm{pF}$, $C_\mu \approx 0$, $r_x = 10\,\Omega$, and $r_o = 100\,\mathrm{k}\Omega$,

(a) Draw two small-signal models for the amplifier, one appropriate for low-frequency analysis and the other for high-frequency analysis. Which capacitors dominate the low- and high-frequency responses?

(b) Set C_π and C_L to open circuits and find the system function for $\mathbf{V}_\pi/\mathbf{V}_s$ at low frequencies.

(c) Set C_S to a short circuit and find the system function for $\mathbf{V}_\pi/\mathbf{V}_s$ at high frequencies.

(d) Draw a magnitude Bode plot for the voltage ratio $\mathbf{V}_\pi/\mathbf{V}_{\mathrm{in}}$.

(e) Find the low- and high-frequency -3-dB midband endpoints of the amplifier.

9.40 Consider the circuit of Fig. 9.20 with a MOSFET substituted for the BJT and a second resistor R_A connected between the gate of the MOSFET and ground, where $R_A = R_B = 500\,\mathrm{k}\Omega$. Ignore the effects of any internal device capacitances, and assume that Q_1 is biased in its constant-current region with $K = 1\,\mathrm{mA/V}^2$

and $I_D = 5\,\mathrm{mA}$. Draw the Bode plot of $|\mathbf{V}_{\mathrm{out}}/\mathbf{V}_{\mathrm{in}}|$ as a function of frequency with and without load elements R_L and C_L connected. What is the midband gain of the amplifier?

9.41 Consider the circuit described in Problem 9.40. If $C_{gs} = 3\,\mathrm{pF}$, $C_{gd} \approx 0$, and $C_L = 2\,\mathrm{pF}$, find the high-frequency -3-dB point of the amplifier. How large can C_L become before it begins to degrade the high-frequency performance?

9.3.2 The Dominant-Pole Concept

9.42 ⚡ An *npn* BJT is connected in the circuit of Fig. 9.76 with $V_{CC} = 15\,\mathrm{V}$, $V_{EE} = -15\,\mathrm{V}$, $R_S = 2\,\mathrm{k}\Omega$, $R_L = 10\,\mathrm{k}\Omega$, and $C_S = 1\,\mu\mathrm{F}$. The BJT has a β of 50.

(a) Use an approximate bias-calculation method (e.g., ignore I_B compared to I_1 and I_C) to choose values of R_1 through R_4 so that $I_C = 2\,\mathrm{mA}$. The bias voltage of the BJT emitter should be set to zero, midway between V_{CC} and V_{EE}.

(b) Find the midband gain and small-signal input and output resistances of the amplifier. Load resistor R_L is considered part of the amplifier.

(c) Find the low-frequency breakpoint due to C_S.

(d) Find C_E such that the emitter is bypassed to signal ground for all frequencies above $20\,\mathrm{Hz}$.

9.43 ⚡ ○ An *npn* BJT is connected in the feedback-bias configuration of Fig. 9.76 with $R_S = 0$, $R_L = 10\,\mathrm{k}\Omega$, $V_{CC} = 12\,\mathrm{V}$, and $V_{EE} = 0$.

(a) Choose values of R_1 through R_4 so that $I_C \approx 1.4\,\mathrm{mA}$ and the midband-gain is about -200 if β_F lies somewhere in the range 50 to 200. Compute values for g_m and r_π.

(b) Choose capacitor values so that the breakpoint frequency of C_S occurs at $0.5\,\mathrm{Hz}$ and the breakpoint frequency of C_C at $5\,\mathrm{Hz}$. Choose C_E so that the emitter is effectively bypassed to ground at about $60\,\mathrm{Hz}$.

(c) Draw the magnitude Bode plot of the amplifier gain.

(d) Modify your choice of capacitors so that only standard 20% values are used. Recalculate the modified breakpoint frequencies of the Bode plot.

9.44 ⚡ ◐ A *pnp* BJT is to be connected into a circuit similar to that of Fig. 9.76. If $V_{CC} = 10\,\mathrm{V}$, $V_{EE} = 0$, and $R_S = 50\,\Omega$, choose a suitable topology and resistor and capacitor values such that the amplifier has a low-frequency -3-dB endpoint at about $65\,\mathrm{Hz}$. Assume Q_1 to have a β_F somewhere in the range 50 to 200.

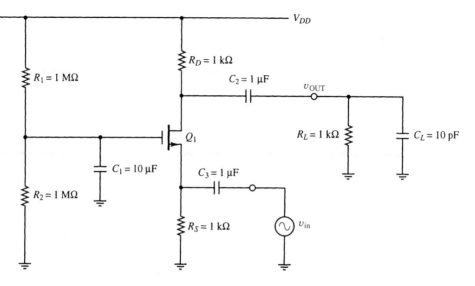

Figure P9.48

9.45 Consider the BJT amplifier of Fig. 9.30 with $R_1 = 200\,\text{k}\Omega$, $R_2 = 100\,\text{k}\Omega$, $C_E = 10\,\mu\text{F}$, and $C_S = 100\,\mu\text{F}$.

(a) Compute the contribution of each low-frequency capacitor to the -3-dB point of the amplifier.

(b) Identify the dominant low-frequency capacitor.

(c) Show that the true pole of the nondominant low-frequency capacitor is erroneously computed using the dominant-pole technique because the dominant capacitor actually behaves as an open circuit at the frequency of the nondominant pole.

9.46 ⚙ ◐ An n-channel enhancement-mode MOSFET is connected in the circuit configuration of Fig. 9.76. If $V_{CC} = 10\,\text{V}$, $V_{EE} = 0$, and $R_S = 50\,\Omega$, choose resistor and capacitor values such that the MOSFET is biased in the constant-current region and the amplifier has a low-frequency -3-dB endpoint at about 100 Hz. Assume Q_1 to have parameters $K = 0.5\,\text{mA}/\text{V}^2$ and $V_{TR} = 2\,\text{V}$.

9.47 ⚙ ◐ Using the feedback-bias configuration, design a MOSFET amplifier with a midband gain of at least 10 and a midband region that extends from no more than 100 Hz to at least 1 MHz. The MOSFET has manufacturer-specified parameters $C_{gs} \approx 10\,\text{pF}$, $C_{gd} \approx 0$, $C_{ds} \approx 1\,\text{pF}$, $0.1 < K < 0.8\,\text{mA}/\text{V}^2$, and $0.5\,\text{V} < V_{TR} < 2\,\text{V}$.

9.48 In the circuit of **Fig. P9.48**, capacitor C_1 grounds the gate of the MOSFET with respect to signals and places the transistor in the current-follower configuration. Estimate the midband gain and the high- and low-frequency -3-dB endpoints of the midband using the dominant-pole

technique if $I_D = 2\,\text{mA}$, $K = 1\,\text{mA}/\text{V}^2$, $C_{gs} = 2\,\text{pF}$, $C_{gd} = 0.5\,\text{pF}$, and $C_{ds} \approx 0$.

9.49 ⚙ ○ Design a BJT amplifier that can be used as part of an old-style phonograph and audio amplifier system. Your amplifier should have a gain of about 100, and a bandwidth limited to the human hearing range of 20 Hz to 20 kHz.

9.50 ⚙ ● Design a BJT amplifier that can be used with a wire coil to measure the ac magnetic fields from power lines. When used to intercept the field to be measured, the coil produces an ac voltage of magnitude 0 to 10 mV at the same frequency as the exciting field. The output of the amplifier must drive a simple half-wave diode rectifier and needle-type mechanical meter movement that requires 1 mA of current for full-scale deflection. Since only power line frequencies and their nearby harmonics are of interest, the amplifier must have a bandwidth limited to the range 50 Hz to 1 kHz.

9.51 ⚙ ◐ Design a MOSFET amplifier with a midband gain of 10 that can be used as part of a multiplexed telephone amplification system. Since only the primary voice channel is to be amplified, the midband region of your amplifier should be confined to the range 200 Hz to 5 kHz. Specify values for the MOSFET parameters.

9.3.3 Effect of Transverse Capacitance on Amplifier Response

9.52 A BJT with $\beta = 100$ and $r_x = 50\,\Omega$ is connected in the circuit of Fig. 9.76 with $V_{CC} = 12\,\text{V}$, $V_{EE} = 0$, $R_1 = 100\,\text{k}\Omega$, $R_2 = 50\,\text{k}\Omega$, $R_3 = 6\,\text{k}\Omega$, $R_4 = 3.3\,\text{k}\Omega$,

$R_S = 4\,\text{k}\Omega$, and $R_L = 4\,\text{k}\Omega$. Assume that all low-frequency capacitors behave as short circuits in and above the midband. Draw the high-frequency magnitude Bode plot of $\mathbf{V}_{\text{out}}/\mathbf{V}_s$ if $C_\pi = 14\,\text{pF}$ and $C_\mu = 2\,\text{pF}$.

9.53 🔊 ○ An n-channel enhancement-mode MOSFET with parameters $K = 0.1\,\text{mA/V}^2$ and $V_{\text{TR}} = 2\,\text{V}$ is connected in the circuit of Fig. 9.76. The MOSFET's internal capacitances are equal to $C_{gs} = 3\,\text{pF}$, $C_{ds} = 0.5\,\text{pF}$, and $C_{gd} = 2.5\,\text{pF}$. The value of r_o for the MOSFET is large enough to be considered infinite.

(a) Choose values for R_1 through R_4 such that the MOSFET is biased with $I_D \approx 1\,\text{mA}$ when $V_{CC} = 20\,\text{V}$ and $V_{EE} = 0$. What is the MOSFET transconductance g_m at this bias current?

(b) Determine the high-frequency -3-dB endpoint of the midband region. Modify your choices in part (a) as required so that the midband extends to at least 10 MHz.

9.54 🔊 ◐ Design a dc-coupled, capacitor-free BJT amplifier using the feedback-bias scheme of Fig. 9.76. In order to achieve dc coupling, v_s must be connected directly to the base of Q_1. The BJT has parameters $f_T = 400\,\text{MHz}$, $C_\mu = 1\,\text{pF}$, and $r_x = 50\,\Omega$. The circuit should be capable of driving a 100-pF load capacitance in parallel with a 1-MΩ load resistance up to a frequency of 200 kHz. Choose appropriate values for R_1, R_2, R_3, R_4, V_{CC}, and V_{EE}.

9.55 🔊 ◐ Design a dc-coupled, capacitor-free MOSFET amplifier using the feedback-bias scheme of Fig. 9.76. In order to achieve dc coupling, v_s must be connected directly to the base of Q_1. The MOSFET has parameters $C_{gs} = 2.5\,\text{pF}$, $C_{gd} = 0.5\,\text{pF}$, and $C_{ds} = 0.1\,\text{pF}$. The circuit should be capable of driving a 1-nF load capacitance in parallel with a 1-MΩ load resistance up to a frequency of at least 100 kHz. Choose appropriate values for all resistors and low-frequency capacitors in the circuit.

9.56 An npn BJT is connected in the circuit of Fig. 9.76. The poles of C_S, C_E, and C_C are to be set to 0.5 Hz, 20 Hz, and 2 Hz, respectively. If $R_1 = 47\,\text{k}\Omega$, $R_2 = 4.7\,\text{k}\Omega$, $R_3 = 5.6\,\text{k}\Omega$, $R_4 = 1\,\text{k}\Omega$, and $R_S = 50\,\Omega$, choose values for C_S, C_C, and C_E such that these poles are realized for $R_L = 10\,\text{k}\Omega$.

9.3.4 Miller's Theorem and Miller Multiplication

9.57 A 10-kΩ transverse capacitance of value $1\,\mu\text{F}$ is driven by a 1-kHz sinusoidal voltage source of amplitude $\mathbf{V}_1 = 10\,\text{V}$ to ground from the left-hand side and a sinusoidal source of amplitude $\mathbf{V}_2 = -20\,\text{V}$ to ground

from the right-hand side ($\mathbf{V}_2$ is 180° out of phase with respect to $\mathbf{V}_1$). Apply Miller's theorem to find a single shunt-capacitor equivalent for the circuit seen by the $\mathbf{V}_1$ source. What happens if the amplitude of $\mathbf{V}_2$ is changed to $+20\,\text{V}$ (in phase with $\mathbf{V}_1$)?

9.58 Consider the MOSFET amplifier of Fig. 9.27. Suppose that a *resistance* of value $R_f = 100\,\text{k}\Omega$ is connected from the gate of the MOSFET to its drain. Use Miller's theorem to determine the input resistance of the modified amplifier in the midband.

9.59 Find the value of the equivalent Miller capacitance C_A for the circuit of Fig. 9.33 if $C_\mu = 0.5\,\text{pF}$, $C_\pi = 8\,\text{pF}$, and $g_m = 150\,\text{mA/V}$. All resistor values are as indicated in the figure. What is the pole frequency for the voltage ratio $\mathbf{V}_\pi/\mathbf{V}_{\text{in}}$?

9.60 Compute the dominant-pole frequency of the circuit of Example 9.7 if $f_T = 600\,\text{MHz}$.

9.61 In the MOSFET circuit of Fig. 9.27, assume that the dominant high-frequency pole occurs in the input loop of the circuit's small-signal model. Use Miller's theorem to find the value of the pole if $C_{gs} = 7\,\text{pF}$ and $C_{gd} = 1\,\text{pF}$. Assume the MOSFET to have parameters $K = 1\,\text{mA/V}^2$ and $V_{\text{TR}} = 4\,\text{V}$.

9.62 In Example 9.8, Miller's theorem was used to estimate the upper -3-dB point of the amplifier of Fig. 9.30. In Example 9.7, the endpoint was found using the Thévenin-resistance method. In this problem, you will show that the endpoint computed by the former method can be made to more closely match the latter if the effect of C_μ on the ratio $\mathbf{V}_{\text{out}}/\mathbf{V}_\pi$ is taken into account.

(a) Use Miller's theorem to model the effect of C_μ as seen by the $g_m v_\pi$ source near the pole frequency of $(C_\pi + C_A)$. Specifically, determine the value of an equivalent parallel capacitance C_B on the output side of the amplifier by shorting the base of the BJT to ground and applying Miller's theorem while exciting the output port of the amplifier with a test source.

(b) Now use superposition of poles to include the contribution of C_B to the upper -3-dB midband endpoint. Compare the result to that obtained in Example 9.7.

As explained in the text, Miller's theorem cannot be used to evaluate the higher-order poles of the amplifier. It is only useful in assessing the effect of C_μ on the ratio $\mathbf{V}_{\text{out}}/\mathbf{V}_\pi$ when computing the value of the dominant Miller input capacitance C_A.

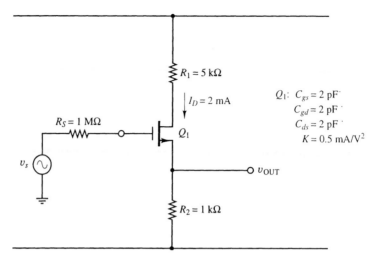

Figure P9.63

9.3.5 High-Frequency Poles with Feedback Resistor

9.63 Find the upper -3-dB point of the MOSFET amplifier of **Fig. P9.63** if Q_1 has parameters $C_{gs} = C_{gd} = C_{ds} = 2\,\text{pF}$ and $K = 0.5\,\text{mA/V}^2$.

9.64 In this problem, you will show how the gain and bandwidth of an amplifier vary with feedback resistance. Consider the small-signal amplifier model of Fig. 9.38 with transistor parameters $C_\mu = 2\,\text{pF}$, $f_T = 400\,\text{MHz}$, and $r_x = 10\,\Omega$. For the purpose of illustration, assume that $\beta_o = 100$ and $I_C = 2.5\,\text{mA}$, as in Example 9.10, so that $g_m = 100\,\text{mA/V}$ and $r_\pi = 1\,\text{k}\Omega$. Compute the high-frequency -3-dB point and the midband gain for values of R_E in the range $0 < R_E < 10\,\text{k}\Omega$. For the purpose of

illustration, ignore the change in bias current (and g_m) that occurs as R_E is changed. Draw the magnitude Bode plot of the amplifier for each value of R_E selected and compare the product of midband gain and -3-dB frequency for each plot.

9.65 ◑ Determine the frequency-response characteristics of the circuit of **Fig. P9.65**.

9.66 ⬛ ○ Design a BJT voltage follower in the feedback-bias configuration for which the upper -3-dB frequency lies at 1 MHz. Transistors are available with parameters $C_\mu = 2\,\text{pF}$, $f_T = 500\,\text{MHz}$, and $r_x = 50\,\Omega$, with $50 < \beta_o < 150$.

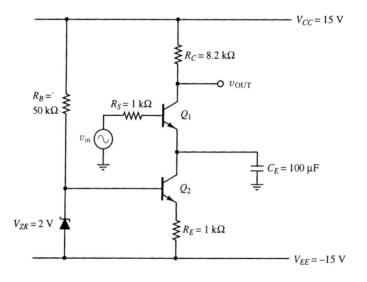

Figure P9.65

9.67 ⅀ ○ Design an *n*-channel enhancement-mode MOSFET inverter in the feedback-bias configuration for which the gain is −4 and the upper −3-dB frequency lies at about 400 kHz. Transistors are available with parameters $C_{gs} = 1.5\,\text{pF}$, $C_{gd} = 1\,\text{pF}$, $C_{ds} = 0.2\,\text{pF}$, and $r_x = 10\,\Omega$, with $V_{TR} = 2\,\text{V}$ and $0.5\,\text{mA/V}^2 < K < 2\,\text{mA/V}^2$.

9.3.6 Frequency Response with Bypass Capacitor

9.68 Consider the circuit of Fig. 9.47 with $\beta_o = 90$. Draw the low-frequency end of the Bode plot of $\mathbf{V}_{\text{out}}/\mathbf{V}_s$ if $R_C = 2.2\,\text{k}\Omega$, $C_S = 150\,\mu\text{F}$, and $C_E = 10\,\mu\text{F}$.

9.69 ⅀ For the amplifier shown in Fig. 9.47, choose values for C_S and C_E such that the low-frequency −3-dB point of the midband is no higher than 20 Hz. Assume that β of the BJT is unknown but lies somewhere between 50 and 200.

9.70 A BJT is connected in the circuit of Fig. 9.76 with $R_S = 1\,\text{k}\Omega$, $R_2 = 5\,\text{k}\Omega$, $R_3 = 2\,\text{k}\Omega$, $R_4 = 500\,\Omega$, $C_S = 100\,\mu\text{F}$ and $C_E = 10\,\mu\text{F}$. Suppose that C_C is replaced by a short circuit and R_L by a capacitance of value $C_L = 4\,\text{nF}$.

(a) If $V_{CC} = 10\,\text{V}$ and $V_{EE} = 0$, find an approximate expression for I_C. What value of R_1 will yield the value $I_C = 2.8\,\text{mA}$ if $V_f = 0.6\,\text{V}$?

(b) Draw a small-signal-circuit model valid in the midband region and find an expression for the midband-gain.

(c) Draw a Bode plot that shows the midband and low-frequency behavior of the amplifier.

9.71 An *n*-channel depletion-mode MOSFET is connected in the circuit of Fig. 9.76 with $V_{CC} = 20\,\text{V}$, $V_{EE} = 0$, $C_S = 0.1\,\mu\text{F}$, $C_E = 10\,\mu\text{F}$, $R_1 = 1.4\,\text{M}\Omega$, $R_2 = 0.6\,\text{M}\Omega$, $R_3 = 5\,\text{k}\Omega$, $R_4 = 3.5\,\text{k}\Omega$, and $R_S = 100\,\text{k}\Omega$. The load components C_C and R_L are absent from the circuit. The MOSFET has parameters $V_{TR} = -2\,\text{V}$, $K = 2\,\text{mA/V}^2$, $C_{gs} = 3\,\text{pF}$, and $C_{gd} = C_{ds} = 1\,\text{pF}$. Draw the magnitude Bode plot of the amplifier gain including the midband-, low-, and high-frequency regions.

9.72 A JFET is connected to the circuit of Fig. 9.76 with $R_S = 100\,\text{k}\Omega$, $R_1 = \infty$, $R_2 = 1\,\text{M}\Omega$, $R_3 = 3\,\text{k}\Omega$, $R_4 = 1\,\text{k}\Omega$, $C_S = 1\,\mu\text{F}$, and $C_E = 100\,\mu\text{F}$. The load elements C_C and R_L are absent from the circuit. The JFET has parameters $I_{DSS} = 8\,\text{mA}$, $V_P = -4\,\text{V}$, $C_{gs} = 5\,\text{pF}$, $C_{gd} = 5\,\text{pF}$, $C_{ds} \approx 0$, and $r_o \approx \infty$.

(a) Find I_D and V_{DS} if $V_{CC} = 15\,\text{V}$ and $V_{EE} = 0$.

(b) Draw a small-signal model of the amplifier valid in the midband. What is the value of the small-signal JFET parameter g_m?

(c) Find and evaluate an expression for the midband-gain.

(d) As the driving frequency is increased or decreased from the midband region, at what frequencies does the gain fall by −3 dB from the midband value?

9.73 An *n*-channel enhancement-mode MOSFET is connected in the circuit of Fig. 9.76 with $C_S = 10\,\mu\text{F}$, $C_E = 1\,\mu\text{F}$, $R_1 = R_2 = 2\,\text{M}\Omega$, $R_3 = 2.5\,\text{k}\Omega$, $R_4 = 1\,\text{k}\Omega$, and $R_S = 50\,\Omega$. The load components C_C and R_L are absent from the circuit. If $g_m = 4\,\text{mA/V}$ at the bias point of Q_1, draw the Bode plot of the amplifier gain up to and including the midband region.

9.74 ○ A BJT with $\beta_o = 100$ and $r_x = 50\,\Omega$ is connected in the circuit of Fig. 9.76 with $V_{CC} = 12\,\text{V}$, $V_{EE} = 0$, $R_1 = 80\,\text{k}\Omega$, $R_2 = 40\,\text{k}\Omega$, $R_3 = 6\,\text{k}\Omega$, $R_4 = 3.3\,\text{k}\Omega$, $R_S = 4\,\text{k}\Omega$, and $R_L = 4\,\text{k}\Omega$. The capacitors have values $C_S = C_C = 1\,\mu\text{F}$ and $C_E = 10\,\mu\text{F}$.

(a) With C_E disconnected and C_S assumed to behave as a short, what is the small-signal Thévenin-resistance seen between the emitter lead of Q_1 and ground?

(b) At what frequency does C_E completely bypass the emitter of Q_1 to ground with respect to incremental signals?

(c) Draw the magnitude and angle Bode plots of $\mathbf{V}_{\text{out}}/\mathbf{V}_s$. Include the effects of the high-frequency capacitors $C_\pi = 14\,\text{pF}$ and $C_\mu = 2\,\text{pF}$ as well as the low-frequency capacitors C_S, C_C, and C_E.

9.75 Find values for the low-frequency zero and pole of the depletion-mode MOSFET inverter with bypass capacitor shown in Fig. 9.50 if $K = 0.8\,\text{mA/V}^2$, $V_{TR} = -1\,\text{V}$, $R_D = 4.7\,\text{k}\Omega$, $R_E = 500\,\Omega$, and $C_E = 100\,\mu\text{F}$.

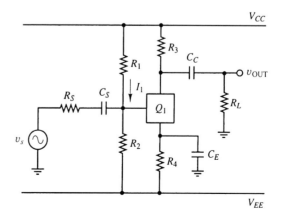

Fig. 9.76

9.76 ⅀ ○ Design a circuit based on the BJT for which the midband-gain magnitude is at least 100, the dc-gain magnitude is no more than 10, and the midband region begins at a frequency no higher than about 10 Hz nor lower than 1 Hz.

9.77 ⅀ ○ Design a circuit based on the enhancement-mode MOSFET for which the midband-gain magnitude is at least 15, the dc-gain magnitude is no more than 2, and the midband region begins at a frequency no higher than about 50 Hz nor lower than 10 Hz.

9.4 Frequency Response of the Differential Amplifier

9.4.1 Differential-Mode Frequency Response

9.78 A BJT differential amplifier of the type shown in Fig. 9.51 is fabricated with BJTs having parameters $\beta = 130$, $f_T = 500$ MHz, $C_\mu = 0.2$ pF, and $r_x = 30\,\Omega$. Suppose that v_1 and v_2 are connected via series resistors of value $R_S = 50\,\Omega$. If $R_{C1} = R_{C2} = 10\,\text{k}\Omega$, find the dc gain and upper -3-dB frequency of the amplifier under differential-mode excitation for $I_o = 2$ mA.

9.79 ○ Consider the BJT differential amplifier of Fig. 9.51, but with resistors R_{E1} and R_{E2} connected between the emitter terminals and node E.

(a) Derive an expression for the midband-gain v_{OUT1}/v_{idm}.

(b) Evaluate the expression of part (a) for $I_o = 5$ mA, $R_{C1} = R_{C2} = 4.7\,\text{k}\Omega$, $R_{E1} = R_{E2} = 1\,\text{k}\Omega$, and $\beta_o = 150$. What is the bias voltage of node E?

(c) Determine the high-frequency -3-dB point of the amplifier under differential-mode excitation if $f_T = 500$ MHz, $C_\mu = 0.5$ pF, and $r_x = 45\,\Omega$.

9.80 An NMOS differential-amplifier is shown in **Fig. P9.80**. Find the midband gain and -3-dB frequency if $I_o = 2$ mA, $Q_A = Q_B$, and the other MOSFETS have parameters $K_1 = K_2 = 5$ mA/V^2, $K_3 = K_4 = 0.5 = $ mA/V^2, $C_{gs1} = C_{gs2} = 25$ pF, $C_{gd1} = C_{gd2} = 5$ pF, $C_{gs3} = C_{gs4} = 1$ pF, $C_{gd3} = C_{gd4} = 0.2$ pF, and $C_{ds} \approx$ 0 for all transistors. Ignore the body effect (assume that the substrate of each MOSFET is connected to its source).

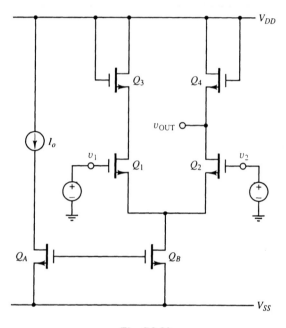

Fig. P9.80

9.81 ○ Determine the differential-mode response of the CMOS amplifier of **Fig. P9.81** if $I_o = 2$ mA, and if $|K| = 1$ mA/V^2, $r_o = 30\,\text{k}\Omega$, $C_{gs} = 2$ pF, and $C_{gd} = 1$ pF for all devices.

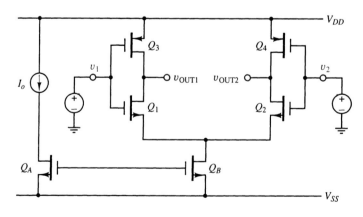

Figure P9.81

9.82 Consider the JFET differential amplifier of Fig. 8.33 (Chapter 8). Find the differential-mode high-frequency -3-dB point if $C_{gs} = 2\,\text{pF}$, $C_{gd} = 1\,\text{pF}$, and $C_{ds} \approx 0$. In this case, input sources v_1 and v_2 are connected to the gates of Q_1 and Q_2 via 50-Ω series resistances.

9.4.2 Common-Mode Frequency Response

9.83 Determine the common-mode gain and lowest common-mode breakpoint frequency of the BJT differential amplifier of Fig. 9.51 if $R_{C1} = R_{C2} = 5\,\text{k}\Omega$, $I_o = 2\,\text{mA}$, $r_n = 50\,\text{k}\Omega$, and a capacitance $C_n = 2\,\text{pF}$ appears in parallel with r_n.

9.84 ○ A BJT differential amplifier of the type shown in Fig. 9.51 is fabricated with $R_{C1} = R_{C2} = 5\,\text{k}\Omega$ and $I_o = 4\,\text{mA}$ using BJTs with parameters $\beta = 170$, $f_T = 450\,\text{MHz}$, $C_\mu = 1.5\,\text{pF}$, $r_x = 30\,\Omega$, and $r_n = 1\,\text{M}\Omega$. A small-signal capacitance $C_n = 5\,\text{pF}$ also appears in parallel with r_n. Suppose that the v_2 input is connected to ground. Determine the magnitude of the output $\mathbf{V}_{\text{out}}/\mathbf{V}_{\text{in}}$ as a function of frequency.

9.85 ◑ A BJT differential amplifier of the type shown in Fig. 9.51 is fabricated with $R_{C1} = 10\,\text{k}\Omega$, $R_{C2} = 0$, and $I_o = 1\,\text{mA}$. The BJTs have parameters $\beta = 120$, $f_T = 650\,\text{MHz}$, $C_\mu = 0.5\,\text{pF}$, $r_x = 50\,\Omega$, and $r_n = 1\,\text{M}\Omega$. A small-signal capacitance $C_n = 5\,\text{pF}$ also appears in parallel with r_n. If the v_2 input is grounded, find the magnitude of the output $\mathbf{V}_{\text{out}}/\mathbf{V}_{\text{in}}$ as a function of frequency.

9.86 ◑ Consider the differential amplifier of Fig. 9.51 with $R_{C1} = R_{C2} = 1\,\text{k}\Omega$, $I_o = 10\,\text{mA}$, $r_n = 100\,\text{k}\Omega$,

$\beta_o = 100$, $C_\mu = 1\,\text{pF}$, $f_T = 400\,\text{MHz}$, $r_x = 50\,\Omega$, and $r_o = 100\,\text{k}\Omega$. Input signals equal to

$$v_1 = 0.01 \sin \omega_x t + 0.005 \cos \omega_x t$$
$$= 0.011 \cos(\omega_x t - 0.35\pi)$$

and $\quad v_2 = 0.01 \sin \omega_x t - 0.005 \cos \omega_x t$
$$= 0.011 \cos(\omega_x t - 0.65\pi)$$

are applied to the amplifier. If $f_x = 8\,\text{MHz}$, find the resulting output voltage v_{out2}.

9.87 ◐ Determine the differential- and common-mode responses of the MOSFET amplifier of **Fig. P9.80** if the MOSFETs have dimensions $W_1 = W_3 = 80\,\mu\text{m}$ and $W_2 = W_4 = 10\,\mu\text{m}$, with $L = 10\,\mu\text{m}$. The MOSFET transconductance parameters are given by $K = K_o(W/L)$, where $K_o = 1\,\text{mA/V}^2$, and the MOSFET capacitances are given by $C_{gs} = C_{gso}(WL)$ and $C_{gd} = C_{gdo}(WL)$, where $C_{gso} = 0.05\,\text{pF}/\mu\text{m}^2$ and $C_{gdo} = 0.02\,\text{pF}/\mu\text{m}^2$. Assume Q_B to have an r_o of $20\,\text{k}\Omega$, and ignore the body effect.

9.4.3 Frequency Response of the Cascode Configuration

9.88 ∑ ◐ The common-base cascode amplifier of **Fig. P9.88** is formed by two BJTs connected in the current-follower configuration, whereby Q_1 serves as the current source driving Q_2. The circuit is often used in applications where the signal-source resistance R_S is small and must be matched by an amplifier input resistance of equally small value.

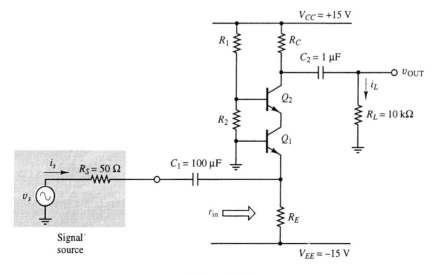

Figure P9.88

(a) Draw the small-signal model of the amplifier in the midband region. Show that the small-signal input resistance r_{in} is equal to $R_E \| [r_{\pi 1}/(\beta_{o1} + 1)]$.

(b) Find the current I_C required to set r_{in} to 50 Ω. It may be helpful to use the approximation $R_E \gg r_{\pi 1}/(\beta_{o1} + 1)$.

(c) Find values for R_E, R_C, R_1, and R_2 such that V_{CE1} and V_{CE2} are both set to about 5 V at the value of I_C found in part (b).

(d) Find the midband-gain v_{out}/v_s.

(e) Find the low-frequency -3-dB endpoint of the midband voltage gain.

(f) If $f_T = 200$ MHz, $r_x \approx 0$, and $C_\mu = 2$ pF, estimate the high-frequency -3-dB point of the midband voltage gain.

9.89 ○ Draw the differential-mode Bode plot (magnitude and angle) for the circuit of Fig. 9.54.

9.90 Consider the small-signal half-circuit of the cascode amplifier shown in Fig. 9.56. Find an expression for the small-signal Thévenin resistance seen by r_{o3} of Q_3 under differential-mode excitation. This expression may be used to assess the conditions under which r_{o3} can be neglected in evaluating amplifier performance.

9.91 ● The circuit of **Fig. P9.91** illustrates a two-BJT version of the *Darlington* connection. Cascading the two BJTs in series results in a very large effective β.

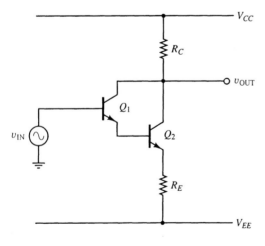

Fig. P9.91

(a) Analyze the circuit and find expressions for the midband gain and the dominant high-frequency pole of the amplifier.

(b) In an alternative connection, the collector of Q_1 is connected to V_{CC}, rather than to the collector of Q_2. Find expressions for the midband gain and the dominant high-frequency pole for this revised connection and compare to part (a).

9.5 Time Response of Electronic Circuits

9.92 A step-voltage source of magnitude $V_o = 5$ V feeds a 10-kΩ load resistor in parallel with a 10-μF capacitor via a 5-kΩ series resistor. Sketch the load voltage as a function of time.

9.93 Find the response of each of the circuits of **Fig. P9.9** to a 5-V step input.

9.94 (a) Find the approximate output of the circuit of **Fig. P9.12** if the input is a 5-V step function. Use engineering approximations where appropriate.

(b) Choose new element values for the circuit, as needed, such that the output droops by no more than 0.1 V for $0 < t < 1$ s.

9.95 ⚡ Redesign the circuit of **Fig. P9.12** so that the rising waveform in response to a step input has a time constant no longer than 1 ns, and the output decays to no more than 10% of its initial stepped value after 100 ms.

9.96 Find the step response of the circuit of **Fig. P9.12** if C_1 is replaced by an inductor of value $L_1 = 10$ mH and if C_2 is removed from the circuit.

9.97 Find the step response of the circuit of **Fig. P9.12** if C_2 is replaced by an inductor of value $L_1 = 100 \mu$H and if C_1 is replaced by a short circuit.

9.98 ○ Consider the diode circuit of Fig. 9.59 with $R_L = 100$ kΩ, $k_j = 400$ pF-V$^{1/2}$, $\Psi_o = 0.9$ V, and $r_d = 10 \Omega$. The input v_{IN} consists of a ± 5-V, 100-Hz square wave.

(a) Approximate the diode depletion capacitance as being constant and estimate its value.

(b) Model the diode diffusion capacitance as a 10-pF capacitor connected across both V_f and r_d in Fig. 9.59(b). Plot the transitions of v_{OUT} as functions of time for both forward- and reverse-biased conditions.

9.99 ○ A bridge rectifier is made from diodes having parameters $k_j = 400$ pF-V$^{1/2}$, $\Psi_o = 0.9$ V, and $r_d = 0$. The bridge is driven by a 10-V peak, 100-Hz square wave and drives a 1-MΩ load. Plot the load voltage versus time. Approximate the diode depletion capacitances as being constant.

9.100 Consider the diode circuit of Fig. 9.59. The undesirable negative peak under reverse-bias diode conditions, observed in the waveform of Fig. 9.60, can be reduced by connecting a small capacitance in parallel with the load R_L. For the element values specified in the analysis of Section 9.5.1, estimate the value of C_L required to reduce the negative peak by half. What is the disadvantage of this method of reducing the reverse-bias voltage peak?

9.101 Find v_{out} in the circuit of Example 9.13 if v_{in} has a magnitude of $\pm 15\,\text{mV}$ and a period of 50 ms, and if $C_S = 3.3\,\mu\text{F}$.

9.102 Find v_{out} in the circuit of Example 9.13 if v_{in} has a magnitude of $\pm 5\,\text{mV}$ and a period of 1 ms, and if $R_B = 220\,\text{k}\Omega$.

9.103 Consider the circuit of Fig. 9.62 with $R_B = 220\,\text{k}\Omega$, $V_{CC} = 5\,\text{V}$, $C_L = 1\,\text{nF}$, $C_S = 3.3\,\mu\text{F}$, and $\beta = 50$.

(a) Choose a standard value for R_C such that v_{OUT} is biased as close as possible to $V_{CC}/2$.

(b) Plot the output versus time if v_{in} is a ± 1-mV, 1-kHz square wave.

9.104 Consider the circuit of Fig. 9.62. Suppose that the BJT is replaced by a MOSFET having parameters $K = 0.02\,\text{mA/V}^2$ and $V_{TR} = 2\,\text{V}$.

(a) Choose values for R_B and R_C such that v_{OUT} is biased at about half of $V_{CC} = 10\,\text{V}$.

(b) If v_{in} is a ± 15-mV, 50-ms square wave, and if $C_S = 3.3\,\mu\text{F}$, plot the output voltage versus time. Ignore the effects of internal MOSFET capacitance.

9.105 Consider the circuit of Fig. 9.62 with $R_B = 1.2\,\text{M}\Omega$, $V_{CC} = 6\,\text{V}$, $C_L = 5\,\text{nF}$, $C_S = 10\,\mu\text{F}$, and $\beta = 50$. Suppose that a resistor of value $R_S = 1\,\text{k}\Omega$ is connected in series between v_{in} and node B.

(a) Choose a standard value for R_C such that v_{OUT} is biased as close as possible to 2 V.

(b) Plot the output voltage versus time if v_{in} is a ± 1-mV, 100-Hz square wave.

9.106 ⊠ Consider the circuit of Fig. 9.62. Choose component values such that the time constant of the rising waveform after an input step function has a maximum value of 1 ms and the droop results in an output decay of no more than 10% in 100 ms. The amplifier should have a midband gain of -50. For the purpose of this problem, assume $\beta_F = 100$, and ignore internal BJT capacitances.

9.107 ⊠ ○ Consider the circuit of Example 9.13. Let the input be a symmetrical square wave of period 10 ms

and peak value 10 mV. If $\beta_F = 100$, $V_{CC} = 10\,\text{V}$, $\eta = 1$, and $V_f = 0.6\,\text{V}$,

(a) Choose R_B and R_C such that the output has a bias value of $V_{OUT} = 5\,\text{V}$ and the amplifier has a gain of -200.

(b) Choose a reasonable value for C_S such that the output signal will be a faithful reproduction of the input signal with minimal rise time and droop.

(c) Find the peak magnitude of the output signal.

9.108 ◑ Consider the circuit of Example 9.13. If $T = 10\,\text{ms}$, the time constant $r_{in}C_S$ will be comparable to the period of the square wave. The response due to a given transition will decay considerably before the next transition arrives, but will not decay to zero. Find a general expression for v_{out} as a function of time by superimposing successive transitions of the form (9.209). Use this relation to plot the output for $0 < t < 25\,\text{ms}$ if $V_s = 5\,\text{mV}$.

9.109 ⊠ ○ Design a MOSFET circuit based on the topology of Fig. 9.62. Assume MOSFET parameters $K = 0.5\,\text{mA/V}^2$ and $V_{TR} = 1\,\text{V}$. Choose component values such that the time constant of the rising waveform after an input step function has a maximum value of 1 ms and the output droops by no more than 10% in 100 ms after reaching its maximum value. The amplifier should have a midband gain of about -4. For the purpose of this problem, ignore the internal MOSFET capacitances C_{gs} and C_{gd}.

9.110 Consider the circuit of Fig. 9.62 with $V_{CC} = 5\,\text{V}$, $R_B = 1\,\text{M}\Omega$, $R_C = 3.3\,\text{k}\Omega$, $C_S = 10\,\mu\text{F}$, and $C_L = 2\,\text{pF}$. Determine the time constant associated with rising and falling output signals if Q_1 has parameters $\beta = 100$, $C_{ob} = C_\mu = 2\,\text{pF}$, and $f_T = 400\,\text{MHz}$.

9.111 Consider the circuit of Fig. 9.62 with $V_{CC} = 12\,\text{V}$, $R_B = 1\,\text{M}\Omega$, $R_C = 5.1\,\text{k}\Omega$, $C_S = 10\,\mu\text{F}$, and $C_L = 7\,\text{pF}$. Determine the fall time of the circuit if Q_1 has parameters $\beta_o = 100$, $C_{ob} = C_\mu = 3\,\text{pF}$, and $f_T = 350\,\text{MHz}$. The fall time is defined as the time required for v_{OUT} to fall from 90% to 10% of its initial value after the application of an input voltage step.

9.112 Consider the circuit of Fig. 9.62 in which Q_1 is a MOSFET with parameters $K = 1\,\text{mA/V}^2$, $V_{TR} = 2\,\text{V}$, $C_{gs} = 2\,\text{pF}$, and $C_{gd} = 1\,\text{pF}$.

(a) For $V_{CC} = 10\,\text{V}$, add a second resistor R_A between the gate and ground, and then choose R_A and R_B such that v_{OUT} is biased at about $V_{CC}/2$ with $R_C = 5\,\text{k}\Omega$.

(b) Suppose that $C_S = 33\,\mu\text{F}$ and $C_L = 8\,\text{pF}$. Determine the time constant associated with rising and falling output signals when the input is a square wave.

9.113 ▣ ◐ Using the circuit of Fig. 9.62 as a guideline, design a BJT circuit that can amplify a ±10-mV, 1-kHz square wave to ±1 V with no more than a 200-ns rise time (defined here as the $1/e$ time constant) and no more than 1% droop. Assume open-circuit load conditions ($C_L = 0$). Specify reasonable BJT parameters in your design.

9.114 ▣ ◐ Design a MOSFET circuit that can amplify

a ±0.2-V, 500-Hz square wave to ±2 V with no more than a 250-ns rise time (defined here as the $1/e$ time constant). Assume open-circuit load conditions. Specify reasonable MOSFET parameters in your design.

9.115 ▣ ◐ An AM detector circuit of the type described in Section 4.4.5 produces a 0.1 V peak digital signal at transmission rates of up to 100 kbits/s. This signal is to be fed as serial input to a microprocessor for decoding. Design a one- or two-transistor circuit that will convert the incoming 0- to 0.1-V signal into a 0- to 5-V signal suitable for digital processing.

◆ SPICE PROBLEMS

9.116 Simulate each of the passive RC circuits of **Fig. P9.9** on SPICE. Find the frequency response of each circuit and confirm the results of Problem 9.9.

9.117 Find the frequency response of the circuit of **Fig. P9.12** by simulating it on SPICE. Show that it has a midband "gain" value of about 0.96, and find the upper and lower -3-dB points.

9.118 Consider the $10\times$ oscilloscope probe of **Fig. P9.17**.

(a) Use SPICE to find the step response of the circuit of **Fig. P9.17(b)** with the resistor and capacitor values found in Problem 9.17, but with C_X absent. Repeat the analysis with C_X in place and confirm that the signal fed to the vertical amplifier of the oscilloscope replicates voltage v_1.

(b) Use SPICE to find the magnitude frequency response of the probe both with and without C_X in place. Confirm that the signal fed to the vertical amplifier of the oscilloscope is not a function of frequency if C_X is in place.

9.119 Simulate the BJT amplifier of Fig. 9.20 on SPICE. Set C_π to the value specified in Example 9.4 and use the .AC command to obtain a system function plot of the output from 0.001 Hz to 1 kHz. Compare the results with those shown in Fig. 9.24.

9.120 Simulate the MOSFET amplifier of Fig. 9.27 on SPICE and confirm the results of Example 9.5.

9.121 Obtain the magnitude Bode plot of the BJT amplifier of Fig. 9.30 by simulating the circuit on SPICE. Estimate the new value of C_E required to extend the low-frequency end of the midband down to 10 Hz. Test your choice of capacitor by again simulating the circuit on SPICE.

9.122 Use SPICE to assess the frequency-response characteristics of the Darlington connection of **Fig. P9.91** for the case $V_{CC} = 20\,\text{V}$, $V_{EE} = -5\,\text{V}$, $R_C = 10\,\text{k}\Omega$, $R_E = 3.6\,\text{k}\Omega$, $\beta_F = 100$, $f_T = 350\,\text{MHz}$, and $C_\mu = 2\,\text{pF}$.

9.123 Use SPICE to determine the frequency response of the circuit of **Fig. P9.65**. Assume transistor parameters $\beta_F = 100$, $f_T = 350\,\text{MHz}$, and $C_\mu = 2\,\text{pF}$.

9.124 Simulate the MOSFET follower of **Fig. P9.63** on SPICE. Find the magnitude and frequency range of the midband gain. Set the MOSFET parameters to the indicated values using the .MODEL statement, with $V_{\text{TR}} = 2\,\text{V}$.

9.125 The frequency response of the emitter-bypassed BJT amplifier of Fig. 9.47 was analyzed as part of the design in Example 9.11. Use SPICE to confirm the plot of Fig. 9.49. Obtain a revised estimate of the low-frequency -3-dB midband endpoint.

9.126 Use SPICE to find the frequency response of the NMOS differential amplifier of **Fig. P9.80** if the MOSFETs have parameters as described in Problem 9.87. In this case, do not ignore the body effect. Assume the substrate of each n-channel MOSFET to be connected to the V_{SS} bus with $V_{DD} = -V_{SS} = 15\,\text{V}$ and $I_o = 2\,\text{mA}$.

9.127 Simulate the BJT cascode circuit of **Fig. P9.88** on SPICE with $R_C = 5\,\text{k}\Omega$, $R_E = 10\,\text{k}\Omega$, $R_1 = R_2 = 500\,\Omega$, and $C_\pi = C_\mu = 1\,\text{pF}$ for both transistors.

(a) Find the magnitude and frequency range of the midband gain.

(b) Excite the amplifier with a 0.1 V, 1-kHz sinusoid. Use the .DISTO command to investigate the harmonic content of the resulting output signal.

9.128 Simulate the BJT cascode differential-amplifier circuit of Fig. 9.54 on SPICE and find the extent of the midband region under differential-mode excitation. The transistors all have parameters $\beta_o = 120$, $C_\mu = 0.5\,\text{pF}$, and $f_T = 350\,\text{MHz}$.

9.129 Repeat Problem 9.128 if enhancement mode MOS-FETs with parameters $K = 0.2\,\text{mA/V}^2$, $V_{TR} = 4\,\text{V}$, $C_{gs} = 1.5\,\text{pF}$, and $C_{gd} = 0.8\,\text{pF}$ are substituted for the two BJTs.

9.130 Use SPICE to evaluate the differential- and common-mode gains of the CMOS differential amplifier of **Fig. P9.81** as functions of frequency if $I_o = 1\,\text{mA}$, $|K| = 2\,\text{mA/V}^2$, $r_o = 20\,\text{k}\Omega$, $C_{gs} = 0.8\,\text{pF}$, and $C_{gd} = 0.2\,\text{pF}$ for all devices, and $V_{DD} = -V_{SS} = 15\,\text{V}$.

9.131 Use SPICE to simulate the half-wave rectifier of Fig. 9.59(a). Excite the circuit with a -10-V step function and plot the resulting output voltage versus time. Compare with the result shown in Fig. 9.61. Include the effects of diode capacitance.

9.132 Use SPICE to find the step response of the MOS-FET amplifier of Fig. 9.27 to a 10-mV step function.

9.133 Simulate the circuit of Fig. 9.62 on SPICE with $V_{CC} = 5\,\text{V}$, $C_S = 1\,\mu\text{F}$, $R_B = 1\,\text{M}\Omega$, $R_C = 3.3\,\text{k}\Omega$, and $\beta_F = \beta_o = 210$. These values were used in Example 9.13. Excite the circuit with a ±5-mV square wave, and find the output for square-wave periods of 100 ms and 0.1 ms. Compare the results with those shown in Figs. 9.65 and 9.66.

9.134 Use SPICE to find the rise time of the amplifier of Fig. 9.62 with $V_{CC} = 5\,\text{V}$, $C_S = 10\,\mu\text{F}$, $R_B = 820\,\text{k}\Omega$, $R_C = 4.7\,\text{k}\Omega$, $C_\pi = 2\,\text{pF}$, and $C_\mu = 1\,\text{pF}$.

9.135 Consider the circuit of Fig. 9.62. Excite the circuit with a step function using the small-signal transistor parameters specified in Example 9.14. Include the "stray" capacitance $C_{be} = 1\,\text{nF}$. As suggested in Exercise 9.48, repeat the simulation with C_{be} omitted so that the rise time of the amplifier will not be dominated by any one high-frequency capacitor, and compare results.

Chapter *10*

Feedback and Stability

T his chapter formally addresses the issues of feedback and feedback stability. Feedback plays a major role in real-life circuits and is a subject worthy of detailed study. Indeed, it is hard to think of any practical circuit or system that does not incorporate some sort of feedback. Feedback can be applied on a small scale or on a large scale and appears in both analog and digital systems. As we shall show, feedback allows circuit characteristics such as gain, input impedance, output impedance, and bandwidth to be precisely controlled while making these parameters insensitive to variations in individual transistor parameters. The latter can be extremely sensitive to fabrication processes, temperature changes, and other environmental factors.

Feedback is not confined to transistor circuits, but can be found in many other engineering and biological systems. Mechanical feedback was first used by inventors in the 1800s to make the speed of steam and internal combustion engines insensitive to mechanical loading. Electrical feedback made possible the first vacuum-tube radio receivers and transmitters. A form of biological feedback is used by the human body to control body temperature, body movement, and heart rate. Feedback is even used in manufacturing systems to make production lines less sensitive to external supply-and-demand variations.

In this chapter, we shall deal almost exclusively with electrical feedback as used in the design of analog amplifiers. Within this limited focus, however, we shall discuss general concepts relevant to feedback of all types. As an engineering tool, feedback has broad relevance and widespread applicability.

10.1 THE NEGATIVE-FEEDBACK LOOP

An amplifier without feedback can generally be represented by the simple block diagram of Fig. 10.1. An input signal x_{IN} feeds the amplifier that processes the signal and produces the output x_{OUT}. The signals x_{IN} and x_{OUT} can represent voltages, currents, or even optical signals.

Figure 10.1
Block diagram of an open-loop amplifier without feedback.

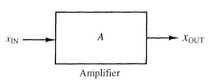

$x_{IN} \longrightarrow$ A $\longrightarrow x_{OUT}$

Amplifier

Indeed, any circuit that consists of a single forward path for the transmission of signals, including virtually all of the amplifiers discussed in the book thus far, can be described by the diagram of Fig. 10.1. If the amplification factor A is a constant or a function only of frequency, the amplifier is said to be linear. More generally, A might represent some nonlinear function that relates x_{OUT} to x_{IN}.

Figure 10.2
Block diagram of a system incorporating feedback. The amplifier output x_{OUT} is fed to the feedback network, which produces the signal x_F and returns it to the amplifier output.

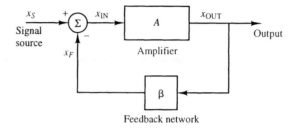

The diagram of Fig. 10.2 represents a system in which a feedback network has been added to the basic amplifier. The amplifier output x_{OUT} is still equal to Ax_{IN}, but in this case, x_{IN} represents only the signal appearing directly at the amplifier input terminals. The variable x_S represents the input signal applied to the entire system by the user. The feedback network accepts x_{OUT} as its input and produces a signal x_F, called the "feedback" signal. The latter is subtracted from x_S at the summation node Σ to produce x_{IN}. Specifically,

$$x_{IN} = x_S - x_F \tag{10.1}$$

Our objective is to find the overall response of the circuit to the applied signal x_S. The most common relationship between x_F and x_{OUT} consists of the simple linear equation

$$x_F = \beta x_{OUT} \tag{10.2}$$

where the feedback factor β is a constant.[1] Determining the amplifier output for such a feedback condition is straightforward. Equation (10.1) can be multiplied by A to yield

$$x_{OUT} = Ax_{IN} = A(x_S - x_F) \tag{10.3}$$

Equation (10.2) for x_F can then be substituted into Eq. (10.3), resulting in

$$x_{OUT} = A(x_S - \beta x_{OUT}) \tag{10.4}$$

As this equation shows, x_{OUT} depends upon itself—a property intrinsic to the nature of a feedback path that connects the output back to the input. Equation (10.4) can be rearranged:

$$x_{OUT}(1 + A\beta) = Ax_S \tag{10.5}$$

and finally put in the form

$$A_{fb} = \frac{x_{OUT}}{x_S} = \frac{A}{1 + A\beta} \tag{10.6}$$

The factor A_{fb} is called the *closed-loop gain* (gain with feedback) of the circuit. It represents the net ratio of x_{OUT} to x_S when a feedback network described by Fig. 10.2 is connected.

---

[1] The symbol β used to describe feedback has nothing to do with the parameter β of the BJT. Both are used by coincidental convention only.

As Eq. (10.6) shows, the basic amplifier gain A in the numerator is divided by a denominator consisting of the large factor $(1 + A\beta)$, resulting in a greatly reduced overall gain A_{fb}. Although it may seem undesirable, this gain reduction actually produces a very desirable effect. If A is large, such that the product $A\beta$ greatly exceeds unity ($A\beta \gg 1$), Eq. (10.6) approaches the limit

$$A_{fb} \approx \frac{A}{A\beta} = \frac{1}{\beta} \tag{10.7}$$

The closed-loop gain A_{fb} becomes independent of A in the limit $A\beta \gg 1$, and depends only on the feedback factor β. This feature is an important one that allows A_{fb} to be precisely set regardless of the exact value of A. Because the feedback network is generally made from passive (and easy-to-control) circuit elements, the many factors that affect A, including component variations, temperature, and circuit nonlinearity, become much less important to the closed-loop circuit. This benefit is generally worth the price of reduced gain, especially because A can usually be made much larger than the required closed-loop gain factor.

Figure 10.3
Noninverting op-amp configuration.

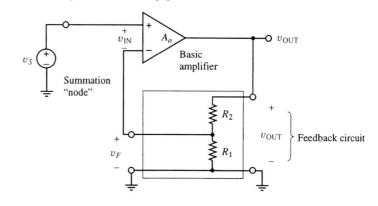

EXAMPLE 10.1

The gain-reduction principle discussed in the preceding paragraph is exemplified by the non-inverting op-amp configuration of Chapter 2. As illustrated in Fig. 10.3, this circuit can indeed be described by the general feedback diagram of Fig. 10.2. In this case, each of the various signals is a voltage. The voltage v_S in Fig. 10.3 functions as the input signal x_S, the op-amp output v_{OUT} functions as the output signal x_{OUT}, and the voltage labeled v_F in the figure serves the role of the feedback signal x_F. The open-loop gain A_o of the op-amp functions as the gain factor A. To the extent that the op-amp is ideal, with $i_- = i_+ \approx 0$, the feedback factor β, defined as the ratio v_F/v_{OUT}, can be determined from the voltage-divider relation:

$$v_F = \frac{R_1}{R_1 + R_2} v_{OUT} = \beta v_{OUT} \tag{10.8}$$

where $\beta = R_1/(R_1 + R_2)$.

The summation function Σ in Fig. 10.2 is performed at the op-amp input terminals, where, by KVL,

$$v_{IN} = v_+ - v_- = v_S - v_F \tag{10.9}$$

For this v_{IN}, the op-amp output becomes

$$v_{OUT} = A_o(v_+ - v_-) = A_o(v_S - v_F) = A_o(v_S - \beta v_{OUT}) \tag{10.10}$$

Solving for v_{OUT} results in

$$v_{OUT} = \frac{A_o}{1 + A_o\beta} v_S \tag{10.11}$$

The typical op-amp has a very large open-loop gain A_o, hence the condition $A\beta \gg 1$ will be met for all reasonable values of R_1 and R_2. The closed-loop gain of the feedback circuit thus can be expressed by the reciprocal of the feedback factor β:

$$\frac{v_{OUT}}{v_S} \approx \frac{1}{\beta} = \frac{R_2 + R_1}{R_1} \tag{10.12}$$

This result is identical to the one found in Chapter 2 using the ideal op-amp approximation and basic circuit theory principles. The result illustrates the utility and simplicity of the general feedback formulation presented in this chapter.

EXERCISE 10.1 An amplifier having an open-loop gain of 5000 is connected in a negative feedback network with a feedback factor of 0.1. What is the gain of the overall amplifier? **Answer:** $A_{fb} = 9.98 \approx 10$

10.2 A noninverting op-amp circuit is made with $R_2 = 100\,\text{k}\Omega$ and $R_1 = 8.2\,\text{k}\Omega$, where R_2 is the feedback resistor. What are the values of the feedback factor and the closed-loop gain?
Answers: 0.076; 13.2

10.2 GENERAL REQUIREMENTS OF FEEDBACK CIRCUITS

The feedback diagram of Fig. 10.2 is a general one that can be applied to many feedback amplifiers. If the summation node is to be physically realizable, x_{IN}, x_S, and x_F must all be of the same signal type. Specifically, the three signals must either be all voltages or all currents. The amplifier output x_{OUT}, however, need not be of the same signal type as its input. It is possible, for example, to have an amplifier with an output signal that is a current and an input signal that is a voltage. In general, the amplification factor A can have dimensional units of $A_v = $ volts/volt, $A_i = $ amperes/ampere, $A_r = $ volts/ampere, or $A_g = $ amperes/volt. The feedback function β must have units that are reciprocal to those of A, such that the product $A\beta$ is dimensionless. This condition ensures that x_F is of the same signal type as x_S and x_{IN}.

For all the circuits discussed in this chapter, the feedback network will be made from passive components only. For such circuits, the feedback factor β can never exceed unity. More generally, feedback can be provided by an active circuit, resulting in larger values of β. Such circuits, however, are beyond the scope of this chapter.

In the feedback loop of Fig. 10.2, x_F is *subtracted* from x_S, making the feedback *negative*. If x_F is *added* to x_S at the summation node, the feedback becomes *positive*. All of the circuits discussed in this chapter, and by far the more common in electronics, are of the negative-feedback type. *Positive* feedback is used in circuits called *oscillators* and also in a class of circuits called *active filters*. These circuits are discussed in Chapter 13.

10.3 EFFECTS OF FEEDBACK ON AMPLIFIER PERFORMANCE

Feedback affects the properties of all amplifiers, regardless of type. Negative feedback reduces amplifier nonlinearity, improves input and output resistance, extends amplifier bandwidth, stabilizes gain, and reduces amplifier sensitivity to transistor parameters. These features are usually desirable ones in amplifier design. In this section, the numerous benefits and effects of feedback are explored in more detail.

Figure 10.4
Nonlinear amplifier transfer characteristic.

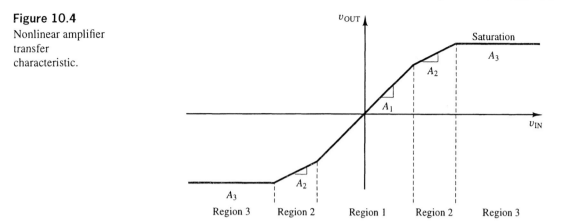

10.3.1 Effect of Feedback on Amplifier Linearity

The ideal analog amplifier has constant gain regardless of signal magnitude. In practice, this ideal is unreachable. As the transfer characteristic of Fig. 10.4 suggests, the gain of most amplifiers changes with signal magnitude, arriving at some saturation region (gain equal to zero) for very large input signals. Negative feedback can do nothing to eliminate saturation, but it can greatly reduce the effects of varying slope. Suppose that an amplifier having the transfer characteristic of Fig. 10.4 is connected in the negative-feedback loop of Fig. 10.2 with constant feedback factor β and voltage input v_S. Over region 1 of the transfer characteristic, the gain with feedback becomes

$$A_{fb1} \triangleq \frac{dv_{OUT}}{dv_S}\bigg|_{\text{region1}} = \frac{A_1}{1 + A_1\beta} \tag{10.13}$$

where $A_1 = dv_{OUT}/dv_{IN}$ over region 1. Similarly, over region 2, the gain becomes

$$A_{fb2} \triangleq \frac{dv_{OUT}}{dv_S}\bigg|_{\text{region2}} = \frac{A_2}{1 + A_2\beta} \tag{10.14}$$

For large A_1 and A_2, the ratio A_{fb2}/A_{fb1} approaches unity, indicating that the two closed-loop gains are close in value:

$$\frac{A_{fb2}}{A_{fb1}} = \frac{A_2}{1 + A_2\beta} \frac{1 + A_1\beta}{A_1} \approx \frac{A_2}{A_2\beta} \frac{A_1\beta}{A_1} \approx 1 \tag{10.15}$$

If A_1 and A_2 are not large, the closed-loop gains will still be much closer in value than the unmodified open-loop gains A_1 and A_2.

Note that A_{fb} falls to zero in the saturation region, where the open-loop gain $A_3 = dv_{OUT}/dv_{IN}$ falls to zero:

$$A_{fb3}\big|_{\text{region3}} = \frac{A_3}{1 + A_3\beta} = \frac{0}{1 + 0 \cdot \beta} = 0 \tag{10.16}$$

Feedback can do nothing to eliminate the effects of amplifier saturation.

EXAMPLE 10.2   A basic amplifier has the open-loop transfer characteristic of Fig. 10.4 with slopes $A_1 = 500$, $A_2 = 300$, and $A_3 = 0$. The amplifier is connected in a negative-feedback loop with $\beta = 0.1$. Find the slope of the closed-loop transfer characteristic over all regions of operation.

Solution

The closed-loop gain over region 1 becomes

$$A_{fb1} = \frac{500}{1 + 500(0.1)} = 9.80 \qquad (10.17)$$

Similarly, the closed-loop gain over region 2 becomes

$$A_{fb2} = \frac{300}{1 + 300(0.1)} = 9.68 \qquad (10.18)$$

The closed-loop gain approaches the value $1/\beta = 10$ in both cases. Although the slopes in regions 2 and 1 have the ratio $300/500 = 0.6$, the slopes with feedback have a ratio of $9.68/9.80 = 0.988$, showing that they are nearly the same. Over the saturation region depicted in Fig. 10.4, where $A_3 = 0$, A_{fb3} equals zero.

DESIGN

EXAMPLE 10.3   Consider the amplifier of Example 10.2, for which the gain ratio A_2/A_1 without feedback equals 0.6. Choose a value for β such that the closed-loop gain falls by no more than 10% from its value in region 1 when the amplifier operates in region 2.

Solution

With the ratio of the two closed-loop gains A_{fb2}/A_{fb1} expressed as m, Eq. (10.15) can be solved for β:

$$m = \frac{A_2(1 + A_1\beta)}{A_1(1 + A_2\beta)} \implies \beta = \frac{A_1 m - A_2}{A_1 A_2(1 - m)} \qquad (10.19)$$

Substituting the specified values of A_1, A_2, and $m = 0.9$ results in

$$\beta = \frac{(500)(0.9) - 300}{(500)(300)(1 - 0.9)} = 0.01 \qquad (10.20)$$

For this value of β, the closed-loop gains become

$$A_{fb1} = \frac{500}{1 + 500(0.01)} = \frac{500}{6} = 83.3 \qquad (10.21)$$

and $$A_{fb2} = \frac{300}{1 + 300(0.01)} = \frac{300}{4} = 75 \qquad (10.22)$$

where A_{fb2} indeed falls by no more than 10% of A_{fb1}. Note that both closed-loop gains approach the value $1/\beta = 100$, which is the gain limit for infinite open-loop gain A.   🔊

We end this section with a formal definition of linearity that can be applied to all feedback amplifiers. The slope dv_{OUT}/dv_{IN}, or gain, of an amplifier transfer characteristic without feedback is equivalent to the open-loop gain A; the gain with feedback is given by $A_{fb} = A/(1 + A\beta)$. The sensitivity of A_{fb} to changes in A can be computed by taking the derivative of A_{fb} with respect to A:

$$\frac{dA_{fb}}{dA} = \frac{(1 + A\beta) - A\beta}{(1 + A\beta)^2} = \frac{1}{(1 + A\beta)^2} \qquad (10.23)$$

If the open-loop gain A undergoes a change ΔA, the change in A_{fb} becomes

$$\Delta A_{fb} = \frac{dA_{fb}}{dA}\Delta A = \frac{\Delta A}{(1+A\beta)^2} \qquad (10.24)$$

The fractional change in A_{fb} can thus be expressed as

$$\frac{\Delta A_{fb}}{A_{fb}} = \frac{\Delta A/(1+A\beta)^2}{A/(1+A\beta)} = \frac{\Delta A}{A}\frac{1}{1+A\beta} \qquad (10.25)$$

The change in A_{fb} is smaller than the change in A by a factor of $1/(1+A\beta)$.

EXERCISE 10.3   A transistor amplifier has a gain of 100 for small v_{OUT} that falls to 20% of this value when v_{OUT} is large. What is the minimum value of β for a negative feedback network that will ensure a linear transfer characteristic to within 2%?   **Answer:**   0.39

     10.4   An operational amplifier with a manufacturer-specified gain of 10^4 for small outputs has a gain that actually varies by as much as 50% over the full range of output voltage. What is the maximum closed-loop gain for negative feedback that will ensure a linear transfer characteristic to within 0.1%?   **Answer:** ≈ 20

10.3.2 Effect of Feedback on Amplifier Bandwidth

In Chapter 2, the effect of feedback on frequency response was examined in the context of the operational amplifier. In this section, we review this topic and discuss it from a more general point of view. Our discussion of feedback has thus far assumed the open-loop amplifier gain A and feedback factor β to be independent of frequency. More generally, these quantities are frequency-dependent functions of the form $A(j\omega)$ and $\beta(j\omega)$. Direct substitution into the closed-loop gain equation (10.6) results in

$$A_{fb}(j\omega) = \frac{\mathbf{X}_{out}}{\mathbf{X}_s} = \frac{A(j\omega)}{1+A(j\omega)\beta(j\omega)} \qquad (10.26)$$

where $\mathbf{X}_{out}$ and $\mathbf{X}_s$ are phasors that represent either voltage or current signals in the sinusoidal steady state. Although frequency-dependent β functions are common in electronics, we wish here to focus specifically on the effect of a frequency-dependent $A(j\omega)$. We thus assume β to be a constant (created by a purely resistive feedback network). For constant β, the gain expression (10.26) simplifies to

$$\frac{\mathbf{X}_{out}}{\mathbf{X}_s} = \frac{A(j\omega)}{1+A(j\omega)\beta} \qquad (10.27)$$

The typical open-loop gain $A(j\omega)$ has several poles that reduce its magnitude at increased frequencies. Suppose that one of these poles is strongly dominant, such that $A(j\omega)$ can be approximated over much of its range by

$$A(j\omega) = \frac{A_o}{1+j\omega/\omega_p} \qquad (10.28)$$

This expression has a single pole at ω_p and reverts to its dc value of A_o well below ω_p. Substituting Eq. (10.28) into Eq. (10.27) results in

$$\frac{\mathbf{X}_{out}}{\mathbf{X}_s} = \frac{A_o/(1+j\omega/\omega_p)}{1+A_o\beta/(1+j\omega/\omega_p)} \qquad (10.29)$$

Multiplying numerator and denominator by $(1 + j\omega/\omega_p)$ results in

$$\frac{\mathbf{X}_{\text{out}}}{\mathbf{X}_s} = \frac{A_o}{(1 + j\omega/\omega_p) + A_o\beta} \equiv \frac{A_o}{(1 + A_o\beta) + j\omega/\omega_p} \qquad (10.30)$$

Finally, dividing numerator and denominator by $(1 + A_o\beta)$ produces the result

$$A_{\text{fb}}(\omega) = \frac{\mathbf{X}_{\text{out}}}{\mathbf{X}_s} = \frac{A_o}{1 + A_o\beta} \frac{1}{1 + j\omega/\omega_p(1 + A_o\beta)} \qquad (10.31)$$

This expression for the closed-loop gain contains a constant factor $A_o/(1 + A_o\beta)$, which we recognize as the closed-loop gain at dc, where $A(j\omega) = A_o$. The second term in Eq. (10.31) describes the frequency response of the closed-loop gain caused by the pole of $A(j\omega)$. Specifically, the denominator of the closed-loop gain has a single pole at

$$\omega_{\text{fb}} = \omega_p(1 + A_o\beta) \qquad (10.32)$$

The closed-loop pole frequency ω_{fb} is much higher in value than the open-loop pole frequency ω_p. The effect of negative feedback is to multiply the dominant pole frequency ω_p by the factor $(1 + A_o\beta)$. This result can be explained qualitatively in the following way: Suppose that an amplifier has a large open-loop gain A_o, such that the condition $A_o\beta \gg 1$ is met. Under these conditions, the closed-loop gain at low frequencies will approach the limit $A_{\text{fb}} = 1/\beta$. As the frequency of excitation is increased above the open-loop pole ω_p, the gain $A(j\omega)$ will diminish in magnitude, but as long as the condition

$$|A(j\omega)\beta| \gg 1 \qquad (10.33)$$

is met, the closed-loop gain will still equal $1/\beta$, even though $A(j\omega)$ itself is reduced in value. Only when $|A(j\omega)|$ falls so low that $|A(j\omega)\beta| \gg 1$ is no longer valid will the closed-loop gain begin to fall with frequency. This transition occurs at the frequency $\omega_p(1 + A_o\beta)$.

The typical $A(j\omega)$ actually has a number of poles that contribute to its open-loop response. It can be shown that Eq. (10.32) applies to the dominant pole even when the higher-order pole terms are added to Eq. (10.28).

Note that the product of gain and bandwidth, where bandwidth is determined by the upper -3-dB point of the frequency response, is constant for any feedback factor β. This conclusion was formally derived in Chapter 2 in the context of operational amplifiers (see Section 2.6.7), but applies to any amplifier incorporating feedback. The *unity-gain frequency* is defined as the value of ω at which the magnitude of the open-loop response falls to 1. Since the gain-bandwidth product of an amplifier is constant regardless of any feedback network, the unity-gain frequency is equivalent to the bandwidth when $\beta = 1$. It represents the largest bandwidth that the amplifier can have with a passive-feedback network.

EXERCISE 10.5 An amplifier with a midband gain of 200 and high-frequency poles at 50 kHz and 4 MHz is connected in a negative feedback loop with $\beta = 0.02$. What is the high-frequency -3-dB point of the response? What is the closed-loop gain?   **Answer:**   250 kHz; 40

10.6 Show that the closed-loop pole value (10.32) can be derived by setting the magnitude of the frequency-dependent gain expression (10.31) to $1/\sqrt{2}$, or -3 dB.

10.7 An op-amp with $A_o = 10^5$ and a dominant pole at 4 Hz is connected as a unity-gain voltage follower. What is the closed-loop bandwidth of the circuit?   **Answer:**   400 kHz

10.8 The op-amp of Exercise 10.7 is connected as an inverting amplifier with a gain of -50. What is the closed-loop bandwidth?   **Answer:**   8 kHz

10.9 For the op-amp described in Exercise 10.7, determine the closed-loop bandwidth of a noninverting amplifier for which $R_1 = R_2$.   **Answer:**   200 kHz

10.4 THE FOUR BASIC AMPLIFIER TYPES

A circuit used for electronic amplification can be designed to respond to either voltage or current as its primary input signal. Similarly, the circuit can be designed to supply either a voltage or a current as its primary output signal. Depending on its mix of input and output signals, an amplifier can be classified into one of the four basic types summarized by Fig. 10.5. A *voltage amplifier* with gain A_v accepts a voltage as it input signal and provides a voltage as its output signal. A *current amplifier* with gain A_i has input and output signals that are both currents.

Figure 10.5
The four basic amplifier types.

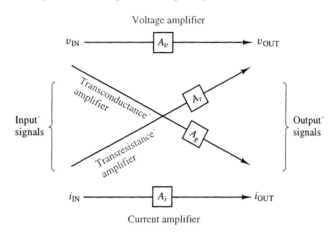

A circuit in which the input signal is a voltage and the output signal a current is called a *transconductance amplifier*, or sometimes a *voltage-to-current converter*. The amplification factor A_g for a transconductance amplifier, defined as the ratio i_{OUT}/v_{IN}, has the units of amperes per volt, or conductance.

A *transresistance amplifier* with gain A_r accepts a current as its input signal and provides a voltage as its output signal. The amplification factor A_r of a transresistance amplifier, sometimes called a *current-to-voltage converter*, is defined as the ratio v_{OUT}/i_{IN} and has the units of volts per ampere, or resistance.

Figure 10.6
(a) Ideal voltage-input amplifier has infinite input resistance. (b) Real voltage-input amplifier has finite input resistance, resulting in a reduction in the actual signal voltage applied to the amplifier input port.

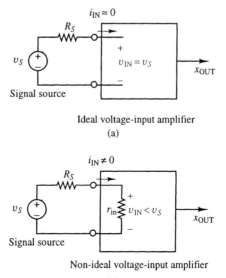

10.4.1 Amplifier Port Characteristics

An amplifier that accepts a voltage signal as its input must have a very high input resistance so that the current drawn from the signal source is minimal. The need for this requirement is illustrated in Fig. 10.6, where a voltage source and Thévenin resistance R_S drive a voltage-input amplifier. If the amplifier has a large input resistance ($r_{in} = \infty$), as in Fig. 10.6(a), v_{IN} will equal the open-circuit voltage v_S, regardless of the value of R_S. If r_{in} is not large, as in Fig. 10.6(b), v_{IN} will be less than open-circuit voltage v_S, and a current $i_{IN} = v_S/(R_S + r_{in})$ will flow into the amplifier. Thus, a condition for proper design of a voltage-input amplifier becomes $r_{in} \gg R_S$.

Figure 10.7
(a) Ideal voltage-output amplifier has zero output resistance.
(b) Real voltage-output amplifier has nonzero output resistance, resulting in a reduction in the actual signal voltage applied to the load.

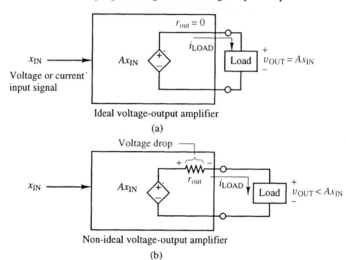

A similar phenomenon occurs at the output port of an amplifier designed for voltage output. Such an amplifier must have a small output resistance so that the open-circuit voltage it produces will appear entirely across the load, regardless of load impedance. This concept is illustrated in Fig. 10.7(a), where $r_{out} = 0$. If r_{out} is nonzero and the load draws current, as in Fig. 10.7(b), the load voltage will be reduced from the open-circuit value Ax_{IN} as current is drawn through r_{out}. A properly designed voltage-output amplifier meets the condition $i_{LOAD}r_{out} \ll v_{OUT}$. If the load is resistive, this condition can be expressed as $r_{out} \ll R_{LOAD}$.

Figure 10.8
(a) Ideal current-input amplifier has zero input resistance.
(b) Real current-input amplifier has nonzero input resistance, resulting in the shunting of some current into the Norton resistance of the current signal source.

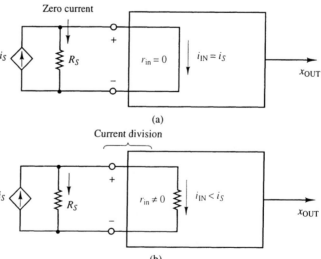

Similar statements can be made about amplifiers designed for current input and output. An amplifier designed to accept current signals ideally should have zero input resistance so as to develop minimal voltage drop across its input terminals. As illustrated in Fig. 10.8(a), this condition allows the amplifier input port to draw all the current available from the current signal source. A nonzero amplifier input resistance, such as the one in Fig. 10.8(b), causes voltage to develop across the Norton resistance R_S of the input source, resulting in the division of current between r_{in} and R_S. A well-designed current-input amplifier will meet the condition $r_{in} \ll R_S$.

An amplifier designed for current output must have a large output resistance so as to maintain its output current regardless of load. The ideal current-output amplifier, depicted in Fig. 10.9(a), has infinite output resistance. If its Norton resistance r_{out} is finite, as in Fig. 10.9(b), any voltage developed across the load will cause some of the dependent source current Ax_{IN} to flow through r_{out}. A well-designed current-output amplifier will meet the condition $v_{OUT}/r_{out} \ll i_{LOAD}$. If the load is resistive, this condition becomes $r_{out} \gg R_{LOAD}$. Note that i_{OUT} in a current-output amplifier is defined by convention as the current flowing *into* the output port, as indicated in Fig. 10.9.

Figure 10.9
(a) Ideal current-output amplifier has infinite output resistance.
(b) Real current-output amplifier has finite output resistance, resulting in a reduction in the actual signal current fed to the load.

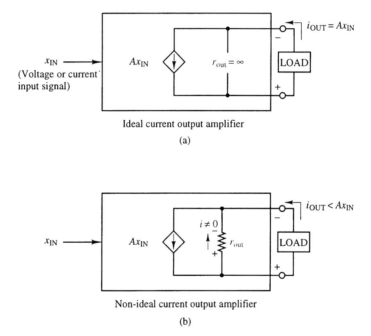

Ideal current output amplifier

(a)

Non-ideal current output amplifier

(b)

The types of input and output signals for which an amplifier is designed completely determine the way in which it must be connected in a feedback loop. If the feedback is to operate successfully, the input and output ports of the feedback network must be compatible with the port characteristics of the amplifier. These issues are discussed in detail in the next two sections.

10.4.2 Output Sampling

When feedback is applied to an amplifier, the feedback network must sense the output in order to produce the feedback signal x_F. This sensing operation is called *sampling*.[2] A voltage-output

---

[2] Continuous, analog sampling in the context of feedback should not be confused with digital sampling, which consists of capturing the value of an analog signal at discrete moments in time. This latter operation is discussed in Chapter 15.

Figure 10.10
Voltage sampling at the output port of a voltage-output amplifier. This connection is also called "shunt sampling." In the ideal case, $r_{in-\beta}$ is infinite.

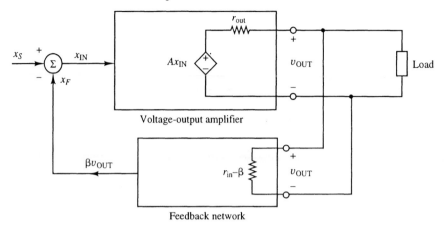

amplifier is readily sampled by tapping the output port in parallel, as illustrated in Fig. 10.10. This parallel topology is sometimes called the "shunt" connection because of its similarity to the railroad tracks used to shunt railroad cars off the line. The shunt connection is appropriate for a voltage-output amplifier because the sampled quantity v_{OUT} can be applied directly to the input port of the feedback network. The input resistance $r_{in-\beta}$ of the feedback circuit must be large (ideally infinite) so as to minimally load the amplifier. This requirement can be expressed quantitatively as $r_{in-\beta} \gg r_{out}$.

If the amplifier is designed for current output, the series connection of Fig. 10.11 is more appropriate for output sampling. The series connection allows the output current i_{OUT} to also flow directly into the input port of the feedback circuit. In this case, $r_{in-\beta}$ must be small (ideally zero) so as to minimize the total series load seen by the amplifier. In quantitative terms, this latter requirement can be expressed as $r_{in-\beta} \ll r_{out}$.

Figure 10.11
Current sampling at the output port of a current-output amplifier. This connection is also called "series sampling." In the ideal case, $r_{in-\beta}$ is zero.

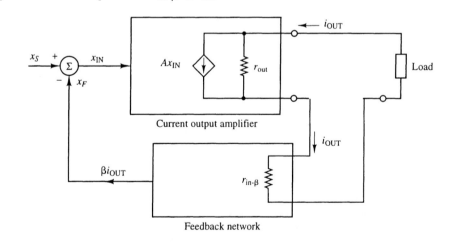

10.4.3 Input Mixing

The negative-feedback diagram of Fig. 10.2 shows the feedback signal x_F being subtracted from the applied signal x_S at the summation node Σ. This simple algebraic subtraction (or addition in the case of positive feedback) is called *input mixing*. In order for input mixing to be possible, the signals x_{IN}, x_F, and x_S must all be of the same type, that is, either all voltages or all currents. If the amplifier is a voltage-input type, the three inputs will be voltages that are most easily mixed using the series connection of Fig. 10.12. If the output resistance of the feedback network, labeled

$r_{\text{out-}\beta}$ in Fig. 10.12, is small, then v_F will be nearly equal to the dependent-source voltage $A\beta v_{\text{IN}}$. Taking KVL around the input loop yields

$$v_{\text{IN}} = v_S - v_F \qquad (10.34)$$

which is equivalent to the algebraic summation depicted in Fig. 10.2. As Eq. (10.34) suggests, series mixing, which derives from KVL, is ideal for adding or subtracting voltage signals. If v_{IN} and v_F were to be connected in parallel, summation of voltage signals would not be possible. A conflicting condition would arise in which v_S and v_F each attempt to establish themselves independently across the same set of terminals.

Figure 10.12
Voltage mixing at the input port of a voltage-input amplifier. The three signals x_{IN}, x_F, and x_S must all be voltages.

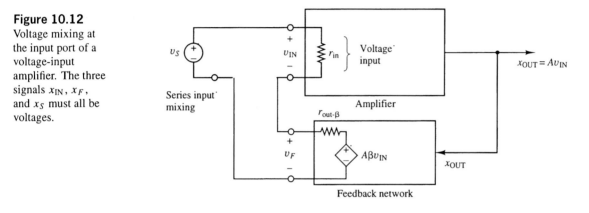

If x_{IN}, x_S, and x_F are currents, input mixing is more appropriately accomplished using the parallel, or shunt-mixing, topology of Fig. 10.13. If $r_{\text{out-}\beta}$ of the feedback network is large, then i_F will be nearly equal to the dependent-source current $A\beta i_{\text{IN}}$. Application of KCL to node X yields

$$i_{\text{IN}} = i_S - i_F \qquad (10.35)$$

Note that i_F is defined as positive *into* the output port of the feedback network. As Eq. (10.35) shows, shunt mixing, which derives from KCL, is ideal for adding or subtracting current signals. In this case, a series connection would lead to a conflict in which i_S and i_F attempt to establish themselves independently around the same loop.

Figure 10.13
Current mixing at the input port of a current-input amplifier. The three signals x_{IN}, x_F, and x_S must all be currents.

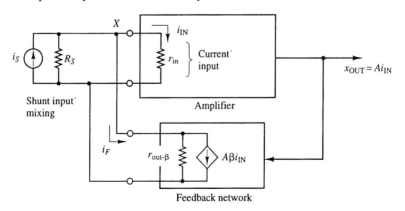

We note with curiosity that sampling of output voltage signals involves a parallel connection, whereas mixing of input voltage signals requires a series connection. Conversely, sampling of output current signals involves a series connection, but the mixing of input current signals requires a parallel connection. The paradox is resolved by noting the characteristics of the amplifier port to which the feedback network is connected. A parallel connection is used whenever the amplifier port involved, whether input or output, has a low impedance. Conversely, a series connection is used when the amplifier port involved has a high impedance.

10.5 THE FOUR FEEDBACK TOPOLOGIES

Each of the four basic amplifier types—voltage, current, transresistance, and transconductance—has its own appropriate feedback topology determined entirely by the amplifier's input and output signal types. A voltage amplifier, for which the input, and output signals are both voltages, requires voltage, or series, mixing at the input and voltage, or shunt sampling, at the output. This feedback topology, called the series-mixing/shunt-sampling (series/shunt) connection, is illustrated in Fig. 10.14. The feedback network must be designed for small $r_{\text{out-}\beta}$ and large $r_{\text{in-}\beta}$ if the feedback is to operate successfully.

Figure 10.14
The series-mixing/shunt-sampling (series/shunt) feedback topology appropriate for a voltage amplifier.

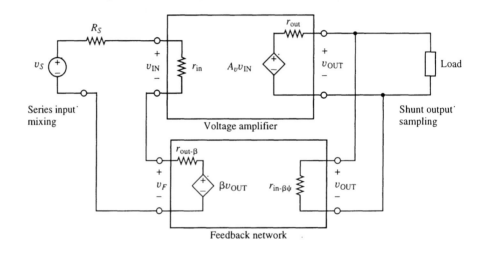

A current amplifier, for which the input and output signals are both currents, requires the shunt-mixing/series-sampling (shunt/series) feedback topology of Fig. 10.15. The feedback network must be designed for large $r_{\text{out-}\beta}$ and small $r_{\text{in-}\beta}$ if this scheme is to operate successfully.

Figure 10.15
The shunt-mixing/series-sampling (shunt/series) feedback topology appropriate for a current amplifier.

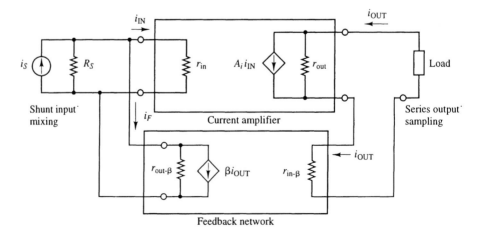

Figure 10.16
The shunt-mixing/shunt-sampling (shunt/shunt) feedback topology appropriate for a transresistance amplifier.

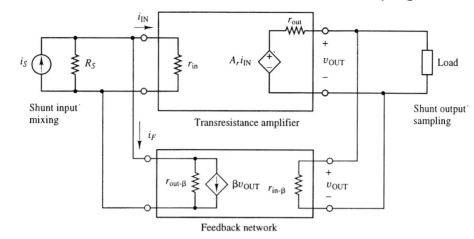

If the feedback loop is built around a transresistance amplifier (input current; output voltage), the shunt-mixing/shunt-sampling feedback topology of Fig. 10.16 must be used. Conversely, if the feedback loop is built around a transconductance amplifier (input voltage; output current), the series-mixing/series-sampling feedback topology of Fig. 10.17 must be used.

Figure 10.17
The series-mixing/series-sampling (series/series) feedback topology appropriate for a transconductance amplifier.

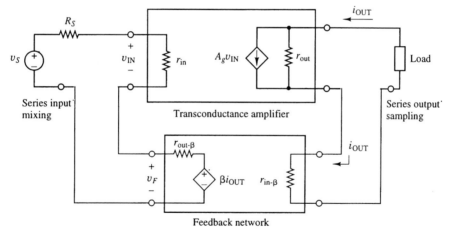

Amplifier Type		Input Signal	Output Signal	Feedback Topology	
				Input Mixing	Output Sampling
Voltage	(A_v)	Voltage	Voltage	Series	Shunt
Current	(A_i)	Current	Current	Shunt	Series
Transresistance	(A_r)	Current	Voltage	Shunt	Shunt
Transconductance	(A_g)	Voltage	Current	Series	Series

Table 10.1. Summary of the Four Basic Feedback Topologies

The four basic feedback topologies described above are summarized in Table 10.1. Successful implementation of these topologies requires that the ports of the amplifier and feedback network have the impedance characteristics summarized in Table 10.2. For the purpose of this second table, an input port is defined as a set of terminals, whether on the amplifier or the feedback network, that *accepts* a signal. An output port is defined as a set of terminals that *provides* a signal. By these definitions, the output port of the feedback network is connected at the point of input signal mixing, and the input port of the feedback network is connected at the point of output signal sampling.

Port Type	Input-Mixing Connection			Output-Sampling Connection		
	Amplifier Input Port	Feedback-Circuit Output Port	Signal Type	Amplifier Output Port	Feedback-Circuit Input Port	Signal Type
Series-connected	High Z	Low Z	Voltage	High Z	Low Z	Current
Shunt-connected	Low Z	High Z	Current	Low Z	High Z	Voltage

Table 10.2. Required Port-Impedance Characteristics

EXERCISE 10.10   For the output-voltage sampling connection of Fig. 10.14, find an expression for the voltage ratio v_{OUT}/Av_{IN} if the load consists of a resistance of value R_L.

10.11   For the output-current sampling connection of Fig. 10.15, find an expression for the current ratio i_{OUT}/Ai_{IN} if the load consists of a resistance of value R_L.

10.12   Draw the voltage- (series-) mixing connections required for positive feedback with $v_{IN} = v_S + v_F$.

10.13   Draw the current- (shunt-) mixing connections required for positive feedback with $i_{IN} = i_S + i_F$.

10.6   EFFECT OF FEEDBACK CONNECTIONS ON AMPLIFIER PORT RESISTANCE

The series and shunt input-mixing connections have a profound effect on the overall input resistance R_{in} of an amplifier with feedback. The series mixing connection, used for voltage-input amplifiers, significantly *increases* R_{in}, causing it to approach the ideal value of infinity desired of a voltage-input amplifier. Conversely, the shunt mixing connection, used for current-input amplifiers, significantly *reduces* R_{in}, causing it to approach the ideal value of zero desired of a current-input amplifier. These impedance-improving features at the input port are the result of negative feedback.

On the output side of the amplifier, the type of feedback connection also alters the overall output-port resistance R_{out}. The shunt sampling connection, used for voltage-output amplifiers, significantly *decreases* R_{out}, causing it to approach the ideal value of zero desired of a voltage-output amplifier. Conversely, the series sampling connection, used for current-output amplifiers, significantly *increases* R_{out}, causing it to approach the ideal value of infinity desired of a current-output amplifier. These impedance-improving features at the output port are again the result of negative feedback. In the sections that follow, we examine the effective resistance created by each of these port connections in more detail.

10.6.1 Input Resistance of the Series Input-Mixing Connection

We illustrate the effect of negative feedback on input resistance by first examining the series input-mixing connection of Fig. 10.18. The input resistance of the basic amplifier is represented by r_{in}. The output port of the feedback network is represented by a dependent voltage source of value $A\beta v_{IN}$ in series with a Thévenin resistance $r_{out\text{-}\beta}$. Such a dependent source appropriately models the output of the feedback circuit for either a voltage amplifier, for which

$$v_F = \beta v_{OUT} = \beta A_v v_{IN} \tag{10.36}$$

(β and A_v *both* have units of volts/volt) or a transconductance amplifier, for which

$$v_F = \beta i_{OUT} = \beta A_g v_{IN} \tag{10.37}$$

(β has units of volts/ampere; A_g has units of amperes/volt). The value of A in Eqs. (10.36) and (10.37) is assumed to include the effects of any output loading caused by the feedback circuit or amplifier load.

Figure 10.18
Determining the input resistance presented to a voltage signal source by the series input-mixing connection.

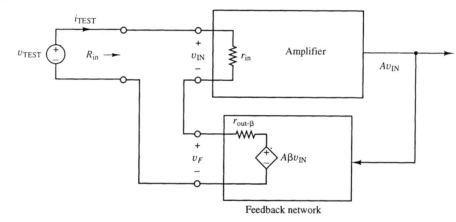

The overall input resistance R_{in} can be found by applying a v_{TEST} voltage source to the overall amplifier, as in Fig. 10.18, and computing the resulting input current i_{TEST}. The latter can be determined from the net voltage drop across the series combination of r_{in} and $r_{out\text{-}\beta}$:

$$i_{TEST} = \frac{v_{TEST} - A\beta v_{IN}}{r_{in} + r_{out\text{-}\beta}} = \frac{v_{TEST} - A\beta(i_{TEST}r_{in})}{r_{in} + r_{out\text{-}\beta}} \tag{10.38}$$

where $v_{IN} = i_{TEST}r_{in}$. Combining terms of i_{TEST} in Eq. (10.38) results in

$$i_{TEST}(r_{in} + r_{out\text{-}\beta} + A\beta r_{in}) = v_{TEST} \tag{10.39}$$

or
$$R_{in} = \frac{v_{TEST}}{i_{TEST}} = r_{in}(1 + A\beta) + r_{out\text{-}\beta} \tag{10.40}$$

For very large loop gain $A\beta$, such that $A\beta \gg 1$, $r_{out\text{-}\beta}$ becomes an insignificant term, and the value of R_{in} approaches the limit

$$R_{in} \approx r_{in}(1 + A\beta) \approx A\beta r_{in} \tag{10.41}$$

By making A arbitrarily large, R_{in} can be made to approach the limit of infinity desired of an ideal voltage-input amplifier. The action of negative feedback is solely responsible for this increase in overall input resistance. It can be explained by considering the effect of the negative feedback connection on the amplifier input port. When a v_{TEST} is applied, the amplifier output increases, causing the feedback network to produce a voltage v_F that acts to limit the current flow through r_{in}. This effect causes r_{in} to appear to have a larger value than it actually does.

10.6.2 Input Resistance of the Shunt Input-Mixing Connection

A similar analysis yields the overall input resistance of the shunt mixing connection of Fig. 10.19. The input resistance of the basic amplifier is again represented by r_{in}, but in this case, the output port of the feedback network is represented by a dependent *current* source of value $A\beta i_{IN}$ in parallel with a Norton resistance $r_{out\text{-}\beta}$. Such a dependent source appropriately models the output of the feedback network for either a current amplifier or a transresistance amplifier. Note that i_F is defined as positive *into* the output port of the feedback network, so that i_F shunts off some of the current from the input source i_S. This condition is necessary if the feedback is to be negative.

Figure 10.19
Determining the input resistance presented to a current signal source by the shunt input-mixing connection.

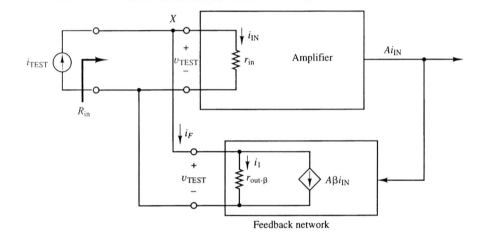

The overall input resistance R_{in} can be determined by connecting a current source i_{TEST} and finding the voltage v_{TEST} developed across the amplifier input port. For the circuit of Fig. 10.19, v_{TEST} is equal to $i_{IN}r_{in}$ and also to $i_1 r_{out\text{-}\beta}$. Taking KCL at node X results in

$$i_{IN} = i_{TEST} - i_F = i_{TEST} - (i_1 + A\beta i_{IN}) \tag{10.42}$$

Substituting $i_{IN} = v_{TEST}/r_{in}$ and $i_1 = v_{TEST}/r_{out\text{-}\beta}$ into Eq. (10.42) results in

$$\frac{v_{TEST}}{r_{in}} = i_{TEST} - \frac{v_{TEST}}{r_{out\text{-}\beta}} - A\beta\frac{v_{TEST}}{r_{in}} \tag{10.43}$$

Collecting terms in v_{TEST} and solving for v_{TEST}/i_{TEST} results in

$$R_{in} = \frac{v_{TEST}}{i_{TEST}} = \left[\frac{1 + A\beta}{r_{in}} + \frac{1}{r_{out\text{-}\beta}}\right]^{-1} \equiv r_{out\text{-}\beta}\left\|\frac{r_{in}}{1 + A\beta}\right. \tag{10.44}$$

If the condition $A\beta \gg 1$ is met, the effect of $r_{out\text{-}\beta}$ becomes negligible, and R_{in} approaches the value $r_{in}/A\beta$. By making A arbitrarily large, R_{in} can be made to approach the ideal value of zero desired of a current-input amplifier.

The reduction in input resistance again can be explained by considering the action of negative feedback. When a current i_{TEST} is applied, the amplifier output increases, causing the feedback circuit to produce the current i_F. When the feedback is negative, i_F shunts most of i_{TEST} away from r_{in}, thereby reducing the voltage rise across r_{in} and making r_{in} appear to be much smaller in value than it actually is.

EXERCISE 10.14 The increase in input resistance realized with the series mixing connection occurs only when the feedback loop is closed and properly working. If the feedback loop is disconnected at the output of the amplifier, determine the resulting value of R_{in}.   **Answer:** $r_{in} + r_{out-\beta}$

10.15 Find R_{in} of the shunt mixing connection if the feedback loop is disconnected at the output of the amplifier.   **Answer:** $r_{in} \| r_{out-\beta}$.

10.16 For the shunt input-mixing configuration of Fig. 10.19, prove that i_F has the units of current regardless of whether a current amplifier or a transresistance amplifier is represented.

10.17 Derive Eq. (10.44) from Eq. (10.43).

10.6.3 Output Resistance of the Shunt Output-Sampling Connection

The shunt-sampling connection used for voltage-output amplifiers can be represented by the diagram of Fig. 10.20. The basic amplifier has Thévenin output resistance r_{out}, and the input port of the feedback circuit has resistance $r_{in-\beta}$. This representation ignores the feedforward contribution of the feedback network to the amplifier output (e.g., the portion of x_S that arrives at v_{OUT} via transmission from left to right through the feedback network). The output resistance is measured with the input signal x_S set to zero. With x_S set to zero, the value of the dependent source Ax_{IN} in Fig. 10.20 becomes $-Ax_F \equiv -A\beta v_{OUT}$ for either a voltage-input ($x_S = v_S$) or a current-input ($x_S = i_S$) amplifier.

Figure 10.20
Determining the output resistance of the shunt output-sampling connection with the input signal x_S set to zero.

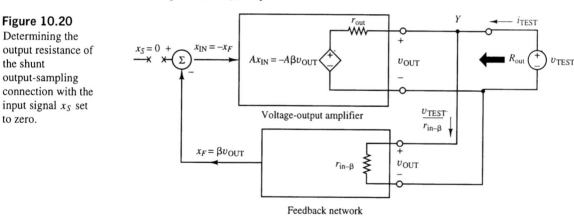

Voltage-output amplifier

Feedback network

The output resistance can be determined by applying a v_{TEST} source to the amplifier output terminals with $x_S = 0$ and computing i_{TEST}. Application of KCL to node Y results in

$$i_{TEST} = \frac{v_{TEST} - (-A\beta v_{TEST})}{r_{out}} + \frac{v_{TEST}}{r_{in-\beta}} \qquad (10.45)$$

where $v_{OUT} = v_{TEST}$ appears across $r_{in-\beta}$, so that the dependent source becomes $-A\beta v_{TEST}$. Solving Eq. (10.45) for the ratio v_{TEST}/i_{TEST} results in

$$R_{out} = \frac{v_{TEST}}{i_{TEST}} = \left[\frac{1 + A\beta}{r_{out}} + \frac{1}{r_{in-\beta}}\right]^{-1} = \frac{r_{out}}{1 + A\beta} \| r_{in-\beta} \qquad (10.46)$$

In the limit of large A, the effect of $r_{in-\beta}$ becomes negligible, and R_{out} approaches the value $r_{out}/A\beta$, which is inversely proportional to A. For very large A, R_{out} approaches the ideal value of zero desired of a voltage-output amplifier. By increasing A, the output impedance R_{out} can be reduced to an arbitrarily small value without altering the closed-loop amplifier gain, which approaches the constant value $1/\beta$ as A is increased.

The reduction in apparent output resistance again can be explained by considering the action of negative feedback. R_{out} is really a measure of how much the output voltage decreases as the load current is increased. Suppose that a signal x_S drives the amplifier without feedback. As current is drawn by the load, the output voltage will be reduced by the voltage drop across r_{out}. If the same amplifier is connected in a feedback loop, the voltage that is actually delivered to the load will also be the voltage applied to the feedback circuit. Any reduction in v_{OUT} caused by the load will thus result in a reduced value of x_F, which will cause $x_{IN} = x_S - x_F$ to increase, leading to an increase in the dependent-source voltage Ax_{IN}. This increase in open-circuit voltage helps to compensate for the reduction in v_{OUT} caused by the output loading. The "tighter" the feedback loop, that is, the larger the loop gain $A\beta$, the better the correction to the output loading and the smaller the apparent output resistance R_{out}.

10.6.4 Output Resistance of the Series Output-Sampling Connection

A similar analysis yields the apparent output resistance of the series output-sampling connection of Fig. 10.21. The Norton output resistance of the open-loop amplifier is represented by r_{out}, and the input port of the feedback circuit by the resistance $r_{in-\beta}$. This representation again ignores the feedforward contribution of the feedback network to the amplifier output. With x_S set to zero, the dependent source Ax_{IN} in Fig. 10.21 takes on the value $-Ax_F = -A\beta i_{OUT}$.

Figure 10.21
Determining the output resistance of the series output-sampling connection with the input signal x_S set to zero.

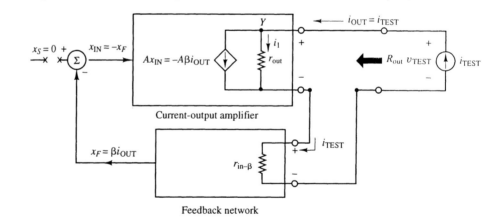

The apparent output resistance R_{out} can be found by applying an i_{TEST} current source and computing the voltage developed across the entire series-connected output port. Taking KVL around the output loop in Fig. 10.21 results in

$$v_{TEST} = i_1 r_{out} + i_{TEST} r_{in-\beta} \tag{10.47}$$

where i_1, the current flowing through the Norton resistance r_{out}, can be found by applying KCL to node Y:

$$i_1 = i_{TEST} + Ax_{IN} \equiv i_{TEST} + A\beta i_{OUT} \tag{10.48}$$

With i_{OUT} equal to i_{TEST}, Eq. (10.48) becomes

$$i_1 = i_{TEST} + A\beta i_{TEST} = (1 + A\beta)i_{TEST} \tag{10.49}$$

Substituting this equation into (10.47) results in

$$v_{\text{TEST}} = (1 + A\beta)i_{\text{TEST}}r_{\text{out}} + i_{\text{TEST}}r_{\text{in-}\beta} \tag{10.50}$$

or
$$R_{\text{out}} = \frac{v_{\text{TEST}}}{i_{\text{TEST}}} = r_{\text{out}}(1 + A\beta) + r_{\text{in-}\beta} \tag{10.51}$$

If the condition $A\beta \gg 1$ is met, $r_{\text{in-}\beta}$ becomes an insignificant term, and R_{out} approaches the limit $R_{\text{out}} \approx A\beta r_{\text{out}}$. By increasing the loop gain $A\beta$, R_{out} can be made to approach the value of infinity desired of a current-output amplifier. This increase in output resistance again can be realized without altering the closed-loop amplifier gain, which is determined for large A solely by the feedback function β.

The large value of R_{out} can be explained by first considering the amplifier without feedback. For a current-output amplifier, R_{out} is a measure of how much i_{OUT} will be decreased as the load impedance is increased. If a load of nonzero resistance is connected to the output, a voltage will develop across the load, causing a voltage to also develop across r_{out}. The resulting current flow into r_{out} will cause less current from the dependent source to flow as i_{OUT}. With feedback connected, the reduction in i_{OUT} will also cause a reduction in the current to $r_{\text{in-}\beta}$, and hence a reduction in the feedback signal x_F. With smaller x_F, the net signal $x_{\text{IN}} = x_S - x_F$ to the amplifier will be larger, resulting in an increase in the dependent source current. This increase helps to compensate for the current being shunted through the Norton resistance r_{out}, leading to less reduction in i_{OUT} and a larger apparent R_{out}. The "tighter" the feedback loop, that is, the larger the loop gain $A\beta$, the better the correction to the output loading and the larger the apparent R_{out}.

EXERCISE 10.18 The reduction in output resistance provided by the shunt-sampling connection is realized only when the feedback loop is properly closed. Find the value of R_{out} if the feedback loop is broken.
Answer: $r_{\text{out}} \| r_{\text{in-}\beta}$

10.19 Find the value of R_{out} for the series output sampling connection if the feedback loop is broken.
Answer: $r_{\text{out}} + r_{\text{in-}\beta}$

10.7 EXAMPLES OF REAL FEEDBACK AMPLIFIERS

The previous sections of this chapter have dealt with feedback amplifiers from an abstract, theoretical point of view. Beginning with this section, we examine real amplifiers representing each of the four basic feedback topologies. These amplifiers are each discussed in the context of the theoretical principles of closed loop feedback. For some of these examples, the feedback approach may seem cumbersome, and direct analysis via KVL and KCL may appear to be an easier way of predicting circuit behavior. For such circuits, one may ask why we bother at all with the formal feedback approach. The answer to this question is two-fold. First of all, an understanding of feedback plays in important role in the design process itself. By understanding the effects of feedback, the designer can make intelligent choices in the selection of component values and circuit topologies. Secondly, an understanding of feedback greatly simplifies the analysis of circuits with nearly ideal behavior. The operational amplifier circuits of Chapter 2, for example, are good examples of circuits more easily analyzed using the formal feedback approach.

10.7.1 Op-Amp Voltage Amplifier (Series/Shunt Feedback)

As previously discussed in Example 10.1, the noninverting op-amp circuit of Fig. 10.3 is really a voltage amplifier connected in the series-input/shunt-output feedback configuration. These connections are illustrated by the equivalent diagram of Fig. 10.22, where the nonideal op-amp has finite r_{in} and nonzero r_{out}. The equivalent circuit shown for the feedback network can be derived by finding the Thévenin equivalent seen looking into each port of the network consisting of R_1 and R_2 in Fig. 10.3. In this case, the value of $r_{in-\beta}$, evaluated with the feedback connection to v_- disconnected, becomes $R_1 + R_2$. Similarly, the value of $r_{out-\beta}$, evaluated with the op-amp output shorted to ground (v_{OUT} set to zero) becomes $R_1 \| R_2$. The parallel shunt connection at the output may be difficult to recognize due to the ground connections, but it is present.

Figure 10.22
Series-input/shunt-output feedback representation of the noninverting amplifier with $\beta = R_1/(R_1 + R_2)$.

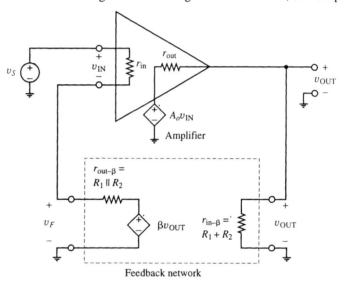

Feedback network

EXAMPLE 10.4

An op-amp with open-loop parameters $r_{in} = 1\,\text{M}\Omega$, $r_{out} = 100\,\Omega$, and $A_o = 10^5$ is connected as a noninverting amplifier with $R_1 = 10\,\text{k}\Omega$ and $R_2 = 50\,\text{k}\Omega$. Find the resulting closed-loop parameters R_{in}, R_{out}, and A_{fb} using the formal feedback representation of Fig. 10.22.

Solution

For the indicated resistor values, the feedback factor becomes

$$\beta = \frac{R_1}{R_1 + R_2} = \frac{10\,\text{k}\Omega}{10\,\text{k}\Omega + 50\,\text{k}\Omega} = \frac{1}{6} \approx 0.17 \qquad (10.52)$$

The condition $A_o\beta \gg 1$ is certainly met for the large open-loop gain of 10^5, hence the closed-loop gain A_{fb} approaches the value $1/\beta = 6$. Substitution of appropriate values into Eq. (10.40) for the input resistance of the series-mixing connection yields

$$R_{in} = r_{in}(1 + A_o\beta) + r_{out-\beta} = 1\,\text{M}\Omega[1 + (10^5)(0.17)] + (10\,\text{k}\Omega)\|(50\,\text{k}\Omega) = 16.7\,\text{G}\Omega \quad (10.53)$$

Similarly, substitution of appropriate values into Eq. (10.46) for the output resistance of the shunt-sampling connection results in

$$\begin{aligned} R_{out} &= \frac{r_{out}}{1 + A\beta}\bigg\|r_{in-\beta} \\ &= \frac{100\,\Omega}{1 + 10^5(0.17)}\bigg\|(10\,\text{k}\Omega + 50\,\text{k}\Omega) = 0.006\,\Omega \end{aligned} \qquad (10.54)$$

Note that the approximations $R_{in} \approx A\beta r_{in}$ and $R_{out} \approx r_{out}/A\beta$ are valid in this case.

10.7.2 MOSFET Transconductance Amplifier with Feedback Resistor (Series/Series Feedback)

The MOSFET in Fig. 10.23(a) functions as a transconductance amplifier. It is driven by an input voltage source v_S and pulls output current i_D down through an arbitrary load element. Feedback is provided by resistor R_F, which functions in the same way as the similarly positioned resistor of the feedback-bias configuration. As a current-output amplifier, the output must be series sampled. In this case, i_{OUT}, which flows through the load, can be sampled by sensing the current through R_F, because the i_D flowing into the drain of the MOSFET also flows out the source.

Figure 10.23
MOSFET
transconductance
amplifier.
(a) Actual
circuit. (b) Small-
signal equivalent
with two-port
feedback model.

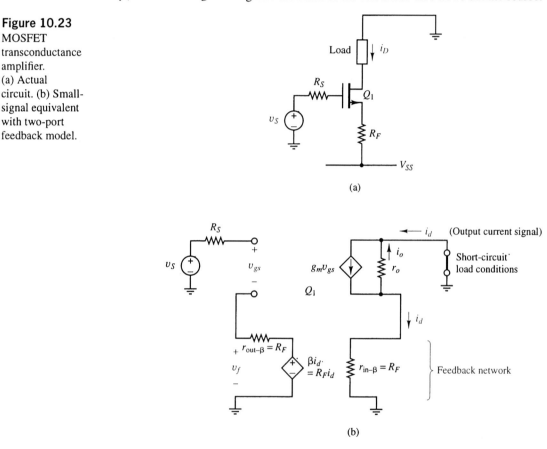

(a)

(b)

The small-signal transconductance gain i_d/v_s can be found using the small-signal analysis techniques of Chapter 7, but it can also be found using the feedback principles of this chapter. Consider the small-signal model of Fig. 10.23(b), for example. The basic MOSFET amplifier has infinite input resistance, an r_{out} equal to r_o of the MOSFET, and a transconductance gain g_m. The feedback network formed by R_F is represented as a two-port network having input resistance $r_{in-\beta} = R_F$, output resistance $r_{out-\beta} = R_F$, and a dependent-source output voltage $v_f = \beta i_d$, where $\beta = R_F$. This two-port model is easily verified by evaluating the Thévenin equivalent of the feedback network as seen by the input loop and output loop of the MOSFET circuit. The feedforward component of v_s is zero in this case because the MOSFET gate prevents direct current flow from v_s through R_F.

The open-loop transconductance gain of the small-signal amplifier is evaluated with the input-mixing disabled (β set to zero) and the current-output terminal (drain lead of Q_1) shorted to ground. This short-circuit output condition is analogous to the open-circuit condition used to evaluate voltage-output amplifiers. In evaluating the open-loop gain, it is important to include the loading effects of the feedback network. The presence of $r_{\text{out-}\beta}$ does not load down the v_S source because the gate and source terminals form an open circuit. Loading by $r_{\text{in-}\beta}$ does affect the output, however, if the condition $r_o \gg R_F$ is not met. With this loading effect included, the output current in Fig. 10.23(b) becomes, via current division,

$$i_d = g_m v_{gs} \frac{r_o}{r_o + R_F} \tag{10.55}$$

where i_d flows through R_F, and where $v_{gs} = v_s$ when $\beta = 0$. Given Eq. (10.55), the open-loop transconductance gain, as loaded by $r_{\text{in-}\beta}$, becomes

$$A_g = \frac{i_d}{v_s} = \frac{g_m r_o}{r_o + R_F} = \frac{g_m}{1 + R_F/r_o} \tag{10.56}$$

Note that this gain reverts to the value $A_g = g_m$ in the limits $r_o = \infty$ or $R_F = 0$.

As previously noted, the small-signal feedback factor β for this circuit, determined by the voltage drop across R_F, is given by

$$\beta = \frac{v_f}{i_d} = R_F \tag{10.57}$$

For the A_g and β given by Eqs. (10.56) and (10.57), the closed-loop gain becomes

$$A_{\text{fb}} = \frac{A_g}{1 + A_g \beta} = \frac{g_m/(1 + R_F/r_o)}{1 + g_m R_F/(1 + R_F/r_o)} \tag{10.58}$$

Multiplying through by $(1 + R_F/r_o)$ results in

$$A_{\text{fb}} = \frac{g_m}{1 + R_F/r_o + g_m R_F} \tag{10.59}$$

EXERCISE 10.20   For the amplifier of Fig. 10.23, derive the small-signal open-loop gain directly. Show that it is given by the same expression as Eq. (10.59).

10.7.3   Single-Transistor Transresistance Amplifier (Shunt/Shunt Feedback)

The BJT circuit of Fig. 10.24(a) is a transresistance amplifier (current input, voltage output), for which shunt-mixing/shunt-sampling (shunt/shunt) feedback is appropriate. This circuit typically might be used to convert the current signal from a high-impedance sensor into a voltage signal. Such a transformation allows a low-impedance load to be driven at a high power level without drawing significant power from the sensor. Without the feedback resistor R_F, the open-loop gain, as determined from the small-signal model of Fig. 10.24(b), would be just $v_{\text{out}}/i_s = -\beta_o R_C$, where $i_b = i_s$. Note that the Greek letter beta is used to describe the BJT parameter β_o and the feedback factor β by coincidence only.

The details of the shunt/shunt feedback topology are more easily recognized if the circuit's small-signal behavior is modeled as in Fig. 10.25, where the transresistance amplifier is represented as a two-port network with transresistance gain $A_r = v_{\text{out}}/i_s = -\beta_o R_C$, input resistance r_π, and output resistance R_C. This model for the amplifier block does not include the effects of input and output loading by the feedback network. The r_o of the BJT, which appears directly in parallel with R_C in the small-signal model, is presumed large and is omitted. The feedback resistor R_F is shown in its shunt/shunt connection topology.

Figure 10.24
(a) Transresistance
amplifier made
from a single BJT.
The feedback
network is formed
by resistor R_F.
(b) Small-signal
representation.

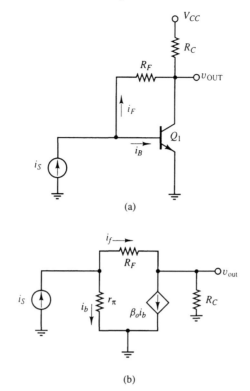

(a)

(b)

The circuit can be further refined if each port of the feedback network is represented by an appropriate equivalent circuit, as in Fig. 10.26. In this case, $r_{in\text{-}\beta} = R_F$ represents the resistance presented to the v_{out} terminal with the ouput terminals of the feedback network shorted together. The resistance $r_{out\text{-}\beta} = R_F$ and dependent source v_{out}/R_F represent the Norton equivalent seen at the output port of the feedback network with a voltage source of value v_{out} connected to the input port of the feedback network. Note that the feedback factor β, which is defined as the ratio i_f/v_{out} with a short circuit applied across the output port of the feedback network, becomes $\beta = -1/R_F$.

Figure 10.25
Feedback
representation of
the small-signal
behavior of the
circuit of
Fig. 10.24.

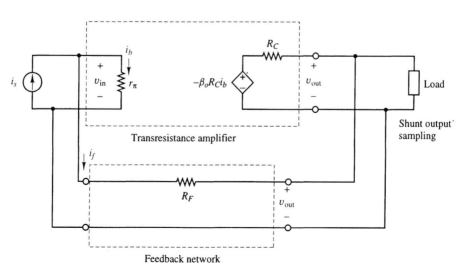

Transresistance amplifier

Shunt output
sampling

Feedback network

Figure 10.26
Feedback network
of Fig. 10.25
represented by an
equivalent circuit
taken at each of its
ports. The
feedforward
contribution at the
input port of the
feedback network
has not been
included.

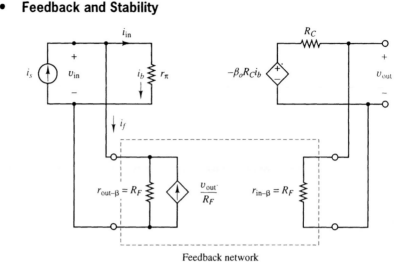

Feedback network

The circuit of Fig. 10.26 does not model the phenomenon of *feedforward* through R_F which accounts for the component of input signal that travels from the amplifier input terminals to the output via the feedback network (e.g., from left to right in Fig. 10.26). Because this signal path does not include the amplifier block, the contribution of the feedforward signal to v_{out} is small and usually can be ignored.

The role of feedback in establishing the closed-loop gain, including the effects of input and output loading by the feedback network, can be determined by analyzing Fig. 10.26. Direct application of current division to the amplifier input port results in

$$i_b = \left(i_s + \frac{v_{out}}{R_F}\right) \frac{R_F}{R_F + r_\pi} \tag{10.60}$$

This equation reflects the loading of the input port by $r_{out\text{-}\beta} = R_F$. The first term in parentheses consists of the sum of i_s and the current v_{out}/R_F from the feedback network.

The output voltage of the amplifier can be determined using voltage division between R_C and $r_{in\text{-}\beta}$:

$$v_{out} = -\beta_o R_C i_b \frac{R_F}{R_F + R_C} = -\beta_o R_C \frac{R_F}{R_F + R_C} \left(i_s + \frac{v_{out}}{R_F}\right) \frac{R_F}{R_F + r_\pi} \tag{10.61}$$

where $r_{in\text{-}\beta} = R_F$, and i_b is given by Eq. (10.60). Solving this equation for v_{out}/i_s leads to, after some algebra,

$$\frac{v_{out}}{i_s} = \frac{-\beta_o R_C \frac{R_F}{R_F + r_\pi} \frac{R_F}{R_F + R_C}}{1 + \frac{\beta_o R_C}{R_F} \frac{R_F}{R_F + r_\pi} \frac{R_F}{R_F + R_C}} \tag{10.62}$$

This equation has the form of the feedback equation $A_{fb} = A_r/(1 + A_r\beta)$, where $\beta = -1/R_F$, and A_r now represents the open-loop amplifier gain with the effects of input and output loading by the feedback network taken into account:

$$A_r = -\beta_o R_C \underbrace{\frac{R_F}{R_F + r_\pi}}_{\text{input loading}} \underbrace{\frac{R_F}{R_F + R_C}}_{\text{output loading}} \tag{10.63}$$

In the limits $R_F \gg r_\pi$ and $R_F \gg R_C$, which occur when the effects of input and output loading are insignificant, Eq. (10.62) reduces to

$$\frac{v_{\text{out}}}{i_s} = \frac{-\beta_o R_C}{1 + \frac{\beta_o R_C}{R_F}} \qquad (10.64)$$

Equation (10.64) again has the form of the classic feedback equation (10.6), with open-loop (and unloaded) gain $v_{\text{out}}/i_{\text{in}} = -\beta_o R_C$ and feedback factor $\beta = -1/R_F$. In the limit of large $\beta_o R_C$, Eq. (10.64) produces the result $A_{\text{fb}} \approx 1/\beta = -R_F$.

EXERCISE 10.21 For the circuit of Fig. 10.24, derive an expression for $A_r = v_{\text{out}}/i_s$ directly from the small-signal model. Show that A_r equals the expression obtained in Eq. (10.63), and also approaches the value $A_{\text{fb}} \approx -1/R_F$ in the limit of large R_F and $\beta_o R_C$.

10.22 Using the formulations of Section 10.6, find expressions for the overall input and output resistances of the BJT circuit of Fig. 10.24 in the limit of large $\beta_o R_C$ and R_F.

Answer: $R_{\text{in}} \approx R_F/g_m R_C$; $R_{\text{out}} \approx R_F/\beta_o$

DESIGN

EXAMPLE 10.5 Design a circuit based on the BJT transresistance amplifier of Fig. 10.24 that has a gain of -10 V/mA, an input resistance of less than 50 Ω, and an output resistance of less than 200 Ω. The circuit should operate from a single 12 V supply and be biased so that v_{OUT} is set to approximately half of V_{CC}. For the purpose of illustration, assume that a BJT with $\beta_F = \beta_o = 100$ and $V_f = 0.7$ V is available.

Solution

• Assess the goals of the problem

The circuit configuration, supply voltage, and transistor parameters are already specified. The problem therefore reduces to choosing values for R_C and R_F.

• Choose a design strategy

The problem statement specifies that v_{OUT} be biased at $V_{CC}/2$. An analysis of the bias configuration of this circuit with $i_s = 0$ reveals that

$$v_{\text{OUT}} = V_{CC} - (\beta_F + 1)I_B R_C = V_{CC} - (\beta_F + 1)\frac{V_{\text{OUT}} - V_f}{R_F}R_C \qquad (10.65)$$

where currents I_C and $I_B = (V_{\text{OUT}} - V_f)/R_F$ both flow through R_C. Applying the condition $V_{\text{OUT}} \approx V_{CC}/2$ to Eq. (10.65) leads, after some algebra, to

$$\frac{R_C}{R_F} = \frac{V_{CC}/2}{(\beta_F + 1)(V_{CC}/2 - V_f)} = \frac{6 \text{ V}}{101(6 \text{ V} - 0.7 \text{ V})} \approx 0.011 \qquad (10.66)$$

This ratio can be used to select values for R_C and R_F.

• Choose values for the resistors in the circuit

We begin the procedure by assuming the approximation $A_{\text{fb}} \approx -R_F$ to apply. The closed-loop transresistance gain of the circuit must equal -10 V/mA, hence the value $R_F = 10 \text{ k}\Omega$ seems appropriate. As suggested by Eq. (10.46), the output resistance for large β_o becomes

$$R_{\text{out}} \approx \frac{r_{\text{out}}}{A\beta} \approx \frac{R_C}{-\beta_o R_C(-1/R_F)} = R_F/\beta_o = (10 \text{ k}\Omega)/100 = 100 \text{ }\Omega \qquad (10.67)$$

This value is less than the maximum specified value of 200 Ω.

Achieving the ratio (10.66) with the chosen value of R_F requires a collector resistor of value

$$R_C = (R_C/R_F)(R_F) = (0.011)(10\,\text{k}\Omega) = 110\,\Omega \qquad \textbf{(10.68)}$$

For this value of R_C, the bias value of collector current becomes, assuming $I_B \ll I_C$,

$$I_C = \frac{V_{CC} - V_{OUT}}{R_C} = \frac{12\,\text{V} - 6\,\text{V}}{110\,\Omega} \approx 54\,\text{mA} \qquad \textbf{(10.69)}$$

so that the amplifier input resistance, as suggested by Eq. (10.44), becomes

$$R_{\text{in}} \approx \frac{r_{\text{in}}}{A\beta} = \frac{r_\pi}{-\beta_o R_C(-1/R_F)} = \frac{R_F}{g_m R_C} = \frac{R_F}{R_C}\frac{\eta V_T}{I_C} = \frac{10\,\text{k}\Omega}{110\,\Omega}\frac{(1)(0.025\,\text{V})}{54\,\text{mA}} \approx 42\,\Omega \quad \textbf{(10.70)}$$

This value is less than the maximum specified value of 50 Ω.

• Evaluate the design and revise if necessary

The computed values of R_{in}, R_{out}, and gain A_{fb} all meet the designated specifications, but in low-power applications, the bias current of 54 mA might be considered excessive for a single-BJT circuit. A smaller bias current can be realized by relaxing the constraint on R_{out}, so that a larger value of R_C can be chosen. Increasing R_C by a factor of about 5 to the value $R_C = 500\,\Omega$ results in a larger open-loop gain $-\beta_o R_C$, and a closed-loop gain that even more closely approaches the -10 V/mA design specification, but reduces the bias current considerably. This design modification is explored in more detail in SPICE Example 10.12.

EXERCISE 10.23 Estimate the gain, input resistance, and output resistance of the transresistance amplifier of Fig. 10.24 if $R_F = 100\,\text{k}\Omega$, $R_C = 5\,\text{k}\Omega$, and $\beta = 250$. **Answer:** -100 V/mA; $240\,\Omega$; $400\,\Omega$

Figure 10.27
Current amplifier with feedback.

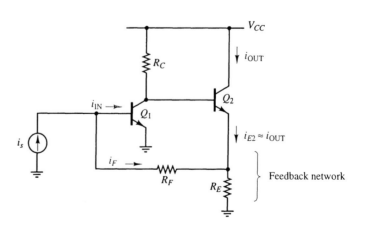

10.7.4 BJT Current Amplifier with Feedback (Shunt/Series Feedback)

The circuit of Fig. 10.27 consists of a current amplifier to which a shunt/series feedback network formed by R_F and R_E has been added. Shunt mixing of currents at the input port occurs at the base of Q_1, as it does in the circuit of the previous example. Output-current sampling of i_{OUT} is performed at the emitter of Q_2, where the current $i_{E2} \approx i_{OUT}$ is fed to the feedback network. The feedback network diverts a fraction of i_{E2} to the base of Q_1. The feedback factor β can be evaluated by temporarily grounding R_F where it connects to the base of Q_1, as in Fig. 10.28, and computing i_F via current division:

$$i_F = -i_{OUT}\frac{R_E}{R_E + R_F} = \beta i_{OUT} \tag{10.71}$$

where
$$\beta = \frac{-R_E}{R_E + R_F} \tag{10.72}$$

Figure 10.28
Feedback network of Fig. 10.27 with output port connected to ground. Feedback current i_F is determined by current division.

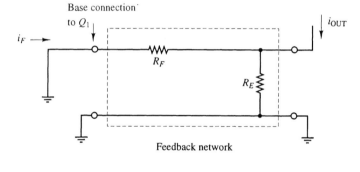

The open-loop gain of the amplifier, including the effects of loading by $r_{in\text{-}\beta}$ and $r_{out\text{-}\beta}$, can be found from the model of Fig. 10.29, where the ports of the feedback network are represented by appropriate Norton equivalent circuits. For simplicity, the output resistance $r_{out} = r_{o2}$ of Q_2 has been assumed large enough compared to $r_{in\text{-}\beta}$ to be ignored. The open-loop current gain, including the loading effects of the feedback network, but computed with the feedback factor β set to zero, is easily shown to be

$$A_i = \frac{i_{out}}{i_s} = -\beta_{o1}\beta_{o2}\underbrace{\frac{R_E + R_F}{r_{\pi 1} + R_E + R_F}}_{\text{input loading factor}}\underbrace{\frac{R_C}{R_C + r_{\pi 2} + (\beta_{o2} + 1)(R_E \| R_F)}}_{\text{output loading factor}} \tag{10.73}$$

Figure 10.29
Model of the current amplifier of Fig. 10.27 with feedback.

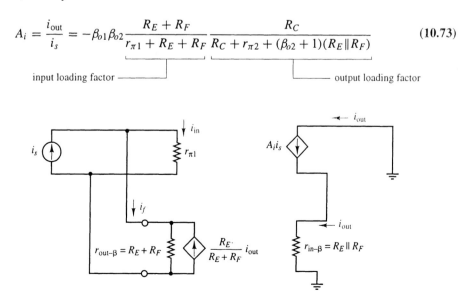

where β_{o1} and β_{o2} are the small-signal current-gain parameters of Q_1 and Q_2. With the direction of i_f defined as in Fig. 10.29, the extraction of i_f from the summation node at the base of Q_1 constitutes negative feedback. In the limit of large R_F (small feedback factor β), A_i approaches the value

$$A_i \approx -\beta_{o1}\beta_{o2}\frac{R_C}{R_C + r_{\pi 2} + (\beta_{o2} + 1)R_E} \tag{10.74}$$

which is the open-loop gain of the two BJT circuits in cascade. With an expression found for A_i, the overall gain of the circuit with the feedback connected can be found using the standard feedback formula (10.6):

$$A_{fb} = \frac{A_i}{1 + A_i\beta} \tag{10.75}$$

EXAMPLE 10.6
Find the gain of the circuit of Fig. 10.27 if $\beta_{o1} = \beta_{o2} = 200$, $r_{\pi 1} = r_{\pi 2} \approx 10\,\mathrm{k\Omega}$, $R_E = 100\,\Omega$, $R_F = 5\,\mathrm{k\Omega}$, and $R_C = 1\,\mathrm{k\Omega}$.

Solution

For the indicated resistor values, the feedback factor becomes

$$\beta = -\frac{R_E}{R_E + R_F} = -\frac{100\,\Omega}{100\,\Omega + 5\,\mathrm{k\Omega}} \approx -0.02 \tag{10.76}$$

The open-loop gain, computed with the loading by $r_{\mathrm{in}\text{-}\beta}$ and $r_{\mathrm{out}\text{-}\beta}$ taken into account, becomes, from Eq. (10.73),

$$A_i \approx -(200)(200)\frac{5.1\,\mathrm{k\Omega}}{10\,\mathrm{k\Omega} + 5.1\,\mathrm{k\Omega}}\frac{1\,\mathrm{k\Omega}}{1\,\mathrm{k\Omega} + 10\,\mathrm{k\Omega} + (201)[(100\,\Omega)\|(5\,\mathrm{k\Omega})]} \approx -440 \tag{10.77}$$

Applying Eq. (10.75) with this value of A_i results in

$$A_{fb} = \frac{A_i}{1 + A_i\beta} = \frac{-440}{1 + (-440)(-0.02)} \approx -44.9 \tag{10.78}$$

EXERCISE 10.24
Compute the gain A_{fb} directly from the small-signal model of the circuit of Fig. 10.27. Compare with the approximate result (10.78).

10.25 Find the value of R_{in} for the amplifier described in Example 10.6. **Answer:** $R_{in} \approx 850\,\Omega$

10.8 FEEDBACK-LOOP STABILITY

When feedback is applied to an amplifier, an unstable condition can occur that results in a phenomenon called *oscillation*. Oscillation is generally undesirable, because it produces an output that exists independently of any applied input signal and carries no signal information. When oscillation due to instability does occur, its magnitude is usually large, swamping out any amplified signals that may be present. In this section, the issues surrounding oscillation and the unstable feedback condition are explored in more detail.

10.8.1 Amplifier Phase Shift

As the analysis of Chapter 9 shows, an amplifier's output signal acquires an additional phase shift of $-90°$ relative to the input signal whenever the input frequency is increased beyond a pole of the amplifier's system function. If the circuit has three or more poles, as do many multistage amplifiers, a frequency will exist between the second and third poles at which the total additional phase shift contributed by the poles becomes exactly $-180°$.[3] Such a phase shift can cause negative feedback to behave like positive feedback at the $-180°$ phase-shift frequency, leading to instability and unwanted oscillations.

Figure 10.30
Bode plot of the open-loop gain of a typical multistage amplifier.

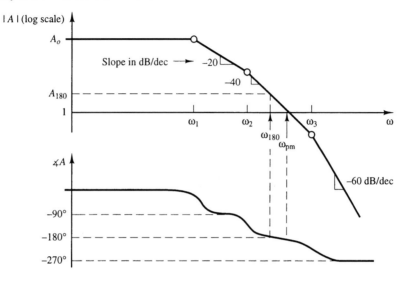

The stability issue can best be illustrated by considering the Bode plot of Fig. 10.30, which describes the open-loop magnitude and phase response of a typical multistage amplifier. Although a voltage amplifier is depicted, the analysis applies equally well to any of the four principal amplifier types discussed in this chapter.

The Bode plot of Fig. 10.30 has a dominant pole at ω_1 and two more poles at higher frequencies. The frequency at which $\angle A$ becomes $-180°$, labeled ω_{180}, lies somewhere between ω_2 and ω_3, as discussed above. Exactly at the frequency ω_{180}, $A(j\omega)$ can be expressed as

$$A(j\omega)\big|_{\omega_{180}} = -A_{180} \tag{10.79}$$

where A_{180} is a positive number less than the midband gain A_o. The cumulative phase shift of $-180°$ that occurs at ω_{180} becomes equivalent to multiplication by a factor of -1.

If the amplifier is connected in a negative-feedback loop, the closed-loop gain becomes

$$A_{\text{fb}} = \frac{V_{\text{out}}}{V_{\text{in}}} = \frac{A(j\omega)}{1 + A(j\omega)\beta} \tag{10.80}$$

where β is assumed to be independent of frequency. If V_{in} is a sinusoid of frequency ω_{180}, the response at $\omega = \omega_{180}$ predicted by Eq. (10.80) becomes

$$\frac{V_{\text{out}}}{V_{\text{in}}} = \frac{-A_{180}}{1 - A_{180}\beta} \tag{10.81}$$

---

[3] In a system function with only two poles, the phase shift will approach $-180°$ asymptotically only at "infinite" frequency.

If the product $A_{180}\beta$ happens to equal unity, the denominator of this equation becomes zero and the closed-loop gain becomes infinite. Such a condition enables an output signal of frequency ω_{180} to exist even if the input signal is zero. This phenomenon is called *oscillation*. Oscillation at ω_{180} can be started in an unstable circuit by extraneous noise or transient signals. Once initiated, an output oscillation at ω_{180} becomes self-sustaining, because no input signal is required to drive it. In designing a feedback circuit for amplification purposes, steps must be taken to ensure that self-sustained oscillation at ω_{180} does not occur. Note that oscillation is sometimes desirable and can be used to generate signals at specific frequencies. These applications are covered in Chapter 13.

It can be shown that the less-stringent inequality $A_{180}\beta \geq 1$ is sufficient to cause oscillation at ω_{180}. If $A_{180}\beta$ exceeds unity, any oscillation initiated by noise or transient signals will increase in magnitude until amplifier nonlinearity (output saturation, for example) reduces the value of A_{180} and limits further growth of the output signal.

10.8.2   Evaluating Feedback Stability Using the Nyquist Plot

The stability of a feedback loop is easily evaluated if the feedback factor β and the amplifier's open-loop transfer function $A(j\omega)$ are known. The latter can be determined by calculation, data sheet, or direct measurement. One convenient method for evaluating stability involves the use of the *Nyquist plot*. A Nyquist diagram is made by plotting the real and imaginary parts of the loop gain $A(j\omega)\beta$ in the complex plane using frequency as a plotting parameter. At any given frequency, the length and angle of the radial vector extending from the origin to the curve become equivalent to the magnitude and phase angle, respectively, of the product $A(j\omega)\beta$.

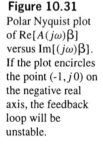

Figure 10.31
Polar Nyquist plot of $\text{Re}[A(j\omega)\beta]$ versus $\text{Im}[(j\omega)\beta]$. If the plot encircles the point $(-1, j0)$ on the negative real axis, the feedback loop will be unstable.

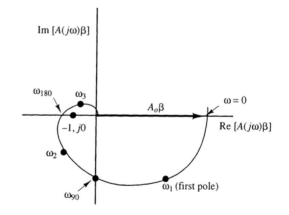

The typical open-loop amplifier approaches a constant gain A_o and zero phase angle as ω approaches zero. If such an amplifier is connected in a feedback loop with feedback factor β, the Nyquist plot of its loop gain, shown in Fig. 10.31, will begin on the positive real axis, where $\text{Re}[A(j\omega)\beta] = A_o\beta$ and $\text{Im}[A(j\omega)\beta] = 0$. As the frequency is increased through the first pole, the imaginary part of the denominator of $A(j\omega)\beta$ will increase, causing $\text{Im}[A(j\omega)\beta]$ to become negative, and causing the Nyquist plot to head toward the negative imaginary axis. As the frequency parameter passes beyond the first pole, the angle of $A(j\omega)\beta$ will reach $-90°$ at the frequency ω_{90}, where the real part of $A(j\omega)\beta$ becomes zero.

As ω passes beyond the amplifier's second pole, the angle of $A(j\omega)\beta$ will reach $-180°$ at ω_{180}, where the imaginary part of $A(j\omega)\beta$ will become zero and the real part of $A(j\omega)\beta$ negative. If unwanted oscillations are to be prevented, the stability condition $|A(j\omega)\beta| < 1$ must be met at ω_{180}. This condition is tantamount to the requirement that the magnitude of $\text{Re}[A(j\omega)\beta]$ be less than unity at ω_{180}. The circuit will thus be stable only if its Nyquist plot does not encircle the point $(-1, j0)$ on the negative-real axis. If the plot encircles this critical point, the circuit will be unstable and will oscillate at ω_{180}.

10.8.3 Bode-Plot Analysis of Feedback Stability

The stability of a feedback loop can also be assessed by examining its Bode plot and evaluating one of two related parameters. One parameter, called the *gain margin*, is defined as the difference between unity and the magnitude of $A(j\omega)\beta$ at ω_{180}:

$$\text{Gain margin} \triangleq 1 - |A(j\omega)\beta|_{\omega_{180}} \equiv 1 - A_{180}\beta \tag{10.82}$$

where $A_{180} = |A(j\omega_{180})|$. If the stability condition $A_{180}\beta < 1$ is to be met, the gain margin must be positive. If the gain margin is negative, the circuit will be unstable and will oscillate at ω_{180}. In a stable feedback amplifier, the gain margin represents the amount by which the loop gain can be increased before instability occurs. Note that the closed-loop gain $A_{fb} = 1/\beta$ becomes smaller as β is increased, hence the stability problem becomes worse for *smaller* closed-loop gains.

The gain margin (10.82) is sometimes evaluated with each of the terms expressed in decibels:

$$\text{Gain margin} \triangleq 0\,\text{dB} - 20\,\log_{10}(A_{180}\beta) \tag{10.83}$$

This alternative representation is often useful for evaluating the stability condition, as we shall see shortly.

A second useful parameter, called the *phase margin*, is evaluated at the frequency ω_u where the magnitude of $A(j\omega)\beta$ reaches unity. The phase margin is defined as the difference between the angle of $A(j\omega)\beta$ at ω_u and the angle $-180°$:

$$\text{Phase margin} \triangleq \sphericalangle A(j\omega)\beta\Big|_{\omega_u} - (-180°) = 180° + \sphericalangle A(j\omega_u)\beta \tag{10.84}$$

Note that $\sphericalangle A(j\omega_u)\beta$ is negative for the typical multipole system function.

A positive phase margin indicates that $|A(j\omega)\beta|$ is reduced to unity *before* its angle becomes $-180°$, implying a stable amplifier. A negative phase margin indicates that $|A(j\omega)\beta|$ is reduced to unity *after* its angle reaches $-180°$, implying that $|A(j\omega)\beta|$ will exceed unity *at* the frequency ω_{180}. This latter condition, of course, results in instability.

The gain and phase-margin parameters are not independent, and can be derived in terms of each other. In general, doubling the gain margin of a given amplifier, for example, will result in a doubling of its phase margin, and vice versa.

DESIGN

EXAMPLE 10.7

The manufacturer of a particular operational amplifier states that it has a three-pole open-loop transfer function given at room temperature by

$$A(j\omega) = \frac{V_{\text{out}}}{V_{\text{in}}} = \frac{A_o}{(1 + j\omega/\omega_1)(1 + j\omega/\omega_2)(1 + j\omega/\omega_3)} \tag{10.85}$$

where $A_o = 10^6$, $\omega_1 = 10\,\text{rad/s}$, and $\omega_2 = \omega_3 = 10^6\,\text{rad/s}$. In real use, the poles of the op-amp are as specified, but temperature and fabrication variations may cause A_o to vary by as much as $\pm 10\,\text{dB}$. Design an amplifier using this op-amp and an appropriate frequency-independent feedback network. The closed-loop amplifier should have a bandwidth of at least 25 kHz and be stable for all expected values of A_o. What range of closed-loop gain is possible given these specifications?

Solution

• Assess the goals of the problem

The primary specifications are on bandwidth and stability, both of which depend on the choice of feedback factor β. Because the closed-loop gain will be essentially equal to $1/\beta$, the range of possible gain values becomes a secondary byproduct of the design process. The principal focus is on finding the range of possible values of β to meet bandwidth and stability constraints.

• Choose a design strategy

We first evaluate the manufacturer-specified transfer function (10.85) to find its value of A_{180}. Given the expected variations in A_o, we may then find the minimum gain margin required for stability. This minimum gain margin will, in turn, determine the maximum allowed value of β. Note that a temperature-related decrease in A_o will serve to improve the gain margin of the closed-loop amplifier, but an increase in A_o may cause A_{180} to rise above unity, leading to instability.

We next address the bandwidth specification by computing the op-amp's gain–bandwidth product. The latter applies to the op-amp whether under open-loop or closed-loop conditions. The maximum possible gain, and hence the minimum possible β, can be determined from the bandwidth specification.

• Find the value of A_{180} for the specified transfer function:

Finding the value A_{180} first requires knowledge of the frequency ω_{180} at which the angle of $A(j\omega)$ reaches $-180°$. Because β is frequency independent, ω_{180} is determined solely by $A(j\omega)$ and can be found by setting the expression for $\angle A(j\omega)$ to the value $-180°$:

$$\angle A(j\omega_{180}) = -\tan^{-1}\frac{\omega_{180}}{\omega_1} - \tan^{-1}\frac{\omega_{180}}{\omega_2} - \tan^{-1}\frac{\omega_{180}}{\omega_3} = -180° \qquad (10.86)$$

Solving Eq. (10.86) for ω_{180} (e.g., by trial and error) using the pole values specified in Eq. (10.85) results in $\omega_{180} \approx 10^6$ rad/s. The magnitude $A_{180} = |A(j\omega_{180})|$ at this frequency is found by substituting ω_{180} into (10.85) and taking the magnitude:

$$A_{180} = \left| \frac{10^6}{[1 + (j10^6/10)][1 + (j10^6/10^6)]^2} \right| \approx \frac{10^6}{(10^5)(\sqrt{2})^2} = 5 \qquad (10.87)$$

For this value of A_{180}, stability of the feedback loop, as determined by the condition $A_{180}\beta < 1$, requires that β be less than $1/5 = 0.2$. Note that the smaller the value of β (and the *larger* the closed-loop gain), the more stable the amplifier.

• Determine the minimum gain margin for stability

As stated at the outset, the value of A_o for this amplifier, and hence the A_{180} computed in (10.87), may increase by as much as 10 dB. A minimum gain margin of 10 dB, evaluated with $A_o = 10^6$, is thus required if the feedback loop is to be stable. This requirement can be expressed, via Eq. (10.83), as $-20 \log_{10}(A_{180}\beta) \geq -10$ dB, or

$$\beta \leq \frac{10^{(-10\,\text{dB}/20\,\text{dB})}}{A_{180}} = \frac{0.316}{5} = 0.063 \qquad (10.88)$$

This feedback factor represents the largest that can be tolerated if stability is to be ensured over the complete range of A_o. A feedback factor of this value corresponds to a minimum gain of $1/\beta = 15.8$.

• Find the gain–bandwidth product of the amplifier

Since ω_1 is much smaller than ω_2 and ω_3, the dominant pole of $A(j\omega)$ occurs at $f_L = \omega_1/2\pi$, and the bandwidth of the op-amp becomes $A_o f_L = (10^6)(10\,\text{rad/s})(2\pi)^{-1} \approx 1.6\,\text{MHz}$. If the minimum bandwidth of the closed-loop amplifier is to be 25 kHz, the maximum possible closed-loop gain becomes

$$A_{\text{fb}} = \frac{A_o f_L}{\text{BW}} = \frac{1.6\,\text{MHz}}{25\,\text{kHz}} = 64 \tag{10.89}$$

for which the minimum feedback factor is $\beta = 1/64 = 0.016$

Figure 10.32

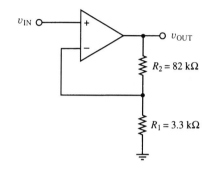

• Choose a topology and resistor values for the feedback network

A feedback factor between the extremes 0.063 and 0.016 can be easily achieved using a resistive divider connected in the series/shunt feedback configuration, as in Fig. 10.32. We recognize this amplifier as the noninverting op-amp circuit of Chapter 2. For the indicated resistor values, the feedback factor becomes $\beta = R_1/(R_1 + R_2) = 0.039$, which clearly lies between the computed minimum and maximum limits. At this value of β, the gain becomes $1/\beta = 25.8 \equiv 28.2\,\text{dB}$ and the bandwidth becomes $A_o f_L/A_{\text{fb}} = (1.6\,\text{MHz})/(25.8) \approx 62\,\text{kHz}$. Evaluating the gain margin in dB results in

$$\begin{aligned}
\text{Gain margin} &= 0\,\text{dB} - 20\log_{10}(A_{180}\beta) \\
&= 0\,\text{dB} - 20\log_{10} A_{180} - 20\log_{10}\beta \\
&= 0\,\text{dB} - 20\log_{10} 5 - 20\log_{10} 0.039 \\
&= 0\,\text{dB} - 14\,\text{dB} - (-28\,\text{dB}) = 14\,\text{dB} \tag{10.90}
\end{aligned}$$

EXAMPLE 10.8

The voltage amplifier of Example 10.7 is connected in a series/shunt feedback loop with $\beta = 0.1$. This value of β results in a stable feedback circuit when $A_o = 10^6$. Find its gain and phase margins.

Solution

If A_o remains fixed at 10^6, the feedback circuit is indeed stable, because the value $\beta = 0.1$ yields

$$\text{Gain margin} = 1 - |A(j\omega)\beta|_{\omega_{180}} = 1 - (5)(0.1) = 0.5 \tag{10.91}$$

which is less than unity. Expressed in decibels, the gain margin becomes $0\,\text{dB} - 20\log_{10} 0.5 = 6\,\text{dB}$.

Computing the phase margin requires that ω_u, the frequency at which $|A(j\omega)\beta|$ equals unity, be found. Because β equals 0.1, the frequency at which $|A(j\omega)\beta| = 1$ is also the frequency at which $|A(j\omega)| = 10$:

$$
|A(j\omega_u)| = \left| \frac{10^6}{(1 + j\omega_u/10)(1 + j\omega_u/10^6)^2} \right|
$$
$$
= \frac{10^6}{\left(1 + \omega_u^2/100\right)^{1/2} (1 + \omega_u^2/10^{12})} = 10 \tag{10.92}
$$

Equation (10.92) can be solved for ω_u in closed form or by iteration on a computer to obtain the value $\omega_u = 0.682 \times 10^6$ rad/s. The angle of $A(j\omega)$ at this frequency becomes

$$
\angle A(j\omega) \equiv -\tan^{-1}\frac{\omega_u}{10} - 2\tan^{-1}\frac{\omega_u}{10^6} \approx -90° - 2 \times 34.3° = -158.6° \tag{10.93}
$$

Using this value of ω_u, the phase margin can be calculated:

$$
\text{Phase margin} = 180° - 158.6° = 21.4° \tag{10.94}
$$

Both the gain and phase margins of this amplifier are well above the zero values that mark the stability limit.

EXERCISE 10.26 Find the values of gain and bandwidth at each of the β limits found in Example 10.7.

10.27 Consider the amplifier of Example 10.7. If connected in a feedback circuit with $\beta = 1$ and $A_o = 10^6$, the circuit will be unstable. What will be the gain margin, phase margin, and frequency of oscillation in this case?   **Answer:** -4; $-36.9°$; 160 kHz

10.28 An op-amp having the transfer function of Example 10.7 is connected in the inverting configuration, where R_2 is the feedback resistor and R_1 the input resistor. What values of R_2 and R_1 result in a minimally stable circuit?   **Answer:** $R_2/R_1 = 5$

10.29 Write a computer program that will precisely yield the value of $\omega_u = 0.682 \times 10^6$ rad/s given in Example 10.8.

10.8.4 Frequency Compensation

Unwanted oscillation in a feedback circuit can be prevented by the use of *frequency compensation*. Compensation consists of altering the loop gain $A(j\omega)\beta$ of an unstable configuration so that the stability condition $A_{180}\beta < 1$ is met. This alteration consists of adding another pole to the amplifier's transfer function, by either modifying its internal circuitry or by adding an external network to it.

The concept of frequency compensation can be illustrated by considering an amplifier with transfer function

$$
A_1(j\omega) = \frac{A_o}{(1 + j\omega/\omega_1)(1 + j\omega/\omega_2)(1 + j\omega/\omega_3)} \tag{10.95}
$$

where $\omega_1 < \omega_2 < \omega_3$. As previously discussed, the phase angle of this three-pole transfer function will reach $-180°$ at some frequency between ω_2 and ω_3. Suppose that an additional *compensation* pole ω_c is added to the amplifier's transfer function, causing an additional binomial term $(1 + j\omega/\omega_c)$ to appear in its denominator. The resulting transfer function becomes

$$A_2(j\omega) = A_1(j\omega)\frac{1}{1 + j\omega/\omega_c}$$

$$= \frac{A_o}{(1 + j\omega/\omega_1)(1 + j\omega/\omega_2)(1 + j\omega/\omega_3)(1 + j\omega/\omega_c)} \tag{10.96}$$

If ω_c is chosen to be smaller than the original ω_{180}, the new ω_{180} will be smaller in frequency, since ω_c will contribute its negative phase shift before the original ω_{180} is reached. If ω_c is appropriately placed at a low enough frequency, it can cause the magnitude of $A(j\omega)\beta$ to fall below unity before the new ω_{180} is reached, resulting in a stable circuit. This concept is illustrated in the next example.

EXAMPLE 10.9

An amplifier having the three-pole open-loop response of Fig. 10.33, with a dc gain of 100 dB and poles at 10^5, 10^6, and 10^7 rad/s is connected in a negative-feedback loop. The poles are spaced closely enough so that the phase angle decreases continuously between the first and third poles.

(a) Show that the maximum allowed β to ensure stability is 0.001.

(b) To what value of closed-loop gain does this β correspond?

(c) Determine the placement of a fourth pole ω_c such that the amplifier will be stable when β is increased to 0.02, for a closed-loop gain of 50.

Figure 10.33
Bode plot of the transfer function described in Example 10.9.

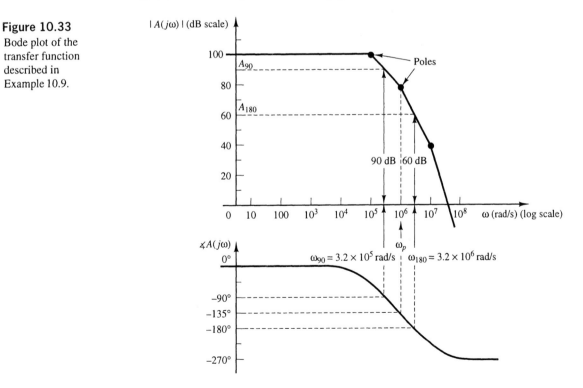

Solution

Without the added pole ω_c in place, the value $\omega_{180} \approx 3.2 \times 10^6$ rad/s is graphically obtained from the angle plot of Fig. 10.33. The magnitude of $A(j\omega)$ at this frequency as obtained from the magnitude plot is seen to be $A_{180} = 60\,\text{dB} \equiv 1000$. Meeting the stability condition $A_{180}\beta < 1$ with an A_{180} of 1000 requires that β be no greater than 0.001. This value corresponds to a minimum closed-loop gain of $1/\beta = 1000$. If connected in a feedback loop with $\beta = 0.02$, for a gain of $1/\beta = 50$, the amplifier will be unstable.

For the purpose of identifying the required location of the pole ω_c that must be added to the response, we identify the frequency $\omega_{90} = 3.2 \times 10^5\,\text{rad/s}$ in Fig. 10.33 at which $\measuredangle A(j\omega) = -90°$. If ω_c is located well below ω_{90}, the phase contribution of ω_c at ω_{90} will be an additional $-90°$, so that the total phase shift at ω_{90} will become $-180°$. With the added pole ω_c in place, the stability condition becomes

$$A_{90}\beta < 1 \qquad (10.97)$$

where A_{90} is the gain at the $-90°$ phase-shift frequency ω_{90} *before* the addition of the pole ω_c. Equation (10.97) can also be expressed in decibel form as

$$|A_{90}|_{\text{dB}} + |\beta|_{\text{dB}} < 0 \qquad (10.98)$$

The uncompensated transfer function of Fig. 10.33 has a gain magnitude of 90 dB at ω_{90}. If a feedback amplifier with a closed-loop gain of 50, or 34 dB, is to be made, β must be set to 0.02, or -34 dB. If the revised stability condition (10.97) is to be met with ω_c in place, A_{90} must be reduced via ω_c from 90 dB to a maximum of 34 dB, so that $A_{90}\beta$ will not exceed unity.

The required maximum location of ω_c can be found by noting that it will initiate a slope of $-20\,\text{dB/decade}$ in $A(j\omega)$ at frequencies well above ω_c. This decline will be superimposed on any gain reduction introduced by the other poles of $A(j\omega)$. At ω_{90}, the gain reduction produced by ω_c will be subtracted from the preexisting A_{90}, as depicted in Fig. 10.34. The gain reduction

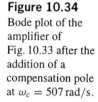

Figure 10.34

Bode plot of the amplifier of Fig. 10.33 after the addition of a compensation pole at $\omega_c = 507\,\text{rad/s}$.

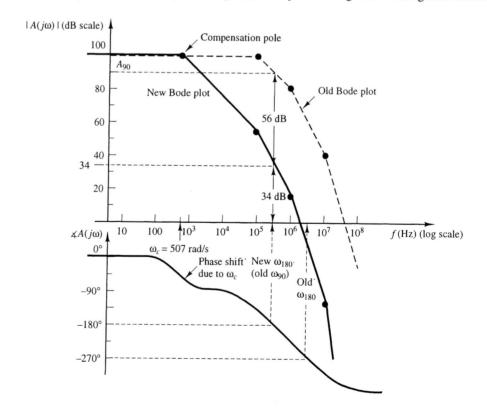

must be at least $-(90\,\text{dB} - 34\,\text{dB}) = -56\,\text{dB}$, hence ω_c must be located below ω_{90} by at least

$$\frac{-56\,\text{dB}}{-20\,\text{dB/decade}} = 2.8\,\text{decades} \tag{10.99}$$

in frequency. Expressed in radians/second, the highest possible ω_c for compensation therefore becomes

$$\omega_c = \frac{\omega_{90}}{10^{2.8}} = \frac{3.2 \times 10^5\,\text{rad/s}}{631} = 507\,\text{rad/s} \tag{10.100}$$

A plot of the revised transfer function with ω_c set to the maximum allowed frequency of $507\,\text{rad/s}$ is shown in Fig. 10.34.

EXAMPLE 10.10   In practice, a phase margin of at least $45°$ is desirable to avoid peaks in response near the dominant pole. Determine the placement of ω_c for the amplifier of the previous example such that the phase margin equals $45°$ for the largest possible feedback factor $\beta = 1$.

Solution

With $\beta = 1$, the magnitude of $A(j\omega)\beta$ becomes identical to $|A(j\omega)|$ alone. Let ω_p be the frequency at which the angle of $A(j\omega)$ equals $-90° - 45° = -135°$. On the plot of Fig. 10.33, it can be seen that $\omega_p = 10^6\,\text{rad/s}$. If ω_c is located well below ω_p, the angle contribution of ω_c at ω_p will be $-90°$, increasing the total phase shift at ω_p to $-90° - 135° = -225°$. If the magnitude of $A(j\omega)$ can be reduced to unity at ω_p, this revised phase shift at ω_p will result in the desired phase margin of $180° - 225° = -45°$. As shown in Fig. 10.33, the magnitude of $A(j\omega)$ equals $80\,\text{dB}$ at ω_p without the compensation pole ω_c. The latter must be placed so as to reduce $|A(j\omega_p)\beta|$ by $-80\,\text{dB}$. The gain reduction introduced by ω_c will proceed at $-20\,\text{dB/decade}$, hence ω_c must be located a distance of $-80\,\text{dB}/(-20\,\text{dB/decade}) = 4\,\text{decades}$ below ω_p:

$$\omega_c = \frac{\omega_p}{10^4} = 100\,\text{rad/s} \tag{10.101}$$

10.8.5 External Compensation

The method by which ω_c can be added to the loop gain $A(j\omega)\beta$ may involve any of several alternatives. The method of *internal compensation*, examined in detail in Chapter 12, is generally the preferred method. Internal compensation involves modifying the internal structure of the open-loop amplifier so that an additional compensation pole ω_c is introduced to guarantee stability. When internal compensation is used, the value of ω_c becomes independent of the external feedback network. In an op-amp, internal compensation is set by the manufacturer at fabrication time, hence designing for stability involves choosing the right op-amp subject to the constraints of the required gain. Manufacturers's specifications for a given op-amp indicate the maximum permitted feedback factor β_{max} if the circuit is to be stable. The reciprocal of β_{max} determines the minimum closed loop gain for which stability can be guaranteed. The popular LM741 op-amp, for example, is fully compensated so as to be stable for all values of β up to 1. For this particular op-amp, even a unity-gain amplifier will be stable. The undercompensated LF357 amplifier will only be stable for $\beta < 0.2$; any feedback configuration using the LF357 must have a closed-loop gain greater than 5. In general, the smaller the value of β_{max}, the greater the gain-bandwidth product of the amplifier.

Figure 10.35
External compensation by adding a simple RC circuit between v_{OUT} and the op-amp output terminal.

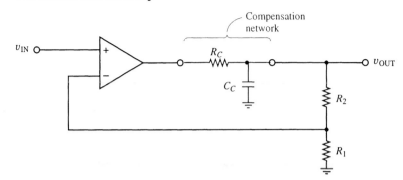

An amplifier can also be compensated using *external compensation*, in which the compensation pole ω_c is introduced via an external RC network. In principle, the connection illustrated in Fig. 10.35, in which a simple RC filter has been cascaded with the amplifier's output port and included inside the feedback loop, can be used to introduce the required pole. If the RC filter is not significantly loaded by the R_2–R_1 feedback network, elements are simply chosen so that $1/R_C C_C = \omega_c$. This method of compensation is seldom used in practice, however, because it substantially reduces the gain-bandwidth product of the op-amp, and thus reduces the bandwidth of the closed-loop amplifier.

Figure 10.36
External compensation network for an undercompensated op-amp in a unity-gain configuration.

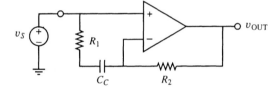

If an undercompensated op-amp *must* be used with $\beta > \beta_{max}$, the external compensation technique illustrated in Fig. 10.36 may also be employed. Connections are shown here for a unity-gain amplifier. At low frequencies, C_C behaves as an open-circuit, causing the circuit to acquire the topology of a unity-gain voltage follower in which R_2 is the sole feedback element. At higher frequencies, C_C behaves as a short circuit, adding R_1 to the feedback network. In this latter case, β is reduced to $R_1/(R_1 + R_2)$. If the revised β is reduced below β_{max} for $\omega > \omega_c$, the resulting amplifier will be stable. At high frequencies where C_C acts as a short, v_S is applied as a simultaneous input to the inverting and noninverting op-amp configurations formed by R_1 and R_2. Use of superposition leads to

$$v_{out} = \frac{R_2 + R_1}{R_1} v_S - \frac{R_2}{R_1} v_S = \frac{R_1}{R_1} v_S = v_S \quad \textbf{(10.102)}$$

which implies unity gain.

EXAMPLE 10.11 A unity gain amplifier is to be constructed from an op-amp for which $f_{180} = 1\,\text{MHz}$ and $\beta_{max} = 0.2$. Choose suitable resistor and capacitor values using the arrangement of Fig. 10.36 such that the amplifier will be stable at all frequencies.

Solution

The value of β at high frequencies must be reduced to a maximum of 0.2, hence the resistors must be chosen so that $R_1/(R_1 + R_2) < 0.2$. Arbitrarily selecting an R_2 of $10\,\text{k}\Omega$ yields

$$R_1 < \frac{0.2 R_2}{1 + 0.2} = 0.17(10\,\text{k}\Omega) = 1.7\,\text{k}\Omega \quad \textbf{(10.103)}$$

Choosing a value of $R_1 = 1\,\text{k}\Omega$ leaves ample gain and phase margins. The frequency at which C_C undergoes the transition from open-circuit to short-circuit behavior can be computed by finding the Thévenin resistance seen by C_C. Given that the v_+ and the v_- terminals of the op-amp form a virtual short circuit, this resistance is equal simply to R_1. Setting the compensation frequency to a value below $f_{180} = 1\,\text{MHz}$ requires that C_C be greater than $1/(2\pi f_{180} R_1) = 1/(2\pi)(1\,\text{MHz})(1\,\text{k}\Omega) \approx 160\,\text{pF}$.

EXERCISE 10.30 Consider the amplifier described by the open-loop response of Fig. 10.33. By how much must the gain at ω_{90} be reduced if the amplifier is to be stable in all negative-feedback configurations, including those with $\beta = 1$? Below what frequency must ω_c be located?   **Answer:** $-90\,\text{dB}$; $10.1\,\text{rad/s}$

10.31 Select the value of R_1, R_2 and C_C in Example 10.11 if $\beta_{\text{max}} = 0.1$ and $f_{180} = 2\,\text{MHz}$.

SUMMARY

- Feedback is an important tool used in many circuit design applications.
- Negative feedback occurs when sampled output is used to reduce the input.
- Feedback improves the linearity, input resistance, output resistance, and frequency response of an amplifier at the expense of reduced gain magnitude.
- The open-loop gain of an amplifier used in a feedback loop is usually made very large so that the closed-loop gain will depend only on the values of the feedback elements.
- An amplifier may take the form of a voltage amplifier, current amplifier, transresistance amplifier, or transconductance amplifier. Each of these amplifier types requires a different feedback configuration.
- The closed-loop gain of a feedback amplifier with large loop gain $A\beta$ is approximately equal to $1/\beta$.
- The feedback factor β describes the portion of the output signal that is fed back for input mixing.
- Series input mixing is used when the input to the amplifier is a voltage. Shunt input mixing is used when the input is a current.
- Shunt output sampling is used when the output of the amplifier is a voltage. Series output sampling is used when the output is a current.
- Negative feedback reduces the resistance of a shunt-connected port and increases the resistance of a series-connected port.
- A negative-feedback amplifier can become unstable if the loop gain $A\beta$ undergoes a phase shift of $-180°$ with increasing frequency before its magnitude falls below unity.
- The gain and phase margins describe how close an amplifier is to its stability limit. A stable amplifier has positive gain and phase margins; an unstable amplifier has negative margins.
- Analysis using Nyquist and Bode plots can determine the state of stability of a feedback amplifier.
- A feedback amplifier can be made stable via the method of frequency compensation, in which a low-frequency pole is introduced into the loop gain. A compensation pole reduces the magnitude of $A\beta$ to unity before the $-180°$ phase-shift frequency is reached.

◆ **SPICE EXAMPLES**

EXAMPLE 10.12   Example 10.5 concerned the design of a BJT transresistance amplifier having the layout shown in Fig. 10.24. Hand calculation yielded the resistor values $R_C = 110 \, \Omega$ and $R_F = 10 \, k\Omega$ as reasonable choices for meeting the specifications $A_{fb} = -10 \, V/mA$, $R_{in} < 50 \, \Omega$, and $R_{out} < 200 \, \Omega$. Use SPICE to verify the validity of the design with $R_C = 110 \, \Omega$ and $R_F = 10 \, k\Omega$. Vary R_C over the range $100 \, \Omega$ to $1 \, k\Omega$ and observe the effects of R_C on the actual amplifier parameters.

Solution

The circuit contains only a few components and is easily simulated in SPICE. In the input listing that follows, the BJT is connected with its collector at node 2, its base at node 3, and its emitter at node 0 (ground). The small-signal amplifier parameters are determined using the .TF (small-signal transfer function) command for each value of R_C. The value of R_C is stepped from $100 \, \Omega$ to $1 \, k\Omega$ in increments of $100 \, \Omega$ using the .STEP command.

Input File

```
TEST OF TRANSRESISTANCE AMPLIFIER DESIGN of EXAMPLE 10.5

*Enter the supply voltage and input current source:
      VCC 1 0 12V
      iS  0 3 0

*Specify a transistor with βF = 100:
      Q1 2 3 0 BJT
      .MODEL BJT NPN(BF=100)

*Specify the resistor values:
      RF 2 3 10k
*Vary RC from 100Ω to 500Ω in steps of 100Ω using a .MODEL statement,
*a .STEP command, and the variable parameter RES (Rstep is the name of the
*.MODEL statement):
      RC 1 2 Rstep 1
      .MODEL Rstep RES
      .STEP RES Rstep(R) 100 500 100

*Compute the small-signal transfer function with iS as the input and V(2)
*as the output variable (analysis will be performed for each
value of RC):
      .TF V(2) iS
      .END
```

Results. The output listing produced by SPICE for the fixed resistor values $R_C = 110 \, \Omega$ and $R_F = 10 \, k\Omega$ (.STEP command omitted from the simulation) follows. Note that the values of A_{fb}, R_{in}, and R_{out} calculated by SPICE differ significantly from those targeted in Example 10.5. These discrepancies occur because the approximations $A_{fb} \approx 1/\beta$, $R_{in} \approx r_{in}/\beta$, and $R_{out} \approx r_{out}/\beta$ used in Example 10.5 are really only valid when the loop gain $A_r\beta$ is very much larger than unity. For the one-stage BJT amplifier of Fig. 10.24, this condition is only weakly met.

　　The results for values of R_C from $100 \, \Omega$ to $1 \, k\Omega$ are summarized in Table 10.3. As shown by these results, increasing R_C increases the value of A_r, which also increases the loop gain. Larger A_r makes the actual closed-loop gain more closely approach the ideal of $1/\beta = -10 \, mA/V$. Increasing R_C also reduces the bias current through Q_1. At $R_C = 100 \, \Omega$, I_C is equal to almost 56 mA. At $R_C = 1 \, k\Omega$, I_C is reduced to about 10 mA. Note that R_C cannot be increased indefinitely because the bias value of V_{OUT} decreases with increasing R_C. At some value of R_C, the BJT will saturate and the BJT will no longer operate in its active region. ∎

$R_C\,(\Omega)$	$V_{OUT}\,(V)$	$I_C\,(mA)$	$R_{in}\,(\Omega)$	$R_{out}\,(\Omega)$	$A_{fb}\,(V/mA)$
100	6.4	55.8	23.4	49.9	−4.96
200	4.6	37.2	23.6	66.5	−6.61
300	3.6	27.9	23.9	75.0	−7.43
'400	3.1	22.3	24.1	80.1	−7.92
500	2.7	18.6	24.3	83.6	−8.24
600	2.4	16.0	24.5	86.2	−8.48
700	2.2	14.0	24.7	88.2	−8.65
800	2.1	12.4	25.0	89.8	−8.98
900	1.9	11.2	25.2	91.1	−8.90
1000	1.8	10.2	25.4	92.2	−8.99

Table 10.3. Results of SPICE Simulation for Values of R_C from $100\,\Omega$ to $1\,k\Omega$

Output Listing (for $R_C = 110\,\Omega$)

```
************* Evaluation PSpice (January 1993) ************
TEST OF TRANSRESISTANCE AMPLIFIER DESIGN of EXAMPLE 10.5

****   SMALL-SIGNAL-BIAS SOLUTION  ** TEMPERATURE =   27.000 DEG C
NODE   VOLTAGE    NODE   VOLTAGE    NODE   VOLTAGE
( 1)   12.0000    ( 2)   6.1459     (3 )   .8767

VOLTAGE SOURCE CURRENTS
     NAME           CURRENT
     VCC          -5.322E-02

****      SMALL-SIGNAL CHARACTERISTICS

     V(2)/iS = -5.198E+03    (transresistance gain in V/A)
     INPUT RESISTANCE AT iS =  2.345E+01     (23.4 Ω)
     OUTPUT RESISTANCE AT V(2) =  5.224E+01  (52.2 Ω)

     JOB CONCLUDED
```

EXAMPLE 10.13   In this example, the effect of feedback on amplifier linearity is demonstrated using SPICE. The circuit of Fig. 10.37 shows an amplifier connected in the noninverting feedback configuration. With no feedback connected, the amplifier has a highly nonlinear open-loop gain. The dependent polynomial source E1 samples v_{IN} and produces a voltage $v(3) = 10v_{IN} - 0.5v_{IN}^2$ at node 3. The linear dependent source E2 multiplies this voltage by a gain factor of $A_o = 100$ to produce v_{OUT}. When the feedback network is connected, the relationship between v_{OUT} and v_S becomes much less nonlinear.

Figure 10.37
Nonlinear amplifier
in a feedback loop.

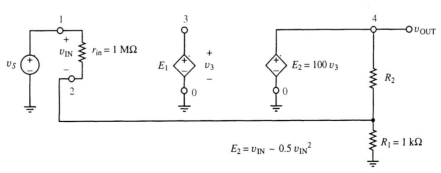

$$E_2 = v_{IN} - 0.5\,v_{IN}^2$$

Simulate the circuit in SPICE with the feedback network disconnected. Plot the dc transfer characteristic for $0 < v_S < 0.1$ V and observe its nonlinearity. Next connect the feedback network and plot the transfer characteristic using feedback factors of 0.001, 0.01, and 0.1. Compare the results in each case.

Solution

An appropriate input file for the simulation follows. The feedback factor β, which is equal to $R_1/(R_1 + R_2)$, can be set to zero, 0.001, 0.01, and 0.1 using resistor values $R_1 = 1 \, \text{k}\Omega$ and $R_2 = \infty$, $999 \, \text{k}\Omega$, $99 \, \text{k}\Omega$, and $9 \, \text{k}\Omega$, respectively.

Input File

```
EFFECT of FEEDBACK ON NONLINEAR AMPLIFIER

*Enter the input source and input resistance of the amplifier:
      vS   1 0 0V
      rin 1 2 1MEG

*Make the amplifier nonlinear using a dependent polynomial source:
      E1   3 0 poly(1) 1,2  0  1  -0.5
*Set the open-loop gain Ao of the amplifier (last coefficient of E2):
      E2   4 0 3 0 100
*                     ^-------- Gain Ao

*Connect the feedback network in the noninverting configuration:
      R2   4 2 9k
      R1   2 0 1k
*(These resistor values are for β = 0.1.  Use R2 = 99k for β = 0.01,
*R2 = 999k for β = 0.001, and R2 = ∞ for β = 0).

*Plot the dc transfer characteristic for positive vS:
      .DC vS  0 1 0.01
      .PROBE v(4)
      .END
```

Results. A plot of each dc transfer characteristic, obtained using the .PROBE utility, is shown in Fig. 10.38. The top curve shows the nonlinear open-loop gain of the amplifier with no feedback ($\beta = 0$). As β is increased, the curve becomes more linear, but the overall gain goes down. By the time β reaches 0.1, the plot is essentially a straight line with slope $A_{\text{fb}} = 1/\beta = 10$. ∎

Figure 10.38

Plot of v_{OUT} versus v_{IN} for the circuit of Fig. 10.37 for varying values of feedback factor β.

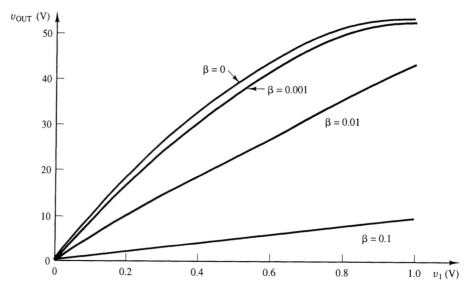

EXAMPLE 10.14   Simulate the op-amp circuit of Example 10.7 on SPICE. Show that for β < 0.2, the circuit is stable, whereas for β > 0.2, the circuit is unstable.

Solution

The op-amp analyzed in Example 10.7 has a dc gain of 10^6, one pole at 10 rad/s, and two poles at 10^6 rad/s. These characteristics can be simulated using the op-amp subcircuit shown in Fig. 10.39. Each of the intermediate stages formed by E1, E2, and E3 is responsible for creating one amplifier pole, with $1/R_A C_A = 10$ rad/s and $1/R_B C_B = 1/R_C C_C = 10^6$ rad/s. The large dc gain is produced by E4, for which the dependent-source coefficient is 10^6.

In the input file to follow, the subcircuit in Fig. 10.39 is used to simulate two circuits, one with β = 0.1 (just under the stability limit) and one with β = 0.25 (just over the stability limit). The input to the circuit consists of a short, 0.1-μs, 1-V pulse that momentarily excites the circuit.

Input File

```
*FEEDBACK STABILITY OF THE NONINVERTING AMPLIFIER of EXAMPLE 10.7

*Define an op-amp with a pole at 10 rad/s and two poles at 10⁶ rad/s:
 .SUBCKT OP-AMP 1 2 9

*First stage:
        E1 3 0 1 2 1
        RA 3 4 1k
        CA 4 0 100uF
*Second stage:
        E2 5 0 4 0 1
        RB 5 6 1k
        CB 6 0 0.001uF
*Third stage:
        E3 7 0 6 0 1
        RC 7 8 1k
        CC 8 0 0.001uF
*Fourth stage:
        E4 9 0 8 0 1e6
        .ENDS OP-AMP
*End of subcircuit

*Build a stable circuit with β = 0.1:
        XOP-AMP1  1 2 3  OP-AMP
        R1 2 0 1k
        R2 3 2 9k

*Build an unstable circuit with β = 0.25:
        XOP-AMP2 1 4 5 OP-AMP
        R3 4 0 1k
        R4 5 4 3k

*Excite both circuits with an "impulse" function
        vs 1 0 PWL(0 0  0.1u 1  0.2u 0)

*Observe the output of each circuit for 20 μs:
        .TRAN 0.1u 20u  0  0.1u
        .PROBE v(1) v(3) v(5)
        .END
```

Figure 10.39
Op-amp subcircuit
that simulates the
amplifier of
Example 10.7 with
a dc gain of 10^6,
one pole at
$1/R_A C_A =$
$10\,\text{rad/s}$, and two
poles at $1/R_B C_B =$
$1/R_C C_C =$
$10^6\,\text{rad/s}$. As
shown in
Example 10.7,
stability requires a
β less than $1/5 =$
0.2.

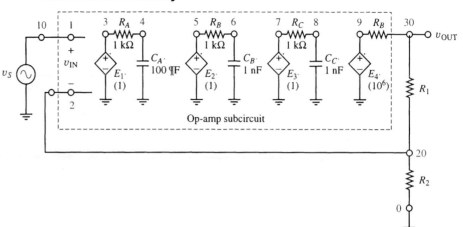

Results. The output for $\beta = 0.25$ depicted in Fig. 10.40, shows oscillations of growing amplitude and frequency of about 160 kHz. This output is consistent with a feedback instability at $f_{180} = \omega_{180}/2\pi$. Because the simulated op-amp has no saturation limits, the magnitude of these oscillations increases without bound as time progresses.

When β is reduced to 0.1, the circuit is technically stable but is so near its stability limit that the output still exhibits oscillatory behavior. In this case, however, the magnitude of the oscillations decays with time. Only for much larger gain margins will the oscillations be of negligible magnitude so that only an amplified version of the input pulse is produced at the output. ∎

Figure 10.40
Results of the
SPICE simulation
of Example 10.14.
For $\beta < 0.2$, the
circuit is stable and
the oscillation
initiated by the
impulse v_S decays
toward zero. For
$\beta > 0.2$, the
magnitude of the
oscillation grows.

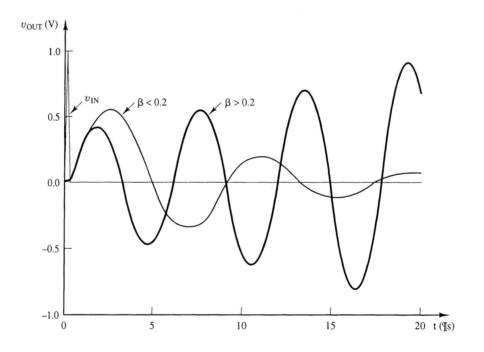

PROBLEMS

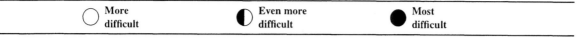

| ○ More difficult | ◐ Even more difficult | ● Most difficult |

10.1 The Negative-Feedback Loop

10.1 An amplifier with large but varying open-loop gain A_o is connected in a negative feedback loop with $\beta = 0.01$, so that the closed-loop gain approaches the value $1/\beta = 100$. Over what range of A_o will the actual closed-loop gain fall by no more than 1% of this ideal value?

10.2 A negative feedback circuit is formed from an amplifier with open-loop gain $A_o = 10^5$ using a feedback factor of $\beta = 0.001$. What is overall gain A_{fb}? Calculate the change in A_{fb} if A_o increases by 10%.

10.3 A noninverting amplifier is made from an op-amp with $A_o = 10^6$, $R_2 = 47\,\text{k}\Omega$, and $R_1 = 5.6\,\text{k}\Omega$. What are the values of the feedback factor β and the closed-loop gain A_{fb}?

10.4 ○ Draw the feedback-loop block diagram of a two-input op-amp summation amplifier. In this case, three signals must enter the summation node of the feedback diagram. Find an expression for the feedback factor β. Show that the closed-loop gain function (10.6) approaches the two-input gain function of Chapter 2 in the limit of large open-loop op-amp gain.

10.5 ○ Draw the feedback-loop block diagram of a two-input op-amp difference amplifier of the type shown in Fig. 2.13. Find an expression for the feedback factor β. Show that the closed-loop gain function (10.6) approaches the two-input gain function of Chapter 2 in the limit of large open-loop op-amp gain.

10.2 General Requirements of Feedback Circuits

10.6 Show that the Schmitt trigger of Section 2.5.2 incorporates positive feedback. What is the output of the circuit for zero input?

10.7 Prove that the signal x_{IN} in Fig. 10.2 will always be less than x_S as long as the product $A\beta$ is positive.

10.8 How must the feedback diagram of Fig. 10.2 be modified if β is a positive number less than unity and the gain A is negative?

10.9 It is possible to create a circuit in which feedback with $\beta > 1$ is provided by active devices. Show that the combination of small gain A and large feedback factor β does not lead to large amplification in the feedback loop of Fig. 10.2.

10.3 Effects of Feedback on Amplifier Performance

10.3.1 Effect of Feedback on Amplifier Linearity

10.10 An amplifier with gain $A_o = 1000$ is connected in a negative feedback loop such that $A_{fb} = 50$. By how much will A_{fb} change if A_o changes by 10%?

10.11 An amplifier is connected in a negative feedback circuit with $\beta = 0.01$. Plot the closed-loop gain as a function of A_o for $1 < A_o < 1000$.

10.12 An amplifier has the open-loop transfer characteristic of Fig. 10.4 with slope $A_1 = 2 \times 10^3$ for $0 < v_{OUT} < 5\,\text{V}$, and slope $A_2 = 0.5 \times 10^3$ for $5\,\text{V} < v_{OUT} < 12\,\text{V}$. The amplifier is connected in a negative feedback loop with $\beta = 0.05$. Plot the resulting closed-loop transfer characteristic v_{OUT}/v_S.

10.13 An amplifier having the transfer characteristic of Fig. 10.4 with $A_1 = 1200$ and $A_2 = 400$ is connected in a negative feedback loop. Choose a value for β such that the closed-loop gain falls by no more than 5% when the amplifier operates in region 2. What is the closed-loop gain for this value of β?

10.14 A voltage amplifier has an open-loop transfer characteristic that has three regions with different slopes. Its transfer characteristic begins at the origin, and its output is equal to 2 V at $v_{IN} = 0.2\,\text{V}$, to 4 V at $v_{IN} = 0.6\,\text{V}$, and to 5 V for $v_{IN} \geq 1\,\text{V}$. The amplifier is connected in a negative feedback loop with $\beta = 0.1$. Plot v_{OUT} versus v_S over the range $0 < v_{OUT} < 5\,\text{V}$.

10.15 An amplifier has the open-loop transfer characteristic of Fig. 10.4 with slope $A_1 = 5 \times 10^4$ for $0 < v_{OUT} < 7\,\text{V}$, and slope $A_2 = 2 \times 10^4$ for $7\,\text{V} < v_{OUT} < 15\,\text{V}$.

(a) If v_{IN} consists of a ± 2-V peak triangular waveform, plot the resulting output as a function of time.

(b) The amplifier is connected in a negative feedback loop with $\beta = 0.1$. Plot the output if the input v_S to the feedback amplifier is again a ± 2-V triangular waveform.

10.16 ⚏ Design an amplifier with an overall gain of $50 \pm 1\%$. Assume that a number of amplifier stages exist for which the open-loop gain is not precisely known, but lies somewhere in the range 1000 to 2000. Your finished design may consist of a cascade of one or more similar stages. For the purpose of this problem, assume that the feedback factor β of any single stage can be known to arbitrary accuracy.

10.17 ⚙ A BJT inverter with a collector resistor of $5\,k\Omega$, a base resistor of $10\,k\Omega$, and an emitter that is grounded with respect to small signals is biased at $I_B = 10\,\mu A$. The circuit is made from a BJT whose β varies over the range 150 to 250. Specify the requirements of a feedback circuit that will result in a closed-loop gain that varies by no more than 1%. What is the value of A_{fb} for such a circuit?

10.18 ⚙ ○ An enhancement-mode NMOS inverter with enhancement-mode NMOS pull-up load has its source grounded and is biased in the constant-current region. The W/L ratio of the pull-up load is known to be 0.1, but due to a masking error, the W/L ratio of the driven MOSFET may lie anywhere in the range 0.2 to 10. Specify the requirements of a feedback circuit that will result in a closed-loop gain that varies by no more than 5%. What is the value of A_{fb} for such a circuit? You may not be able to meet the design specifications in this case.

10.3.2 Effect of Feedback on Amplifier Bandwidth

10.19 An op-amp with a gain–bandwidth product of 4×10^5 is connected in the noninverting configuration with a closed-loop gain of 50. What is its closed-loop bandwidth?

10.20 An op-amp has an open-loop gain of 10^6 and a dominant open-loop pole at 5 Hz. If a noninverting amplifier with a gain of 100 is made from this op-amp, what is the resulting amplifier bandwidth? What is the gain–bandwidth product?

10.21 An op-amp has an open-loop gain of 10^4 and a dominant open-loop pole at 4 Hz. If an inverting amplifier with a gain of -100 is made from this op-amp, what is the resulting amplifier bandwidth? What is the gain–bandwidth product?

10.22 A noninverting amplifier with a gain of 25 and a bandwidth of at least 50 kHz is to be made from an op-amp. Specify the minimum open-loop gain for the op-amp if its dominant pole frequency is located at 4 Hz.

10.23 An inverting amplifier with a gain of -50 and a bandwidth of at least 10 kHz is to be made from an op-amp. Specify the minimum open-loop gain for the op-amp if its dominant pole frequency is located at 5 Hz.

10.24 Find the unity-gain frequency of an op-amp with an open loop gain of 10^6 and a dominant pole at 4 Hz.

10.25 ⚙ Design a cascade of two or more noninverting amplifiers such that the overall cascade has a gain of 1000 and a bandwidth of at least 10 kHz. The available op-amps each have an open-loop gain of 10^5 and a dominant pole at 4 Hz.

10.26 ⚙ Design an inverting amplifier that can amplify the voice signal from a dynamic microphone and deliver the resulting signal to a power amplifier. The power amplifier requires a signal on the order of 1 V peak to be driven to full power output. The dynamic microphone has an internal series resistance of $10\,k\Omega$ and produces a 10-mV peak signal when excited by a normal speaking voice. The amplifier should respond to at least the normal range of human hearing (about 20 Hz to 15 kHz).

10.27 ⚙ Design an amplifier system that can add together the signals from two dynamic microphones (see Problem 10.26) and deliver the resulting signal to a power amplifier and loudspeaker. The power amplifier requires a signal of about 0.5 V peak to be driven to full-power output. The entire system should employ separate volume controls for each input and should respond to at least the normal range of human hearing (about 20 Hz to 15 kHz).

10.4 The Four Basic Amplifier Types

10.28 The output from a transconductance amplifier feeds the input to a transresistance amplifier. What is the resulting amplifier type of the overall cascade?

10.29 Show that each of the four amplifier systems depicted in Fig. 10.5 can provide power gain, regardless of amplifier type.

10.5 The Four Feedback Topologies

10.30 An ideal voltage amplifier is connected in the feedback loop of Fig. 10.14. Find values for A_v and β if $v_S = 50\,mV$, $v_F = 45\,mV$, and $v_{OUT} = 5\,V$.

10.31 An ideal current amplifier is connected in the feedback loop of Fig. 10.15. Find values for A_i and β if $i_S = 10\,mA$, $i_F = 9\,mA$, and $i_{OUT} = 1\,A$.

10.32 An ideal transresistance amplifier is connected in the feedback loop of Fig. 10.16. Find values for A_r and β if $i_S = 20\,\mu A$, $i_F = 18\,\mu A$, and $v_{OUT} = 12\,V$.

10.33 An ideal transconductance amplifier is connected in the feedback loop of Fig. 10.17. Find values for A_g and β if $v_S = 50\,mV$, $v_F = 49.5\,mV$, and $i_{OUT} = 5.2\,mA$.

10.34 An ideal current amplifier is connected in the feedback loop of Fig. 10.15. If $i_S = 10\,\text{mA}$, $A_i = 8000$, and $\beta = 0.1$, find values for i_F and i_{OUT}.

10.35 An ideal current amplifier is connected in the feedback loop of Fig. 10.15. If $A_i = 20{,}000$, $i_S = 20\,\text{mA}$, and $i_F = 8\,\text{mA}$, find i_{OUT} and β.

10.36 The feedback network β of a voltage amplifier described by the feedback diagram of Fig. 10.14 consists of a simple voltage divider formed by $R_1 = 100\,\Omega$ and $R_2 = 10\,\text{k}\Omega$, where the feedback signal is tapped from R_1. Find numerical values for β, $r_{\text{in--}\beta}$, and $r_{\text{out-}\beta}$.

10.37 The feedback function β of a current amplifier described by the feedback diagram of Fig. 10.15 consists of a current divider formed by $R_1 = 100\,\Omega$ and $R_2 = 10\,\Omega$, where the feedback current signal through R_1 is shunted from the amplifier input terminals. Find numerical values for β, $r_{\text{in-}\beta}$, and $r_{\text{out-}\beta}$.

10.38 ○ The feedback network to a voltage amplifier consists of a resistor and a forward-biased diode, as depicted in **Fig. P10.38**. The diode is kept in forward-bias conditions at all times by a dc bias component V_1 of v_{OUT}. Choose R_1 and V_1 such that $\beta \approx 0.005$ and $r_{\text{in-}\beta} = 10\,\text{k}\Omega$. What is the resulting value of $r_{\text{out-}\beta}$?

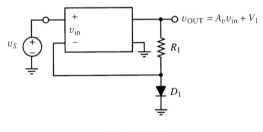

Fig. P10.38

10.6 Effect of Feedback Connections on Amplifier Port Resistance

10.39 A voltage amplifier with $A_v = 10^4$, $r_{\text{in}} = 1\,\text{k}\Omega$, and $r_{\text{out}} = 100\,\Omega$ is connected in a negative feedback loop with $\beta = 0.1$. What are the resulting values of A_{fb}, R_{in}, and R_{out} if $r_{\text{in-}\beta} = \infty$ and $r_{\text{out-}\beta} = 0$?

10.40 A voltage amplifier with $A_v = 5000$, $r_{\text{in}} = 3\,\text{k}\Omega$, and $r_{\text{out}} = 20\,\Omega$ is connected in a negative feedback loop with $\beta = 0.02$. What are the resulting values of A_{fb}, R_{in}, and R_{out} if $r_{\text{in-}\beta} = 100\,\text{k}\Omega$ and $r_{\text{out-}\beta} = 10\,\Omega$?

10.41 A current amplifier with $A_i = 100$, $r_{\text{in}} = 5\,\Omega$, and $r_{\text{out}} = 1\,\text{M}\Omega$ is connected in a negative feedback loop with $\beta = 0.05$. What are the resulting values of A_{fb}, R_{in}, and R_{out} if $r_{\text{in-}\beta} = 0$ and $r_{\text{out-}\beta} = \infty$?

10.42 A current amplifier with $A_i = 4000$, $r_{\text{in}} = 100\,\Omega$, and $r_{\text{out}} = 100\,\text{k}\Omega$ is connected in a negative feedback loop with $\beta = 0.1$. What are the resulting values of A_{fb}, R_{in}, and R_{out} if $r_{\text{in-}\beta} = 15\,\Omega$ and $r_{\text{out-}\beta} = 1\,\text{M}\Omega$?

10.43 A transresistance amplifier with $A_r = 500\,\text{V/mA}$, $r_{\text{in}} = 15\,\Omega$, and $r_{\text{out}} = 10\,\Omega$ is connected in a negative feedback loop with $\beta = 0.05$. What are the resulting values of A_{fb}, R_{in}, and R_{out} if $r_{\text{in-}\beta} = r_{\text{out-}\beta} = \infty$?

10.44 A transresistance amplifier with $A_r = 1200\,\text{V/mA}$, $r_{\text{in}} = 50\,\Omega$, and $r_{\text{out}} = 80\,\Omega$ is connected in a negative feedback loop with $\beta = 0.001$. What are the resulting values of A_{fb}, R_{in}, and R_{out} if $r_{\text{in-}\beta} = r_{\text{out-}\beta} = 1\,\text{M}\Omega$?

10.45 A transconductance amplifier with $A_g = 100\,\text{mA/V}$, $r_{\text{in}} = 500\,\text{k}\Omega$, and $r_{\text{out}} = 1\,\text{M}\Omega$ is connected in a negative feedback loop with $\beta = 0.5$. What are the resulting values of A_{fb}, R_{in}, and R_{out} if $r_{\text{in-}\beta} = r_{\text{out-}\beta} = 0$?

10.46 A transconductance amplifier with $A_g = 6000\,\text{mA/V}$, $r_{\text{in}} = 10\,\text{M}\Omega$, and $r_{\text{out}} = 500\,\text{k}\Omega$ is connected in a negative feedback loop with $\beta = 0.5$. What are the resulting values of A_{fb}, R_{in}, and R_{out} if $r_{\text{in-}\beta} = r_{\text{out-}\beta} = 50\,\Omega$?

10.7 Examples of Real Feedback Amplifiers

10.7.1 Op-Amp Voltage Amplifier (Series/Shunt Feedback)

10.47 An op-amp has an open-loop gain of 10^6, an open-loop output resistance of $10\,\Omega$, and an open-loop input resistance of $2\,\text{M}\Omega$. If the op-amp is used to make a noninverting amplifier with a gain of 100, find the overall circuit input and output resistances.

10.48 ▣ Design a voltage amplifier using the μA741 op-amp, for which $A_v \approx 200{,}000$, $r_{\text{in}} \approx 2\,\text{M}\Omega$, and $r_{\text{out}} \approx 75\,\Omega$. Your amplifier should have an input resistance of at least $1000\,\text{M}\Omega$ and an output resistance of less than $0.1\,\Omega$. What is the largest gain achievable for these specifications?

10.49 ▣ Suppose that an op-amp voltage amplifier is required with a gain of 100 and a minimum input resistance of $1\,\text{G}\Omega$. Choose a real op-amp with sufficient open-loop gain and input resistance to meet these specifications. Possibilities to investigate include the μA741, LF411, and LM101A operational amplifiers.

10.50 ◐ Consider a voltage amplifier with a gain of 10 made from an op-amp with parameters $A_o = 5 \times 10^4$,

$r_{in} = 50\,\text{k}\Omega$, and $r_{out} = 500\,\Omega$. Suppose that the op-amp has a single pole that causes the open-loop gain to roll off at $-20\,\text{dB/decade}$, reaching unity gain at 4 MHz. Find equivalent circuit representations for the frequency-dependent input and output impedances $\mathbf{Z}_{in}$ and $\mathbf{Z}_{out}$ of the circuit.

10.51 ⚡ ◑ The differential amplifier of **Fig. P10.51** incorporates series/shunt feedback.

  (a) If $R_B \gg R_3$, find an approximate expression for the closed-loop gain v_{OUT}/v_S.

  (b) Choose I_o, V_{CC}, V_{EE}, and all resistor values such that the closed-loop gain is approximately 10.

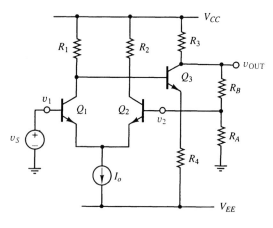

Fig. P10.51

10.52 ⚡ ◯ Design a modification to the circuit of **Fig. P10.51** so that the constraint $R_B \gg R_3$ is not needed. This design change will enable the conditions $r_{in\text{-}\beta} \gg r_{out}$ and $r_{out\text{-}\beta} \ll r_{in}$ to be more closely met.

10.7.2 MOSFET Transconductance Amplifier with Feedback Resistor (Series/Series Feedback)

10.53 Consider the MOSFET amplifier of Fig. 10.23. Derive expressions for the closed-loop gain and overall output resistance R_{out} directly from the circuit's small-signal model. Include the effects of the small-signal MOSFET output resistance r_o. Compare to the results obtained using feedback analysis.

10.54 ◑ In this problem, the overall output resistance R_{out} of the MOSFET transconductance amplifier of Fig. 10.23 is examined.

  (a) Derive an expression for R_{out} directly from the small-signal amplifier model.

  (b) Attempt to evaluate R_{out} by substituting the open-loop gain (10.56) into Eq. (10.51) and compare with the correct answer from part (a).

The results of parts (a) and (b) show an inconsistency because the open-loop gain substituted into Eq. (10.51) must be the *unloaded* value. The analysis leading to Eq. (10.51) already accounts for the loading of the output port by R_F. Also including output loading in the expression for the open-loop gain in Eq. (10.51) is tantamount to accounting for the loading effect twice.

  (c) Compute the open-loop gain of the amplifier with $r_{in\text{-}\beta} = R_F$ replaced by a short circuit in the output loop. Retain $r_{out\text{-}\beta} = R_F$ in the input loop. Show that substituting this open-loop gain into Eq. (10.51) results in the correct expression for R_{out}.

10.55 ⚡ ◯ Consider the MOSFET transconductance amplifier of Fig. 10.23. Suppose that the MOSFET has parameters $K = 2\,\text{mA/V}^2$, $V_{TR} = 2\,\text{V}$, and $V_A = 20\,\text{V}$. Choose values for V_{SS} and R_F such that a closed-loop gain of approximately 2 mA/V is realized.

10.56 ◑ The circuit of **Fig. P10.56**, called the *series/series triple*, utilizes series input mixing and series output sampling. Suppose that $R_{C1} = 9\,\text{k}\Omega$, $R_{C2} = 5\,\text{k}\Omega$, $R_{C3} = 600\,\Omega$, $R_{E1} = R_{E2} = 100\,\Omega$, and $R_F = 640\,\Omega$.

  (a) Find an expression and value for the open-loop gain with the loading by the feedback circuit included.

  (b) Find an expression and value for the feedback function $\beta = v_F/i_{E3}$. Note that R_F is much larger in value than R_{E1} and R_{E2}.

  (c) What is the closed-loop gain of the amplifier?

  (d) Can such an amplifier be made from two stages?

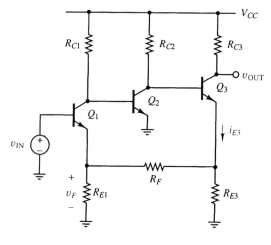

Fig. P10.56

10.57 ⟳ ○ The series/series feedback circuit of **Fig. P10.57** can be used to produce large output currents from a signal voltage v_S.

(a) What is the open-loop transconductance gain A_g of the op-amp–transistor–resistor combination? What practical limitation exists to this circuit?

(b) If the feedback loop is closed, what is the value of the feedback factor $\beta = v_F/i_{OUT}$? (Assume that $i_E \approx i_{OUT}$ and use the ideal op-amp approximation.)

(c) Choose a value for R_F such that $A_g = 100\,\text{mA/V}$.

(d) Now suppose that the op-amp has parameters $r_{in} = 50\,\text{k}\Omega$, $A_v = 500$, and $r_{out} = 1\,\text{k}\Omega$. If $R_S = 1\,\text{k}\Omega$, what is new value of A_{fb}?

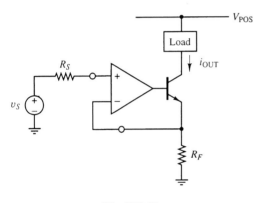

Fig. P10.57

10.58 ⟳ Design a simple modification to the circuit of **Fig. P10.57** that will permit currents of both positive and negative polarities to be produced by v_S.

10.7.3 Single-Transistor Transresistance Amplifier (Shunt/Shunt Feedback)

10.59 The op-amp inverting amplifier of Chapter 2 is really a feedback amplifier in which shunt mixing is employed at the input and shunt sampling at the output. Use feedback analysis to determine the feedback factor, input resistance, output resistance, and closed-loop gain of an inverting amplifier with $r_{in} = 10\,\text{M}\Omega$, $r_{out} = 10\,\Omega$, $R_1 = 10\,\text{k}\Omega$, and $R_2 = 100\,\text{k}\Omega$, where R_2 is the feedback resistor.

10.60 The current-to-voltage converter of **Fig. P10.60** is made from an op-amp with open-loop parameters $A_o = 2 \times 10^5$, $r_{in} = 2\,\text{M}\Omega$, and $r_{out} = 75\,\Omega$. Suppose that $R_F = 1\,\text{M}\Omega$, and that the circuit drives an $R_L = 10\,\text{k}\Omega$ load resistor to ground. Find the transresistance gain, input resistance, and output resistance of the amplifier with feedback.

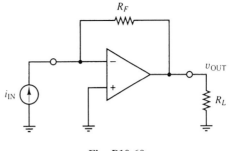

Fig. P10.60

10.61 The simple op-amp circuit of **Fig. P10.60** functions as a transresistance amplifier that converts a current signal into a voltage signal. If the op-amp has parameters $A_o = 500{,}000$, $r_{in} = 1\,\text{M}\Omega$, and $r_{out} = 25\,\Omega$, find the values of the closed-loop A_r, R_{in}, and R_{out}.

10.62 Draw the feedback-loop block diagram of an op-amp integrator made from one feedback capacitor and one input resistor. This circuit incorporates shunt/shunt feedback with an op-amp. Find an expression for the frequency-dependent feedback function β in the sinusoidal steady state. Model the open-loop op-amp gain as a constant A_o. Show that the feedback function (10.6) approaches the integrator transfer function $\mathbf{V}_{out}/\mathbf{V}_{in} = -1/j\omega RC$ in the limit of large open-loop op-amp gain.

10.63 Repeat Problem 10.62 if a resistor R_F is connected in parallel with the feedback capacitor, so as to make a modified op-amp integrator, or low-pass filter.

10.64 Draw the feedback-loop block diagram of an op-amp differentiator made from one input capacitor and one feedback resistor. Find an expression for the frequency-dependent feedback function β in the sinusoidal steady state. Show that the feedback function (10.6) approaches the differentiator transfer function $\mathbf{V}_{out}/\mathbf{V}_{in} = -j\omega RC$ in the limit of large open-loop op-amp gain.

10.65 Consider the analysis of the BJT transresistance amplifier of Figs. 10.24 to 10.26. Transform the output port of the feedback circuit in Fig. 10.26 into its Thévenin equivalent circuit. Use this new representation of the circuit to derive the open-loop gain A_r, the feedback function $\beta = i_F/v_{OUT}$, and the amplifier input and output resistances.

10.66 The transresistance amplifier of Fig. 10.24 is made with $R_C = 5\,\text{k}\Omega$, $R_F = 10\,\text{k}\Omega$, $\beta_o = 100$, and $I_C = 1\,\text{mA}$. Find the transresistance gain A_r, input resistance R_{in}, and output resistance R_{out}.

10.67 ⟳ ○ Design a circuit based on the BJT transresistance amplifier of Fig. 10.24 that has a gain magnitude of 5 V/mA, an input resistance smaller than 500 Ω, and an output resistance smaller than $20\,\Omega$. The circuit should operate from a single 15-V supply and be biased so that Q_1 operates well into the constant-current region. For the purpose of this problem, assume a BJT with $\beta_F = \beta_o = 200$ and $V_f = 0.7\,\text{V}$.

10.68 ○ Use feedback analysis to find an expression for the closed-loop gain of the shunt/shunt circuit of Fig. 10.24 if a MOSFET is substituted for the BJT and if i_S is replaced by a voltage source v_S in series with a resistance R_S. Your answer can be confirmed by direct calculation.

10.69 A piezoelectric pressure transducer with a peak output voltage of 10 V and a series output resistance of $10\,\text{M}\Omega$ can be modeled as a current source that provides a peak short-circuit current of $1\,\mu\text{A}$. It is desired to use the sensor output to drive a 10-kΩ load.

(a) Compute the peak power delivered to the load if the sensor is connected directly to it.

(b) Compute the peak power delivered to the load if the sensor feeds a transresistance amplifier that has parameters $R_{\text{in}} = 10\,\text{k}\Omega$, $A_r = 10\,\text{V/A}$, and $R_{\text{out}} = 10\,\text{k}\Omega$.

10.70 Consider the current-to-voltage converter of **Fig. P10.70**. For the condition $R_F \ll r_{\text{in}}$, find the following:

(a) The open-loop gain A_r with the op-amp input and output ports loaded by the resistance R_F.

(b) The feedback factor $\beta = i_F/v_{\text{OUT}}$.

(c) The closed-loop gain of the overall circuit.

(d) The overall input resistance R_{in} seen by the i_{IN} source.

(e) Derive the gain $v_{\text{OUT}}/i_{\text{IN}}$ using circuit theory principles and the ideal op-amp approximation of Chapter 2. Compare the result with the closed-loop gain derived from the reciprocal of the β found in part (c).

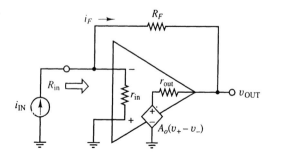

Fig. P10.70

10.71 ◖ Consider the three-stage feedback amplifier of **Fig. P10.71** with $R_{C1} = R_{C2} = R_{C3} = 5\,\text{k}\Omega$ and $R_B = 1\,\text{k}\Omega$. Identify the feedback function β, then find the closed-loop gain A_r, the input resistance R_{in}, and output resistance R_{out}.

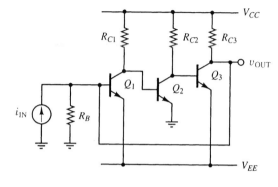

Fig. P10.71

10.7.4 BJT Current Amplifier with Feedback (Shunt/Series Feedback)

10.72 Find the gain of the circuit of Fig. 10.27 if $\beta_F = \beta_o = 75$ for both transistors, $r_{\pi 1} = r_{\pi 2} = 5\,\text{k}\Omega$, $R_E = 220\,\Omega$, $R_F = 4.7\,\text{k}\Omega$, and $R_C = 1.5\,\text{k}\Omega$.

10.73 ⅀ ○ Design a practical bias network for the current amplifier of Fig. 10.27 such that Q_1 and Q_2 are both biased in the active region when $V_{CC} = 12\,\text{V}$, $R_C = 1\,\text{k}\Omega$, and an approximate feedback factor of $\beta = -0.02$ is desired. For the purpose of this problem, assume that $\beta = 75$ for both transistors.

10.74 Consider the circuit of Fig. 10.27. This circuit was analyzed in Example 10.6 assuming $\beta_1 = \beta_2 = 200$. Suppose that component variations results in beta values in the range $50 < \beta < 250$. Find the minimum and maximum possible closed-loop gain of the circuit.

10.75 ⅀ Consider the two-stage BJT circuit shown in Fig. 10.27. Suppose that a collector resistor R_C is connected between Q_2 and the V_{CC} supply bus, and that the right-hand side of R_F is connected to the collector of Q_2, rather than to the emitter of Q_2.

(a) Show that the feedback for this new configuration is positive.

(b) Design a modification to the new circuit that will allow negative feedback to be successfully implemented.

10.8 Feedback-Loop Stability

10.8.1 Amplifier Phase Shift

10.76 The measured open-loop transfer function of a particular op-amp can be approximated by the expression $A(j\omega) = 10^3/[(1 + j\omega/\omega_1)(1 + j\omega/\omega_2)^2]$, where $\omega_1 = 10^4\,\text{rad/s}$ and $\omega_2 = 10^5\,\text{rad/s}$. The op-amp is connected as a noninverting amplifier with a feedback function β that is not a function of frequency.

(a) Find the value of ω at which the loop gain undergoes a $-180°$ phase shift, thereby creating positive feedback.

(b) Find the maximum value of β before the circuit begins to oscillate.

(c) If the circuit does become unstable due to an overly large value of β, what will be the frequency of oscillation?

10.8.3 Bode-Plot Analysis of Feedback Stability

10.77 A particular op-amp has an open-loop dc gain of 80 dB and open-loop poles at 1, 10^5, and 10^7 Hz. If the op-amp is connected as a unity-gain follower, the circuit will be stable. Find the gain and phase margins.

10.78 Consider an op-amp with an open-loop dc gain of 80 dB and open-loop poles at 1, 10^5, and 10^7 Hz. The internal circuitry of the op-amp is redesigned so that the open-loop dc gain is increased to 160 dB without changing the pole frequencies.

(a) How small must the feedback function β be if a circuit made from the op-amp is to be stable?

(b) Implement a negative-feedback circuit with a gain margin of $+10\,\text{dB}$ using the inverting-amplifier configuration.

10.79 An amplifier with a dc gain of 100 dB and poles at $f_1 = 10^5\,\text{Hz}$, $f_2 = 10^6\,\text{Hz}$, and $f_3 = 10^7\,\text{Hz}$ is connected in the inverting-amplifier configuration. What is the minimum inverter gain required to ensure circuit stability?

10.80 An op-amp with a dc gain of 80 dB and poles at $f_1 = 10^4\,\text{Hz}$, $f_2 = 10^5\,\text{Hz}$, and $f_3 = 10^6\,\text{Hz}$ is connected as an inverting amplifier with a gain of -100. The resulting circuit meets the stability condition. What are its gain and phase margins?

10.81 An op-amp with a dc gain of 80 dB has poles at $f_1 = 10^5\,\text{Hz}$, $f_2 = 10^6\,\text{Hz}$, and $f_3 = 10^7\,\text{Hz}$. As designed, this op-amp is not stable in all feedback configurations. It is desired to stabilize the op-amp in the *inverting-amplifier* configuration by the addition of a simple RC filter cascaded with the v_{out} terminal, as in Fig. 10.35.

(a) Suppose the values of R_1, R_2, and R_C are chosen so that the voltage-divider relation can be approximately applied to R_C and C_C. Under such conditions, derive an expression for the open-loop gain of the op-amp with R_C and C_C included but with R_1 and R_2 of the inverting-amplifier feedback network disconnected.

(b) Find the minimum value of C_C that will guarantee the stability of the circuit if $R_C = 1\,\text{k}\Omega$ when $R_1 = R_2$.

10.82 ⓢ ◯ Design a series cascade of amplifiers that has an overall cascade gain of $+30\,\text{dB}$ and a bandwidth of at least 500 Hz. Each stage is to be built using the op-amp described in Problem 10.81. The overall circuit must be stable.

10.8.4 Frequency Compensation

10.83 An amplifier with a dc gain of 80 dB and a dominant pole at 2 MHz must be compensated for all possible closed-loop gains. Discuss the constraints that will determine the frequency at which a compensation pole should be placed.

10.84 Consider the amplifier described by the open-loop response of Fig. 10.33. By how much must the gain at ω_{90} be reduced if the amplifier is to be stable for feedback factors in the range $0.1 < \beta < 0$? Below what frequency must ω_c be located?

10.85 Choose the value of C in Example 10.11 if the amplifier must have a closed-loop gain of -15 and a phase margin of $50°$.

10.86 Find the bandwidth in hertz of the compensated amplifier of Example 10.11 if $A_{\text{fb}} = 300$ and $A_o = 10^5$.

10.87 An amplifier with a dc gain of 75 dB and three open-loop poles at 2×10^4, 10^5, and 3×10^6 Hz is connected in a negative feedback configuration.

(a) Find the maximum allowed β and minimum A_{fb} that will ensure stability.

(b) Determine the placement of a fourth pole f_c such that the amplifier will be stable for all values of β.

10.88 An amplifier with a dc gain of 60 dB and four open-loop poles at 10^3, 5×10^4, 2×10^5, and 10^7 Hz is connected in a negative feedback configuration.

(a) What are the maximum β and minimum A_{fb} that will ensure stability?

(b) How might this amplifier be modified if it is to be stable for gains between 50 and 200 with a gain margin of at least 5 dB?

10.89 ⅀ ○ An inverting amplifier with a dc gain of -20 is to be constructed from an op-amp. In order to ensure stability, a compensation pole at $f_c = 5\,Hz$ is required.

(a) Choose suitable resistor values, and use the arrangement of Fig. 10.36 to implement the required compensation pole.

(b) Design an alternative compensation network that does not make use of scheme shown in Fig. 10.36. You might try adding a capacitor in parallel with the feedback resistor R_F of the inverting amplifier to introduce the required pole.

10.90 ⅀ An op-amp has the open-loop frequency response shown in Fig. 10.33. Design a noninverting amplifier with a closed-loop gain of 40 dB. Find the location of the compensation pole ω_c required to guarantee stability with a phase margin of $45°$. Implement ω_c using an external compensation capacitor.

10.91 ⅀ An inverting amplifier with a dc gain of -10 is to be made from a particular op-amp. Design the circuit such that a compensation pole at $\omega_c = 10\,rad/s$ is added to the closed-loop response.

◆ SPICE PROBLEMS

10.92 Use SPICE to find the exact value of the input resistance of a noninverting amplifier with a gain of $31\,(R_2 = 30\,k\Omega, R_1 = 1\,k\Omega)$ made from an op-amp having parameters $r_{in} = 10\,M\Omega$, $r_{out} = 10\,\Omega$, and $A_o = 10^6$.

10.93 Use SPICE to find the exact value of the output resistance of an inverting amplifier with a gain of $-20\,(R_2 = 200\,k\Omega, R_1 = 10\,k\Omega)$ made from an op-amp having parameters $r_{in} = 10\,M\Omega$, $r_{out} = 10\,\Omega$, and $A_o = 10^6$.

10.94 Use SPICE to simulate a noninverting amplifier with $R_1 = 10\,k\Omega$ and $R_2 = 100\,k\Omega$, where R_2 is the feedback resistor. Assume the op-amp to have internal parameters $r_{in} = 10\,M\Omega$, $r_{out} = 10\,\Omega$, and $A_o = 10^6$. Find the closed-loop parameters A_{fb}, R_{in}, and R_{out}.

10.95 Use SPICE to find the gain, input resistance, and output resistance of the closed-loop amplifier of Fig. 10.24 if $R_F = 220\,k\Omega$, $R_C = 3.9\,k\Omega$, and $\beta_o = 175$.

10.96 A voltage amplifier with open-loop parameters $A_v = 3 \times 10^4$, $r_{in} = 2\,k\Omega$, and $r_{out} = 200\,\Omega$ is connected in a negative feedback loop with $\beta = 0.1$. Use SPICE to determine the resulting closed-loop parameters A_{fb}, R_{in}, and R_{out}.

10.97 A current amplifier with $A_i = 1000$, $r_{in} = 25\,\Omega$, and $r_{out} = 500\,k\Omega$ is connected in a feedback loop with $\beta = 0.05$. Use SPICE to determine the resulting closed-loop parameters A_{fb}, R_{in}, and R_{out}.

10.98 Confirm the results of Example 10.6 by simulating the circuit's small-signal model on SPICE.

10.99 Simulate the circuit of **Fig. P10.51** on SPICE with $V_{CC} = 10\,V$, $V_{EE} = -10\,V$, $I_o = 1\,mA$, $R_1 = R_2 = 5\,k\Omega$, $R_3 = 1\,k\Omega$, $R_4 = 5\,k\Omega$, $R_A = 100\,k\Omega$, and $R_B = 820\,k\Omega$. Find the resulting closed-loop gain, input resistance, and output resistance.

10.100 Simulate the series/series triple of **Fig. P10.56** on SPICE with $V_{CC} = 7\,V$, $R_{C1} = 9\,k\Omega$, $R_{C2} = 5\,k\Omega$, $R_{C3} = 600\,\Omega$, $R_{E1} = R_{E2} = 100\,\Omega$, and $R_F = 640\,\Omega$. Find the resulting closed-loop gain.

10.101 Simulate the MOSFET transconductance amplifier of Fig. 10.23 on SPICE with $K = 0.5\,mA/V^2$, $V_{TR} = 1.5\,V$, $V_A = 35\,V$, $V_{SS} = -6\,V$, $R_S = 50\,\Omega$, and $R_F = 560\,\Omega$. Find the resulting transconductance gain i_{OUT}/v_S, where i_{OUT} is the current i_D flowing into the MOSFET.

10.102 The purpose of this problem is to construct a subcircuit definition for use in SPICE that will represent an op-amp with the magnitude response of Fig. 10.33

(a) Specify a subcircuit op-amp definition that has an open-loop dc gain of 100 dB, as in Fig. 10.33.

(b) Modify the subcircuit so that it has the configuration shown in **Fig. P10.102(a)**. Set the parameters of this new model so that the op-amp has the same dc gain as that of part (a).

(c) Add a simple capacitor to the circuit, as in Fig. **P10.102(b)** so as to create a simple RC filter. The values of R_1 and C_1 should be chosen so that the modified op-amp has a single dominant pole at $10^5\,Hz$.

(d) Now add two more cascaded dependent sources and RC filters to the model so that the op-amp response has the three poles shown in Fig. 10.33. Test your op-amp model by plotting its magnitude response using SPICE.

10.103 Use the op-amp subcircuit developed in Problem 10.102 to test the circuit of Fig. 10.36 on SPICE (see

Example 10.11). Simulate the circuit both with and without the compensation circuit and compare results. Set v_{IN} to a 1-μs, 1-V pulse and observe v_{OUT} as a function of time.

10.104 Use the technique of Prob. 10.102 to model an op-amp with an open-loop dc gain of 80 dB and open-loop poles at 1, 10^5, and 10^7 Hz. Plot the open-loop magnitude and angle responses, then verify that a unity-

gain follower made from this op-amp will be stable.

10.105 Use the technique of Problem 10.102 to model an amplifier with an open-loop dc gain of 200, an infinite r_{in}, zero r_{out}, and high-frequency poles at 100 kHz and 5 MHz. Simulate the circuit on SPICE and find its gain–bandwidth product.

10.106 Simulate the amplifier described in Problem 10.20 on SPICE and determine its closed-loop bandwidth.

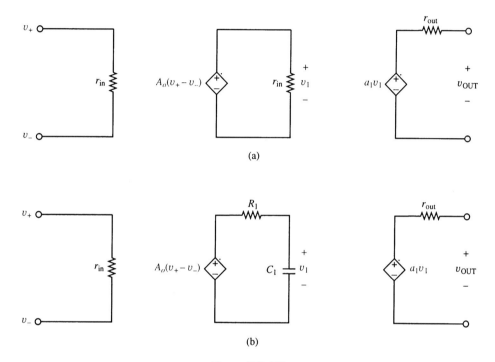

Figure P10.102

Chapter 11

Multistage and Power Amplifiers

U p to this point in the book, our treatment of amplifiers has centered primarily around circuits containing only one or two active devices. Such circuits have limited gain, input resistance, output resistance, and power-handling capabilities, and cannot be used to fulfill every amplification need. Fortunately, the limitations of single-stage amplification can be overcome by the use of multistage amplification. As shown in this chapter, the use of multiple cascaded stages can produce circuits with large overall gain, large input resistance, and small output resistance. The implementation of such circuits requires design and analysis techniques that extend beyond the principles of single-stage design.

The multistage design techniques discussed in this chapter are oriented primarily toward integrated circuits. In an integrated-circuit environment, device parameters that depend on variations in the fabrication process can be controlled only to within a specified range of values. All the devices on an integrated circuit, however, undergo virtually identical fabrication processes. Consequently, devices with closely matched, if not perfectly predictable, parameters are readily available. This characteristic of integrated circuits motivates many of the design techniques that are introduced in this chapter and expanded upon in Chapter 12.

This chapter also contains material on power-amplification stages and power devices. These topics are included here because the last stage of a multistage-amplifier cascade is often a power stage that enables the circuit to supply large amounts of current to its load. Even if the power stage is a stand-alone circuit that is not part of a complete integrated circuit, as in an audio power-amplification system, for example, the power stage must still be preceded by one or more amplification stages, and thus becomes part of a multistage cascade. In this latter situation, many of the multistage design considerations covered in this chapter, including input loading, output loading, and interstage loading, still apply. Given this context, it is appropriate to consider power-amplification stages, and the power devices used to make them, within the same chapter as integrated multistage amplifiers.

11.1 INPUT AND OUTPUT LOADING

In this section, the limitations of single-stage amplification, as typified by the two-port representations of the inverter and voltage-follower of Chapter 7, are examined. The typical inverter has a moderately large gain and has input and output resistances in the kilohm range. The voltage follower has a much higher input resistance and lower output resistance, but has only unity gain. An amplifier is often required that combines the desirable features of both the inverter and the follower: large input resistance, small output resistance, and large voltage gain.

As an example of such a requirement, suppose that a microphone signal, which we shall represent as a 10-mV p–p sinusoidal voltage source v_m, is to be linearly amplified by a factor of about 100. Suppose the 10-mV signal source to have an internal series resistance of $R_m = 1\,\text{k}\Omega$. The output of the amplifier is to be delivered to the 8-Ω load of a standard loudspeaker. One might be tempted to accomplish this task with the high-gain capabilities of a BJT inverter, which could easily be designed to the following specifications:

$$r_{\text{in}} = 5\,\text{k}\Omega \qquad r_{\text{out}} = 1\,\text{k}\Omega \qquad a_v = -100 \tag{11.1}$$

Suppose that an inverter built to these specifications is connected to the 10-mV microphone signal source and 8-Ω load previously described. The basic connections between signal source, amplifier, and load, with the amplifier represented by a linear two-port cell, are shown in Fig. 11.1.[1] The inverter responds to the voltage v_{in} measured across its own input terminals. In this case, v_{in} is not equal to v_m of the signal source, but is instead given by

$$v_{\text{in}} = v_m \frac{r_{\text{in}}}{r_{\text{in}} + R_m} = \frac{5\,\text{k}\Omega}{5\,\text{k}\Omega + 1\,\text{k}\Omega} v_m = 0.83 v_m \tag{11.2}$$

As Eq. (11.2) shows, the loading at the input reduces the 10-mV signal source to 8.3 mV at the amplifier's input terminals. When $R_m \gg r_{\text{in}}$, the input loading becomes even more severe.

Figure 11.1
Linear model of a
single-stage
inverter connected
to a signal source
and load.

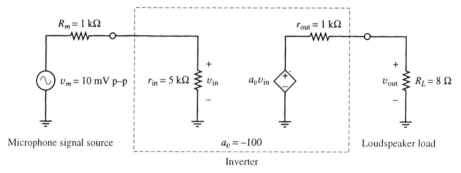

A similar loading process attenuates the signal appearing across the load. In the circuit of Fig. 11.1, the output voltage actually delivered to R_L is given by

$$v_{\text{out}} = a_v v_{\text{in}} \frac{R_L}{R_L + r_{\text{out}}} = a_v v_{\text{in}} \frac{8\,\Omega}{8\,\Omega + 1\,\text{k}\Omega} \approx 0.008 a_v v_{\text{in}} \tag{11.3}$$

Given the loading expressed by Eqs. (11.2) and (11.3), the overall voltage amplification from v_m to v_{out} becomes

$$\frac{v_{\text{out}}}{v_m} = a_v \frac{r_{\text{in}}}{r_{\text{in}} + R_m} \frac{R_L}{R_L + r_{\text{out}}} = -100(0.83)(0.008) \approx -0.66 \tag{11.4}$$

The net amplifier gain (as measured by the amplified version of v_m actually appearing across R_L) is not even close to the desired value of -100, and in fact is less than unity. In this case, the output loading of r_{out} by R_L is so catastrophic that the overall amplification system actually attenuates the input signal.

---

[1] Some confusion may result from the use of the term "load," which is sometimes applied to the pull-up element of an inverter. In this section, the "load" is an external element connected outside the topology of the amplifier.

11.2 TWO-PORT AMPLIFIER CASCADE

The impact of input and output loading can be minimized by creating a cascade of two amplifiers, as depicted in Fig. 11.2. The first stage consists of an inverter with large gain, and the second consists of a voltage follower, or "buffer" stage, with large input resistance and low output resistance. The gains and input voltages of the first and second stages have been labeled a_1, a_2, v_1, and v_2, respectively, and the small-signal input and output resistances labeled r_1 through r_4.

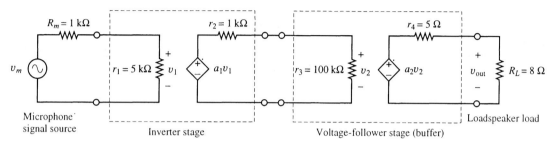

Figure 11.2 Cascaded amplifier consisting of an inverter stage and a voltage-follower stage.

The voltage divider can be applied to each of the coupled ports to obtain the overall gain from v_m to v_{out}. Beginning at the right side of the cascade in Fig. 11.2, v_{out} can be expressed by

$$v_{out} = (a_2 v_2) \frac{R_L}{R_L + r_4}$$

$$= (a_1 v_1) \frac{r_3}{r_3 + r_2} \, a_2 \frac{R_L}{R_L + r_4}$$

$$= v_m \frac{r_1}{r_1 + R_m} \frac{a_1 r_3}{r_3 + r_2} \frac{a_2 R_L}{R_L + r_4} \tag{11.5}$$

A small-signal analysis of the follower configuration reveals that r_{in} and r_{out} of the voltage-follower stage often depend on the r_{in} and r_{out} of the succeeding and preceding stages, respectively. (This problem will be specifically addressed in a later section.) For now, we shall show how the follower alleviates the loading problem by assuming it to have the following typical parameters:

$$r_{in} = r_3 = 100\,\text{k}\Omega \qquad r_{out} = r_4 = 5\,\Omega \qquad a_v = a_2 = 1 \tag{11.6}$$

For these values, the overall cascade gain given by Eq. (11.5) becomes

$$\frac{v_{out}}{v_m} = \left(\frac{5\,\text{k}\Omega}{5\,\text{k}\Omega + 1\,\text{k}\Omega} \right) (-100) \left(\frac{100\,\text{k}\Omega}{100\,\text{k}\Omega + 1\,\text{k}\Omega} \right) (1) \left(\frac{8\,\Omega}{8\,\Omega + 5\,\Omega} \right)$$

$$= \qquad (0.83) \qquad (-100) \qquad (0.99) \qquad (1) \qquad (0.62) \qquad \approx -51 \tag{11.7}$$

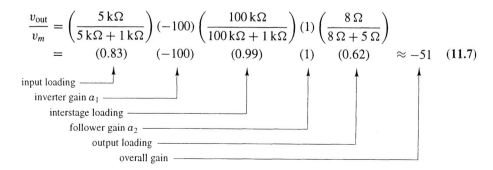

The result (11.7), although only about half the desired overall gain of -100, is much closer to the targeted value than the result (11.4) obtained for the unbuffered inverter.

The technique of multistage cascading is often used to create amplifiers with high input resistance and low output resistance, as well as gains that are much higher than those obtainable from a single stage alone. In the following example, several design and analysis techniques that apply uniquely to multistage amplifiers are illustrated.

EXAMPLE 11.1 The circuit of Fig. 11.3 is a dc-coupled three-stage BJT amplifier cascade made from discrete devices. Estimate the approximate small-signal gain using the two-port cascade technique and suitable engineering approximations.

Figure 11.3
Three-stage
dc-coupled
cascaded amplifier.

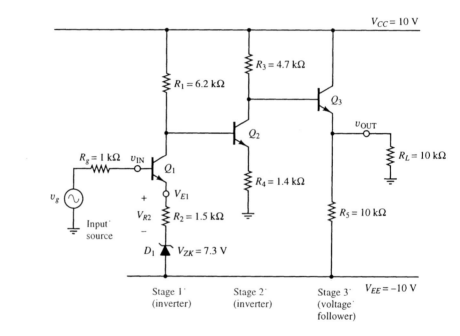

Solution

• **Overview of the amplifier design**

The first two stages are inverters in the feedback-bias configuration, and the third stage is a voltage follower. The zener diode D_1 is included to assist in the biasing of Q_1 and to allow the first stage to have a reasonably high gain. Expressions for the gain, input resistance, and output resistance of each stage, as derived in Chapter 7, are listed in Table 11.1. Also given in the table are approximate expressions for each of these parameters in the limit of large β_o. The incremental r_z of the zener is neglected compared to R_2.

Note that, in general, the input resistance of a follower depends on its load resistance R_L, while the output resistance of a follower depends on the source resistance of the preceding stage. In forming the two-port cascade of Fig. 11.3, however, only one of these dependencies need be included. When the cascade loading factors are computed for the third stage, either dependency will independently account for the feedback coupling introduced by R_5. Including both dependencies in loading-factor expressions would be tantamount to accounting for the effects of the coupling twice. This concept is explored in detail in Problem 11.18. In the case of

Parameter	Stage 1	Stage 2	Stage 3
a_v	$\dfrac{-\beta_{o1}R_1}{r_{\pi 1}+(\beta_{o1}+1)R_2}\approx\dfrac{-R_1}{R_2}$	$\dfrac{-\beta_{o2}R_3}{r_{\pi 2}+(\beta_{o2}+1)R_4}\approx\dfrac{-R_3}{R_4}$	$\dfrac{(\beta_{o3}+1)R_5}{r_{\pi 3}+(\beta_{o3}+1)R_5}\approx 1$
r_{in}	$r_{\pi 1}+(\beta_{o1}+1)R_2\approx\beta_{o1}R_2$	$r_{\pi 2}+(\beta_{o2}+1)R_4\approx\beta_{o2}R_4$	$r_{\pi 3}+(\beta_{o3}+1)R_5\approx\beta_{o3}R_5$
r_{out}	R_1	R_3	$R_5\left\|\dfrac{r_{\pi 3}+R_3}{\beta_{o3}+1}\approx\dfrac{r_{\pi 3}+R_3}{\beta_{o3}}\right.$

* Approximations shown are for large β_o.

Table 11.1. Gain and Resistance Parameters of the Cascade of Fig. 11.3*

Table 11.1, the factor $r_{out2}=R_3$ from stage 2 is included in the expression for r_{out3}, but R_L is omitted from the expression for r_{in3}. As shown in Exercise 11.4, the analysis of the cascade could be approached with equal validity by substituting $R_5\|R_L$ for R_5 in the expression for r_{in3} and by omitting the factor $r_{out2}=R_3$ from the expression for r_{out3}.

• Compute the overall gain of the cascade by considering loading factors

Analysis of the small-signal, two-port cascade model of this amplifier, shown in Fig. 11.4, yields an expression for the overall cascade gain:

$$
\begin{aligned}
\frac{v_{out}}{v_g} &= (a_1)(a_2)(a_3)\frac{r_{in1}}{R_g+r_{in1}}\frac{r_{in2}}{r_{in2}+r_{out1}}\frac{r_{in3}}{r_{in3}+r_{out2}}\frac{R_L}{R_L+r_{out3}}\\
&\approx \frac{-R_1}{R_2}\frac{-R_3}{R_4}(1)\frac{\beta_{o1}R_2}{R_g+\beta_{o1}R_2}\frac{\beta_{o2}R_4}{R_1+\beta_{o2}R_4}\frac{\beta_{o3}R_5}{R_3+\beta_{o3}R_5}\frac{R_L}{R_L+(r_{\pi 3}+R_3)/\beta_{o3}}
\end{aligned}
\tag{11.8}
$$

where large β_o has been assumed. Substitution of resistor values, a typical beta of $\beta_{o1}=\beta_{o2}=\beta_{o3}=100$, and the value $r_{\pi 3}\approx 1.5\,\text{k}\Omega$ yields

$$
\frac{v_{out}}{v_g}=(-4.1)(-3.4)(1)(0.99)(0.96)(0.995)(0.994)\approx 13
\tag{11.9}
$$

Each of the four loading factors in Eq. (11.9) is almost unity—a consequence of the high input resistance of each stage. Note that the overall gain is positive because two negative-gain inverters are cascaded in series.

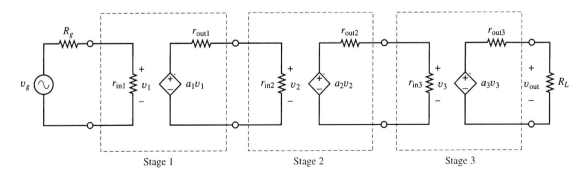

Figure 11.4 Two-port cascade of the amplifier of Fig. 11.3.

EXERCISE 11.1 Show that the "limit of large β_o" applied to the entries in Table 11.1 is tantamount to the conditions $1/g_{m1} \ll R_2$, $1/g_{m2} \ll R_4$, and $1/g_{m3} \ll R_5$.

11.2 For values of β_o in the range 20 to 200 and $g_m = 66.7$ mA/V, compute the gain of the cascade of Fig. 11.3 using the approximate expression (11.8).

11.3 For selected values of β_o in the range 20 to 200 and $g_m = 66.7$ mA/V, compare the results of the approximate expression (11.8) to the cascade gain obtained without engineering approximation. Use the exact expressions in Table 11.1 for a_v, r_{in}, and r_{out}.

11.4 Show that the same approximate gain expression (11.8) results if R_L is included in r_{in3} while r_{out2} is omitted from r_{out3}, so that $r_{in2} = r_{\pi3} + (\beta_o + 1)(R_5 \| R_L)$ and $r_{out3} = R_5 \| [r_{\pi3}/(\beta_o + 1)]$.

11.5 Show that the gain expression (11.8) can be found by analyzing the complete small-signal model of the circuit of Fig. 11.3.

11.3 MULTISTAGE AMPLIFIER BIASING

It is possible to create a multistage cascade in which each stage is separately biased and coupled to adjacent stages via dc blocking capacitors. One possible implementation of a two-stage inverter-follower cascade, for example, is shown in Fig. 11.5. In this circuit, each of the two stages has its own separate feedback bias network. Capacitors C_1 and C_2 isolate the separate bias networks by acting as open circuits to dc. These capacitors allow signals of sufficiently high frequency content to pass through the cascade by acting as short circuits to ac. Similarly, capacitor C_3 provides the small-signal emitter bypass necessary to give the inverter its large gain, but does not affect the bias network. This amplifier configuration has several deficiencies that limit its usefulness as a practical design. First of all, the impedances of the coupling and bypass capacitors become large at low frequencies, thereby reducing the overall amplifier gain. Secondly, as mentioned previously in Chapter 9, the use of discrete capacitors is to be discouraged in modern integrated-circuit design, because capacitors occupy large amounts of valuable "chip" area. The circuit of Fig. 11.5 embodies several principles important to multistage amplification, but does not represent good design practice. It is presented here for illustrative purposes only.

Figure 11.5
Possible implementation of a two-stage inverter-follower cascade. This design technique is not optimal because it uses many discrete capacitors.

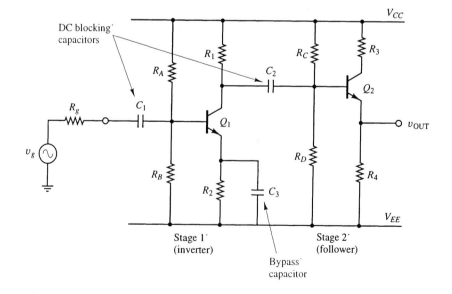

An improved, dc-coupled version of the two-stage amplifier that uses no discrete capacitors is shown in Fig. 11.6. In this circuit, an active bypass device Q_3 replaces C_3, as discussed in Chapter 8 (see Section 8.3.1). The bias configuration also eliminates the need for capacitors C_1 and C_2, because Q_1 is designed to have its base at dc ground, and Stage 1 provides the bias for Stage 2.

Figure 11.6
Two-stage
dc-coupled
cascaded amplifier.

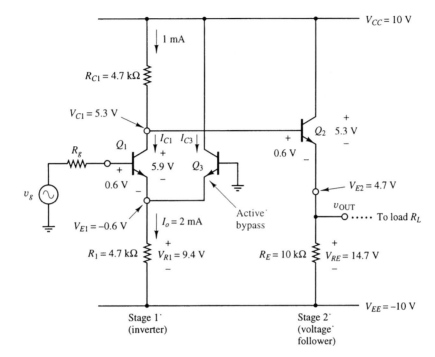

Exact bias calculations of this amplifier are cumbersome because many voltages and currents are involved, and linear cascading principles are not readily applied to biasing. The "constant-voltage" approximation, however, can be used to quickly analyze the bias of the multistage cascade. In the constant-voltage bias approximation, BJT base currents are assumed to be negligibly small compared to collector currents. This assumption allows nodes to which bases are connected to be evaluated under "open-circuit" conditions. The constant-voltage approximation is exact for FETs, for which $I_G = 0$.

EXAMPLE 11.2   Use the constant-voltage approximation to determine the bias voltages and currents in the circuit of Fig. 11.6. Verify the bias currents and node voltages shown in the figure.

Solution

• Find the bias parameters of Q_1

Analysis of the bias network of Fig. 11.6 is simple if the constant-voltage approximation is used. If the base current through R_g is neglected, the dc bias voltage at the base of Q_1 becomes zero, so that

$$V_{E1} = -V_{BE1} \equiv -V_f \tag{11.10}$$

Note that this bias voltage is evaluated with v_g set to zero. The voltage appearing across R_1 becomes

$$V_{R1} = V_{E1} - V_{EE} = (-0.6\,\text{V}) - (-10\,\text{V}) = 9.4\,\text{V} \tag{11.11}$$

where the value $V_f \approx 0.6\,\text{V}$ has been assumed. The resulting current I_o becomes

$$I_o = \frac{V_{R1}}{R_1} = \frac{9.4\,\text{V}}{4.7\,\text{k}\Omega} = 2\,\text{mA} \tag{11.12}$$

The current I_o is split into two equal parts of value $I_{C1} = I_{C3} \approx I_o/2 = 1\,\text{mA}$, so that the collector voltage to ground of Q_1 is given by

$$V_{C1} = V_{CC} - I_{C1}R_{C1} = 10\,\text{V} - (4.7\,\text{k}\Omega)(1\,\text{mA}) = 5.3\,\text{V} \tag{11.13}$$

• **Find the bias parameters of Q_2**

The voltage V_{C1} is fed directly to the base of Q_2, eliminating the need for base resistors to bias the second stage. The voltage across the follower resistor R_E thus becomes

$$V_{RE} = V_{C1} - V_{BE2} - V_{EE} = 5.3\,\text{V} - 0.6\,\text{V} - (-10\,\text{V}) = 14.7\,\text{V} \tag{11.14}$$

Similarly, the bias value of v_{OUT} is set to the value

$$V_{\text{OUT}} = V_{C1} - V_{BE2} = 4.7\,\text{V} \tag{11.15}$$

Finally, the bias parameters of Q_2 are found from

$$V_{CE2} = V_{CC} - V_{\text{OUT}} = 10\,\text{V} - 4.7\,\text{V} = 5.3\,\text{V} \tag{11.16}$$
and
$$I_{C2} = V_{RE}/R_E = 14.7\,\text{V}/10\,\text{k}\Omega = 1.47\,\text{mA} \tag{11.17}$$

Note that if $V_{BE2} = V_f$ is assumed constant, the bias value of V_{OUT} will not change appreciably if a load resistance is connected between the follower output and ground.

EXERCISE 11.6 Find the bias point of transistor Q_3 in Fig. 11.6.   **Answer:** $V_{CE3} = 10.6\,\text{V}$, $I_{C3} = 1\,\text{mA}$

11.7 Find the approximate bias power dissipated in each of the transistors of Fig. 11.6.
Answer: $P_{Q1} = 5.9\,\text{mW}$; $P_{Q2} = 7.8\,\text{mW}$; $P_{Q3} = 10.6\,\text{mW}$

11.8 Analyze the bias components in the circuit of Fig. 11.3 and find the approximate bias point I_C, V_{CE} for each transistor. Use the constant-voltage approximation, and assume the value $V_f = 0.7\,\text{V}$. Also find the approximate value of $r_{\pi 3}$ for $\beta_o = 100$.   **Answer:** $I_{C1} \approx 1.3\,\text{mA}$; $I_{C2} \approx 0.7\,\text{mA}$; $I_{C3} \approx 1.6\,\text{mA}$; $V_{CE1} \approx 2.4\,\text{V}$; $V_{CE2} \approx 5.5\,\text{V}$; $V_{CE3} \approx 4.2\,\text{V}$; $r_{\pi 3} \approx 1.6\,\text{k}\Omega$

11.9 Assess the change in the bias points I_C, V_{CE} for each transistor in Fig. 11.3 if the value $V_f = 0.6\,\text{V}$ is assumed instead of $V_f = 0.7\,\text{V}$. Again use the constant-voltage approximation, as in Exercise 11.8.   **Answer:** $I_{C1} \approx 1.4\,\text{mA}$; $I_{C2} \approx 0.5\,\text{mA}$; $I_{C3} \approx 1.7\,\text{mA}$; $V_{CE1} \approx 1.9\,\text{V}$; $V_{CE2} \approx 6.9\,\text{V}$; $V_{CE3} \approx 3\,\text{V}$

11.4 DC LEVEL SHIFTING

In a dc-coupled multistage cascade, the output bias level of each stage is passed on to the input of the next stage. An *npn* BJT properly biased for adequate swing range must have its collector more positive than its base. Similarly, an *n*-channel FET biased for adequate swing range must have its drain more positive than its gate. A problem arises in large cascades when the bias voltage passed from stage to stage becomes increasingly more positive. If the voltage "stacking" is severe, little swing room is left in the final stages of the cascade.

11.4.1 Level Shifting in BJT Circuits

The problem of level shifting is best illustrated by the BJT circuit of Fig. 11.7. This circuit is the same one examined for its small-signal properties in Example 11.1. In Fig. 11.7, the various bias voltages and currents in the circuit are shown. The base of the *npn* device Q_1 is biased near ground potential by v_g. The collector of Q_1 is biased at a more positive voltage, equal to $V_{C1} = +1.3$ V. The voltage V_{C1} is passed on to the base of Q_2, whose collector is biased at an even higher voltage of $V_{C2} = +7.6$ V. The latter voltage is passed on to the base of Q_3. The base–emitter junction of Q_3 contributes only a small drop of V_f before passing the bias voltage to the v_{OUT} terminal as $+7.0$ V. The resulting bias value of v_{OUT} illustrates the nature of the voltage accumulation problem. Regardless of the exact bias levels chosen in stages 1 and 2, v_{OUT} of this amplifier will always be biased at a positive voltage if the base of Q_1 is biased at ground potential. Such a situation limits the swing range of v_{OUT}, which ideally should be biased midway between V_{CC} and V_{EE}. In the extreme case of many cascaded stages, the bias value of v_{OUT} may even approach the V_{CC} bus.

Figure 11.7
Multistage cascade of Fig. 11.3 with dc bias voltages indicated.

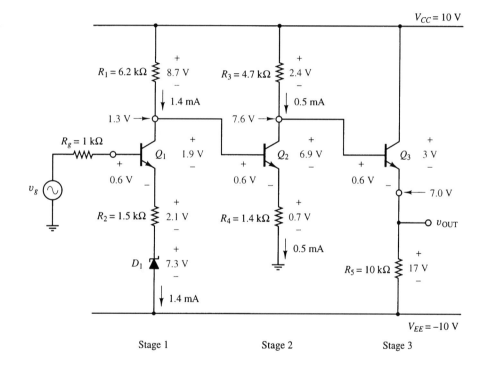

This voltage "stacking" problem can be alleviated by the use of *dc level shifting*. Level shifting alters the bias distribution, but not the gain, of the amplifier cascade. One simple dc level-shifting method involves the insertion of a passive device with a constant dc voltage drop between successive stages of the cascade. The circuit of Fig. 11.8, for example, resembles the circuit of Fig. 11.7, but has a zener diode inserted between the first and second stages. The voltage drop of zener D_2 shifts the bias voltage passed to the base of Q_2 downward by 7.9 V, from 1.3 V to the value -6.6 V. The latter voltage is closer to the negative bus than the voltage at the collector of Q_1. The level shift provided by zener D_2 ultimately allows v_{OUT} to be biased at 0 V, midway between V_{CC} and V_{EE}. To first order, the zener can be modeled as a dc voltage source and as a short circuit to small signals, so that the gain and small-signal behavior of the amplifier are virtually unaffected by the presence of the zener.

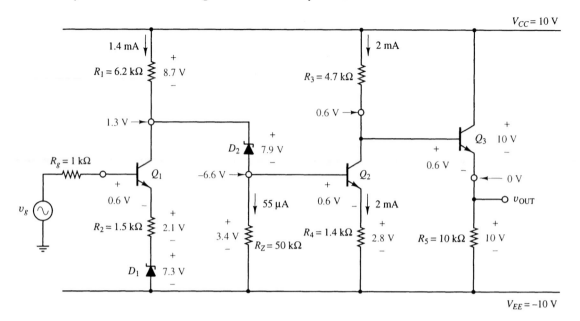

Figure 11.8 Zener diode level shifter added to the circuit of Fig. 11.7. BJT base currents are neglected in computing the indicated bias levels.

EXERCISE 11.10 What value of V_{ZK} should be chosen for zener diodes D_1 and D_2 in Fig. 11.8?
Answer: 7.3 V; 7.9 V

11.11 Analyze the bias of the circuit of Fig. 11.8 and show that v_{OUT} is indeed biased at zero volts relative to ground.

11.12 Redesign the level shifter of Fig. 11.8 by substituting one or more *pn* junction diodes for zener D_2. What are the advantages and disadvantages of such a substitution?

11.13 Redesign the level shifter of Fig. 11.8 by substituting base–emitter junctions of one or more *npn* BJTs for zener D_2.

By using complementary devices, active leveling-shifting techniques can be combined with circuit amplification functions. In the circuit of Fig. 11.9, for example, Q_2 of Fig. 11.7 is replaced by a *pnp* device. This substitution introduces a downward shift in the cascade bias and allows v_{OUT} to be biased midway between V_{CC} and V_{EE}. The shift in bias voltage is made possible because the collector of the *pnp* device must be biased more *negatively* than its base. The bias voltage of 0.4 V passed from stage 2 to stage 3 is therefore closer to V_{EE} than the bias voltage of 6.3 V passed from stage 1 to stage 2. Note that *pnp* and *npn* devices have the same small-signal model; hence the gain of the cascade of Fig. 11.9 can be made the same as that of Fig. 11.7 by choosing the same values of R_1 through R_4.

EXERCISE 11.14 Analyze the bias circuit of Fig. 11.9 and show that v_{OUT} is biased at -0.2 V relative to ground.

11.15 For the circuits of Fig. 11.8 and 11.9, calculate the bias power dissipated in each transistor and resistor.

11.16 Choose a new value of R_5 in Figs. 11.8 and 11.9 so that the sum of bias powers dissipated in devices Q_3 and R_5 will be limited to a total of about 1 mW. **Answer:** $R_5 \geq 196\,k\Omega$

Figure 11.9
Pnp level shifter incorporated into the circuit of Fig. 11.7. BJT base currents are neglected in computing the indicated bias levels.

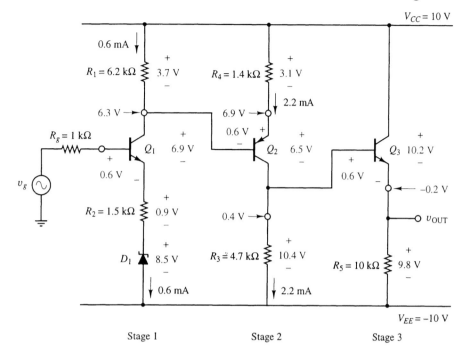

Stage 1   Stage 2   Stage 3

11.4.2 Level Shifting in MOSFET Circuits

If a MOSFET is to be biased with adequate swing range, its drain must be biased at a higher voltage than its gate. A multistage cascade of MOSFET inverters therefore will also experience the "voltage-stacking" problem described in Section 11.4.1. Level shifting can again be introduced into a MOSFET circuit using techniques that are similar to those employed in BJT circuits. Specifically, devices with fixed voltage drops can be inserted between adjacent stages, thus eliminating the voltage-stacking problem. In principle, one could use a series-connected zener diode as a constant-voltage element, as was done with D_2 in the BJT circuit of Fig. 11.8. This method is acceptable for BiCMOS circuits but is undesirable on an NMOS or CMOS integrated circuit, because a separate set of fabrication processes may be required to create the zener. An alternative scheme involving *n*-channel MOSFETs only is illustrated in Fig. 11.10. In this circuit, MOSFETs Q_A and Q_B are biased at a constant-current level by current sources I_1 and I_2. On an actual IC, these sources would be implemented using MOSFET current mirrors, as discussed in Chapter 7. With fixed drain currents I_D, devices Q_A and Q_B sustain constant gate-to-source voltages V_{GS}. These voltage drops are subtracted from V_{D1}, reducing the bias voltage at the gate of Q_3 to the value $V_{D1} - V_{GSA} - V_{GSB}$. The overall small-signal gain of the cascade is virtually unaffected by the presence of Q_A and Q_B, because the gains of these followers approach unity when I_1 and I_2 have large Norton resistances (see Exercise 11.17). In practice, as many level-shifting devices as are required can be inserted into the cascade.

In analog CMOS amplifiers, level shifting is easily accomplished without the need for additional constant-voltage-drop devices. One need only alternate the roles of the *n*- and *p*-channel devices as the cascade progresses. The basic concept, applicable to most CMOS circuits, is illustrated by the simplified amplifier of Fig. 11.11. (A real CMOS amplifier requires additional circuitry to properly establish its bias voltages and currents.) In stage 1, *n*-channel device Q_1 is driven by v_{IN}, and *p*-channel device Q_2 is used as a pull-up load. In stage 2, the roles are reversed—*p*-channel MOSFET Q_3 is driven by the output of the first stage, and *n*-channel device Q_4 is used as a "pull-up" load. With the gate of Q_1 biased at dc ground, proper bias of Q_1 requires

Figure 11.10
Level shifting in a
two-stage NMOS
amplifier. One or
more gate-to-source
voltage drops are
inserted in series
between v_{D1} and
v_{OUT}. Currents I_1
and I_2 are
established by
current mirrors.

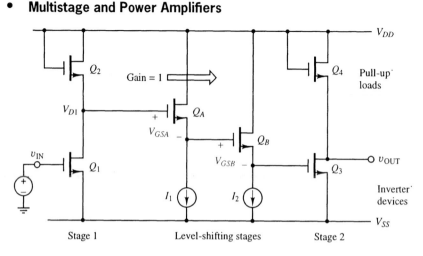

that V_{D1} be biased above ground. The arrangement of Fig. 11.11 causes V_{D1} to be downshifted by the gate-to-drain voltage V_{GD3} of Q_3 before being passed on to V_{OUT}. In this case, V_{GD3} must be positive because the drain of the p-channel device Q_4 must be biased more *negatively* than its gate.

Figure 11.11
Level shifting in
CMOS circuits can
be realized without
additional
components by
alternating the roles
of the n-channel
and p-channel
devices. In stage 1,
n-channel Q_1 is the
driven device. In
stage 2, p-channel
Q_3 is the driven
device.

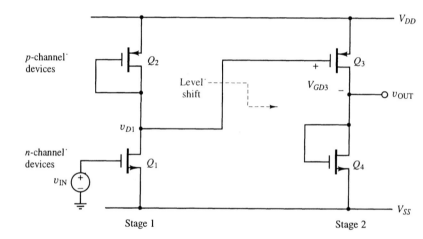

Note that the arrangement of Fig. 11.11 can also be used in BJT circuits by alternating the roles of *npn* and *pnp* devices in successive stages. This level-shifting scheme will be illustrated in the context of the differential amplifier in the next section.

EXERCISE 11.17 Show that the gain of either level-shifting stage formed by Q_A or Q_B in Fig. 11.10 approaches unity in the limit $r_n \to \infty$, where r_n is the Norton resistance of current sources I_1 or I_2. Begin your analysis by computing the gain of Q_A or Q_B with finite r_n, then take the limit of the gain for very large r_n.

11.18 Design a pair of current mirrors using NMOS devices only that will function as current sources I_1 and I_2 in Fig. 11.10.

11.19 Design a cascade made from *npn* and *pnp* BJTs that makes use of the level-shifting topology of Fig. 11.11. How does your circuit differ from the MOSFET version?

11.20 Derive an expression for the overall cascade gain of the CMOS circuit of Fig. 11.11.
Answer: $g_{m1}g_{m3}/g_{m2}g_{m4}$

11.5 DIFFERENTIAL-AMPLIFIER CASCADE

The multistage cascade and dc level-shifting principles of this chapter are readily applied to the differential amplifier circuits of Chapter 8. When two diff-amps are cascaded, the overall gain of the cascade is again subject to input, output, and interstage loading factors. These loading factors may not be the same for differential- and common-mode signals.

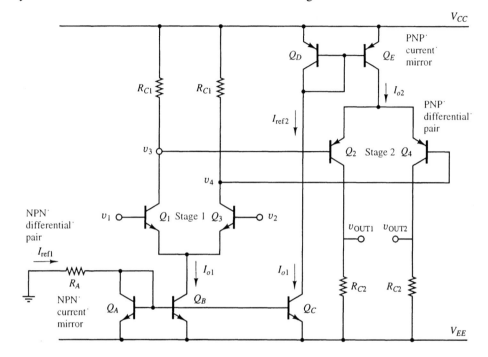

Figure 11.12
Two-stage BJT differential amplifier cascade with complementary *npn* and *pnp* devices.

The circuit of Fig. 11.12 illustrates a typical two-stage BJT diff-amp cascade. The first stage is made from matched *npn* devices and the second from matched *pnp* devices. As suggested in Section 11.4.2, the use of complementary transistors in successive stages provides dc level shifting without the need for additional passive level-shifting devices. Transistors Q_A through Q_E form a pair of complementary current mirrors that set the bias currents of Q_1 through Q_4 to the same value. The reference current I_{ref2} of the *pnp* current mirror is derived from Q_C, which replicates the current I_{ref1} of the *npn* current mirror. Because $g_m = I_C/\eta V_T$ for both *npn* and *pnp* devices, transistors Q_1 through Q_4 have the same value of g_m. Note that although these transistors have the same g_m, the parameters $\beta_{o\text{-NPN}}$ and $\beta_{o\text{-PNP}}$ are seldom equal, so that $\beta_{o\text{-NPN}}/g_m \neq \beta_{o\text{-PNP}}/g_m$, and $r_{\pi N} \neq r_{\pi P}$.

The two stages of the cascade have the same small-signal model, with amplifier parameters given by

$$A_{\text{dm-se}} = \mp \frac{g_m R_C}{2} \tag{11.18}$$

$$A_{\text{cm-se}} = \frac{-\beta_o R_C}{r_\pi + 2(\beta_o + 1)r_o} \approx \frac{-R_C}{2r_o} \tag{11.19}$$

$$r_{\text{in-dm}} = 2r_\pi \tag{11.20}$$

$$r_{\text{in-cm}} = r_\pi + 2(\beta_o + 1)r_o \approx 2\beta_o r_o \tag{11.21}$$

and

$$r_{\text{out-se}} = R_C \tag{11.22}$$

In these expressions, r_o refers to the small-signal output resistance of the current-mirror transistors Q_B and Q_E.

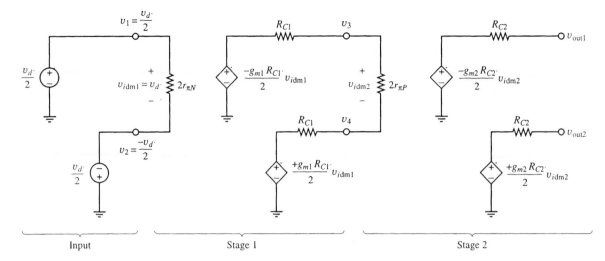

Figure 11.13 Two-port cascade for evaluating the differential-mode performance of the amplifier of Fig. 11.12.

A two-port cascade suitable for evaluating the differential-mode performance of this amplifier is shown in Fig. 11.13. Differential sources of value $\pm v_d/2$ are connected to the v_1 and v_2 inputs, so that $v_{idm1} = v_1 - v_2 = v_d$. Outputs v_{out1} and v_{out2} are measured under open-circuit conditions. Applying the cascading and loading principles to this circuit yields the single-ended differential-mode gains:

$$A_{\text{dm-se1}} = \frac{v_{\text{out1}}}{v_{idm1}} = \frac{-g_{m2}R_{C2}}{2}(-g_{m1}R_{C1})\frac{2r_{\pi P}}{2r_{\pi P} + 2R_{C1}} \tag{11.23}$$

and

$$A_{\text{dm-se2}} = \frac{v_{\text{out2}}}{v_{idm1}} = \frac{+g_{m2}R_{C2}}{2}(-g_{m1}R_{C1})\frac{2r_{\pi P}}{2r_{\pi P} + 2R_{C1}} \tag{11.24}$$

The common-mode performance of the cascade can be evaluated by connecting common-mode inputs to v_1 and v_2, as shown in Fig. 11.14. Applying cascading and loading principles to this circuit yields

$$A_{\text{cm-se1}} = A_{\text{cm-se2}} = \frac{v_{\text{out-se}}}{v_{icm}} = \frac{-R_{C2}}{2r_{oE}}\frac{-R_{C1}}{2r_{oB}}\frac{2\beta_o P r_{oE}}{2\beta_o P r_{oE} + R_{C1}} \tag{11.25}$$

The common-mode rejection ratio of the cascade for single-ended outputs is given by

$$\begin{aligned}
\text{CMRR} &= \frac{|v_{\text{out-se}}/v_{idm}|}{|v_{\text{out-se}}/v_{icm}|} \\
&= \frac{(g_{m2}R_{C2})(g_{m1}R_{C1})/2}{(R_{C2}/2r_{oE})(R_{C1}/2r_{oB})}\frac{r_{\pi P}}{r_{\pi P} + R_{C1}}\frac{2\beta_o P r_{oE} + R_{C1}}{2\beta_o P r_{oE}} \\
&= \frac{g_{m2}g_{m1}r_{oB}r_{\pi P}(2\beta_o P r_{oE} + R_{C1})}{(r_{\pi P} + R_{C1})\beta_o P} = \frac{g_{m1}r_{oB}(2\beta_o P r_{oE} + R_{C1})}{r_{\pi P} + R_{C1}}
\end{aligned} \tag{11.26}$$

where $g_{m2}r_{\pi P} = \beta_o P$. Because $r_{oE} \gg R_{C1}$ and $r_{\pi P} \sim R_{C1}$, the latter expression can be simplified to

$$\text{CMRR} \approx \frac{g_{m1}r_{oB}(2\beta_o P r_{oE})}{r_{\pi P} + R_{C1}} \sim g_{m1}g_{m2}r_{oB}r_{oE} \tag{11.27}$$

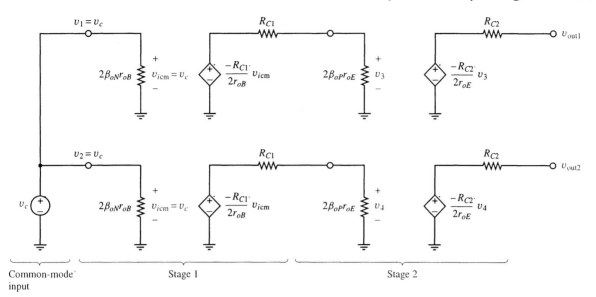

Figure 11.14 Two-port cascade for evaluating the common-mode performance of the amplifier of Fig. 11.12.

Equation (11.27) indicates that the cascaded CMRR is on the order of $(g_m r_o)^2$. Its value should be compared to the CMRR of a single-stage BJT differential amplifier, which is on the order of $g_m r_o$:

$$\text{CMRR} = \frac{r_\pi + 2(\beta_o + 1)r_o}{2r_\pi} \approx \frac{\beta_o r_o}{r_\pi} = g_m r_o \qquad (11.28)$$

A cascade of differential amplifiers has a large differential gain and very large common-mode rejection ratio. Such cascades often form the basis for the internal circuitry of an operational amplifier. In Chapter 12, it will be shown how a diff-amp cascade can be made into an operational amplifier by the addition of middle-gain and output stages.

EXERCISE 11.21 Rederive the differential- and common-mode cascaded gains of the circuit of Fig. 11.12 if the input voltages are connected to the amplifier via series input resistances R_g.

11.22 Rederive the differential- and common-mode cascaded gains of the circuit of Fig. 11.13 for single-ended outputs if a load resistor R_L is connected differentially between the v_{OUT1} and v_{OUT2} terminals.

11.6 POWER-AMPLIFICATION OUTPUT STAGES

A multistage amplifier is often called upon to deliver large amounts of power to a passive load. This power delivery may take the form of a large load current fed to a small resistance or impedance. Alternatively, it may consist of a large-amplitude voltage signal developed across a resistance or impedance of moderate value. In either case, the final stage must be designed to meet the amplifier's output requirements. Although the preceding stages of the amplifier are typically biased for linear operation, using the techniques of Chapter 7, problems may arise if the output stage of the amplifier is biased in the same way. As shown in this section, the power wasted in a linearly biased amplifier stage can be comparable to the power actually delivered to its load.

Consequently, large, oversized transistors capable of safely dissipating this excess power must be used for the output stage. In high-power applications, the power wasted by low amplifier efficiency leads to the generation of heat in the circuit and causes device temperatures to rise.

In this section, several alternatives to the linear amplifier topologies of Chapter 7 are examined for use as output stages. These alternative configurations offer improved efficiency and less stringent device size and power-handling requirements at the expense of true linear operation. Although the devices in a well-designed power-amplification stage are not biased in their active regions, near-linear amplification is nonetheless possible in many situations.

11.6.1 Complementary-Pair (Class B) Output Configuration

When an amplifier is called upon to deliver large load currents, it is desirable to bias the voltage of its output terminal near ground potential. This requirement minimizes the bias power dissipated in both the load element and the active devices of the output stage. Implicit in this design goal is the understanding that the amplifier will operate from bipolar power supplies, so that the output voltage will be capable of swinging in both positive and negative directions. One power amplification stage that meets these needs is called the *complementary-pair*, or "push-pull," follower configuration. The complementary-pair follower configuration is also known as the *class B amplifier*.

Figure 11.15
Complementary-pair "push-pull" amplifier output stage made from BJTs. This circuit is a class B amplifier because neither device is biased in its active region.

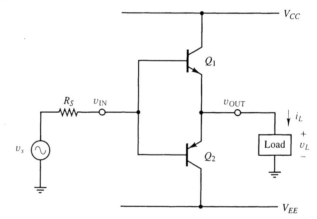

An example of such a circuit is shown in Fig. 11.15. This representative circuit is made from BJTs; similar circuits can be made from MOSFETs or JFETs. The sinusoidal signal source v_s and series resistance R_S represent the output voltage and output resistance of the stage preceding the complementary-pair output stage. If R_S is small, input loading will be minimal and v_{IN} will be approximately equal to v_s.

In the class B amplifier, neither device is biased in the active region. When the input signal is positive, as in Fig. 11.16(a), the upper device is driven into the active region by the input signal, while the lower device remains in cutoff. The large-signal output for positive v_{IN} becomes that of a simple voltage follower, that is,

$$v_{OUT} = v_{IN} - V_f \qquad \text{for } v_{IN} > V_f \qquad (11.29)$$

Similarly, when the input is negative, as in Fig. 11.16(b), the upper device remains in cutoff, while the lower device is driven into the active region by the input signal. The output voltage under such conditions becomes

$$v_{OUT} = v_{IN} + V_f \qquad \text{for } v_{IN} < -V_f \qquad (11.30)$$

If the magnitude of v_{IN} is less than the base–emitter turn-on voltage V_f, v_{OUT} will be equal to zero.

Figure 11.16
Operation of complementary-pair BJT follower for (a) positive $(v_{IN} > V_f)$ and (b) negative $(v_{IN} < -V_f)$ polarities of v_{IN}. When $|v_{IN}| < V_f$, both Q_1 and Q_2 are in cutoff and $v_L = 0$.

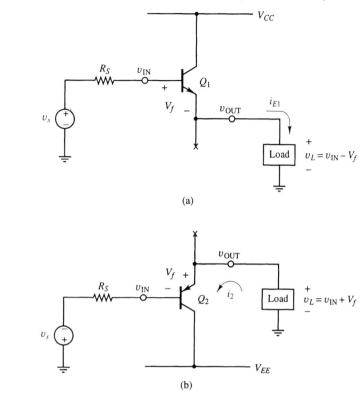

(a)

(b)

The complete transfer characteristic of the output stage, including both positive and negative input voltages, is shown in Fig. 11.17. Near the origin, where $|v_{IN}| < V_f$, the slope changes abruptly. This nonlinearity in the transfer characteristic, called the *crossover distortion* region, represents nonlinear amplifier behavior and results in an output signal that is not an exact replica of the input signal.

Figure 11.17
Large-signal voltage transfer characteristic of the BJT circuit of Fig. 11.15. The crossover distortion region has a width of $2V_f$. The slope of the transfer characteristic outside this region is approximately equal to +1.

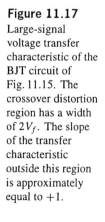
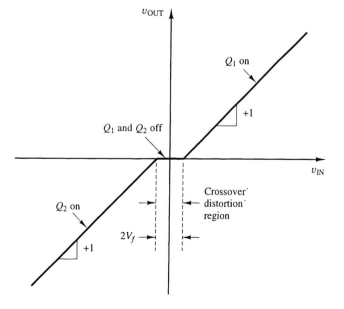

When either device is conducting, the small-signal output resistance of the circuit is given by

$$r_{\text{out}} = \frac{R_S + r_\pi}{\beta_o + 1} \qquad (11.31)$$

Equation (11.31) must be interpreted with care, because the output stage often operates under large-signal conditions where the small-signal model may not apply. The equivalent value of r_π may change as v_{OUT} undergoes large voltage swings. In the case where R_S is large compared to the average r_π, Eq. (11.31) reduces to

$$r_{\text{out}} = \frac{R_S}{\beta_o + 1} \qquad (11.32)$$

which generally has a small value and is independent of r_π.

EXAMPLE 11.3   An inverter stage with an output resistance of $1\,k\Omega$ and a gain of -100 is cascaded with the output stage of Fig. 11.15. Estimate the resulting output resistance and gain of the cascaded combination if $\beta_o = 100$ and if the v_{OUT} terminal feeds a load resistance of value $R_L = 100\,\Omega$.

Solution

If the output stage operates outside the crossover distortion region, the overall gain will be equal to the product of the individual stage gains multiplied by input and output loading factors. The input loading factor will be given by $(\beta_o + 1)R_L/[(\beta_o + 1)R_L + R_S] \approx 0.83$. Assuming r_π to be on the order of $R_S = 1\,k\Omega$, the output resistance of the cascade will be on the order of $(1\,k\Omega + 1\,k\Omega)/50 = 20\,\Omega$, so that the output loading factor will equal $R_L/(R_L + r_{\text{out}}) \approx 0.83$. The approximate value of the gain of the overall cascade thus becomes $(-100)(1)(0.83)(0.83) = -69$.

11.6.2 Linearly Biased (Class A) Output Configuration

For input voltages greater than V_f, the complementary-pair follower of Fig. 11.15 performs the same signal amplification task as the linearly biased voltage follower of Fig. 11.18. In the latter circuit, V_{BB} biases Q_1 in the middle of its active region. The circuit of Fig. 11.18 is an example of a *class A amplifier*. Its output is a linear replica of the input signal and the transistor remains actively biased at all times. The transfer characteristic of the amplifier exhibits no crossover distortion region. Despite these desirable characteristics, the advantage of the class B complementary-pair voltage follower over the simple class A follower becomes evident if their power characteristics are compared under resistive load conditions.

Figure 11.18
One-transistor BJT voltage-follower stage. The voltage V_{BB} represents the dc bias voltage fed from the stage preceding the output stage.

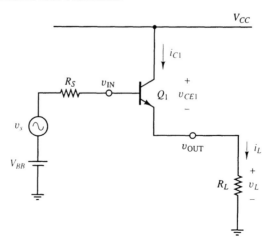

If the amplifier of Fig. 11.18 is biased for maximum symmetrical swing, then the bias value of v_{OUT} will be set to approximately $V_{CC}/2$. If the input signal v_s is a sinusoid that causes a signal component of magnitude V_o to be developed across the load, the total voltage appearing across the load can be written as

$$v_L(t) = \frac{V_{CC}}{2} + V_o \sin \omega t \qquad (11.33)$$

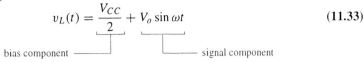

For a $v_L(t)$ of this form, the current flowing through both Q_1 and R_L will be equal to

$$i_L(t) = \frac{V_{CC}}{2R_L} + \frac{V_o}{R_L} \sin \omega t \qquad (11.34)$$

where $i_{C1} \approx i_{E1} = i_L$. For these values of v_L and i_L, the instantaneous power delivered to the load, given by the product of Eqs. (11.33) and (11.34), becomes

$$p_L(t) = v_L(t) i_L(t)$$
$$= \frac{V_{CC}^2}{4R_L} + \frac{V_o V_{CC}}{R_L} \sin \omega t + \frac{V_o^2}{R_L} \sin^2 \omega t \qquad (11.35)$$

Of the three terms on the right-hand side of Eq. (11.35), only the first and third have nonzero time-average values. The time average of $\sin^2 \omega t$ is equal to $1/2$; hence the time-average value of Eq. (11.35) is given by

$$\langle p_L(t) \rangle = \frac{V_{CC}^2}{4R_L} + 0 + \frac{V_o^2}{2R_L} \qquad (11.36)$$

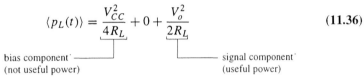

A similar consideration yields the bias and signal components of the power dissipated in the active device Q_1, for which

$$v_{CE1}(t) = V_{CC} - v_L(t) = \frac{V_{CC}}{2} - V_o \sin \omega t \qquad (11.37)$$

and
$$p_Q(t) \approx v_{CE1}(t) i_{C1}(t) \equiv v_{CE1}(t) i_L(t)$$
$$= \frac{V_{CC}^2}{4R_L} - \frac{V_o^2}{R_L} \sin^2 \omega t \qquad (11.38)$$

where the first term in Eq. (11.38) is the bias component, and the second is the signal component. As Eq. (11.38) shows, the bias components of $p_L(t)$ and $p_Q(t)$ are equal when the follower is biased for maximum symmetrical swing with $V_L = V_{CE} = V_{CC}/2$. The time average of $p_Q(t)$ is given by

$$\langle p_Q(t) \rangle = \frac{V_{CC}^2}{4R_L} - \frac{V_o^2}{2R_L} \qquad (11.39)$$

Note that the second term in Eq. (11.39) is negative. The total time average power dissipated in Q_1 is reduced by an amount equal to the time-average signal power delivered to the load. This result can be explained by noting that in the voltage-follower configuration, the voltage across Q_1 is reduced as i_{C1} is increased. Conversely, the current through Q_1 is reduced as v_{CE1} is increased. Hence the total power dissipated in Q_1 is reduced for both positive and negative excursions of $V_o \sin \omega t$.

We next compare the signal power dissipated in the load to the total power drawn from the supply. For a given $v_L(t)$, the power drawn from the V_{CC} supply will be equal to V_{CC} multiplied by the current $i_{C1}(t)$:

$$p_S(t) = V_{CC}\left(\frac{V_{CC}}{2R_L} + \frac{V_o}{R_L}\sin\omega t\right) \tag{11.40}$$

Taking the time-average value of this quantity yields

$$\langle p_S(t)\rangle = \frac{V_{CC}^2}{2R_L} \tag{11.41}$$

Note that $\langle p_S(t)\rangle$ is equal to the sum of $\langle p_L(t)\rangle$ and $\langle p_Q(t)\rangle$.

The efficiency η of a power output stage is defined as the time-average signal power delivered to the load divided by the total time-average power extracted from the supply, that is,

$$\eta = \frac{\langle p_L(t)\rangle|_{\text{signal component}}}{\langle p_S(t)\rangle|_{\text{total}}} \tag{11.42}$$

Applying Eqs. (11.36) and (11.41) to this definition results in

$$\eta = \frac{V_o^2/2R_L}{V_{CC}^2/2R_L} = \frac{V_o^2}{V_{CC}^2} \tag{11.43}$$

If the amplifier is driven to its cutoff and saturation swing limits by the maximum permissible v_s, then the excursion magnitude V_o can be at most equal to $V_{CC}/2$ (if V_{sat} of Q_1 is approximated as zero). For this maximum obtainable magnitude of V_o, the efficiency (11.43) also attains its maximum value, given by

$$\eta_{\max} = \frac{(V_{CC}/2)^2}{V_{CC}^2} = \frac{1}{4} \equiv 25\% \tag{11.44}$$

It can be shown that the theoretical result (11.44) applies generally to all class A amplifiers under sinusoidal excitation, including a linearly biased inverter or current follower. In practice, an efficiency lower than 25% results because the output is seldom driven to the edge of its cutoff and saturation limits in linear applications. Because of its low efficiency, the class A configuration is seldom used as the power output stage of a multistage cascade.

The power characteristics of the class B complementary-pair amplifier of Fig. 11.15 are easily computed for comparison with the class A configuration. For the purpose of this rough calculation, the crossover distortion evident in Fig. 11.17 is ignored. Since no bias current flows through either Q_1 or Q_2, the only power dissipated in the load is that associated with the output signal. If a sinusoidal signal voltage of magnitude V_o is developed across a resistive load of value R_L, the current through the load, carried alternately by Q_1 and Q_2, becomes

$$i_L(t) = \frac{V_o}{R_L}\sin\omega t \tag{11.45}$$

The resulting power dissipated in the load will be equal to the product $v_L(t)i_L(t)$. The time-average value of this product becomes

$$\langle p_L(t)\rangle = \langle v_L(t)i_L(t)\rangle$$
$$= \left\langle V_o\sin\omega t\left(\frac{V_o}{R_L}\sin\omega t\right)\right\rangle = \frac{V_o^2}{2R_L} \tag{11.46}$$

In this case, the time-average sinusoidal signal power dissipated in the load is identical to that of the class A amplifier of Fig. 11.18, as given by Eq. (11.36).

To find the power extracted by the circuit from the power supplies, the current drawn from each voltage source must be computed. The current to the circuit will have two components—one flowing out of V_{CC} when Q_1 conducts, and one flowing into V_{EE} when Q_2 conducts. For a symmetrical input signal, these current components will have the same peak magnitudes. If $V_{EE} = -V_{CC}$, the total power consumed by the circuit will be equal to twice the power drawn from one supply. If the output signal is a sinusoid of magnitude V_o, the time-average power extracted from V_{CC} over one sinusoidal cycle will be equal to

$$
\begin{aligned}
\langle p_{S+}(t) \rangle &= \frac{1}{T} \int_0^{T/2} V_{CC} i_{C1}(t)\, dt \\
&= \frac{1}{T} \int_0^{T/2} V_{CC} \frac{V_o}{R_L} \sin \omega t\, dt
\end{aligned}
\tag{11.47}
$$

where $i_{C1}(t)$ has be equated with the $i_L(t)$ of Eq. (11.45). The upper integration limit in Eq. (11.47) is set to $T/2$ because V_{CC} delivers power only when $i_L(t)$ is positive. The integral of $\sin \omega t$ over the indicated limits is equal to $2/\omega = T/\pi$, so that Eq. (11.47) becomes

$$
\langle p_{S+}(t) \rangle = \frac{V_{CC}}{\pi} \frac{V_o}{R_L}
\tag{11.48}
$$

If $V_{EE} = -V_{CC}$, the time-average power $\langle p_{S-}(t) \rangle$ extracted from the negative supply will also be given by Eq. (11.48). The total power extracted from both supplies therefore becomes

$$
\langle p_S(t) \rangle = \langle p_{S+}(t) \rangle + \langle p_{S-}(t) \rangle = \frac{2 V_{CC}}{\pi} \frac{V_o}{R_L}
\tag{11.49}
$$

For a sinusoidal output signal of magnitude V_o, the efficiency of the amplifier becomes

$$
\eta = \frac{\langle p_L(t) \rangle}{\langle p_S(t) \rangle} = \frac{V_o^2 / 2R_L}{2 V_{CC} V_o / \pi R_L} = \frac{\pi}{4} \frac{V_o}{V_{CC}}
\tag{11.50}
$$

where $\langle p_L(t) \rangle$ is given by Eq. (11.46). If the output is driven to maximum symmetrical swing limits defined by $V_o = \pm V_{CC}$ (which assumes that $V_{\text{sat}} \approx 0$), the efficiency η reaches the value

$$
\eta_{\max} = \frac{\pi}{4} \equiv 78.5\%
\tag{11.51}
$$

Equation (11.51) describes the percentage of power extracted from the supplies that is delivered to the load as signal power. The remaining 21.5% of the extracted power is dissipated in Q_1 and Q_2.

For the same delivered sinusoidal signal power, the complementary-pair class B amplifier of Fig. 11.15 has more than three times the maximum efficiency of the linearly biased class A voltage follower of Fig. 11.18. The class B configuration also dissipates no bias power in the load or in the active devices. For these reasons, an output stage of the class B type is used when efficiency is of prime importance.

EXERCISE 11.23   Show that the maximum attainable efficiency of a class A amplifier under sinusoidal excitation is still equal to 25%, even if the amplifier is operated from symmetrical bipolar power supplies.

11.24   The efficiency calculations performed on the class A voltage follower of Fig. 11.18 also apply to the case where the load resistor appears between the collector of Q_1 and the V_{CC} supply, with the emitter grounded. Show that the efficiency of such an inverting amplifier under sinusoidal excitation is also limited to a maximum of 25%.

11.25   A class A voltage-follower output stage drives an 8-Ω load with a 4-V peak sinusoid at 1 kHz. Find the power delivered to the load. Estimate the power dissipated in the transistors of the output stage.   **Answer:**   1 W; 3 W

11.26   A class B output stage powered by ± 5-V voltage sources drives an 8-Ω load with a 4-V peak sinusoid at 1 kHz. Find the power delivered to the load and the amplifier efficiency. Estimate the power dissipated in the transistors of the output stage.   **Answer:**   1 W; 63%; 1.6 W

11.27   Repeat Exercise 11.26 if the output stage is powered by ± 10-V voltage sources. **Answer:**   1 W; 31%; 3.2 W

Discussion. The concept of maximum obtainable efficiency, as derived in the preceding analysis, assumes that the transistors of the follower stage can be driven to their cutoff and saturation limits without loss of amplifier linearity. Similarly, the derivation assumes that these devices are capable of driving v_{OUT} all the way to the power-supply voltages. In practice, devices are not operated near their cutoff and saturation limits because nonlinear distortion is seldom negligible near these limits. Similarly, the devices used to implement the output stage may not be capable of bringing v_{OUT} close to the power-supply voltages. These limitations are especially true for FET devices, which exhibit square-law behavior and a large v_{DS} at the transition from constant-current to triode-region operation. Unlike the BJT, with its negligible V_{sat}, the voltage across an FET at the limit of constant-current operation is sizable. Even when a class B amplifier is not driven to the maximum theoretical efficiency predicted by Eq. (11.51), however, it generally exhibits an improvement in efficiency over a similar class A configuration. ∎

11.6.3   Minimally Biased (Class AB) Output Configuration

The problem of crossover distortion in a class B push-pull output stage can be alleviated by biasing the complementary-pair devices *slightly* into the active region, just above cutoff. In a BJT amplifier, this task can be accomplished by adding two diode junctions to the base circuit, as shown in Fig. 11.19. In this circuit, the diodes are kept forward biased by the bias network and large-valued resistor R_1. The series combination of D_1 and D_2 is connected in parallel with the base–emitter junctions of Q_1 and Q_2 so that

$$V_{BE1} + V_{EB2} = V_{D1} + V_{D2} = 2V_f \tag{11.52}$$

where V_f is the turn-on voltage of the diodes. If D_1 and D_2 are matched to the base–emitter junctions of Q_1 and Q_2, the emitter currents I_{E1} and I_2 will be equal to the diode current I_D when the load current is zero. The base current to Q_1 must flow from the network that biases D_1 and D_2; the current $-I_{B2}$ flowing out of Q_2 must flow into R_1 (I_{B2} is negative for a *pnp* device).

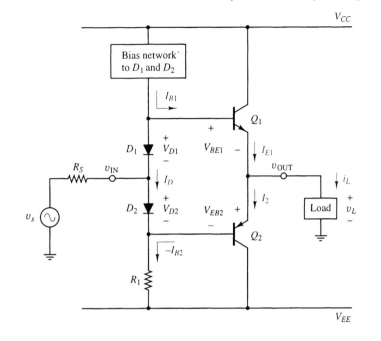

Figure 11.19
Circuit of
Fig. 11.15 with
diodes added to the
base circuit. These
diodes bias Q_1 and
Q_2 slightly in the
active region,
forming a class AB
amplifier.

To first order, D_1 and D_2 will behave as constant-voltage sources of value V_f when an input signal is applied. When v_{IN} is positive, Q_1 will be driven into its active region, with v_{OUT} given by

$$v_{OUT} = v_{IN} + V_{D1} - V_{BE1} = v_{IN} + V_f - V_f = v_{IN} \qquad (11.53)$$

When v_{IN} is negative, Q_2 will be driven into its active region with v_{OUT} given by

$$v_{OUT} = v_{IN} - V_{D2} + V_{EB2} = v_{IN} - V_f + V_f = v_{IN} \qquad (11.54)$$

Equations (11.53) and (11.54) yield the same value $v_{OUT} = 0$ at the crossover point $v_{IN} = 0$. The problem of crossover distortion is virtually eliminated by the use of the two biasing diodes. The price paid is a small decrease in amplifier efficiency, because the complementary-pair devices Q_1 and Q_2 and the diodes dissipate a small amount of bias power.

It is possible to drive the output stage from a node above or below the double diode stack of D_1 and D_2, thereby adding a bias component of value V_f to v_{OUT} (see Exercise 11.32). In integrated-circuit designs, such a connection is often more convenient than the connection shown in Fig. 11.19.

Despite its near-linear input–output characteristic, an amplifier of the type shown in Fig. 11.19 functions in a similar manner to the class B amplifier of Fig. 11.15. Each output device carries signal current to the load for only one polarity of the input signal. Because the output devices are slightly biased into the active region, the circuit is called a *class AB* amplifier.

DESIGN

EXAMPLE 11.4

It is desired to bias the transistors of Fig. 11.19 with a small bias current of 0.5 mA. If the diodes have a scale current of $I_s = 10^{-11}$ mA, find the appropriate value of diode bias current I_D and design a circuit to produce it. Assume that the base–emitter junctions of Q_1 and Q_2 have twice the junction area of the diodes D_1 and D_2 but are otherwise matched. The maximum expected load current is equal to 10 mA. The circuit is operated from ±15-V supplies. For a "worst-case" (minimum) β_F of 50, what is the bias power dissipated in each device when $v_{IN} = 0$?

Solution

• Assess the goals of the problem

The first step involves finding an appropriate bias current I_D for the diodes such that the diode voltages will bias Q_1 and Q_2 at the desired level of 0.5 mA. The second goal involves designing a circuit that can produce the required I_D.

• Choose a design strategy

The scale current parameter of a *pn* junction is proportional to its area, hence I_{EO} of the two transistors will be equal to $2I_S$, where I_S is the scale current of either diode. Given that $|V_{BE}| = V_f$ for each BJT, it thus follows that I_{E1} and I_2 will be equal to $2I_D$. If the former currents are to equal 0.5 mA, I_D must be set to 0.25 mA. This value of I_D can be easily achieved by selecting a second resistor R_2 for the bias network and choosing appropriate values for R_1 and R_2.

• Choose values for the components in the circuit

If the base currents into Q_1 and Q_2 are assumed small, the diode currents through D_1 and D_2 can be expressed approximately by

$$I_D \approx \frac{V_{CC} - V_{D1} - V_{D2} - V_{EE}}{R_1 + R_2} = \frac{V_{CC} - 2V_f - V_{EE}}{R_1 + R_2} \tag{11.55}$$

where $V_{D1} = V_{D2} = V_f$. With $I_D = 0.25$ mA as the desired goal, we next determine the exact value of V_f at this current level. With $I_S = 10^{-11}$ mA, the drop across each diode becomes

$$V_f = \eta V_T \ln \frac{I_D}{I_S} = (1)(0.025) \ln \frac{0.25 \text{ mA}}{10^{-11} \text{ mA}} = 0.60 \text{ V} \tag{11.56}$$

Given that $I_D = 0.25$ mA when $V_f = 0.60$ V, the required values of R_1 and R_2 can be computed. We arbitrarily set R_1 and R_2 to the same value, so that

$$R_1 = R_2 = \frac{V_{CC} - 2V_f - V_{EE}}{2I_D} = \frac{15 \text{ V} - 2(0.60 \text{ V}) - (-15 \text{ V})}{2(0.25 \text{ mA})} = 57.6 \text{ k}\Omega \tag{11.57}$$

• Evaluate the design and revise if necessary

A resistance of 57.6 kΩ is an "oddball" value; a better choice would be the closest standard value of 56 kΩ, leading to diode currents of

$$I_D = \frac{15 \text{ V} - 2(0.6 \text{ V}) - (-15 \text{ V})}{2(56 \text{ k}\Omega)} = 0.257 \text{ mA} \tag{11.58}$$

and transistor bias currents of $2I_D = 0.514$ mA. A quick check reveals that the revised diode current still produces a V_f of about 0.6 V:

$$v_D = \eta V_T \ln \frac{I_D}{I_S} = (1)(0.025 \text{ V}) \ln \frac{0.257 \text{ mA}}{10^{-11} \text{ mA}} = 0.599 \text{ V} \approx 0.6 \text{ V} \tag{11.59}$$

We next test the assumption that the transistor base currents are negligible compared to I_D. For a "worst-case" (smallest expected) β_F of 50, and for $I_{C1} = |I_{C2}| = 0.514$ mA, the base current magnitudes become $(0.514 \text{ mA})/50 \approx 10.3 \text{ } \mu\text{A}$, which is about 25 times smaller than I_D. The actual value of I_D in the circuit becomes $0.257 \text{ mA} - 10.3 \text{ } \mu\text{A} = 0.247 \text{ mA}$. This current is still close to the targeted value of 0.25 mA and will still produce of V_f of 0.6 V, hence the bias currents I_{E1} and I_2 will still be equal to 0.514 mA.

• **Examine the circuit under peak load conditions**

The bias circuit of D_1, D_2, and the two resistors R_1 and R_2 must be examined to ensure that sufficient base current will be available when the load current delivered by Q_1 or Q_2 is a maximum. Under the specified peak load condition $i_L = 10\,\text{mA}$, and again assuming a "worst-case" β_F of 50, the additional peak base current into either device must be $(10\,\text{mA})/50 = 0.2\,\text{mA}$. This current must ultimately flow from the diode bias network, hence i_D will be reduced from its bias value of $I_D \approx 0.25\,\text{mA}$ to $0.05\,\text{mA}$ at the peak of v_{IN}. Note that i_D does not fall to zero at the peak of i_L, hence this bias design should work even at the specified peak load current.

• **Compute the power dissipation in each device with $v_{\text{IN}} = 0$**

When $v_{\text{IN}} = 0$, the power dissipated in each diode is equal to

$$p_D = V_D I_D \approx (0.6\,\text{V})(0.257\,\text{mA}) = 154\,\mu\text{W} \tag{11.60}$$

The bias power dissipated in each transistor is given by

$$\begin{aligned}
p_Q &= V_{CE} I_C + V_{BE} I_B \\
&\approx (15\,\text{V})(0.51\,\text{mA}) + (0.6\,\text{V})\frac{0.51\,\text{mA}}{50} \\
&= 7.7\,\text{mW} + 0.006\,\text{mW} \approx 7.7\,\text{mW}
\end{aligned} \tag{11.61}$$

Note that the $V_{BE} I_B$ term in Eq. (11.61) is negligible compared to the $V_{CE} I_C$ term.

EXERCISE 11.28 For the diode and transistor parameters indicated in Example 11.4, compute the diode voltage v_{D1} when the circuit of Fig. 11.19 delivers a current $i_L = 10\,\text{mA}$ to the load, so that $i_{D1} \approx 0.05\,\text{mA}$. Use a "worst-case" (minimum) value of $\beta_F = 50$ in your calculations. The resulting value of v_{BE1} does not yield an emitter current of $10\,\text{mA}$ in Q_1. From where does the extra required base–emitter voltage come when $i_L = 10\,\text{mA}$?   **Answer:** $0.558\,\text{V}$

11.29 Design a resistive bias network for diodes D_1 and D_2 in Fig. 11.19 such that $I_D = 0.10\,\text{mA}$.

11.30 Using BJTs and other appropriate components, design a current-source bias network for diodes D_1 and D_2 in Fig. 11.19. Set the bias current I_D to $0.25\,\text{mA}$.

11.31 Modify your design of Exercise 11.30 so that $I_D = 0.5\,\text{mA}$. What is the resulting value of bias current through Q_1 and Q_2? The base–emitter junction area of each BJT is twice that of the diodes.

11.32 The output stage of Fig. 11.19 is driven at the node between D_1 and D_2. It is also possible to drive the amplifier from below D_2 or above D_1. If the amplifier is driven from below D_2, follow the analysis of Eqs. (11.52) through (11.54) and show that the output under such conditions is given by $v_{\text{OUT}} = v_{\text{IN}} + V_f$.

11.33 Show that $V_f \approx 0.6\,\text{V}$ in Example 11.4 for $\beta_F = 50$ and $R_1 = R_2 = 56\,\text{k}\Omega$ if base currents are not neglected.

The diode connection of Fig. 11.19 can also be implemented by connecting two BJTs in series, as in Fig. 11.20. This connection is functionally similar to the circuit of Fig. 11.19 but allows for more precise control of the bias current through Q_1 and Q_2. With the bias configuration of Fig. 11.20, it is possible to set the current through Q_1 and Q_2 to a very small value without requiring the use of large-value resistors. This bias configuration is most often found on integrated circuits and in operational amplifiers. We shall study it in more detail in Chapter 12 (see Section 12.2.3.)

Figure 11.20
Biasing diodes are
replaced by BJTs
Q_X and Q_Y.

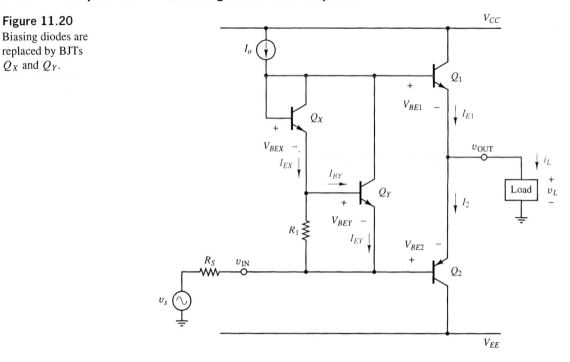

11.7 INTEGRATED-CIRCUIT POWER AMPLIFIERS

A *high-power* device is defined as one that can safely dissipate the heat generated by a large amount of electrical power. The output stages discussed in this chapter are often made as discrete, stand-alone amplifiers using high-power devices. If a class B or class AB output stage made from high-power devices is cascaded with a high-gain operational amplifier, the resulting combination forms a high-power "super" op-amp. An op-amp buffered by a high-power output stage can be connected in any of the usual op-amp feedback configurations and used to supply large load currents. If the components of both the op-amp and the output stage are fabricated together on the same integrated circuit, or "chip," the resulting circuit is called an *integrated-circuit power amplifier*. Low-power devices and high-power devices can be fabricated simultaneously on the same substrate using *hybrid thick-film* technology.

The basic layout of a typical IC power amp is shown in Fig. 11.21. The circuit shown is a simplified version of the commercially available LH0101 power IC op-amp. The principal advantage of the IC power amp of Fig. 11.21 over a similar circuit made from a low-power op-amp and discrete power BJTs is its small size and simplicity of use.

The output stage in Fig. 11.21 consists of a class AB configuration formed by the complementary pair Q_3 and Q_4. The minimal bias required of the class AB configuration is set by the combination of R_1 and R_2, and the forward-biased base–emitter junctions of Q_1 and Q_2. These junctions perform the function of the diodes found in the class AB amplifier of Section 11.6.3. As discussed previously, the presence of these two *pn* junctions acts to minimize the crossover distortion inherent to class B amplifiers.

Analysis of the circuit of Fig. 11.21 shows that when i_O is zero, the bias current flowing out of the base of Q_1 flows into the base of Q_2. When v_O becomes positive, i_O becomes positive and is split between the bases of Q_1 and Q_2, thereby reducing i_1 and increasing i_2. The corresponding decrease in $-i_{E1}$ causes some of the current through R_1 to be diverted to the base of Q_3, where it is amplified by β_{F3} and delivered to the load. When i_O is negative, i_2 and i_{E2} are reduced. Some

Figure 11.21
Simplified version of the LH0101 IC power amplifier. All the components shown would be contained on a single IC chip. (Reprinted with permission of National Semiconductor.)

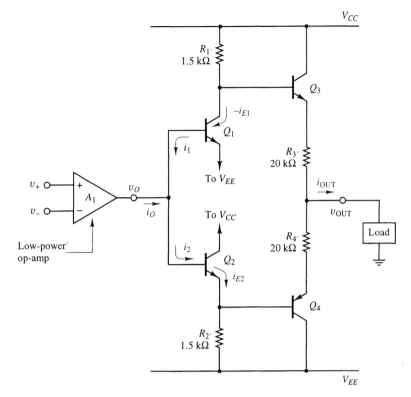

of the current through R_2 is taken out the base of Q_4, where it is amplified by β_{F4} and pulled up through the load. The buffering of i_O by two cascaded BJTs substantially reduces the current drawn from the low-power op-amp A_1.

EXAMPLE 11.5

Consider the circuit of Fig. 11.21. Find the output current i_O of the low-power op-amp A_1 as a function of its output voltage v_O. Assume that $V_{EE} = -V_{CC}$.

If β_F for each of the power transistors in Fig. 11.21 is equal to 20 and the v_{OUT} terminal feeds a 50-Ω load at a voltage of 5 V, estimate the load current i_{OUT} and the current i_O drawn from amplifier A_1.

Solution

• Find an expression for i_O in terms of v_O

Application of KCL to the output node of A_1 results in $i_O = i_2 - i_1$. Currents i_1 and i_2 are determined by R_1 and R_2, and the base–emitter junctions of Q_1 and Q_2. Specifically, i_O can be expressed by the equation

$$i_O = i_2 - i_1 \approx \underbrace{\frac{i_{E2}}{\beta_{F2}}}_{} - \underbrace{\frac{i_{E1}}{\beta_{F1}}}_{} = \underbrace{\frac{v_O - V_f - V_{EE}}{\beta_{F2} R_2}}_{\text{current } i_2} - \underbrace{\frac{V_{CC} - (v_O + V_f)}{\beta_{F1} R_1}}_{\text{current } i_1} \qquad \textbf{(11.62)}$$

where factors of V_f are contributed by the base–emitter junctions of Q_1 and Q_2.

For the conditions $V_{EE} = -V_{CC}$, $R_1 = R_2$, and $\beta_{F1} = \beta_{F2}$, Eq. (11.62) reduces to

$$i_O = \frac{2v_O}{\beta_{F1} R_1} \tag{11.63}$$

• Compute an approximate value for i_O from the given information

If v_{OUT} is equal to 5 V and the load consists of a 50-Ω resistor, the output current i_{OUT} will be equal to 100 mA. The voltage v_O will be approximately equal to v_{OUT}, because the base–emitter junctions of Q_1 and Q_3 contribute opposite and nearly equal voltage drops of value V_f between v_O and v_{OUT}. Consequently, the current i_O given by Eq. (11.63) becomes

$$i_O = \frac{2v_O}{\beta_{F1} R_1} \approx \frac{2(5\,\text{V})}{(20)(1.5\,\text{k}\Omega)} = 0.33\,\text{mA} \tag{11.64}$$

The overall current gain provided by the class AB output stage is seen to be $(100\,\text{mA})/(0.33\,\text{mA})$, or about 300.

EXERCISE 11.34 Show that Eq. (11.62) reduces to Eq. (11.63) when $V_{EE} = -V_{CC}$, $R_1 = R_2$, and $\beta_{F1} = \beta_{F2}$.

11.35 Construct a simplified model for v_O, R_1, R_2, and the base–emitter junctions of Q_1 and Q_2 in Fig. 11.21. Show that when $v_O = 0$, the emitter current $-i_{E1}$ into Q_1 can be expressed by[2]

$$-i_{E1} = \frac{V_{CC} - V_{EE} - 2V_f}{R_1 + R_2} \tag{11.65}$$

11.36 Use superposition in the circuit of Fig. 11.21 to show that

$$i_O = \frac{v_O}{\beta_{F1}(R_1 \| R_2)} \tag{11.66}$$

when $V_{EE} = -V_{CC}$ and $\beta_{F1} = \beta_{F2}$ but $R_1 \neq R_2$. Compare with the result (11.63).

The output stage shown in Fig. 11.21 is but one example of IC power-amplifier construction; many other output configurations are possible. Regardless of its specific output circuitry, however, an IC power op-amp can be connected in any one of the usual negative-feedback op-amp configurations. The output of the "super" op-amp, taken at the output of the high-power stage, functions as the driving terminal for the negative-feedback network and load. If the overall circuit contains negative feedback, the op-amp A_1 will adjust the voltage v_{OUT} until v_+ is approximately equal to v_-. A high-power op-amp thus functions in every respect like a regular op-amp, except that its output terminal is capable of supplying larger amounts of current.

---

[2] Q_1 is a *pnp* device, hence $-i_{E1}$ is positive.

11.8 POWER DEVICES

The class A, class B, class AB and power IC amplifiers of the previous two sections have the capability of delivering large load currents at moderate voltages and thus can deliver large amounts of power. When substantial power is delivered to an amplifier load, some power will always be dissipated in the transistors of the amplifier, even when a higher efficiency class B or class AB configuration is used. A well-designed high-power output stage must utilize specially fabricated power transistors capable of safely handling the electrical power dissipated as heat. The typical power device has a large surface area and is mounted in good thermal contact with its package and ambient surroundings. The transistor cases shown in Fig. 11.22 are typical of those used to package discrete high-power transistors. When power devices are fabricated on an integrated circuit, as in the IC power amplifier of Section 11.7, the entire IC chip is usually packaged in one of the configurations of Fig. 11.22.

Figure 11.22
Typical high-power device packages.

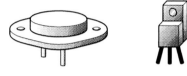

11.8.1 Heat Sinks

A high-power device package is often mounted on a metal *heat sink*, which enhances the overall thermal contact between the device case and the surrounding air. This improved thermal contact facilitates the removal of heat from the device. Heat removal in important, because excess heat can cause a catastrophic rise in device temperature and permanent device failure. A thermal heat sink like the one depicted in Fig. 11.23 draws heat from the device via thermal conduction, then expels the heat into the ambient air via thermal convection and heat radiation.

Figure 11.23
Metallic heat sink used to improve the thermal conductivity between a power device and the surrounding air.

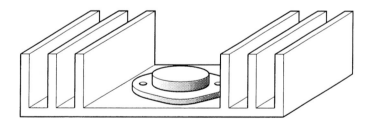

A given heat sink is characterized by a heat-transfer coefficient, or *thermal resistance* Θ (capital greek theta), which describes the flow of heat from the sink to the ambient air for a given rise in heat-sink temperature. The air around the heat sink is assumed to remain at constant temperature. This thermal flow of heat, expressed as an energy flow per unit time, has the units of watts and is governed by the equation

$$P_{\text{therm}} = \frac{T_{\text{sink}} - T_{\text{air}}}{\Theta} \tag{11.67}$$

The thermal resistance Θ has the units of degrees per watt. Equation (11.67) has the functional form of the v–i equation of a resistor, with P_{therm} analogous to current, T_{sink} and T_{air} to voltages, and Θ to electrical resistance.

In a typical heat-sink application, the power device to be cooled is mounted in thermal contact with the heat sink. The imperfect mating of adjacent surfaces introduces additional thermal resistance. The latter is made as small as possible by coating the mated surfaces of the transistor and sink with a thermally conducting compound.

If the overall thermal resistance between the transistor case and the heat sink is designated $\Theta_{\text{case-sink}}$, and the thermal resistance between the heat sink and the ambient air is designated $\Theta_{\text{sink-air}}$, the heat-flow equation (11.67) becomes

$$P_{\text{therm}} = \frac{T_{\text{case}} - T_{\text{air}}}{\Theta_{\text{case-sink}} + \Theta_{\text{sink-air}}} \tag{11.68}$$

Equation (11.68) describes the heat flow from the transistor case, via the heat sink, to the surrounding air.

The temperature of the actual semiconductors inside the case will be higher than the external case temperature, because the semiconductors are separated from the case by an additional component of thermal resistance. Equation (11.68) can be further modified to express the heat flow as a function of the actual device temperature:

$$P_{\text{therm}} = \frac{T_{\text{device}} - T_{\text{air}}}{\Theta_{\text{device-case}} + \Theta_{\text{case-sink}} + \Theta_{\text{sink-air}}} \tag{11.69}$$

In thermal equilibrium, the electrical power P_{elec} dissipated in the device will equal the power P_{therm} flowing out of the device as heat. Equation (11.69) can be used to find the device operating temperature at thermal equilibrium by setting P_{therm} to P_{elec}.

Note that many power devices utilize the case itself as one of the principal current-carrying terminals. Such a connection minimizes the thermal resistance between the working semiconductors and the transistor case. The case is typically assigned to the collector in a power BJT and to the drain in a power MOSFET. In situations where the metallic heat sink must be grounded for mechanical reasons, the transistor case must be insulated from the heat sink by a very thin, electrically insulating, but thermally conducting spacer. This spacer allows the device terminal represented by the case to operate at voltages other than ground. Despite its good thermal conductivity, such a spacer introduces an additional component of thermal resistance between the transistor case and the body of the heat sink.

EXAMPLE 11.6

A power BJT, for which $\Theta_{\text{device-case}} = 4\,°\text{C/W}$, is mounted on a heat sink with $\Theta_{\text{sink-air}} = 5\,°\text{C/W}$, as shown in Fig. 11.24. The mounting method utilizes a 0.2-mm-thick mica spacer, which introduces an additional thermal resistance of $1\,°\text{C/W}$ between the transistor case and the heat sink. If the BJT carries an average current of $i_C = 1\,\text{A}$ at an average voltage of $v_{CE} = 10\,\text{V}$, determine the operating temperature of the transistor substrate, the transistor case, and the heat sink for $T_{\text{air}} = 25°\text{C}$. Neglect the power dissipated in the base–emitter junction of the BJT.

Figure 11.24
Heat flow from a power device to ambient air via a heat sink.

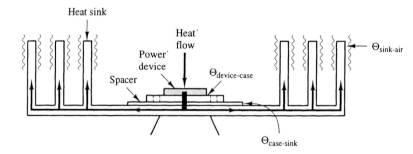

Solution

• Find the power dissipated in the transistor

If the power input $i_B V_{BE}$ is neglected, the electrical power into the BJT becomes

$$P_{\text{elec}} = i_C v_{CE} = (1\,\text{A})(10\,\text{V}) = 10\,\text{W} \tag{11.70}$$

• **Find the temperature of the transistor substrate (T_{device}), the transistor case (T_{case}), and the heat sink (T_{sink})**

The power (11.70) is dissipated as heat in the transistor and must be conducted away by the heat sink if the device temperature is to be kept from increasing without bound. Since $P_{elec} = P_{therm}$ in thermal equilibrium, Eq. (11.69) yields

$$\begin{aligned}
T_{device} &= T_{air} + P_{therm}(\Theta_{device-case} + \Theta_{case-sink} + \Theta_{sink-air}) \\
&= 25°C + (10\,W)(4\,°C/W + 1\,°C/W + 5\,°C/W) \\
&= 25°C + 100°C = 125°C
\end{aligned} \tag{11.71}$$

Similarly, the temperature of the transistor case is given by

$$\begin{aligned}
T_{case} &= T_{air} + P_{therm}(\Theta_{case-sink} + \Theta_{sink-air}) \\
&= 25°C + (10\,W)(1\,°C/W + 5\,°C/W) = 85°C
\end{aligned} \tag{11.72}$$

and the temperature of the heat sink by

$$\begin{aligned}
T_{sink} &= T_{air} + P_{therm}\Theta_{sink-air} \\
&= 25°C + (10\,W)(5\,°C/W) = 75°C
\end{aligned} \tag{11.73}$$

The device and heat-sink system of Fig. 11.24 can be modeled by the electrical analog of Fig. 11.25, where the heat flow is represented by a current source of value P_{therm}. The voltage drop across a given sequence of thermal resistances represents the temperature rise above ambient air (ground) for the corresponding set of thermal elements.

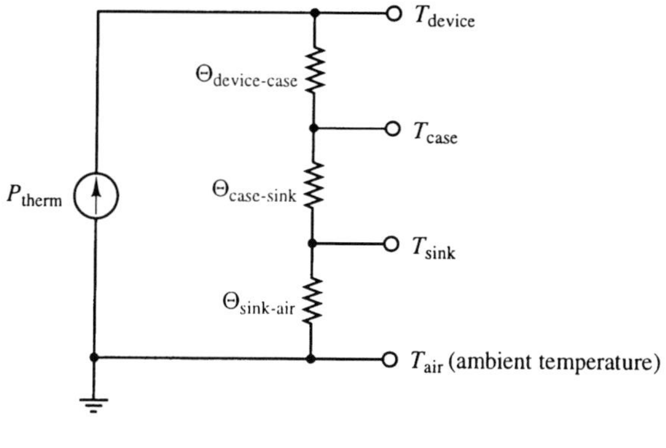

Figure 11.25
Resistive model of the heat flow in Fig. 11.24.

11.8.2 Power BJT

The fabrication geometry of a typical discrete power BJT is shown in Fig. 11.26. The "star"-shaped pattern of the transistor causes the collector–base junction to have a large surface area. Since most of the electrical power in a BJT is dissipated in the base–collector junction, this design maximizes the contact area over which the dissipated heat can be conducted via the substrate to the external case. To facilitate heat conduction, the collector substrate of the BJT is metallurgically bonded to the metallic transistor case.

Figure 11.26
Physical geometry
of a typical discrete
power BJT.

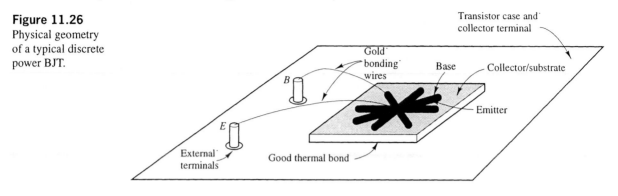

In principle, a given BJT can be operated at any power level, provided that the temperature of the base–collector junction does not exceed a specified maximum value $T_{j\text{-max}}$. This temperature is usually specified by the device manufacturer. If the device temperature rises above $T_{j\text{-max}}$, permanent damage to the device can occur.

As discussed in Section 11.8.1, the flow of heat from the semiconductors of a BJT to the packaging case is determined by the thermal resistance coefficient $\Theta_{\text{device–case}}$. For a given case temperature, the maximum safe operating power of the BJT can thus be expressed by

$$P_{\text{elec-max}} = \frac{T_{j\text{-max}} - T_{\text{case}}}{\Theta_{\text{device–case}}} \tag{11.74}$$

The manufacturer may also specify a maximum safe emitter current, above which damage will occur to the fine gold bonding wire connecting the emitter region to its external terminal.

A plot of $P_{\text{elec-max}}$ versus T_{case} is called the *power derating curve* of the transistor. The power derating curve of a typical power BJT is shown in Fig. 11.27. The *rated power* of the device is defined as the safe electrical power input when the case temperature is equal to room temperature, which is usually defined as $T_0 = 25°C$. In practice, a case temperature of T_0 can be achieved only by using a heat sink with an extremely low thermal resistance to air. The temperature at which the power derating curve crosses the horizontal axis corresponds to $T_{j\text{-max}}$. If the case temperature is at $T_{j\text{-max}}$, no electrical power can be fed to the transistor; otherwise, the transistor temperature will rise above $T_{j\text{-max}}$ and damage will occur.

Figure 11.27
Power transistor
derating curve.

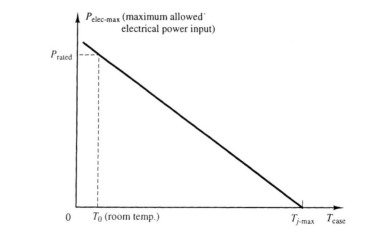

EXAMPLE 11.7

A BJT with a rated power of 25 W and a maximum junction temperature of 200°C is to be operated in air with an ambient temperature of 25°C. The transistor is mounted on a heat sink with $\Theta_{\text{sink-air}} = 2\,°\text{C/W}$. The thermal resistance between the transistor case and the heat sink is equal to 0.5 °C/W. Find the power that can be safely dissipated in the transistor.

Solution

The stated power rating of the BJT assumes a case temperature of $T_0 = 25°\text{C}$. The thermal resistance between the semiconductor substrate and the case can thus be computed:

$$\Theta_{\text{device-case}} = \frac{T_{j\text{-max}} - T_0}{P_{\text{rated}}} = \frac{200°\text{C} - 25°\text{C}}{25\,\text{W}} = 7\,°\text{C/W} \tag{11.75}$$

If the BJT is installed on the specified heat sink, the maximum safe electrical power input becomes

$$\begin{aligned} P_{\text{elec-max}} &= \frac{T_{j\text{-max}} - T_{\text{air}}}{\Theta_{\text{device-case}} + \Theta_{\text{case-sink}} + \Theta_{\text{sink-air}}} \\ &= \frac{200°\text{C} - 25°\text{C}}{7\,°\text{C/W} + 0.5\,°\text{C/W} + 2\,°\text{C/W}} = 18.4\,\text{W} \end{aligned} \tag{11.76}$$

EXAMPLE 11.8

A power BJT is rated at 50 W and has a device-to-case thermal resistance of 4 °C/W. Find the maximum allowed junction temperature $T_{j\text{-max}}$ of the device.

Solution

The rated power of the transistor assumes a case temperature of $T_0 = 25°\text{C}$, that is,

$$P_{\text{rated}} = \frac{T_{j\text{-max}} - T_0}{\Theta_{\text{device-case}}} = 50\,\text{W} \tag{11.77}$$

This equation can be rearranged to yield

$$\begin{aligned} T_{j\text{-max}} &= P_{\text{rated}}\Theta_{\text{device-case}} + T_0 \\ &= (50\,\text{W})(4\,°\text{C/W}) + 25°\text{C} = 225°\text{C} \end{aligned} \tag{11.78}$$

The maximum safe operating power hyperbola, as determined by thermal considerations, is defined for the BJT by the relation $P_{\text{elec-max}} = i_C v_{CE}$. The maximum power hyperbola constitutes but one limit to the allowed operating region of a BJT. Even if the power level is small enough such that the device temperature remains below $T_{j\text{-max}}$, the device current must never exceed the manufacturer-specified value $I_{C\text{-max}}$. As previously mentioned, currents above this value can melt the wires connecting the device terminals to the external case terminals or can break the bonding points between the wires and the device.

A second limit to BJT operation is related to the maximum value of allowed collector–emitter voltage. If v_{CE} exceeds the manufacturer-specified value BV_{CEO}, the electric field within the base–collector depletion region will become large enough to initiate avalanche breakdown, effectively shorting the collector to the base. If the absolute current and voltage limits $I_{C\text{-max}}$ and BV_{CEO} are considered, the safe operating region of a BJT becomes a truncated version of the safe power hyperbola, as shown in Fig. 11.28.

Figure 11.28
Safe operating region of a BJT is defined by the maximum power hyperbola, the maximum collector current $I_{C\text{-max}}$, and the collector-emitter breakdown voltage BV_{CEO}.

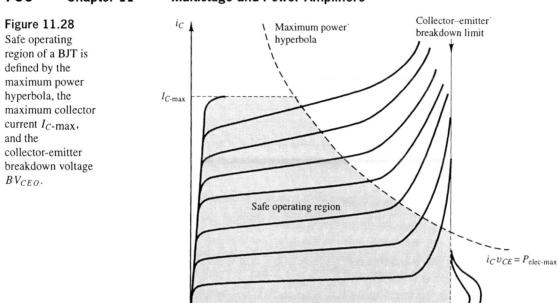

11.8.3 Power MOSFET

Like the power BJT, the power MOSFET is capable of dissipating large amounts of power and can handle large currents and voltages. The planar MOSFET geometry described in Chapter 5 is unsuitable for power MOS devices, however. At large values of v_{DS}, the short channel length in a planar device causes the electric field of the depletion region surrounding the drain to extend all the way to the source. The short channel also causes the drain-to-channel avalanche breakdown voltage to be too low for most high-power applications. The problem is exacerbated by the desire to have an especially short channel in a power device, so that large drain currents can be produced.

A common form of alternative MOSFET, suitable for power applications, is shown in n-channel form in Fig. 11.29. A comparable p-channel device can also be made. This device configuration is called the *double-diffused vertical* MOSFET, or DMOS. The drain contact in a DMOS device has a large surface area and is located at the bottom of a lightly doped n-type region. This construction has several effects on transistor behavior. First, the light doping of the drain region causes the drain-to-channel depletion region to extend primarily into the drain area, rather than across the channel and toward the source. This wider drain-to-channel depletion region results in a larger drain-to-source breakdown voltage than can be obtained in a low-power planar device. At the same time, the short channel necessary for large current conduction is preserved, because the width of the channel region between the source and the active portion of the drain (located just beneath the gate) can be made very small. Finally, the large collection area of the n^+ drain contact at the bottom of the MOSFET facilitates the flow of large currents without appreciable ohmic heating in the drain region.

Although the physical appearance of a DMOS transistor differs significantly from that of a low-power planar MOSFET, its v–i characteristics are essentially the same in both the constant-current and triode regions of operation. The principal difference occurs at very high currents, where a phenomenon known as *velocity saturation* limits the flow of current through the device. In this high-current region of operation, the velocity of electrons is no longer linearly proportional to the electric field produced by v_{DS} in the channel region. This effect causes the usual square-law be-

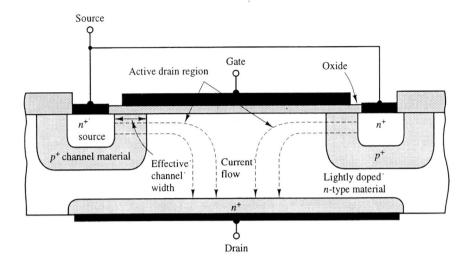

Figure 11.29
Physical geometry of a high-power double-diffused vertical MOSFET (DMOS) device.

havior of the MOSFET in the constant-current region to be replaced by the more complex equation[3]

$$i_D = \frac{\mu_e \epsilon_{ox}}{2t_{ox}} \frac{W}{L} \frac{(v_{GS} - V_{TR})^2}{1 + a(v_{GS} + V_{TR})} \tag{11.79}$$

where a is a constant, and $K = (\mu_e \epsilon_{ox}/2t_{ox})W/L$, as before.

The temperature dependence of the DMOS transistor also resembles that of a low-power planar MOSFET. Specifically, a DMOS transistor exhibits a zero-temperature-coefficient (ZTC) point in its i_D–v_{GS} transconductance curve, as discussed in Section 5.6.1.

A DMOS transistor can be used in any of the power-amplifier configurations discussed in this chapter. In practice, DMOS transistors are biased below the ZTC point, so that an increase in device temperature leads to a decrease in current. This choice of operating region prevents a phenomenon known as *thermal runaway*, in which an increase in device temperature causes an increase in device current, which, in turn, causes a further rise in device temperature.

A Final Note.

The issues of power dissipation, heat conduction, and safe operating region were introduced in this chapter in the context of high-power devices. These same concepts also apply to medium-power, low-power, and micropower circuit applications. The issues of power dissipation and heat flow become particularly important in the environment of a very-large-scale (VLSI) or ultra-large-scale (ULSI) integrated circuit. Although the power dissipated in a single device on a VLSI or ULSI chip is miniscule, the extremely small device dimensions, large number of devices, and crowded packing densities make the removal of excess heat critical to the design of certain types of integrated circuits. In the environment of discrete low-power and medium-scale integrated (MSI) circuits, the issues of power dissipation and heat conduction can become equally important. If circuits made from low-power devices are to be pushed to their operation limits, heat sinking and adherence to safe operating-region criteria must be carefully observed. Although the heat sinks found in low-power discrete and MSI designs are much less obtrusive than the large, bulky heat sinks of high-power BJT and MOSFET circuits, they are no less crucial to successful circuit design and operation.

---

[3] See, for example, P. R. Grey and R. G. Meyer, *Analysis and Design of Analog Integrated Circuits*, 3rd Ed, New York: John Wiley, 1993, pp. 73–76. Or see C. G. Fonstad, *Microelectronic Devices and Circuits*, New York: McGraw-Hill, 1994, pp. 281–285.

SUMMARY

⦿ Multistage cascading permits several single-stage amplifiers to be combined into one circuit.

⦿ Multistage cascading can produce an amplifier with large gain, high input resistance, and low output resistance.

⦿ Input loading, output loading, and interstage loading all affect the performance of a multistage amplifier.

⦿ The small-signal behavior of a multistage amplifier can be modeled by cascading an appropriate number of small-signal two-port amplifier modules.

⦿ Dc level shifting improves the swing range of a dc-coupled multistage amplifier.

⦿ The cascading principle can be applied to a multistage differential amplifier.

⦿ The last stage of a multistage amplifier is often designed for power amplification.

⦿ A power-amplification stage has a voltage gain of unity but provides substantial current gain.

⦿ In a class A power amplifier, a single device drives the load during both positive and negative excursions of the output signal.

⦿ The output device in a class A amplifier is biased in the middle of its swing range. The class A configuration requires substantial bias power and is not very efficient.

⦿ In a class B power amplifier, complementary output devices drive the load. One device carries current to the load during positive output excursions and the other device carries current during negative excursions.

⦿ The output devices in a class B amplifier are biased in the cutoff region so that no power is wasted by the bias configuration. The efficiency of a class B amplifier depends on the magnitude and shape of the output waveform but is considerably higher than the efficiency of a comparable class A amplifier.

⦿ A class B power amplifier suffers from crossover distortion.

⦿ In a class AB amplifier, the output devices are biased slightly out of the cutoff region. In all other respects, a class AB amplifier functions like a class B amplifier. The biasing in a class AB amplifier helps to reduce crossover distortion at the expense of some wasted bias power.

⦿ A power output stage is typically made from special transistors with large power-dissipation capabilities.

⦿ The temperature of a device depends on the total electrical power dissipated in the device, the ambient air temperature, and the thermal resistance between the device and the ambient.

⦿ The maximum power that a device can safely dissipate is determined by the maximum temperature that its semiconductors can sustain without damage.

⦿ A device can safely dissipate its rated power if its case is held at room temperature.

⦿ The maximum permissible power dissipation for a device will be less than its rated power if its case is held above room temperature.

⦿ The power-dissipation capability of a device can be improved by mounting the device on a thermal heat sink. A good heat sink holds the device's case near room temperature.

⦿ Power BJT and power MOS devices utilize special geometries to achieve their high power-dissipation capabilities.

⦿ A double-diffused vertical power MOSFET permits large currents to flow at large values of v_{DS}.

◆ **SPICE EXAMPLE**

EXAMPLE 11.9 | Verify the bias design of Example 11.4 by simulating the circuit of Fig. 11.19 on SPICE. Use SPICE to assess the extent of crossover distortion when the class AB amplifier drives a 1-kΩ load. Compare to the class B amplifier of Fig. 11.15 under similar conditions. Plot the transfer characteristic in each case for $-10\,\text{V} < v_{\text{IN}} < 10\,\text{V}$ with power-supply voltages of $\pm15\,\text{V}$ and an R_S of zero.

Figure 11.30
Class AB amplifier with diode-connected BJTs Q_A and Q_B.

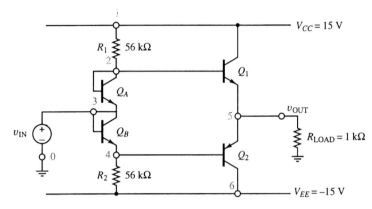

Solution

EXAMPLE 11.10 | For this simulation, we shall implement the diodes of the class AB amplifier in the form of BJTs with base connected to collector, as shown in Fig. 11.30. In this way, the diodes and the BJT base–emitter junctions will have identical parameters, and the junction area of the diodes can be set to half that of the BJTs using the AREA parameter of the element statements. A suitable SPICE file listing follows.

Input File

```
TEST OF CLASS B and CLASS AB AMPLIFIERS
*Set the power-supply voltages:
        VCC 1 0    15V
        VEE 6 0   -15V
*Specify the transistors in the class A amplifier:
        QA 2 2 3 lowpower 100
        QB 3 3 4 lowpower 100
        Q1 1 2 5 Noutput 200
        Q2 6 4 5 Poutput 200
        .MODEL lowpower NPN(BF=100  IS=1e-14)
        .MODEL Noutput  NPN(BF=100  IS=1e-14)
        .MODEL Poutput  PNP(BF=100  IS=1e-14)
*Add the rest of the elements:
        R1 1 2   56k
        R2 4 6   56k
        Rload 5 0 1k
        vIN 3 0 0V
*Specify bias point and dc transfer characteristic analyses:
        .OP
        .DC vIN  -10V 10V 0.1V
        .PROBE v(5) v(15)
*Specify the class B circuit with QA and QB absent:
        Q11 1 3 15 Noutput 200
        Q22 6 3 15 Poutput 200
        Rload2 15 0 1k
        .END
```

Results. As shown in the abridged and annotated output listing that follows, the value of $I_{C1} = 0.503$ mA is reasonably close to the value of 0.5 mA targeted in Example 11.4. As indicted in the node-voltage table of the output listing, the bias value of v_{OUT} is essentially zero.

Figure 11.31
Plot of transfer characteristics of class AB and class B amplifiers with 1-kΩ load.

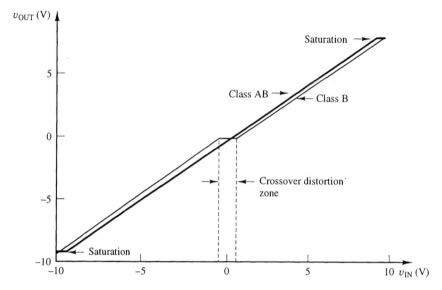

The two transfer characteristics are shown in Fig. 11.31. The class B amplifier exhibits a crossover distortion region of approximately ± 0.6 V. This distortion region is considerably reduced in the class AB circuit. The total bias power dissipated in the latter, however, is approximately 23 mW. Note that the class AB amplifier saturates when v_{OUT} reaches approximately ± 9 V. This condition occurs when all the current available from R_1 or R_2 is diverted to the bases of the output transistors, leaving Q_A or Q_B in cutoff. ∎

Output Listing

```
**** Evaluation PSpice ************
TEST OF CLASS B and CLASS AB AMPLIFIERS
****   SMALL-SIGNAL BIAS SOLUTION    TEMPERATURE =   27.000 DEG C
   NODE     VOLTAGE    NODE     VOLTAGE    NODE     VOLTAGE    NODE    VOLTAGE
 (   1)     15.0000  (    2)      .5003  (    3)     0.0000  (    4)   -.5003
 (   5)   -107.1E-18 (    6)   -15.0000  (   15)   201.9E-27
VOLTAGE-SOURCE CURRENTS
NAME           CURRENT
VCC          -7.617E-04
VEE           7.617E-04
vIN          -3.289E-18
TOTAL POWER DISSIPATION   2.29E-02  WATTS
****   OPERATING-POINT INFORMATION *****
       BIPOLAR JUNCTION TRANSISTORS
NAME        Q1            Q2            QN            QP
MODEL       lowpower      lowpower      Noutput       Poutput
IB          2.51E-06      2.51E-06      5.03E-06      -5.03E-06
IC          2.51E-04      2.51E-04      5.03E-04      -5.03E-04
VBE         5.00E-01      5.00E-01      5.00E-01      -5.00E-01
VBC         0.00E+00      0.00E+00      -1.45E+01     1.45E+01
VCE         5.00E-01      5.00E-01      1.50E+01      -1.50E+01
BETADC      1.00E+02      1.00E+02      1.00E+02      1.00E+02
JOB CONCLUDED
```

PROBLEMS

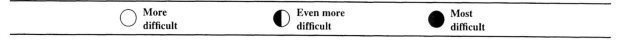

More difficult	Even more difficult	Most difficult

11.1 Input and Output Loading

11.1 The signal from a video recorder is to be fed to the input terminals of a TV monitor. The video recorder output can be modeled as a 0.25-V peak, time-varying voltage source in series with a 10-kΩ resistor. If the TV monitor requires a signal on the order of 0.1 V in order to properly operate, what must be its minimum input resistance?

11.2 A public address amplifier has an output resistance of 3.9 Ω. The circuit can deliver up to 10 W of power to a matched load if appropriately driven. The amplifier must drive eight 8-Ω loudspeakers at as high a power level as possible. Devise a suitable connection scheme and determine the power delivered to each loudspeaker if the amplifier is driven to its maximum possible output voltage.

11.3 A multistage amplifier is formed by cascading two stages. Each stage has a small-signal input resistance of 2 kΩ, an output resistance of 1 kΩ, and a gain of −10. If the input is driven by a voltage source of value 1 mV p-p via a 3-kΩ Thévenin source resistance, find the peak-to-peak output voltage. Evaluate the output under open-circuit load conditions.

11.4 An op-amp is connected as an inverting amplifier with input resistor $R_1 = 2.2\,\text{k}\Omega$ and feedback resistor $R_2 = 100\,\text{k}\Omega$. The circuit is driven by a voltage signal source having parameters $v_{\text{Th}} = 2.5\cos\omega t$ and $R_{\text{Th}} = 600\,\Omega$. Find the resulting output voltage.

11.5 🔢 An amplifier is required that can accept the audio signal from a compact disc (CD) player and deliver an amplified version to an 8-Ω loudspeaker. The CD player can be modeled as a voltage signal source of up to ±1 V peak magnitude in series with a 10-kΩ output resistance. Specify the parameters of an amplifier designed to deliver 10 W of peak output power when driven by the maximum output of the CD player. In a normal "stereo" system, two such amplifiers would be required, one for the left channel and one for the right channel.

11.2 Two-Port Amplifier Cascade

11.6 A three-stage amplifier is made by cascading three inverters. The small-signal behavior of each inverter can be represented by a two-port cell having an input resistance of 2.5 kΩ, an output resistance of 500 Ω, and a gain of −80. Find the input resistance, output resistance, and gain of the overall cascade.

11.7 A three-stage amplifier is made by cascading three inverters. The small-signal behavior of the first inverter can be modeled by a two-port cell with parameters $r_{\text{in1}} = 17\,\text{k}\Omega$, $r_{\text{out1}} = 2.2\,\text{k}\Omega$, and $a_{v1} = -65$. The other two stages can be represented by similar models with $r_{\text{in2}} = 10\,\text{k}\Omega$, $r_{\text{out2}} = 150\,\Omega$, $a_{v2} = -45$, and $r_{\text{in3}} = 5.8\,\text{k}\Omega$, $r_{\text{out3}} = 1.5\,\text{k}\Omega$, and $a_{v3} = -130$. Find the input resistance, output resistance, and gain of the overall cascade.

11.8 Three inverting amplifier stages having the parameters listed in the following table are to be cascaded together to form a three-stage amplifier. Choose the ordering of stages so that the gain will be a maximum when the output stage drives a 100-Ω load and the input is driven by a voltage signal source that has a 10-kΩ series Thévenin resistance.

	Inverter 1	Inverter 2	Inverter 3
Gain	−85	−150	−140
Input resistance	12 kΩ	18 kΩ	45 kΩ
Output resistance	1.1 kΩ	130 kΩ	80 Ω

11.9 Each stage of a three-stage amplifier is represented by a two-port cell with the following parameters. Find the gain, input resistance, and output resistance of the complete amplifier cascade.

	Stage 1	Stage 2	Stage 3
Gain	10	−100	2
Input resistance	10 kΩ	1 MΩ	100 kΩ
Output resistance	1 kΩ	100 Ω	100 Ω

11.10 A three-stage amplifier is modeled as a cascade of two-port cells. The parameters of each cell are indicated in the following table. The input is driven by a voltage source of magnitude v_g via a 500-Ω series resistance. The load consists of another circuit that can be represented as a 50-Ω resistance. Draw a diagram of the two-port cascade, and find an expression and a value for the overall gain v_{out}/v_g.

	Stage 1	Stage 2	Stage 3
r_{in}	5 kΩ	10 kΩ	1 kΩ
r_{out}	1 kΩ	1 kΩ	100
a_v	−6	−110	1

11.11 ⚡ A three-stage BJT amplifier is to be designed in which each stage is represented by a two-port cell. Specify the input resistance, output resistance, and gain of each cell. The overall cascade should have a gain of +400, but no one stage should have a gain larger than 60. The overall input resistance should be at least 10 kΩ and the overall output resistance no greater than 100 Ω. Each cell in your cascade should have parameters typical of an inverter or follower made from BJTs.

11.12 ⚡ A three-stage MOSFET amplifier is to be designed in which each stage is represented by a two-port cell. Specify the input resistance, output resistance, and gain of each cell. The overall cascade should have a gain of ±100, but no one stage should have a gain larger than 10. The overall input resistance should be at least 1 MΩ and the overall output resistance no greater than 100 Ω. Each cell in your cascade should have parameters typical of either an inverter or follower made from MOSFETs.

11.13 ⚡ Design a two-stage amplifier in block diagram form that can amplify the signal from a voltage signal source and 5-kΩ series Thévenin resistance. The amplifier must deliver its output to a 1-kΩ load. Two basic amplifier modules are available, one with parameters $r_{inA} = 1\,\text{k}\Omega$, $r_{outA} = 10\,\text{k}\Omega$, and $a_{vA} = 100$, and the other with parameters $r_{inB} = 10\,\text{k}\Omega$, $r_{outB} = 1\,\text{k}\Omega$, and $a_{vB} = 10\,\text{k}\Omega$. Calculate the overall cascade gain including the various loading factors.

11.14 ⚡ Design a three-stage amplifier in block diagram form that has an overall gain magnitude of 900, an input resistance of at least 5 kΩ, and an output resistance of no more than 100 Ω. Each stage is to consist of either a BJT inverter with a gain of no more than −40 or a BJT voltage follower. Specify the parameters of each stage, using values typical for the chosen stage configuration.

11.15 ○ Analyze the circuit of Fig. 11.3. Find the bias values of I_C and V_{CE} for each transistor in the circuit. What are the positive and negative swing limits of voltage v_{OUT}?

11.16 Consider the two-stage BJT amplifier of **Fig. P11.16**. Suppose that $R_1 = 10\,\text{k}\Omega$, $R_2 = 1\,\text{k}\Omega$, $R_S = 10\,\text{k}\Omega$, $R_3 = 10\,\text{k}\Omega$, and $V_{CC} = 15\,\text{V}$.

(a) Choose V_{EE} such that Q_1 is biased with $I_C \approx 0.5\,\text{mA}$.

(b) Find the bias current through Q_2.

(c) Estimate the small-signal input resistance seen by v_s, the small-signal Thévenin output resistance presented by the v_{OUT} terminal, and the overall cascade gain. Assume a reasonable value for β if necessary.

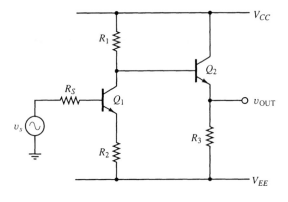

Fig. P11.16

11.17 ⚡ In analog BJT integrated circuits, *npn* and *pnp* devices are often used on the same chip. This characteristic greatly expands the range of possible design parameters. In this problem, the concept is illustrated using the two-stage BJT amplifier of **Fig. P11.16**. Redesign the circuit such that the first stage is made from a *pnp* BJT. The small-signal properties of the overall cascade should remain approximately the same.

11.18 ○ In this problem, the input and output resistances of a BJT voltage follower are examined. As discussed in Section 11.3, the input resistance of a follower is dependent on its load, and the output resistance is dependent on the series resistance of the source driving the follower. As the results of this problem show, only one of these dependencies need be included in evaluating a multistage cascade.

(a) Consider the small-signal model of a BJT follower shown in **Fig. P11.18**. Find an expression for the overall amplifier gain v_{out}/v_s with the load R_L connected.

(b) Represent the circuit between the v_{out} terminal and ground by its Thévenin equivalent with the load R_L disconnected. Compute the open-circuit voltage gain without approximation. Include the source resistance R_S in computing the output resistance r_{out}. Reconnect R_L and use the principle of output loading to find an expression for v_{out}/v_s. Your result should agree with that of part (a).

(c) Now find the gain of the circuit with R_L connected and with v_s and R_S replaced by a single voltage source connected between the v_b node and ground. Include the load resistance R_L in computing the input resistance r_{in}. Reconnect v_s and R_S and use the principle of input loading to find an expression for v_{out}/v_s. Your result should agree with that of part (a).

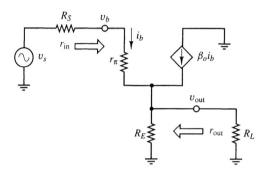

Fig. P11.18

11.19 ⯀ ◑ It is possible to create a multistage cascade in which each stage is separately biased and coupled to adjacent stages by dc blocking capacitors. One possible implementation of such a two-stage amplifier is illustrated in **Fig. P11.19**. In this circuit, each BJT has its own separate feedback-bias network. Capacitors C_1 and C_2 isolate the bias networks by acting as open circuits to dc, but as low-impedance paths to ac signal components of sufficiently high frequency. The presence of C_3 enables the feedback-bias resistor R_2 to properly bias Q_1 while grounding the emitter of Q_1 with respect to signals of sufficiently high frequency. This feature results in a larger midband gain.

(a) Find an expression for the signal gain v_{out}/v_s of the two-stage cascade in terms of resistor values and other parameters. Represent each stage by an appropriate linear two-port model.

(b) For the case $V_{CC} = 12\,\text{V}$, $V_{EE} = -12\,\text{V}$, and $R_S = 50\,\Omega$, choose values for the resistors in the circuit such that both transistor stages are properly biased somewhere in the middle of the active region. Limit your resistor choices to 5% standard values. Aim for an approximate cascade gain of about 100, using your answer to part (a) as a guide. Note that the output signal voltage is measured with no external load resistance connected to Q_2, hence output loading is not a consideration.

(c) Now determine the actual gain of the amplifier by analyzing the small-signal model of the cascade as one entity. Compare to the approximate result obtained in part (b).

11.20 ⯀ ◑ Consider the two-stage resistive-load MOSFET amplifier of **Fig. P11.20**.

(a) Find expressions for the input resistance seen at the v_G node, the output resistance seen at the v_{OUT} node, and the small-signal amplification factor v_{out}/v_g. The bypass capacitor C_1 should be set to a short circuit in the amplifier's small-signal model.

(b) Choose values for all resistors such that the two-stage cascade has an overall gain factor v_{out}/v_g of about -5.

(c) Modify the circuit so that an input voltage source can be connected directly between the v_G terminal and ground without upsetting the bias point of Q_1. The gain of the modified amplifier should be approximately the same as that obtained in part (b).

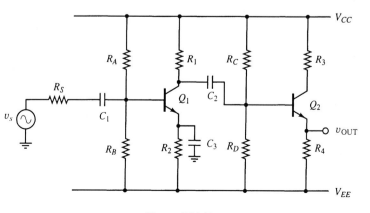

Figure P11.19

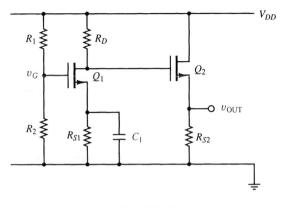

Fig. P11.20

11.21 Consider the two-stage CMOS amplifier of **Fig. P11.21**.

(a) Find an expression for the input resistance, output resistance, and small-signal gain v_{out}/v_s if all devices are biased in the constant-current region.

(b) Find the bias values of all gate-to-source voltages and the output voltage v_{OUT} for the case $V_{DD} = 3\,V$, $V_{SS} = -3\,V$, $K_1 = |K_3| = 4\,mA/V^2$, and $K_2 = |K_4| = 2\,mA/V^2$. Assume all devices to have a threshold voltage of $|V_{TR}| = 2\,V$.

(c) Evaluate the overall gain for the bias parameters determined in part (b).

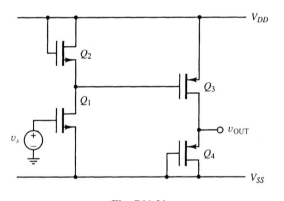

Fig. P11.21

11.22 Consider the circuit of **Fig. P11.16** with a capacitor C_2 connected in parallel with R_2.

(a) Choose resistor values so that v_{OUT} is biased midway between $V_{CC} = 12\,V$ and $V_{EE} = -12\,V$. Aim for a midband-gain magnitude of at least 50.

(b) Estimate the values of amplifier input and output resistance in the midband.

(c) Estimate the value of C_2 required to set the lower limit of the midband region to a maximum of 100 Hz.

11.23 ◑ Consider the two-stage amplifier of **Fig P11.19** with C_2, R_C, and R_D omitted, and with the collector of Q_1 dc coupled to the base of Q_2. Suppose that $V_{CC} = 12\,V$, $V_{EE} = 0$, $R_A = 30\,k\Omega$, $R_B = 10\,k\Omega$, $R_1 = 10\,k\Omega$, $R_2 = 3\,k\Omega$; $R_3 = 0$; $R_4 = 4.3\,k\Omega$, $R_g = 50\,\Omega$, $C_1 = 1\,\mu F$, and $C_3 = 10\,\mu F$. The object of this problem is to study the effect of the internal device capacitances on the high-frequency -3-dB point ω_H of the amplifier. If $C_{\pi 1} = C_{\pi 2} = 11\,pF$ and $C_{\mu 1} = C_{\mu 2} = 1\,pF$, find ω_H. Use the superposition-of-poles technique and the principles of multistage loading. Assume $r_x = 0$. Comment on why $C_{\mu 1}$ and $C_{\pi 1}$ have such different effects on ω_H.

11.3 Multistage Amplifier Biasing

11.24 Find the bias voltages and currents in the circuit of Fig. 11.3 if $R_1 = 10\,k\Omega$, $R_2 = 680\,\Omega$, $R_3 = 3.3\,k\Omega$, $R_4 = 4.7\,k\Omega$, $R_5 = 120\,\Omega$, $R_L = \infty$, and $V_{ZK} = 9\,V$. Assume that $V_f = 0.7\,V$. What is the approximate small-signal gain of the cascade?

11.25 Use the constant-voltage approximation to find the bias voltages and currents in the circuit of Fig. 11.6 if $R_{C1} = 8.2\,k\Omega$, $R_1 = 12\,k\Omega$, and $R_E = 15\,k\Omega$. Assume that $V_f = 0.7\,V$.

11.26 ⅀ ◑ **Figure P11.26** illustrates one possible topology for a three-stage BJT amplifier. The cascaded inverters provide large voltage gain, and the follower stage provides a small output resistance.

(a) Derive an approximate expression for the midband voltage gain using either the two-port-cell cascade method or the complete small-signal circuit model.

(b) Choose values for all resistors so that Q_1, Q_2, and Q_3 are appropriately biased and the cascade has a midband gain of about 50. Include the effects of interstage loading.

(c) Predict the maximum expected positive and negative swings of each stage.

(d) Predict the low-frequency -3-dB point of the amplifier's frequency response.

11.27 ⅀ ○ Using the layout of **Fig. P11.26** as a guide, design an amplifier that has a dc coupled input signal. Assume that a $V_{EE} = -15$-V power supply is also available. The amplifier should have a gain of approximately 50.

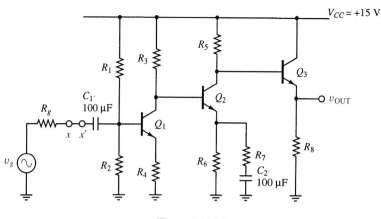

Figure P11.26

11.28 ⚡ ◑ Design a capacitor-free amplifier suitable for integrated-circuit fabrication that has the same small-signal topology as the discrete-component amplifier of **Fig. P11.26**. Make use of the active bypass technique in the second stage of the amplifier to achieve the necessary gain. Assume that a $V_{EE} = -15$-V power supply is also available. The amplifier should have a gain of approximately 75.

11.29 In the circuit of **Fig. P11.29**, two BJT voltage followers are cascaded to produce a circuit with very high input resistance. Determine the power dissipation in each transistor if $V_f = 0.7$ V. What is the total bias power dissipated in the circuit?

11.4 DC Level Shifting

11.30 Find the bias value of the output voltage in the circuit of Fig. 11.8 if V_{ZK1} of D_1 is changed to 6 V. Can the value of V_{ZK2} be changed to compensate?

11.31 Find the bias value of v_{OUT} in the circuit of Fig. 11.8 if the supplies are changed to ± 12 V.

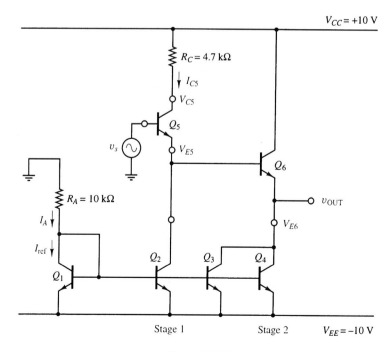

Figure P11.29

11.32 Redesign the circuit of Fig. 11.8 so that zener D_2 is placed between the second and third stages.

11.33 For the *pnp* level-shifted circuit of Fig. 11.9, determine the bias voltages and currents if the circuit is connected to ± 15-V power supplies.

11.34 ⚡ Consider the amplifier layout of **Fig. P11.26**. Redesign the circuit so that the input capacitor C_1 and bypass capacitor C_2 are eliminated, and the output voltage is biased near ground potential. In redesigning the circuit, incorporate some form of level shifting to establish proper bias levels. Assume that a second $V_{EE} = -15$-V power supply and *pnp* transistors are available.

11.35 ⚡ Consider the three-stage BJT amplifier with a *pnp* level shifter shown in Fig. 11.9. Redesign the circuit so that v_{OUT} is still biased near ground potential but the amplifier gain is increased to at least 20.

11.36 Consider the MOSFET amplifier with level shifting shown in Fig. 11.10. Suppose that the devices have parameters $K = 0.5$ mA/V^2 and $V_{TR} = 1$ V, except for Q_1 and Q_3, for which $K_1 = K_3 = 16$ mA/V^2. Choose appropriate values for I_1 and I_2 such that v_{OUT} is biased at a voltage of zero when $V_{DD} = -V_{SS} = 15$ V.

11.37 ⚡ ○ Using the level-shifting method illustrated in Fig. 11.10, design a three-stage MOSFET amplifier with a gain magnitude of about 100 and a dc output bias

voltage close to zero. Set the power-supply voltages to ± 15 V and assume that the MOSFETs have parameters $V_{TR} = 1.5$ V and $\mu_e \epsilon_{ox}/t_{ox} = 0.2$ mA/V^2.

11.5 Differential-Amplifier Cascade

11.38 A BJT differential amplifier with collector resistors of 20 kΩ is dc cascaded with a second differential amplifier in which the collector resistors are equal to 3 kΩ. Each stage is biased by a separate 0.5-mA current source. Find the midband differential-mode and common-mode gains of the resulting multistage amplifier with the output taken in single-ended fashion from one of the collector terminals of the second stage. Assume the value $\beta \approx 100$, if required.

11.39 ⚡ ○ Consider the circuit of **Fig. P11.39**.

(a) Choose values for I_o, R_1, R_2, V_{CC}, V_{EE}, and R_4 such that the collector voltages of Q_3 and Q_4 have maximum differential-mode swing range. Your choices should result in equal bias values for v_{C3} and v_{C4}. (Why?)

(b) Design a biasing current source I_o using as many transistors and resistors as required.

(c) Bias the v_{OUT} terminals at approximately ground potential by adding level shifting to the second stage.

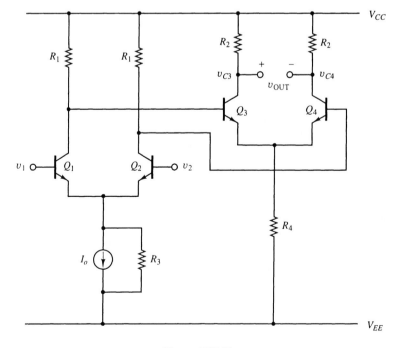

Figure P11.39

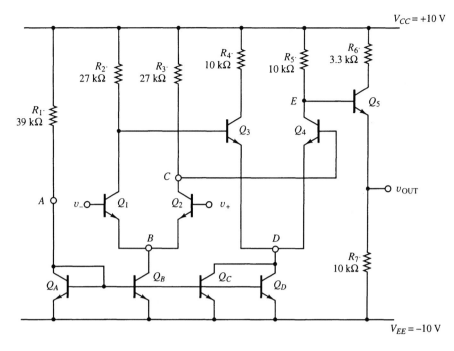

Figure P11.41

11.40 ○ The circuit of **Fig. P11.40** depicts a simple two-stage BJT operational amplifier. All *pnp* transistors are matched and all *npn* transistors are matched, but *npn* ≠ *pnp*. Assume that $\beta = 50$ for all transistors.

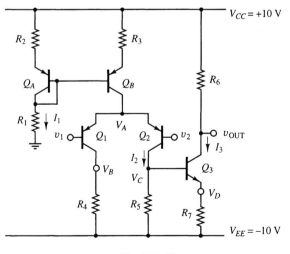

Fig. P11.40

(a) Choose values of R_1 through R_7 such that the small-signal differential gain is greater than 100. Resistors R_4 and R_5 should have equal values. Specify which input is v_+ and which is v_-.

(b) Determine the bias values of node voltages V_A, V_B, V_C, V_D, and V_{OUT}. Determine the bias currents I_1, I_2, and I_3.

(c) What is the small-signal output resistance of the amplifier?

(d) What is the differential-mode input resistance of the amplifier?

11.41 ◐ The circuit of **Fig. P11.41** represents the internal circuitry of a simple BJT operational amplifier. In answering the following questions, use appropriate engineering approximations where required. Assume the BJTs in the circuit to be matched and to have the following parameters: $C_\mu = 1\,\text{pF}$, $V_A \approx \infty$, $V_{BE} = 0.7\,\text{V}$, $V_{sat} \approx 0.2\,\text{V}$, $\beta_o = 200$, and $\eta = 1$. Find the bias currents through each BJT and the bias voltages of nodes A, B, C, D, E, and v_{OUT}. Compute the midband gain, small-signal output resistance, and differential-mode input resistance of the amplifier.

11.42 ○ Suppose that the BJTs in the circuit of **Fig. P11.41** have parameters $f_T = 100\,\text{MHz}$ and $r_x = 2\,\Omega$. Find the upper −3-dB point of the amplifier.

11.43 Σ Using the BJT op-amp circuit of **Fig. P11.41** as a guide, design a modification for the circuit so that the bias value of v_{OUT} is equal to zero relative to ground. The bias level of v_{OUT} in the existing circuit is approximately 4.3 V.

11.44 Using the techniques discussed Section 11.5, derive an expression for the differential- and common-mode gains and the CMRR of the CMOS differential amplifier cascade shown in **Fig. P11.44**.

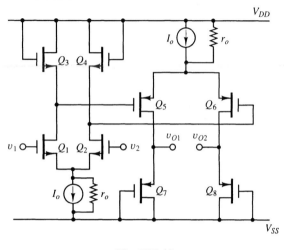

Fig. P11.44

11.45 🔁 ● The four-stage amplifier of **Fig. P11.45** must be designed to have a gain of 2×10^6 when $R_S = 1\,k\Omega$. The output must be biased such that $v_{OUT} = 0$ when $v_{idm} = 0$.

(a) Temporarily ignore loading factors, and find values for bias voltages V_{C2}, V_{C3}, and V_{C5} such that the product of the stage gains is about 10 times as large as the required value. Note that this design step can be performed without calculating any current or resistor values.

(b) Now include the effect of loading factors. Find appropriate values for R_{ref}, R_2, R_3, and R_5, as well as the currents I_{o2}, I_{o3}, and I_{o5}, such that the overall gain requirement is met.

(c) In order for the amplifier to approximate linear behavior, the incremental change in $v_{BE} \equiv v_\pi$ in each device must be small. What is the approximate limit on v_π, and in which transistors of your design is this criteria most likely to be violated?

11.6 Power-Amplification Output Stages

11.6.1 Complementary-Pair (Class B) Output Configuration

11.46 Suppose that the class B output stage of Fig. 11.15 is made from complementary p- and n-channel enhancement-mode MOSFETs. Draw the large-signal v_{IN}–v_{OUT} transfer characteristic if the load is a resistor of value R_{LOAD}.

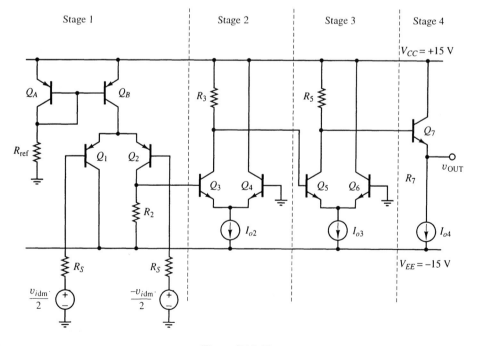

Figure P11.45

11.47 A class B BJT power amplifier is made by incorporating a power amplifier stage into an op-amp voltage-follower feedback loop, as shown in **Fig. P11.47**. The circuit is driven by a sinusoidal voltage source of 13 V peak. Find the time-average power dissipated in the load, the time-average power dissipated in the transistors, the approximate power drawn from the power supplies, and the amplifier efficiency.

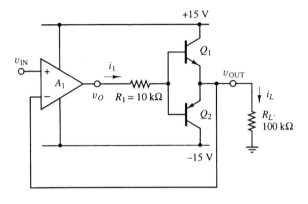

Fig. P11.47

11.48 Using the circuit of **Fig. P11.47** as a guide, design a class B MOSFET power amplifier with op-amp feedback. Draw the transfer characteristics of v_{OUT} and v_O versus v_{IN}.

11.49 A power amplifier is needed to drive a dc servo motor in its forward and reverse directions. The speed of the motor is controlled by varying the voltage applied to its terminals. For the purposes of this problem, model the motor as a 3-Ω resistive load. Design a circuit based on the class B power amplifier that can deliver up to ±10 V to the motor from ±12-V supply busses. What will be the maximum current flow through the BJTs? What will

be their maximum power dissipation?

11.6.2 Linearly-Biased (Class A) Output Configuration

11.50 The efficiency analysis of the class A and class B amplifiers of Section 11.6.2 focuses exclusively on sinusoidal excitation. Show that the maximum theoretical efficiencies of these amplifier types under square-wave excitation become 50% and 100%, respectively.

11.51 Show that the maximum theoretical efficiency of a class A amplifier becomes 17% under symmetrical triangle-wave excitation. Show that the efficiency of a class B amplifier becomes 67% under similar conditions.

11.52 ○ The circuit of **Fig. P11.52** illustrates the use of a two-winding magnetic-core transformer to improve the efficiency of the class A amplifier. In this case, a transformer couples the emitter loop of a class A BJT amplifier to a loudspeaker load. The latter can be modeled as a resistor of value $R_{LOAD} = 8\,\Omega$. If the ac current flowing into the primary of the transformer is I_1, the ac current I_2 flowing from the secondary will be equal to nI_1, where n is the transformer turns ratio. Similarly, if a voltage V_2 develops across the secondary, a voltage $V_1 = nV_2$ will develop across the primary. Since the ratio V_2/I_2 defines R_{LOAD}, the latter will appear to have a resistance of value $V_1/I_1 = nV_2/(I_2/n) = n^2 R_{LOAD}$ when viewed from the primary.

Suppose that $n = 8$. If $\beta_F = 100$, find the maximum theoretical efficiency of the amplifier. How large should C_S be made to ensure operation down to at least 20 Hz? Note that the improvement in efficiency of this circuit results not because of the impedance-transforming properties of the transformer, but because the transformer cannot pass *dc* currents. This feature results in zero bias-power dissipation in the load.

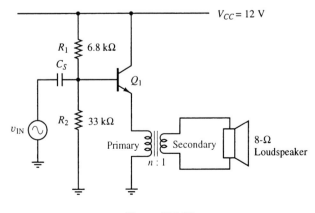

Figure P11.52

11.53 ⚡ ○ Using the circuit of **Fig. P11.52** as a guide (see Problem 11.52), design a transformer-coupled power amplifier that can drive an 8-Ω load if $n = 8$ and $V_{CC} = 20\,\text{V}$, where n is the transformer turns ratio.

11.6.3 Minimally Biased (Class AB) Output Configuration

11.54 Consider the class AB power-amplifier circuit of Fig. 11.19. Suppose that the bias network in the box consists of a single resistor R_2. The diodes have saturation (scale) currents of $10^{-14}\,\text{A}$, and the power-supply voltages are set to $V_{CC} = -V_{EE} = 10\,\text{V}$.

(a) Choose R_1 and R_2 so that the crossover distortion zone is reduced from $v_{IN} = \pm V_f$ to $\pm 0.1\,\text{V}$. Assume a V_f of 0.7 V.

(b) Repeat part (a) for a crossover distortion zone of about 0.01 V.

11.55 ⚡ The class B output stage of Fig. 11.15 is made from complementary p- and n-channel enhancement-mode MOSFETs. Design a modification to the circuit so that it will become a class AB amplifier with reduced crossover distortion.

11.56 Consider the circuit of Fig. 11.20. Find the bias currents in Q_1 and Q_2 when $I_o = 200\,\mu\text{A}$ and $R_1 = 40\,\text{k}\Omega$. For the purpose of this calculation, neglect base currents with respect to emitter currents, and assume all devices to have scale currents of $I_{EO} = 0.8 \times 10^{-11}\,\text{mA}$.

11.57 Choose R_1 and I_o in the circuit of Fig. 11.20 so that Q_1 and Q_2 will have bias currents of about $10\,\mu\text{A}$.

11.58 ⚡ ○ Consider the circuit of Fig. 11.20. Design an I_o current source for Q_X and Q_Y that will produce an I_o of about $200\,\mu\text{A}$. Use either npn or pnp transistors in your design. Assume power supply voltages of $\pm 15\,\text{V}$.

11.59 ⚡ ○ An audio power-amplification stage is required that can deliver up to 0.5 W to an 8-Ω loudspeaker load while operating from a single 12-V supply. To avoid unnecessary power dissipation in the loudspeaker, no dc bias current may flow into it. Design a capacitively coupled, class AB amplifier with a maximum cutoff frequency of 50 Hz. Assume that power BJTs with $\beta_F = 60$ are available. Determine the required power ratings of the transistors.

11.7 Integrated-Circuit Power Amplifiers

11.60 Consider the IC power amplifier of Fig. 11.21. Suppose that the load consists of an 8-Ω resistor and that devices Q_1 through Q_4 have a β_F of 50. A feedback circuit consisting of $R_A = 100\,\text{k}\Omega$ and $R_B = 10\,\text{k}\Omega$

is connected between v_{OUT} and v_-, such that the circuit functions as a noninverting amplifier with a gain of 11.

(a) Estimate the load current i_{OUT} and the current i_o drawn from the amplifier A_1 when $v_{IN} = 1\,\text{V}$.

(b) Now suppose that Q_1 and Q_2 are absent, so that v_O drives the bases of Q_3 and Q_4 directly. Estimate the revised currents i_{OUT} and i_o for $v_{IN} = 1\,\text{V}$.

11.61 Perform the analysis of Example 11.5 without neglecting base currents with respect to emitter currents. What are the values of i_{OUT} and i_o if $\beta_F = 20$, $R_{LOAD} = 50\,\Omega$, and $v_{LOAD} = 5\,\text{V}$?

11.8 Power Devices

11.8.1 Heat Sinks

11.62 A power transistor has an internal thermal resistance of $\Theta_{\text{device–case}} = 3.6\,°\text{C/W}$. The transistor case is mounted on a heat sink with $\Theta_{\text{sink-air}} = 4.5\,°\text{C/W}$. The mounting method utilizes a 0.2-mm-thick plastic spacer with $\Theta_{\text{case-sink}} = 0.5\,°\text{C/W}$. If the transistor dissipates a total average power of 20 W, determine the device temperature inside the case. Assume an ambient air temperature of $T_{\text{air}} = 25°\text{C}$.

11.63 A particular transistor is called upon to dissipate 30 W of power. If the device temperature can be no higher than 200°C, what is the maximum permissible thermal resistance between the ambient air (25°C) and the case? The transistor has a $\Theta_{\text{device–case}}$ of 5 °C/W.

11.64 A power device with a maximum junction temperature of 300°C must dissipate 45 W of power. The device has a $\Theta_{\text{device–case}}$ of 5 °C/W, and a $\Theta_{\text{case-air}}$ of 1.5 °C/W. What is the maximum permissible ambient temperature in which the circuit can be operated? What recommendation would you make to the engineer designing this power circuit?

11.65 A thermal sensor is used to monitor the temperature of the output transistor of a power amplifier. The sensor is bonded to the case of the power transistor using a thin mica spacer having a thermal resistance of 0.2 °C/W. The thermal resistance between the sensor and ambient air (25°C) is 0.5 °C/W. If the maximum allowed transistor case temperature is 170°C, at what temperature should the circuit monitoring the temperature sensor shut down the amplifier?

11.8.2 Power BJT

11.66 A BJT has a rated power of 15 W and a maximum junction temperature of 200°C. The device is to operate in air at an ambient temperature of 25°C. The transistor is mounted on a heat sink with $\Theta_{sink-air} = 4\,°C/W$ and $\Theta_{case-sink} = 1\,°C/W$. Find the actual power that can be safely dissipated in the transistor.

11.67 A BJT has a rated power of 75 W and a maximum junction temperature of 215°C. The device is operated inside a closed compartment in which the air temperature may reach up to 75°C. If $\Theta_{case-sink} = 1\,°C/W$, what is the maximum acceptable $\Theta_{sink-air}$ if the device it to be operated at 35 W?

11.68 A particular device must safely dissipate 10 W of power. The device is to be operated in air with an ambient temperature of 25°C. If $\Theta_{case-air} = 5\,°C/W$ and $T_{j-max} = 200°C$, specify the minimum rated power of the device.

11.8.3 Power MOSFET

11.69 A MOSFET has a rated power of 25 W. Its junc-tion temperature can be no more than 150°C. Find a rela-tionship between the actual operating power and the total $\Theta_{case-air}$ if the junction temperature is not to be exceeded. The ambient air has a temperature of 25°C.

11.70 A power MOSFET has a maximum substrate tem-perature of 200°C. The total thermal resistance between the substrate and the ambient air, including the case and mounting hardware, is 5 °C/W. The air has a temperature of 25°C.

(a) What is the maximum time-average power that can be dissipated in the MOSFET?

(b) The heat sink is improved by the addition of water cooling, which removes heat at the rate of 10 J/s. What is the maximum permissible time-average power dissipation under these conditions?

◆ SPICE PROBLEMS

11.71 Verify the bias calculations of the circuit of Fig. 11.6 under open-circuit load conditions with $\beta_F = 100$ by simulating the circuit on SPICE. Also use SPICE to find the small-signal voltage gain of the circuit when $R_L = 10\,k\Omega$. How does the bias of the circuit change when the load is connected?

11.72 Use SPICE to assess the sensitivity of the bias levels in the circuit of Fig. 11.6 to values of β_F over the range 50 to 200.

11.73 Verify the bias design of the circuit of Fig. 11.7 on SPICE. Show that each BJT operates in the active region. Find the current through and voltage across each transistor.

11.74 Consider the three-stage dc-coupled amplifier of Fig. 11.7.

(a) Estimate the overall cascade gain using the two-port cascade technique.

(b) Obtain a more accurate value for v_{out}/v_g by simu-lating the circuit on SPICE.

(c) Use SPICE to obtain the operating-point values for each of the transistors in the circuit.

(d) What is the total bias power dissipated in the cir-cuit?

11.75 Simulate the level-shifted amplifier circuit of Fig. 11.8 on SPICE. Find the sensitivity of the bias value of v_{OUT} to the V_{ZK} value of D_1.

11.76 Simulate the level-shifted amplifier circuit of Fig. 11.9 on SPICE. Find the change in the bias value of v_{OUT} per volt change in V_{CC}.

11.77 Use SPICE to assess the power efficiency of the circuit of Fig. 11.9 in the case where R_5 is the load resis-tor. Increase the value of v_g until the output reaches either its positive or negative clipping limit. For this maximally driven case, compare the time-average signal power dis-sipated in R_5 to the total power drawn from the V_{CC} and V_{EE} power supplies.

11.78 Consider the cascaded differential-amplifier cir-cuit of Fig. 11.12. If $V_{CC} = -V_{EE} = 15$ V, choose the value of R_A such that $I_{ref1} \approx 1$ mA, then select values for the remaining resistors so that the differential-mode gain measured differentially between v_{OUT1} and v_{OUT2} is approximately 400. Use SPICE to find the actual differential-mode gain of the circuit for $\beta_F = 100$. Also use SPICE to find the common-mode gain of the circuit for single-ended outputs.

11.79 Simulate the class B power amplifier of Fig. 11.15 on SPICE with $R_S = 50\,\Omega$, $V_{CC} = -V_{EE} = 15$ V, and a load resistor $R_{LOAD} = 100\,\Omega$ connected between v_{OUT} and ground. Plot the large-signal transfer characteristic. Over what values of v_{IN} does the crossover distortion re-gion extend? Use SPICE to estimate the overall amplifier efficiency.

11.80 Simulate the class B power amplifier of Fig. 11.15 on SPICE with $R_S = 1\,k\Omega$. In this case, the load consists of a square-law device with parameters $A = 1\,mA/V^2$

and $V_{TR} = 2\,V$, so that $i_L = A(v_L - V_{TR})^2$. Use a MOSFET with its gate connected to its drain to simulate the square-law device. Plot the circuits's large-signal transfer characteristic. Over what values of v_{IN} does the crossover distortion region extend?

11.81 Simulate the class AB power amplifier of Fig. 11.19 on SPICE. Let the bias network consist of a resistor of value $R_2 = 1\,k\Omega$, with $R_1 = 1\,k\Omega$ also. Specify diodes and BJTs with the same junction area. Use SPICE to assess the extent of the crossover distortion region as a function of I_D over the range $1\,\mu A < I_D < 1\,mA$.

11.82 Simulate the class AB power amplifier of Fig. 11.20 on SPICE. Set the bias elements to $I_o = 50\,\mu A$ and $R_1 = 30\,k\Omega$. Use SPICE to assess the extent of the crossover distortion region as a function of I_o over the range $1\,\mu A < I_o < 1\,mA$.

11.83 Simulate the output stage of Fig. 11.21 (not including the op-amp A_1) on SPICE. Set the value of β_F to 20 for each BJT. Plot the transfer characteristic of v_{OUT} versus input voltage v_O over the range $-10\,V < v_O < +10\,V$ for no load and for load resistors of value $100\,\Omega$ and $8\,\Omega$. Also plot the input current i_O and the base currents to Q_3

and Q_4 for each value of R_{LOAD}.

11.84 Model the amplifier circuit of **Fig. P11.47** on SPICE using a linear dependent source for the op-amp. Set the op-amp parameters to $r_{in} = 10\,M\Omega$, $r_{out} = 100\,\Omega$, and $A_o = 10^5$.

(a) Plot the large-signal input–output transfer characteristic over the range $-5\,V < v_{IN} < +5\,V$ when the load is a 1-kΩ resistor.

(b) Rerun the simulation, but ground the v_{in} terminal and drive transistors Q_1 and Q_2 directly via the base resistor R_1. The op-amp output should be disconnected from R_1 for this purpose. Compare the transfer characteristic to that obtained in part (a).

11.85 Model the transformer-coupled circuit of **Fig. P11.52** on SPICE with $C_S = 100\,\mu F$ and $n = 8$. Find the gain as a function of frequency over the range $0.01\,Hz < f < 10\,kHz$.

11.86 Simulate your design of Problem 11.45 (**Fig. P11.45**) on SPICE. Determine the values of differential-mode and common-mode gains.

Chapter 12

Analog Integrated Circuits

M any of the chapters in this book deal with the component parts of modern analog integrated circuits. The basic amplifier topologies of Chapter 6, the bias configurations of Chapter 7, the differential amplifier of Chapter 8, and the multistage considerations of Chapter 11 all play important roles in the synthesis of analog amplification systems. In this chapter, we discuss several design and analysis techniques that belong exclusively to the realm of analog integrated circuits, with particular focus on the monolithic (single-chip) IC operational amplifier. The low cost, widespread availability, and ease of use of the op-amp allow it to be treated as a simple electronic device having only three terminals: v_+, v_-, and v_{OUT}. This approach, which was followed in Chapter 2, enables the novice circuit designer to progress quickly through the fundamentals of electronics and learn basic amplification principles. Despite the usefulness of this simplistic view, the op-amp is really a complex and precision-engineered electronic circuit that has many characteristics and features not described by the simple three-terminal viewpoint. Only through a detailed study of the op-amp can we gain a deeper understanding of the origin of the op-amp's fundamental open-loop parameters as well as its many nonideal characteristics. In this chapter, we examine the internal structure of the op-amp in detail. Our approach shall be one of case studies, first beginning with the popular (and defacto standard) LM741 integrated-circuit operational amplifier, then progressing on to a more generic CMOS configuration. We begin the chapter with a review of the basic component structure of the modern IC op-amp, as originally introduced in Chapter 2.

12.1 BASIC OPERATIONAL-AMPLIFIER CASCADE

As illustrated in Fig. 12.1, the internal circuitry of the typical operational amplifier consists of a dc-coupled multistage cascade that enhances differential-mode performance and minimizes common-mode response. The differential input stage feeds a single-ended middle-gain stage in which the op-amp's frequency-response profile is also set. The final stage, usually preceded by a level shifter, consists of a class B or class AB output buffer stage that enables the op-amp to provide large load currents of both polarities. The differential input stage provides the op-amp with its large input resistance, while the output stage provides the op-amp's low output resistance. The gain of the overall cascade can be as high as 10^6 or more for a BJT design.

Figure 12.1
Operational
amplifier cascade.

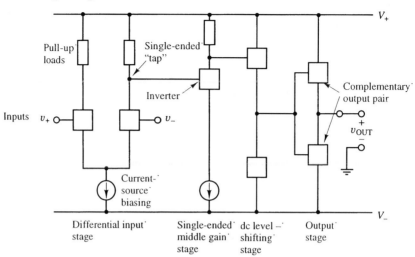

Most op-amps are designed for use with symmetrical bipolar power-supply voltages, although operation with asymmetrical supply voltages is often possible. Some op-amps can also be used with a single-polarity power supply.

When op-amps were first introduced in the mid–1960s, limitations on fabrication technology required them to be made only from BJTs. Modern technologies have also made possible both MOSFET and BiFET (bipolar-FET) op-amp designs. In the latter technology, BJTs and FETs are fabricated simultaneously on the same integrated circuit. The FET input stage of a BiFET op-amp produces an extremely high input resistance, and subsequent BJT stages produce large overall gain. Most recently, CMOS op-amps have become possible due to the development of fabrication techniques for creating reliable *p*-channel and *n*-channel transistors on the same chip. The low power consumption of CMOS circuits, as well as the availability of computer-aided design tools and simple fabrication techniques for CMOS circuits have made CMOS amplifiers an increasingly popular choice among designers of application-specific analog integrated circuits.

12.2 CASE STUDY: THE LM741 BIPOLAR OPERATIONAL AMPLIFIER

The LM741 op-amp was first introduced in the mid–1960s as an improvement on earlier op-amp designs. The "741" has since become a standard electronic "workhorse" suitable for countless op-amp applications. At least one version of the 741 op-amp is now offered by virtually all major manufacturers of analog integrated circuits. Although other op-amps with improved properties have been developed for demanding design applications, the 741 remains the op-amp of choice for students of electronics. Its low cost, stability in most all negative-feedback configurations, and relative immunity to abuse have made it an excellent choice for learning applications. In this section, we study the internal circuitry of the LM741 op-amp, first by dissecting its individual stages, and then by examining the properties of the entire op-amp cascade. A schematic diagram of the entire 741 cascade is shown in Fig. 12.2. We begin by analyzing its differential input stage in the next section.

12.2.1 BJT Input Stage of the LM741 Op-Amp

The key elements of the differential input stage of the LM741 operational amplifier of Fig. 12.2 are shown in Fig. 12.3. The simultaneous use of *npn* and *pnp* devices produces both a large gain

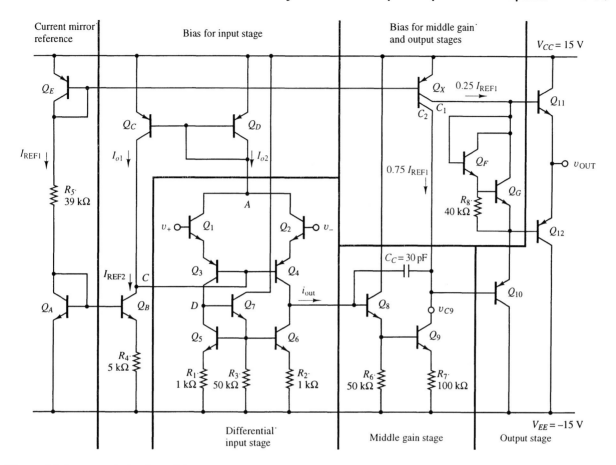

Figure 12.2 Internal circuitry of the LM741 op-amp. The short-circuit-protection components of the output stage have been omitted from the diagram.

and dc level shifting at the same time. As discussed in Chapter 11 (see Section 11.4), the latter is necessary if the output terminal of the op-amp is to be biased midway between its power-supply voltages.

Note that the signal output of the input stage of Fig. 12.3 is a current rather than a voltage. This signal current is fed to the next stage of the op-amp cascade, as we shall see in Section 12.2.2. For now, we shall feed the output signal current i_{out} to an arbitrary load resistance R_{LOAD}, so that we can focus on the properties of the input stage itself. The labeling of the v_+ and v_- terminals in Fig. 12.3 reflects the effect of these voltages on the polarity of the output voltage of the entire op-amp, rather than the voltage developed across R_{LOAD}.

We next find the current flowing through each side of the differential amplifier. As the analysis will show, the current I_{o2} flowing out of Q_D is made equal to I_{REF2} via the combined actions of the Widlar source of Q_A and Q_B and the *pnp* current mirror formed by Q_C and Q_D. The current I_{o2}, which acts as the reference current to the *pnp* current mirror, is set by the voltage applied across the base–emitter junctions of Q_1 through Q_4. The bases of Q_1 and Q_2 are held at dc ground potential by the input voltage sources connected to the v_+ and v_- terminals. The voltage across the series-connected base–emitter junctions of Q_1–Q_3 and Q_2–Q_4 are therefore set by the voltage of node C, which is the output node of the Widlar source.

We note that I_{o2} is duplicated as I_{o1} by the *pnp* current mirror. The latter current then flows into node C. To the extent that base currents I_{B3} and I_{B4} can be neglected compared to I_{o1} and

Figure 12.3
BJT input stage of
the LM741 op-amp.
R_{LOAD} represents
the input resistance
of the next stage.
The signal output
of this circuit is the
current i_{out}.

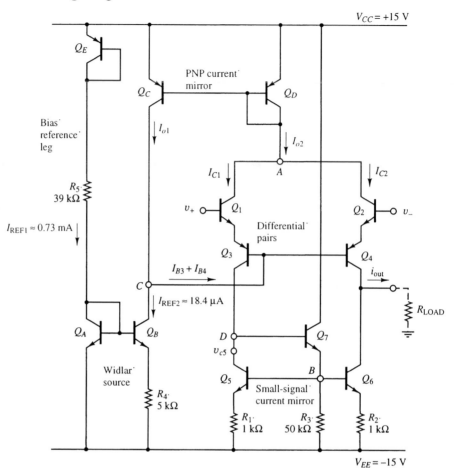

I_{REF2}, the duplicated current I_{o1} must equal I_{REF2}, as can be seen by applying KCL to node C. The nature of a current source is such that its terminal voltage adjusts to match the constraint of its load. In this case, the Widlar source adjusts the voltage of node C until the currents through the complementary pairs Q_1–Q_3 and Q_2–Q_4 allow the condition $I_{o1} = I_{\text{REF2}}$ to be met. (More precisely, the condition $I_{o1} = I_{\text{REF2}} + I_{B3} + I_{B4}$ is met at node C, but we have assumed that I_{B3} and I_{B4} are negligible.) The net result of this bias arrangement is that I_{o2} is set to the value I_{REF2}, causing currents of value $I_{C1} \approx I_{\text{REF2}}/2$ and $I_{C2} \approx I_{\text{REF2}}/2$ to flow through Q_1 and Q_2.

Bias Design

As discussed in the preceding paragraph, bias in the circuit of Fig. 12.3 is set by the Widlar source formed by Q_A and Q_B. Use of the Widlar source allows small bias currents to be established without requiring large-value resistors. As shown in Section 8.3.8, I_{REF2} can be found by iterative solution of the equation

$$I_{\text{REF2}} = \frac{\eta V_T}{R_4} \ln \frac{I_{\text{REF1}}}{I_{\text{REF2}}} \tag{12.1}$$

where $\quad I_{\text{REF1}} = \dfrac{V_{CC} - V_{EE} - 2V_f}{R_5} = \dfrac{15\,\text{V} - (-15\,\text{V}) - 1.4\,\text{V}}{39\,\text{k}\Omega} \approx 0.73\,\text{mA} \tag{12.2}$

The voltage $V_f = 0.7\,\text{V}$ represents the base–emitter turn-on voltage of Q_E and Q_A. Applying the iterative solution method to Eq. (12.1) yields $I_{\text{REF2}} \approx 18.4\,\mu\text{A}$. This result may be checked by direct substitution of I_{REF2} back into the equation.

With dc currents of $I_{REF2}/2 = 18.4\,\mu A/2 = 9.2\,\mu A$ established down the legs of the differential amplifier, the bias value of V_{C5}, measured at the collector of Q_5 (node D), becomes

$$V_{C5} = V_{EE} + I_{E5}R_1 + V_{BE5} + V_{BE7}$$
$$\approx V_{EE} + \frac{I_{REF2}}{2}R_1 + 2V_f = -15\,\text{V} + (9.2\,\mu A)(1\,\text{k}\Omega) + 1.4\,\text{V} = -13.6\,\text{V} \tag{12.3}$$

where V_{BE5} and V_{BE7} are approximately equal to the base–emitter turn-on voltage V_f, and the base current into Q_7 has been neglected. The bias voltage at the collector of Q_6 is determined by its connection to the succeeding stage, as we shall see in Section 12.2.2.

Small-Signal Behavior

An appropriate small-signal model for the input stage is shown in Fig. 12.4. The *pnp* and *npn* transistors do not necessarily have the same β_o, even though they are fabricated on the same integrated circuit. All six of devices Q_1 through Q_6 have the same g_m, however, since all have approximately the same value of collector current. Note that *pnp* and *npn* devices are represented by the same small-signal model, as discussed in Chapter 7. The small-signal output resistances presented to node C by the Widlar source and the *pnp* current mirror are modeled by r_{oW} and r_{oC}, respectively, in Fig. 12.4. Similarly, the small-signal resistances contributed by the current-mirror transistors Q_C and Q_D at node A are each represented by an r_π to ground.

The small-signal behavior of this circuit is dominated by followers Q_1 and Q_2, which feed the emitters of the *pnp* cascode devices Q_3 and Q_4. The latter devices are connected in the current-follower configuration (also known as the "common-base" connection), with signal fed into the emitter and output taken from the collector. Under differential-mode excitation, no signal current flows through r_{oW} and r_{oC}, so that the voltage of node C remains at signal ground.

The output signal current i_{out} at the joined collectors of Q_4 and Q_6 is given by

$$i_{out} = -(i_{c4} + i_{c6}) \tag{12.4}$$

Evaluation of i_{out} requires expressions that relate i_{c4} and i_{c6} to the differential-mode input $(v_+ - v_-)$.

An expression for i_{c4} can be found by evaluating i_{b4}. The value of i_{b4} can be found by taking KVL around the outer loop of the circuit of Fig. 12.4, yielding

$$v_+ - v_- = i_{b1}r_{\pi1} - i_{b3}r_{\pi3} + i_{b4}r_{\pi4} - i_{b2}r_{\pi2} \tag{12.5}$$

The symmetry of the circuit implies that $i_{b1} = -i_{b2}$ and $i_{b3} = -i_{b4}$ under differential-mode excitation, so that Eq. (12.5) becomes

$$v_+ - v_- = i_{b4}(r_{\pi3} + r_{\pi4}) - i_{b2}(r_{\pi1} + r_{\pi2}) \tag{12.6}$$

Transistor pairs Q_1–Q_3 and Q_2–Q_4 will also have equal and opposite emitter currents, so that, for example,

$$(\beta_{o2} + 1)i_{b2} = -(\beta_{o4} + 1)i_{b4} \tag{12.7}$$

Equation (12.6) therefore can be written as

$$v_+ - v_- = i_{b4}(r_{\pi3} + r_{\pi4}) + \frac{\beta_{o4} + 1}{\beta_{o2} + 1}i_{b4}(r_{\pi1} + r_{\pi2}) \tag{12.8}$$

If *npn* parameters are expressed by $\beta_{o1} = \beta_{o2} = \beta_{oN}$ and $r_{\pi1} = r_{\pi2} = r_{\pi N}$, and *pnp* parameters by $\beta_{o3} = \beta_{o4} = \beta_{oP}$ and $r_{\pi3} = r_{\pi4} = r_{\pi P}$, then Eq. (12.8) can be solved for i_{b4} and put in the form

$$i_{b4} = \frac{v_+ - v_-}{[(\beta_{oP} + 1)/(\beta_{oN} + 1)]2r_{\pi N} + 2r_{\pi P}} \approx \frac{v_+ - v_-}{(2\beta_{oP}/\beta_{oN})r_{\pi N} + 2r_{\pi P}} \tag{12.9}$$

Figure 12.4
Small-signal model
of the LM741 input
stage.

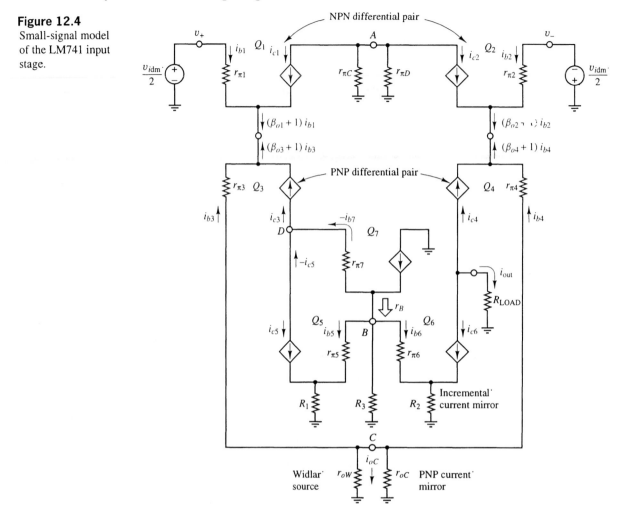

Transistors Q_1 through Q_4 have the same g_m, so the denominator of Eq. (12.9) may be expressed as

$$2\beta_{oP}\frac{r_{\pi N}}{\beta_{oN}} + 2r_{\pi P} = \frac{2\beta_{oP}}{g_m} + 2r_{\pi P} \equiv 4r_{\pi P} \tag{12.10}$$

Equation (12.9) therefore simplifies to

$$i_{b4} \approx \frac{v_+ - v_-}{4r_{\pi P}} = \frac{v_{idm}}{4r_{\pi P}} \tag{12.11}$$

where $v_{idm} = v_+ - v_-$. The resulting i_{c4} component of i_{out} becomes

$$i_{c4} = \beta_{o4}i_{b4} = \frac{\beta_{oP}}{4r_{\pi P}}(v_+ - v_-) = \frac{g_m}{4}(v_+ - v_-) \tag{12.12}$$

Evaluation of i_{c6} as a function of $(v_+ - v_-)$ requires an examination of the role of Q_7 in the circuit. When differential-mode signals are applied to v_+ and v_-, the outer loop of the circuit is balanced, and node C remains at signal ground. Node B, unlike node C, does not remain at signal ground, but responds to the emitter current of Q_7, which flows into $r_{\pi5}$, R_3, and $r_{\pi6}$.

The small-signal resistance r_B between node B and ground, as seen by the emitter of Q_7, is composed of the parallel combination of three resistances and is given by

$$r_B = R_3 \| [r_{\pi 5} + (\beta_{o5} + 1)R_1] \| [r_{\pi 6} + (\beta_{o6} + 1)R_2] \qquad (12.13)$$

Because R_3, β_{o5}, and β_{o6} are large, the value of r_B is also large compared to other small-signal resistances in the circuit. The current $-i_{c5}$ is made approximately equal to i_{c3} by the action of Q_7. For a positive increase in i_{c3}, a portion of i_{c3} is pulled up through $r_{\pi 7}$ as $-i_{b7}$. This negative base current causes Q_7 to pull signal current up through the high resistance r_B, thereby increasing i_{b5} in the negative direction. The voltage of node B and the currents $-i_{b5}$ and $-i_{c5}$ are increased by the action of Q_7 until KCL is satisfied at node D with

$$-i_{b7} - i_{c5} = i_{c3} \qquad (12.14)$$

Because i_{b7} is negligible compared to i_{c5}, Eq. (12.14) reduces to

$$-i_{c5} \approx i_{c3} \qquad (12.15)$$

The small-signal current i_{c5} is thus made into the negative image of i_{c3} by the action of Q_7.

Resistors R_1 and R_2 have the same value, hence transistors Q_5 and Q_6 form a small-signal current mirror that replicates i_{b5} in i_{b6}.[1] The collector current of Q_6 thus becomes a mirror image of i_{c5}. Given the relation (12.15), i_{c6} also becomes a negative mirror image of i_{c3}. Given that $i_{b3} = -i_{b4}$, it follows that $i_{c3} = -i_{c4}$; hence i_{c6} becomes

$$i_{c6} = -i_{c3} = i_{c4} \qquad (12.16)$$

As seen by this analysis, the net function of Q_5, Q_6, and Q_7 is to duplicate the small-signal current i_{c4} in Q_6.

With i_{c4} described by Eq. (12.12), the output current of the amplifier under differential-mode conditions becomes

$$i_{\text{out}} = -(i_{c4} + i_{c6}) = -\frac{g_m}{2}(v_+ - v_-) \equiv -\frac{g_m}{2}v_{i\text{dm}} \qquad (12.17)$$

For bias currents $I_{C1} = I_{C2} = I_{\text{REF2}}/2 = 9.2\,\mu\text{A}$, the transconductance gain of the input stage becomes

$$\frac{i_{\text{out}}}{v_{i\text{dm}}} = \frac{-g_m}{2} = \frac{-I_{C1}}{2\eta V_T} = -\frac{9.2\,\mu\text{A}}{(2)(1)(25\,\text{mV})} = -0.18\,\text{mA/V} \qquad (12.18)$$

When the output terminal is connected to the load resistance R_{LOAD}, the voltage gain becomes

$$\frac{v_{\text{out}}}{v_{i\text{dm}}} = \frac{-g_m}{2}R_{\text{LOAD}} \qquad (12.19)$$

This gain is equivalent to the value achievable from the BJT diff-amp of Section 8.3. The level-shifting function of the *npn–pnp* combination in this circuit makes it preferable for op-amp applications. More importantly, the small-signal reflection of i_{c1} in i_{c6} by the small-signal current mirror substantially reduces the common-mode gain of the circuit. Under common-mode excitation, the current $-i_{c4}$ will essentially be a replica of i_{c6}, so that i_{out} becomes approximately equal to zero.

We note that the small value of I_C in these transistors leads to a correspondingly small value of g_m. The resulting transconductance gain (12.18) may seem low compared to the values obtained in the single-stage BJT amplifiers of Chapter 7. In an op-amp, however, small bias currents are desirable at every stage if the current drawn from the power supplies is to be held to a minimum. Moreover, as a *current* output circuit, the voltage gain of the input stage of Fig. 12.3 can still be made large by choosing a large value for R_{LOAD}. Since R_{LOAD} is really just the input resistance of the next stage of the op-amp, a large voltage gain can be achieved by designing the next stage so that it has a large input resistance.

---

[1] If R_1 is not equal to R_2, Q_3 and Q_4 form a modified Widlar source called the "Wilson" current source.

EXERCISE 12.1 For the circuit of Fig. 12.3, find I_{C1} and I_{C2} in terms of I_{REF2} if BJT base currents are not neglected compared to collector currents. Assume npn devices to be matched to other npn devices, and pnp devices to be matched to other pnp devices; however, npn $\neq$ pnp.
Answer:

$$I_{C1} = I_{C2} = \frac{I_{REF2}}{2} \frac{(\beta_P + 1)(\beta_P + 2)}{(\beta_P + 2)(\beta_N + 1) + \beta_P(\beta_P + 1)}$$

where $\beta_P = \beta_{F\text{-pnp}}$ and $\beta_N = \beta_{F\text{-npn}}$.

12.2 Derive the common-mode gain of the circuit of Fig. 12.3.

12.2.2 Middle-Gain Stage of the LM741 Op-Amp

The circuit of Fig. 12.5 shows the key elements of the middle-gain stage of the LM741 operational amplifier. Its input consists of the output current of the preceding differential stage. The circuit consists of a follower Q_8 cascaded with an inverter Q_9. The follower is responsible for providing the large R_{LOAD} required of the preceding input stage, as described in Section 12.2.1. The follower also provides large current gain. The pull-up load to Q_9 consists of the incremental output resistance r_{oX} presented by the connection to Q_X. The large value of r_{oX} produces an inverter in Q_9 with a very large voltage gain. As a result, the cascaded combination of Q_8 and Q_9 functions as a current-to-voltage converter with an extremely large current-to-voltage conversion ratio v_{out}/i_{in}. This ratio, called the transresistance gain, has units of volts per milliampere. The circuit also can be thought of as a voltage amplifier with a voltage gain of v_{out}/v_{in}, where v_{in} is computed from the product $i_{in}R_{LOAD}$.

Figure 12.5
BJT op-amp middle gain stage consists of follower Q_8 and inverter Q_9. The small-signal output resistance r_{oX} serves as the inverter load to Q_9.

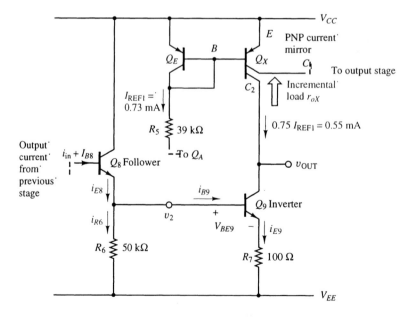

The pnp transistor used for Q_X is actually a "split collector" device that has two parallel collectors and a common base–emitter junction. Such a device can be fabricated in lateral planar geometry, as illustrated in Fig. 12.6. In a lateral pnp BJT, the base region also functions as the local device substrate. In practice, the n-type base region shown in Fig. 12.6 is locally implanted on a global p-type substrate used to fabricate the npn devices of the op-amp (see Appendix B).

Figure 12.6
Layout and circuit symbol of lateral split-collector *pnp* transistor (simplified fabrication diagram): (a) side view; (b) top view; (c) circuit symbol.

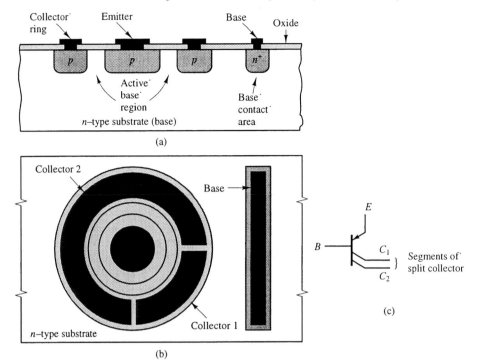

The circular collector ring in Fig. 12.6 is split into two unequal segments, one occupying 3/4 of the circumference and the other 1/4. Each segment captures its proportional share of the total collector current. The larger collector segment is connected to Q_9, so that the bias current fed to Q_9 is equal to $0.75 I_{REF1} = 0.55$ mA, where $I_{REF1} = 0.73$ mA is established by Q_E, Q_A, and R_5, as described in the preceding section. Collector C_2 of Q_X thus forces a current of value $I_{C9} = 0.55$ mA into Q_9 and R_7. The sum of the voltage drops across R_7 and V_{BE9} fixes the voltage drop across R_6, thereby setting the current through Q_8. The current through Q_8 can be set to a very small value using this scheme, with the result that minimal bias current is drawn as I_{B8} from the preceding differential input stage. Specifically, the bias current through R_6 is given by

$$I_{R6} = \frac{V_{BE9} + I_{E9} R_7}{R_6} \tag{12.20}$$

where

$$V_{BE9} = \eta V_T \ln \frac{I_{E9}}{I_{EO9}} \tag{12.21}$$

The bias current through Q_8 becomes

$$I_{E8} = I_{R6} + I_{B9} \tag{12.22}$$

If the condition $I_{E9} \approx I_{C9} = 0.55$ mA is assumed, and the value of I_{EO} set to 10^{-11} mA, Eq. (12.21) yields

$$V_{BE9} = (1)(0.025 \text{ V}) \ln \frac{0.55 \text{ mA}}{10^{-11} \text{ mA}} = 0.618 \text{ V} \tag{12.23}$$

The value $\eta = 1$ has been used in this equation, because I_{E9} is in the milliampere range.

Substituting the value (12.23) into Eq. (12.20) yields the bias current through R_6:

$$I_{R6} = \frac{0.618 \text{ V} + (0.55 \text{ mA})(100 \text{ }\Omega)}{50 \text{ k}\Omega} = \frac{0.618 \text{ V} + 0.055 \text{ V}}{50 \text{ k}\Omega} \approx 13.5 \text{ } \mu\text{A} \tag{12.24}$$

If I_{B9} is assumed negligibly small compared to I_{R6}, then I_{E8} will also be equal to $13.5\,\mu A$. An I_{E8} of $13.5\,\mu A$ will require a bias current I_{B8} that is only in the nanoampere range, hence bias-current loading of the output of the preceding differential stage will be minimal.

EXERCISE 12.3 Using the result (12.24), find I_{E8} and I_{B8} for the case where I_{B9} is not neglected. Assume the typical value $\beta_F = 200$ for both transistors.   **Answer:** $I_{B9} = 2.8\,\mu A$, $I_{E8} = 16.3\,\mu A$, $I_{B8} \approx 81\,nA$

12.4 For a β_F in the range $50 < \beta_F < 250$, find the values of I_{E8} and I_{B9} in the circuit of Fig. 12.5. Use the values of R_6 and R_7 shown in the figure.   **Answer:** $2.2\,\mu A < I_{B9} < 11\,\mu A$; $15.7\,\mu A < I_{E8} < 24.5\,\mu A$

For an inverter with the topology of Q_9 in Fig. 12.5, the bias value of v_{OUT} would normally be determined by the inverter's pull-up resistance. In this case, the load to Q_9 consists only of the incremental output resistance of Q_X. If no load is connected to the v_{OUT} terminal, devices Q_9 and Q_X behave as current sources in series. To first order, the bias value of v_{OUT} under such conditions is indeterminate. A more detailed examination (see Problem 12.19) shows that the bias voltage V_{OUT} is determined by the incremental resistance values seen looking into the collectors of Q_9 and Q_X. In practice, the bias value V_{OUT} is also set by the input conditions of the next stage in the cascade, as discussed in Section 12.2.3.

Figure 12.7
Small-signal model of the circuit of Fig. 12.5; The split collector of Q_X is represented by its output resistance r_{oX2}.

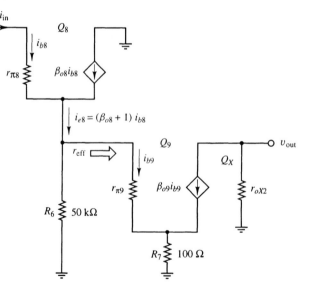

The small-signal behavior of the circuit of Fig. 12.5 can be derived from the model of Fig. 12.7. In this model, the active load Q_X is represented by the small-signal output resistance r_{oX2} seen looking into collector C_2 of Q_X. It can be shown (see Problem 12.23) that r_{o9} of Q_9 has little effect on the small-signal behavior of the circuit; hence r_{o9} is omitted from the circuit model of Fig. 12.7. The use of Q_X as a load to the inverter Q_9 illustrates the concept of an *active load*, whereby a properly biased active device replaces a fixed passive resistor. The active load concept was explored previously in the context of the NMOS and CMOS amplifiers of Chapter 7 and the differential amplifier in Chapter 8.

If the follower Q_8 is to contribute to the gain of the op-amp cascade, most of the signal current i_{e8} must be transferred to the base of the inverter Q_9. This objective is met by making R_6 large, as shown by the following analysis.

The incremental resistance r_{eff} seen looking into the base of Q_9 is given by

$$r_{\text{eff}} = r_{\pi 9} + (\beta_{o9} + 1)R_7 \tag{12.25}$$

The fraction of i_{e8} flowing into r_{eff} can be found using the current-divider relation:

$$i_{b9} = i_{e8}\frac{R_6}{R_6 + r_{\text{eff}}} = (\beta_{o8} + 1)i_{b8}\frac{R_6}{R_6 + r_{\text{eff}}} \tag{12.26}$$

Note that i_{b8} is equal to i_{in}. The output voltage of the circuit is given by

$$v_{\text{out}} = -\beta_{o9}i_{b9}r_{oX2} \tag{12.27}$$

Substituting the appropriate expressions for i_{b9} and r_{eff} into Eq. (12.27) with $i_{b8} = i_{\text{in}}$ results in

$$v_{\text{out}} = -\beta_{o9}(\beta_{o8} + 1)r_{oX2}\underbrace{\frac{R_6}{R_6 + r_{\pi 9} + (\beta_{o9} + 1)R_7}}_{\text{current-division factor}}i_{\text{in}} \tag{12.28}$$

The current-division factor in Eq. (12.28) has a value less than unity, but the effect of this factor on the gain will not be severe if R_6 is much larger than $(\beta_{o9} + 1)R_7$ and $r_{\pi 9}$. This point is illustrated in the following numerical example.

EXAMPLE 12.1 For the circuit of Fig. 12.7, let $R_6 = 50\,\text{k}\Omega$ and $R_7 = 100\,\Omega$, as shown. Suppose that $\beta_o = 150$ for Q_8 and Q_9, and assume that Q_X has an Early voltage of $V_{AX} = 50\,\text{V}$. These transistor parameters are typical of those found inside the typical LM741. If $I_{C9} = 0.55\,\text{mA}$, as previously computed, estimate the transresistance gain $a_r = v_{\text{out}}/i_{\text{in}}$ of this middle-gain stage.

Solution

For the suggested value of $\beta_{o8} = \beta_{o9} = 150$, $r_{\pi 9}$ will have the value

$$r_{\pi 9} = \frac{\beta_{o9}}{g_{m9}} = \beta_{o9}\frac{\eta V_T}{I_{C9}} = \frac{(1)(150)(0.025\,\text{V})}{0.55\,\text{mA}} \approx 6.8\,\text{k}\Omega \tag{12.29}$$

Similarly, for the specified Early voltage, r_{oX2} has the value

$$r_{oX2} \approx \frac{V_{AX}}{I_{C9}} = \frac{50\,\text{V}}{0.55\,\text{mA}} = 90.9\,\text{k}\Omega \tag{12.30}$$

For the value of $r_{\pi 9}$ given by Eq. (12.29), the current-division factor in Eq. (12.28) becomes

$$\frac{R_6}{R_6 + r_{\pi 9} + (\beta_{o9} + 1)R_7} = \frac{50\,\text{k}\Omega}{50\,\text{k}\Omega + 6.8\,\text{k}\Omega + (151)(0.1\,\text{k}\Omega)} \approx 0.70 \tag{12.31}$$

With R_6 set to the large value $50\,\text{k}\Omega$, the attenuation imposed by the current division factor is minimal. When the number given by Eq. (12.31) is multiplied by the product $\beta_{o9}(\beta_{o8} + 1)r_{oX2}$ in Eq. (12.28), the current-to-voltage conversion ratio $v_{\text{out}}/i_{\text{in}}$ becomes

$$a_r \triangleq \frac{v_{\text{out}}}{i_{\text{in}}} = -(150)(151)(90.9\,\text{k}\Omega)(0.70) = -1.44 \times 10^6\,\text{V/mA} \tag{12.32}$$

This value is still a very large number even though the current-division term contributes a factor of 0.70 to the stage gain.

If the circuit of Fig. 12.5 is fed from the LM741 input stage of Fig. 12.3, for which

$$i_{\text{out}} = -\frac{g_{m1}}{2}(v_+ - v_-) \tag{12.33}$$

a large differential-mode voltage gain results, as shown in the next example.

EXAMPLE 12.2 The LM741 BJT differential input stage of Fig. 12.3 is cascaded with the middle-gain stage of Example 12.1. Find the resulting voltage gain of the two cascaded stages. Assume that the differential pair devices Q_1 through Q_4 in Fig. 12.3 are biased with $I_C = 18.4\,\mu A/2 = 9.2\,\mu A$, as in Section 12.2.1, so that $g_{m1} = g_{m2} = g_{m3} = g_{m4} \approx 0.37\,\text{mA/V}$.

Solution

When the two stages are cascaded, the output current of the differential stage becomes the input current to the middle-gain stage. The resulting overall voltage gain can be found by substituting Eq. (12.33) into Eq. (12.28):

$$v_{\text{out}} = \beta_{o9}(\beta_{o8} + 1)r_{oX2}\frac{R_6}{R_6 + r_{\pi 9} + (\beta_{o9} + 1)R_7}\frac{g_{m1}}{2}(v_+ - v_-) \equiv a_r\frac{g_{m1}}{2}(v_+ - v_-) \quad \textbf{(12.34)}$$

Substituting the values from Example 12.1 into Eq. (12.34) results in

$$A_o = \frac{v_{\text{out}}}{v_+ - v_-} = (1.44 \times 10^6\,\text{V/mA})\frac{0.37\,\text{mA/V}}{2} \approx 2.7 \times 10^5 \equiv 109\,\text{dB} \quad \textbf{(12.35)}$$

This value is reasonable for the open-loop gain of an op-amp and is within the typical manufacturer's specified gain range for the LM741. Only a buffering output stage with unity voltage gain need be added to complete the op-amp cascade. As discussed in Chapter 2, a large open-loop op-amp gain, as typified by the value computed in Eq. (12.35), makes possible the use of the ideal op-amp approximation.

Note that the gain expression (12.34) contains the device parameter β_o. When the op-amp is fabricated on a monolithic integrated circuit, β_o can be predicted to within a range of probable values. Under such conditions, a minimum value for the op-amp gain can be calculated.

12.2.3 Output Stage of the LM741 Op-Amp

The cascaded combination of differential input stage and middle-gain inverter stage provides the LM741 with its characteristic large open-loop gain. In order to provide the 741 with the low incremental output resistance characteristic of a well-designed op-amp, the signal from the middle-gain stage of Section 12.2.2 must be buffered by a low-output-resistance circuit before being passed to the op-amp's output terminal. The output stage must also produce an output voltage capable of swinging in both positive and negative directions. The class AB complementary pair follower of Chapter 11 is ideal for this task. The LM741 output stage shown in Fig. 12.8 is a unity-voltage-gain class AB amplifier in which Q_{11} functions as an *npn* follower for positive output currents and Q_{12} as a *pnp* follower for negative output currents. The entire circuit is driven from the base of Q_{12}, where it is buffered by the *pnp* follower Q_{10}. The unity-gain follower formed by Q_{10} acts to increase the input resistance presented to the preceding middle-gain stage.

The series diodes used in Section 11.6.3 (Fig. 11.19) to form the class AB configuration have been replaced here by the base–emitter junctions of Q_F and Q_G. The net voltage drop developed across these junctions establishes a small bias current in the output devices Q_{11} and Q_{12} even when the input signal is zero. As discussed in Section 11.6.3, this small bias current eliminates much of the crossover distortion inherent to BJT push-pull output stages. Transistors Q_F and Q_G are used instead of diodes to allow for more precise control of the voltage drops applied to the base–emitter junctions of Q_{11} and Q_{12}, as we shall see shortly. Because the bias currents in Q_{11} and Q_{12} are set via precise control of their base–emitter voltages, we shall not assume V_{BE} to equal a generic V_f but shall instead compute V_{BE} exactly for each device in the circuit.

Figure 12.8

Output stage of the LM741 op-amp. The base–emitter junctions of Q_F and Q_G replace the diodes used in the class AB amplifier of Section 11.6.3. The entire circuit is driven at the base of Q_{12} via the *pnp* follower Q_{10}.

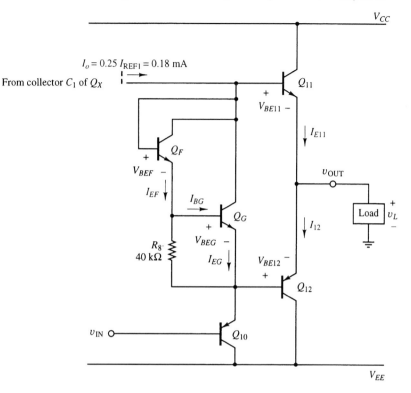

Bias Design

Analysis of the bias design of the output stage of Fig. 12.8 is straightforward if base currents are neglected with respect to emitter currents. We note that a bias current $I_o = 0.25 I_{REF1} \approx 180\,\mu A$ is provided by the split collector C_1 of Q_X, where $I_{REF1} = 0.73\,\text{mA}$ was computed in Eq. (12.2). This current from Q_X divides between Q_F and Q_G, then exits the circuit by way of Q_{10}. The way in which I_o is distributed between Q_F and Q_G will determine the base–emitter junction voltages of these devices, and hence the net voltage applied to the base–emitter junctions of output transistors Q_{11} and Q_{12}. As we now show, the division of I_o between Q_F and Q_G is controlled by the value of R_8.

We initially estimate the base–emitter junction voltage of Q_G to have the value $V_{BEG} \approx 0.6\,\text{V}$. With the base currents into Q_G and Q_F neglected compared to their emitter currents, the value of I_{EF} becomes

$$I_{EF} \approx \frac{V_{BEG}}{R_8} = \frac{0.6\,\text{V}}{40\,\text{k}\Omega} = 15\,\mu A \tag{12.36}$$

The remaining portion of I_o (with the base current into Q_{11} neglected) must flow through Q_G, so that $I_{EG} \approx 165\,\mu A$. For an assumed transistor scale current of $I_{EO} = 10^{-11}\,\text{mA}$, this value of I_{EG} requires a V_{BEG} of

$$V_{BEG} = \eta V_T \ln \frac{I_{EG}}{I_{EO}} = (1)(0.025\,\text{V}) \ln \frac{165\,\mu A}{10^{-11}\,\text{mA}} = 0.588\,\text{V} \tag{12.37}$$

This value is consistent with the value of 0.6 V originally assumed.

The value of V_{BEF} resulting from the emitter current (12.36) is given by

$$V_{BEF} = \eta V_T \ln \frac{I_{EF}}{I_{EO}} = (1)(0.025\,\text{V}) \ln \frac{15\,\mu A}{10^{-11}\,\text{mA}} = 0.528\,\text{V} \tag{12.38}$$

The sum of voltages appearing across the base–emitter junctions of Q_{11} and Q_{12} is equal to the sum $V_{BEF} + V_{BEG}$. Q_{11} and Q_{12} share the same current, hence $V_{BE11} = V_{EB12}$, and it follows that

$$V_{BE11} = V_{EB12} = \frac{V_{BEF} + V_{BEG}}{2} = \frac{0.588 \text{ V} + 0.528 \text{ V}}{2} = 0.558 \text{ V} \qquad \textbf{(12.39)}$$

For this value of base–emitter voltage, the bias current through both Q_{11} and Q_{12} for $v_{\text{IN}} = 0$ becomes

$$I_{E11} = I_{E12} = I_{EO}(e^{v_{BE}/\eta V_T} - 1) \approx (10^{-11} \text{ mA}) \exp\left(\frac{0.558 \text{ V}}{0.025 \text{ V}}\right) \approx 50 \,\mu\text{A} \qquad \textbf{(12.40)}$$

EXERCISE 12.5 Repeat the calculations leading to Eq. (12.39) without neglecting base currents. Assume the BJTs to have parameters $\beta_{\text{npn}} = 150$ and $\beta_{\text{pnp}} = 50$.

12.6 Choose a value for R_8 in Fig. 12.8 such that Q_{11} and Q_{12} have bias currents of about $10 \,\mu\text{A}$.

12.7 What bias current flows through Q_{11} and Q_{12} in Fig. 12.8 if resistor R_8 is removed, so that $R_8 = \infty$?

12.2.4 Complete Op-Amp Cascade

The cascade of amplifier stages described in the preceding sections gives the LM741 its characteristic input resistance, output resistance, gain, and common-mode rejection ratio. In this section, we examine the interconnection of these stages to form the complete LM741 op-amp cascade of Fig. 12.2. Note that this simplified version of the 741 omits several components that protect the op-amp output from short circuits or excessive output currents. This additional circuitry is examined in Problem 12.28.

Bias Summary

As illustrated in the preceding sections, the bias for the entire op-amp originates in the reference leg formed by Q_A and Q_E. This leg drives the current mirrors of each stage in the cascade. The current through Q_A is mirrored in Q_B, and again in Q_C and Q_D. The current through Q_D biases the differential input stage of Q_1–Q_7. The current in the reference leg of Q_A and Q_E is also mirrored in the split-collector device Q_X. One collector of Q_X biases the middle-gain stage of Q_8 and Q_9, and the other collector biases the output stage formed by Q_{11} and Q_{12}. Devices Q_F and Q_G are used to bias Q_{11} and Q_{12} for class AB operation. This bias scheme minimizes crossover distortion in the output stage, as discussed in Section 12.2.3.

Note that the bias value of v_{C9} (at the output of the middle-gain stage) is set by Q_{10} to approximately two V_{BE} drops below v_{OUT}. When the op-amp is connected in a negative-feedback configuration, the feedback network helps to set the bias value of v_{OUT}, and, in turn, the bias value of v_{C9}.

Summary of Gain Characteristics

The output current signal from the differential stage, taken from the collectors of Q_4 and Q_6, is fed directly to the input of the middle-gain stage at the base of Q_8. The latter is designed to have a high input resistance. The combination of these two stages results in the large overall differential-mode gain (12.35). Because the middle-gain stage also has a large output resistance, it cannot drive the output stage directly. To help alleviate this problem, the unity-gain *pnp* follower Q_{10} is connected between Q_9 and Q_{12} to act as an impedance buffer. This buffer minimizes interstage loading between the middle-gain stage and the output stage while not significantly degrading the large gain of the first two stages. The voltage gain of the complementary-pair output stage is also unity, hence the value $A_o = 2.7 \times 10^5$ computed in Eq. (12.35) describes the open-loop gain of the entire op-amp cascade.

Note that a capacitor C_C connects the output of the middle-gain stage back to its input. This connection is responsible for the internal compensation of the op-amp and directly sets the op-amp's open-loop frequency response. A detailed analysis of the function of this capacitor is provided in Section 12.2.5.

Output Saturation Levels

The output saturation levels of the op-amp are principally governed by the various transistors in the output stage. The v_{OUT} terminal will reach $V_{sat\text{-}pos}$ when Q_X saturates, hence the positive output limit is given by

$$
\begin{aligned}
V_{sat\text{-}pos} &= V_{CC} - V_{satX} - v_{BE11} \\
&= 15\,\text{V} - 0.2\,\text{V} - 0.7\,\text{V} \approx 14.1\,\text{V}
\end{aligned}
\tag{12.41}
$$

or about 0.9 V below V_{CC}. Note that $V_f = 0.7\,\text{V}$ is substituted for v_{BE11} in Eq. (12.41), rather than the value 0.558 V computed in Section 12.2.3, because Q_{11} is likely to be heavily conducting when the op-amp reaches its positive saturation limit. In any case, Eq. (12.41) serves as a guideline only.

The v_{OUT} terminal will reach $V_{sat\text{-}neg}$ when Q_{10} saturates. The output voltage under these conditions becomes

$$
\begin{aligned}
V_{sat\text{-}neg} &\approx V_{EE} + V_{sat} + V_{EB12} \\
&= -15\,\text{V} + 0.2\,\text{V} + 0.7\,\text{V} = -14.1\,\text{V}
\end{aligned}
\tag{12.42}
$$

or about 0.9 V above V_{EE}. These output saturation levels lie near the power-supply voltages V_{CC} and V_{EE}, adding credibility to the ideal op-amp approximation, which assumes the op-amp to saturate at V_{CC} and V_{EE}.

12.2.5 Frequency Compensation

Many integrated-circuit op-amps contain an internal capacitance that limits the overall frequency response of the cascade. Such frequency limitation, called *compensation*, is necessary to guarantee the stability of the op-amp under all negative-feedback conditions. The issue of feedback stability was addressed at length in Chapter 10. In this section, we examine the effects of the compensation capacitor in the LM741 and its role in producing a stable frequency-response profile.

Figure 12.9
Compensation capacitor connected between the input and the output of the middle gain stage.

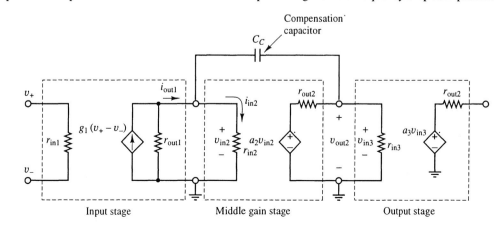

Input stage   Middle gain stage   Output stage

In the LM741 op-amp, the compensation capacitor is connected between the input and output terminals of the middle-gain stage. If the middle-gain stage is represented by a two-port Thévenin cell, the connection of the compensation capacitor appears as in Fig. 12.9, where the input stage and output stage of the op-amp are also represented by two-port cells. The middle-gain stage is an inverter, hence its gain a_2 is negative.

Application of Miller's theorem at the input port of the middle-gain stage yields the modified circuit of Fig. 12.10. In this latter circuit, the transverse capacitance C_C has been replaced by an equivalent parallel capacitance C_1 seen at the input port. As discussed in Chapter 9, the value of this equivalent parallel capacitance is equal to

$$C_1 = C_C \left(1 - \frac{\mathbf{V}_{out2}}{\mathbf{V}_{in2}}\right) \tag{12.43}$$

where $\mathbf{V}_{in2}$ and $\mathbf{V}_{out2}$ are phasors that represent v_{in2} and v_{out2} in the sinusoidal steady state.

Figure 12.10
Circuit as seen from the input port of the middle gain stage. The compensation capacitor C_C is represented by an equivalent capacitance C_1 in parallel with the input port.

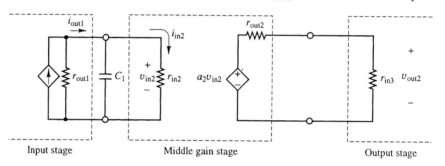

Input stage   Middle gain stage   Output stage

The effect of C_C on the op-amp's frequency response can be assessed via the dominant-pole technique of Section 9.3.2. Assuming that C_1 is the dominant pole, the ratio $\mathbf{V}_{out2}/\mathbf{V}_{in2}$ can be computed without regard for the effect of C_C on $\mathbf{V}_{out2}$. Under such conditions, the value of C_1 in Eq. (12.43) becomes

$$C_1 \approx C_C \left(1 - a_2 \frac{r_{in3}}{r_{out2} + r_{in3}}\right) \tag{12.44}$$

In a well-designed op-amp like the 741, r_{in3} will be much larger than r_{out2}. At the same time, the gain a_2 is negative, hence Eq. (12.44) can be simplified to

$$C_1 \approx C_C(1 + |a_2|) \tag{12.45}$$

Note that Miller multiplication produces a large C_1 without requiring a large C_C. This feature is advantageous in integrated-circuit designs because a small C_C occupies less total chip area. Computing the dominant pole caused by C_C requires knowledge of the small-signal resistance appearing in parallel with C_1 in Fig. 12.10. This resistance is equal to $r_{out1} \| r_{in2}$, so that the computed pole frequency of C_1 becomes

$$\omega_{C1} = \frac{1}{(r_{out1} \| r_{in2})C_1} = \frac{1}{(r_{out1} \| r_{in2})C_C(1 + |a_2|)} \tag{12.46}$$

The dominant pole of C_1 will initiate a -20 dB/decade decline in the op-amp's open-loop frequency response, beginning at the pole frequency ω_{C1}. If ω_{C1} is located well below the poles caused by the internal device capacitances of the circuit, the response will be sufficient to produce stable gain and phase margins under all feedback conditions.

EXAMPLE 12.3

In the LM741, the compensation capacitor C_C is equal to 30 pF and is connected across the middle-gain stage, as in Fig. 12.2. Analyze the circuit and estimate the location of the dominant-pole frequency $f_1 = \omega_{C1}/2\pi$. Assume that r_{in3}, seen looking into the base of Q_{10}, approaches an infinite value. Assign the typical parameters $\beta_o = 150$ and $V_A = 50$ V to the npn BJTs. It can be shown that these parameters yield an output resistance r_{out1} for the differential input stage of 6.7 MΩ (see Problem 12.7).

Solution

As previously shown, the bias current I_{REF1} is given by

$$I_{REF1} = \frac{V_{CC} - V_{EE} - 2V_f}{R_5} = \frac{15\,\text{V} - (-15\,\text{V}) - 1.4\,\text{V}}{39\,\text{k}\Omega} \approx 0.73\,\text{mA} \qquad (12.47)$$

The resulting current I_{C9}, produced by collector C_2 of the current mirror Q_X, becomes

$$I_{C9} \approx 0.75 I_{REF} = 0.55\,\text{mA} \qquad (12.48)$$

For this value of I_{C9}, the value of I_{C8} is equal to about 13.5 μA if base currents are neglected, as computed in Eq. (12.24). For these values of I_{C8} and I_{C9}, the values of $r_{\pi 8}$ and $r_{\pi 9}$ become

$$r_{\pi 8} = \frac{\beta_{o8}}{g_{m8}} = \frac{\beta_{o8}\eta V_T}{I_{C8}} = \frac{(150)(1)(0.025\,\text{V})}{13.5\,\mu\text{A}} \approx 278\,\text{k}\Omega \qquad (12.49)$$

and

$$r_{\pi 9} = \frac{\beta_{o9}}{g_{m9}} = \frac{\beta_{o9}\eta V_T}{I_{C9}} = \frac{(150)(1)(0.025\,\text{V})}{0.55\,\text{mA}} \approx 6.82\,\text{k}\Omega \qquad (12.50)$$

For these values of resistance, the incremental input resistance of the middle-gain stage of Fig. 12.7 becomes

$$r_{in2} = r_{\pi 8} + (\beta_{o8} + 1)\{R_6 \| [r_{\pi 9} + (\beta_{o9} + 1)R_7]\}$$
$$\equiv 278\,\text{k}\Omega + 151[50\,\text{k}\Omega \| (6.8\,\text{k}\Omega + 151 \cdot 100\,\Omega)] \approx 2.58\,\text{M}\Omega \qquad (12.51)$$

If Q_X has an Early voltage of 50 V, the value of r_{oX} seen by Q_9 will be equal to

$$r_{out2} = r_{oX} = \frac{V_A}{I_{C3}} \approx \frac{50\,\text{V}}{0.55\,\text{mA}} = 90.9\,\text{k}\Omega \qquad (12.52)$$

as computed previously.

The current-to-voltage conversion ratio $a_r = v_{out2}/i_{in2}$ of the middle-gain stage is given by Eqs. (12.28) and (12.32) and is equal to -1.44×10^6 V/mA. The current i_{in2} is the signal current flowing into the input terminal of the middle-gain stage. Since $v_{in2} = i_{in2}r_{in2}$, the voltage gain a_2 in Fig. 12.10 can be expressed in terms of a_r in the following way:

$$a_2 = \frac{v_{out2}}{v_{in2}} = \frac{a_r i_{in2}}{i_{in2}r_{in2}} \equiv \frac{a_r}{r_{in2}} = \frac{-1.44 \times 10^6\,\text{V/mA}}{2.58 \times 10^3\,\text{k}\Omega} \approx -558 \qquad (12.53)$$

For this value of a_2, the Miller-multiplied capacitance C_1 becomes

$$C_1 = C_C(1 + |a_2|) = 30\,\text{pF}(1 + 558) \approx 16.8\,\text{nF} \qquad (12.54)$$

The dominant-pole frequency of C_1, as given by Eq. (12.46), becomes

$$f_1 = \frac{\omega_{C1}}{2\pi} = \frac{1}{2\pi}\frac{1}{(r_{out1}\|r_{in2})C_1} = \frac{1}{2\pi(6.7\,\text{M}\Omega\|2.58\,\text{M}\Omega)(16.8\,\text{nF})} \approx 5.1\,\text{Hz} \qquad (12.55)$$

It can be shown (see discussion on pole splitting to follow) that the higher-order poles of the LM741 lie well above the unity-gain frequency f_o, so that the Bode plot of the open-loop response can be drawn as in Fig. 12.11. We note that the gain–bandwidth product of the op-amp must remain constant, hence the unity-gain frequency f_o can be computed from the product $GB = A_o f_1 = (2.7 \times 10^5)(5.1\,\text{Hz}) \approx 1.4\,\text{MHz}$, so that $f_o = GB/A(f_o) = 1.4\,\text{MHz}$.

Figure 12.11

Magnitude Bode plot of the open-loop response of the LM741 op-amp based on the assumed parameters of Example 12.3.

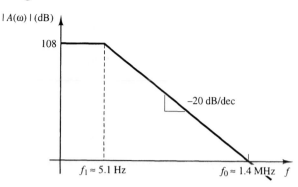

12.2.6 Pole Splitting

In the analysis of the preceding section, the dominant pole of the LM741 created by C_C in the middle-gain stage was computed using Miller's theorem. The higher-order, nondominant poles of this stage also play an important role in determining amplifier stability. As illustrated in Chapter 10, higher-order poles affect the location of ω_{180}, the frequency at which the phase shift of the open-loop gain reaches $-180°$. An exact computation of the nondominant pole frequencies of the middle-gain stage is difficult because they are affected not only by C_C, but also by the internal capacitances of the various BJTs in the circuit. Their locations may be estimated, however, by modeling all device capacitances as two parallel capacitors C_1 and C_2 connected across the input and output ports, respectively, of the middle-gain stage.

Without C_C in place, C_1 and C_2 would cause poles to appear at $\omega_1 = 1/r_1 C_1$ and $\omega_2 = 1/r_2 C_2$, where r_1 and r_2 represent the small-signal Thévenin resistances seen by C_1 and C_2, respectively. A detailed SPICE analysis of the circuit with C_C absent shows the pole of C_2 to lie at about 320 kHz. At this frequency, the gain of the op-amp falls to approximately 80 dB. The latter is much too high to guarantee a stable phase margin for all values of feedback factor β. In the LM741, the compensation capacitor alleviates this situation.

Figure 12.12

Model of the LM741 middle gain stage in which pole splitting is illustrated.

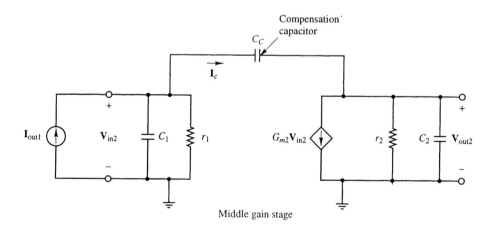

Middle gain stage

For the purpose of estimating the effect of C_C on the poles of the amplifier, it is advantageous to model the output of the middle-gain stage by its small-signal Norton equivalent, as shown in Fig. 12.12. The short-circuit current of the dependent source is set to the value $G_{m2}\mathbf{V}_{in2} = -a_2\mathbf{V}_{in2}/r_{out2}$, so that

$$G_{m2} = -a_2/r_{out2} = -(-558/90.9\,\text{k}\Omega) \approx 6.1\,\text{mA/V} \qquad (12.56)$$

Analysis of this circuit proceeds in much the same way as the that of Example 9.9 in Section 9.3.4, where the pole frequencies caused by the transverse capacitance C_μ in the BJT inverter were computed. In this case, applying KCL to the input node results in

$$\mathbf{I}_{out1} = \mathbf{V}_{in2}(j\omega C_1) + \mathbf{V}_{in2}/r_1 + (\mathbf{V}_{in2} - \mathbf{V}_{out2})(j\omega C_C) \tag{12.57}$$

where the phasor $\mathbf{I}_{out1}$ represents the output current of the preceding differential stage. Similarly, applying KCL to the output node results in

$$(\mathbf{V}_{in2} - \mathbf{V}_{out2})(j\omega C_C) = G_{m2}\mathbf{V}_{in2} + \mathbf{V}_{out2}/r_2 + \mathbf{V}_{out2}(j\omega C_2) \tag{12.58}$$

Solving these equations for the ratio $\mathbf{V}_{out2}/\mathbf{I}_{out1}$ results, after some algebra, in

$$\frac{\mathbf{V}_{out2}}{\mathbf{I}_{out1}} =$$

$$\frac{G_{m2}r_1r_2(1 - j\omega C_C/G_{m2})}{1 + j\omega[r_1(C_1 + C_C) + r_2(C_2 + C_C) + G_{m2}r_1r_2C_C] + (j\omega)^2 r_1 r_2[C_1 C_2 + C_C(C_1 + C_2)]} \tag{12.59}$$

This equation is similar in form to Eq. (9.128) derived in Chapter 9. The poles of $\mathbf{V}_{out2}/\mathbf{I}_{out1}$ are located at the frequencies at which the denominator of Eq. (12.59) equals zero. The latter can be expressed as a product of binomials:

$$\left(1 + \frac{j\omega}{\omega_1}\right)\left(1 + \frac{j\omega}{\omega_2}\right) = 1 + j\omega\left(\frac{1}{\omega_1} + \frac{1}{\omega_2}\right) + \frac{(j\omega)^2}{\omega_1\omega_2} \tag{12.60}$$

Following the procedure of Example 9.9, we next assume the pole ω_1 associated with Miller multiplication to be dominant, so that $\omega_1 \gg \omega_2$. Equation (12.60) can then be written in the approximate form

$$\left(1 + \frac{j\omega}{\omega_1}\right)\left(1 + \frac{j\omega}{\omega_2}\right) \approx 1 + \frac{j\omega}{\omega_1} + \frac{(j\omega)^2}{\omega_1\omega_2} \tag{12.61}$$

Equating ω_1 with the reciprocal of the coefficient multiplying $j\omega$ in the denominator of Eq. (12.59) leads to

$$\omega_1 \approx \frac{1}{r_1(C_1 + C_C) + r_2(C_2 + C_C) + G_{m2}r_1r_2C_C} \approx \frac{1}{G_{m2}r_1r_2C_C} \tag{12.62}$$

where the second approximation is valid because $G_{m2}r_1r_2$ is much larger than either r_1 or r_2. Similarly, equating $\omega_1\omega_2$ with the reciprocal of the coefficient multiplying $(j\omega)^2$ in the denominator of Eq. (12.59) leads to

$$\omega_2 \approx \frac{G_{m2}C_C}{C_1 C_2 + C_C(C_1 + C_2)} \tag{12.63}$$

Without the compensation capacitor C_C in place, the two poles of the middle-gain stage would be located at $\omega_1 = 1/r_1C_1$ and $\omega_2 = 1/r_2C_2$. With C_C in place, the value of ω_1 *decreases* with increasing C_C or G_{m2}. Conversely, the value of ω_2 *increases* with C_C or G_{m2}, reaching a limiting value of $G_{m2}/(C_1 + C_2)$ for very large C_C.

The moving of preexisting poles in opposite directions caused by the addition of a transverse capacitance is known as *pole splitting*. In compensated op-amps like the LM741, pole splitting moves the amplifier's nondominant pole well above the unity-gain frequency, so that the open-loop response of the op-amp can be approximated as having only a single dominant pole at about 5 Hz. In the LM741, this feature results in a response that is stable for all values of β up to unity. A detailed analysis of the LM741 using SPICE shows the second pole frequency $f_2 = \omega_2/2\pi$ to lie at about 100 MHz, which is well above the unity-gain frequency of 1.4 MHz.

12.2.7 Origin of Slew-Rate Limitation

The concept of the slew-rate limit and its effect on op-amp performance was introduced in Chapter 2 (see Section 2.6.4). In this section, the physical origin of the slew-rate limit is examined in the context of the LM741. The slew-rate limit is related to the compensation capacitor C_C connected across the input and output ports of the middle-gain stage. The internal circuitry of the op-amp can charge this capacitor no faster than a fixed, finite rate. If the op-amp is externally driven in a way that would exceed this charging rate, the op-amp output will reach its slew-rate limit.

The origin of the maximum capacitor charging rate can be identified by considering the two-port cell representation of the middle-gain stage shown in Fig. 12.13. For the purpose of discussion, suppose the op-amp to operate under open-loop conditions. If a step voltage is applied to the op-amp input terminals, the output current i_{out} of the differential input stage will also be driven toward a step change in magnitude. The resulting magnitude of i_{out} under such conditions can be no greater than the total current available from the bias network of the input stage. In the case of the 741, this bias network consists of a current mirror of fixed value. If the maximum magnitude of i_{out} is reached, the input stage will function as a fixed current source of value I_{MAX} driving the middle-gain stage. Under such conditions, the combination of capacitor C_C and middle-gain stage will function as an integrator driven by a constant-current source, as depicted in Fig. 12.13. The rate of change of v_{OUT2} for this latter circuit is equal to

$$\left| \frac{dv_{OUT2}}{dt} \right| = \left| \frac{I_{MAX}}{C_C} \right| \tag{12.64}$$

where I_{MAX} can be positive or negative, depending on the polarity of the applied input voltage. The value (12.64) represents the maximum time rate of change of v_{OUT2}, regardless of the external connections to the op-amp. If the input stage is driven in such a way that its i_{out} reaches I_{MAX} or $-I_{MAX}$, the op-amp output will reach its slew-rate limit. This description presents a simplified picture of the slew-rate limiting process, but provides a means of estimating the value of the maximum possible slew rate.

Figure 12.13
Model of the input stage, compensation capacitor, and middle gain stage under slew-rate limited conditions.

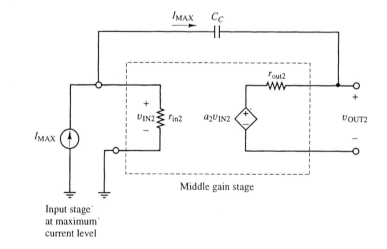

The input stage of the LM741, shown as part of Fig. 12.2, is biased by a current mirror and Widlar source that produces the current I_{o2}. As illustrated in Section 12.2.1 for the component values in Fig. 12.2, I_{o2} will be equal to about 18.4 μA, where $I_{o2} \approx I_{REF2}$. Under the current-limiting conditions described above, transistors Q_2 and Q_4 will supply all of I_{o2} as positive i_{out} for $v_- \gg v_+$. Similarly, Q_1 and Q_3 will supply all of I_{o2} to Q_5 for $v_+ \gg v_-$; this current will be reflected in Q_6 and result in an i_{out} of $-I_{o2}$. In either case, I_{o2} represents the maximum positive or negative value that i_{out} can have when the input is driven by a large signal. If i_{out} is forced to either limiting value of $\pm I_{o2}$ at any time, the output of the middle-gain stage of Fig. 12.13 will be limited to the integration rate:

$$\left|\frac{dv_{OUT2}}{dt}\right|_{max} = \frac{I_{o2}}{C_C} = \frac{18.4\,\mu A}{30\,pF} \approx 0.61\ V/\mu s \qquad (12.65)$$

This value falls within the range of the manufacturer-specified slew-rate for the LM741 op-amp.

12.2.8 Offset-Null Adjustment

The internal circuitry of an op-amp usually contains unavoidable imbalances and asymmetries that lead to nonzero values of input offset voltage and input offset current. The effect of these parameters on op-amp performance was discussed in Sections 2.6.2 and 2.6.3. In this section, one technique for counteracting these imbalances in the LM741 is presented.

Figure 12.14
Connection of R_X to the offset-null terminals of the LM741 op-amp.

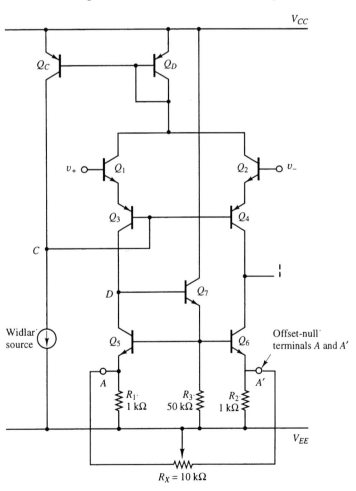

Many op-amps, including the LM741, are equipped with a pair of external *offset-null* terminals that can be used to "fine-tune" the differential input stage. The offset-null terminals in the LM741 input stage are located as shown in Fig. 12.14. If an external potentiometer R_X is connected between these terminals and the V_{EE} supply, as shown in the figure, an extra resistance will appear in parallel with each of the resistors R_1 and R_2. If the arm of R_X is centered, R_1 and R_2 will each be connected to an additional parallel resistance $R_X/2$. If the arm of R_X is adjusted to an off-center position, R_1 and R_2 will see different parallel resistances and an asymmetry will be introduced in the circuit. If the arm is properly adjusted, the asymmetry will produce an artificial offset component that will cancel the effects of the input offset voltage and input bias current produced by the stage itself. In practice, the op-amp output must be observed with a measuring instrument when the op-amp is connected to its feedback network and no input signals are applied. The balancing potentiometer R_X is adjusted until the op-amp output is forced to zero. The value of R_X required for a given op-amp depends on the structure of its differential input stage and is usually specified by the op-amp manufacturer.

> ***Discussion.*** The numerical values computed in this section are based on assumed values for parameters of the various transistors in the LM741 op-amp. The actual gain, frequency response, input resistance, and output resistance of any given 741 are likely to vary from part to part and even from manufacturer to manufacturer. In general, each device manufacturer publishes a data sheet on which the minimum and/or maximum guaranteed values of the various op-amp parameters are indicated. These minimum and maximum values are computed based on the expected range of transistor parameters for the devices in the circuit.
>
> ■

12.3 CASE STUDY: A SIMPLE CMOS OPERATIONAL AMPLIFIER

As discussed in previous chapters, the CMOS process produces circuits made entirely from n- and p-channel MOSFET transistors. The absence of resistors or other semiconductor components on CMOS integrated circuits makes them easy to fabricate and design. In this section, we examine the various construction and operating features of a typical CMOS operational amplifier. The gain of an analog CMOS amplifier is much smaller than that of its BJT counterpart. Several CMOS stages must be cascaded to achieve large overall gain. Interstage loading in multistage CMOS circuits is minimal, however, because each stage in the cascade drives the purely capacitive load of another MOSFET gate. CMOS op-amps are crucial to a class of networks called *switched capacitor filters*, to be discussed in Chapter 13. Note that CMOS op-amps are not generally available as "off-the-shelf" components. Rather, each is custom-designed and fabricated to meet the needs of a specific application.

On a given CMOS integrated circuit, the value of the threshold voltage V_{TR} cannot be accurately predicted, but all devices fabricated on the same IC will have the *same* value of V_{TR}. Similarly, the value of K for a given device cannot be accurately predicted, but the ratio of the K parameters of any two devices on the same IC can be set precisely by adjusting device geometry at fabrication time. Fortunately, most CMOS circuit characteristics depend on ratios of K parameters. The design of a CMOS amplifier thus involves the specification of length-to-width ratios for each device in the circuit.

One very popular CMOS op-amp configuration, based on the CMOS differential amplifier of Section 8.4.3, is illustrated in Fig. 12.15. In the op-amp version of the circuit, the differential input stage is buffered by a simple inverting output stage consisting of Q_5 and Q_C. In the sections that follow, the various properties of the amplifier, including its bias, small-signal gain, and frequency response, are examined in more detail.

Figure 12.15
A simple CMOS op-amp. The differential device Q_2 is actively loaded by r_{o4} of Q_4, and the output device Q_5 is loaded by r_{oC} of Q_C.

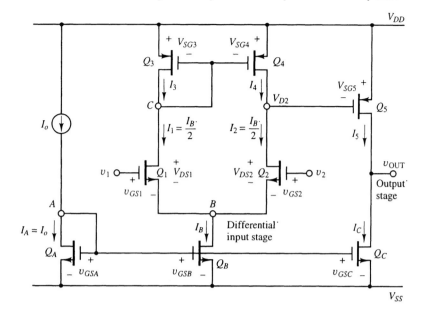

12.3.1 Bias Design

The two stages of the CMOS amplifier of Fig. 12.15 are biased by a current mirror consisting of Q_A, Q_B, and Q_C. The reference current I_o feeding Q_A is created either by an on-chip circuit or fed to Q_A via an external terminal. MOSFET Q_C serves a dual purpose—one as the current mirror biasing Q_5, the other as the active load to Q_5. The n-channel devices Q_A, Q_B, and Q_C are assumed to have the same threshold voltage V_{TRN}, since they undergo the same fabrication process. Similarly, these devices share the same process-dependent parameters μ_e, ϵ_{ox}, and t_{ox}, allowing relative values of the conductance parameter $K = (\mu_e \epsilon_{ox}/2t_{ox})(W/L)$ to be set for each device by choosing its width W and length L.

If the effects of r_{oA} through r_{oC} on the bias currents are neglected, the gate-to-source voltage of each MOSFET can be expressed as $|V_{GS}| = (I_D/|K|)^{1/2} + |V_{TR}|$ (absolute values required for p-channel devices). Given the condition $V_{GSA} = V_{GSB}$, the bias current I_B flowing through Q_B can be found from the equation

$$(I_A/K_A)^{1/2} + V_{TRN} = (I_B/K_B)^{1/2} + V_{TRN} \qquad (12.66)$$

Solving for I_B results in

$$I_B = \frac{K_B}{K_A} I_A = \frac{W_B/L_B}{W_A/L_A} I_o \qquad (12.67)$$

where $I_A = I_o$ is the current through Q_A. Similarly,

$$I_C = \frac{K_C}{K_A} I_A = \frac{W_C/L_C}{W_A/L_A} I_o \qquad (12.68)$$

The bias currents I_B and I_C, set via the W/L ratios of Q_A, Q_B, and Q_C, must be chosen such that MOSFETs Q_1 through Q_5 are biased in the constant-current region. We note that Q_3, with its gate connected to its drain, automatically operates in the constant-current region with $V_{DS3} = V_{GS3}$. (These voltages on a p-channel device are still measured from drain to source and from gate to source, respectively, even though the device appears "upside down" on the circuit diagram.) If the differential stage is to be symmetrical, Q_4 should be biased such that

$V_{DS4} = V_{DS3}$. In this case, V_{DS4} is equal to V_{GS5}, hence the bias current through Q_5 must be adjusted such that $V_{GS5} = V_{GS3}$. As implied by Eqs. (12.66) through (12.68), the requirement $V_{GS5} = V_{GS3}$ can be met by satisfying the condition

$$\frac{I_3}{W_3/L_3} = \frac{I_5}{W_5/L_5} \tag{12.69}$$

If Q_1 and Q_2 are matched (i.e., have the same W/L ratios), the current I_3 will equal $I_B/2$, where I_B is given by Eq. (12.67). Similarly,

$$I_5 = I_C = \frac{W_C/L_C}{W_A/L_A}I_o \tag{12.70}$$

The constraint (12.69) is thus equivalent to

$$\frac{\frac{1}{2}\frac{W_B/L_B}{W_A/L_A}I_o}{W_3/L_3} = \frac{\frac{W_C/L_C}{W_A/L_A}I_o}{W_5/L_5} \tag{12.71}$$

where $I_3 = I_B/2$ is derived from Eq. (12.67). Equation (12.71) can be simplified to the following expression:

$$\frac{W_3}{L_3} = \frac{1}{2}\frac{(W_B/L_B)(W_5/L_5)}{W_C/L_C} \tag{12.72}$$

If the W/L ratios of Q_B, Q_C, Q_3, and Q_5 are chosen to satisfy Eq. (12.72), then V_{GS5} will equal V_{GS3}, and the node voltage V_{D2} will be biased at the value

$$V_{D2} = V_{DD} - V_{SG3} = V_{DD} - (I_B/2|K_3|)^{1/2} + |V_{TRP}| \tag{12.73}$$

The bias values of the various other node voltages in the circuit also can be computed using the relation $|V_{GS}| = (I_D/|K|)^{1/2} + |V_{TR}|$. Specifically, the voltage of node A becomes

$$V_A = V_{SS} + V_{GSA} = V_{SS} + (I_o/K_A)^{1/2} + V_{TRN} \tag{12.74}$$

Similarly, with $v_1 = v_2 = 0$, the bias voltage of node B becomes

$$V_B = 0 - V_{GS1} = -(I_B/2K_1) - V_{TRN} \tag{12.75}$$

The bias value of the v_{OUT} node under open-loop, no-load conditions is determined by the values of v_{DS5} and v_{DSC} at which the constant-current v–i characteristics of Q_5 and Q_C intersect. In general, this no-load bias point is not important, because the load and the feedback connections play a more significant role in determining the bias value of v_{OUT}.

EXAMPLE 12.4

Suppose that the CMOS amplifier of Fig. 12.15 is fabricated with the W/L ratios indicated in Table 12.1. The op-amp is powered by supply voltages $V_{DD} = 5\,\text{V}$ and $V_{SS} = -5\,\text{V}$. Find the bias values of the various voltages and currents in the circuit for the case $I_o = 20\,\mu\text{A}$, $V_{TRN} = |V_{TRP}| = 1\,\text{V}$, $\mu_e C_{\text{ox}} = 20\mu\text{A/V}^2$, $\mu_h C_{\text{ox}} = 10\mu\text{A/V}^2$, and $V_{AN} = V_{AP} = 100\,\text{V}$.

	Q_A	Q_B	Q_C	Q_1	Q_2	Q_3	Q_4	Q_5
$W(\mu\text{m})$	10	20	10	20	20	5	5	90
$L(\mu\text{m})$	10	100	10	5	5	11	11	20

Table 12.1. W and L values for the CMOS circuit of Fig. 12.15

Solution

• Find the bias currents through each leg of the current mirror

Given that $I_A = I_o = 20\,\mu A$, the various bias currents become, via Eqs. (12.67) and (12.68),

$$I_B = \frac{W_B/L_B}{W_A/L_A} I_o = \frac{20/100}{10/10} 20\,\mu A = 4\,\mu A \qquad (12.76)$$

and

$$I_C = \frac{W_C/L_C}{W_A/L_A} I_o = \frac{10/10}{10/10} 20\,\mu A = 20\,\mu A \qquad (12.77)$$

• Find the bias value of node voltage V_C

This node voltage will be equal to $V_{DD} - V_{SG3}$. Computing the gate-to-source voltage $V_{SG3} \equiv -V_{GS3}$ requires knowledge of $|K_3|$. From the given data,

$$|K_3| = \frac{1}{2}\frac{W_3}{L_3}\mu_h C_{ox} = \frac{5}{2(11)}(10\,\mu A/V^2) = 2.27\mu A/V^2 \qquad (12.78)$$

Using this value of $|K_3|$, and the value $I_3 = I_B/2 = 2\,\mu A$, it follows that

$$\begin{aligned} V_{SG3} &= (I_3/|K_3|)^{1/2} + |V_{TRP}| \\ &= [(2\,\mu A)/(2.27\mu A/V^2)]^{1/2} + 1\,V \approx 1.9\,V \end{aligned} \qquad (12.79)$$

and $V_C = V_{DD} - V_{SG3} = 5\,V - 1.9\,V = 3.1\,V$.

• Find the bias value of node voltage V_B

For the W/L values indicated in Table 1, the K parameters of Q_1 and Q_2 become

$$K_1 = K_2 = \frac{1}{2}\frac{W_1}{L_1}\mu_e C_{ox} = \frac{20}{2(5)}(20\,\mu A/V^2) = 40\,\mu A/V^2 \qquad (12.80)$$

With $I_1 = I_2 = I_B/2 = 2\,\mu A$, and with v_1 and v_2 held at dc ground by input signal sources, it follows that

$$\begin{aligned} V_{GS1} &= V_{GS2} = (I_1/K_1)^{1/2} + V_{TRN} \\ &= [(2\,\mu A)/(40\,\mu A/V^2)]^{1/2} + 1\,V \approx 1.2\,V \end{aligned} \qquad (12.81)$$

where $V_B = 0 - V_{GS1} = -1.2\,V$.

• Find the bias value of node voltage V_{D2}

The value of this node voltage is determined by V_{SG5}, which is, in turn, determined by the current

$$I_5 = I_C = \frac{W_C/L_C}{W_A/L_A} I_o = \frac{10/10}{10/10}(20\,\mu A) = 20\,\mu A \qquad (12.82)$$

and the conductance parameter

$$|K_5| = \frac{1}{2}\frac{W_5}{L_5}\mu_h C_{ox} = \frac{90}{2(20)}(10\,\mu A/V^2) = 22.5\,\mu A/V^2 \qquad (12.83)$$

Specifically, from the v–i equation of Q_5, it follows that

$$\begin{aligned} V_{SG5} &= (I_5/|K_5|)^{1/2} + |V_{TRP}| \\ &= [(20\,\mu A)/(22.5\,\mu A/V^2)]^{1/2} + 1\,V \approx 1.9\,V \end{aligned} \qquad (12.84)$$

Note that this voltage is identical to V_{SG3}, as computed in Eq. (12.79), because the W/L ratios of Q_B, Q_C, Q_3, and Q_5 satisfy Eq. (12.72). Finally, from KVL,

$$V_{D2} = V_{DD} - V_{SG5} = 5\,V - 1.9\,V = 3.1\,V \qquad (12.85)$$

which is identical to the previously computed voltage of node C.

12.3.2 Small Signal Differential-Mode Performance

The CMOS amplifier of Fig. 12.15 will respond to both differential- and common-mode signals. Its gain under differential-mode excitation can be easily derived using the results of Section 8.4.3. In this case, the incremental output resistance r_{o4} of Q_4 acts as the pull-up load to Q_2. As noted in Eq. (8.187), the differential-mode gain observed at the drain of Q_2, as enhanced by the current-mirror action of Q_3 and Q_4, becomes

$$A_{\text{dm-se2}} = \frac{v_{d2}}{v_{\text{idm}}} = g_{m2}(r_{o2}\|r_{o4}) \tag{12.86}$$

where $v_{\text{idm}} = v_1 - v_2$. Because r_{o4} is comparable to r_{o2}, both resistances are included in the equation.

The signal v_{d2} from the drain of Q_2 is fed to the gate of Q_5. This latter transistor acts as a p-channel inverter actively loaded by r_{oC} of Q_C. The gain of the second stage thus becomes

$$\frac{v_{\text{out}}}{v_{d2}} = -g_{m5}(r_{oC}\|r_{o5}) \tag{12.87}$$

Note that r_{o5} is included in this equation because it is comparable to r_{oC}.

The loading of the v_{d2} node by the gate of Q_5 is negligible, because the latter presents a pure capacitive load to the former. The overall signal gain of the two-stage cascade thus can be determined by simply multiplying together the individual gain equations (12.86) and (12.87), yielding

$$A_{\text{dm}} = \frac{v_{\text{out}}}{v_{\text{idm}}} = -g_{m2}g_{m5}(r_{o2}\|r_{o4})(r_{oC}\|r_{o5}) \tag{12.88}$$

We note that $g_m = 2\sqrt{KI_D}$ for a MOSFET in the constant-current region, and that $r_o = V_A/I_D$, where V_A is the MOSFET's Early voltage. Given that $I_2 = I_4$ and $I_5 = I_C$, Eq. (12.88) can be expressed in the alternative form

$$\begin{aligned}
A_{\text{dm}} &= -(2\sqrt{K_2 I_2})(2\sqrt{K_5 I_5})\frac{V_{A2}}{I_2}\left\|\frac{V_{A4}}{I_4}\times\frac{V_{AC}}{I_C}\right\|\frac{V_{A5}}{I_5} \\
&= -4\sqrt{K_2 K_5}\sqrt{I_2 I_5}\frac{(V_{AN}/I_2)(V_{AP}/I_2)}{V_{AN}/I_2 + V_{AP}/I_2}\frac{(V_{AN}/I_5)(V_{AP}/I_5)}{V_{AN}/I_5 + V_{AP}/I_5} \\
&= -4\sqrt{K_2 K_5}\sqrt{I_2 I_5}\frac{1}{I_2}\frac{1}{I_5}\left(\frac{V_{AN}V_{AP}}{V_{AN}+V_{AP}}\right)^2 \\
&= -4\left(\frac{K_2 K_5}{I_2 I_5}\right)^{1/2}\left(\frac{V_{AN}V_{AP}}{V_{AN}+V_{AP}}\right)^2
\end{aligned} \tag{12.89}$$

where V_{AN} and V_{AP} are the Early voltages of the n-channel and p-channel devices, respectively. If $V_{AN} = V_{AP} = V_A$, the last term in Eq. (12.89) becomes $(V_A^2/2V_A)^2 = V_A^2/4$, so that the equation reduces to

$$A_{\text{dm}} = -V_A^2\left(\frac{K_2 K_5}{I_2 I_5}\right)^{1/2} \tag{12.90}$$

Note that A_{dm} becomes larger as I_2 and I_5 are reduced. This effect occurs because the incremental output resistances r_{o2}, r_{o4}, r_{oC}, and r_{o5} of Q_2, Q_4, Q_C, and Q_5 become larger with decreasing magnitude of drain current I_D.

As these results show, the circuit designer has several degrees of freedom in setting the gain of a CMOS op-amp. Given knowledge of the process-dependent parameters V_{TR}, V_A, and $\mu\epsilon_{\text{ox}}/t_{\text{ox}} \equiv \mu C_{\text{ox}}$ for n-channel and p-channel devices, where $K = \mu C_{\text{ox}}W/(2L)$, the open-loop gain of the amplifier can be set by appropriately choosing the W/L ratios of each device.

EXAMPLE 12.5

Using the circuit diagram of Fig. 12.15 as a guide, design a CMOS op-amp that has an open-loop gain of approximately $-50{,}000$. Given that the layout of the circuit has already been determined, the design consists of simply specifying I_o and choosing the W/L values for each MOSFET? Suppose that the IC fabrication process yields devices with parameters $V_{TRN} = 1\,\text{V}$, $V_{TRP} = -1\,\text{V}$, $V_{AN} = V_{AP} = V_A = 100\,\text{V}$, $\mu_e \epsilon_{ox}/t_{ox} \equiv \mu_e C_{ox} = 20\,\mu\text{A/V}^2$ (for n-channel devices), and $\mu_h C_{ox} = 10\,\mu\text{A/V}^2$ (for p-channel devices). Note that $\mu_e C_{ox} = 2\mu_h C_{ox}$, because $\mu_e = 2\mu_h$.

Solution

• Assess the goals of the problem

We must choose the W/L values for each MOSFET such that the differential-mode gain, as given by Eq. (12.90), is approximately equal to $-50{,}000$. Many different combinations will do the job, hence some of the choices will be arbitrary.

• Choose a design strategy

Substitution of $K = \mu C_{ox}/2$ into Eq. (12.90) yields the following expression for the gain in terms of V_A and the various W/L ratios:

$$A_{dm} = -V_A^2 \left[\frac{(\mu_e C_{ox}/2)(\mu_h C_{ox}/2)(W_2/L_2)(W_5/L_5)}{I_2 I_5} \right]^{1/2} \qquad (12.91)$$

The required gain can be achieved by choosing appropriate values for W, L, and I_D for each MOSFET.

Substitution of the known values for V_A and μC_{ox} into Eq. (12.91) results in

$$
\begin{aligned}
A_{dm} &= \frac{-(100\,\text{V})^2}{2}[(20\,\mu\text{A/V}^2)(10\,\mu\text{A/V}^2)]^{1/2}\left(\frac{W_2}{L_2}\frac{W_5}{L_5}\frac{1}{I_2 I_5}\right)^{1/2} \\
&= -(7.07 \times 10^4\,\mu\text{A})\left(\frac{W_2}{I_2 L_2}\frac{W_5}{I_5 L_5}\right)^{1/2}
\end{aligned}
\qquad (12.92)
$$

• Choose W/L and I_D for each MOSFET

If the factors of W/L are set to the values listed in Table 12.1, and if I_o is again chosen to be $20\,\mu\text{A}$, the circuit will be properly biased with $I_B = 4\,\mu\text{A}$ and with all devices in the constant-current region. Note that the current I_B in Q_B will divide equally between Q_1 and Q_2, so that $I_2 = I_B/2 = 2\,\mu\text{A}$. Conversely, the current through Q_C will be pulled exclusively through Q_5, so that $I_5 = I_C = 20\,\mu\text{A}$.

• Evaluate the design and revise if necessary

Substitution of the chosen W/L and I_D values into Eq. (12.92) yields a gain of

$$A_{dm} = -(7.07 \times 10^4\,\mu\text{A})\left[\frac{20\,\mu\text{m}}{(2\,\mu\text{A})(5\,\mu\text{m})}\frac{90\,\mu\text{m}}{(20\,\mu\text{A})(20\,\mu\text{m})}\right]^{1/2} = -4.74 \times 10^4 \qquad (12.93)$$

This value of gain falls somewhat short of the design goal of $A_{dm} = -5 \times 10^4$. Decreasing the reference current to $I_o = 19\,\text{mA}$ while preserving all W/L ratios results in $I_2 = I_B/2 = 1.9\,\text{mA}$ and $I_5 = 19\,\text{mA}$, so that

$$A_{dm} = -(7.07 \times 10^4\,\mu\text{A})\left[\frac{20\,\mu\text{m}}{(1.9\,\mu\text{A})(5\,\mu\text{m})}\frac{90\,\mu\text{m}}{(19\,\mu\text{A})(20\,\mu\text{m})}\right]^{1/2} = -5 \times 10^4 \qquad (12.94)$$

12.3.3 Common-Mode Performance

The common-mode gain of the CMOS amplifier of Fig. 12.15 can be found using the analysis technique of Section 8.4.3. In this case, if Q_B has an incremental output resistance of r_{oB}, the common-mode gain of the input stage will become

$$\frac{v_{d2}}{v_{icm}} = \frac{-g_{m2}(r_{o4}\|r_{o2})}{1 + 2g_{m2}r_{oB}} \tag{12.95}$$

Substituting V_A/I_D for each r_o and $2\sqrt{K_2 I_2}$ for g_{m2} results in

$$\frac{v_{d2}}{v_{icm}} = -\frac{2\sqrt{K_2 I_2}\frac{V_{AP}}{I_4}\|\frac{V_{AN}}{I_2}}{1 + 4\sqrt{K_2 I_2}(V_{AN}/I_B)} = -\frac{2\sqrt{K_2 I_2}\frac{V_{AP}}{I_B/2}\|\frac{V_{AN}}{I_B/2}}{1 + 4\sqrt{K_2 I_2}(V_{AN}/I_B)}$$

$$= -\frac{4\sqrt{K_2 I_2}}{I_B}\frac{V_{AP}V_{AN}/(V_{AP} + V_{AN})}{1 + 4\sqrt{K_2 I_2}(V_{AN}/I_B)} \tag{12.96}$$

where $I_2 = I_4 = I_B/2$. Given that $g_{m2}r_{oB} \gg 1$, this equation reduces to the simpler expression

$$\frac{v_{d2}}{v_{icm}} = \frac{V_{AP}V_{AN}/(V_{AP} + V_{AN})}{V_{AN}} = \frac{V_{AP}}{V_{AP} + V_{AN}} \tag{12.97}$$

Multiplying Eq. (12.97) by the gain of the output stage yields the overall common-mode gain of the amplifier:

$$A_{cm} = \frac{v_{out}}{v_{icm}} = \frac{V_{AP}}{V_{AP} + V_{AN}}g_{m5}(r_{o5}\|r_{oC}) \equiv \frac{V_{AP}}{V_{AP} + V_{AN}}\frac{2\sqrt{K_5 I_5}}{I_5}\frac{V_{AP}V_{AN}}{V_{AP} + V_{AN}}$$

$$= \frac{2V_{AP}^2 V_{AN}}{(V_{AP} + V_{AN})^2}\sqrt{\frac{K_5}{I_5}} \tag{12.98}$$

For the case $V_{AN} = V_{AP}$, Eq. (12.98) reduces to

$$A_{cm} = \frac{2V_A^3}{(2V_A)^2}\sqrt{\frac{K_5}{I_5}} = \frac{V_A}{2}\sqrt{\frac{K_5}{I_5}} \tag{12.99}$$

The common-mode rejection ratio (CMRR) for this limiting case becomes

$$\left|\frac{A_{dm}}{A_{cm}}\right| = \frac{V_A^2\left(\frac{K_2 K_5}{I_2 I_5}\right)^{1/2}}{(V_A/2)(K_5/I_5)^{1/2}} = 2V_A(K_2/I_2)^{1/2} \tag{12.100}$$

where A_{dm} is obtained from Eq. (12.90). As Eq. (12.100) shows, the CMRR can be made large by choosing a large W/L ratio for Q_2, thus producing a large K_2, and also by keeping the bias current to Q_2 small.

EXERCISE 12.8   Find the common-mode gain and CMRR for the circuit of Fig. 12.15 using the parameters of Example 12.5.   **Answer:**   75; 1265

12.3.4 Signal Swing Range

Applying a common-mode signal to the amplifier may cause the input stage to leave the region where all devices operate in the constant-current region, thus greatly reducing the differential-mode gain. If the common-mode input voltage v_{icm} becomes too positive, then v_{DS1} and v_{DS2} will no longer exceed $(v_{GS1} - V_{TRN})$ and $(v_{GS2} - V_{TRN})$, and Q_1 and Q_2 will leave the constant-current region. Similarly, if v_{icm} becomes too negative, then the condition $v_{DSB} > v_{GSB} - V_{TRN}$ will no longer be satisfied, and Q_B will leave the constant-current region.

SUMMARY

- An operational amplifier is fabricated by cascading a differential input stage, one or more additional gain stages, and a single-ended output stage.
- A given op-amp stage can be made from BJTs, MOSFETs, or JFETs.
- An op-amp cascade usually employs some form of dc-level shifting.
- An op-amp often contains devices of both polarities, that is, *npn* and *pnp* BJTs or *n*-channel and *p*-channel MOSFETs.
- The differential input stage of an op-amp provides a large incremental input resistance.
- The cascading of stages provides the op-amp with large open-loop gain.
- The power output stage of an op-amp provides a small incremental output resistance.
- The open-loop gain of a frequency-compensated op-amp has a dominant pole at a relatively low frequency.
- An internally compensated op-amp contains an integrated-circuit capacitor.
- An op-amp can be externally compensated by the addition of an external capacitor to the feedback network.
- The slew-rate limitation of an op-amp is related to the maximum rate at which the compensation capacitor or other capacitances can be charged and discharged by the op-amp's internal circuitry.
- Many op-amps contain a set of external offset-null terminals. Proper connection of these terminals to a variable resistor and power-supply bus permits the effects of input offset voltage and input bias current to be corrected.
- The differential- and common-mode gains of a CMOS op-amp are determined by fixing the W/L ratios of the various MOSFETs in the circuit.

◆ SPICE EXAMPLE

Many readers following the SPICE examples are likely to be using the evaluation version of PSPICE freely distributed by the MicroSim Corporation. This version of SPICE, while supporting all the features of the full-scale program, limits the user to a maximum of 10 transistors and 65 nodes. These limits do not permit a simulation of the complete LM741 op-amp described in this chapter, but do allow for portions of the circuit to be simulated. The limitations of Evaluation PSPICE were considered in developing the SPICE problems for this chapter.

Because many students may wish to program the SPICE example to follow, it focuses on the CMOS op-amp of Section 12.3, rather than on the LM741. The former has fewer than 10 transistors and is easily simulated on Evaluation PSPICE. The reader with access to a full-scale version of SPICE can investigate the complete LM741 op-amp via the end-of-chapter SPICE problems specifically set aside for this purpose.

EXAMPLE 12.6

Confirm the CMOS amplifier bias design of Example 12.4 by simulating the circuit on SPICE using the W and L values shown in Table 12.1 and Early voltages of value $V_{AN} = V_{AP} = 100\,\text{V}$. Determine the bias voltages of all nodes, and observe the current flow through each transistor. Also find the small-signal voltage gain and the large-signal swing range of the amplifier under open-loop, differential-mode conditions.

Solution

A version of the circuit with nodes suitably numbered for SPICE is shown in Fig. 12.16. The W/L values for each transistor, as obtained from Table 12.1, are also shown. The MOSFETs are labeled using the symbol M_n so as to be consistent with the use of the letter M to designate MOSFETs in SPICE. The W and L values of the devices are set in the element statements. Note that the parameter KP in each .MODEL statement is equivalent to the quantity μC_{ox}. Differential-mode excitation is produced by driving the v_1 input terminal with a voltage source v_1 and the v_2 terminal with a dependent source of value $E_2 = -v_1$.

Input File

```
CMOS OP-AMP of Fig. 12.16

*Specify the power supplies and other sources:
      VDD 1 0   5V
      VSS 9 0  -5V
      Io   1 7   20u
      v1   4 0   0V
      E2   5 0   4 0 -1 ; This dependent source creates a differential-mode input

*Specify the n-channel MOSFETs and their W and L values:
      M1   2 4 6 6 nchan W=20u L=  5u
      M2   3 5 6 6 nchan W=20u L=  5u
      MA   7 7 9 9 nchan W=10u L= 10u
      MB   6 7 9 9 nchan W=20u L=100u
      MC   8 7 9 9 nchan W=10u L= 10u
      .MODEL nchan NMOS(VTO=1V KP=20u LAMBDA=0.01); (VA = 100 V)

*Specify the p-channel MOSFETs and their W and L values:
      M3   2 2 1 1 pchan W= 5u L=11u
      M4   3 2 1 1 pchan W= 5u L=11u
      M5   8 3 1 1 pchan W=90u L=20u
      .MODEL pchan PMOS(VTO=-1V KP=10u LAMBDA=0.01); (VA = 100 V)

      .OP                  ;Print the bias voltages and currents
      .TF v(8) v1          ;Find the small-signal, differential-mode gain
      .DC v1 -1m 1m 0.01m  ;Plot the large-signal dc transfer characteristic
      .PROBE v(8)
      .END
```

Results. The various node voltages and device currents are shown in the abbreviated output listing that follows. The device currents agree reasonably well with those found in Example 12.5, but differ slightly due to the effect of the V_A parameter, which was assumed to be infinite ($r_o = \infty$) in the hand calculations. Similarly, the node voltage $V_{D2} = 3.06\,\text{V}$ (node 3) computed by SPICE differs slightly from the value of 3.1 V found in Example 12.4 (see Eq. (12.85)). Nevertheless, the simulation shows the design method based on the hand calculations of Example 12.4 to be reasonably accurate.

Figure 12.16
CMOS op-amp of Fig. 12.15 with nodes suitably numbered for SPICE. Differential-mode excitation is produced by driving the input v_2 with a dependent voltage source of value $E2 = -v_1$.

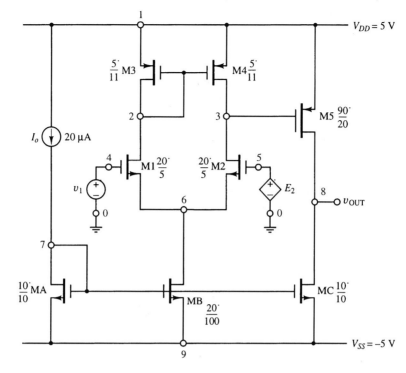

The results of the .TF analysis show the small-signal voltage gain to be approximately -1.05×10^5, with an r_{in} of $10^{20}\ \Omega$ (essentially infinite) and an r_{out} of $2.55\ \text{M}\Omega$. The plot of v_{OUT} versus v_1, obtained using the .PROBE utility and shown in Fig. 12.17, indicates large-signal op-amp limits of approximately $+4.7\,\text{V}$ and $-5\,\text{V}$. Note that the transfer characteristic does not cross through the origin, but instead shows a dc output-offset voltage of approximately $1.5\,\text{V}$ when $v_1 = 0$. ■

Figure 12.17
Plot of the large-signal v_1–v_{OUT} voltage transfer characteristic for the CMOS op-amp of Example 12.6.

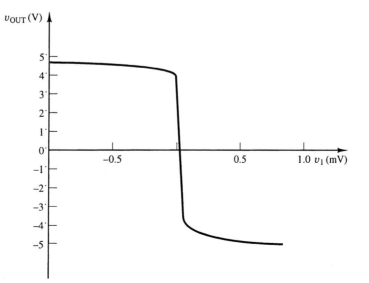

Output Listing

```
*********** Evaluation PSpice ************

****      MOSFET MODEL PARAMETERS
          nchan      pchan
          NMOS       PMOS
  LEVEL   1          1
      L   100.E-06   100.E-06; (Default values of L and W that would be used were
      W   100.E-06   100.E-06;  L and W not specified in each element statement)
    VTO   1          -1
     KP   20.E-06    10.E-06
  GAMMA   0          0
    PHI   0.6        0.6
 LAMBDA   0.01       0.01
   PBSW   0.8        0.8
    TOX   0          0


****      SMALL-SIGNAL BIAS SOLUTION        TEMPERATURE =   27.000 DEG C

  NODE   VOLTAGE    NODE   VOLTAGE    NODE   VOLTAGE    NODE   VOLTAGE
(   1)   5.0000   (   2)   3.0646   (   3)   3.0646   (   4)   0.0000
(   5)   0.0000   (   6)  -1.2204   (   7)  -2.6024   (   8)    .4137
(   9)  -5.0000
**** MOSFET OPERATING-POINT INFORMATION   TEMPERATURE =   27.000 DEG C

NAME      M1         M2         MA         MB         MC
MODEL     nchan      nchan      nchan      nchan      nchan
ID        2.03E-06   2.03E-06   2.00E-05   4.05E-06   2.06E-05
VGS       1.22E+00   1.22E+00   2.40E+00   2.40E+00   2.40E+00
VDS       4.29E+00   4.29E+00   2.40E+00   3.78E+00   5.41E+00
VBS       0.00E+00   0.00E+00   0.00E+00   0.00E+00   0.00E+00
VTH       1.00E+00   1.00E+00   1.00E+00   1.00E+00   1.00E+00
VDSAT     2.20E-01   2.20E-01   1.40E+00   1.40E+00   1.40E+00
GM        1.84E-05   1.84E-05   2.86E-05   5.80E-06   2.95E-05
GDS       1.94E-08   1.94E-08   1.95E-07   3.91E-08   1.95E-07

NAME      M3         M4         M5
MODEL     pchan      pchan      pchan
ID       -2.03E-06  -2.03E-06  -2.06E-05
VGS      -1.94E+00  -1.94E+00  -1.94E+00
VDS      -1.94E+00  -1.94E+00  -4.59E+00
VBS       0.00E+00   0.00E+00   0.00E+00
VTH      -1.00E+00  -1.00E+00  -1.00E+00
VDSAT    -9.35E-01  -9.35E-01  -9.35E-01
GM        4.33E-06   4.33E-06   4.40E-05
GDS       1.99E-08   1.99E-08   1.97E-07

****      SMALL-SIGNAL CHARACTERISTICS

      V(8)/v1 = -1.048E+05
      INPUT RESISTANCE AT v1     = 1.000E+20
      OUTPUT RESISTANCE AT V(8) = 2.550E+06

      JOB CONCLUDED
```

◆

PROBLEMS

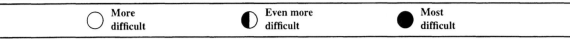

	More difficult		Even more difficult		Most difficult

12.2 Case Study: The LM741 Bipolar Operational Amplifier

12.2.1 BJT Input Stage of the LM741 Op-Amp

12.1 The LM741 op-amp is designed to operate over a wide range of supply voltages, with ±15 V being the optimum. Compute the bias currents I_{REF1} and I_{REF2} in the LM741 input stage of Fig. 12.3 if the supply voltages are set to ±10 V instead of ±15 V.

12.2 Compute the value of the current I_{REF2} in the LM741 input stage of Fig. 12.3 without neglecting the base currents flowing into Q_A, Q_B, Q_C, and Q_D. Assume that $\beta_N = \beta_P = 200$.

12.3 The fabrication process used by a particular manufacturer of the LM741 yields resistance values that may vary by as much as ±20% from specified values. Compute the possible range of I_{REF1} and I_{REF2} in the input stage of Fig. 12.3 under these conditions.

12.4 ○ Consider the LM741 input stage of Fig. 12.3. Suppose that R_{LOAD} consists of a short circuit to ground. Compute the bias voltages of the nodes in the circuit under these conditions. Use the approximate values of I_{REF1} and I_{REF2} indicated in the figure for your calculations. Assume transistor parameters $I_{EO} = 10^{-11}$ mA, $\eta = 1$, $\beta_N = 250$, and $\beta_P = 200$. Do not assume a value of V_f

for any base–emitter junction other than those of Q_E and Q_A, for which $V_f \approx 0.7$ V.

12.5 Compute the small-signal transconductance gain i_{out}/v_{idm} of the LM741 input stage of Fig. 12.3 if R_5 is changed from 39 kΩ to 33 kΩ.

12.6 Consider the small-signal representation of the LM741 input stage shown in Fig. 12.4. For the case $R_1 \neq R_2$, show that Q_5 and Q_6 form a small-signal current source of the modified Widlar, or Wilson, type. Find an expression for the resulting transconductance gain i_{out}/v_{idm}.

12.7 ○ Consider the LM741 input stage of Fig. 12.3 with the value of I_{REF2} shown in the figure.

(a) Compute the bias currents through Q_1, Q_2, Q_3, and Q_4.

(b) Assume the BJTs to have Early voltages of $V_A = 50$ V. Estimate the value of r_o for Q_4 and Q_6 given the bias currents found in part (a).

(c) Beginning with the small-signal circuit model, use the test current-source method to find the output resistance seen at the collector terminals of Q_4 and Q_6. It may be helpful to divide the output resistance into two parallel pieces—one contributed by Q_4 and its associated circuitry, and one contributed by Q_6.

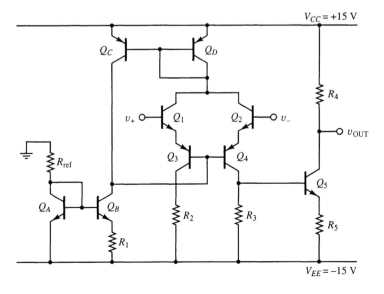

Figure P12.8

12.8 The circuit of **Fig. P12.8** illustrates a simple BJT op-amp that resembles the input stage of the LM741. It consists of a differential input stage cascaded with a class A voltage-follower output stage formed by Q_5.

(a) Choose values for R_{ref} and R_1 such that the bias currents through R_2 and R_3 are both set to approximately 50 μA.

(b) Choose values for R_2, R_3, R_4, and R_5 such that v_{OUT} is biased at approximately zero.

12.9 Consider the circuit of **Fig. P12.8** that depicts a simple two stage BJT op-amp. Suppose that R_1 and R_{ref} are chosen so that the current through Q_B is about 100 μA. If Q_3, Q_4, R_2, and R_3 are omitted from the circuit, and the emitters of Q_1 and Q_2 connected directly to the collector of Q_B, find the resulting bias voltage measured at the collector of Q_B.

12.10 ⚡ ○ Consider the two-stage BJT op-amp shown in **Fig. P12.8**. Choose appropriate values for all resistors to yield a differential gain of at least +60 dB with a bias value for v_{OUT} of approximately zero.

12.11 ◖ The BiFET (bipolar-FET) input stage of **Fig. P12.11** uses BJTs and JFETs together in the same circuit.

(a) Find expressions for I_{o1}, I_{D1}, and I_{D2}.

(b) Find expressions for the voltages of nodes D and E in terms of V_f and other parameters.

(c) Use the expression for the voltage of node E to find an approximate expression for I_{C4}.

(d) Apply KCL at the output node to find an expression for the bias value of v_{OUT}.

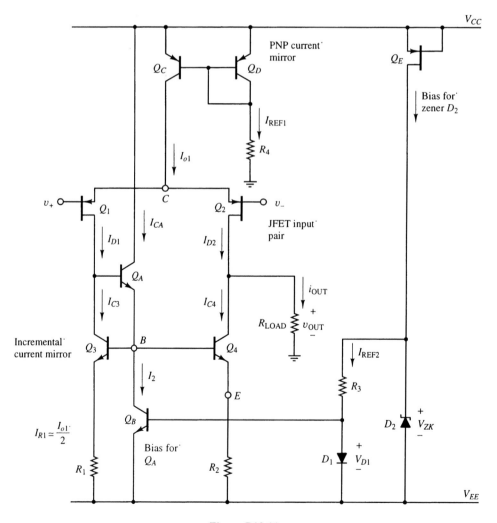

Figure P12.11

12.12 ⚡ Consider the BiFET differential input stage of **Fig. P12.11**. Choose values of V_{ZK2} and R_3 so that I_{REF2} is set to $100\,\mu\text{A}$ for a range of V_{CC} and V_{EE} values from ± 3 to $\pm 15\,\text{V}$, where $V_{EE} = -V_{CC}$. What range of parameter values must the JFET Q_E have if the chosen bias design is to be valid?

12.13 ○ Consider the BiFET differential input stage shown in **Fig. P12.11**. Use small-signal modeling to find an expression for the gain $v_{\text{out}}/v_{i\text{dm}}$. Show that the common-mode gain is approximately zero. Choose resistor values to yield a small-signal transconductance gain $i_{\text{out}}/v_{i\text{dm}}$ of about $50\,\text{mA/V}$. What transistor parameters limit the swing range of v_{OUT}?

12.14 Modify the bias circuitry of the BiFET input stage of **Fig. P12.11** so that bias currents I_{REF1} and I_{REF2} are both made independent of V_{CC} and V_{EE}.

12.2.2 Middle-Gain Stage of the LM741 Op-Amp

12.15 Find the bias current flowing through Q_8 in the LM741 middle-gain stage of Fig. 12.5 if R_7 is changed from $100\,\Omega$ to $200\,\Omega$. Assume that $I_{EO} = 10^{-11}\,\text{mA}$.

12.16 Find I_{E8} and I_{B8} in the LM741 middle-gain stage of Fig. 12.5 if I_{B9} is not neglected compared to I_{R6} and if $\beta_F = 150$ for the two *npn* transistors.

12.17 Suppose that the fraction of current captured by collector C_2 in the LM741 middle-gain stage of Fig. 12.5 is equal to 0.5 instead of 0.75. Find the resulting bias current through Q_8.

12.18 ○ Consider the LM741 middle-gain stage of Fig. 12.5. In this problem, the bias currents through each of the transistors in the circuit are examined. For the purpose of computation, assume the base of Q_8 to be grounded via a 5-MΩ resistance.

(a) Neglect base currents with respect to collector currents and find the voltage drop across R_7. Next estimate the bias value of I_{R6} by assuming a value for V_{BE9}.

(b) Use the iterative technique with $I_{EO} = 10^{-12}\,\text{A}$ to compute V_{BE9} exactly. How does your answer to part (a) change?

(c) Now modify your results to include the base current into Q_9. Assume β_F to be equal to 200. Compare with the results of parts (a) and (b).

12.19 In this problem, the bias value of v_{OUT} for the LM741 middle-gain stage of Fig. 12.5 is computed. For the purpose of this problem, assume Q_X and Q_9 to have Early voltages of value $V_A = 50\,\text{V}$.

(a) If $\beta_F = 200$ for all transistors, draw the v–i characteristic of Q_9 for the appropriate value of I_{B9}.

(b) Now consider R_7, Q_X, V_{CC}, and V_{EE} to be the load circuit for Q_9 and plot the resulting load curve over the graph of part (a). For the purpose of this plot, assume that $I_{E9} \approx I_{C9}$.

(c) Obtain the bias point of v_{OUT} by locating the point of intersection of the load curve of part (b) with the v–i characteristic of part (a).

12.20 ○ Consider the LM741 middle-gain stage of Fig. 12.5. In this problem, we shall examine the effect on the middle-gain stage of output resistances r_{o4} and r_{o6} from the input stage of Fig. 12.3.

(a) Represent the middle-gain stage by a small-signal model. Its input should be driven by a voltage source in series with r_{out} from the previous input stage. The latter should include the effects of r_{o4} and r_{o6}.

(b) Find the gain $v_{\text{out}}/v_{\text{in}}$ with r_{o9} included in the model for Q_9. Compare with the result obtained in Section 12.2.2.

12.21 Determine the small-signal transresistance gain a_r of the LM741 middle-gain stage of Fig. 12.5 if $R_6 = 25\,\text{k}\Omega$ and $R_7 = 200\,\Omega$. Assume the BJTs to have parameters $\beta_o = 200$ and $V_A = 30\,\text{V}$.

12.22 ○ Compute the input resistance of the LM741 middle-gain stage of Fig. 12.5. Use this value of input resistance as the load to the input stage of Fig. 12.3. Treat the cascade of the two stages as a single voltage amplifier and compute the gain $v_{\text{out}}/v_{i\text{dm}}$.

12.23 ◉ The small-signal behavior of the LM741 middle-gain stage of Fig. 12.5 can be derived from the small-signal model of Fig. 12.7. In this model, the active load presented to Q_9 by Q_X is represented by the small-signal output resistance r_{oX2}. In this problem, we show that r_{o9} of Q_9 has little effect on circuit behavior and can be omitted from the small-signal circuit model.

(a) Find an expression for i_{b9} in the circuit of Fig. 12.7.

(b) From the expression of part (a), find the small-signal gain $v_{\text{out}}/i_{\text{in}}$ without including the effect of r_{o9}.

(c) Redraw the circuit to include r_{o9} between the collector and emitter of Q_9. Use KVL and KCL to find an expression for v_{out} as a function of i_{b9} with r_{o9} included in the circuit. Note that the presence of r_{o9} will affect the value of i_{b9} computed in part (a).

(d) Use the result of part (c) to find the gain $v_{\text{out}}/i_{\text{in}}$ with r_{o9} included. Compare to the result of part (a).

12.2.3 Output Stage of the LM741 Op-Amp

12.24 Compute the bias currents in Q_{11} and Q_{12} in the LM741 output stage of Fig. 12.8 if R_8 is changed from $40\,\text{k}\Omega$ to $25\,\text{k}\Omega$. Assume the BJTs to have an I_{EO} of 10^{-11} mA, as in the analysis leading to Eq. (12.40).

12.25 Choose a new value for R_8 in the LM741 output stage of Fig. 12.8 such that Q_{11} and Q_{12} have bias currents of about $150\,\mu\text{A}$.

12.26 Compute the bias currents in Q_{11} and Q_{12} in the LM741 output stage of Fig. 12.8 if the BJTs have an I_{EO} of 10^{-10} mA.

12.27 The circuit shown in **Fig. P12.27** is used in the design of integrated-circuit output stages to magnify the value of a forward-biased base–emitter junction voltage.

(a) Express the current through R_2 as a function of V_{BE1}.

(b) Express the current through R_1 as a function of the overall circuit voltage V_{BB}. Assume the base current of Q_1 to be negligibly small.

(c) Combine the results of parts (a) and (b) to find an expression for V_{BB} in terms of R_1 and R_2 when I_1 is positive.

(d) Evaluate V_{BB} for the case $R_1 = R_2 = 10\,\text{k}\Omega$ and draw the v–i characteristic of I_{BB} versus V_{BB} as measured at the circuit terminals.

(e) Choose values for R_1 and R_2 such that $V_{BB} = 1.1\,\text{V}$.

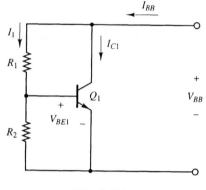

Fig. P12.27

12.28 The simplified version of the LM741 output stage shown in Fig. 12.8 omits several components designed to protect the output from short circuits or excessive output currents. The additional components that appear in the actual LM741 are shown in **Fig. P12.28**. Note that most of the output current flows through R_9 for positive i_{OUT} and through R_{10} for negative i_{OUT}.

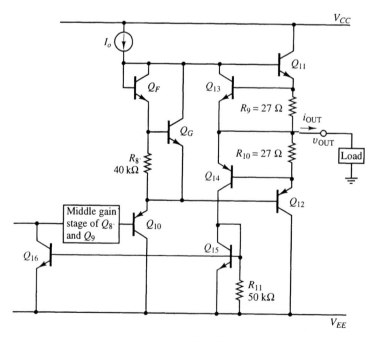

Figure P12.28

(a) Assume the turn-on voltage of Q_{13} and Q_{14}, which have large cross-sectional areas, to be about 0.6 V. At what values of i_{OUT} will Q_{13} or Q_{14} begin to conduct?

(b) If i_{OUT} is equal to one of the values found in part (a), estimate the current through Q_{11} and Q_{12}.

(c) When current flows through device Q_{14}, it will also flow through R_{11}. How large must the current through these components become before Q_{15} begins to conduct?

(d) Explain the function of Q_{16} in protecting the devices in the output stage in the event that the output terminal is short circuited to ground.

(e) Why do R_9 and R_{10} have an insignificant effect on op-amp performance under normal operating conditions?

12.29 For the LM741 output stage of Fig. 12.8, estimate the values of $V_{sat-pos}$ and $V_{sat-neg}$ if the op-amp is powered by ±10-V supplies.

12.2.5 Frequency Compensation

12.30 Show that the result (12.44), which describes Miller multiplication, can also be found if the output port of the middle-gain stage in Fig. 12.9 is represented by a dependent current source and a parallel Norton resistance.

12.31 Estimate the effect on the dominant-pole frequency of the LM741 if C_C is changed from 30 pF to 50 pF.

12.32 The location of the dominant pole of the LM741 was computed in Example 12.3 using the value $\beta_o = 150$. Suppose that $\beta_o = 250$. Estimate the effect on the dominant-pole frequency.

12.33 Suppose that the fraction of total collector current captured by C_2 of Q_X in the LM741 is equal to 0.5 instead of 0.75. Determine the resulting effect on the dominant-pole frequency initiated by C_C.

12.2.6 Pole Splitting

12.34 Show that Eq. (12.59) follows from Eqs. (12.57) and (12.58).

12.35 Compute the value of the non-dominant pole ω_2 for the LM741, as represented by the circuit of Fig. 12.12, if $r_1 = 6\,M\Omega$, $r_2 = 100\,k\Omega$, $C_1 = 10\,pF$, $C_2 = 20\,pF$, and $C_C = 30\,pF$.

12.36 By how much must ω_1 be larger than ω_2 if Eq. (12.62) is to accurately predict the dominant amplifier pole to within 20%?

12.37 Show that the product of the two split poles computed in Eqs. (12.62) and (12.63) is equal to the product of the pole computed at the input port with the output grounded times the pole computed at the output port with the input grounded.

12.38 Prove that the LM741 will be stable for all feedback factors β if its dominant pole lies at 5 Hz and its second pole at 1.4 MHz.

12.2.7 Origin of Slew-Rate Limitation

12.39 Compute the slew-rate of the LM741 if the compensation capacitor is changed from 30 to 50 pF.

12.40 Compute the slew-rate of the LM741 if the bias resistor R_5 is changed from 39 to 25 kΩ.

12.41 What value of resistor R_5 is required if the slew-rate of the LM741 is to be increased to 1.5 V/μs?

12.3 Case Study: A Simple CMOS Operational Amplifier

12.3.1 Bias Design

12.42 The current source I_o in the CMOS amplifier of Fig. 12.15 is set to 10 μA. Choose W and L values for Q_A, Q_B, and Q_C such that $I_B = 7.5\,mA$ and $I_C = 16\,\mu A$.

12.43 Find the bias voltage of node C in the CMOS amplifier of Fig. 12.15 if $I_o = 15\,\mu A$, $V_{DD} = -V_{SS} = 7\,V$, $W_A = L_A = 15\,\mu m$, $W_B = 8\,\mu m$, $L_B = 80\,\mu m$, $W_3 = 10\,\mu m$, $L_3 = 14\,\mu m$, and $\mu_h C_{ox} = 12\,\mu A/V^2$.

12.44 ○ Find the bias voltage of node C in the CMOS amplifier of Fig. 12.15 if $I_o = 20\,\mu A$, $V_{DD} = -V_{SS} = 10\,V$, $W_A = 15\,\mu m$, $L_A = 12\,\mu m$, $W_B = 12\,\mu m$, $L_B = 65\,\mu m$, $W_3 = L_3 = 10\,\mu m$, and $\mu_h C_{ox} = 15\,\mu A/V^2$. What will be the bias values of node B and V_{D2} if Q_5 is matched to Q_3, Q_C is matched to Q_A, and Q_1 and Q_2 have dimensions $W = 25\,\mu m$ and $L = 5\,\mu m$?

12.45 ◑ Find the bias values of the various voltages and currents in the CMOS amplifier of Fig. 12.15 if the W values in Table 12.1 are all multiplied by a factor of 2.

12.46 ○ The circuit of **Fig. P12.46** is an example of NMOS technology, in which all devices are *n*-channel, enhancement-mode MOSFETs. Find expressions for the node voltages V_A and V_B. Show that the total voltage drop $(V_{DD} - V_{SS})$ is divided into thirds if Q_A, Q_B, and Q_C are matched devices (same K and V_{TR}).

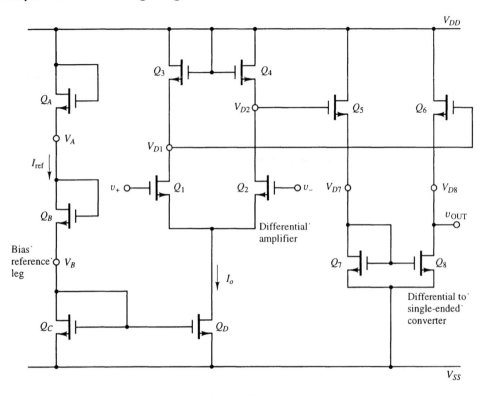

Figure P12.46

12.47 Consider the NMOS amplifier of **Fig. P12.46**. For W/L ratios of $K_A/K_B = K_A/K_C = 1/4$, find the node voltage V_A when $V_{SS} = -V_{DD}$.

12.48 ◐ Consider the NMOS amplifier of **Fig. P12.46**. If Q_A, Q_B, and Q_C are matched devices (same K and V_{TR}), then V_{DSA} and V_{DSB} will equal $(V_{DD} - V_{SS})/3$. Suppose that $K_3 = K_4 = K_A/2$ and the ratios K_B/K_C and K_5/K_7 are equal. Find expressions for the bias voltages V_{D1}, V_{D2}, V_{D7}, and V_{D8}.

12.49 Find expressions for the small-signal differential- and common-mode gains of the NMOS amplifier of **Fig. P12.46** as measured at the drains of Q_1 and Q_2.

12.50 ● Consider the NMOS amplifier of **Fig. P12.46**. Assume the voltages v_{D1} and v_{D2} appearing at the drains of Q_1 and Q_2 to have a differential-mode component only. Find an expression for the small-signal gain v_{out}/v_{d1} of the differential-to-single-ended converter, where $v_{d2} = -v_{d1}$. In performing this analysis, include the output resistance r_{o8} of Q_8 but assume the r_o of all other MOSFETs in the circuit to be infinite. (*Hint*: Apply the principle of superposition.)

12.3.2 Small-Signal Differential-Mode Performance

12.51 Find the differential-mode gain of the CMOS amplifier of Fig. 12.15 if the devices have the parameters listed in Table 12.1 and Early voltages of $V_A = 100\,\mathrm{V}$.

12.52 Find the differential-mode gain of the CMOS amplifier of Fig. 12.15 if the devices have the parameters listed in Table 12.1, with $V_{AN} = 50\,\mathrm{V}$ and $V_{AP} = 75\,\mathrm{V}$.

12.53 ⬛ Design a CMOS op-amp based on the circuit of Fig. 12.15 that has an open-loop gain of approximately $-30{,}000$. The devices have parameters $V_{TRN} = 2\,\mathrm{V}$, $V_{TRP} = -1\,\mathrm{V}$, $V_{AN} = V_{AP} = 50\,\mathrm{V}$, $\mu_e \epsilon_{ox}/t_{ox} = 15\,\mu\mathrm{A/V^2}$, and $\mu_h \epsilon_{ox}/t_{ox} = 7.5\,\mu\mathrm{A/V^2}$.

12.54 ⬛ Design a CMOS op-amp based on the circuit of Fig. 12.15 that has an open-loop gain of approximately $-75{,}000$. The devices have parameters $V_{TRN} = 1.5\,\mathrm{V}$, $V_{TRP} = -2\,\mathrm{V}$, $V_{AN} = 30\,\mathrm{V}$, $V_{AP} = 25\,\mathrm{V}$, $\mu_e \epsilon_{ox}/t_{ox} = 24\,\mu\mathrm{A/V^2}$, and $\mu_h \epsilon_{ox}/t_{ox} = 12\,\mu\mathrm{A/V^2}$.

12.55 An example of a revised CMOS output stage is shown in **Fig. P12.55**. The follower Q_2 is a p-channel device with the n-channel device Q_3 as its active load. The combination of Q_2 and Q_3 drives a complementary-pair follower formed by Q_4 and Q_5. Derive an expression for the gain v_{out}/v_{in}.

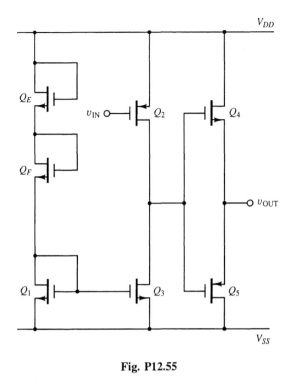

V_{DD}

Q_E

v_{IN} — Q_2

Q_4

Q_F

v_{OUT}

Q_1

Q_3

Q_5

V_{SS}

Fig. P12.55

12.56 Consider the CMOS output stage of **Fig. P12.55**. Choose values of W and L for all devices such that the output terminal is biased at ground potential when $v_{IN} = 0$. The smallest device dimension is to be $2\,\mu m$. The voltage sources V_{DD} and V_{EE} have equal magnitudes.

12.3.3 Common-Mode Performance

12.57 Find the common-mode gain of the CMOS amplifier of Fig. 12.15 if the devices have the parameters listed in Table 12.1 and Early voltages of value $V_A = 80\,V$.

12.58 Find the common-mode gain of the CMOS amplifier of Fig. 12.15 if the devices have the parameters listed in Problem 12.44, with $V_A = 40\,V$ for all devices.

12.59 Find the common-mode rejection ratio of the CMOS amplifier of Fig. 12.15 if it has the W and L values listed in Table 12.1 and $V_A = 50\,V$ for all devices.

12.60 Suggest a modification to the CMOS circuit of Fig. 12.15 such that the common-mode gain is negative rather than positive.

◆ SPICE PROBLEMS

12.61 Confirm the bias analysis of the input stage of the LM741 performed in Section 12.2.1 by simulating the circuit on SPICE. Verify the voltages and currents found in Eqs. (12.1) to (12.3).

12.62 Use SPICE to compute the small-signal transconductance gain of the input stage of the LM741 by simulating the circuit on SPICE. Use the component values given in Section 12.2.1. Compare the results of the simulation to Eq. (12.18).

12.63 Use SPICE to compute the bias currents through Q_8 and Q_9 in the middle-gain stage of Fig. 12.5.

(a) For $\beta_F = 200$, adjust I_{EO} over the range 10^{-11} to 10^{-14} A in multiples of 10 and assess the effect on the circuit.

(b) For $I_{EO} = 10^{-12}$ A, adjust β_F over the range 50 to 250 and assess the effect on the circuit.

12.64 Simulate the middle-gain stage of the LM741 op-amp on SPICE. Determine the circuit's small-signal gain. The split-collector device Q_X can be simulated by connecting two *pnp* devices in parallel—one with a relative base–emitter junction area of 0.25, the other with a relative junction area of 0.75.

12.65 Confirm the analysis of the output stage of the LM741 performed in Section 12.2.3 by simulating the circuit on SPICE. Find the bias levels and small-signal gain.

12.66 Simulate the V_{BE} multiplier circuit of **Fig. P12.27** with $R_2 = 1\,k\Omega$ and R_1 equal to several values over the range $100\,\Omega$ to $1\,M\Omega$. Find the value of V_{BB} and the small-signal resistance looking into the I_{BB} node for each value of R_1.

12.67 Simulate the output stage of the LM741 shown in **Fig. P12.28** on SPICE. This version of the circuit includes the short-circuit protection components Q_{13}–Q_{16}. For the purpose of this problem, model the middle-gain stage as a unity-gain dependent source, and drive the circuit at the collector of Q_{16} using a voltage source v_{IN} in series with a 1-$M\Omega$ resistor. Plot the dc transfer characteristic for $-10\,V < v_{IN} < 10\,V$ using various load resistors in the range $10\,k\Omega$ to $1\,\Omega$.

12.68 The evaluation version of PSPICE used by many students limits simulations to 10 transistors. If your version of SPICE permits larger circuits, simulate the complete LM741 BJT op-amp of Fig. 12.2 on SPICE. Determine the differential- and common-mode open-loop gains as well as the amplifier's input and output resistances. Assume that all *npn* devices have parameters $\beta_F = 200$ and $V_A = 150\,\text{V}$; assume that all *pnp* devices have parameters $\beta_F = 50$ and $V_A = 50\,\text{V}$. All BJTs in the circuit are fabricated so that $I_{EO} = 10^{-11}\,\text{mA}$. The split-collector device Q_X can be simulated by connecting two *pnp* devices in parallel—one with a relative base–emitter junction area of 0.25, the other with a relative junction area of 0.75.

12.69 If your version of SPICE permits a sufficient number of transistors, simulate the entire LM741 op-amp of Fig. 12.2 on SPICE. Assume the *npn* devices to have parameters $\beta_F = 200$ and $V_A = 150\,\text{V}$; assume the *pnp* devices to have parameters $\beta_F = 50$ and $V_A = 50\,\text{V}$. All BJTs in the circuit are fabricated so that $I_{EO} = 10^{-11}\,\text{mA}$. Connect the op-amp as a noninverting amplifier with a gain of 2. Find the positive and negative saturation limits of v_{OUT} and compare with the output saturation values V_{CC} and V_{EE} assumed in the ideal op-amp approximation.

12.70 If your version of SPICE permits it, simulate the entire LM741 op-amp of Fig. 12.2 on SPICE. Set *npn* parameters to $\beta_F = 200$ and $V_A = 150\,\text{V}$, and *pnp* parameters to $\beta_F = 50$ and $V_A = 50\,\text{V}$, with $I_{EO} = 10^{-11}\,\text{mA}$ in both cases. Apply a step function to the v_+ input with the v_- input grounded and determine the slew rate of the amplifier.

12.71 If your version of SPICE permits it, simulate the entire LM741 op-amp of Fig. 12.2 on SPICE. Set *npn* parameters to $\beta_F = 200$ and $V_A = 150\,\text{V}$, and *pnp* parameters to $\beta_F = 50$ and $V_A = 50\,\text{V}$, with $I_{EO} = 10^{-11}\,\text{mA}$ in both cases. By making appropriate connections to the op-amp, determine the values of its input offset current, input bias current, and input offset voltage.

12.72 If your version of SPICE permits it, simulate the entire LM741 op-amp of Fig. 12.2 on SPICE. Set *npn* parameters to $\beta_F = 200$ and $V_A = 150\,\text{V}$, and *pnp* parameters to $\beta_F = 50$ and $V_A = 50\,\text{V}$, with $I_{EO} = 10^{-11}\,\text{mA}$ in both cases. Connect the op-amp as a unity-gain follower and determine its unity-gain frequency.

12.73 Simulate the BiFET input stage of **Fig. P12.11** on SPICE with $R_1 = R_2 = 1\,\text{k}\Omega$, $R_3 = 50\,\text{k}\Omega$, and

$R_{\text{LOAD}} = 1\,\text{M}\Omega$. Set V_{ZK2} to 5 V, β_F to 150, and set the *p*-channel JFET parameters to $I_{DSS} = -4\,\text{mA}$ and $V_P = 4\,\text{V}$.

12.74 Simulate the NMOS circuit of **Fig. P12.46** using the values $KP = 0.4\,\text{mA/V}^2$, and $W = L = 2\,\mu\text{m}$ for all devices except Q_1 and Q_2, for which $W = 40\,\mu\text{m}$, and Q_A, for which $W = 4\,\mu\text{m}$. Assess the role of the body effect on the amplifier's large-signal transfer characteristic.

12.75 Assess the effect on the bias levels in the CMOS amplifier of Example 12.4 if the supplies are changed to $V_{DD} = 15\,\text{V}$ and $V_{SS} = -15\,\text{V}$. Simulate the circuit on SPICE using the W and L values listed in Table 12.1.

12.76 Confirm the differential-mode gain found in Example 12.5 for the CMOS amplifier of Fig. 12.15 by simulating the circuit on SPICE. Also determine the value of the common-mode gain.

12.77 Simulate the CMOS amplifier of Fig. 12.15 on SPICE using the W and L values listed in Table 12.1. Determine the gain and large-signal swing range measured at the v_{OUT} terminal for several values of Early voltage V_A over the range $20\,\text{V} < V_A < \infty$.

12.78 Simulate the CMOS amplifier of Fig. 12.15 on SPICE using the W and L values listed in Table 12.1 and Early voltages of $V_A = 50\,\text{V}$. Assess the effect of including the body voltage v_{BS} in the *n*-channel devices Q_1 and Q_2. The substrates of these devices should be connected to the V_{SS} bus.

12.79 Simulate the CMOS amplifier of Example 12.4 for the case in which the V_{SS} supply is omitted, so that $V_{SS} = 0$. Simulate the circuit on SPICE using the W and L values listed in Table 12.1, with $V_A = 50\,\text{V}$, and plot the amplifier's large-signal transfer characteristic.

12.80 Simulate the CMOS amplifier of Fig. 12.15 on SPICE using the W and L values listed in Table 12.1 and Early voltages of value $V_A = 50\,\text{V}$. Also include device capacitances $C_{GSo} = 1\,\text{pF}$ and $C_{GDo} = 0.5\,\text{pF}$ for all MOSFETs, where C_{GSo} and C_{GDo} describe the capacitances of a device for which $W = L = 1\,\mu\text{m}$. Use SPICE to determine the amplifier's open-loop frequency response.

Chapter 13

Active Filters and Oscillators

$\mathbf{E}$lectrical elements have been used to make frequency-selective filters since the early part of the 20th century. These early filters, which utilized only passive inductors, capacitors, and resistors, helped foster the development of the first radio transmitters and receivers by providing circuits with frequency-selective capabilities. The limited range and selectivity of passive RLC circuits were improved somewhat by the invention of the vacuum tube, which permitted the design of filter circuits with feedback. Modern filter design really began with the arrival of high-quality integrated-circuit operational amplifiers in the early 1960s. Modern filters utilize op-amps in combination with RC feedback networks to provide countless filter functions with a wide range of frequency-selective properties.

As demonstrated in Chapter 10, the frequency response of an op-amp feedback circuit can be dramatically changed by the addition of capacitors to its feedback network. This property can be exploited to produce op-amp circuits with well-defined and controllable frequency-response characteristics. Such circuits are part of a family of stable analog feedback circuits called *active filters*. An analog feedback circuit that is intentionally operated outside its stability limit is called an *oscillator*. In this chapter, the characteristics and properties of several active filter and oscillator circuits are examined in detail. The functions performed by these circuits are important in many signal- and information-processing applications. As we shall see, an active op-amp filter can achieve all of its desired properties without the use of inductors. This result is fortunate, because the inductors needed for filter circuits below about 1 MHz tend to be large, difficult to produce in ideal form, and unsuitable for fabrication on an integrated circuit. Filter circuits made solely from op-amps, resistors, and capacitors are readily fabricated in an integrated-circuit environment.

13.1 A SIMPLE FIRST-ORDER ACTIVE FILTER

As a prelude to a general discussion of active filters, we first illustrate the basic concepts of active filtering using the circuit of Fig. 13.1(a). This simple filter is a low-pass variety that passes all frequency components below its cutoff frequency and attenuates all frequency components above. (We also recognize this circuit as the modified op-amp integrator of Chapter 2.) Because we are interested in the behavior of the circuit under sinusoidal steady-state, rather than transient, conditions, the circuit is best analyzed in the frequency domain.

Figure 13.1

A simple active filter example. (a) Inverting amplifier with feedback "element" $\mathbf{Z}_2 = R_2 \| \mathbf{Z}_C$; (b) equivalent topology of the inverting-amplifier configuration.

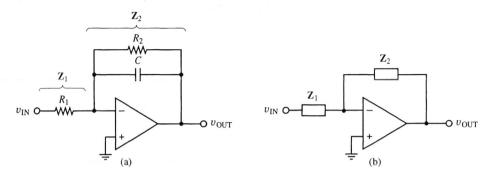

The circuit has the same basic topology as the inverting amplifier of Fig. 13.1(b), but in this case, the parallel combination $R_2 \| C$ is used as a feedback element. In the frequency domain, the capacitor behaves as an impedance element of value $\mathbf{Z}_C = 1/j\omega C$. The output of the filter of Fig. 13.1(a) can be found by first expressing $R_2 \| \mathbf{Z}_C$ as a single feedback impedance element of value

$$\mathbf{Z}_2 = R_2 \| \frac{1}{j\omega C} = \frac{R_2}{1 + j\omega R_2 C} \tag{13.1}$$

By analogy to the inverting-amplifier topology of Fig. 13.1(b), the output of the filter becomes

$$\frac{\mathbf{V}_{out}}{\mathbf{V}_{in}} = -\frac{\mathbf{Z}_2}{\mathbf{Z}_1} = -\frac{R_2}{R_1} \frac{1}{1 + j\omega R_2 C} \tag{13.2}$$

where $\mathbf{Z}_1 = R_1$, and where v_{OUT} and v_{IN} have been represented in sinusoidal phasor form as $\mathbf{V}_{out}$ and $\mathbf{V}_{in}$.

Figure 13.2

Magnitude Bode plot of the active filter of Fig. 13.1(a). The filter's "cutoff" frequency is designated ω_o.

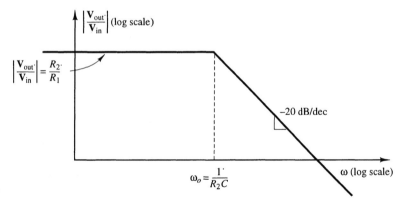

The transfer function (13.2) has a single pole at $\omega_o = 1/R_2 C$ and a gain of $-R_2/R_1$ well below ω_o. The magnitude Bode plot of this transfer function, shown in Fig. 13.2, has the basic form of a single-pole low-pass filter. As the frequency of the input signal is increased above the "cutoff" frequency ω_o, the filter output decreases at the rate of $-20\,\text{dB}$ per decade.

The frequency dependency described by the transfer function (13.2) can also be synthesized using the passive RC circuit of Fig. 13.3, for which

$$\frac{\mathbf{V}_{out}}{\mathbf{V}_{in}} = \frac{\mathbf{Z}_C}{R_2 + \mathbf{Z}_C} = \frac{1}{1 + j\omega R_2 C} \tag{13.3}$$

The Bode plot of the latter circuit's response has the same shape as the Bode plot of Fig. 13.2. The advantage of the active filter version of Fig. 13.1(a) over the passive version of Fig. 13.3 is twofold. First, the dc gain of the active filter can be adjusted by changing the ratio R_2/R_1. The dc response of the passive circuit of Fig. 13.3 has a fixed value of unity. Second, the output impedance of the active filter of Fig. 13.1(a) is negligibly small; the op-amp functions as a voltage source that drives the output terminal. This feature allows the active filter to drive a load impedance or another stage in a multistage filter cascade without changing the filter characteristics. In contrast, the output impedance of the passive circuit of Fig. 13.3 is equal to $R_2 \| (1/j\omega C)$. This relatively high impedance causes the circuit's output voltage and frequency response to be affected by the characteristics of its load.

Figure 13.3
Passive RC circuit having the same general frequency response as the active filter of Fig. 13.1(a).

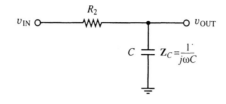

In principle, the transfer function of any passive filter can be synthesized in active form to realize the advantages stated above. Additionally, passive filter circuits that would normally require inductors can be made in active form without the use of inductors. As discussed previously, high-quality inductors are difficult to make in both discrete and integrated environments and are usually avoided in modern active circuit design.

13.2 IDEAL FILTER FUNCTIONS

The low-pass filter function described in Section 13.1 is but one of a class of analog filter functions that also includes high-pass, band-pass, and band-reject filters. As its name implies, the high-pass filter passes only those frequency components that lie above some designated cutoff frequency.

Figure 13.4
Ideal "brick-wall" responses of (a) low-pass; (b) high-pass; (c) band-pass; and (d) band-reject filters.

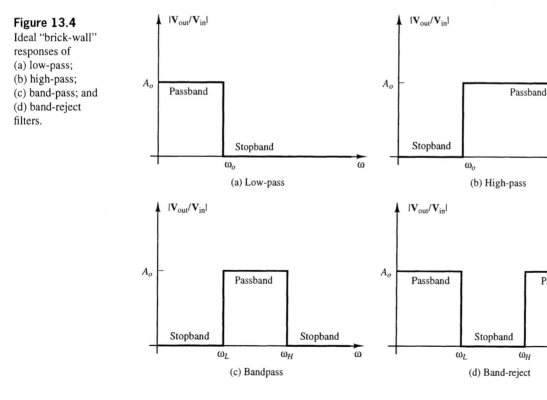

The band-pass filter transmits only those frequency components lying within a range specified by upper and lower cutoff limits. The band-reject filter is the inverse of the band-pass filter; it passes only those frequency components lying *outside* some specified frequency range.

Figure 13.5
Filter function definitions shown for (a) low-pass filter and (b) band-pass filter.

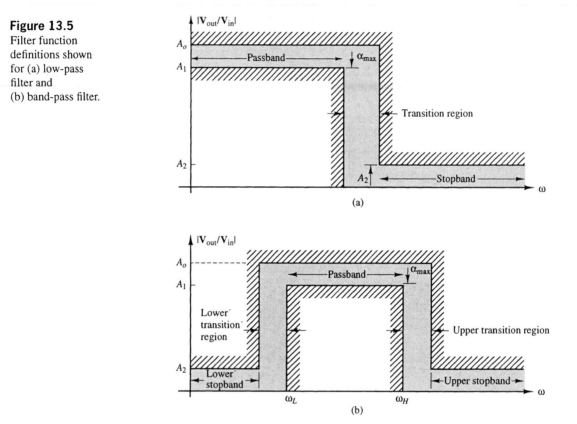

The basic forms of the transfer function for each of the various filter types are depicted in Fig. 13.4. These perfect, boxlike plots are sometimes called *brick-wall* responses. Each one represents an ideal case in which the filter gain remains constant over frequency regions where signal transmission is desired and falls to zero otherwise. Much of filter design is concerned with approaching these ideal responses as closely as possible while remaining within the practical constraints of part count, cost limitations, and filter complexity. As an example of this concept, consider the simple low-pass op-amp filter of Fig. 13.1(a). Its −20-dB per decade rolloff above ω_o provides only a very crude approximation to the ideal brick-wall low-pass response of Fig. 13.4(a). The filter is inexpensive and easy to build, however, and is adequate for many applications. A more complex op-amp circuit involving many more components could be constructed to provide a response more closely approaching the ideal, but this choice would result in a larger number of parts, and hence a greater cost per filter.

In order to quantify the degree to which any given filter approaches the ideal "brick-wall" response, it is helpful to define several quantities related to the filter's response curve. These quantities are summarized in Fig. 13.5 using the low-pass and band-pass filters as examples. Similar definitions exist for the high-pass and band-reject filters. The filter's *passband* is defined as the frequency region over which signal transmission is desired. The largest response occurring anywhere within the passband is designated A_o. In the ideal case, the filter gain would be equal to A_o throughout the passband. A real filter will always have a gain that changes with frequency, hence the parameter A_1 is used to define the lowest value to which the passband gain can fall and still be acceptable to the designer. Any departure from the ideal of constant passband gain may also be expressed as a maximum acceptable attenuation within the passband, defined by the factor $\alpha_{max} = A_o - A_1$.

If an ideal filter could be constructed, its signal transmission would immediately fall to zero outside the passband. In any real filter, some signal transmission always occurs outside the passband. The quantity A_2 defines the maximum signal transmission acceptable to the designer outside of the passband. The frequency at which signal transmission first falls to A_2 defines the beginning of the filter's *stopband*; the region between the passband and stopband is called the *transition* region. Note that the band-pass filter has two transition regions and two stopbands. Similarly, the band-reject filter has two transition regions and two passbands.

In general, the gain of a filter may lie anywhere between the limits A_o and A_1 in the passband; similarly, the gain may lie anywhere below the value A_2 within the stopband. The plot of Fig. 13.6(a) shows a low-pass filter response that decreases monotonically from its value of A_o at $\omega = 0$ and reaches the value A_1 only once before leaving the passband. The filter response of Fig. 13.6(b) cycles between A_o and A_1 several times within the passband and also cycles between zero and A_2 within the stopband. The peak passband gain A_o is reached at some frequency other than zero in this second example. Both plots in Fig. 13.6 represent valid low-pass filter responses and reasonable approximations to ideal brick-wall behavior. Each type of response can be produced by an appropriately designed filter circuit.

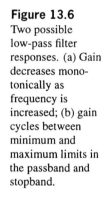

Figure 13.6
Two possible low-pass filter responses. (a) Gain decreases monotonically as frequency is increased; (b) gain cycles between minimum and maximum limits in the passband and stopband.

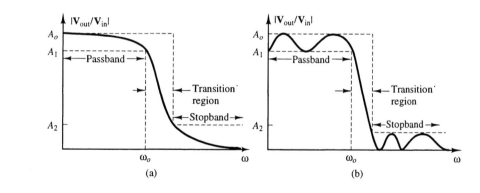

13.3 SECOND-ORDER FILTER RESPONSES

The low-pass filter of Section 13.1 is an example of a first-order filter. Its single pole in the denominator causes the magnitude $|\mathbf{V}_{out}/\mathbf{V}_{in}|$ to fall off as $1/\omega$, or -20 dB/decade, at frequencies well above ω_o. The steep walls of the ideal response of Fig. 13.4(a) are only weakly approximated by the -20 dB/decade slope of a first-order filter. A better approximation can be realized by using filters of higher order. The order of a filter is formally defined as the number of poles in the denominator of the transfer function. As a general rule, filters of higher order will have steeper transition region slope(s). The transfer function of a second-order low-pass filter, for example, falls off as $1/\omega^2$, or at -40 dB/decade, at frequencies well above its poles. Its slope will be twice as steep as that of a first-order filter, making it a better approximation to the ideal brick-wall response. Transfer functions of even higher order will produce steeper transition-region slopes. In this section, we examine the properties of several second-order filter configurations. In Section 13.4, these filters are used as basic building blocks to synthesize filters of higher order using the technique of *cascading*.

13.3.1 The Biquadratic Filter Function

The transfer function of a second-order filter can be described in terms of a ratio two of quadratic polynomials

$$H(j\omega) = \frac{\mathbf{V}_{\text{out}}}{\mathbf{V}_{\text{in}}} = A_o \frac{(1 + j\omega/\omega_1)(1 + j\omega/\omega_2)}{(1 + j\omega/\omega_3)(1 + j\omega/\omega_4)} \qquad (13.4)$$

In both the numerator and denominator, the quadratic polynomial has been expressed as the product of two binomials, as in Chapter 9. If the filter is of order 2 or higher, the poles and zeros are generally complex numbers. A transfer function with complex poles and zeros is more readily described using the s-plane representation in the sinusoidal steady-state, where $s = j\omega$. The s-plane is defined by a set of real and imaginary axes that are used to plot the real and imaginary components of each pole and zero in the system. In the s-plane, the sinusoidal driving frequency of the filter is equivalent to the imaginary-axis variable $s = j\omega$. If complex numbers $s_1, \cdots, s_n$ are used to describe the poles and zeros, the biquadratic transfer function (13.4) takes on the form

$$H(s) = A_o \frac{(1 + s/\omega_1)(1 + s/\omega_2)}{(1 + s/\omega_3)(1 + s/\omega_4)} \equiv A \frac{(s + s_1)(s + s_2)}{(s + s_3)(s + s_4)} \qquad (13.5)$$

Equation (13.5) can also be expressed in the general form

$$H(s) = \frac{a_2 s^2 + a_1 s + a_0}{b_2 s^2 + b_1 s + b_0} \qquad (13.6)$$

where the coefficients $a_1, \cdots, a_n$ and $b_1, \cdots, b_n$ include combinations of the poles $s_1, \cdots, s_4$. This ratio of quadratic polynomials is sometimes called the *biquadratic* transfer function, or simply "biquad." It can be used to describe virtually any second-order filter by appropriate selection of the a and b coefficients. The denominator of the transfer function describing a second-order filter must introduce a factor of $1/\omega^2$ at high frequencies; this criterion can be met by adjusting the coefficients b_0, b_1, and b_2 in (13.6) so that the s^2 factor in the denominator dominates at high frequencies. The filter's overall behavior—that is, whether it will be a low-pass, high-pass, band-pass, or band-reject filter—is established by adjusting the numerator coefficients a_0, a_1, and a_2.

13.3.2 Second-Order Active Low-Pass Filter

If the coefficients a_2 and a_1 in Eq. (13.6) are set to zero, the transfer function acquires the form

$$H(s) = \frac{a_0}{b_2 s^2 + b_1 s + b_0} \qquad (13.7)$$

Figure 13.7
Second-order active
low-pass filter of
the Sallen–Key
type.

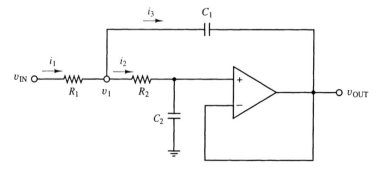

We recognize this function as that of a second-order low-pass filter. At frequencies near $s = j\omega \approx 0$, the response approaches the constant value $H(s) = a_0/b_0$. At very high frequencies, the response approaches the limit $H(s) = a_0/b_2s^2$. Because $s = j\omega$, this limit falls off as $1/\omega^2$ with a slope of -40 dB/decade, as required of a second-order low-pass filter. One filter circuit that has a transfer function of the form (13.7) is shown in Fig. 13.7. The circuit is sometimes called a Sallen–Key filter after its original inventors.[1] Its output as a function of frequency can be found by direct analysis using KVL and KCL. Applying KCL to the v_1 node with all currents represented as phasors yields

$$\mathbf{I}_1 = \mathbf{I}_2 + \mathbf{I}_3 \tag{13.8}$$

If the impedance of each capacitor is represented by $\mathbf{Z}_n = 1/j\omega C_n$, Eq. (13.8) can be expressed as

$$\frac{\mathbf{V}_{in} - \mathbf{V}_1}{R_1} = \frac{\mathbf{V}_1}{R_2 + \mathbf{Z}_2} + \frac{\mathbf{V}_1 - \mathbf{V}_{out}}{\mathbf{Z}_1} \tag{13.9}$$

The op-amp voltage v_+ can be found in terms of $\mathbf{V}_1$ from the complex form of the voltage divider:

$$\mathbf{V}_+ = \mathbf{V}_1 \frac{\mathbf{Z}_2}{R_2 + \mathbf{Z}_2} \tag{13.10}$$

The op-amp output is connected directly to the v_- terminal, thereby forming a voltage follower between v_+, v_-, and v_{OUT}. This connection forces v_{OUT} to have the same value as v_+, so that Eq. (13.10) becomes

$$\mathbf{V}_{out} = \mathbf{V}_1 \frac{\mathbf{Z}_2}{R_2 + \mathbf{Z}_2} \tag{13.11}$$

Rearranging Eq. (13.11) results in

$$\mathbf{V}_1 = \mathbf{V}_{out} \frac{R_2 + \mathbf{Z}_2}{\mathbf{Z}_2} \tag{13.12}$$

Combining Eqs. (13.9) and (13.12) leads to an expression for $\mathbf{V}_{out}$ as a function of $\mathbf{V}_{in}$:

$$\frac{\mathbf{V}_{in}}{R_1} - \mathbf{V}_{out} \frac{R_2 + \mathbf{Z}_2}{R_1 \mathbf{Z}_2} = \frac{\mathbf{V}_{out}}{\mathbf{Z}_2} + \mathbf{V}_{out} \frac{R_2 + \mathbf{Z}_2}{\mathbf{Z}_1 \mathbf{Z}_2} - \frac{\mathbf{V}_{out}}{\mathbf{Z}_1} \tag{13.13}$$

Equation (13.13) can be solved for $\mathbf{V}_{out}$, resulting in

$$\mathbf{V}_{out} \left(\frac{1}{\mathbf{Z}_2} + \frac{R_2 + \mathbf{Z}_2}{\mathbf{Z}_1 \mathbf{Z}_2} - \frac{1}{\mathbf{Z}_1} + \frac{R_2 + \mathbf{Z}_2}{R_1 \mathbf{Z}_2} \right) = \frac{\mathbf{V}_{in}}{R_1} \tag{13.14}$$

or

$$\mathbf{V}_{out} = \mathbf{V}_{in} \frac{\mathbf{Z}_1 \mathbf{Z}_2}{\mathbf{Z}_1 \mathbf{Z}_2 + \mathbf{Z}_1(R_1 + R_2) + R_1 R_2} \tag{13.15}$$

Substitution of $1/j\omega C_1$ and $1/j\omega C_2$ for $\mathbf{Z}_1$ and $\mathbf{Z}_2$ in Eq. (13.15) results in

$$H(j\omega) = \frac{\mathbf{V}_{out}}{\mathbf{V}_{in}} = \frac{1}{1 - \omega^2(R_1 R_2 C_1 C_2) + j\omega C_2(R_1 + R_2)} \tag{13.16}$$

It is possible to factor the denominator of this frequency-dependent transfer function into the standard "product of binomials" form of Chapter 9. For all but a few values of R_1, R_2, C_1,

---

[1] R. P. Sallen and E. L. Key, "A Practical Method of Designing RC Active Filters," *IRE Transactions on Circuit Theory*, Vol. CT–2, 74–85, March 1955.

and C_2, this factoring reveals poles in the denominator that are complex numbers. The transfer function (13.16) can also be represented as $H(s)$ in the s-plane if the following substitutions are made:

$$s = j\omega \tag{13.17}$$

$$\omega_o = \frac{1}{\sqrt{R_1 R_2 C_1 C_2}} \tag{13.18}$$

and
$$Q = \frac{\omega_o R_1 R_2 C_1 C_2}{C_2 (R_1 + R_2)} = \sqrt{\frac{C_1}{C_2}} \left(\frac{\sqrt{R_1 R_2}}{R_1 + R_2} \right) \tag{13.19}$$

Note that the parameter Q, called the "quality factor" of the filter, is dimensionless.

By using these substitutions, Eq. (13.16) can be expressed as

$$H(s) = \frac{\mathbf{V}_{out}}{\mathbf{V}_{in}} = \frac{1}{1 + s^2/\omega_o^2 + s/\omega_o Q} = \frac{\omega_o^2}{s^2 + s(\omega_o/Q) + \omega_o^2} = \frac{\omega_o^2}{(s - s_1)(s - s_2)} \tag{13.20}$$

On the right-hand side of Eq. (13.20), the denominator has been factored into two complex binomials $(s - s_1)$ and $(s - s_2)$, where

$$s_1 = -\frac{\omega_o}{2Q} + \left[\left(\frac{\omega_o}{2Q} \right)^2 - \omega_o^2 \right]^{1/2} \tag{13.21}$$

and
$$s_2 = -\frac{\omega_o}{2Q} - \left[\left(\frac{\omega_o}{2Q} \right)^2 - \omega_o^2 \right]^{1/2} \tag{13.22}$$

We recognize Eq. (13.20) as a biquad transfer function in which the s^2 and s coefficients in the numerator are set to zero. This feature causes the response to be unity at dc ($s = 0$) and to fall off as $1/s^2$ at high frequencies. As the expressions (13.21) and (13.22) indicate, s_1 and s_2 are complex conjugates with equal real parts and with imaginary parts of the same magnitude but opposite sign.

The factors s_1 and s_2 represent the poles of the transfer function (13.16). For the case where the R and C values yield a Q less than 0.5, the factor in brackets in Eqs. (13.21) and (13.22) will be positive, so that s_1 and s_2 will be real and Eq. (13.20) can be written in the form

$$\frac{\mathbf{V}_{out}}{\mathbf{V}_{in}} = \frac{\omega_o^2/s_1 s_2}{[(j\omega/\omega_1) + 1][(j\omega/\omega_2) + 1]} = \frac{1}{(1 + j\omega/\omega_1)(1 + j\omega/\omega_2)} \tag{13.23}$$

where $\omega_1 = -s_1$ and $\omega_2 = -s_2$. Equation (13.23) is produced by dividing the numerator and denominator on the right-hand side of Eq. (13.20) by $s_1 s_2$. The right-hand side of Eq. (13.23) is in the standard product of binomials form of Chapter 9, wherein the frequency response is described by two simple, real poles at ω_1 and ω_2. We recognize this transfer function as that of a low-pass filter of second order. At frequencies well below ω_1 and ω_2, its gain is unity. At frequencies well above ω_1 and ω_2, its gain falls off as $1/\omega^2$, that is, at -40 dB per decade in frequency.

If $Q = 0.5$ exactly, the factor inside the brackets in Eqs. (13.21) and (13.22) becomes zero. For this case, the poles of the filter coincide at ω_o, reducing the transfer function to

$$\frac{\mathbf{V}_{out}}{\mathbf{V}_{in}} = \frac{1}{(1 + j\omega/\omega_o)^2} \tag{13.24}$$

If Q is larger than 0.5, the square-root terms in Eqs. (13.21) and (13.22) become imaginary, and the poles become complex-conjugate numbers $\mathbf{s}_1$ and $\mathbf{s}_2$ with real part equal to

$$s_R = -\frac{\omega_o}{2Q} \tag{13.25}$$

and imaginary parts equal to

$$\pm js_I = \pm j[\omega_o^2 - (\omega_o/2Q)^2]^{1/2} \equiv \pm j[\omega_o^2 - s_R^2]^{1/2} \tag{13.26}$$

These complex poles $\mathbf{s}_1$ and $\mathbf{s}_2$ are located to the left of the imaginary s-axis in the s-plane, as shown in Fig. 13.8. Their placement in the left-half plane results because the s_R given by Eq. (13.25) is negative. The radial distance d from the origin to each of the poles $\mathbf{s}_1$ and $\mathbf{s}_2$ in Fig. 13.8 is given by

$$\begin{aligned} d &= |s_R + js_I| \\ &= \{(\omega_o/2Q)^2 + [\omega_o^2 - (\omega_o/2Q)^2]\}^{1/2} = \omega_o \end{aligned} \tag{13.27}$$

As this equation shows, the poles $\mathbf{s}_1$ and $\mathbf{s}_2$, when complex, lie on a circle of radius ω_o at an angle determined by the value of Q. The real and imaginary parts s_R and s_I are not independent. For a given ω_o, specifying s_R automatically specifies Q and s_I.

When the poles $\mathbf{s}_1$ and $\mathbf{s}_2$ are imaginary, the magnitude of the transfer function must be expressed as

$$|H(s)| = \left| \frac{\mathbf{V}_{\text{out}}}{\mathbf{V}_{\text{in}}} \right| = \frac{\omega_o^2}{|s - \mathbf{s}_1||s - \mathbf{s}_2|} \tag{13.28}$$

Figure 13.8
Location of poles $\mathbf{s}_1$ and $\mathbf{s}_2$ for $Q > 0.5$ in the s-plane for the low-pass filter transfer function (13.20).

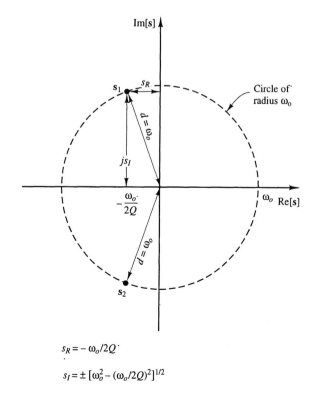

$s_R = -\omega_o/2Q$

$s_I = \pm [\omega_o^2 - (\omega_o/2Q)^2]^{1/2}$

Figure 13.9
Poles of the
low-pass filter of
Fig. 13.7 in the
s-plane. The
magnitude of
$\mathbf{V}_{out}/\mathbf{V}_{in}$ is
proportional to the
reciprocal of the
product $d_1 d_2$.

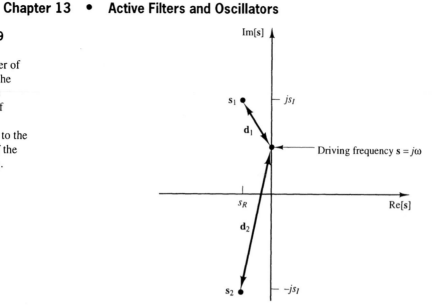

where $s = j\omega$. In the s-plane representation, the magnitudes $|s - s_1|$ and $|s - s_2|$ are determined by the distances between the poles s_1 and s_2 and the location $s = j\omega$ on the imaginary s-axis. As depicted in Fig. 13.9, the magnitude $|\mathbf{V}_{out}/\mathbf{V}_{in}|$ becomes

$$|H(s)| = \left|\frac{\mathbf{V}_{out}}{\mathbf{V}_{in}}\right| = \frac{\omega_o^2}{|s - s_1||s - s_2|} \equiv \frac{\omega_o^2}{d_1 d_2} \tag{13.29}$$

where d_1 and d_2 are the lengths of vectors $\mathbf{d}_1$ and $\mathbf{d}_2$, respectively.

For small frequencies, such that the driving point $s = j\omega$ in Fig. 13.8 is located near the origin, the vectors $\mathbf{d}_1$ and $\mathbf{d}_2$ have approximately the same length ω_o, and Eq. (13.29) yields $|\mathbf{V}_{out}/\mathbf{V}_{in}| \approx 1$. This situation is depicted in Fig. 13.10(a).

Figure 13.10
Low-pass filter
transfer function
(13.20) in the
s-plane. The
lengths of vectors
$\mathbf{d}_1$ and $\mathbf{d}_2$ are
shown at three
different driving
frequencies:
(a) $\omega = 0$;
(b) $0 < \omega < s_I$;
(c) $\omega > s_I$.

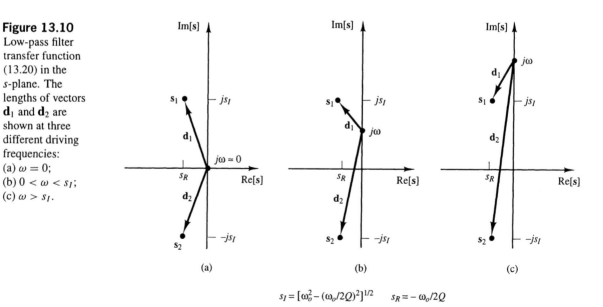

$$s_I = [\omega_o^2 - (\omega_o/2Q)^2]^{1/2} \qquad s_R = -\omega_o/2Q$$

As the frequency is raised and $j\omega$ rises up the imaginary s-axis, as in Fig. 13.10(b), $\mathbf{d}_2$ becomes longer and $\mathbf{d}_1$ shorter by approximately the same amount, so that $|\mathbf{V}_{out}/\mathbf{V}_{in}|$ remains approximately constant. As ω exceeds the value s_I, both $\mathbf{d}_1$ and $\mathbf{d}_2$ increase in length, and $|\mathbf{V}_{out}/\mathbf{V}_{in}|$ begins to decrease as $1/\omega^2$. The latter case is depicted in Fig. 13.10(c).

We now consider the case of large Q. If Q exceeds the value $1/\sqrt{2} = 0.707$, $\mathbf{s}_1$ will lie closer to the imaginary axis than to the real axis, because s_I will be larger than s_R. For this case, the product $d_1 d_2$ will be smaller than ω_o^2 as the driving frequency $j\omega$ passes the value js_I. The resulting Bode plot of $|\mathbf{V}_{out}/\mathbf{V}_{in}|$ will thus display a rise at the frequency $j\omega = js_I$. At very large values of Q (values of 10 or more), the Bode plot will actually peak sharply as $\mathbf{s} = j\omega$ passes through js_I. Note that js_I will be approximately equal to $j\omega_o$ as Q becomes very large.

Figure 13.11 shows the magnitude Bode plot of the second-order low-pass filter function (13.20) for several values of Q, including small and large values. In all cases, the roll-off at high frequencies proceeds at -40 dB/decade because $|\mathbf{V}_{out}/\mathbf{V}_{in}|$ decreases as $1/\omega^2$. When Q is less than $1/\sqrt{2} = 0.707$, the filter response decreases gradually as the driving frequency ω passes through ω_o. When Q is greater than 0.707, the filter response peaks above unity as ω passes through ω_o. When Q is equal to $1/\sqrt{2} = 0.707$, the horizontal portion of the plot extends as far as possible to the right without rising, and the filter's -3-dB point lies exactly at ω_o. This condition is sometimes called the *maximally flat* response. The maximally flat transfer function represents a good approximation to the ideal brick-wall response when the filter of Fig. 13.7 is used in stand-alone fashion. When several circuits are cascaded, so as to produce a maximally-flat overall filter response of higher order, the poles of each second-order section are sometimes located so as to produce values of Q other than $1/\sqrt{2}$. This concept is explored in detail in Section 13.4.

Figure 13.11
Magnitude plot of the second-order low-pass filter transfer function (13.20) for several values of Q. The slope of the Bode plot above ω_o is equal to -40 dB/decade. When $Q = 1/\sqrt{2} = 0.707$, the plot is said to be maximally flat. As Q becomes large, the pole frequency s_1 begins to approach the value $j\omega_o$.

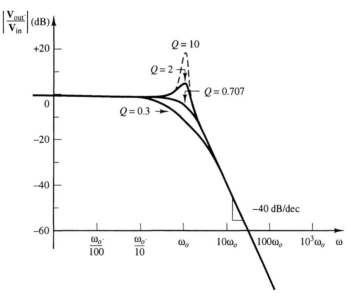

DESIGN

EXAMPLE 13.1

An amplitude-modulated (AM) radio transmission consists of a 530-kHz carrier modulated by an audio signal with frequency components from 300 Hz to 10 kHz. (See Section 4.4.5 for a discussion of amplitude modulation.) The signal is passed through a diode detector, which produces an output consisting of the desired audio signal plus unwanted frequency components at 530 kHz and above. Design a second-order analog filter that will pass the desired audio signal while attenuating the unwanted signals by at least -60 dB.

Solution

• Assess the goals of the problem

The filter must have a relatively constant gain up to 10 kHz and reduced gain at higher frequencies. The response above 10 kHz must fall with frequency at a rate sufficient to produce the required minimum attenuation of -60 dB at 530 kHz.

• Choose a design strategy

The upper frequency end of the filter's passband should be set to at least 10 kHz to guarantee full signal transmission of the desired audio signal. The frequency 530 kHz lies 1.72 decades above 10 kHz (10 kHz $\times 10^{1.72} \approx 530$ kHz), hence the required per-decade attenuation of the filter becomes $(-60$ dB$)/(1.72$ decade$) \approx -35$ dB/decade. This requirement is more than met by using a Sallen–Key second-order low-pass filter with a cutoff frequency (-3-dB point) of 10 kHz and a maximally-flat Q of 0.707. Such a filter will have a gain of unity in its passband and a high-frequency rolloff of -40 dB/decade above its cutoff frequency $f_o = \omega_o/2\pi$. This rolloff is larger than the minimum required value of -35 dB/decade.

• Choose values for all components in the circuit

The selection of R_1, R_2, C_1, and C_2 is not unique; many values will yield the required values of $Q = 0.707$ and $\omega_o = 2\pi f_o = 2\pi (10$ kHz$) = 6.28 \times 10^4$ rad/s. If R_1 and R_2 are arbitrarily chosen to have the same value R, Eqs. (13.18) and (13.19) yield

$$\omega_o = \frac{1}{R\sqrt{C_1 C_2}} \tag{13.30}$$

and

$$Q = \frac{1}{2}\sqrt{\frac{C_1}{C_2}} \tag{13.31}$$

These equations can also be expressed in the form:

$$C_1 C_2 = \frac{1}{(\omega_o R)^2} \tag{13.32}$$

and

$$\frac{C_1}{C_2} = 4Q^2 \tag{13.33}$$

The resistance value R is arbitrarily chosen as 10 kΩ. Combining Eqs. (13.32) and (13.33) and substituting the desired values of ω_o and Q yields

$$C_1 = \frac{2Q}{\omega_o R} = \frac{2(0.707)}{(6.28 \times 10^4 \text{ rad/s})(10 \text{ k}\Omega)} \approx 2.3 \text{ nF} \tag{13.34}$$

and

$$C_2 = \frac{1}{2Q\omega_o R} = \frac{1}{2(0.707)(6.28 \times 10^4 \text{ rad/s})(10 \text{ k}\Omega)} \approx 1.1 \text{ nF} \tag{13.35}$$

• Evaluate the design and revise if necessary

The choice of $\omega_o = 2\pi (10$ kHz$)$ places the -3-dB point exactly at 10 kHz. The components of the incoming signal that lie near 10 kHz will thus be attenuated slightly by the filter. Moving ω_o to the highest value possible will ensure that the entire input spectrum is passed with nearly the same gain. Since the filter has a rolloff of -40 dB/decade, an attenuation of -60 dB at 530 kHz could still be realized by placing the cutoff frequency only 1.5 decades below 530 kHz, that is, at 530 kHz$/10^{1.5} \approx 16.8$ kHz. This design change would require new capacitor values of $C_1 = 1.3$ nF and $C_2 = 670$ pF.

Discussion. The capacitor values determined in this example are not standard values, hence one additional design modification might be to change C_1 and C_2 to their nearest respective "off-the-shelf" component values, with appropriate changes in R_1 and R_2. Alternatively, if the capacitors are to be fabricated on an integrated circuit, these nonstandard values can be chosen at fabrication time. ∎

EXERCISE 13.1 Redesign the filter of Example 13.1 so that $Q = 0.6$. This choice of Q will result in a filter for which the response is not maximally flat. **Answer:** (one possible design) $R_1 = R_2 = 10\,k\Omega$; $C_1 = 1.9\,nF$; $C_2 = 1.3\,nF$

13.2 Design a second-order filter with a -3-dB cutoff frequency of 1 kHz and a Q of 1.3. This choice of Q will result in a filter with a peak in its response near ω_o. **Answer:** (one possible design) $R_1 = R_2 = 5\,k\Omega$; $C_1 = 83\,nF$; $C_2 = 12\,nF$

13.3 Show that Eq. (13.13) leads to Eq. (13.16).

13.4 For the pole frequencies defined by Eqs. (13.21) and (13.22), show that $s_1 s_2 = \omega_o^2$. Express s_1 and s_2 as $s_R \pm j s_I$, where s_R and s_I are given by Eqs. (13.25) and (13.26).

13.5 Plot the magnitude and angle of the low-pass filter transfer function (13.20) as a function of frequency for several values of Q. Write a computer program to help with these calculations.

13.3.3 Second-Order Active High-Pass Filter

The active circuit of Fig. 13.12 is a second-order high-pass filter of the Sallen–Key type. A high-pass filter transmits frequency components above its cutoff frequency ω_o and attenuates frequency components below ω_o. The circuit of Fig. 13.12 is the *dual* of the low-pass filter of Fig. 13.7; the locations of all capacitors and resistors are exchanged. Because the basic circuit topology is preserved, the output can be found by exchanging the R and $j\omega C$ terms in the transfer function for the low-pass filter. Performing this operation on Eq. (13.15) yields

$$\mathbf{V}_{out} = \mathbf{V}_{in} \frac{R_1 R_2}{R_1 R_2 + R_1(\mathbf{Z}_1 + \mathbf{Z}_2) + \mathbf{Z}_1 \mathbf{Z}_2} \tag{13.36}$$

Substitution of $1/j\omega C_1$ for $\mathbf{Z}_1$ and $1/j\omega C_2$ for $\mathbf{Z}_2$ in Eq. (13.36) and some manipulation results in

$$\mathbf{V}_{out} = \mathbf{V}_{in} \frac{(j\omega)^2 C_1 C_2 R_1 R_2}{1 - \omega^2 (C_1 C_2 R_1 R_2) + j\omega R_1 (C_1 + C_2)} \tag{13.37}$$

Figure 13.12
Second-order
Sallen–Key active
high-pass filter.
The slope of the
Bode plot below ω_o
is equal to
$+40\,dB/decade$.

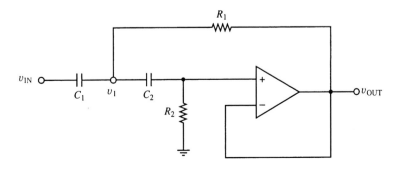

The transfer function (13.37) can again be represented in the s-plane by making the following substitutions:

$$s = j\omega \tag{13.38}$$

$$\omega_o = \frac{1}{\sqrt{R_1 R_2 C_1 C_2}} \tag{13.39}$$

and

$$Q = \frac{\omega_o R_1 R_2 C_1 C_2}{R_1(C_1 + C_2)} = \sqrt{\frac{R_2}{R_1}}\left(\frac{\sqrt{C_1 C_2}}{C_1 + C_2}\right) \tag{13.40}$$

With these substitutions, Eq. (13.37) becomes

$$H(s) = \frac{\mathbf{V}_{\text{out}}}{\mathbf{V}_{\text{in}}} = \frac{s^2}{s^2 + s(\omega_o/Q) + \omega_o^2} = \frac{s^2}{(s - s_1)(s - s_2)} \tag{13.41}$$

The roots of the denominator of this expression are again given by Eqs. (13.21) and (13.22), respectively, and Eq. (13.41) again has the form of a biquadratic transfer function. In this case, the numerator consists of a single factor of s^2. At high frequencies, the denominator approaches a limit consisting of a single factor s^2, but this factor is canceled by the factor of s^2 in the numerator. Hence $|\mathbf{V}_{\text{out}}/\mathbf{V}_{\text{in}}|$ approaches a limit of unity gain at high frequencies. As the driving frequency is reduced well below ω_o, the denominator in (13.41) approaches the constant value ω_o^2. In this case, the factor of s^2 in the numerator causes $|\mathbf{V}_{\text{out}}/\mathbf{V}_{\text{in}}|$ to fall toward zero at the rate of 40 dB/decade.

Figure 13.13
The lengths of vectors $\mathbf{d_1}$, $\mathbf{d_2}$, and $\mathbf{d_3}$ shown at three different frequencies ω for the second-order high-pass filter function (13.41) with $Q > 0.5$:
(a) $\omega \approx 0$;
(b) $0 < \omega < s_I$;
(c) $\omega > s_I$.

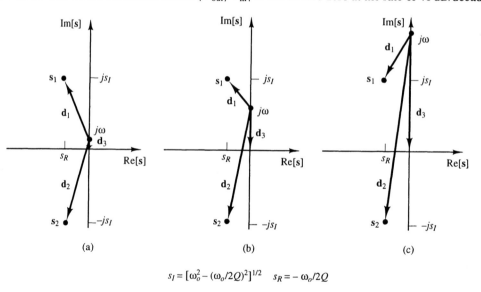

$$s_I = [\omega_o^2 - (\omega_o/2Q)^2]^{1/2} \quad s_R = -\omega_o/2Q$$

The magnitude of $\mathbf{V}_{\text{out}}/\mathbf{V}_{\text{in}}$ as a function of input frequency can be determined by again examining the vectors $\mathbf{d}_1$ and $\mathbf{d}_2$ in the s-plane, as in Fig. 13.13. An additional vector $\mathbf{d}_3$, which extends from the location of the driving frequency $j\omega$ to the origin, is needed to represent the factor of s^2 in the numerator of Eq. (13.41). This vector has a length $d_3 = \omega$. When $j\omega$ lies near zero, as in Fig. 13.13(a), vectors $\mathbf{d}_1$ and $\mathbf{d}_2$ have nearly the same length, and $\mathbf{d}_3$ has approximately zero length. The magnitude of the transfer function (13.41) in the limit $j\omega \sim 0$ thus becomes

$$\left|\frac{\mathbf{V}_{\text{out}}}{\mathbf{V}_{\text{in}}}\right| = \frac{d_3^2}{d_1 d_2} \approx 0 \tag{13.42}$$

where d_1, d_2, and d_3 represent the lengths of $\mathbf{d}_1$, $\mathbf{d}_2$, and $\mathbf{d}_3$, respectively. If $j\omega$ is increased above zero, d_2 will increase by about the same amount that d_1 decreases, as in the low-pass filter case. Were it not for d_3, this relationship would again keep $|\mathbf{V}_{out}/\mathbf{V}_{in}|$ constant for $\omega < s_I$. The length d_3 increases with $j\omega$, however, so that $|\mathbf{V}_{out}/\mathbf{V}_{in}|$ for the high-pass filter increases as the square of ω. As ω approaches s_I, d_1 reaches a minimum. Well above $\omega = s_I$, the lengths d_1, d_2, and d_3 all approach the same value, so that the magnitude of $|\mathbf{V}_{out}/\mathbf{V}_{in}|$ approaches unity. Note that if $s_R \ll s_I$, s_I will be approximately equal to ω_o, as seen from Eq. (13.26).

Figure 13.14
Magnitude plot of the second-order high-pass filter transfer function (13.41) for several values of Q. When $Q = 1/\sqrt{2} = 0.707$, the plot is said to be maximally flat.

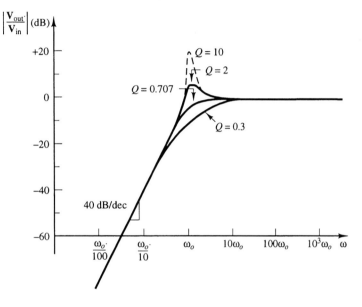

The magnitude Bode plots of this filter for several values of Q are shown in Fig. 13.14. Each plot has a low-frequency $+40$ dB/decade slope (the response increases as ω^2), a corner frequency near ω_o, and a flat region above ω_o. As in the low-pass filter case, if $Q < 0.5$, the poles s_1 and s_2 become real and the filter function (13.41) can be expressed by

$$H(j\omega) = \frac{\mathbf{V}_{out}}{\mathbf{V}_{in}} = \frac{(j\omega)^2/\omega_1\omega_2}{(1 + j\omega/\omega_1)(1 + j\omega/\omega_2)} \tag{13.43}$$

where $\omega_1 = s_1$, $\omega_2 = s_2$, and $\omega_1\omega_2 = \omega_o^2$. Equation (13.43) is in the product-of-binomials form introduced in Chapter 9. For the case $Q = 0.5$, the poles ω_1 and ω_2 coincide at ω_o.

We note in Fig. 13.14 that when Q is small, the magnitude of the response increases gradually as ω pass through ω_o. When Q is equal to $1/\sqrt{2} = 0.707$, the response becomes maximally flat. For larger values of Q, a peak appears in the response at $\omega = \omega_o$.

EXERCISE 13.6

Design a second-order high-pass filter with parameters $\omega_o = 1$ kHz and $Q = 0.707$.

13.7   Draw the angle plot $\measuredangle \mathbf{V}_{out}/\mathbf{V}_{in}$ as a function of frequency for the high-pass filter transfer function (13.41) for several values of Q.

13.8   Show that Eq. (13.36) leads to Eq. (13.37).

13.3.4 Second-Order Active Band-Pass Filter

The op-amp circuit shown in Fig. 13.15 is a *second-order active band-pass* filter. It transmits only those frequency components contained within its passband and attenuates both low-frequency and high-frequency components that lie outside this range. The band-pass behavior of the circuit can be confirmed qualitatively by examining the circuit in the limits of zero and infinite frequency. In the limit $j\omega \to 0$, capacitors C_1 and C_2 behave as open circuits, and v_{IN} is effectively disconnected from the op-amp terminals. The remaining dc portion of the circuit has the form of a follower with zero input. In the limit $j\omega \to \infty$, capacitors C_1 and C_2 behave as short circuits, and the circuit functions as an inverting amplifier with a feedback impedance of zero. Because the gain of an op-amp inverter is proportional to its feedback impedance, the resulting output equals zero regardless of input.

Figure 13.15
Second-order active bandpass filter of the Sallen–Key type.

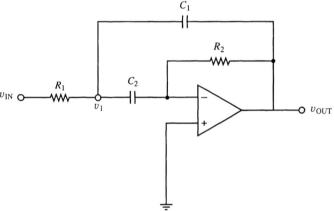

A mathematical expression for the transfer function of this circuit can be derived from KVL and KCL. The op-amp, C_2, and R_2 form an inverting amplifier between v_{OUT} and v_1. The voltage v_{OUT} can thus be expressed in the frequency domain by

$$\mathbf{V}_{\text{out}} = -\frac{R_2}{\mathbf{Z}_2}\mathbf{V}_1 \tag{13.44}$$

where $\mathbf{Z}_2 = 1/j\omega C_2$. Both $\mathbf{V}_{\text{in}}$ and $\mathbf{V}_{\text{out}}$ contribute to $\mathbf{V}_1$; hence $\mathbf{V}_1$ can be found using superposition and the complex form of the voltage divider. Alternately setting $\mathbf{V}_{\text{out}}$ and $\mathbf{V}_{\text{in}}$ to zero with v_- assumed to be at virtual ground potential yields a value for $\mathbf{V}_1$:

$$\mathbf{V}_1 = \mathbf{V}_{\text{in}}\frac{\mathbf{Z}_1\|\mathbf{Z}_2}{R_1 + \mathbf{Z}_1\|\mathbf{Z}_2} + \mathbf{V}_{\text{out}}\frac{R_1\|\mathbf{Z}_2}{\mathbf{Z}_1 + R_1\|\mathbf{Z}_2} \tag{13.45}$$

Substituting Eq. (13.45) for $\mathbf{V}_1$ into Eq. (13.44) results in

$$\mathbf{V}_{\text{out}} = -\frac{R_2}{\mathbf{Z}_2}\left(\mathbf{V}_{\text{in}}\frac{\mathbf{Z}_1\|\mathbf{Z}_2}{R_1 + \mathbf{Z}_1\|\mathbf{Z}_2} + \mathbf{V}_{\text{out}}\frac{R_1\|\mathbf{Z}_2}{\mathbf{Z}_1 + R_1\|\mathbf{Z}_2}\right) \tag{13.46}$$

Equation (13.46) can be simplified by substituting appropriate expressions for each parallel combination of impedances and by moving the factor of $R_2/\mathbf{Z}_2$ inside the parentheses:

$$\begin{aligned}
\mathbf{V}_{\text{out}} &= -\left[\mathbf{V}_{\text{in}}\frac{\mathbf{Z}_1 R_2/(\mathbf{Z}_1 + \mathbf{Z}_2)}{R_1 + \mathbf{Z}_1\mathbf{Z}_2/(\mathbf{Z}_1 + \mathbf{Z}_2)} + \mathbf{V}_{\text{out}}\frac{R_1 R_2/(R_1 + \mathbf{Z}_2)}{\mathbf{Z}_1 + R_1\mathbf{Z}_2/(R_1 + \mathbf{Z}_2)}\right] \\
&= -\left(\frac{\mathbf{V}_{\text{in}}\mathbf{Z}_1 R_2 + \mathbf{V}_{\text{out}}R_1 R_2}{R_1\mathbf{Z}_1 + R_1\mathbf{Z}_2 + \mathbf{Z}_1\mathbf{Z}_2}\right)
\end{aligned} \tag{13.47}$$

Solving Eq. (13.47) for $\mathbf{V}_{\text{out}}$ results, after some algebra, in

$$\mathbf{V}_{\text{out}} = \frac{-\mathbf{Z}_1 R_2}{R_1 \mathbf{Z}_1 + R_1 \mathbf{Z}_2 + \mathbf{Z}_1 \mathbf{Z}_2 + R_1 R_2} \mathbf{V}_{\text{in}} \tag{13.48}$$

Substituting $1/j\omega C_1$ for $\mathbf{Z}_1$ and $1/j\omega C_2$ for $\mathbf{Z}_2$ into Eq. (13.48) produces the desired transfer function:

$$\begin{aligned} H(j\omega) = \frac{\mathbf{V}_{\text{out}}}{\mathbf{V}_{\text{in}}} &= \frac{-R_2/j\omega C_1}{R_1/j\omega C_1 + R_1/j\omega C_2 + (1/j\omega C_1)(1/j\omega C_2) + R_1 R_2} \\ &= \frac{-R_2(j\omega C_2)}{j\omega R_1(C_1 + C_2) + 1 + (j\omega C_1)(j\omega C_2)R_1 R_2} \end{aligned} \tag{13.49}$$

Dividing the numerator and denominator of Eq. (13.49) by $R_1 R_2 C_1 C_2$ and expressing $j\omega$ as s results in

$$H(s) = \frac{\mathbf{V}_{\text{out}}}{\mathbf{V}_{\text{in}}} = \frac{-j\omega R_2 C_2/R_1 R_2 C_1 C_2}{\frac{(j\omega C_1)(j\omega C_2)R_1 R_2 + j\omega R_1(C_1+C_2) + 1}{R_1 R_2 C_1 C_2}} = \frac{-\omega_o^2 R_2 C_2 s}{s^2 + s(\omega_o/Q) + \omega_o^2} \tag{13.50}$$

where

$$\omega_o = \frac{1}{\sqrt{R_1 R_2 C_1 C_2}} \tag{13.51}$$

and

$$Q = \frac{\omega_o R_1 R_2 C_1 C_2}{R_1(C_1 + C_2)} = \sqrt{\frac{R_2}{R_1}} \left(\frac{\sqrt{C_1 C_2}}{C_1 + C_2} \right) \tag{13.52}$$

Equation (13.50) has the form of a biquad transfer function in which the numerator contains a single factor of s. As the frequency is reduced well below ω_o, the denominator approaches a constant value of ω_o^2. The factor of s in the numerator thus causes $|\mathbf{V}_{\text{out}}/\mathbf{V}_{\text{in}}|$ to fall toward zero at the rate of 20 dB/decade as the frequency is reduced. As the frequency is increased well above ω_o, the denominator of $H(s)$ approaches the limit s^2. The factor of s in the numerator cancels one factor of s in the denominator, leaving a single factor of s in the denominator that causes $|\mathbf{V}_{\text{out}}/\mathbf{V}_{\text{in}}|$ to be reduced as $1/s$ (i.e., at the rate of -20 dB/decade) as the frequency is increased. At the passband center frequency ω_o, the magnitude $|\mathbf{V}_{\text{out}}/\mathbf{V}_{\text{in}}|$ can be found by substituting $\omega = \omega_o$ into Eq. (13.49):

$$\begin{aligned} \left| \frac{\mathbf{V}_{\text{out}}}{\mathbf{V}_{\text{in}}} \right|_{\omega=\omega_o} &= \left| \frac{-j\omega_o R_2 C_2}{j\omega_o R_1(C_1 + C_2) + 1 - \omega_o^2 R_1 R_2 C_1 C_2} \right| \\ &= \frac{R_2 C_2}{R_1(C_1 + C_2)} \end{aligned} \tag{13.53}$$

where $\omega_o = 1/\sqrt{R_1 R_2 C_1 C_2}$. Given the expression (13.52) for Q, Eq. (13.53) becomes

$$\left| \frac{\mathbf{V}_{\text{out}}}{\mathbf{V}_{\text{in}}} \right|_{\omega=\omega_o} = \sqrt{\frac{R_2 C_2}{R_1 C_1}} Q \tag{13.54}$$

As Eq. (13.54) suggests, $|\mathbf{V}_{\text{out}}/\mathbf{V}_{\text{in}}|$ at $\omega = \omega_o$ is dependent on Q. Analysis of Eq. (13.50) in the s-plane yields the magnitude plots of Fig. 13.16, shown for several values of Q and values of R and C such that $R_2 C_2/R_1 C_1 = 100$.

Figure 13.16
Magnitude plot of
the band-pass filter
transfer function
(13.50) for several
values of Q. The
components $R_1, C_1,$
$R_2,$ and C_2 have
been chosen so that
$R_2C_2/R_1C_1 = 100$.
The locations of the
passband limits ω_1
and ω_2 are shown
for the case
$Q = 0.01$.

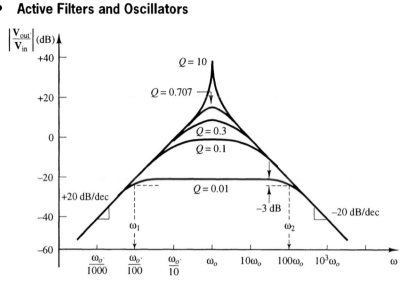

For any value of Q, the two frequencies ω_1 and ω_2 at which the output falls by a factor of $1/\sqrt{2} \equiv -3$ dB from its value at $\omega = \omega_o$ are called the *half-power* frequencies of the band-pass filter. The difference $\omega_2 - \omega_1$ is called the *bandwidth* of the filter. The relationship between ω_o, ω_1, ω_2, and the bandwidth of the filter for the specific case $Q = 0.01$ is illustrated in Fig. 13.16. For large Q, analysis of the transfer function in the s-plane shows the half-power frequencies to be located at $\omega_o \pm \omega_o/2Q$ on either side of ω_o, as we now show.

The denominator of Eq. (13.50) can be factored into two binomials $(\mathbf{s} - \mathbf{s}_1)$ and $(\mathbf{s} - \mathbf{s}_2)$ and the magnitudes of these binomials represented by the vectors $\mathbf{d}_1$ and $\mathbf{d}_2$ shown in Fig. 13.17. The vector $\mathbf{d}_3$ in Fig. 13.17 represents the factor of $\mathbf{s}$ in the numerator of Eq. (13.50). As shown in Fig. 13.17(a), the vector length d_1 reaches its minimum value at the frequency $\omega = s_I$. If s_R is small compared to s_I (i.e., large Q), then

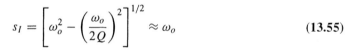

$$s_I = \left[\omega_o^2 - \left(\frac{\omega_o}{2Q} \right)^2 \right]^{1/2} \approx \omega_o \qquad (13.55)$$

Figure 13.17
The s-plane
representation of
the poles of the
second-order
band-pass filter of
Fig. 13.15:
(a) $\omega = s_I$;
(b) $\omega = s_I - |s_R|$.

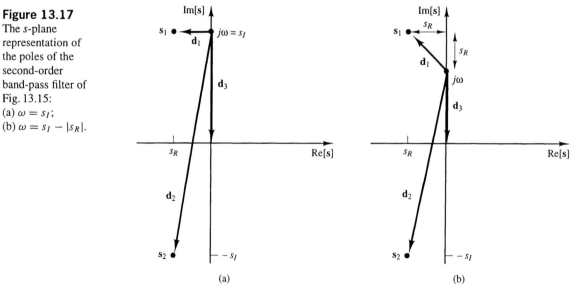

At this frequency, the magnitude of $\mathbf{V}_{out}/\mathbf{V}_{in}$, which is proportional to d_3/d_1d_2, reaches its maximum value. If ω is decreased to the value $(s_I - s_R)$, as in Fig. 13.17(b), d_1 will increase by a factor of $\sqrt{2}$. If s_R is small compared to s_I, the vector lengths d_2 and d_3 will be decreased only slightly at this new frequency. As a result of these combined effects, $|\mathbf{V}_{out}/\mathbf{V}_{in}|$ will fall by a factor of $1/\sqrt{2}$ from its value at $\omega = s_I \approx \omega_o$. The half-power frequency ω_1, therefore, is equal to

$$\omega_1 = s_I - s_R \tag{13.56}$$

Note from the denominator of Eq. (13.50) that $|s_R|$ is equal to $\omega_o/2Q$.

Given the result (13.55), the half-power frequency becomes

$$\omega_1 \approx \omega_o - \frac{\omega_o}{2Q} \tag{13.57}$$

This same reasoning can be applied to the case $\omega_2 = (s_I + s_R)$ to yield

$$\omega_2 \approx \omega_o + \frac{\omega_o}{2Q} \tag{13.58}$$

The resulting bandwidth of the filter for the case $s_R \ll s_I$ is thus equal to

$$\text{BW} = \omega_2 - \omega_1 = \frac{\omega_o}{Q} \tag{13.59}$$

Note that a small s_R is equivalent to a large Q, since $Q = \omega_o/2s_R$.

EXERCISE 13.9   Derive the result (13.54) by substituting $s = j\omega_o$ into Eq. (13.50).

13.10   Synthesize a second-order band-pass filter with a Q of 100 and a center frequency of 10 kHz. What is the magnitude of the response at $\omega = \omega_o$? What is the bandwidth of the filter?

13.11   Design a band-pass filter using the specifications of Exercise 13.10 so that the gain is equal to $+60\,\text{dB}$ at ω_o. What is the bandwidth of your new design?

13.12   By analyzing the transfer function (13.50) in the s-plane, show that the band-pass filter of Fig. 13.15 produces responses with the magnitude plots of Fig. 13.16.

13.13   Draw the angle plot $\angle \mathbf{V}_{out}/\mathbf{V}_{in}$ as a function of frequency for the band-pass filter transfer function (13.49) for several values of Q. Write a computer program to help with these calculations.

13.14   Show that Eq. (13.48) follows from Eq. (13.47).

13.15   Arrive at the result (13.48) using KCL at the v_1 node, rather than the superimposed complex voltage-divider expression (13.45).

13.4 ACTIVE FILTER CASCADING

The usefulness of active filters becomes apparent when two or more are cascaded together to produce filter transfer functions of increased order or complexity. As mentioned in Section 13.1, the output of any one active op-amp filter stage will appear as a voltage source to the input of the next stage, so that interstage loading problems are virtually nonexistent. The overall transfer function of an active filter cascade thus will be equal to the simple product of the transfer functions of each of its individual stages. A filter of order higher than 2 is easily synthesized by simply cascading several one- or two-pole filters in series. The poles of each component filter are appropriately chosen such that the desired overall response is achieved.

In general, the higher the order of a filter cascade, the more closely its transfer function can be made to approach one of the ideal "brick-wall" responses of Fig. 13.4. In this section, we examine the techniques required to accomplish this task. Although we shall focus primarily on the low-pass filter, the concepts introduced apply equally well to high-pass, band-pass, and band-reject filters.

13.4.1 Low-Pass Butterworth Response

The key to designing any filter cascade is the selection of the pole frequencies for each of its component sections. The various poles in a cascade act jointly to produce its overall response. In analyzing the requirements for optimal pole placement, it is convenient to describe a filter's response in terms of the magnitude squared of its transfer function $H(s)$. In the sinusoidal steady-state, where $s = j\omega$, $|H(s)|^2$ becomes equivalent to $|H(j\omega)|^2$, which is a function of ω^2.

The ideal "brick-wall" low-pass filter response has a magnitude of A_o in the passband and a magnitude of zero outside the passband. The magnitude-squared response $|H(s)|^2 = |H(j\omega)|^2$ of such a filter becomes

$$|H(j\omega)|^2 = A_o^2 \qquad 0 < \omega < \omega_o$$
$$= 0 \qquad \omega > \omega_o \qquad (13.60)$$

This ideal response can be closely approximated by the realizable function

$$|H(j\omega)|^2 = \frac{A_o^2}{1 + f(\omega^2)} \qquad (13.61)$$

The function $f(\omega^2)$ must be chosen such that $f(\omega^2) \ll 1$ for $\omega < \omega_o$, and $f(\omega^2) \gg 1$ for $\omega > \omega_o$. Such an $f(\omega^2)$ will cause $|H(j\omega)|^2$ to quickly approach the limits A_o^2 for $\omega < \omega_o$ and zero for $\omega > \omega_o$.

One possible choice for $f(\omega^2)$, generally attributed to the British engineer S. Butterworth,[2] consists of the simple function $f(\omega^2) = (\omega^2/\omega_o^2)^n$, where n is a positive integer called the *order* of the filter cascade. This choice of $f(\omega^2)$ results in a magnitude-squared function of

$$|H(j\omega)|^2 = \frac{A_o^2}{1 + (\omega/\omega_o)^{2n}} \qquad (13.62)$$

Figure 13.18
Plot of $|H(j\omega)|$ given by Eq. (13.62) for several values of n.

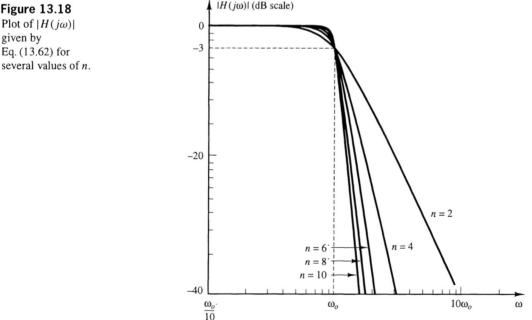

---

[2] S. Butterworth, "On the Theory of Filter Amplifiers," *Wireless Engineer*, Vol. 7, 536–541, October 1930.

The larger the order n, the more rapidly $|H(j\omega)|^2$ will fall toward zero when ω exceeds ω_o; similarly, the larger the value of n, the more rapidly $|H(j\omega)|^2$ will approach A_o^2 as ω falls below ω_o. It can be shown that this choice for $f(\omega^2)$ produces the flattest possible transfer function in the passband. The first $2n - 1$ derivatives of $|H(j\omega)|$ will contain a factor of ω, and thus will be equal to zero at $\omega = 0$. This feature is sometimes called the *maximally flat* condition. A plot of $|H(j\omega)|$ for various values of n is shown in Fig. 13.18.

The magnitude-squared function (13.62) may also be described in the s-plane, where the filter input frequency becomes $\mathbf{s} = j\omega$ and the filter poles are complex. Substituting $\mathbf{s}/j$ for ω in Eq. (13.62) and expressing the function as $|H(s)|$ results in

$$|H(s)| = \left| \frac{A_o}{\sqrt{1 + [(s/j)/\omega_o]^{2n}}} \right| \equiv \left| \frac{A_o}{\sqrt{1 + (-1)^n (s/\omega_o)^{2n}}} \right| \tag{13.63}$$

where $1/j^2 = -1$. We assume that $H(s)$ can be factored into the product of binomials form, so that the magnitude function (13.63) can also be expressed as

$$|H(s)| = \left| \frac{A_o}{(1 - s/s_1)(1 - s/s_2) \cdots (1 - s/s_n)} \right| \tag{13.64}$$

The poles of $H(s)$ expressed in Eq. (13.64) can be determined by finding the values of $\mathbf{s}$ at which the denominator of Eq. (13.63) goes to zero:

$$\sqrt{1 + (-1)^n (s/\omega_o)^{2n}} = 0 \tag{13.65}$$

Equation (13.65) can also be expressed as

$$(s/\omega_o)^{2n} = \frac{-1}{(-1)^n} \equiv (-1)^{n-1} \tag{13.66}$$

When the order n of the filter is *odd*, such that $(-1)^{n-1} = 1$, Eq. (13.66) becomes

$$\mathbf{s} = \omega_o (1)^{1/2n} \tag{13.67}$$

The root of 1 has been explicitly written as $(1)^{1/2n}$ to remind us that $\mathbf{s}$ is complex, hence $(1)^{1/2n}$ will have $2n$ possible values in the complex plane. Each value of $(1)^{1/2n}$ will be a complex vector that, when multiplied by itself $2n$ times, yields a factor of 1. If the factor 1 is expressed as $e^{j2k\pi}$, where k is any integer, then Eq. (13.67) becomes

$$\mathbf{s} = \omega_o e^{jk\pi/n} \tag{13.68}$$

Note that $e^{jk\pi/n}$ is a unit vector of angle $k\pi/n$. Raising it to the $2n$ power yields $e^{j2k\pi}$, which is equal to 1 regardless of the value of k. The values of $\mathbf{s}$ that satisfy Eq. (13.65) thus will consist of vectors of magnitude ω_o and angle $k\pi/n$, where k is an integer and n is the order of the filter. A little thought will reveal that there are exactly $2n$ unique vectors described by Eq. (13.68), one for each of the values $k = 0$ through $k = 2n - 1$; higher values of k simply yield the first $2n$ vectors multiplied by factors of $e^{j2\pi} = 1$.

Figure 13.19
Vectors indicating the s-values that satisfy Eq. (13.65). The s-values that lie in the left-half plane are chosen as the poles for the nth-order Butterworth filter.

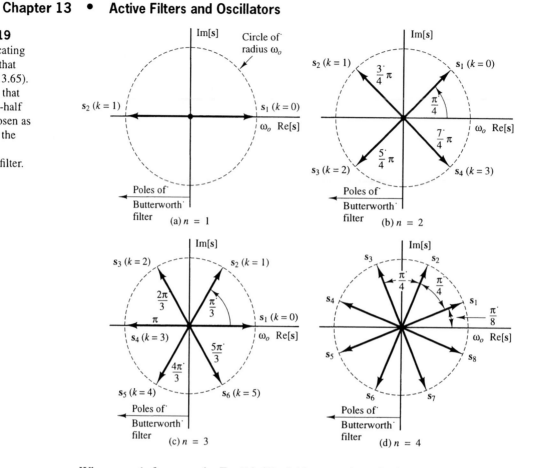

When $n = 1$, for example, Eq. (13.68) yields two value of **s** that satisfy Eq. (13.65)—one at $s_1 = \omega_o e^{j0} \equiv \omega_o$ (for $k = 0$), and one at $s_2 = \omega_o e^{j\pi} \equiv -\omega_o$ (for $k = 2n - 1 = 1$). These s-value locations are illustrated by vectors in Fig. 13.19(a). Because $H(s)$ must be a stable filter function with all its poles in the left half of the s-plane, those s-values having negative real parts are associated with $H(s)$; those having positive real parts are associated with the negative root of the denominator of Eq. (13.63). The circuit that produces the first-order Butterworth response must contain a pole at the location $s_1 = -\omega_o$.

When $n = 3$, the s-values satisfying Eq. (13.65) have magnitude ω_o and angles of 0, $\pi/3$, $2\pi/3$, π, $4\pi/3$, and $5\pi/3$ (for $k = 0, \cdots, 5$), as illustrated in Fig. 13.19(c). Those s-values lying to the left of the Im[s] axis constitute the desired poles of $H(s)$ and are chosen to obtain a third-order Butterworth response.

When the order n of the filter is *even*, such that $(-1)^{n-1} = -1$, Eq. (13.66) becomes

$$\mathbf{s} = \omega_0(-1)^{1/2n} \tag{13.69}$$

In this case, the factor (-1) can be expressed as $e^{j\pi}e^{j2k\pi}$, where k is again any integer. Equation (13.69) then becomes

$$\mathbf{s} = \omega_0 e^{j(\pi+2k\pi)/2n} \tag{13.70}$$

In this case, raising the unit vector $e^{j(\pi+2k\pi)/2n}$ to the $2n$ power yields a value of -1, regardless of the value of k. The values of **s** that satisfy Eq. (13.69) now consist of vectors of magnitude ω_o and angle $\pi(1 + 2k)/2n$. Multiplying the angle of any of these vectors by $2n$ yields a net angle of π.

When $n = 2$, for example, Eq. (13.70) yields s-values of magnitude ω_o and angles $\pi/4$, $3\pi/4$, $5\pi/4$, and $7\pi/4$, as shown in Fig. 13.19(b). These angles correspond to those obtained for $k = 0$ through 3. Similarly, when $n = 4$, Eq. (13.70) yields eight s-values of magnitude ω_o and angles spaced $\pi/4$ apart, beginning at $n = \pi/8$, as illustrated in Fig. 13.19(d). These angles correspond to those obtained for $k = 0$ through 7. In both cases, those s-values lying to the left of the Im[s] axis are chosen for the poles of $H(s)$ and must be produced by the filter cascade in order to obtain the second- and fourth-order Butterworth responses, respectively.

The preceding analysis is easily generalized. The $2n$ s-values that satisfy Eqs. (13.68) or (13.70) can be expressed by the equations

$$\left.\begin{array}{l} s_R = \omega_o \cos k\pi/n \\ s_I = j\omega_o \sin k\pi/n \end{array}\right\} \quad (n \text{ odd}) \tag{13.71}$$

and

$$\left.\begin{array}{l} s_R = \omega_o \cos (1 + 2k)\pi/2n \\ s_I = j\omega_o \sin (1 + 2k)/2n \end{array}\right\} \quad (n \text{ even}) \tag{13.72}$$

where $k = 0, \cdots, 2n - 1$ in each case. The n poles selected for a stable nth-order Butterworth filter will be those values given by Eqs. (13.71) and (13.72) for which s_R is negative.

The locations of the poles found in this way for order up to $n = 9$ are summarized in Table 13.1. The pole values in the table are computed for a filter cutoff frequency of $\omega_o = 1$ rad/s; Butterworth pole locations for other desired cutoff frequencies are obtained by multiplying the values in the table by the desired ω_o. Note that of the $2n$ poles that satisfy Eqs. (13.68) or (13.70), there will always be exactly n poles for which s_R is negative.

In practice, the n required poles of an nth-order Butterworth response are produced by cascading filter sections of smaller order. When the cascade order is even, such sections may consist solely of second-order Sallen–Key biquad sections of the type introduced in Section 13.3. If the cascade order is odd, an additional first-order filter of the type introduced in Section 13.1 must also be included. In either case, if all the poles in the table are represented somewhere in the cascade, they will act in concert to produce an overall response of the Butterworth type.

	$n = 2$	$n = 3$	$n = 4$	$n = 5$	$n = 6$	$n = 7$	$n = 8$	$n = 9$
s_R	-0.707	-0.500	-0.924	-0.809	-0.259	-0.901	-0.195	-0.940
s_I	$\pm j0.707$	$\pm j0.866$	$\pm j0.383$	$\pm j0.588$	$\pm j0.966$	$\pm j0.431$	$\pm j0.981$	± 0.342
s_R		-1.000	-0.383	-0.309	-0.707	-0.223	-0.556	-0.174
s_I		0.000	$\pm j0.924$	$\pm j0.951$	$\pm j0.707$	$\pm j0.975$	$\pm j0.831$	± 0.985
s_R				-1.000	-0.966	-0.623	-0.831	-0.500
s_I				0.000	$\pm j0.259$	$\pm j0.782$	$\pm j0.556$	$\pm j0.866$
s_R						-1.000	-0.981	-0.766
s_I						0.000	$\pm j0.195$	$\pm j0.643$
								-1.000
								0.000

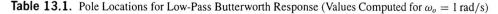

Table 13.1. Pole Locations for Low-Pass Butterworth Response (Values Computed for $\omega_o = 1$ rad/s)

DESIGN

EXAMPLE 13.2   A low-pass audio filter is required for a telephone filtering operation. The filter's passband must extend to 1 kHz. When the frequency is increased to 5 kHz, the response must fall by at least -50 dB from the response at dc. Design an appropriate active filter based on the Butterworth response.

Solution

• Assess the goals of the problem

A filter cascade must be designed such that the -3-dB endpoint ω_o of the Butterworth passband lies at $2\pi \times 1$ kHz $= 6.28 \times 10^3$ rad/s. The response must also be no greater than -50 dB at 5 kHz $\equiv 3.14 \times 10^4$ rad/s. The frequency 5 kHz lies 0.7 decade above 1 kHz [number of decades $= \log_{10}(5\,\text{kHz}/1\,\text{kHz}) = 0.7$], hence the slope of the filter response must roll off by at least -50 dB/0.7 decade $= -71.5$ dB/decade above the passband. Each filter pole adds -20 dB/decade to the rolloff, hence a four-pole filter with a total rolloff of -80 dB/decade is required.

• Choose a design strategy

A fourth-order Butterworth filter is easily made by cascading two second-order low-pass filter sections of the Sallen–Key type. The poles must be chosen such that all entries under the $n = 4$ column in Table 13.1 are included. As indicated by Eq. (13.25), the poles of the Sallen–Key filter of Fig. 13.7 have a real part $s_R = -\omega_o/2Q$. This equation can be solved for the Q required to achieve a desired s_R:

$$Q = \frac{\omega_o}{-2s_R} \tag{13.73}$$

As indicated in Table 13.1, one pair of poles in the Butterworth response for $n = 4$ must have $s_{R1} = -0.924\omega_o$, hence one Sallen–Key section in the cascade should be designed for $Q_1 = 1/(2)(0.924) = 0.54$. Similarly, another pair of poles must have $s_{R2} = -0.383$, hence the other filter section should be designed for $Q_2 = 1/(2)(0.383) = 1.31$. As we now show, these Q choices automatically yield the required values of s_I listed in Table 13.1. From Eq. (13.26) for the Sallen–Key filter, we obtain:

$$\pm js_{I1} = \pm j(\omega_o^2 - s_{R1}^2)^{1/2} = \pm j[1 - (0.924)^2]^{1/2}\omega_o = \pm j0.383\omega_o \tag{13.74}$$

and

$$\pm js_{I2} = \pm j[1 - (0.383)^2]^{1/2}\omega_o = \pm j0.924\omega_o \tag{13.75}$$

Figure 13.20
Two-stage
fourth-order
low-pass
Butterworth filter
cascade.

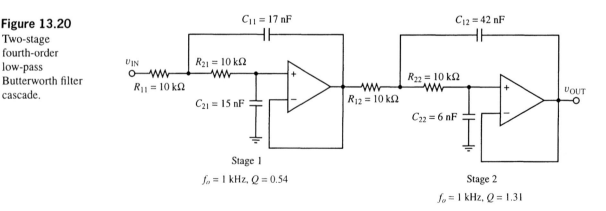

$C_{11} = 17$ nF $C_{12} = 42$ nF

v_{IN} $R_{21} = 10$ kΩ $R_{22} = 10$ kΩ v_{OUT}

$R_{11} = 10$ kΩ $R_{12} = 10$ kΩ

$C_{21} = 15$ nF $C_{22} = 6$ nF

Stage 1
$f_o = 1$ kHz, $Q = 0.54$

Stage 2
$f_o = 1$ kHz, $Q = 1.31$

This result is expected, because both the Butterworth pole locations and those of the Sallen–Key filter section lie on circles of radius ω_o. Setting the latter to the former simply involves the proper selection of Q.

• Specify values for all resistors and capacitors in the circuit

With Q_1 and Q_2 selected, the problem reduces to the selection of values for all the resistors and capacitors in the cascade. One possible selection of R and C values is shown in Fig. 13.20. For simplicity, the resistors in each stage have been chosen to have the same value $R_1 = R_2 = 10\,\mathrm{k\Omega}$. Under these conditions, Equations (13.18) and (13.19) relating ω_o and Q of the Sallen–Key section to the R and C values become

$$\omega_o = \frac{1}{R\sqrt{C_1 C_2}} \tag{13.76}$$

and

$$Q = \frac{1}{2}\sqrt{\frac{C_1}{C_2}} \tag{13.77}$$

These equations can be solved for the required values of C_1 and C_2. For stage 1, the solution of Eqs. (13.76) and (13.77) yields

$$C_{11} = \frac{2Q_1}{\omega_o R} = \frac{2(0.54)}{(2\pi)(1\,\mathrm{kHz})(10\,\mathrm{k\Omega})} = 17.2\,\mathrm{nF} \approx 17\,\mathrm{nF} \tag{13.78}$$

and

$$C_{21} = \frac{1}{2Q_1\omega_o R} = \frac{1}{2(0.54)(2\pi)(1\,\mathrm{kHz})(10\,\mathrm{k\Omega})} = 14.7\,\mathrm{nF} \approx 15\,\mathrm{nF} \tag{13.79}$$

Similarly, the capacitor values for stage 2 become

$$C_{12} = 41.7\,\mathrm{nF} \approx 42\,\mathrm{nF} \tag{13.80}$$

and

$$C_{22} = 6.07\,\mathrm{nF} \approx 6\,\mathrm{nF} \tag{13.81}$$

Because interstage loading between stage 1 and stage 2 is negligible, the transfer function of the overall cascade is equal to the product of the transfer functions of each stage:

$$\frac{\mathbf{V}_{out}}{\mathbf{V}_{in}} = \underbrace{\frac{\omega_{o1}^2}{[s^2 + s(\omega_{o1}/Q_1) + \omega_{o1}^2]}}_{\text{stage 1}} \underbrace{\frac{\omega_{o2}^2}{[s^2 + s(\omega_{o2}/Q_2) + \omega_{o2}^2]}}_{\text{stage 2}}$$

$$= \frac{\omega_{o1}^2 \omega_{o2}^2}{(s - s_1)(s - s_2)(s - s_3)(s - s_4)} \tag{13.82}$$

The magnitude plot of the overall transfer function is shown in Fig. 13.21. Also shown in the figure are the magnitude plots of each individual stage. We note that the response of the first Sallen–Key stage, with Q less than 0.707, begins to fall off well below ω_o. Conversely, the response of the second stage, with Q greater than 0.707, peaks as ω approaches ω_o. Combining these responses yields the maximally flat response of the overall cascade. The second stage essentially "pulls up" the sagging response of the first.

Figure 13.21
Magnitude plot of
the two-stage
low-pass filter
cascade of
Fig. 13.20. Dashed
lines show the
magnitude plots of
the individual
stages.

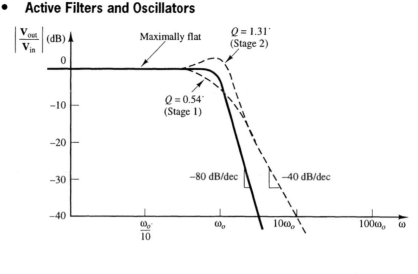

EXERCISE 13.16 In the cascade of Fig. 13.20, the low-Q stage is placed before the high-Q stage. Will the same result be obtained if the order of the stages is reversed?

13.17 Add a third stage to the cascade of Fig. 13.20 so that the cascade has a midband gain of 20 dB. Can this midband gain be obtained without adding a third cascaded stage?

13.18 Add a third stage to the filter cascade of Fig. 13.20 to produce a low-pass Butterworth filter with a -120 dB/decade high-frequency roll-off. Given the discussion of Chapter 10, what instability problems might arise in such a filter design? (*Hint*: Consider stray capacitance.)

13.4.2 Low-Pass Chebyshev Filter Response

The Butterworth response of Section 13.4.1 is noted for its very flat response in the passband region. The maximally flat Butterworth passband is acquired at the expense of a comparatively weak rolloff with frequency outside the passband. One alternative to the Butterworth response, called the *Chebyshev response*, produces a much sharper roll-off for the same number of poles. The price paid is a cyclic variation in gain with frequency inside the passband. Like the Butterworth filter, the Chebyshev filter attempts to approximate the ideal brick-wall response via the magnitude-squared function

$$|H(j\omega)|^2 = \frac{A_o}{1 + f(\omega^2)} \tag{13.83}$$

To obtain the Chebyshev response, the function chosen for $f(\omega^2)$ becomes $\varepsilon^2 [C_n(\omega/\omega_c)]^2$, where the parameter ε ($\varepsilon \leq 1$) is called the *ripple factor*, ω_c is the filter *cutoff frequency*, and the function C_n is an nth-order Chebyshev polynomial defined by $C_n(x) = \cos[n(\cos^{-1}x)]$. The magnitude-squared function (13.83) thus becomes

$$|H(j\omega)|^2 = \frac{A_o^2}{1 + \varepsilon^2 \cos^2[n(\cos^{-1}\omega/\omega_c)]} \tag{13.84}$$

Figure 13.22
Chebyshev
responses for $n = 2$
and $n = 5$ with
$\varepsilon = 1$. The value
$\varepsilon = 1$ causes
$|H(j\omega)|$ to cycle
between A_o and
$A_o/\sqrt{2}$ a total of n
times between the
frequencies $\omega = 0$
and $\omega = \omega_c$. For
$\omega > \omega_c$, the
response falls as the
hyperbolic cosine.

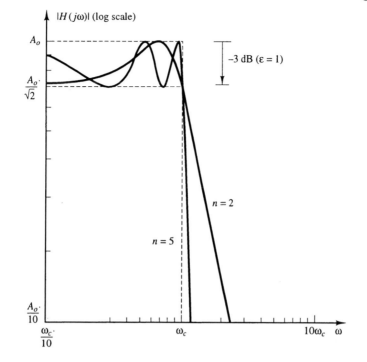

For a given filter order n, the function $[C_n(\omega/\omega_c)]^2$ will oscillate between 0 and 1 exactly n times as the frequency is changed from $\omega = 0$ to $\omega = \omega_c$. This property causes $|H(j\omega)|$ to oscillate between the limits A_o and $A_o/\sqrt{(1 + \varepsilon^2)}$ within the passband. The magnitude plots of two representative Chebyshev filter responses, one for $n = 2$ and one for $n = 5$, are illustrated for the case $\varepsilon = 1$ in Fig. 13.22. The factor $\varepsilon = 1$ causes the filter gain to dip to $A_o/\sqrt{2} \equiv -3\,\text{dB}$ within the passband, and also at $\omega = \omega_c$. This variation in gain is sometimes referred to as *passband ripple*. The magnitude of the passband ripple, expressed as a loss coefficient α, can be computed for any value of ε by evaluating Eq. (13.84) at those frequencies where $C_n(\omega/\omega_c) = 1$, yielding

$$\dot{\alpha} = 20 \log_{10} \left| \frac{H(j\omega)}{A_o} \right| = 20 \log_{10} \frac{1}{\sqrt{1 + \varepsilon^2}} \tag{13.85}$$

Conversely, Eq. (13.85) can be solved for ε in terms of a desired α. Exponentiating (13.85) with respect to power 10 and solving for ε leads to

$$\varepsilon = \sqrt{10^{-\alpha/10} - 1} \tag{13.86}$$

When ω exceeds ω_c, the argument of the Chebyshev polynomial in (13.84) becomes imaginary, yielding the result

$$C_n(\omega/\omega_c) = \cos[n(\cos^{-1}\omega/\omega_c)] \equiv \cos[n(j\cosh^{-1}\omega/\omega_c)] \equiv \cosh[n(\cosh^{-1}\omega/\omega_c)] \tag{13.87}$$

The magnitude-squared function $|H(j\omega)|^2$ thus falls off as the reciprocal of a hyperbolic cosine of order n, which is, in essence, a decaying exponential in frequency. This exponential nature of the Chebyshev function for $\omega > \omega_c$ causes it to fall off much more rapidly than a Butterworth function of similar order. As shown in the previous section, the Butterworth response falls off only as $1/\omega^n$ above its critical frequency ω_o. If the Chebyshev function is to be realized, the poles of the filter cascade must be located at the roots of the denominator of Eq. (13.84). These roots are those values of **s** that make the denominator of Eq. (13.84) go to zero:

$$1 + \varepsilon^2 \cos^2[n(\cos^{-1}(s/j)/\omega_c)] = 0 \tag{13.88}$$

It can be shown[3] that the left half-plane roots of this equation are n in number and lie at locations

$$s_R = -\omega_c \sin\left(\frac{2k+1}{2n}\pi\right)\sinh\gamma \tag{13.89}$$

and

$$js_I = j\omega_c \cos\left(\frac{2k+1}{2n}\pi\right)\cosh\gamma \tag{13.90}$$

where

$$\gamma = \frac{1}{n}\sinh^{-1}\left(\frac{1}{\varepsilon}\right) \tag{13.91}$$

and $k = 0, 1, \cdots, n-1$. The values of s_R and s_I for various values of ε and n are summarized in Table 13.2 for $\omega_c = 1\,\text{rad/s}$. Poles for higher values of n are easily computed using Eqs. (13.89) and (13.90). Note that the poles of a Chebyshev filter do not lie on a circle of radius ω_c in the complex plane. It can be shown that they lie on an ellipse with foci at $\pm j\omega_c$, an imaginary axis of length $\omega_c \cosh\gamma$, and a real axis of length $\omega_c \sinh\gamma$.

	$\varepsilon = 1$ ($\alpha = -3\,\text{dB}$)		$\varepsilon = 0.51$ ($\alpha = -1\,\text{dB}$)		$\varepsilon = 0.35$ ($\alpha = -0.5\,\text{dB}$)	
n	s_R	s_I	s_R	s_I	s_R	s_I
1	-1.002	0.000	-1.965	0.000	-2.863	0.000
2	-0.322	$\pm j0.777$	-0.549	$\pm j0.895$	-0.713	$\pm j1.004$
3	-0.149	$\pm j0.904$	-0.247	$\pm j0.966$	-0.313	$\pm j1.022$
	-0.299	0.000	-0.494	0.000	-0.627	0.000
4	-0.085	$\pm j0.947$	-0.140	$\pm j0.983$	-0.175	$\pm j1.016$
	-0.206	$\pm j0.392$	-0.337	$\pm j0.407$	-0.423	$\pm j0.421$
5	-0.055	$\pm j0.966$	-0.090	$\pm j0.990$	-0.112	$\pm j1.012$
	-0.144	$\pm j0.597$	-0.234	$\pm j0.612$	-0.293	$\pm j0.625$
	-0.178	0.000	-0.290	0.000	-0.362	0.000

Table 13.2. Pole Locations for the Low-Pass Chebyshev Response (Values computed for $\omega_c = 1\,\text{rad/s}$)

DESIGN

EXAMPLE 13.3

Redesign the low-pass Butterworth cascade of Example 13.2 so that it produces a fourth-order Chebyshev response with a passband ripple of $-3\,\text{dB}$. The filter cutoff frequency should still be $1\,\text{kHz}$. Compare its magnitude response plot to that obtained for the Butterworth filter in Example 13.2.

---

[3] See, for example, M. E. Van Valkenburg, *Analog Filter Design*. New York: Holt, Rinehart and Winston, 1982.

Solution

The pole locations for the revised filter can be determined from the $n = 4$ entries under the $\varepsilon = 1$ ($\alpha = -3\,\text{dB}$) column in Table 13.2. In this case, the four pole values must lie at $(0.085 \pm j0.947)\omega_c$ and $(0.206 \pm j0.392)\omega_c$. As before, each pair of poles can be produced by a second-order Sallen–Key low-pass filter section. The required ω_o for each individual section can be determined from the expressions for s_R and s_I for the second-order Sallen–Key filter. These expressions were previously derived in Section 13.3.2:

$$s_R = -\omega_o/2Q \tag{13.92}$$

and
$$s_I = [\omega_o^2 - (\omega_o/2Q)^2]^{1/2} = (\omega_o^2 - s_R^2)^{1/2} \tag{13.93}$$

From the second equation, we note that $\omega_o^2 = s_R^2 + s_I^2$. The corner frequency ω_o of the first Sallen–Key section should thus be set to the value computed using one set of required Chebyshev poles:

$$\omega_{o1} = [(-0.085)^2 + (0.947)^2]^{1/2}\omega_c = 0.951\omega_c \tag{13.94}$$

Similarly, the corner frequency of the second Sallen–Key section should be set to the value

$$\omega_{o2} = [(-0.206)^2 + (0.392)^2]^{1/2}\omega_c = 0.443\omega_c \tag{13.95}$$

The corresponding values of Q are obtained from Eq. (13.92):

$$Q_1 = -\omega_{o1}/2s_R = -0.951\omega_c/2(-0.085)\omega_c = 5.59 \tag{13.96}$$

and
$$Q_2 = -\omega_{o2}/2s_R = -0.443\omega_c/2(-0.206)\omega_c = 1.075 \tag{13.97}$$

These values of ω_o and Q are set by choosing proper R and C values for each filter section. Expressions for the latter quantities were also derived previously in Section 13.3.2:

$$\omega_o = (R_1 R_2 C_1 C_2)^{-1/2} \tag{13.98}$$

and
$$Q = (C_1/C_2)^{1/2} \times (R_1 R_2)^{1/2}/(R_1 + R_2) \tag{13.99}$$

If we again choose all resistors to be $R = 10\,\text{k}\Omega$, as in Example 13.2, Eqs. (13.98) and (13.99) can be solved for C_1 and C_2:

$$C_1 = 2Q/\omega_o R \tag{13.100}$$

and
$$C_2 = 1/(2Q\omega_o R) \tag{13.101}$$

For $\omega_c = 2\pi \times 1\,\text{kHz}$, the required capacitor values for the first section of the Chebyshev cascade become

$$C_{11} = \frac{2Q_1}{\omega_{o1} R} = \frac{2(5.59)}{(2\pi)(0.951 \times 1\,\text{kHz})(10\,\text{k}\Omega)} = 187\,\text{nF} \tag{13.102}$$

and
$$C_{21} = \frac{1}{2Q_1\omega_{o1} R} = \frac{1}{2(5.59)(2\pi)(0.951 \times 1\,\text{kHz})(10\,\text{k}\Omega)} = 1.5\,\text{nF} \tag{13.103}$$

Similarly, the capacitor values for the second section of the Chebyshev cascade become

$$C_{12} = \frac{2Q_2}{\omega_{o2} R} = \frac{2(1.075)}{(2\pi)(0.443 \times 1\,\text{kHz})(10\,\text{k}\Omega)} = 77.2\,\text{nF} \tag{13.104}$$

and
$$C_{22} = \frac{1}{2Q_2\omega_{o2} R} = \frac{1}{2(1.075)(2\pi)(0.443 \times 1\,\text{kHz})(10\,\text{k}\Omega)} = 16.7\,\text{nF} \tag{13.105}$$

The magnitude plot of the new Chebyshev response is shown in Fig. 13.23. Also shown for comparison is the magnitude plot of the original Butterworth response. Note that the Chebyshev response has a much steeper rolloff above ω_c than does the Butterworth response.

Figure 13.23

Comparison of fourth-order Butterworth and Chebyshev ($\varepsilon = 1$) responses.

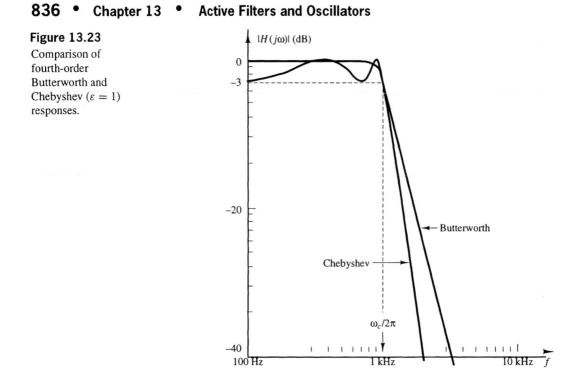

Note that the two Sallen–Key filter sections have different corner frequencies ω_o, where neither is equal to the cutoff frequency ω_c of the Chebyshev function. This situation arises because the poles of the Chebyshev response lie on an ellipse, whereas those of the Sallen–Key sections lie on a circle of radius ω_o. The only way that the required poles can be produced by Sallen–Key sections is to set ω_o of the latter to values that are different from ω_c. This relationship is illustrated in Fig. 13.24, where the loci of possible pole locations of the two Sallen–Key sections in this problem are superimposed on the loci of poles for a Chebyshev function with ripple factor $\varepsilon = 1$ and order $n = 4$.

Figure 13.24

Loci of Chebyshev poles and the poles of the Sallen–Key filter sections for the circuit of Example 13.3. $\omega_{o1} = 0.951\omega_c$, $\omega_{o2} = 0.443\omega_c$, $s_1 = (-0.085 - j0.947)\omega_c$, and $s_2 = (-0.206 - j0.392)\omega_c$.

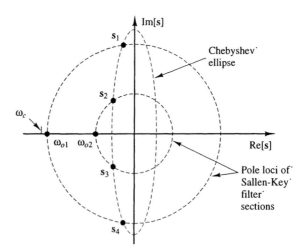

13.4.3 High-Pass and Band-Pass Cascades

The discussions of the previous two sections have focused exclusively on the low-pass filter. The techniques used to synthesize low-pass Butterworth and Chebyshev filters can also be used to create high-pass and band-pass filters. Such a transformation is readily accomplished by simply substituting second-order Sallen–Key sections of the high-pass or band-pass type. One need only choose R and C values for each section such that the required poles for a Butterworth or Chebyshev response are represented. The high-pass or band-pass topologies will then ensure that the resulting filter cascade will be of the proper type and will have the same corner frequency as its low-pass counterpart. The principal difference is that the equations relating ω_o and Q to the resistor and capacitor values vary for each of the various Sallen–Key filter types.

EXAMPLE 13.4

Using the design of Example 13.2 as a guideline, specify the R and C values of a fourth-order high-pass Butterworth filter cascade with a corner frequency of 1 kHz.

Solution

The desired response can be realized by cascading two high-pass filter sections of the Sallen–Key type, each with a corner frequency of $\omega_o = 2\pi \times 1\,\text{kHz}$. As specified in Table 13.1, and as discussed in Example 13.2, the poles of a fourth-order Butterworth cascade must be located at $(-0.924 \pm j0.383)\omega_o$ and $(-0.383 \pm j0.924)\omega_o$, where $s_R = -\omega_o/2Q$ and $s_I = (\omega_o^2 - s_R^2)^{1/2}$. These s_R and s_I values are the same ones used in the design of the low-pass filter, hence the high-pass sections again should be designed such that $Q_1 = 0.54$ and $Q_2 = 1.31$. For the high-pass filter Sallen–Key section, however, Q is given by $(R_2/R_1)(C_1C_2)^{1/2}/(C_1 + C_2)$, as previously derived in Section 13.3.3. If all capacitors are set to a common value C, the expressions for ω_o and Q become

$$\omega_o = 1/C\sqrt{R_1 R_2} \tag{13.106}$$

and

$$Q = \sqrt{\frac{R_2}{R_1}}\left(\frac{\sqrt{C_1 C_2}}{C_1 + C_2}\right) = \frac{1}{2}\sqrt{\frac{R_2}{R_1}} \tag{13.107}$$

These equations may be solved for R_1 and R_2 in terms of Q and ω_o:

$$R_1 = 1/2Q\omega_o C \tag{13.108}$$

and

$$R_2 = 2Q/\omega_o C \tag{13.109}$$

If C is arbitrarily chosen as 10 nF, the required resistor values for the first high-pass section with $Q_1 = 0.54$ thus become

$$R_{11} = \frac{1}{2Q_1\omega_o C} = \frac{1}{2(0.54)(2\pi \times 1\,\text{kHz})(10\,\text{nF})} = 14.7\,\text{k}\Omega \tag{13.110}$$

and

$$R_{21} = \frac{2Q_1}{\omega_o C} = \frac{2(0.54)}{(2\pi \times 1\,\text{kHz})(10\,\text{nF})} = 17.2\,\text{k}\Omega \tag{13.111}$$

Similarly, the required resistor values for the second section become

$$R_{12} = \frac{1}{2Q_2\omega_o C} = \frac{1}{2(1.31)(2\pi \times 1\,\text{kHz})(10\,\text{nF})} = 6.07\,\text{k}\Omega \tag{13.112}$$

and

$$R_{22} = \frac{2Q_2}{\omega_o C} = \frac{2(1.31)}{(2\pi \times 1\,\text{kHz})(10\,\text{nF})} = 41.7\,\text{k}\Omega \tag{13.113}$$

A complete circuit diagram of the filter is shown in Fig. 13.25.

Figure 13.25
Fourth-order high-pass Butterworth cascade with $\omega_o = 1$ kHz.

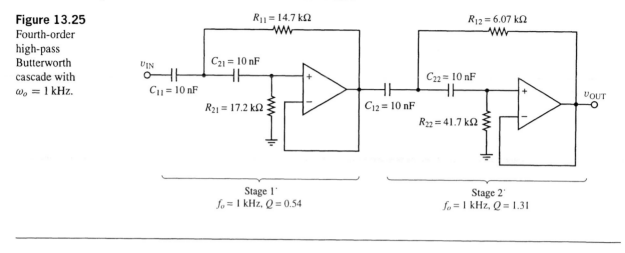

Stage 1
$f_o = 1$ kHz, $Q = 0.54$

Stage 2
$f_o = 1$ kHz, $Q = 1.31$

EXERCISE 13.19 Show that Eqs. (13.108) and (13.109) can be derived from Eqs. (13.106) and (13.107).

13.20 Design a high-pass filter version of the Chebyshev filter of Example 13.3. The filter should have a corner frequency of 1 kHz and a high-frequency passband ripple of -3 dB.

13.4.4 Other Filter Types

The discussions of this chapter might seem to imply that all filter cascades are best designed using Sallen–Key sections, and that all filtering needs can be met using the Butterworth and Chebyshev responses. Actually, many other types of filters and response functions exist. The inverse-Chebyshev filter function produces a response with ripple in the stopband, rather than in the passband. The passband gain of an inverse-Chebyshev filter is a monotonically decreasing function of frequency. The elliptical filter function results in ripple in both the passband and stopband while producing a very narrow transition region. The Bessel filter results in a cascade with constant time delay, regardless of frequency. In many of these filter applications, Sallen–Key sections are better replaced by *multiple-feedback* sections of various types. The reader with an interest in these topics should consult one or more of the excellent references on active filters cited in the Bibliography at the end of the book.

13.5 MAGNITUDE AND FREQUENCY SCALING

The computed values of resistance and capacitance required of a given filter design may not always be convenient. If the filter is made from discrete components, resistors and capacitors may be available only in certain standard values. At other times, it may be necessary to preserve the shape of a filter's response while altering the location of its passband. One useful technique for dealing with such situations is called *scaling*.

The scaling principle can be understood by way of an example. Suppose that the computed resistor and capacitor values required for a particular filter design do not coincide with available component values. The values of the filter parameters ω_o and Q can still be preserved by choosing scaled resistor and capacitor values. For each of the second-order filter sections examined in Section 13.3, the value of ω_o is equal to $1/\sqrt{R_1 R_2 C_1 C_2}$, as indicated in Table 13.3. If R_1 and R_2 are both multiplied by a factor M, and if C_1 and C_2 are divided by the same factor M, then ω_o

will stay the same. Similarly, multiplying R_1 and R_2 by a factor M and dividing C_1 and C_2 by the same factor M will preserve the value of Q. A scaling operation of this type that preserves both ω_o and Q is called *magnitude scaling*.

Filter	ω_o	Q	Reference Equation for Q
Low-pass	$\dfrac{1}{\sqrt{R_1 R_2 C_1 C_2}}$	$\sqrt{\dfrac{C_1}{C_2}} \dfrac{\sqrt{R_1 R_2}}{R_1 + R_2}$	(13.19)
High-pass	$\dfrac{1}{\sqrt{R_1 R_2 C_1 C_2}}$	$\sqrt{\dfrac{R_2}{R_1}} \dfrac{\sqrt{C_1 C_2}}{C_1 + C_2}$	(13.40)
Band-pass	$\dfrac{1}{\sqrt{R_1 R_2 C_1 C_2}}$	$\sqrt{\dfrac{R_2}{R_1}} \dfrac{\sqrt{C_1 C_2}}{C_1 + C_2}$	(13.52)

Table 13.3. Values of ω_o and Q for Second-Order Sallen–Key Filter Sections

---

EXAMPLE 13.5   A second-order Sallen–Key type low-pass filter with parameters $f_o = \omega_o/2\pi = 1\,\text{kHz}$ and $Q = 0.6$ can be made using the values $R_1 = R_2 = 10\,\text{k}\Omega$, $C_1 = 18.3\,\text{nF}$, and $C_2 = 13\,\text{nF}$. Redesign the filter so that only standard 5% tolerance values of components are used.

Solution

Dividing each capacitance by the factor $M = 0.39$ yields the values

$$C_1' = \frac{C_1}{M} = \frac{18.3\,\text{nF}}{0.39} \approx 47\,\text{nF} \tag{13.114}$$

and

$$C_2' = \frac{C_2}{M} = \frac{13\,\text{nF}}{0.39} \approx 33\,\text{nF} \tag{13.115}$$

These values of capacitance are available as 5% stock values. The values of ω_o and Q of the filter will be preserved if R_1 and R_2 are multiplied by $M = 0.39$:

$$R_1' = R_2' = MR = (0.39)(10\,\text{k}\Omega) = 3.9\,\text{k}\Omega \tag{13.116}$$

These resistor values are also available as 5% stock values.

EXERCISE 13.21   Use magnitude scaling to redesign the filter of Example 13.5 so that $R_1' = R_2' = 5\,\text{k}\Omega$.
Answer: $C_1 = 36.6\,\text{nF}$; $C_2 = 26\,\text{nF}$

13.22   Use magnitude scaling to redesign the high-pass filter of Fig. 13.25 so that all capacitors are 47 nF.
Answer: $R_{11} = 3.1\,\text{k}\Omega$, $R_{21} = 3.7\,\text{k}\Omega$, $R_{12} = 1.3\,\text{k}\Omega$, $R_{22} = 8.9\,\text{k}\Omega$

The scaling principle also can be used to change the pole frequencies of a filter while preserving the overall shape of its response. The latter is specifically dependent on the dimensionless parameter Q. If all resistors *and* capacitors in one of the filters of Table 13.3 are divided by the same factor M, then ω_o will be multiplied by a factor of M^2:

$$\omega_o' = \frac{1}{\sqrt{(R_1 R_2/M^2)(C_1 C_2/M^2)}} = \frac{M^2}{\sqrt{R_1 R_2 C_1 C_2}} = M^2 \omega_o \qquad (13.117)$$

An examination of Table 13.3 will yield the expression for Q to be preserved in each case. For the second-order low-pass filter, for example, dividing each R and C by M results in a new Q that is identical to the original:

$$Q' = \sqrt{\frac{C_1/M}{C_2/M}} \frac{\sqrt{R_1 R_2/M^2}}{(R_1 + R_2)/M} = \sqrt{\frac{C_1}{C_2}} \frac{\sqrt{R_1 R_2}}{R_1 + R_2} = Q \qquad (13.118)$$

EXAMPLE 13.6

For the active low-pass filter of Example 13.5, choose new values of R_1, R_2, C_1, and C_2 such that $f_o = \omega_o/2\pi$ is changed from 1 to 5 kHz. The value of Q should remain at $Q = 0.6$.

Solution

The following values of resistance and capacitance will achieve the desired results. Each is found by dividing the original value by the factor $M = \sqrt{5}$:

$$R_1 = R_2 = \frac{10\,\text{k}\Omega}{\sqrt{5}} \approx 4.5\,\text{k}\Omega \qquad (13.119)$$

$$C_1 = \frac{18.3\,\text{nF}}{\sqrt{5}} \approx 8.2\,\text{nF} \qquad (13.120)$$

$$C_2 = \frac{13\,\text{nF}}{\sqrt{5}} \approx 5.8\,\text{nF} \qquad (13.121)$$

The resulting f_o will be equal to $f_o' = (\sqrt{5})^2(1\,\text{kHz}) = 5\,\text{kHz}$.

EXERCISE 13.23

Redesign the filter of Example 13.5 so that f_o is changed from 1 to 10 kHz.
Answer: (one possible design): $R_1 = R_2 \approx 3.2\,\text{k}\Omega$; $C_1 = 5.8\,\text{nF}$; $C_2 = 4.1\,\text{nF}$

13.24

Redesign the filter of Example 13.5 so that f_o is changed from 1 to 10 kHz in such a way that only standard 5% tolerance values of components are used.

The scaling principle can also be applied to cases where the critical frequency ω_o of a filter is to be preserved, but the shape of the filter response, as determined by the value of Q, is to be changed. Since each Q in Table 13.3 depends on ratios of R and C, the value of ω_o can be preserved by multiplying and dividing appropriate resistors and capacitors by the same numerical factor. The expressions for Q vary with the type of filter section, hence each case must be examined individually.

EXAMPLE 13.7

For the second-order low-pass filter of Example 13.5, choose new component values such that f_o remains at 1 kHz, but Q is changed from 0.6 (below maximally flat) to 1.2 (above maximally flat).

Solution

For the second-order low-pass filter entry of Table 13.3, Q is given by

$$Q = \sqrt{\frac{C_1}{C_2}} \frac{\sqrt{R_1 R_2}}{R_1 + R_2} \tag{13.122}$$

Multiplying C_1 by $M = 2$ and dividing C_2 by $M = 2$ while keeping the resistor values the same will preserve the original value of ω_o, but will multiply the original Q by the required factor of 2. The new values of capacitance should be set to

$$C_1' = MC_1 = 2(18.3\,\text{nF}) = 36.6\,\text{nF} \tag{13.123}$$

and

$$C_2' = \frac{C_2}{M} = \frac{13\,\text{nF}}{2} = 6.5\,\text{nF} \tag{13.124}$$

EXERCISE 13.25

Redesign the low-pass filter of Example 13.5 so that Q is changed from 0.6 to 0.2. The value of ω_o should stay the same.   **Answer:** $C_1' = 6.1\,\text{nF}$; $C_2' = 39\,\text{nF}$

13.26 Redesign the low-pass filter of Example 13.5 so that f_o is changed from 1 kHz to 500 Hz and Q is changed from 0.6 to 0.2.   **Answer:**   (one possible design) $R_1' = R_2' = 20\,\text{k}\Omega$; $C_1' = 6.1\,\text{nF}$; $C_2' = 39\,\text{nF}$

13.27 Determine whether or not the design requirements of Example 13.7 can be met by changing resistor values only.

13.6 SWITCHED-CAPACITOR NETWORKS AND FILTERS

Many of the filtering functions discussed in this chapter can be synthesized using switched-capacitor networks. A switched-capacitor network is built entirely from operational amplifiers, MOSFETs, and capacitors of very small value. If the op-amps are also made from MOSFETs only, the entire switched-capacitor network will contain no resistors whatsoever and can be easily built on a single integrated circuit. The pole and zero frequencies of a switched-capacitor circuit can be controlled externally via the frequency of a single synchronizing "clock." This feature constitutes one of the principal advantages of such circuits over conventional analog filter circuits.

Figure 13.26
Basis of the switched-capacitor filter: (a) circuit that simulates a resistor of value T_c/C; (b) switch timing diagram.

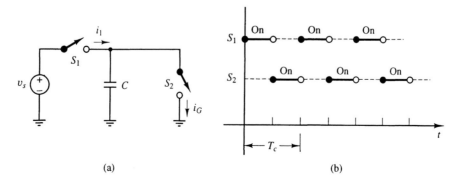

The basis of the switched-capacitor network lies in the simple circuit of Fig. 13.26(a), which simulates the function of a resistor. The synchronized switches S_1 and S_2 open and close every half-cycle of the clock interval T_c, as illustrated by the timing diagram of Fig. 13.26(b), and periodically charge and discharge the capacitor. For this structure to behave as a resistor, the following assumptions must be valid:

1. At any given moment, only one switch is closed; both switches are never closed simultaneously.

2. The interval during which *neither* switch is closed is very short compared to the clock period T_c.

3. If S_1 is closed, the capacitor charges instantly to the signal voltage v_s.

4. If S_2 is closed, the capacitor discharges instantly to ground.

During the interval when S_1 is closed, the charge on the capacitor reaches the value

$$q = Cv_s \tag{13.125}$$

This charge is completely discharged to ground when S_2 is closed. The energy stored in the capacitor is dissipated in the contact resistance of the switch. Over a single clock interval T_c, the total charge per unit time leaving the v_s source is equal to

$$\frac{\Delta q}{\Delta t} = \frac{Cv_s}{T_c} \tag{13.126}$$

This charge must be continuously supplied to the capacitor by the v_s source, since the charge is discharged to ground after every clock cycle. When averaged over many clock cycles, the flow of charge from v_s corresponds to an average current of value

$$\langle i_1 \rangle = \left\langle \frac{dq}{dt} \right\rangle = \frac{Cv_s}{T_c} \tag{13.127}$$

The quantity T_c/C has the units of resistance (second/coulomb/volt $\equiv$ ohm), hence Eq. (13.127) has the form of Ohm's law

$$\langle i_1 \rangle = \frac{v_s}{R} \tag{13.128}$$

The discharge mechanism of the switched capacitor circuit of Fig. 13.26(a) can thus be modeled as a resistance of value T_c/C.

Figure 13.27
(a) Practical implementation of the switched-capacitor circuit of Fig. 13.26(a).
(b) Timing diagram of voltages ϕ_1 and ϕ_2. The ϕ_1 and ϕ_2 pulses do not overlap.

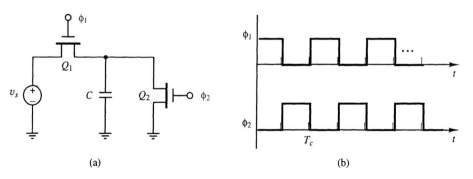

(a)   (b)

A practical version of the simulated switched-capacitor resistor, in which the switches are replaced by enhancement-mode MOSFETs, is shown in Fig. 13.27(a). In this circuit, the MOSFETs are fabricated with geometric symmetry, so that the drain and source terminals are interchangeable. Consequently, neither side is specified as the source on the MOSFET circuit symbols. The voltages ϕ_1 and ϕ_2 are nonoverlapping clock signals of period T_c. As shown in Fig. 13.27(b), they are never high simultaneously. The magnitudes of ϕ_1 and ϕ_2 exceed the MOS-FET threshold voltage V_{TR}. Like the switched circuit of Fig. 13.26(a), the circuit of Fig. 13.27(a) is virtually indistinguishable from a fixed resistor of value T_c/C if the time scale of interest is much longer than T_c.

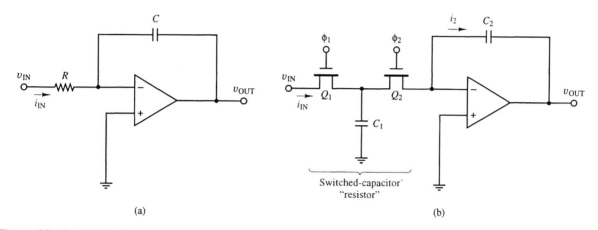

(a)   (b)

Figure 13.28   (a) Simple op-amp integrator; (b) switched-capacitor equivalent.

In Fig. 13.28, the switched-capacitor "resistor" of Fig. 13.27 is used to build a simple op-amp integrator. The negative-feedback connection holds the v_- terminal at the ground potential of v_+, so that the time average of i_{IN} in Fig. 13.28(b) is given by

$$\langle i_{IN} \rangle = \frac{v_{IN}}{T_c/C_1} \tag{13.129}$$

Because i_{IN} must ultimately flow through C_2 as i_2, the output of the circuit is given by

$$\frac{dv_{OUT}}{dt} = \frac{-\langle i_2 \rangle}{C_2} = -\frac{v_{IN}}{T_c}\frac{C_1}{C_2} \tag{13.130}$$

This equation may be integrated with respect to time to yield

$$v_{OUT} = -\frac{C_1}{C_2}\frac{1}{T_c}\int_0^t v_{IN}\, dt \tag{13.131}$$

As Eq. (13.131) shows, the gain of the integrator depends on the clock frequency T_c, which can be adjusted to any value that is shorter than the shortest time span of interest of the input signal. The integrator gain is also proportional to the ratio of C_1 to C_2. Note that the gain does not depend on the actual values of these capacitors, but only on their ratio. In theory, C_1 and C_2 can be made arbitrarily small. This feature allows switched-capacitor filter networks to be built on integrated circuits, where small capacitors that occupy a minimal amount of surface area are available.

Figure 13.29
Modified
switched-capacitor
network allows a
connection between
ungrounded nodes.

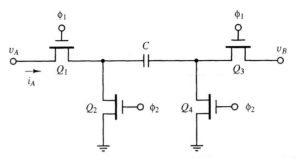

The simulated switched-capacitor resistor of Fig. 13.27 has two major limitations. First, it can only simulate a resistor that has one side connected to ground potential. It also suffers from susceptibility to the stray capacitance that may exist between the terminals of the MOSFETs and ground. The modified network of Fig. 13.29 overcomes these limitations. During the ϕ_1 interval, the capacitor is charged to the voltage $v_A - v_B$ via the series MOSFETs Q_1 and Q_3. During the ϕ_2 interval, the capacitor is discharged through the closed loop formed by the parallel MOSFETs Q_2 and Q_4. The energy stored in the capacitor is dissipated in these devices. The charge stored in the capacitor during ϕ_1 is equal to

$$q = C(v_A - v_B) \tag{13.132}$$

The time-average current from v_A to v_B therefore is equal to

$$\langle i_A \rangle = \left\langle \frac{dq}{dt} \right\rangle = \frac{C}{T_c}(v_A - v_B) \tag{13.133}$$

If T_c is short compared to time scales of interest, this network once again simulates a resistance of value T_c/C. Note that if v_B is held at ground potential, the entire network reduces to that of Fig. 13.27(a), because the parallel combination of Q_3 and Q_4 then behaves as a continuous connection to ground.

In Fig. 13.30(b), the modified switched-capacitor filter network of Fig. 13.29 is used to simulate the simple low-pass filter of Fig. 13.30(a). The frequency domain transfer function of the latter is given by

$$\frac{\mathbf{V}_{out}}{\mathbf{V}_{in}} = -\frac{R_2}{R_1}\left(\frac{1}{1 + j\omega R_2 C_X}\right) \tag{13.134}$$

By analogy, the transfer function of the switched-capacitor filter version is given by

$$\frac{\mathbf{V}_{out}}{\mathbf{V}_{in}} = -\frac{T_c/C_2}{T_c/C_1}\left[\frac{1}{1 + (j\omega T_c/C_2)C_X}\right] \\ = -\frac{C_1}{C_2}\left[\frac{1}{1 + j\omega T_c(C_X/C_2)}\right] \tag{13.135}$$

The dc gain of this filter depends only on the ratio C_1/C_2 of two fixed capacitors, which, in theory, can be made arbitrarily small. Similarly, the pole frequency of the filter response, given by

$$\omega_o = \frac{C_2}{C_X T_c} \tag{13.136}$$

can be adjusted via the clock frequency T_c. This latter feature is one of the more useful characteristics of switched-capacitor filters. Once again, the switched-capacitor circuit will imitate the RC version of the filter only if T_c is much shorter than the shortest time span of interest of the input signal.

Figure 13.30
(a) Simple low-pass
filter; (b) switched-
capacitor
equivalent.

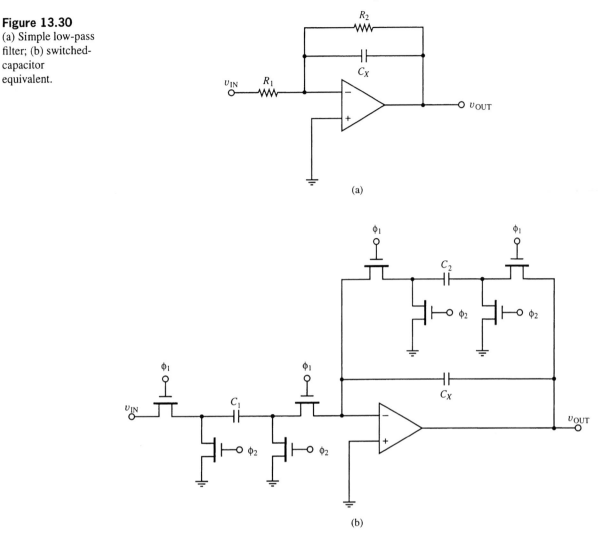

(a)

(b)

EXERCISE 13.28 In theory, the capacitances of a switched-capacitor network can be made arbitrarily small if the proper capacitance ratios are preserved. Discuss the practical limitations to the minimum attainable capacitor size, which is generally limited to the tenths-of-picofarad range or larger. Also discuss the limitations to the maximum practical size of the capacitors in a switched-capacitor network.

13.29 The switched-capacitor network concept can be applied to each of the active filters examined in Section 13.3. Draw a switched-capacitor version of the second-order band-pass filter of Section 13.3.4. Show that its parameters can be expressed by $\omega_o = (C_A C_B)^{1/2}/T_c(C_1 C_2)^{1/2}$ and $Q = (C_A C_1 C_2)^{1/2}/\sqrt{C_B}(C_1 + C_2)$, where C_A and C_B are the capacitors of the switched-capacitor resistors.

13.30 Show that the quantity T_c/C has the units of resistance.

13.31 Draw the circuit diagram of a switched-capacitor version of the second-order low-pass filter of Fig. 13.7.

13.32 Draw the circuit diagram of a switched-capacitor version of the second-order high-pass filter of Fig. 13.12.

13.33 Consider the second-order Sallen–Key band-pass filter of Fig. 13.15 with $C_1 = 19\,\text{nF}$ and $C_2 = 13\,\text{nF}$. If the filter is made in a switched-capacitor version, choose C_A, C_B, and T_c so that the filter has parameters $\omega_o = 2\pi \times 1\,\text{kHz}$ and $Q = 0.49$. **Answer:** (one possible design) $C_A = C_B = 100\,\text{pF}$; $T_c \approx 1\,\mu s$

13.7 OSCILLATORS

In Chapter 10, the role of the stability condition in preventing unwanted oscillations was made evident. In this section, we examine circuits in which the stability condition is intentionally violated as a means of *creating* oscillation. Appropriate selection of feedback-loop parameters in an active filter circuit can lead to oscillation at a fixed, determinable frequency. Oscillators similar to those discussed in this section are important to a variety of analog and digital signal-processing applications.

13.7.1 Wien-Bridge Oscillator

An example of a simple oscillator circuit called the Wien-bridge oscillator is shown in Fig. 13.31. This circuit consists of a noninverting amplifier with an additional RC filter network connected between the output terminal, the v_+ terminal, and ground. This filter network applies an attenuated, phase-shifted version of v_{OUT} to the noninverting amplifier input v_+. The circuit is designed to have a self-sustaining output, hence no input terminal is indicated.

Figure 13.31
Wien-bridge oscillator consists of a noninverting amplifier and a phase-shift network formed by $\mathbf{Z}_1$ and $\mathbf{Z}_2$.

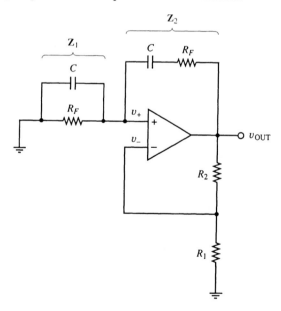

In the frequency domain, the impedances $\mathbf{Z}_1$ and $\mathbf{Z}_2$ indicated in Fig. 13.31 can be expressed by

$$\mathbf{Z}_1 = R_F \left\| \frac{1}{j\omega C} \right. = \frac{R_F/j\omega C}{R_F + 1/j\omega C} = \frac{R_F}{1 + j\omega R_F C} \tag{13.137}$$

and

$$\mathbf{Z}_2 = \frac{1}{j\omega C} + R_F = \frac{1 + j\omega R_F C}{j\omega C} \tag{13.138}$$

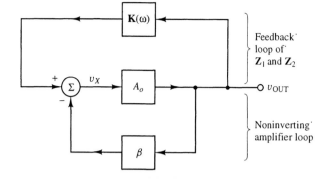

The circuit is best analyzed by considering the feedback-loop diagram of Fig. 13.32. The negative-feedback function β represents the fraction of v_{OUT} applied to the v_- terminal by R_1 and R_2:

$$v_- = \beta v_{OUT} \equiv \frac{R_1}{R_1 + R_2} v_{OUT} \tag{13.139}$$

Similarly, the positive-feedback function $\mathbf{K}$ represents the fraction of v_{OUT} applied to the v_+ terminal by $\mathbf{Z}_1$ and $\mathbf{Z}_2$. If v_{OUT} and v_+ are expressed as phasors, this second feedback relationship can be expressed by

$$\mathbf{V}_+ = \mathbf{K}(\omega)\mathbf{V}_{out} = \frac{\mathbf{Z}_1}{\mathbf{Z}_1 + \mathbf{Z}_2}\mathbf{V}_{out} \tag{13.140}$$

where the frequency-dependent feedback function is given by

$$\begin{aligned}\mathbf{K}(\omega) &= \frac{\mathbf{Z}_1}{\mathbf{Z}_1 + \mathbf{Z}_2} = \frac{R_F/(1 + j\omega R_F C)}{[R_F/(1 + j\omega R_F C)] + [(1 + j\omega R_F C)/j\omega C]} \\ &= \frac{j\omega R_F C}{j\omega R_F C + (1 + j\omega R_F C)^2} = \frac{j\omega R_F C}{[1 - (\omega R_F C)^2] + 3j\omega R_F C}\end{aligned} \tag{13.141}$$

It can be shown that the oscillation frequency depends only on $\mathbf{Z}_1$ and $\mathbf{Z}_2$, and not on the internal frequency-response properties of the op-amp, as long as the bandwidth of the noninverting amplifier embedded within the circuit is much larger than the frequency of oscillation. Subject to this constraint, we model the op-amp as having a constant open-loop gain A_o.

The output of the combined feedback loops in Fig. 13.32 is given in the frequency domain by

$$\mathbf{V}_{out} = A_o\mathbf{V}_x$$

where
$$\mathbf{V}_x = \mathbf{V}_+ - \mathbf{V}_- = \mathbf{K}(\omega)\mathbf{V}_{out} - \beta\mathbf{V}_{out} \tag{13.142}$$

Combining Eqs. (13.141) and (13.142) results in

$$\mathbf{V}_{out} = A_o\left[\mathbf{K}(\omega) - \beta\right]\mathbf{V}_{out} \tag{13.143}$$

Equation (13.143) can be satisfied for $A_o \to \infty$ only if $\left[\mathbf{K}(\omega) - \beta\right] = 0$, that is, if

$$\frac{\mathbf{Z}_1}{\mathbf{Z}_1 + \mathbf{Z}_2} = \frac{R_1}{R_1 + R_2} \tag{13.144}$$

The expression on the left side of Eq. (13.144) is complex, hence the real and imaginary parts of the equation must be satisfied independently. The ratio $R_1/(R_1 + R_2)$ is entirely real.

The imaginary part of $\mathbf{Z}_1/(\mathbf{Z}_1 + \mathbf{Z}_2)$ must therefore equal zero if Eq. (13.144) is to be satisfied. An examination of Eq. (13.141) reveals that the imaginary part of $\mathbf{Z}_1/(\mathbf{Z}_1 + \mathbf{Z}_2)$ will be zero if

$$1 - (\omega R_F C)^2 = 0 \qquad (13.145)$$

Under this condition, the factors of j in the numerator and denominator of Eq. (13.141) cancel, making the remaining expression completely real. Equation (13.145) will be satisfied if

$$\omega = \omega_F = \frac{1}{R_F C} \qquad (13.146)$$

At this frequency, the real part of $\mathbf{K}(\omega) = \mathbf{Z}_1/(\mathbf{Z}_1 + \mathbf{Z}_2)$ becomes

$$\mathrm{Re}\left[\frac{\mathbf{Z}_1}{\mathbf{Z}_1 + \mathbf{Z}_2} \right]_{\omega=1/R_F C} = \frac{j\omega R_F C}{3j\omega R_F C} = \frac{1}{3} \qquad (13.147)$$

The quantity $\mathbf{K}(\omega) - \beta$ will thus be equal to zero at the frequency ω_F if $\beta = 1/3$.

If R_1 and R_2 are chosen such that $\beta = 1/3$, the circuit of Fig. 13.31 will be capable of sustaining a nonzero v_{OUT} at the frequency $\omega_F = 1/R_F C$. Such an output is equivalent to sinusoidal oscillation at ω_F. It can be shown that if R_1 and R_2 are chosen so that $\beta < 1/3$ (higher noninverting amplifier gain), the oscillation frequency of the circuit will still be $\omega_F = 1/R_F C$, but the amplitude of the oscillation will increase with time. In practice, the saturation limits of the op-amp will limit the magnitude of the oscillating v_{OUT}, so that the output becomes a sinusoid of fixed frequency and amplitude. In essence, the incremental gain of the noninverting amplifier is reduced as the op-amp approaches saturation, thus reducing the loop gain.

EXERCISE 13.34　Repeat the analysis of the Wien-bridge oscillator of Fig. 13.31 for the case in which the resistors in the positive-feedback network have different values R_{F1} and R_{F2}. Show that the frequency of oscillation for the case $\beta \leq 1/3$ is equal to $\omega_F = 1/\sqrt{R_{F1} R_{F2}} C$.

13.35　Repeat the analysis of the Wien-bridge oscillator of Fig. 13.31 for the case in which the resistors in the positive-feedback network have different values R_{F1} and R_{F2} and the capacitors have different values C_1 and C_2. Show that the frequency of oscillation for the case $\beta \leq 1/3$ is equal to $\omega_F = 1/\sqrt{R_{F1} R_{F2} C_1 C_2}$.

13.7.2 Phase-Shift Oscillator

A second useful op-amp oscillator circuit, called the *phase-shift oscillator*, is shown in Fig. 13.33. The heart of the circuit consists of a frequency-independent inverting amplifier A_3 designed for a gain of $K = -R_4/R_3$. The output of this amplifier is connected to a three-stage RC filter in which intermediate stages are buffered by unity-gain op-amp followers. This three-stage RC network is used to add a 180° phase shift to v_{OUT} before returning it to the amplifier input at the v_3 terminal. Note that the input resistance of the inverting amplifier, equal to R_3, appears as a resistance to ground to the capacitor C_3. These two components function as one of the RC phase-shift stages.

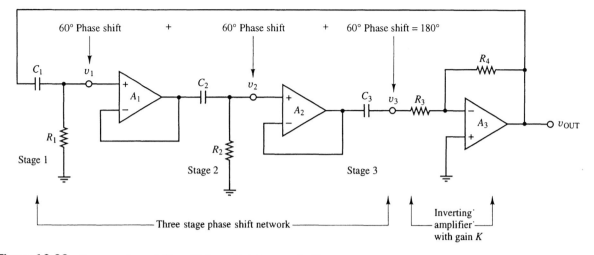

Figure 13.33 Op-amp phase-shift oscillator with interstage buffers.

The three-stage phase-shift network shown in Fig. 13.33 is more complicated than necessary. In a practical version of the oscillator, the unity-gain followers can be omitted. Including them here, however, simplifies circuit analysis and highlights the role of the phase-shift network in sustaining oscillation. Without the unity-gain followers present, the loading of each stage in the phase-shift network by its successor must be taken into account.

The circuit of Fig. 13.33 is designed to oscillate below any of the high-frequency poles of the op-amp's internal closed-loop gain. The circuit thus can be modeled by the simplified feedback diagram of Fig. 13.34, in which the inverting amplifier is represented by the constant $K = -R_4/R_3$. The frequency-dependent feedback function $\beta(\omega)$ of the phase-shift network can be expressed by

$$\beta(\omega) = \frac{j\omega R_1 C_1}{(1 + j\omega R_1 C_1)} \frac{j\omega R_2 C_2}{(1 + j\omega R_2 C_2)} \frac{j\omega R_3 C_3}{(1 + j\omega R_3 C_3)} \qquad (13.148)$$

With interstage loading eliminated by the unity-gain buffers, Eq. (13.148) consists of the simple product of the transfer functions of each individual stage in the phase-shift network.

The sole input to the summation node in Fig. 13.34 is the output of the phase-shift feedback network; hence the output v_{OUT} becomes

$$v_{OUT} = K v_3 = K\beta(\omega)v_{OUT} \qquad (13.149)$$

This equation will yield a nonzero v_{OUT} if $K\beta(\omega) = 1$. Since K is negative, this condition can be met only if $\beta(\omega)$ contributes a 180° phase shift, in effect multiplying K by (-1).

Figure 13.34
Feedback
representation of
the phase-shift
oscillator of
Fig. 13.33.

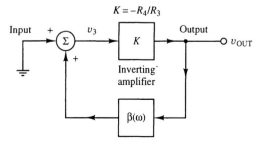

For the simplified case $C_1 = C_2 = C_3 = C$ and $R_1 = R_2 = R_3 = R$, the condition $K\beta(\omega) = 1$ becomes

$$K\beta(\omega) = \frac{(-R_4/R_3)(j\omega RC)^3}{(1+j\omega RC)^3} = \frac{(R_4/R_3)(j\omega RC)(\omega RC)^2}{(1+j\omega RC)[1+2j\omega RC - (\omega RC)^2]} \quad (13.150)$$

Further simplification of the denominator in Eq. (13.150) results in

$$K\beta(\omega) = \frac{(R_4/R_3)(j\omega RC)(\omega RC)^2}{[1 - 3(\omega RC)^2] + j\omega RC[3 - (\omega RC)^2]} \quad (13.151)$$

If the condition $K\beta(\omega) = 1$ is to be met, the imaginary component of Eq. (13.151) must equal zero. The numerator of Eq. (13.151) is purely imaginary, hence the entire equation will become real if the denominator is also purely imaginary, that is, if $1 - 3(\omega RC)^2 = 0$, or

$$\omega = \frac{1}{\sqrt{3}RC} \quad (13.152)$$

At this frequency, Eq. (13.151) becomes

$$K\beta(\omega) = \frac{(R_4/R_3)(j/\sqrt{3})(1/3)}{0 + (j/\sqrt{3})(3 - 1/3)} = \frac{(R_4/R_3)(1/3)}{8/3} = \frac{1}{8}\frac{R_4}{R_3} \quad (13.153)$$

If $K = R_4/R_3 = 8$, then $K\beta(\omega)$ will equal 1 at $\omega = 1/(\sqrt{3}RC)$, and oscillation will be possible at this frequency.

Note that if K is greater than 8, oscillation will still occur at the frequency $\omega = 1/(\sqrt{3}RC)$, but the amplitude of the oscillations will increase without bound until the saturation limits of the op-amp are reached. In general, a practical oscillator must always utilize an output-limiting network (in essence, a voltage-dependent gain) if oscillations of stable amplitude are desired. The saturation limits of an op-amp are sufficient for this purpose.

Given the result (13.152), the role of the three-stage phase-shift network becomes evident. At the frequency $\omega = 1/(\sqrt{3}RC)$, where $\omega RC = 1/\sqrt{3} = 0.577$, the angle between v_{OUT} and the output of the first stage, taken at v_1, becomes

$$\measuredangle\frac{j\omega RC}{1+j\omega RC} = 90° - \tan^{-1}\omega RC = 90° - \tan^{-1}0.577 = 90° - 30° = 60° \quad (13.154)$$

The angles contributed by each of the remaining two stages in the phase-shift network are also equal to 60°; hence the three-stage phase-shift network contributes a total angle of 180° at the frequency $\omega = 1/(\sqrt{3}RC)$. This condition turns the negative-feedback connection of the inverting amplifier into a purely real positive-feedback connection at the frequency of oscillation.

Figure 13.35
Practical version of
the op-amp
phase-shift
oscillator.

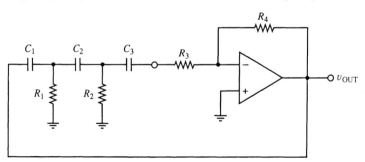

A practical version of the phase-shift oscillator, in which the unity-gain buffers are absent, is shown in Fig. 13.35. Analysis of this circuit is complicated by the loading between successive stages in the phase-shift network. Despite the tedious algebra involved, it is possible to show that the phase-shift network in Fig. 13.35 has a $\beta(\omega)$ given by

$$\beta(\omega) = \frac{(j\omega RC)(j\omega RC)^2}{[1 - 6(\omega RC)^2] + j\omega RC[5 - (\omega RC)^2]} \tag{13.155}$$

Oscillation occurs at $\omega = 1/(\sqrt{6}RC)$, where $\beta(\omega)$ becomes purely real. At this frequency, each stage of the phase-shift network contributes a phase shift of 60°, so that

$$K\beta(\omega) = \frac{(R_4/R_3)(1/6)}{5 - 1/6} = \frac{1}{29}\frac{R_4}{R_3} \tag{13.156}$$

As Eq. (13.156) shows, the condition $K\beta(\omega) = 1$ will be met if $R_4/R_3 = 29$.

EXERCISE 13.36 For the phase-shift oscillator of Fig. 13.33, use Eq. (13.150) to arrive at the result (13.151).

13.37 The phase-shift oscillator need not be made from an operational amplifier. Show that the MOSFET circuit of Fig. 13.36 will oscillate at $\omega = 1/(\sqrt{6}RC)$ if $2\sqrt{KI_D}R_D \geq 29$, where K is the conductance parameter of Q_1.

13.38 Design a phase-shift oscillator using a noninverting amplifier as the basic building block.

Figure 13.36
MOSFET
phase-shift
oscillator.

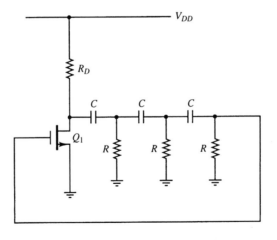

13.7.3 Tuned LC Oscillators

An oscillator can be made using a single amplification stage and an inductor–capacitor (LC) network to provide the necessary $-180°$ phase shift between the output and the input. The oscillation frequency of such a circuit is easily adjusted, or *tuned*, over a wide range of frequencies by changing the value of a single capacitor or inductor. Tuned LC oscillators are used in many applications, including radio transmitters, AM and FM receivers, and sinusoidal function generators.

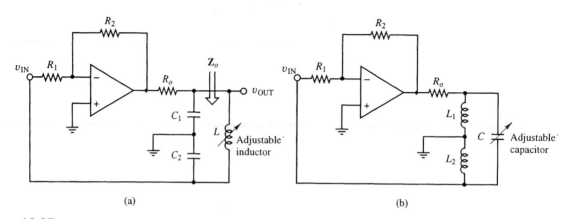

(a)   (b)

Figure 13.37 (a) Colpitts tuned LC oscillator configuration; (b) Hartley oscillator configuration.

One popular tuned LC oscillator configuration, shown in Fig. 13.37(a), is called the *Colpitts* oscillator. The resonant combination of C_1, C_2, and L is called the *tank* circuit. The oscillation frequency is changed via the adjustable inductor L. An analogous circuit called the *Hartley* oscillator, shown in Fig. 13.37(b), is formed by changing the capacitors in the Colpitts tank circuit to inductors and by changing the tunable element to an adjustable capacitor. Both the Colpitts and Hartley oscillator configurations require an amplifier with inverting (negative) gain in order to sustain oscillations. In the circuits of Fig. 13.37, this gain is provided by an operational amplifier connected in the inverting configuration with feedback resistors R_1 and R_2.

We now examine the Colpitts circuit to determine the conditions for oscillation. A similar analysis can be performed on the Hartley oscillator. The analysis is most easily performed if the Colpitts circuit is redrawn as in Fig. 13.38, where the components of the inverting amplifier are represented as a single block having an amplification factor $A = -R_2/R_1$. If amplification is provided by a single- or multiple-transistor circuit, rather than by an IC op-amp, a suitable expression for the gain is used instead.

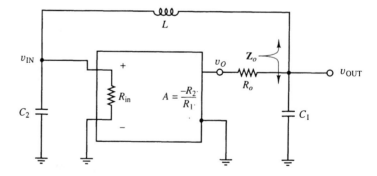

Figure 13.38 Alternative representation of the Colpitts oscillator of Fig. 13.37(a). The amplification block A could represent any amplifier with an inverting gain. To simplify the calculations, the amplifier input resistance R_{in} is assumed to be much larger than the reactance of C_2.

The significance of R_o is twofold. It represents the output resistance of the amplifier, and also accounts for the losses that occur in the LC tank circuit. In the ideal (but unreachable) case where $R_o = 0$, the tank circuit would be capable of sustaining oscillation without the regenerative effect of the inverting amplifier.

The analysis begins by expressing v_{OUT} via voltage division as a fraction of the amplifier output v_O. With v_{IN} and v_{OUT} expressed as phasors, this relationship becomes

$$\mathbf{V}_{\text{out}} = \frac{\mathbf{Z}_o}{R_o + \mathbf{Z}_o}\mathbf{V}_o = \frac{\mathbf{Z}_o}{R_o + \mathbf{Z}_o}A\mathbf{V}_{\text{in}} \equiv \mathbf{K}_1 A\mathbf{V}_{\text{in}} \tag{13.157}$$

where $\mathbf{Z}_o$ is the impedance seen looking into the LC tank circuit, A is equal to $-R_2/R_1$, and the ratio $\mathbf{K}_1 = \mathbf{Z}_o/(R_o + \mathbf{Z}_o)$ is frequency-dependent. The input voltage $\mathbf{V}_{\text{in}}$ can also be determined by voltage division. If R_{in} is assumed to be much larger than the reactance $1/\omega C_2$, it follows that

$$\mathbf{V}_{\text{in}} = \frac{\mathbf{Z}_{C2}}{\mathbf{Z}_{C2} + \mathbf{Z}_L}\mathbf{V}_{\text{out}} = \frac{1/j\omega C_2}{1/j\omega C_2 + j\omega L}\mathbf{V}_{\text{out}} = \frac{1}{1 - \omega^2 LC_2}\mathbf{V}_{\text{out}} \equiv \mathbf{K}_2\mathbf{V}_{\text{out}} \tag{13.158}$$

where $\mathbf{K}_2 = 1/(1 - \omega^2 LC_2)$ is also frequency-dependent. The analysis can be performed without assuming large R_{in}, but the algebra becomes more complicated. Since we wish to focus on the key features that allow the circuit to oscillate, we assume R_{in} to be large without loss of generality.

With the voltage-division factors expressed as $\mathbf{K}_1$ and $\mathbf{K}_2$, the overall output becomes

$$\mathbf{V}_{\text{out}} = \mathbf{K}_1 A\mathbf{V}_{\text{in}} = \mathbf{K}_1 A\mathbf{K}_2\mathbf{V}_{\text{out}} \tag{13.159}$$

As this equation suggests, the circuit is capable of sustaining oscillations at any frequency for which $A\mathbf{K}_1\mathbf{K}_2 = 1$.

We now determine an expression for $\mathbf{K}_1$. If R_{in} is again assumed very large, the impedance $\mathbf{Z}_o$ seen looking into the tank circuit from the v_{OUT} terminal can be expressed as

$$\mathbf{Z}_o = \left(\frac{1}{j\omega C_1}\right) \Big\| \left(j\omega L + \frac{1}{j\omega C_2}\right) = \frac{(1/j\omega C_1)(j\omega L + 1/j\omega C_2)}{1/j\omega C_1 + j\omega L + 1/j\omega C_2} \tag{13.160}$$

After some algebra, this expression reduces to

$$\mathbf{Z}_o = \frac{1 - \omega^2 LC_2}{j\omega(C_1 + C_2 - \omega^2 LC_1C_2)} \tag{13.161}$$

With $\mathbf{Z}_o$ expressed in the form (13.161), the voltage-division ratio $\mathbf{K}_1$ becomes

$$\begin{aligned}
\mathbf{K}_1 &= \frac{\mathbf{Z}_o}{R_o + \mathbf{Z}_o} = \frac{(1 - \omega^2 LC_2)/j\omega(C_1 + C_2 - \omega^2 LC_1C_2)}{R_o + (1 - \omega^2 LC_2)/j\omega(C_1 + C_2 - \omega^2 LC_1C_2)} \\
&= \frac{1 - \omega^2 LC_2}{(1 - \omega^2 LC_2) + j\omega R_o(C_1 + C_2 - \omega^2 LC_1C_2)}
\end{aligned} \tag{13.162}$$

With $\mathbf{K}_2 = 1/(1 - \omega^2 LC_2)$, the product $A\mathbf{K}_1\mathbf{K}_2$ can be computed:

$$\begin{aligned}
A\mathbf{K}_1\mathbf{K}_2 &= -\frac{R_2}{R_1}\frac{1 - \omega^2 LC_2}{(1 - \omega^2 LC_2) + j\omega R_o(C_1 + C_2 - \omega^2 LC_1C_2)}\frac{1}{1 - \omega^2 LC_2} \\
&= -\frac{R_2}{R_1}\frac{1}{(1 - \omega^2 LC_2) + j\omega R_o(C_1 + C_2 - \omega^2 LC_1C_2)}
\end{aligned} \tag{13.163}$$

The circuit will sustain oscillations with no external input if Eq. (13.163) is equal to unity with zero imaginary part. This condition requires that the factor multiplying $j\omega$ in the denominator become zero, which will occur at the frequency $\omega = \omega_o$ where $\omega_o^2 LC_1C_2 = C_1 + C_2$, or

$$\omega_o = \left(\frac{C_1 + C_2}{LC_1C_2}\right)^{1/2} \tag{13.164}$$

At this frequency, the remaining real part of Eq. (13.163) will become unity when

$$-\frac{R_2}{R_1}\frac{1}{1-\omega_o^2 LC_2}=1 \tag{13.165}$$

or

$$-\frac{R_2}{R_1}\frac{1}{1-(C_1+C_2)/C_1}\equiv\frac{R_2}{R_1}\frac{C_1}{C_2}=1 \tag{13.166}$$

When Eq. (13.166) is satisfied, the circuit will be capable of sustaining oscillations at the frequency ω_o. We note that (13.166) requires the amplifier to have a gain magnitude R_2/R_1 that is proportional to the capacitance feedback ratio C_2/C_1.

In practice, the gain of the amplifier is made larger than the value required by Eq. (13.166). This feature is necessary if the oscillations, started by noise or other random voltage fluctuations, are to grow to a usable magnitude. When the amplifier gain exceeds the ratio (13.166), the magnitude of the oscillations will increase until circuit nonlinearities, such as the saturation of the op-amp, reduce the incremental gain at large output amplitudes and limit oscillation growth.

13.7.4 Crystal Oscillators

An oscillator in which the Colpitts tank circuit contains a piezoelectric crystal element is called a *crystal oscillator*. A piezoelectric element, such as the one depicted in Fig. 13.39(a), is formed by placing capacitive electrodes on either side of a crystalline material such as quartz. This arrangement produces a circuit element with the equivalent electrical network of Fig. 13.39(b), where C_P is the capacitance formed by the crystal's electrodes, R_S represents the losses in the crystal, and L and C_S reflect the electromechanical coupling between the electrodes and the crystal lattice. In the typical piezoelectric crystal, C_S is very small (in the femtofarad range), L very large (as high as tens to hundreds of henries), and C_P of moderate size (in the picofarad range). Because the loss resistance R_S is very small, the crystal possesses a resonant terminal impedance with an extremely high Q.

Figure 13.39
(a) Piezoelectric crystal element;
(b) equivalent circuit diagram;
(c) reactance as a function of frequency. Circuit appears inductive only between ω_s and ω_p.

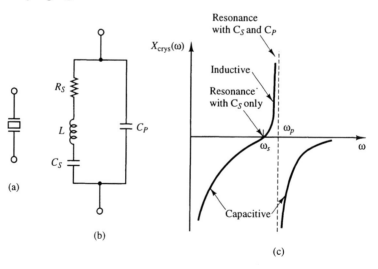

An expression for the terminal impedance can be found by combining in parallel the impedances of the two legs. With capacitive impedance expressed as $1/j\omega C$, inductive impedance as $j\omega L$, and R_S assumed negligibly small, the net impedance becomes

$$\mathbf{Z}_{\text{crys}}=\frac{1}{j\omega C_P}\left\|\left(j\omega L+\frac{1}{j\omega C_S}\right)=\frac{1}{j\omega C_P}\frac{j\omega L+1/j\omega C_S}{1/j\omega C_P+j\omega L+1/j\omega C_S}\right. \tag{13.167}$$

Algebraic manipulation of this equation leads to

$$\mathbf{Z}_{\text{crys}} = \frac{1}{j\omega C_P}\left[\frac{1/LC_S - \omega^2}{(C_S + C_P)/LC_S C_P - \omega^2}\right] \equiv \frac{1}{j\omega C_P}\left[\frac{\omega_s^2 - \omega^2}{\omega_p^2 - \omega^2}\right] \tag{13.168}$$

where $\omega_s = (1/LC_S)^{1/2}$ reflects the series resonance of L with C_S only, and $\omega_p = [(C_S + C_P) /LC_S C_P]^{1/2}$ reflects the resonance of L with C_P *and* C_S in series. Note that ω_p will always be larger than ω_s for nonzero C_S and C_P.

For ω below both ω_s and ω_p, and for ω above both ω_s and ω_p, the expression in brackets in Eq. (13.168) remains positive, so that $\mathbf{Z}_{\text{crys}}$ becomes a capacitive impedance of the form $1/j\omega C$. For a very narrow region between ω_s and ω_p, however, the expression in brackets becomes negative, transforming the leading factor of $1/j$ in Eq. (13.168) into $-1/j \equiv j$. Over this range of frequency, $\mathbf{Z}_{\text{crys}}$ becomes an inductive impedance of value

$$jX_{\text{crys}}(\omega) = j\frac{1}{\omega C_P}\left|\frac{\omega_s^2 - \omega^2}{\omega_p^2 - \omega^2}\right|_{\omega_s < \omega < \omega_p} \tag{13.169}$$

where $\mathbf{Z}_{\text{crys}} = jX_{\text{crys}}$. This phenomenon is illustrated in Fig. 13.39(c), where the reactance $X_{\text{crys}}(\omega)$ is plotted as a function of frequency. A negative reactance implies capacitive behavior, and a positive reactance indicates inductive behavior.

If the crystal is substituted for the tank inductor L in a Colpitts oscillator, oscillation can occur under the conditions outlined in Section 13.7.3 only at frequencies where the crystal appears inductive. With the crystal connected, the capacitances C_P and C_S also will become part of the tank circuit. Because C_S is much smaller than either C_P or the external capacitances of the Colpitts oscillator, and because the these capacitances all resonate in series with the inductor in the resonant tank circuit, the tank circuit will be dominated by C_S, and oscillation will occur at a frequency that is independent of external elements. One of the key advantages of a piezoelectric crystal oscillator is its frequency stability. A frequency of oscillation that is constant to within fractional parts per million is possible under ideal conditions. The principal disadvantage of a crystal oscillator is the fixed nature of its frequency, which cannot be adjusted.

Figure 13.40
Simple
one-transistor
crystal oscillator
based on the
Colpitts
configuration.

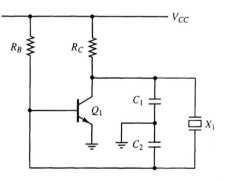

An example of a Colpitts crystal oscillator based on a single-transistor amplifier is illustrated in Fig. 13.40. In this circuit, the BJT inverter, which has gain $-g_m R_C$, provides the gain necessary to overcome the tank circuit losses and sustain oscillation. The resistor R_B biases Q_1 into its active region. Many other crystal oscillator configurations are possible. A crystal oscillator could be made using the Colpitts op-amp oscillator configuration of Fig. 13.37(a), for example.

13.7.5 Schmitt-Trigger Oscillator

One final oscillator circuit to be studied in this section utilizes the op-amp Schmitt-trigger circuit of Chapter 2. If a Schmitt trigger is combined with a simple RC network, the resulting circuit will produce square-wave oscillations at a fixed frequency. Such a circuit represents a type of oscillator known as the *astable multivibrator*. Unlike the linear oscillator circuits examined so far in this chapter, astable circuits specifically rely on the nonlinear nature of electronic devices to produce periodic signals. We study the Schmitt-trigger oscillator here, rather than in Chapter 2, because it need not be limited to op-amp form. It can be made from a variety of active devices, including discrete BJTs and MOSFETs. We shall study some of these latter astable oscillators in the context of digital circuits in Chapter 15. The analysis presented here is a most general one that applies to astable circuits of all types.

Figure 13.41
Schmitt-trigger RC oscillator, or "op-amp clock."

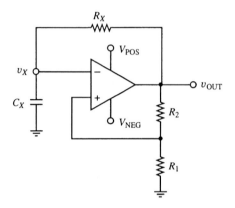

Consider the Schmitt-trigger RC oscillator shown in Fig. 13.41. This circuit is sometimes called the *op-amp clock*. In the analysis that follows, R_1 and R_2 are set to the same value, and the op-amp is assumed to be ideal. To understand the operation of the circuit, suppose that the op-amp has just switched from its negative saturation limit of V_{NEG} to its positive saturation limit of V_{POS}. Moreover, suppose the capacitor voltage to be equal to $V_{NEG}/2$ at the time of switching. Under these conditions, the circuit composed of R_X and C_X will see a step change in voltage, from V_{NEG} to V_{POS}, subject to the initial condition $v_X = V_{NEG}/2$. The capacitor will begin to charge toward V_{POS} according to the equation

$$v_X = V_{POS}(1 - e^{-t/R_X C_X}) + \frac{V_{NEG}}{2}e^{-t/R_X C_X} \qquad (13.170)$$

This expression can be derived from the differential equation of the $R_X C_X$ circuit. Direct substitution of $t = 0$ shows Eq. (13.170) to indeed satisfy the initial condition $v_X(0) = V_{NEG}/2$ and the final condition $v_X(\infty) = V_{POS}$. Note that V_{NEG}, as defined, is a negative number.

When the output of the Schmitt trigger switches to V_{POS} at $t = 0$, the voltage at the v_+ terminal becomes $V_{POS}/2$. As the capacitor charges toward V_{POS}, it will surpass the voltage of the v_+ terminal some time t_1 later, causing the Schmitt trigger to switch from its positive saturation state to its negative saturation state. The output of the Schmitt trigger will thus switch from V_{POS} to V_{NEG}. This transition reexcites the $R_X C_X$ network with a new step change in voltage from V_{POS} to V_{NEG}, subject to the new initial condition $v_X(t_1) = V_{POS}/2$. The charging of the capacitor will then be governed by the equation

$$v_X = V_{NEG}\left[1 - e^{-(t-t_1)/R_X C_X}\right] + \frac{V_{POS}}{2}e^{-(t-t_1)/R_X C_X} \qquad (13.171)$$

When the output of the Schmitt trigger switches to V_{NEG}, the voltage at the v_+ terminal becomes $V_{NEG}/2$. As the capacitor charges toward V_{NEG}, it will again surpass the voltage of the v_+ terminal, this time causing the output of the Schmitt trigger to switch from V_{NEG} to V_{POS} when v_X reaches $V_{NEG}/2$. This transition is identical to the originally assumed transition from V_{NEG} to V_{POS}; hence, the process as described above repeats itself indefinitely.

The time delays between Schmitt trigger transitions can be computed by solving Eq. (13.170) for the condition $v_X = V_{POS}/2$ and Eq. (13.171) for the condition $v_X = V_{NEG}/2$. Time t_1, which describes the time at which v_X reaches $V_{POS}/2$ and v_{OUT} switches to V_{NEG}, is reached when

$$v_X = V_{POS}(1 - e^{-t_1/R_X C_X}) + \frac{V_{NEG}}{2}e^{-t_1/R_X C_X} = \frac{V_{POS}}{2} \tag{13.172}$$

Rearranging this equation results in

$$\left(\frac{V_{NEG}}{2} - V_{POS}\right)e^{-t_1/R_X C_X} = -\frac{V_{POS}}{2} \tag{13.173}$$

The latter equation can be solved for t_1 to yield

$$t_1 = R_X C_X \ln \frac{V_{POS} - V_{NEG}/2}{V_{POS}/2} \tag{13.174}$$

Because V_{NEG} is a negative number, the quantity $(V_{POS} - V_{NEG}/2)$ is larger than V_{POS}.

A similar analysis yields the time t_2 at which v_X reaches $V_{NEG}/2$ and v_{OUT} is switched from V_{NEG} to V_{POS}:

$$t_2 = t_1 + R_X C_X \ln \frac{V_{NEG} - V_{POS}/2}{V_{NEG}/2} \tag{13.175}$$

For the case $V_{NEG} = -V_{POS}$, Eqs. (13.174) and (13.175) yield the results

$$t_1 = R_X C_X \ln 3 = 1.1 R_X C_X \tag{13.176}$$

and
$$t_2 = t_1 + 1.1 R_X C_X = 2.2 R_X C_X \tag{13.177}$$

Figure 13.42
Voltages v_X and v_{OUT} versus time for the Schmitt-trigger oscillator circuit of Fig. 13.41 for the case $V_{NEG} = -V_{POS}$ and $R_1 = R_2$.

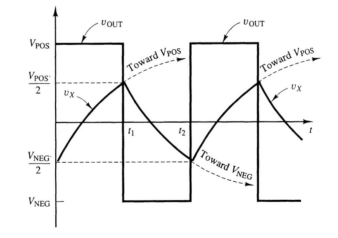

The waveforms v_{OUT} and v_X computed for the case $R_1 = R_2$ and $V_{NEG} = -V_{POS}$ are shown in Fig. 13.42. The frequency of oscillation, which is equal to $1/t_2$, can be fixed by selecting an appropriate value for R_X. Alternatively, the frequency can be made adjustable by choosing a variable resistor for R_X. The *duty cycle* of the oscillator, defined as the percentage of time that v_{OUT} spends in its positive state, will be equal to 50% if $V_{NEG} = -V_{POS}$. If V_{NEG} and V_{POS} are set to different values, the duty cycle can be made asymmetrical.

EXERCISE	13.39	Sketch the v_{OUT} and v_X waveforms for the circuit of Fig. 13.33 when $R_2 = 2R_1$.
	13.40	Sketch the v_{OUT} and v_X waveforms for the circuit of Fig. 13.33 when $V_{NEG} = -2V_{POS}/3$.
	13.41	Sketch the v_{OUT} and v_X waveforms for the circuit of Fig. 13.33 when $V_{NEG} = -2V_{POS}$.

SUMMARY

- Active filters are made from op-amps, resistors, and capacitors.
- An active filter has higher input resistance and lower output resistance than a comparable passive filter. An active filter can provide gain, but a passive filter cannot.
- The frequency response of a first-order low- or high-pass filter has a single pole.
- The frequency response of a second-order band-pass, low- or high-pass filter has two poles.
- Active filters can be cascaded to produce filters with a variety of frequency-response properties.
- A Butterworth filter has a maximally flat, monotonically decreasing response in its passband.
- A Chebyshev filter has a cyclic variation of its gain with frequency in the passband, but has a much steeper roll-off than a Butterworth filter of comparable order.
- The Butterworth and Chebyshev responses are produced by cascading first- and second-order sections in series. The required poles of the overall response must be provided by the various filter sections. Sallen–Key filter sections are often used.
- Magnitude scaling preserves the shape and pole locations of a filter's response while allowing its resistor or capacitor values to be changed.
- Frequency scaling preserves the overall shape of a filter's response while allowing its poles and zeros to be changed.
- Switched capacitor networks can be used to produce resistor-free active filters.
- A switched-capacitor filter is made from integrated-circuit capacitors and MOSFET op-amps.
- The response of a switched-capacitor filter depends on capacitance ratios but not on absolute capacitor values.
- The response of a switched-capacitor filter depends on the period of its driving clock.
- The period of the clock that drives a switched-capacitor filter must be much shorter than the smallest time interval of interest.
- An active filter in which the stability criteria is intentionally violated functions as an oscillator.
- Wien-bridge and phase-shift oscillators produce sinusoidal oscillations.
- Colpitts and Hartley oscillators use resonant LC "tank" circuits to produce oscillations of adjustable frequency.
- A crystal oscillator uses a resonant piezoelectric element in the tank circuit of an LC oscillator.
- A crystal oscillator produces oscillations at a fixed, very stable frequency.
- A Schmitt trigger with an added RC network functions as a square-ware oscillator.

◆ SPICE EXAMPLES

EXAMPLE 13.8   Suppose that the fourth-order Butterworth cascade of Fig. 13.20 is built for a particular telephone application where the circuit will be housed outdoors (e.g., on a utility pole). If the resistors have a temperature coefficient of 1% per °C and the capacitors 2% per °C, assess the performance of the filter over the temperature range $-15°C$ to $45°C$. This temperature range is typical for worst-case operating conditions in a temperate climate zone.

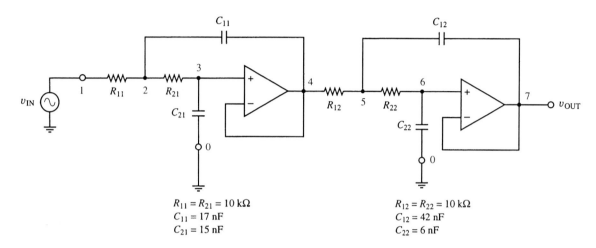

$$R_{11} = R_{21} = 10 \text{ k}\Omega$$
$$C_{11} = 17 \text{ nF}$$
$$C_{21} = 15 \text{ nF}$$

$$R_{12} = R_{22} = 10 \text{ k}\Omega$$
$$C_{12} = 42 \text{ nF}$$
$$C_{22} = 6 \text{ nF}$$

Figure 13.43   Fourth-Order Butterworth filter of Fig. 13.20 with nodes suitably numbered for SPICE.

Solution

The circuit is easily simulated using the circuit layout and node numbers of Fig. 13.43. In the input file to follow, each op-amp is defined using the subcircuit definition of Fig. 13.44. The temperature dependence of the resistors is modeled by simply adding a linear temperature-coefficient parameter $TC = 0.01$ to each element statement. An alternative method based on the .MODEL statement is used to specify the temperature dependence of the capacitors. In this latter method, a base capacitance of 1 nF is declared in the .MODEL statement TCAP and an appropriate corresponding multiplier appended to each capacitor element statement. Meanwhile, a common temperature coefficient $TC1 = 0.02$ is included in the capacitor .MODEL statement. The temperature is varied from $-15°C$ to $45°C$ in steps of $15°C$ using a .STEP command in the LIST mode, and the frequency response is determined using a .AC command.

Figure 13.44
Subcircuit op-amp
definition with
$A = 10^5$,
$R_{in} = 1 \text{ M}\Omega$, and
$R_{out} = 100 \, \Omega$.

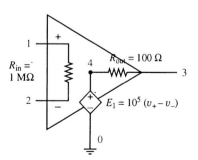

Input File

```
LOW-PASS BUTTERWORTH FILTER WITH TEMPERATURE DEPENDENCE

*Define the op-amp using a subcircuit definition:
*(No temperature dependence is included for the op-amp)
        .SUBCKT OP-AMP  1 2 3
        Rin  1 2 1meg
        E1   4 0 1 2 1e5
        Rout 4 3 100
        .ENDS OP-AMP

*Construct the first Sallen-Key section using the R and C values from
*Fig. 13.20:
        X1   3 4 4 OP-AMP
        R11 1 2 10k  TC=0.01
        R21 2 3 10k  TC=0.01
        C11 2 4 TCAP 17
        C21 3 0 TCAP 15

*Construct the second Sallen-Key section:
        X2   6 7 7 OP-AMP
        R12 4 5 10k  TC=0.01
        R22 5 6 10k  TC=0.01
        C12 5 7 TCAP 42
        C22 6 0 TCAP 6

*Model the temperature dependence of the capacitors in the .MODEL statement:
        .MODEL TCAP CAP(C = 1nF; TC1 = 0.02)

*Specify the input source:
        Vin 1 0 AC 1V 0

*Specify the parameters of the frequency analysis:
        .AC  DEC  50 100Hz 100kHz

*Step the temperature over the range -15°C to 45°C using the
*listed values:
        .STEP TEMP LIST -15 0 15 30 45

        .PROBE V(7)
        .END
```

Results. The results of this simulation are shown in Fig. 13.45. As these magnitude plots show, the overall shape of the response is preserved with temperature, but the corner frequency shifts from about 1.7 kHz at −15°C to 845 Hz at 45°C. The decrease in ω_o with increasing temperature is expected because the resistors and capacitors increase with temperature, and ω_o is equal to $1/(R_1 R_2 C_1 C_2)^{1/2}$. Similarly, with all resistors equal to the same value, the Q parameters of each stage, which determine the shape of the overall response, are dependent solely on the capacitance ratios C_{11}/C_{21} and C_{12}/C_{22}. The capacitor temperature dependence is simulated in the program by multiplying each capacitor by the same factor $(1 + TC_1 \Delta T)$, hence the parameter Q becomes insensitive to temperature.

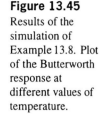

Figure 13.45
Results of the simulation of Example 13.8. Plot of the Butterworth response at different values of temperature.

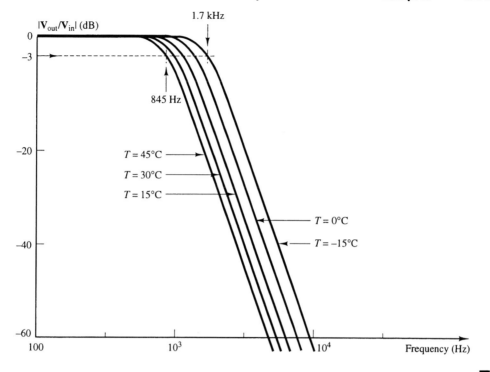

EXAMPLE 13.9

Simulate the switched-capacitor filter of Fig. 13.30(b) on SPICE. Show that it has same response as the analog filter of Fig. 13.30(a). Choose element values such that each circuit has a dc gain of -5 and a cutoff frequency of $f_o = \omega_o/2\pi = 1\,\text{kHz}$.

Solution

The analog filter will meet the required specifications if the elements are set to the values $R_1 = 2\,\text{k}\Omega$, $R_2 = 10\,\text{k}\Omega$, and $C_X = 15.9\,\text{nF}$. These values will indeed yield a dc gain of -5 and a cutoff frequency of $f_o = \omega_o/2\pi = 1\,\text{kHz}$.

The clock period of the switched-capacitor filter is next chosen (somewhat arbitrarily) to be $10\,\mu\text{s}$. This value, which corresponds to a clock frequency of $100\,\text{kHz}$, lies well above the expected f_o of the filter and is consistent with the requirements of switched-capacitor filtering. It is also slow enough that it can be easily simulated within a reasonable time on SPICE.

The switched-capacitor version of the feedback capacitor is next set to the value $C_X = 15.9\,\text{nF}$, which is identical to its counterpart in the analog filter. According to Eq. (13.136), the cutoff frequency ω_o of the switched-capacitor filter will be $C_2/C_X T_c$. Achieving an f_o of $1\,\text{kHz}$ therefore requires that C_2 be set to the value

$$C_2 = 2\pi f_o C_X T_c = 2\pi (1\,\text{kHz})(15.9\,\text{nF})(10\,\mu\text{s}) = 1\,\text{nF} \qquad (13.178)$$

Note that this capacitance value yields an equivalent resistance of

$$R_2 = T_c/C_2 = (10\,\mu\text{s})/(1\,\text{nF}) = 10\,\text{k}\Omega \qquad (13.179)$$

which is consistent with the value of R_2 in the analog filter. Similarly, achieving an equivalent value for R_1 of $2\,\text{k}\Omega$ requires that C_1 be set to the value

$$C_1 = T_c/R_1 = (10\,\mu\text{s})/(2\,\text{k}\Omega) = 5\,\text{nF} \qquad (13.180)$$

These elements are labeled in the circuit diagram of Fig. 13.46, which depicts single-pole switched-capacitor and analog filters driven from the same source v_{IN}.

Figure 13.46

Switched-capacitor
and analog versions
of the low-pass
filter of Fig. 13.30.

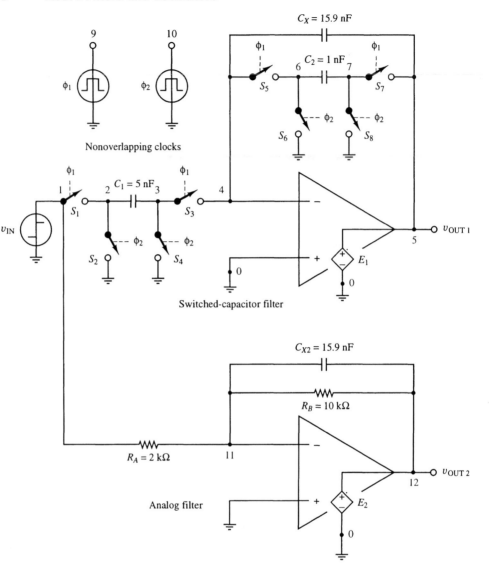

The operation of a switched-capacitor network is most easily modeled in SPICE using voltage-controlled switches to simulate the MOSFETs. This method is preferred because SPICE has difficulty obtaining an initial bias solution if actual MOSFET devices are used. In the circuit of Fig. 13.46, the transverse switches are actuated by voltage ϕ_1, and the shunt switches are actuated by ϕ_2. As indicated in the input file that follows, all switches are set to the closed position when the control voltage rises to 4 V; conversely, switches are set to the open position when the control voltage falls to 1 V. These staggered levels help to maintain the nonoverlapping status of ϕ_1 and ϕ_2. The on resistance RON of each switch is set to 1 Ω, yielding a maximum switched-capacitor decay time of $(5\,\text{nF})(2\,\Omega) = 10\,\text{ns}$, which is much shorter than any time interval of interest. The nonoverlapping clocks ϕ_1 and ϕ_2 are generated using identical 0- to 5-V pulsed periodic waveforms with rise and fall times of 1 μs, and with the source ϕ_2 delayed by a half period (5 μs).

Ideally, the frequency analysis of a .AC command would provide the most appropriate simulation, but SPICE cannot execute a .AC command using transient waveforms such as ϕ_1 and ϕ_2. The two circuits therefore are analyzed and compared in the time domain using a 1-V step input. An annotated input file for the simulation follows.

Input File

```
COMPARISON OF SWITCHED-CAPACITOR and ANALOG FILTERS
*Enter the voltage sources in the circuit:
     vIN 1 0 dc 1
*Specify two 0- to 5-V, 100-kHz, nonoverlapping clocks by delaying φ₂ by 50μs:
     VPHI1  9 0 PULSE(0 5  0 1u 1u 3u 10u)
     VPHI2 10 0 PULSE(0 5 5u 1u 1u 3u 10u)
*Enter the switches of the switched-capacitor filter:
     S1 1 2  9 0 switch
     S2 2 0 10 0 switch
     S3 3 4  9 0 switch
     S4 3 0 10 0 switch
     S5 4 6  9 0 switch
     S6 6 0 10 0 switch
     S7 7 5  9 0 switch
     S8 7 0 10 0 switch
     .MODEL switch VSWITCH( VON=4V VOFF=1V RON=1 )
*Enter the capacitor values:
     CX 4 5 15.9nF
     C1 2 3  5nF
     C2 6 7  1nF
*Use a dependent source with gain of 10⁶ to simulate the op-amp:
     E1 5 0 0 4 1e6
*Add the analog filter:
     RA   1 11 2k
     RB  11 12 10k
     CX2 11 12 15.9nF
     E2  12 0 0 11 1e6
*Perform the transient analysis:
     .TRAN 1u 1m
     .PROBE V(5) V(12)
     .END
```

Results. The two clock voltages ϕ_1 and ϕ_2 produced by the simulation are shown in Fig. 13.47. The output of both filters is shown in Fig. 13.48. As these plots indicate, the output of the switched-capacitor filter closely resembles that of the analog version. Both exhibit a dc output of -5 V after about 1 ms, an exponential rise toward -5 V after $t = 0$, and a time constant of $1/R_2C_X = 0.16$ ms. Such a step response is to be expected from a 1-kHz, single-pole RC filter. ■

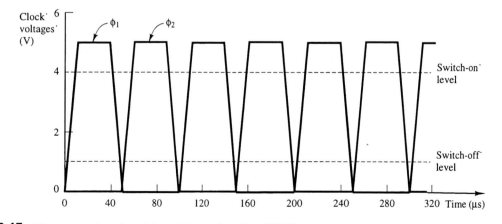

Figure 13.47 Nonoverlapping clocks ϕ_1 and ϕ_2 produced by SPICE.

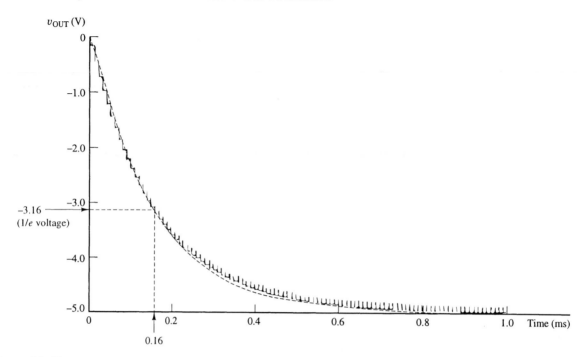

Figure 13.48 Output produced by SPICE for the filters of Fig. 13.46 in response to a 1-V step input. Solid line: switched-capacitor filter; dashed line: analog filter.

◆

PROBLEMS

13.1 A Simple First-Order Active Filter

13.1 ⟫ Using the circuit configuration of Fig. 13.1(a) as a guideline, design a low-pass filter that attenuates all frequency components above 1 kHz. What advantages does this circuit have over a passive low-pass filter made from resistors and capacitors alone?

13.2 The passive RC filter of Fig. 13.3 is constructed with $R_2 = 5.6\,k\Omega$ and $C = 0.033\,\mu F$. The filter feeds the input of a noninverting op-amp amplifier for which $R_2 = 81\,k\Omega$ (feedback resistor) and $R_1 = 4.7\,k\Omega$. Draw the magnitude Bode plot of the resulting overall response.

13.3 Show that the op-amp circuit of **Fig. P13.3** functions as a first-order high-pass filter. Find the cutoff frequency in Hz, the passband gain in dB, and the slope outside the passband in dB/decade for the case $R_1 = 2\,k\Omega$, $R_2 = 10\,k\Omega$, and $C = 0.05\,\mu F$.

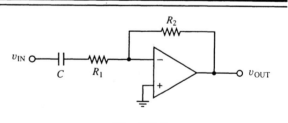

Fig. P13.3

13.4 ⟫ The circuit of **Fig. P13.3** functions as a simple high-pass filter. Choose values for the three passive elements such that the filter has a gain of 40 dB in the passband and a cutoff frequency at 15 kHz. If the filter is to pass signals with spectral content between 15 and 50 kHz, what must be the minimum unity-gain frequency of the op-amp used to construct the filter?

13.5 ⟫ ○ Consider the high-pass filter of **Fig. P13.3**. Transform this filter into a band-pass type by adding one additional capacitor to the circuit. Derive an expression for passband gain and the high- and low-frequency cutoff frequencies of your filter.

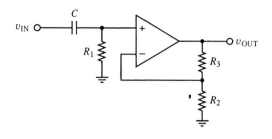

Fig. P13.6

13.6 Show that the circuit of **Fig. P13.6** functions as a single-pole high-pass filter of first order. Find an expression for the cutoff frequency and gain in the passband region. Evaluate these expressions for $R_1 = R_2 = 2.7\,\text{k}\Omega$, $R_3 = 18\,\text{k}\Omega$, and $C = 0.33\,\mu\text{F}$.

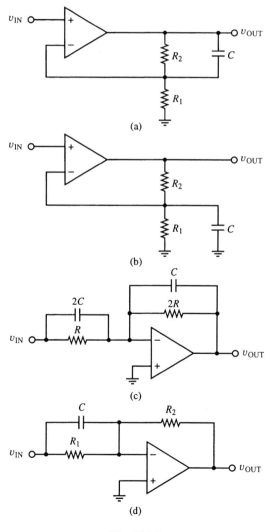

Fig. P13.7

13.7 For each of the op-amp circuits of **Fig. P13.7**, find an expression for the closed-loop gain as a function of frequency.

13.8 Design a first-order active filter that has a gain magnitude of 20 between 0 Hz and 5 kHz and a gain magnitude of 1 at higher frequencies. These design objectives can be achieved using a single capacitor only. What is the width of the transition region for your filter?

13.9 Design a first-order active filter that has a gain magnitude of 2 at frequencies well below about 1 kHz and a gain magnitude of 10 at frequencies well above 1 kHz. What is the width of the transition region for your filter? You can achieve these design goals using a single capacitor only.

13.10 An active filter can be made by cascading an integrator with a differentiator. Derive an expression for the overall transfer function of such a cascade.

13.11 The circuit of **Fig. P13.11** functions as a bandpass filter in which R_1 and C_1 set the high-frequency endpoint of the passband, and R_2 and C_2 set the low-frequency endpoint.

(a) Derive an expression for the transfer function of the filter in the sinusoidal steady state.

(b) Choose values for the resistors and capacitors such that the passband endpoints are set to 100 Hz and 10 kHz, with a passband gain of 40 dB.

(c) What is the input resistance of the filter in the passband region?

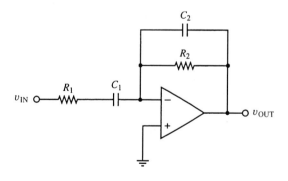

Fig. P13.11

13.12 Show that the largest possible value of Q for the low-pass filter of Fig. 13.7 occurs at the capacitance ratio $C_1/C_2 = 1$.

13.13 Find the transfer function of the circuit of **Fig. P13.13** in the sinusoidal steady state if $R_1 = R_2 = R_3 = R_4 = R$. This circuit functions as a noninverting integrator.

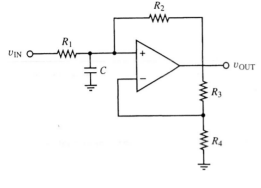

Fig. P13.13

13.2 Ideal Filter Functions

13.14 Suppose that a low-pass filter is required with a cutoff frequency of 20 kHz. The gain of the filter must lie between 0 dB and −3 dB within this passband. At 200 kHz, the gain must be reduced by at least −12 dB. Determine the lowest-order filter transfer function that can meet these specifications.

13.15 A high-pass filter is required for which the pass-band "gain" is at least −6 dB, but no more than 0 dB, above 10 kHz. At 5 kHz, the gain must be reduced by at least −10 dB from the passband value. Determine the lowest-order filter transfer function that can meet these specifications.

13.16 A low-pass filter is used to approximate an ideal brick-wall response. What must be the minimum order of the filter if its gain is to be reduced by −40 dB at $\omega = 5\omega_o$, where ω_o denotes the edge of the passband?

13.17 A high-pass filter is used to approximate an ideal brick-wall response. What must be the minimum order of the filter if its gain is to be reduced by −35 dB at $\omega = 0.8\omega_o$, where ω_o denotes the edge of the passband?

13.18 A low-pass filter has a cutoff frequency of ω_o. Which of the following most closely approaches the ideal low-pass brick-wall response:

(a) A filter that reduces the gain by −20 dB at $\omega = 4\omega_o$.

(b) A filter that reduces the gain by −40 dB at $\omega = 8\omega_o$.

(c) A filter that reduces the gain by −10 dB at $\omega = 2\omega_o$.

13.19 A band-pass filter is required such that the gain falls by at least −15 dB when ω is increased by a factor of 10 above the upper edge of the passband or decreased by a factor of 10 below the lower edge of the passband. What is the minimum required order of the filter?

13.20 A band-pass filter is required such that the gain falls by at least −35 dB when ω is increased by a factor of 5 above the upper edge of the passband. Similarly, the gain must fall by at least a factor of −25 dB when ω is decreased by a factor of 4 below the lower edge of the passband. What is the minimum required order of the filter?

13.3 Second-Order Filter Responses

13.3.1 The Biquadratic Filter Function

13.21 A low-pass filter has a transfer function given by $H(s) = 200/(s + 10)(s + 200)$.

(a) What is the gain in the passband?

(b) What are the biquad coefficients $a_0, \cdots, a_2$ and $b_0, \cdots, b_2$ of $H(s)$?

(c) Sketch the magnitude plot of the transfer function.

13.22 A high-pass filter has a transfer function given by $H(s) = s^2/(s + 10^3)(s + 2 \times 10^3)$.

(a) What is the gain in the passband?

(b) What are the biquad coefficients $a_0, \cdots, a_2$ and $b_0, \cdots, b_2$ of $H(s)$?

(c) Sketch the magnitude plot of the transfer function.

13.23 A low-pass filter has a biquadratic transfer function given by $H(s) = (s + 3)(s + 4)/(s + 0.5)(s + 2)$.

(a) What is the dc gain of the filter?

(b) What are the biquad coefficients $a_0, \cdots, a_2$ and $b_0, \cdots, b_2$ of $H(s)$?

(c) Sketch the magnitude plot of the transfer function.

(d) What is the roll-off at high frequencies?

13.24 A high-pass filter has a transfer function given by $H(s) = s^2/(s + 10)(s + 2 \times 10^4)$.

(a) What is the gain in the passband?

(b) What are the biquad coefficients $a_0, \cdots, a_2$ and $b_0, \cdots, b_2$ of $H(s)$?

(c) Sketch the magnitude plot of the transfer function.

13.25 A second-order filter has poles at $s = -0.5 \pm j0.866$. The response goes to zero at $\omega = 5$ rad/s, but becomes a constant value of 5 as ω approaches zero. Find the biquadratic transfer function of the filter.

13.26 A filter has a biquadratic transfer function given by $H(j\omega) = 10^4/(10^3 - \omega^2 + j100w)$.

(a) What is the filter type?

(b) What is the gain at $\omega = 0$?

(c) Identify the biquad coefficients $a_0, \cdots, a_2$ and $b_0, \cdots, b_2$ of $H(s)$.

(d) What is the cutoff frequency of the filter?

(e) What is the Q of the filter?

13.27 A low-pass filter has a transfer function given by $H(s) = 1/(s + 1)(s^2 + s + 1)$.

(a) What is the filter order?

(b) Find the location of the filter poles.

(c) Express $|H(s)|$ as a function of $j\omega$. Show that $|H(j\omega)|$ reduces to a simple function of ω.

13.28 A second-order filter has poles $s_1 = j500$ rad/sec and $s_2 = -j500$ rad/sec, plus a zero at $s_3 = 300$ rad/sec. What is the filter's biquadratic transfer function? Identify the values of the biquad coefficients $a_0, \cdots, a_2$ and $b_0, \cdots, b_2$ of $H(s)$.

13.29 A second-order filter has a biquadratic transfer function given by $H(s) = 100s/(s^2 - 200s + 300)$.

(a) What is the behavior of the filter for $\omega \to 0$ and for $\omega \to \infty$?

(b) From the answer to part (a), determine whether the filter is low-pass, high-pass, or band-pass.

(c) Do the poles of the filter lie in the left-half or right-half plane? If the latter condition is true, the filter will be unstable and will oscillate at ω_o.

13.30 A biquad filter has poles located at $\mathbf{s} = -288 \pm j408$, a dc gain of 2, and no zeros.

(a) What are the values of ω_o and Q for the filter?

(b) What is the complete transfer function of the filter?

13.31 A second-order biquad filter has no zeros and two poles located at angles $\pm 1.7\pi$ relative to the positive real axis.

(a) What type of filter is this?

(b) What are the real and imaginary parts of the poles expressed as functions of ω_o?

(c) What is the Q of the filter?

13.3.2 Second-Order Active Low-Pass Filter

13.32 A low-pass filter of the type shown in Fig. 13.7 is constructed with $C_1 = 0.03\,\mu F$, $C_2 = 0.015\,\mu F$, and $R_1 = R_2 = 7.5\,k\Omega$.

(a) What is the filter cutoff frequency ω_o?

(b) At what frequency will the filter response fall by -3 dB from its low-frequency value.

13.33 A low-pass filter of the type shown in Fig. 13.7 is constructed with $C_1 = C_2 = 2.2\,\mu F$, $R_1 = 15\,k\Omega$, and $R_2 = 42\,k\Omega$.

(a) What is the filter cutoff frequency ω_o?

(b) At what frequency will the filter response fall by -3 dB from its low-frequency value.

13.34 Consider the second-order low-pass filter of Fig. 13.7. If $C_2 = 0.01\,\mu F$, choose the values of the remaining elements if the cutoff frequency is to be 1 kHz and $Q = 1/\sqrt{2} = 0.707$.

13.35 Consider the second-order low-pass filter of Fig. 13.7. If $C_1 = 300\,nF$, choose the values of the remaining elements if the cutoff frequency is to be 100 kHz and $Q = 1/\sqrt{2} = 0.707$.

13.36 Σ Suppose that the only capacitors available to you have values of $0.005\,\mu F$. Use them to design a second-order low-pass filter with $Q = 0.707$ and a cutoff frequency of 3000 Hz.

13.37 Σ Design a low-pass filter that meets the following specifications: passband gain = 0 dB, cutoff frequency = 40 kHz, $Q = 0.707$, high-frequency roll-off = -20 dB/decade.

13.38 Σ Design a low-pass filter that meets the following specifications: passband gain = 0 dB, cutoff frequency = 10 Hz, $Q = 0.707$, and high-frequency roll-off = -40 dB/decade.

13.39 Σ Design a low-pass filter that meets the following specifications: passband gain = 0 dB, cutoff frequency = 1.5 kHz, $Q = 0.5$, and high-frequency roll-off = -40 dB/decade.

13.40 Σ Design a low-pass filter that meets the following specifications: passband gain = $+20$ dB, cutoff frequency = 250 kHz, $Q = 1.2$, and high-frequency roll-off = -40 dB/decade.

13.41 Σ An amplitude-modulated (AM) radio transmission consists of a 1.03-MHz carrier modulated by a 100-Hz to 15-kHz audio signal. The received signal is passed through a diode detector that produces an output consisting of the desired audio signal plus unwanted carrier signal components at 1.03 MHz and above. Design a second-order analog filter that will pass the desired audio signal while attenuating the unwanted signals by at least -60 dB.

13.3.3 Second-Order Active High-Pass Filter

13.42 Σ Design a first-order high-pass filter that meets the following specifications: passband gain = 0 dB, cutoff frequency = 5 kHz, $Q = 0.707$, and low-frequency slope = $+20$ dB/decade.

13.43 A high-pass filter of the type shown in Fig. 13.12 is constructed with $C_1 = 0.05\,\mu F$, $C_2 = 0.022\,\mu F$, and $R_1 = R_2 = 5.6\,k\Omega$.

(a) What is the filter cutoff frequency ω_o?

(b) At what frequency will the filter response fall by -3 dB?

13.44 A high-pass filter of the type shown in Fig. 13.12 is constructed with $C_1 = C_2 = 68\,\text{nF}$, $R_1 = 39\,\text{k}\Omega$, and $R_2 = 33\,\text{k}\Omega$.

(a) What is the filter cutoff frequency ω_o?

(b) At what frequency will the filter response fall by $-3\,\text{dB}$?

13.45 Consider the second-order high-pass filter of Fig. 13.12. If $C_2 = 0.05\,\mu\text{F}$, choose the values of the remaining elements if the cutoff frequency is to be 2.5 kHz and $Q = 1/\sqrt{2} = 0.707$.

13.46 Consider the second-order high-pass filter of Fig. 13.12. If $C_1 = 180\,\text{nF}$, choose the values of the remaining elements if the cutoff frequency is to be 120 kHz and $Q = 1/\sqrt{2} = 0.707$.

13.47 Show that a second-order high-pass filter of the type shown in Fig. 13.12 can be constructed with a Q of $1/\sqrt{2} = 0.707$ by choosing values $C_1 = 2C$, $C_2 = C$, and $R_1 = R_2 = R$, where C_1 is the feedback capacitor, and C_2 is the input capacitor. Express the -3-dB cutoff frequency in terms of R and C.

13.48 ⅀ Design a high-pass filter that has a Q of 0.707, a passband gain of 0 dB, a cutoff frequency of 10 kHz, and a low-frequency slope of $+40\,\text{dB/decade}$.

13.49 ⅀ Design a high-pass filter that meets the following specifications: passband gain = 0 dB, cutoff frequency = 20 Hz, $Q = 0.707$, and low-frequency slope = $+40\,\text{dB/decade}$.

13.50 ⅀ Design a high-pass filter that meets the following specifications: Passband gain = 0 dB, cutoff frequency = 1.5 kHz, $Q = 0.5$, and low-frequency slope = $+40\,\text{dB/decade}$.

13.51 ⅀ Design a high-pass filter that meets the following specifications: Passband gain = $+20\,\text{dB}$, cutoff frequency = 4.5 kHz, $Q = 1.2$, and low-frequency slope = $+40\,\text{dB/decade}$.

13.3.4 Second-Order Active Band-Pass Filter

13.52 A bandpass filter has a center frequency of 2800 Hz and a bandwidth of 900 Hz.

(a) What is the Q of this filter?

(b) What are the upper and lower cutoff frequencies f_1 and f_2?

13.53 A bandpass filter has a Q of 2.5 and a center frequency of 300 Hz.

(a) What are the upper and lower cutoff frequencies f_1 and f_2?

(b) What is the bandwidth of the filter?

13.54 The second-order bandpass filter of Fig. 13.15 is constructed with $R_1 = 2.6\,\text{k}\Omega$, $R_2 = 180\,\text{k}\Omega$, and $C_1 = C_2 = 10\,\text{nF}$.

(a) Find the center passband frequency ω_o.

(b) Find the upper and lower endpoints ω_1 and ω_2 of the passband.

(c) Draw the magnitude Bode plot of the filter response.

13.55 Consider the second-order bandpass filter of Fig. 13.15. Suppose that the filter has a passband gain magnitude of $A = 0.5 \equiv -6\,\text{dB}$ and a quality factor of $Q = 0.5$.

(a) What will be the filter bandwidth expressed in units of ω_o?

(b) Suggest a simple modification to the filter such that the passband gain magnitude is set to unity without changing Q or the bandwidth.

13.56 Choose resistor values for the bandpass filter of Fig. 13.15 such that $f_o = 1590\,\text{Hz}$ and $Q = 10$. Let $C_1 = C_2 = 0.001\,\mu\text{F}$.

13.57 ⅀ Using the band-pass filter of Fig. 13.15 as a guide, design a narrow-passband filter that has a peak response at $f_o = 10.6\,\text{kHz}$ and a Q of 25. Draw a magnitude Bode plot for the response of your filter. What is the magnitude of the response in dB at f_o?

13.58 ⅀ Design a second-order band-pass filter based on the circuit of Fig. 13.15 such that the center of the passband is at 5 kHz and the bandwidth is 1 kHz. What should be the Q of the filter? Choose appropriate values for the resistors and capacitors in the circuit.

13.59 A second-order band-pass filter is required with $f_o = 10\,\text{kHz}$ and $|V_{\text{out}}/V_{\text{in}}| = +60\,\text{dB}$ at f_o. If the filter is implemented using the circuit of Fig. 13.15, choose appropriate values for R_1, R_2, C_1, and C_2. What are the resulting values of Q and bandwidth for this filter?

13.60 ⅀ A second-order band-pass filter is required with a center frequency of $f_o = 54\,\text{kHz}$ and a passband gain of $+50\,\text{dB}$. If the filter is implemented using the circuit of Fig. 13.15 with $C_1 = C_2$, choose appropriate values for R_1 and R_2. What is the resulting value of Q for the filter? What is its bandwidth?

13.61 The circuit of **Fig. P13.61** is called an *all-pass* filter. Its transfer function has constant magnitude with frequency, but the filter introduces a 90° phase shift at the filter frequency f_o. Use KVL and KCL to derive the transfer function of this filter. Draw the magnitude and angle Bode plots of the filter.

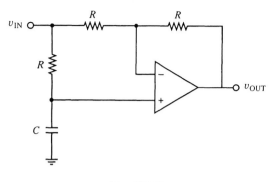

Fig. P13.61

13.62 ⅀ ◑ Design a modem receiver capable of decoding digital information transmitted over the telephone. In this case, assume that tone frequencies of 3 kHz and 4.5 kHz are used to represent logic **0** and logic **1**, respectively. Your system should provide the necessary digital outputs at a 0- to 5-V level. Estimate the maximum BAUD rate of your receiver system.

13.4 Active Filter Cascading

13.4.1 Low-Pass Butterworth Response

13.63 A low-pass Butterworth filter has order $n = 5$. Determine the output attenuation in dB at $\omega = 5\omega_o$.

13.64 Find the signal attenuation (in decimal *and* dB) of a fourth-order low-pass Butterworth filter at $\omega = 1.5\omega_o$.

13.65 Determine the order n of a low-pass Butterworth filter that must provide a minimum of -40 dB attenuation at $\omega = 2\omega_o$.

13.66 Prove that the first $(2n - 1)$ derivatives of a Butterworth filter of order n are equal to zero at $\omega = 0$.

13.67 ⅀ Design a low-pass Butterworth filter that meets the following specifications: passband gain $= 0$ dB, cutoff frequency $= 1$ Hz, and high-frequency roll-off $= -40$ dB/decade.

13.68 ⅀ Use cascaded Sallen–Key sections to synthesize a sixth-order, low-pass Butterworth filter with a cutoff frequency of 1 kHz.

13.69 ⅀ ○ Design a low-pass filter cascade based on the Butterworth response that has a -60 dB/decade roll-off outside the passband. The passband gain at dc should be $+10$ dB and the cutoff frequency 15 kHz.

13.70 ⅀ A low-pass audio filter is required for a video communication operation. The filter's passband must extend from dc to 5 kHz and the filter must attenuate the signal by at least -50 dB at 20 kHz. Design an appropriate active low-pass Butterworth filter.

13.71 ⅀ ◑ Not-so-famous engineer I. M. A. Bozo has recorded a live rock concert using a tape recorder and microphones. The tape is to be played into a digitizing unit that will make the master for compact-disc versions of the concert. Unfortunately, Bozo has failed to connect one of the microphone ground shields, so that the recording contains a considerable amount of 60-Hz noise picked up from power-line interference. The production manager has asked you, as chief engineer, to salvage the master recording. You have decided to build a circuit that can filter out the unwanted noise when the tape recording is played into the digitizing unit. Design such a filter by cascading a low-pass filter with a high-pass filter. Your circuit must attenuate the 60-Hz noise component by at least -80 dB while allowing as much of the remaining spectrum as possible to pass with zero attenuation.

13.72 ◑ A particular application requires a maximally flat filter with a maximum passband gain of 0 dB and a passband width from 0 to 80 Hz with passband attenuation equal to no more than -0.5 dB at 80 Hz. If the filter's half-power frequency ω_o is $2\pi (100$ Hz$)$, specify the required order and pole locations of an appropriate Butterworth filter. Note that ω_o is defined as the frequency at which the Butterworth response falls by $1/\sqrt{2} \equiv -3$ dB. In this case, the usable passband is defined as the frequency range over which the gain falls by a maximum of -0.5 dB. Also find the frequency at which the response falls by -20 dB.

13.4.2 Low-Pass Chebyshev Response

13.73 A low-pass Chebyshev filter has order $n = 4$. Determine the output attenuation in dB at $\omega = 4\omega_o$.

13.74 Find the signal attenuation (in decimal *and* dB) of a sixth-order low-pass Chebyshev filter at $\omega = 2\omega_o$.

13.75 Determine the order n of a low-pass Chebyshev filter that must provide a minimum of -40 dB attenuation at $\omega = 2\omega_o$.

13.76 ⅀ Design a third-order Chebyshev filter that has a maximum passband gain of unity, -3-dB passband ripple, and a cutoff frequency of 5 kHz.

13.77 ⅀ Design a fourth-order Chebyshev filter that has a maximum passband gain of unity, -1-dB passband ripple, and a cutoff frequency of 10 kHz.

13.78 ⅀ ○ Design a low-pass filter cascade based on the Chebyshev response that has a -60 dB/decade roll-off outside the passband and a maximum passband gain of $+10$ dB. The filter cutoff frequency should be 20 kHz and the passband ripple -3 dB.

13.79 ⚡ A low-pass audio filter is required for a filtering operation in a wireless communication system. The filter's −3-dB passband endpoint must extend to 2740 Hz and the filter must attenuate its input signal by at least −60 dB at 5400 Hz. Design an appropriate active filter cascade based on the Chebyshev response.

13.4.3 High-Pass and Band-Pass Cascades

13.80 ⚡ Cascade a second-order low-pass filter with a second-order high-pass filter to create a band-pass filter with a +40 dB/decade slope below the passband and a −40 dB/decade slope above the passband. The passband should have a gain of 0 dB, a center frequency of 1 kHz, and lower and upper corner frequencies of 100 Hz and 10 kHz, respectively. Set the Q of each filter to $1/\sqrt{2} = 0.707$.

13.81 A fourth-order high-pass filter is constructed by cascading two second-order filters in series. Each stage has a corner frequency of 1 kHz, and Q values of 1.31 and 0.541, respectively. Show that this filter combination produces a high-pass Butterworth response. Draw the Bode plot of the overall transfer function.

13.82 A fourth-order high-pass filter is made by cascading two second-order high-pass filters each having a Q of 0.707. What is the response at ω_o? Draw the Bode plot of the cascade response and compare to that of a fourth-order Butterworth response.

13.83 ⚡ ○ Design a high-pass filter cascade with +60 dB/decade roll-off below its cutoff frequency. The filter should have maximally flat response in the passband, a passband gain of 20 dB, and a cutoff frequency of 10 Hz.

13.84 ⚡ ◑ Mobile radio communication from cars, trucks, and boats often requires the use of land-based repeater stations. A repeater receives the relatively weak signal from a mobile transmitter and rebroadcasts it at a much higher power level from a tall antenna. In this way, the communication range is greatly extended.

To avoid the use of the repeater by unauthorized personnel, or the keying of the repeater transmitter from extraneous interference, some repeater transmitters will activate only while the mobile vehicle transmits a subaudible sinusoidal tone simultaneously with the transmitted speech information. Typical subaudible tones lie in the frequency range 67 to 180 Hz.

In this problem, you are to design a filter circuit capable of detecting a 100.5-Hz subaudible tone and sending a digital activation signal to the repeater transmitter. Assume that the repeater's receiver has demodulated the signal sent by the mobile vehicle, so that its radio-frequency carrier has been eliminated. The input to your circuit will thus consist only of the subaudible tone plus voice information lying in the frequency band 200 Hz to 5 kHz. You must design your circuit such that any voice information does not activate the transmitter if the correct subaudible tone is not present.

13.85 ⚡ ● Design a crude speech-recognition system capable of providing a five-bit digital output in response to voice information lying in the following five frequency bands: 100–300 Hz; 300–500 Hz; 500 Hz–1 kHz; 1–2.5 kHz; and 2.5–4 kHz.

13.86 ⚡ ● An infrared LED–phototransistor system is required by a mobile robot to help it detect wall obstructions. The light emitted by the LED will reflect off walls and be received by the phototransistor. To avoid interference from ambient room light powered at 50 Hz, the LED is to be energized by a 1-kHz square wave. The signal received from the phototransistor will then be fed to a filter to attenuate any 100 Hz signals present from ambient light. (Ambient light from bulbs powered at 50 Hz has a frequency of 100 Hz, rather than 50 Hz, because the bulbs blink for both the positive and negative portions of the ac power-line cycle.)

Design a system that can provide the necessary detection and filtering operation. Specifically, you should design circuits for the LED emitter and photodetector, and the required filter circuit tuned to 1 kHz. Your system should attenuate any 100-Hz signals detected by the phototransistor circuit by at least −60 dB, and it should pass any 1-kHz signals with a gain of +20 dB.

13.5 Magnitude and Frequency Scaling

13.87 Use magnitude scaling to redesign the filter of Example 13.5 such that $C_1 = 0.1\,\mu F$ and $C_2 = 0.068\,\mu F$.

13.88 The paper design of a second-order low-pass filter of the Sallen–Key type yields the following component values: $R_1 = R_2 = 26.3\,k\Omega$, $C_1 = 106\,nF$, and $C_2 = 70.2\,nF$. Specify alternative values for the resistors and capacitors such that only standard component values are used.

13.89 A low-pass Sallen–Key filter of the type shown in Fig. 13.7 is made with $R_1 = 15\,k\Omega$, $R_2 = 33\,k\Omega$, $C_1 = 0.022\,\mu F$, and $C_2 = 0.031\,\mu F$.

(a) What are ω_o and Q of this filter?

(b) Choose new values for the resistors and capacitors such that only standard component values are used.

13.90 The corner frequency f_o of the low-pass filter described in Problem 13.89 is equal to about 274 Hz. Use frequency scaling to move f_o to 2.5 kHz.

13.91 The all-pass filter of **Fig. P13.61** has a gain magnitude that is constant with frequency but a phase that shifts by 90° at ω_p (see Problem 13.61). The filter is constructed with $R = 15\,\text{k}\Omega$ and $C = 0.015\,\mu\text{F}$, so that the phase-shift frequency occurs at about 700 Hz.

(a) Use magnitude scaling to change R to 100 kΩ while preserving the original phase-shift frequency of the filter.

(b) Use magnitude scaling to instead change the capacitor value to 1 nF while preserving the original phase-shift frequency of the filter.

(c) Use frequency scaling to find the new phase-shift frequency if R is kept at 15 kΩ but C is changed to 10 nF.

(d) Use frequency scaling to change the phase-shift frequency to 10 kHz when $C = 1$ nF.

13.6 Switched-Capacitor Networks and Filters

13.92 Draw a switched-capacitor version of each of the filters shown in **Fig. P13.7**.

13.93 Draw the switched-capacitor version of the bandpass filter of Fig. 13.15.

13.94 A basic switched-capacitor element representing a 10 kΩ resistor is required for processing signals in the audio range (dc to 20 kHz). Choose appropriate values for the switched-capacitor C and clock frequency T_c.

13.95 🔢 Using the topology of Fig. 13.30, choose a clock frequency T_c and values for C_1, C_2, and C_X such that the circuit forms a low-pass filter with a dc gain magnitude of 5 and a cutoff frequency of 10 kHz. What is the input resistance of your filter?

13.96 ◯ Using the topology of Fig. 13.30 as a guide, design an analogous switched-capacitor high-pass filter. Choose a clock frequency T_c and values for all capacitors such that the filter has a passband gain magnitude of 10 and a cutoff frequency of 20 kHz.

13.97 Consider the basic switched-capacitor element shown in Fig. 13.27. This circuit simulates a resistor connected to ground. If Q_1 and Q_2 have parameters $K = 0.5\,\text{mA/V}^2$ and $V_{TR} = 1$ V, estimate the charge and discharge times of the capacitor if $C = 50$ pF. Assume that v_s, ϕ_1, and ϕ_2 are voltages of 5-V magnitude. What practical limitations does your result impose on the operation of the circuit?

13.98 The switched-capacitor element of Fig. 13.27 is constructed from MOSFETs with parameters $K = 0.5\,\text{mA/V}^2$ and $V_{TR} = 1$ V, and with $C = 50$ pF. Estimate the time-average power dissipation in Q_1 and Q_2 if v_s is a dc voltage of magnitude 5 V and $T_c = 1$ ms.

13.99 Find an equivalent circuit for the switched-capacitor network of **Fig. P13.99**.

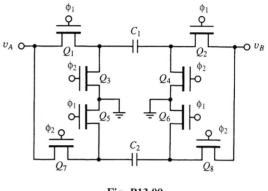

Fig. P13.99

13.100 ◐ Consider the switched-capacitor network of **Fig. P13.100** with $C_1 = C_2 = 1$ pF and $T_c = 10\,\mu\text{s}$.

(a) Plot v_{OUT} as a function of time if v_{IN} is a 1-V step function.

(b) Repeat part (a) for the case $C_1 = 1$ pF and $C_2 = 3$ pF.

(c) What general class of filter is simulated by this circuit?

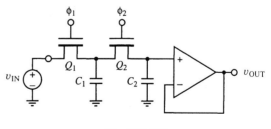

Fig. P13.100

13.101 Consider the circuit of **Fig. P13.100**. Using switched-capacitor elements only, modify the circuit so that its output is amplified by a factor of 10.

13.7 Oscillators

13.7.1 Wien-Bridge Oscillator

13.102 Specify the R and C values of a Wien-bridge oscillator designed to oscillate at 5 kHz.

13.103 The positive-feedback network of the Wien-bridge oscillator circuit of Fig. 13.31 has the form of a band-pass filter. Find the transfer function that relates V_+ to V_{out}. Identify ω_o and Q of this band-pass filter function.

13.104 Consider the Wien-bridge oscillator of Fig. 13.31 in which the two resistors R_F have unequal values R_A and R_B, and the two capacitors C have unequal values C_A and C_B, where R_A and C_A form the parallel-connected input devices and R_B and C_B the series-connected feedback devices.

(a) What is the criterion for oscillation in this case?

(b) Derive an expression for the frequency of oscillation.

13.105 🔊 🌑 Using a 60-Hz Wien-bridge oscillator as a basis, design a variable-amplitude portable ac power source capable of producing a 0- to 120-V, 60-Hz sinusoidal voltage from two 12-V storage batteries. The amplitude of the output should be adjustable via a single variable-resistor control. The circuit must include a suitable transformer and driving circuit.

13.7.2 Phase-Shift Oscillator

13.106 The block diagram of a positive-feedback oscillator is shown in **Fig. P13.106**, where $\mathbf{P} = 1 ∡ 45°$ for all ω. Find the minimum gain A necessary to sustain oscillation and find the frequency of oscillation.

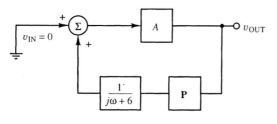

For all ω: $|\mathbf{P}(j\omega)| = 1$; $∡\,\mathbf{P}(j\omega) = +45°$

Fig. P13.106

13.107 🌑 The circuit diagram of **Fig. P13.107** represents one form of phase-shift oscillator. Suppose that all resistors are equal to R and all capacitors to C.

(a) Show that oscillations occur at $\omega = \sqrt{3}/RC$ if each amplifier stage has a gain of -2.

(b) Show that oscillations will still occur at $\omega = \sqrt{3}/RC$ if any one stage has a gain of -2 and the other two stages have a gain of $+2$.

(c) Show that oscillations will still occur at $\omega = \sqrt{3}/RC$ as long as the product $A_1A_2A_3$ is equal

to -8, regardless of the values of the individual stage gains.

How would you actually implement such an oscillator using operational amplifiers?

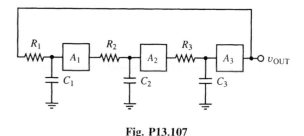

Fig. P13.107

13.108 Find the frequency of oscillation for the phase-shift oscillator of **Fig. P13.107** if all the resistors and capacitors have different values.

13.109 In the phase-shift oscillator of **Fig. P13.107**, each resistor is replaced by an inductor of value L and each capacitor by a resistor of value R. Show that the circuit will oscillate at a constant frequency. Find the frequency of oscillation.

13.110 ● Analyze the circuit of **Fig. P13.110** and show that it produces a sinusoidal voltage of fixed frequency. What is the frequency of oscillation?

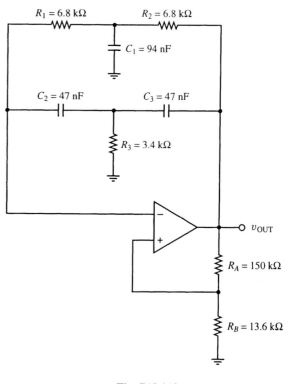

Fig. P13.110

13.7.3 Tuned *LC* Oscillators

13.111 The Colpitts oscillator of Fig. 13.37(a) is constructed with $C_1 = C_2 = 0.1\,\mu F$ and $L = 100\,mH$. Choose appropriate values for R_1 and R_2, and determine the frequency of oscillation.

13.112 The Colpitts oscillator of Fig. 13.37(a) is constructed with $R_1 = 100\,k\Omega$ and $R_2 = 5\,k\Omega$. Choose appropriate values for C_1, C_2, and L so that the circuit oscillates at 100 kHz.

13.113 ⬛ ○ Design a one-transistor version of the Colpitts oscillator of Fig. 13.37(a). The amplifier in your circuit should consist of a single BJT properly biased in the active region.

13.114 ⬛ ○ Design a one-transistor version of the Colpitts oscillator of Fig. 13.37(a). The amplifier in your circuit should consist of a single MOSFET properly biased in the constant-current region.

13.115 Derive the criteria for oscillation for the Hartley oscillator of Fig. 13.37(b). Your analysis should parallel that of Eqs. (13.157) through (13.166) performed for the Colpitts oscillator.

13.7.4 Crystal Oscillators

13.116 Show that Eq. (13.168), which describes the terminal impedance of a piezoelectric crystal, can be derived from the circuit representation of Fig. 13.39(b).

13.117 Find the exact values of $f_s = \omega_s/2\pi$ and $f_p = \omega_p/2\pi$ of a quartz crystal for which $L = 2\,mH$, $C_S = 1\,pF$, and $C_P = 200\,pF$.

13.118 The Colpitts oscillator of Fig. 13.40 is constructed from a crystal with parameters $L = 0.1H$, $C_S = 0.1\,pF$, and $C_P = 100\,pF$. Assume that R_C and R_B are properly chosen so that the BJT is biased in the active region and oscillation occurs. What is the frequency of oscillation?

13.119 A Colpitts crystal oscillator of the type shown in Fig. 13.40 oscillates at 4.1 MHz. If the crystal has

capacitances $C_S = 0.8\,pF$ and $C_P = 340\,pF$, estimate the value of the inductance L in the crystal's equivalent circuit model.

13.120 A Colpitts crystal oscillator of the type shown in Fig. 13.40 oscillates at 100 MHz. If the crystal has an inductance of $L = 5\,\mu H$, estimate the value of the two capacitances in the crystal's equivalent circuit model.

13.7.5 Schmitt-Trigger Oscillator

13.121 The Schmitt-trigger circuit of Fig. 13.41 is constructed with $V_{POS} = 10\,V$, $V_{NEG} = -10\,V$, $R_1 = R_2 = 10\,k\Omega$, $R_X = 12\,k\Omega$, and $C_X = 0.33\,\mu F$. What is the period of the resulting square-wave oscillation?

13.122 Choose all R and C values for the Schmitt-trigger oscillator of Fig. 13.41 such that the frequency of oscillation is 3.3 kHz.

13.123 ○ The Schmitt-trigger circuit of Fig. 13.41 is constructed with $V_{POS} = 15\,V$, $V_{NEG} = -15\,V$, $R_1 = 10\,k\Omega$, $R_2 = 27\,k\Omega$, $R_X = 56\,k\Omega$, and $C_X = 0.15\,\mu F$. What is the period of the resulting square-wave oscillation? Note that R_1 and R_2 do not have the same value.

13.124 Find the output versus time and the period of the Schmitt-trigger clock of Fig. 13.41 if $R_1 = R_2$, $V_{POS} = 15\,V$, and $V_{NEG} = -10\,V$. Note that V_{POS} and V_{NEG} have different magnitudes.

13.125 The circuit of Fig. 13.41 is used to create a square-wave oscillator. Suppose that $V_{POS} = 15\,V$, $V_{NEG} = -15\,V$, and that the op-amp has a slew rate of 10 V/μs.

(a) What is the maximum possible frequency of oscillation that will permit v_{OUT} to reach its V_{POS} and V_{NEG} saturation limits?

(b) Sketch v_{OUT} versus time if R_X and C_X are chosen in an attempt to create an oscillation frequency that exceeds this limit.

◆ SPICE PROBLEMS

13.126 Simulate the low-pass filter of Fig. 13.1(a) on SPICE with $R_1 = 10\,k\Omega$, $R_2 = 100\,k\Omega$, and $C = 0.01\,\mu F$. Use the .AC analysis command to find the circuit's magnitude response. Verify that the circuit has a dc gain magnitude of 10, a cutoff frequency of $1/2\pi R_2 C \approx 160\,Hz$, and a high-frequency roll-off of $-20\,dB/decade$. Assume the op-amp to have parameters

$r_{in} = 10\,M\Omega$, $r_{out} = 5\,\Omega$, and $A_o = 10^6$.

13.127 Simulate the high-pass filter of **Fig. P13.3** on SPICE with $R_1 = 27\,k\Omega$, $R_2 = 470\,k\Omega$, and $C = 0.005\,\mu F$. Use .AC analysis to find the circuit's magnitude response. Verify that the passband gain magnitude is equal to R_2/R_1, the cutoff frequency is $1/2\pi R_1 C \approx 1180\,Hz$, and the low-frequency slope is $+20\,dB/decade$.

Assume the op-amp to have the parameters of an LM741 op-amp.

13.128 Simulate each of the circuits of **Fig. P13.7** on SPICE for the case $R_1 = 1\,k\Omega$, $R_2 = 10\,k\Omega$, and $C = 1\,\mu F$. Assume the op-amp to be an LM741. Plot the magnitude response in each case using the .AC command.

13.129 Consider the active low-pass Sallen–Key filter circuit of Fig. 13.7 with component values $R_1 = R_2 = 27\,k\Omega$, $C_1 = 0.27\,\mu F$, and $C_2 = 0.015\,\mu F$. Compute the cutoff frequency ω_o by hand, and then verify it by simulating the circuit on SPICE. Also verify that the magnitude Bode plot has the same general shape as that depicted in Fig. 13.11 for the value of $Q \approx 2$. Assume the op-amp to have the parameters of an LM741.

13.130 Simulate the low-pass Sallen–Key filter circuit of Fig. 13.7 on SPICE with component values $R_1 = R_2 = 10\,k\Omega$, $C_1 = 33\,\mu F$, and $C_2 = 0.1\,\mu F$. Assume the op-amp to have parameters $r_{in} = 20\,M\Omega$, $r_{out} = 10\,\Omega$, and $A_o = 10^5$. The resulting response should have a Q of about 9.1. Now simulate the circuit while varying the ratio C_1/C_2 but keeping their product constant. In this way, the cutoff frequency ω_o will be preserved while the value of Q is changed. Verify that the shape of the magnitude plot produced by SPICE is altered, but that the cutoff frequency ω_o remains unchanged.

13.131 Simulate the all-pass filter of **Fig. P13.61** on SPICE for the case $R = 10\,k\Omega$ and $C = 0.01\,\mu F$. Verify that $|\mathbf{V}_{out}/\mathbf{V}_{in}|$ does not change with frequency, but $\angle\,\mathbf{V}_{out}/\mathbf{V}_{in}$ undergoes a 90° phase shift at $f_o = 1/2\pi RC$.

13.132 Simulate the high-pass Sallen–Key filter circuit of Fig. 13.12 on SPICE with component values $R_1 = R_2 = 100\,k\Omega$, $C_1 = 0.27\,\mu F$, and $C_2 = 0.1\,\mu F$. Assume the op-amp to have parameters $r_{in} = 20\,M\Omega$, $r_{out} = 10\,\Omega$, and $A_o = 10^5$. Use .AC analysis to find the circuit's magnitude-response plot. Compare results to those predicted by hand calculation.

13.133 Use SPICE to obtain the plots of Fig. 13.14 in which the frequency response of the high-pass Sallen–Key filter of Fig. 13.12 is plotted for various values of Q.

13.134 Simulate the band-pass filter of Fig. 13.15 on SPICE using the component values $R_1 = R_2 = 100\,k\Omega$ and $C_1 = C_2 = 0.1\,\mu F$, for a Q of 0.5. Assume the op-amp to have parameters $r_{in} = 10\,M\Omega$, $r_{out} = 100\,\Omega$, and $A_o = 10^6$. Verify that the magnitude Bode plot obtained with $Q = 0.5$ has a shape that lies below the maximally

flat, $Q = 0.707$ response.

13.135 Simulate the two-stage Butterworth cascade of Example 13.2 (see Fig. 13.20) on SPICE. Verify that the circuit produces the desired maximally flat response with a cutoff frequency of 1 kHz. Also obtain the output of each stage separately so as to reproduce the plot of Fig. 13.21.

13.136 Use SPICE to simulate the two-stage Butterworth cascade of Fig. 13.20 in which the positions of the resistors and capacitors in each stage are reversed so as to form two Sallen–Key sections of the high-pass type.

13.137 Simulate the two-stage Chebyshev filter of Example 13.3 on SPICE. Verify that the plot of Fig. 13.23 is obtained from the redesigned filter.

13.138 Simulate the fourth-order high-pass Butterworth filter of Example 13.4 (see Fig. 13.25). Verify that the response is maximally flat with a cutoff frequency of 1 kHz and that the slope of the low-frequency response is +80 dB/decade.

13.139 In Example 13.5, magnitude scaling is used to modify the resistor and capacitor values in a filter circuit without changing its overall response. Use SPICE to simulate the original and modified circuits. Verify that each has the same frequency-response characteristics.

13.140 Simulate the original and modified circuits described in Example 13.6. Verify that the corner frequency of the low-pass filter is changed from 1 to 5 kHz.

13.141 Simulate the original and modified circuits described in Example 13.7. Verify that the corner frequency of the low-pass filter remains unchanged in the modified filter, but that the response changes from a Q of 0.6 (below maximally flat) to a Q of 1.2 (above maximally flat). As a reference point, you should also simulate a filter having a maximally flat ($Q = 1/\sqrt{2}$) response.

13.142 Simulate the switched-capacitor element of Fig. 13.27 on SPICE with $C = 100\,pF$, $v_S = 5\,V$, and $T_c = 0.1\,ms$. Use the default SPICE parameters for the MOSFETs. Find the average current drawn from the v_S source.

13.143 Simulate the switched-capacitor integrator of Fig. 13.28(b) on SPICE with $C_1 = 10\,pF$, $C_2 = 0.1\,\mu F$, and $T_c = 0.1\,ms$, and with v_{IN} set to a 5-V step function. Assume the op-amp to have infinite input resistance, zero output resistance, and a gain of 2×10^5. Plot the response of the integrator as a function of time for $0 < t < 20\,ms$. Compare to that predicted by substituting a resistor of equivalent value for the switched-capacitor element formed by C_1.

13.144 Simulate the switched-capacitor integrator of Fig. 13.28(b) on SPICE with $C_1 = 50\,\text{pF}$, $C_2 = 100\,\text{pF}$, and $T_c = 100\,\mu\text{s}$. Assume the op-amp to have infinite input resistance, zero output resistance, and a gain of 10^6. Find the response of the circuit if v_{IN} is a 5-V peak, 500-Hz sinusoid.

13.145 Simulate the Wien-bridge oscillator circuit of Fig. 13.31 on SPICE with $R_F = 100\,\text{k}\Omega$, $R_1 = 10\,\text{k}\Omega$, $R_2 = 22\,\text{k}\Omega$, and $C = 1\,\mu\text{F}$.

(a) Assume the op-amp to be linear with infinite input resistance, zero output resistance, and a gain of 10^5. Plot v_{OUT} versus time. You may have to excite the grounded input terminal with a short voltage pulse in order to initiate oscillation.

(b) Now assume the op-amp to have nonlinear limits. Limit the output of the simulated op-amp using a circuit made from 1-kΩ resistors and diodes with reverse-breakdown voltages set to $-15\,\text{V}$. Plot v_{OUT} versus time for the revised circuit.

13.146 Simulate the circuit of **Fig. P13.110** on SPICE. Assume an op-amp with the parameters of an LM741.

Plot the voltage v_{OUT} versus time. It may be necessary to include voltage as an initial condition on one of the capacitors in order to get the circuit to oscillate.

13.147 Simulate the circuit of **Fig. P13.107** on SPICE using dependent sources for blocks A_1 through A_3. Show that oscillation will occur at $f = \sqrt{3}/2\pi RC$ for the case $R_1 = R_2 = R_3 = R$, $C_1 = C_2 = C_3 = C$, and $A_1 A_2 A_3 \geq 8$. Choose any reasonable values for R and C to run your simulation.

13.148 Simulate the phase-shift oscillator of Fig. 13.33 on SPICE using the values $R_1 = R_2 = R_3 = 10\,\text{k}\Omega$ and $C_1 = C_2 = C_3 = 0.01\,\mu\text{F}$. Assume the op-amps to have infinite input resistance, zero output resistance, and gains of 5×10^5. Plot v_{OUT} versus time. You may have to seed one of the capacitors with an initial condition or excite the circuit with a short voltage or current pulse in order to initiate oscillation.

13.149 Simulate the MOSFET phase-shift oscillator of Fig. 13.36 with $R_D = 5\,\text{k}\Omega$, $C = 0.1\,\mu\text{F}$, and $R = 10\,\text{k}\Omega$. Experiment with different values of K for the MOSFET and determine the minimum K value for sustained oscillation. Excite the gate of Q_1 with a short voltage pulse to initiate oscillation.

Chapter 14

Digital Circuits

I n this chapter, the fundamentals of digital circuits, originally introduced in Chapter 6, are explored in more detail. Although much of this book has focused on analog circuits and applications, the realm of digital circuits is equally important. Many data and signal processing applications involve the manipulation of data in digital form. Probing the physical world requires analog sensors and circuitry, but the information obtained is often converted to digital form for processing. Digital systems have higher noise immunity than equivalent analog systems; hence many electronic communication systems use digital rather than analog signals. Computers, which are based on binary logic, are made almost entirely from digital devices. The subject of digital circuits and systems can rightfully occupy an entire book. In the limited space of this chapter, the discussion focuses on the basic electronic building blocks that form the foundations of digital systems. Our starting point for this discussion will be the BJT and MOSFET inverter circuits introduced in Chapter 6. The mathematical language of digital electronics is a system of logical operations and functions called Boolean algebra. The reader is presumed to have had some exposure to Boolean algebra and to have some familiarity with the basic concepts of binary numbers and logic gates.

14.1 FUNDAMENTAL CONCEPTS OF DIGITAL CIRCUITS

14.1.1 Scale of Integration

Digital systems typically require a large number of active devices. Consequently, digital circuits are usually implemented in a integrated-circuit environment. A digital system can be synthesized by interconnecting several smaller integrated circuits; alternatively, the system can be fabricated entirely on one IC. The former method, called *off-the-shelf* design, involves the selection of standard integrated circuits packaged as stock, multipin components made by one or more commercial IC manufacturers. In contrast, fabricating an entire digital system on one IC usually involves the design of a custom-made integrated circuit.

The more complex a given logic system, the more densely packed are its devices on the integrated circuit. Off-the-shelf components are usually fabricated with the least dense packaging and are classified as small-scale integrated (SSI) or medium-scale integrated (MSI) circuits. In SSI circuits, up to 10 gates appear on a single IC, while between 10 and 100 appear on MSI circuits. A circuit containing between 100 and 1000 gates is classified as a large-scale integrated

(LSI) circuit; a circuit with more than 1000 gates (up to 100,000 or more) is classified as a very-large-scale integrated (VLSI) circuit. LSI and VLSI circuits are frequently used in digital microprocessor and memory circuits. Processes that produce more than 10^6 devices on an IC are classified as ultralarge-scale integrated (ULSI) circuits.

The task of fabricating a reliable VLSI circuit is a formidable one. It has been said, for example, that a 4-megabit random-access memory (RAM) chip, now routinely found in desktop computers, can be likened to a map of the entire country showing all highways, major access roads, side streets, and alleys. The presence of one pothole on the map will make the chip unusable.

14.1.2 Logic Families

Each of the various logical functions performed by a digital circuit can be implemented in a number of ways. It is customary to limit the various methods of implementation to one of several *logic families*. Within a given logic family, logic gates are synthesized from one type of active device only. The choice of device is determined by the fabrication technology inherent to the logic family. Within a logic family, gates are easily interconnected to logic elements of the same family.

In the NMOS logic family, logic gates are made entirely from n-channel enhancement-mode and depletion-mode MOSFETs. Resistors are not used at all in NMOS logic circuits. Because n-channel MOSFETs can be fabricated easily and with very small size, the "packing density" of NMOS transistors can be made very high. Consequently, the NMOS family is well suited to the design of VLSI microprocessor and memory circuits, where large numbers of devices are required. NMOS logic elements are also capable of producing circuits with moderately high computational speed. Unfortunately, NMOS circuits have a low load-current driving capability. Use of the NMOS family is dominant in many VLSI microprocessor and random-access memory applications. In these circuits, where logic gates drive only the inputs of other MOSFET gates, the load-current limitation of NMOS is not a significant problem. Conversely, many NMOS VLSI circuits contain special driving devices for interfacing with the outside world. The low load-driving capability of NMOS circuits makes them unsuitable as off-the-shelf components, however. For this reason, NMOS logic gates are not available in standard logic packages as are TTL and ECL gates.

In the similar CMOS (complementary MOS) family, all logic elements are implemented only from n-channel and p-channel enhancement-mode MOSFETs. Like NMOS gates, CMOS gates are easy to fabricate and simple to design. In addition, CMOS circuits consume very little power. CMOS circuits are an obvious choice for battery-powered circuits such as digital watches, hand-held calculators, cellular telephones, and portable computers. The speed of CMOS circuits has become competitive with that of other logic families, so that CMOS has become a general-purpose digital logic family encompassing the advantages of both low power *and* high speed. The BiCMOS family is also rapidly becoming an important logic family. In the BiCMOS family, MOSFETs and BJTs are fabricated on the same integrated circuit and are combined within the same logic gate. MOSFETs are used as input devices to produce large input impedances, and BJTs are used to achieve high speed and large load-driving capability.

Compared to logic circuits made from BJTs, circuits in the CMOS, BiCMOS, and NMOS families are more susceptible to damage from electrostatic overstress (EOS) and electrostatic discharge (ESD). Improper handling of MOS components results in static charge transfer from personnel or objects to the input leads of the IC packages. These discharges can cause the gate oxide layers of MOS devices to be punctured. As a result, unconnected MOS devices require special "antistatic" packaging and careful handling procedures when digital systems are assembled.

In the TTL (transistor–transistor logic) family, all logic elements are made from *npn* BJTs, diodes, and resistors. Devices in the TTL family are driven to cutoff and saturation to achieve high and low logic states. TTL was one of the first digital technologies to be developed and has been a traditional choice in systems where moderately high computational speed is required but power consumption is not a main concern. Most TTL logic gates are available as off-the-shelf components packaged as multipin LSI level integrated circuits. Compared with other MOSFET-based logic families, TTL components are far less susceptible to physical damage from the electrostatic discharge (ESD) that occurs during manual handling of components. For these reasons, the TTL logic family is ideal for laboratory "breadboarding" of digital system designs.

In the ECL (emitter-coupled logic) family, logic elements are made from BJTs and resistors connected in the differential-amplifier topology. Unlike the TTL family, the active devices in the ECL family are not driven to their saturation limits. The signal swings in ECL logic circuits are small, requiring less time to charge and discharge the various circuit capacitances. As a result, digital systems implemented in the ECL family are capable of very fast operating speeds. The price paid is the large power consumption required to keep the BJTs of an ECL gate biased in their active regions. The ECL logic family is used in the design of ultrahigh-speed "supercomputers" and digital processing circuits.

From a binary point of view, each of the various logic families mentioned above produces gates that are functionally identical. From an electronic point of view, each family has different properties and characteristics. In the sections of this chapter, these properties and characteristics are examined and discussed in detail.

14.1.3 Definition of Logic Levels

The basic features of a digital system were previously introduced in Chapter 6 (see Section 6.4), where the concepts of the digital signal, digital regime, and high and low logic levels were all defined. In this section, we examine the definitions of high and low logic levels in more detail.

A large-scale digital system usually consists of an interconnection of many gates from the same logic family. Some gates act as input ports that receive digital information. These input gates pass their output signals on to other gates, which may feed still more gates down the line. If the digital character of the signals is to be preserved as they are passed from one gate to another, the voltage levels $V(1)$, representing logic **1**, and $V(0)$, representing logic **0**, must be consistently reproduced by each gate. By convention, the voltage levels $V(1)$ are $V(0)$ are designated V_{OH} and V_{OL}, respectively. They are illustrated on the generic transfer characteristic of Fig. 14.1.

Figure 14.1
Definitions of V_{OH} (logic **1**) and V_{OL} (logic **0**) voltage levels. These voltage levels must self-consistently reproduce each other.

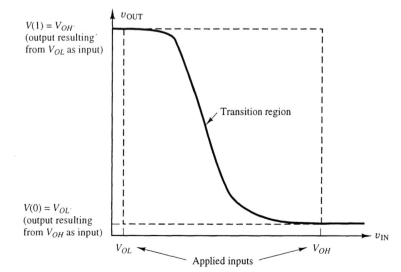

The values of V_{OH} and V_{OL} must always be defined in a self-consistent way such that an inverter receiving V_{OL} as its input from some other gate will produce V_{OH} as its output, while an inverter receiving V_{OH} as its input will produce V_{OL} as its output. Briefly stated, a logic inverter must be designed such that V_{OH} and V_{OL} self-consistently reproduce each other. This relationship between V_{OH} and V_{OL} places in important constraint on the inverter transfer characteristic. As illustrated in Fig. 14.1, a transfer characteristic in which V_{OL} and V_{OH} reproduce each other must traverse from the upper left-hand corner to the lower right-hand corner of a square of height and width $V_{OH} - V_{OL}$. As a consequence, the slope of the transfer characteristic's transition region must be at least as steep as -1. A transfer characteristic with slope everywhere less than -1 would be incapable of traversing the diagonal of the square. In practice, a real logic inverter is subject to an even more stringent slope condition. Because the extremities of the typical transfer characteristic have slopes less than -1, as does the one depicted in Fig. 14.1, the slope in the transition region must actually be steeper than -1 if the output is to fall to V_{OL} when the input is equal to V_{OH}.

The flat extremes of a transfer characteristic, particularly near the low input end, are actually advantageous from a digital point of view. The value of V_{OH} will generally be equal to the open-circuit output voltage obtained with the input set to zero. In most logic families of practical importance, V_{OH} will persist as the output even when the input is made slightly nonzero, such as when V_{OL} is applied. In the simple BJT inverter, for example, the value $v_{OUT} = V_{CC}$ is obtained when the transistor is in cutoff with $v_{IN} = 0$. The output will not begin to fall from V_{CC} until v_{IN} rises above the threshold value V_f, which is larger than $V_{OL} \approx V_{sat}$. This feature of the inverter transfer characteristic results in a V_{OH} that is easy to identify and becomes equivalent to the inverter's open-circuit output voltage. A similar situation exists for MOSFET inverters, for which the input threshold voltage is V_{TR} and the logic level V_{OL} is smaller than V_{TR}.

14.1.4 Noise Margins

In practice, the actual input voltage received by one gate from another gate may fall below V_{OH} or lie above V_{OL}. Voltage fluctuations can occur due to the coupling of electromagnetic interference into interconnect paths, the "ringing" of parasitic LC elements excited by digital

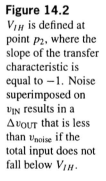

Figure 14.2
V_{IH} is defined at point p_2, where the slope of the transfer characteristic is equal to -1. Noise superimposed on v_{IN} results in a Δv_{OUT} that is less than v_{noise} if the total input does not fall below V_{IH}.

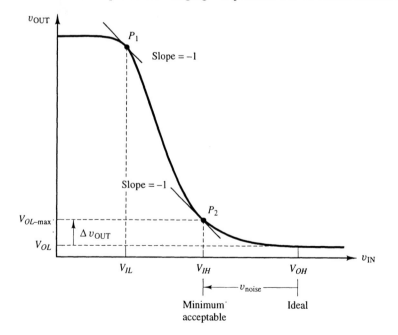

switching transients, or simply the loading of a given gate output by the other gates that it drives. As a consequence, a gate must often process less-than-perfect input signals. If the digital system is to work, deviations from V_{OH} and V_{OL} produced by one gate must be attenuated, rather than amplified, by succeeding gates. Amplification of input noise or voltage fluctuations would cause logic levels at gates somewhere down the line to enter their undefined, analog regions of operation. The goal of attenuating, rather than amplifying, input noise fluctuations is accomplished by defining two input voltage levels V_{IL} and V_{IH} at the points p_1 and p_2, respectively, where the slope of the inverter transfer characteristic equals -1. These input voltage values, illustrated in Fig. 14.2, represent logic levels that are barely acceptable as low and high inputs, respectively, under worst-case conditions. Specifically, V_{IL} is defined as the highest voltage that a given gate can accept as a logic **0** and still produce a logic **1** output acceptable to other gates. Similarly, V_{IH} is defined as the smallest voltage that a given gate can accept as a logic **1** and still produce a logic **0** output acceptable to other gates. The voltage-level definitions for V_{IH} and V_{IL} are arrived at by the following consideration. Suppose that the input signal to a gate contains an unwanted noise component $\Delta v_{IN} = v_{\text{noise}}$. The gate output will contain a corresponding noise component Δv_{OUT} determined by the inverter's transfer characteristic. As illustrated in Fig. 14.2, applying a logic high input signal with noise component v_{noise} will yield an output fluctuation Δv_{OUT} that is *less* in magnitude than v_{noise} as long as the total input signal does not fall below V_{IH}. This result is guaranteed because an input signal more positive than V_{IH} will always fall over a portion of the transfer characteristic where the magnitude of the slope, or gain factor, is less than unity. Similarly, as illustrated in Fig. 14.3, applying a low input signal with noise component v_{noise} will also yield an output fluctuation Δv_{OUT} that is *less* in magnitude than v_{noise} if the total input signal does not exceed V_{IL}. This result is again guaranteed because an input signal less positive than V_{IL} will also fall over a portion of the transfer characteristic where the gain factor is less than unity. The chosen values of V_{IH} and V_{IL} thus will yield an output noise component that is always less than the input noise component. In this way, if Δv_{OUT} is passed on to another gate as an input signal, the noise component Δv_{OUT} of the output of the latter gate will be even smaller than the original v_{noise}. The noise disturbance will be attenuated as it passes from gate to gate.

Figure 14.3

V_{IL} is defined at point p_1, where the slope of the transfer characteristic is also equal to -1. Noise superimposed on v_{IN} results in a Δv_{OUT} that is less than v_{noise} if the total input does not rise above V_{IL}.

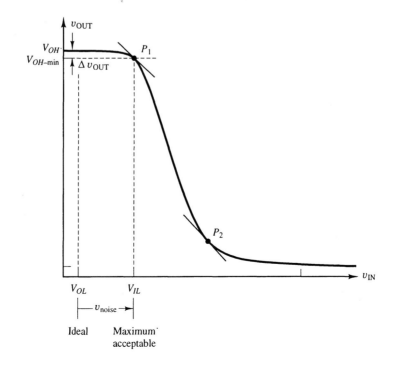

Discussion. The preceding discussion focuses on noise that causes a gate's input signal to move toward the analog region, such that $v_{IN} < V_{OH}$ or $v_{IN} > V_{OL}$. It is equally possible for noise to cause the input to be larger than V_{OH} or smaller than V_{OL}. In these cases, the noise components appearing at the gate output again will be attenuated, but in any case will be less problematic because the gate's operating point will be moved *away* from the analog region. ∎

Figure 14.4
Ideal inverter output levels, worst-case output levels, and noise margins.

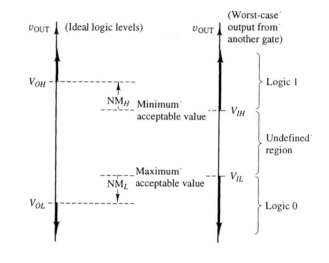

When the output of one logic gate feeds the input to another, the relationship between V_{OH}, V_{IH}, V_{OL}, and V_{IL} becomes significant. Two figures of merit, called the *noise margins*, are defined in Fig. 14.4. The noise margins indicate the relative immunity of a given logic family to noise-induced excursions into the analog region. A logic family's noise margins are evaluated from the point of view of a single inverter operating somewhere within a digital system. The input to the inverter consists of the output of some other digital gate of the same family. Suppose that the signal supplied by the other gate is a logic **1**. Ideally, this input should be equal to V_{OH}, but if its value is as low as V_{IH}, it will still represent a valid high signal. Noise fluctuations superimposed on the logic **1** input will become a problem only if they cause the input to fall below V_{IH}. The difference between V_{OH} and V_{IH} thus represents a margin of safety called the *high noise margin*, defined by

$$NM_H = V_{OH} - V_{IH} \qquad (14.1)$$

ideal logic **1** level ⌐   ⌐ minimum acceptable logic **1** to guarantee noise attenuation

The parameter NM_H applies to the case of a high input voltage. In a working digital system, NM_H must be positive, so that V_{IH} is less than V_{OH}. The larger the value of NM_H, the more immune the inverter will be to noise fluctuations superimposed on a logic **1** input signal.

Now suppose that the signal supplied by the previous gate is a logic **0**. Ideally, its value should be V_{OL}, but it could be as high as V_{IL} and still represent a valid low signal. Noise fluctuations superimposed on the logic **0** input will become a problem only if they cause the input to rise above V_{IL}. The difference between V_{IL} and V_{OL} thus represents a second margin of safety called the *low noise margin*, defined by

$$NM_L = V_{IL} - V_{OL} \qquad (14.2)$$

maximum acceptable logic **0** ⌐   ⌐ ideal logic **0** to guarantee noise attenuation

In a working digital system, NM_L must also be positive, with V_{OL} less than V_{IL}. The larger the value of NM_L, the more immune the inverter will be to noise fluctuations superimposed on a logic 0 input signal. In general, large values of NM_H and NM_L result in a gate that can process noise fluctuations without losing the digital character of the logic levels.

EXERCISE 14.1 The transition voltages of a logic inverter are measured on a curve tracer and are found to be $V_{IH} = 2\,\text{V}$, $V_{IL} = 0.8\,\text{V}$, $V_{OH} = 2.8\,\text{V}$, and $V_{OL} = 0.3\,\text{V}$. Find the high and low noise margins of the logic gate. **Answer:** $NM_H = 0.8\,\text{V}$, $NM_L = 0.5\,\text{V}$

14.2 A BJT inverter with resistive load is made with $V_{CC} = 5\,\text{V}$, $R_C = 1\,\text{k}\Omega$, $R_B = 10\,\text{k}\Omega$, $\beta_F = 100$, $V_f = 0.7\,\text{V}$, and $V_{\text{sat}} = 0.2\,\text{V}$. Find the values of V_{OH}, V_{OL}, V_{IH}, and V_{IL}. **Answer:** $V_{OH} = 5\,\text{V}$; $V_{OL} = 0.2\,\text{V}$; $V_{IH} = 1.18\,\text{V}$; $V_{IL} = 0.7\,\text{V}$

14.1.5 Fan-Out and Fan-In

When a logic inverter or gate is called upon to drive a load, output loading may cause its output voltage to fall into the analog region. The term *fan-out* refers to the number of other gates that a logic gate can drive if its output is to remain above V_{IH} when high and below V_{IL} when low. If the gate's fan-out is kept below the maximum permissible number, its output voltage will always fall within acceptable logic limits when the gate is driven by inputs of acceptable logic level. The fan-out of a gate is one of the parameters that characterizes its performance capabilities. The larger the fan-out, the greater the load-driving capability of the gate.

The term *fan-in* refers to the number of inputs that a single gate can accommodate. A three-input AND gate, for example, has a fan-in of 3. The technology of a logic family determines the maximum fan-in that is practical for a given gate configuration. A gate is never fabricated with a fan-in that exceeds this number. The issues of a fan-out and fan-in will be addressed in the context of specific logic families in later sections.

14.1.6 Propagation Delay

The speed of operation of a logic family is characterized by three quantities called the rise time, fall time, and propagation delay. These parameters describe the time delay that results when a logic gate makes the transition from one state to another. Time delays occur because the capacitances associated with the gate circuit cannot be charged and discharged instantaneously. Capacitance is present in the electronic devices of the gate and also in the interconnecting paths between devices.

The rise time t_r of a logic gate is defined as the time required for v_{OUT} to change from 10 to 90% of the difference between $V(0)$ and $V(1)$. Similarly, the fall time t_f is defined as the time required for v_{OUT} to fall by the same amount. These definitions are depicted in Fig. 14.5(a). Note that the rise and fall times are not necessarily equal to each other. The symbols t_r and t_f are sometimes replaced by the symbols t_{TLH} and t_{THL}, signifying the *Low-to-High* and *High-to-Low* transition times, respectively.

The rise time and fall time are related to two propagation-delay parameters that describe the time lag between an input signal transition and the corresponding output transition. As shown in Fig. 14.5(b), the propagation delay is measured between the points on the input and output waveforms where the voltage lies halfway between $V(0)$ and $V(1)$. The propagation delay during a low-to-high output transition is designated t_{PLH}, and the delay during a high-to-low output transition is designated t_{PHL}. Like rise and fall times, propagation delays are caused by the capacitances associated with the devices in the gate circuit. These delays are related to each

Figure 14.5

Definition of time delays in digital circuits: (a) rise and fall time; (b) propagation delay time.

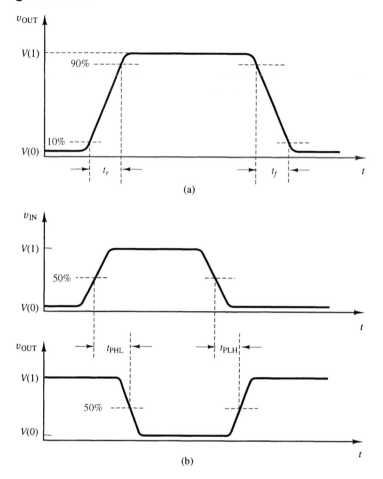

other by the same capacitive phenomena. The rise- and fall-time parameters are used to describe the time characteristics of a single gate-output signal. The propagation-delay parameters apply to the relationship between input and output signals.

14.1.7 Power Dissipation

The power required by a digital circuit is often a quantity of interest to the circuit designer. Modern digital circuits incorporate large numbers of devices, and the demands on the voltage supply required to power the circuit must always be considered. This issue becomes particularly important when considering the digital circuits that go into laptop computers, cellular telephones, calculators, and other battery-powered devices. The power required by some logic families, including TTL, NMOS, and ECL, can be substantial; conversely, the CMOS and BiCMOS logic families require much less power for circuit operation.

The power required by each gate in a digital circuit ultimately translates into current drawn from the circuit's power supply. The power consumed by each gate also contributes to the total amount of heat that must be dissipated by the circuit and its immediate environment. Knowledge of both of these quantities is important to the digital designer.

The power dissipated by an individual gate has both a static component and a dynamic component. The former is related to the power drawn when the gate is not changing logic states; the latter is related to the power dissipated only during transitions between logic states. To illustrate the origins of static and dynamic power dissipation, consider the waveforms shown in

Fig. 14.6. These waveforms represent the voltage measured at the output of a logic inverter and the current drawn by the inverter from its power supply. The waveforms depicted in Fig. 14.6(a) represent those of an ideal inverter with an infinite transfer characteristic slope, and rise and fall times of zero. The static power consumption of the ideal inverter is zero because either v_{OUT} or i_S is zero at all times. Similarly, the transitions between high and low logic states are instantaneous; hence the dynamic power consumption is also zero.

Figure 14.6
Illustration of static (p_S) and dynamic (p_D) power dissipation:
(a) ideal case:
$p_S = p_D = 0$;
(b) nonideal case:
$p_S \neq 0$ and
$p_D \neq 0$. Current i_S is the power-supply current.

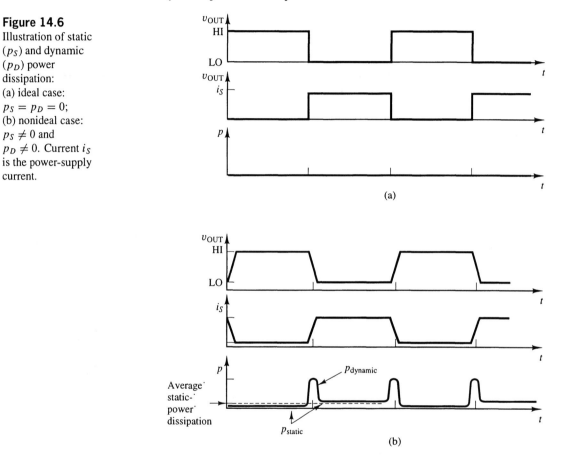

The nonideal waveforms of Fig. 14.6(b) are more typical of those found in a real inverter. In this case, the static power consumption is not zero because neither v_{OUT} nor i_S ever reach values of zero; hence the product $v_{OUT}i_S$ is always nonzero. A gate is presumed to spend about half its time in its high and low output states; hence the average static power dissipation is equal to the average of the static power dissipated when the output is high and that dissipated when the output is low.

The dynamic power dissipation associated with the waveforms of Fig. 14.6(b) is not zero because the transitions between high and low states do not occur instantaneously. During transition periods, both v_{OUT} and i_S reach values midway between their minimum and maximum levels, and the product $v_{OUT}i_S$ is nonzero. This product represents power dissipated in the devices of the gate circuit and constitutes one possible component of the dynamic power consumption. A second component of dynamic power consumption is associated with the charging of any load capacitance fed by the gate output. During every output transition to the high state, a load capacitance connected to the gate output must be charged to the value $V(1)$. This charging operation stores energy equal to $C[V(1)]^2/2$ and requires that an approximately equal amount of energy be dissipated in the charging circuit. When the output falls to $V(0)$ during the next transition, the energy stored in the load capacitance is discharged and dissipated as heat. This energy must be replaced the next time the load capacitance is charged to $V(1)$. The larger the load capacitance and the more frequently the gate output is switched, the larger the dynamic power dissipation. Note that a digital gate can draw power during a transition even if its static power dissipation is zero.

Both static and dynamic power dissipation contribute to the total power consumed by a gate. In some logic families, such as TTL and ECL, static power dissipation constitutes the major component. In other logic families, most typically the CMOS, BiCMOS, and certain VLSI NMOS families, dynamic power dissipation is the dominant component of the total power consumption. The latter quantity is proportional to the rate at which the gate is switched, because power is dissipated only during logic transitions.

EXERCISE 14.3 A step function in voltage is applied to a simple RC circuit. Prove that the energy dissipated in the resistor between $t = 0$ and $t = \infty$ is equal to the energy stored in the capacitor.

14.1.8 Delay–Power Product

An ideal digital circuit is instantaneously fast and requires minimal power. In real life, it is seldom possible to produce a digital circuit that has both short transition times and low power consumption. Fast logic circuits usually require the expenditure of significant amounts of static and dynamic power. One figure of merit that indicates the degree to which a logic family approaches the ideal is called the delay–power product. The quantity DP, expressed in joules, is defined as the product of the average propagation delay multiplied by the time-average power dissipation:

$$\text{DP} = t_D p_D \tag{14.3}$$

where

$$t_D = \frac{t_{\text{PHL}} + t_{\text{PLH}}}{2} \tag{14.4}$$

and

$$p_D = \langle p_{\text{dynamic}} + p_{\text{static}} \rangle \tag{14.5}$$

The smaller the DP product of a logic family, the closer its characteristics approach those of an ideal logic element.

EXAMPLE 14.1 The propagation delay times of a particular TTL logic gate are equal to $t_{\text{PLH}} = 15\,\text{ns}$ and $t_{\text{PHL}} = 8\,\text{ns}$. The gate draws 12 mA from a +5-V supply when the output is low and draws 4 mA when the output is high. The gate has logic levels $V(0) \approx 0.3\,\text{V}$ and $V(1) \approx 3.6\,\text{V}$.

(a) Find the average static power dissipation. Assume that the gate spends approximately half of its time with the output high and half with the output low.

(b) Estimate the value of the delay–power product if the gate is switched at a 1-MHz rate. The gate drives an external load capacitance of about 30 pF.

Solution

• Compute the average static power dissipation

Because the gate spends equal amounts of time in its high- and low-output states, the average static power dissipation can be computed from the product of the power-supply voltage and the average power-supply current:

$$\langle p_{\text{static}} \rangle = (+5\,\text{V}) \left(\frac{12\,\text{mA} + 4\,\text{mA}}{2} \right) = 40\,\text{mW} \tag{14.6}$$

• **Estimate the dynamic power dissipation**

The dynamic power can be estimated by assuming the load capacitance to be charged to $V(1) \approx$ 3.6 V after every output transition from low to high and to be discharged to $V(0) \approx 0.3$ V after every transition from high to low. The indicated values of $V(1)$ and $V(0)$ produce stored capacitive energies $CV^2/2$ of 194 pJ (picojoules) when the output is high and 1.4 pJ when the output is low. The average dynamic energy loss associated with the charging of the capacitor will be equal to twice the net energy lost in the capacitor after each discharge cycle. The factor of 2 accounts for the energy dissipated in the gate circuit during the charging of the capacitor.

If the gate is switched at 1 MHz, the time between capacitor chargings will be equal to 1 μs. Taking all these factors into account yields an average dynamic power dissipation of

$$\langle p_{\text{dynamic}} \rangle = 2 \left(\frac{194\,\text{pJ} - 1.4\,\text{pJ}}{1\,\mu\text{s}} \right) = 0.39\,\text{mW} \tag{14.7}$$

• **Multiply the total power dissipation (static plus dynamic) by the average propagation delay to obtain the delay–power product**

The delay–power product for this gate becomes

$$\text{DP} = p_D \left(\frac{t_{\text{PLH}} + t_{\text{PHL}}}{2} \right) = (40\,\text{mW} + 0.39\,\text{mW}) \left(\frac{15\,\text{ns} + 8\,\text{ns}}{2} \right) \approx 464\,\text{pJ} \tag{14.8}$$

In this case, the static power dissipation represents the dominant energy-loss mechanism.

EXERCISE 14.4 Compute the delay–power product for the gate of Example 14.1 if the gate draws 15 mA from its +5-V supply in the high-output state and 7 mA in the low-output state. **Answer:** 637 pJ

14.5 Compute the delay–power product for the gate of Example 14.1 if the gate spends 2/3 of its time in the high-output state and 1/3 of its time in the low-output state. **Answer:** 541 pJ

14.6 Suppose that the TTL gate of Example 14.1 is redesigned to reduce static power dissipation. If the currents drawn from the +5-V supply in the high- and low-output states are maintained in the same proportion, at what value of high-output-state current will the static power dissipation just equal the dynamic power dissipation? Assume that the gate spends half its time in the high-output state and half in the low-output state. **Answer:** 117 μA

14.2 CMOS LOGIC FAMILY

In the CMOS logic family, gates are made entirely from n-channel and p-channel enhancement-mode MOSFETs using the basic inverter configuration of Fig. 14.7. CMOS circuits are readily designed using automated and computer-aided design techniques, and are relatively easy to fabricate. These features, combined with low cost, low-power consumption, and moderately fast switching speeds, have made CMOS the logic family of choice in many applications. In addition, the geometric simplicity of CMOS integrated circuits facilitates the use of computer-aided design, layout, and fabrication tools.

Figure 14.7
Basic CMOS logic inverter with capacitive load C_L. Q_P charges C_L during a low-to-high output transition; Q_N discharges C_L during a high-to-low output transition.

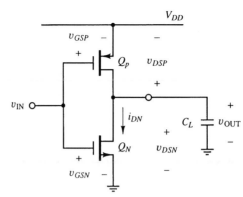

A given CMOS inverter generally drives only the inputs of other CMOS gates, hence the typical CMOS inverter drives a purely capacitive load. In the CMOS inverter, the p-channel and n-channel devices Q_P and Q_N are driven simultaneously by an input voltage v_{IN} derived from the output of some previous gate. When v_{IN} is high, Q_N is made to conduct while Q_P is forced into cutoff. This action discharges the load capacitance C_L to ground. When v_{IN} is low, Q_N is forced into cutoff while Q_P is made to conduct. This action charges the load capacitance C_L to the supply voltage V_{DD}. If Q_N and Q_P are fabricated to have the same value of K, the circuit will be a symmetrical one with equal rise and fall times. Additionally, the gate will have equal current driving capabilities in the high and low logic states. Because the n-channel device is in cutoff when the input is low (logic **0**), and the p-channel device is in cutoff when the input is high (logic **1**), the CMOS configuration consumes essentially zero static power. Power is consumed by the inverter only when the output makes a transition between its high and low states.

14.2.1 CMOS Inverter Transfer Characteristic

The basic form the voltage transfer characteristic of the CMOS inverter, shown here in Fig. 14.8, was obtained previously in Chapter 6 using graphical construction. In this section, we examine the transfer characteristic in more detail. Specifically, we wish to find V_{OH} and V_{OL} and to derive the points p_1 and p_2 at which the slope of the transfer characteristic equals -1. These transition points define the values of V_{IL} and V_{IH}, and thus determine the inverter's noise margins.

Figure 14.8
Transfer characteristic of the CMOS inverter of Fig. 14.7. If $K_N = K_P$, the transition voltage v_{IC} will be symmetrically located at $V_{DD}/2$.

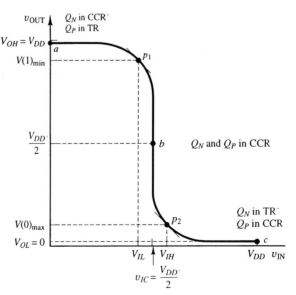

For this particular gate, Q_N is in cutoff when the output is high, hence $V_{OH} = V_{DD}$. Similarly, Q_P is in cutoff when the output is low, hence $V_{OL} = 0$. Points p_1 and p_2 can be found by considering the v–i characteristics of Q_N and Q_P. At p_2, the n-channel device Q_N will be in the triode region and the p-channel device Q_P in the constant-current region, as illustrated in Fig. 14.9(a). Because both devices share the same current, their v–i characteristics will be related by the equation

$$K_N[2(v_{GSN} - V_{TRN})v_{DSN} - v_{DSN}^2] = -K_P(v_{GSP} - V_{TRP})^2 \qquad (14.9)$$

where K_P is negative, and i_{DP} is defined as positive up into the drain of Q_P. In this equation, V_{TRN} and V_{TRP} are the threshold voltages of the n-channel and p-channel devices, respectively. For enhancement-mode devices, V_{TRN} is positive and V_{TRP} is negative. The various device terminal voltages indicated in Fig. 14.7 are related to v_{IN} and v_{OUT} by

$$v_{GSN} = v_{IN} \qquad (14.10)$$

$$v_{GSP} = -(V_{DD} - v_{IN}) \qquad (14.11)$$

and $\qquad v_{DSN} = v_{OUT} \qquad (14.12)$

Substituting these equivalent voltages into Eq. (14.9) results in

$$K_N[2(v_{IN} - V_{TRN})v_{OUT} - v_{OUT}^2] = -K_P(V_{DD} - v_{IN} + V_{TRP})^2 \qquad (14.13)$$

Taking the derivative of this equation with respect to v_{IN} yields factors of dv_{OUT}/dv_{IN}:

$$2K_N(v_{IN} - V_{TRN})\frac{dv_{OUT}}{dv_{IN}} + 2K_N v_{OUT} - 2K_N v_{OUT}\frac{dv_{OUT}}{dv_{IN}} = 2K_P(V_{DD} - v_{IN} + V_{TRP}) \quad (14.14)$$

Figure 14.9
Intersection of v–i characteristics of Q_N and Q_P at
(a) point p_2;
(b) point p_1;
(c) point b in Fig. 14.8.

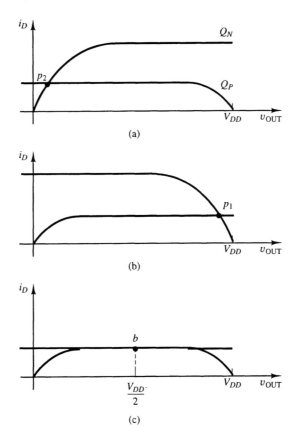

At point p_2, v_{IN} is equal to V_{IH}. Substituting this value in Eq. (14.14), together with the condition $dv_{OUT}/dv_{IN} = 1$, which is also true at p_2, results in

$$-2K_N(V_{IH} - V_{TRN}) + 4K_N v_{OUT} = 2K_P(V_{DD} - V_{IH} + V_{TRP}) \tag{14.15}$$

or

$$v_{OUT}\Big|_{v_{IN}=V_{IH}} = \frac{1}{2}(V_{IH} - V_{TRN}) + \frac{1}{2}\frac{K_P}{K_N}(V_{DD} - V_{IH} + V_{TRP}) \tag{14.16}$$

If the gate is symmetric, with $K_P = -K_N$, Eq. (14.16) becomes

$$v_{OUT}\Big|_{v_{IN}=V_{IH}} = V_{IH} - \frac{V_{DD}}{2} - \frac{V_{TRN} + V_{TRP}}{2} \tag{14.17}$$

Note that the condition $K_N = -K_P$ can be established by setting W_P/L_P to the value mW_N/L_N, where $m = \mu_e/\mu_h$ and $|K| = \frac{1}{2}\mu(\epsilon_{ox}/t_{ox})(W/L)$. (For silicon, $m \approx 2$.)

Equation (14.17) provides a relationship between v_{OUT} and V_{IH} at point p_2. Substituting Eq. (14.17) for v_{OUT} and the value V_{IH} for v_{IN} into Eq. (14.13) yields an equation valid at point p_2 that can be used to find V_{IH}:

$$K_N\left[2(V_{IH} - V_{TRN})\left(V_{IH} - \frac{V_{DD}}{2} - \frac{V_{TRN} + V_{TRP}}{2}\right) - \left(V_{IH} - \frac{V_{DD}}{2} - \frac{V_{TRN} + V_{TRP}}{2}\right)^2\right]$$
$$= -K_P(V_{DD} - V_{IH} + V_{TRP})^2 \tag{14.18}$$

For $K_N = -K_P$ and $V_{TRP} = -V_{TRN} = V_{TR}$, this equation can be solved, after some algebra, to yield

$$V_{IH} = \frac{5}{8}V_{DD} - \frac{V_{TR}}{4} \tag{14.19}$$

In this equation, V_{TR} is a positive number. Note that it is possible to build complementary n-channel and p-channel MOSFETs such that $V_{TRP} = -V_{TRN}$.

The value of V_{IL} at point p_1 can be found by similar algebraic calculations. Alternatively, when $K_N = -K_P$ and $V_{TRP} = -V_{TRN}$, the symmetry of the transfer characteristic can be exploited. In the latter method, V_{IL} can be found by noting that V_{IL} and V_{IH} lie equidistant from the transition voltage $v_{IC} = V_{DD}/2$, that is,

$$\frac{V_{IH} + V_{IL}}{2} = \frac{V_{DD}}{2} \tag{14.20}$$

Equation (14.20) can be solved for V_{IL} in terms of V_{IH}, and Eq. (14.19), for V_{IH} can be substituted to yield

$$V_{IL} = V_{DD} - V_{IH}$$
$$= V_{DD} - \left(\frac{5}{8}V_{DD} - \frac{V_{TR}}{4}\right) = \frac{3}{8}V_{DD} + \frac{V_{TR}}{4} \tag{14.21}$$

The noise margins of the inverter when $K_N = -K_P$ and $V_{TRN} = -V_{TRP}$ can be calculated from the expressions for V_{OH}, V_{OL}, V_{IH}, and V_{IL}. Specifically, with $V_{OH} = V_{DD}$ and $V_{OL} = 0$, the noise margins become

$$NM_H = V_{OH} - V_{IH} = V_{DD} - \left(\frac{5}{8}V_{DD} - \frac{V_{TR}}{4}\right) = \frac{3}{8}V_{DD} + \frac{V_{TR}}{4} \tag{14.22}$$

and

$$NM_L = V_{IL} - V_{OL} = \left(\frac{3}{8}V_{DD} + \frac{V_{TR}}{4}\right) - 0 = \frac{3}{8}V_{DD} + \frac{V_{TR}}{4} \tag{14.23}$$

As Eqs. (14.22) and (14.23) show, the symmetry of the transfer characteristic for the case $K_N = -K_P$ and $V_{TRP} = -V_{TRN}$ yields identical high and low margins. If $K_N \neq -K_P$ or $V_{TRP} - V_{TRN}$, the transfer characteristic will not be symmetric and the noise margins will not be equal. The results (14.22) and (14.23) illustrate another advantage of CMOS logic circuits. For a given V_{TRN} and V_{TRP}, the high and low noise margins can be improved simply by increasing the value of V_{DD}.

EXERCISE 14.7 Find the value of v_{OUT} at points p_1 and p_2 in Fig. 14.8 for the case $K_N = -K_P$ and $V_{TRN} = -V_{TRP}$.   **Answer:** p_2: $V_{DD}/8 - V_{TR}/4$, p_1: $7V_{DD}/8 + V_{TR}/4$

14.8 Solve Eq. (14.18) to obtain Eq. (14.19).

14.9 Find the value of V_{IL} at point p_1 via direct analysis of the MOSFET $v-i$ equations, rather than by the symmetry argument. Your calculations should parallel those of Eqs. (14.13) to (14.19).

14.2.2 Dynamic Behavior of CMOS Inverter

The dynamic behavior of a CMOS inverter is influenced by both internal device and external load capacitances. The load capacitance to a particular gate is generally dominated by the gate-to-source input capacitances of the gates it drives, but contributions also are made by the interconnect paths connecting one gate to another. Together these capacitances determine the rise time, fall time, and propagation time of a CMOS gate.

An estimate of the general dynamic performance of CMOS circuits can be made by examining the case of a CMOS inverter driving a load capacitance, as illustrated in Fig. 14.10. In this case, C_L represents all relevant capacitances appearing between v_{OUT} and ground. For the purpose of illustration, we assume the input v_{IN} to be an ideal voltage step function that makes transitions between 0 and V_{DD} with zero rise and fall times. This assumption simplifies the calculation of the rise time, fall time, and propagation delay of the gate output v_{OUT}.

Figure 14.10
CMOS inverter connected to capacitive load C_L. The input v_{IN} is assumed to be an ideal digital signal that makes step transitions between 0 and V_{DD} with zero rise and fall times.

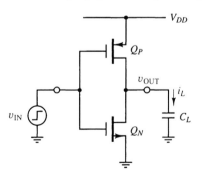

We first examine the discharging of C_L when v_{OUT} undergoes a high-to-low transition. This transition occurs in response to a step change in v_{IN} from 0 to V_{DD}. Prior to the jump in the input signal, v_{IN} will be equal to zero. The load capacitance C_L will be charged to V_{DD}, with Q_N in cutoff and Q_P deep into its triode region at zero current. When v_{IN} makes its transition to V_{DD}, Q_P will be driven into cutoff, and Q_N will be driven into the constant current region with v_{DS} equal to the initial capacitor voltage $v_{OUT} = V_{DD}$. The current drawn out of the capacitor by Q_N will discharge C_L, causing its voltage to drop toward zero, as illustrated in Fig. 14.11. Although Q_N will eventually enter its triode region as $v_{DSN} = v_{OUT}$ falls, an estimate of the time required to discharge C_L can be made by assuming Q_N to conduct in the constant-current region for the duration of the propagation delay time t_{PHL}. This assumption is reasonable, because Q_N will operate in the constant-current region until v_{OUT} falls below the value $v_{GSN} - V_{TRN} \equiv V_{DD} - V_{TRN}$. Because t_{PHL} is defined as the time required for v_{OUT} to fall to 50% of its initial value, and because V_{TRN} is generally a voltage on the order of one-fourth of V_{DD} or less, the amount of time spent by Q_N in the triode region will be small compared to the time spent in the constant-current region.

Figure 14.11
Discharge of C_L by the current through Q_N. The input v_{IN} drives Q_P into cutoff. Q_N will operate in the constant-current region over most of the propagation delay time t_{PHL}.

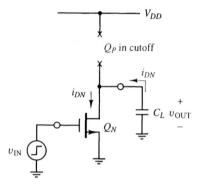

For a constant i_{DN}, the capacitance C_L will discharge according to the equation

$$\frac{\Delta v_{OUT}}{\Delta t} = \frac{i_L}{C_L} \tag{14.24}$$

where $i_L = -i_{DN}$. With i_{DN} equal to $K_N(v_{GSN} - V_{TRN})^2 = K_N(V_{DD} - V_{TRN})^2$, the time t_{PHL} required for v_{OUT} to fall from V_{DD} to $V_{DD}/2$ becomes

$$t_{PHL} = \Delta t = \frac{C_L \Delta v_{OUT}}{-i_{DN}} = \frac{C_L(-V_{DD}/2)}{-K_N(V_{DD} - V_{TRN})^2} \tag{14.25}$$

EXAMPLE 14.2   The CMOS inverter of Fig. 14.7 drives a 2-pF load capacitance. For the typical case $V_{DD} = 5\,\text{V}$, $V_{TRN} = 0.2V_{DD} = 1\,\text{V}$, and $K_N = 0.2\,\text{mA/V}^2$, find the propagation delay time t_{PHL}.

Solution
Substitution of the value $V_{TRN} = 0.2V_{DD}$ into Eq. (14.25) results in

$$t_{PHL} = \frac{C_L(-V_{DD}/2)}{-K_N(V_{DD} - V_{TRN})^2} = \frac{C_L(V_{DD}/2)}{K_N(V_{DD} - 0.2V_{DD})^2} \approx \frac{0.8C_L}{K_N V_{DD}} \tag{14.26}$$

or $t_{PHL} = 0.8(2\,\text{pF})/(0.2\,\text{mA/V}^2)(5\,\text{V}) = 1.6\,\text{ns}$

When v_{IN} undergoes a high-to-low transition, the load capacitance will be *charged* from zero to V_{DD} via the current through Q_P. In this case, Q_N will be in cutoff, and Q_P will operate in its constant-current region for most of the transition. If the gate is symmetrical with $K_P = -K_N$ and $V_{TRP} = -V_{TRN}$, substitution of the voltage change $\Delta v_{OUT} = V_{DD}/2$ into Eq. (14.25) with K_P substituted for K_N yields the same Δt given by Eq. (14.26). Because a CMOS gate with $K_P = -K_N$ and $V_{TRP} = -V_{TRN}$ is functionally symmetrical, t_{PLH} becomes equal to t_{PHL}.

EXERCISE 14.10   Find the values of t_{PHL} and t_{PLH} for the CMOS inverter of Fig. 14.7 if $V_{DD} = 5\,\text{V}$, $K_N = 0.3\,\text{mA/V}^2$, $K_P = -0.3\,\text{mA/V}^2$, and $V_{TRN} = -V_{TRP} = 1.5\,\text{V}$. The inverter drives a 0.7-pF load capacitance.   **Answer:** 0.48 ns

14.2.3 CMOS Logic Gates

The logic functions NAND and NOR are readily implemented using the CMOS inverter topology of the previous section. A two-input NOR gate, for example, is depicted in Fig. 14.12. The load capacitance C_L represents the input to another CMOS gate. In the lower half of the circuit, logic inputs A and B drive the parallel-connected n-channel transistors Q_1 and Q_2. In the upper half of the circuit, these same logic inputs also drive the series-connected p-channel transistors Q_3 and Q_4. When both inputs are low, Q_1 and Q_2 become open circuits. At the same time, Q_3 and Q_4 are forced into conduction and provide a conducting path between the load and V_{DD}. Conversely, when either input A or B is high, the series connection of Q_3 and Q_4 is broken by the high input, but a conduction path from C_L to ground is provided by Q_1 or Q_2.

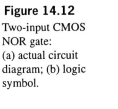

Figure 14.12

Two-input CMOS NOR gate: (a) actual circuit diagram; (b) logic symbol.

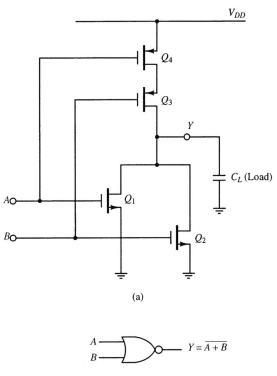

The circuit shown in Fig. 14.13 is a two-input CMOS NAND gate. In contrast to the NOR gate of Fig. 14.12, the roles of n-channel and p-channel devices are reversed. The parallel connection of devices in the upper half of the circuit is complemented by a series connection of devices in the lower half. When either input A or B is low, the conduction path provided by Q_1 and Q_2 is broken. The parallel connection of Q_3 and Q_4 ensures that a conducting path is provided between the load capacitance C_L and V_{DD}. Similarly, when both A and B are high, C_L is connected to ground via Q_1 and Q_2. For this case, both Q_3 and Q_4 become open circuits, and C_L is disconnected from V_{DD} and discharged to ground.

EXERCISE 14.11   Draw the circuit diagrams for a three-input NAND gate and a three-input NOR gate in the CMOS configuration.

Figure 14.13
Two-input CMOS
NAND gate:
(a) circuit diagram;
(b) logic symbol.

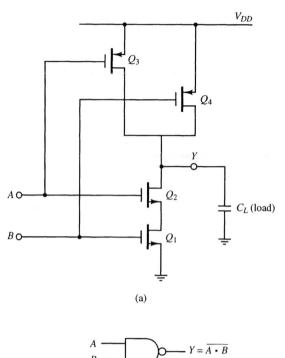

(a)

(b)

Discussion. The inverter symmetry requirement $K_N = |K_P|$ must be modified in a CMOS gate with multiple inputs. When devices of geometrical length L_1 and L_2 are connected in series, the conduction path provided by them becomes equivalent to that of one device of length $L_1 + L_2$. For this reason, series devices in a two-input CMOS circuit must be fabricated with twice the width of a single device acting alone. Circuits with more than two inputs must have series-connected devices of proportionately larger width. The width multiplication factor for series-connected devices must be combined with the mobility-related requirement that a silicon p-channel device, for which $\mu_h = \mu_e/2$, have twice the width of an equivalent n-channel device. In the two-input NOR gate of Fig. 14.12, for example, Q_3 and Q_4 are p-channel devices and are connected in series. If all devices have the same length L, Q_3 and Q_4 must be fabricated with four times the width of Q_1 and Q_2. Similarly, in the two-input NAND gate of Fig. 14.13, Q_1 and Q_2 are connected in series and must have twice the width of an n-channel device acting alone. Hence, if all devices have the same length L, Q_1 and Q_2 must be fabricated with the *same* width as the p-channel devices Q_3 and Q_4.

The device geometry requirements discussed in the preceding paragraph can be generalized to apply to a CMOS gate with an arbitrary number of inputs. For a silicon NOR gate with N inputs, the device width-to-length ratios should be designed so that

$$\frac{W_P}{L_P} = 2\,\mathrm{N}\frac{W_N}{L_N} \tag{14.27}$$

Similarly, for a NAND gate with N inputs, the width-to-length ratios should be designed so that

$$\frac{W_N}{L_N} = \frac{N}{2}\frac{W_P}{L_P} \tag{14.28}$$

where, in both cases, $\mu_e = 2\mu_h$, and $|K| = \mu\epsilon_{\mathrm{ox}}/2t_{\mathrm{ox}}$. ∎

EXERCISE 14.12   A three-input CMOS NAND gate is fabricated so that the smallest device dimension is $1\,\mu$m. If the width-to-length ratios of the n-channel devices are equal to 2, find the dimensions of all devices in the circuit and the approximate total surface area occupied by the gate.

Body Effect in CMOS Gates

The effect of nonzero source-to-substrate voltage v_{SB} (the body effect) was shown to modify the transfer characteristic of the NMOS inverter in Chapter 6. In CMOS circuits, the substrates of all n-channel devices are connected to the most negative voltage in the circuit, and the substrates of all p-channel devices are connected to the most positive voltage. As illustrated by the circuit diagrams of Fig. 14.14, which show the substrate connections, the body effect need not be considered in a CMOS inverter consisting of one n-channel and one p-channel device. For the circuit shown, the source of the n-channel MOSFET is connected to ground, and the source of the p-channel MOSFET is connected to V_{DD}, so that $v_{SB} = 0$ in each case.

In the more complex CMOS logic gates described in this section, the source terminals of all n-channel devices are not connected to ground; similarly, the source terminals of all p-channel devices are not connected to V_{DD}. Consequently, v_{SB} will not be zero for all devices, and the modified formula for V_{TR} introduced in Chapter 6,

$$\Delta V_{\text{TR}} = \gamma \left[(v_{SB} + 2\phi_F)^{1/2} - (2\phi_F)^{1/2} \right] \tag{14.29}$$

must be considered in deriving the gate transfer characteristics.

Figure 14.14
Substrate connections in a CMOS logic inverter. The representations (a) and (b) are equivalent.

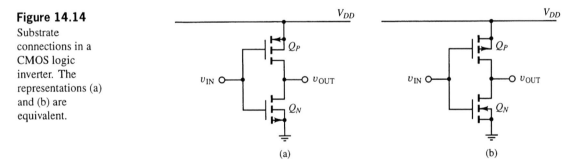

(a)   (b)

14.2.4 CMOS Transmission Gate

The availability of n-channel and p-channel devices in CMOS circuits makes it possible to fabricate the *bidirectional transmission gate* shown in Fig. 14.15(a). The function of this circuit is to connect terminal A to terminal B whenever the control input C is high. The transmission gate must also be fed an input signal $\overline{C}$ equal to the complement of C. When C is low (and $\overline{C}$ is high), the n-channel and p-channel devices are both forced into cutoff. In this state, the transmission gate becomes an open circuit and terminal A is disconnected from terminal B. When C is high (and $\overline{C}$ is low), Q_1 and Q_2 both enter a conducting state. The use of complementary devices permits current flow in either direction—hence the term *bidirectional*. The logic symbol of the bidirectional transmission gate is shown in Fig. 14.15(b).

The availability of a transmission gate in the CMOS family adds another degree of freedom to the digital design process. The transmission gate makes possible a configuration called the *tristate output* in CMOS circuits. It also makes possible a CMOS version of a logic element called the *type D flip-flop*, to be discussed in Chapter 15.

Figure 14.15
CMOS
transmission gate:
(a) basic circuit;
(b) logic symbol.

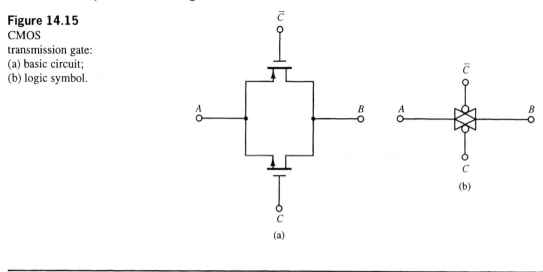

(a)

(b)

EXERCISE 14.13 Draw the geometrical layout diagram of the transmission gate of Fig. 14.15.

14.14 Design a CMOS circuit that will function as a bidirectional transmission gate controlled by a single input signal. Your circuit should be similar to that of Fig. 14.14, but should also contain components to generate the required $\overline{C}$ signal from a single control input C.

14.3 NMOS LOGIC FAMILY

In the following sections, the basic characteristics of the NMOS inverter introduced in Chapter 6 are examined from a digital point of view. In the NMOS logic family, digital gates are made entirely from n-channel MOSFETs. The relatively small size of NMOS devices makes them ideal for use in VLSI applications. In addition, the geometric simplicity of NMOS integrated circuits facilitates automated and computer-aided design techniques. Like their CMOS counterparts, NMOS circuits can be designed without the use of resistors, which are difficult to fabricate and occupy large amounts of surface area on an integrated circuit. Since the inverter forms the fundamental building block of more complex NMOS circuits, its properties typify those of the entire NMOS family. A digital NMOS inverter can be made using either an enhancement-mode or a depletion-mode pull-up load. The device accepting the input is always an enhancement-mode device.

14.3.1 NMOS Inverter with Enhancement Load

The circuit diagram of an enhancement-load inverter is shown in Fig. 14.16. If devices Q_1 and Q_2 undergo the same fabrication process, they will have the same value of threshold voltage V_{TR}. As discussed previously, the K parameter of each device depends on its geometrical dimensions, hence the logic gate characteristics that depend on K can be set at fabrication time by choosing appropriate W and L values. The gate of the pull-up load Q_2 is connected to its drain, so that $v_{DS2} = v_{GS2}$. This connection ensures that Q_2 will always operate in its constant-current region with $v_{DS2} > (v_{GS2} - V_{TR})$. As discussed in Chapter 5, an enhancement-mode MOSFET with its gate connected to its drain behaves like an ideal square-law device. An NMOS gate with the configuration of Fig. 14.16 is sometimes called the *saturated-load* inverter.

Figure 14.16
NMOS inverter with enhancement-mode pull-up load.

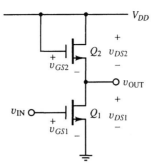

The transfer characteristic of the inverter of Fig. 14.16 was previously studied in Chapter 6. As the analysis of Section 6.4.5 shows, if v_{IN} exceeds V_{TR} and if Q_1 remains in the constant-current region, the relationship between v_{OUT} and v_{IN} becomes

$$v_{OUT} = V_{DD} + \left(\sqrt{\frac{K_1}{K_2}} - 1\right) V_{TR} - \sqrt{\frac{K_1}{K_2}} v_{IN} \tag{14.30}$$

The slope of Eq. (14.30) is constant and given by

$$\frac{dv_{OUT}}{dv_{IN}} = -\sqrt{\frac{K_1}{K_2}} \tag{14.31}$$

The slope of the transfer characteristic of a digital inverter must exceed -1 over its transition region; hence the ratio K_1/K_2 must have a magnitude greater than unity.

When the NMOS gate of Fig. 14.16 makes a transition between its output states, Q_1 will operate for part of the time in the triode region when v_{IN} is small. Deriving an expression for the portion of the transfer characteristic over which Q_1 operates in the triode region requires simultaneous solution of the equations:

$$i_{D1} = K_1[2(v_{IN} - V_{TR1})v_{OUT} - v_{OUT}^2] \tag{14.32}$$

and
$$i_{D2} = K_2(V_{DD} - v_{OUT} - V_{TR2})^2 \tag{14.33}$$

where $i_{D1} = i_{D2}$. This mathematical solution involves considerable algebraic manipulation. Alternatively, this portion of the transfer characteristic can be obtained graphically, as was done in Section 6.4.4. Either method will yield the complete transfer characteristic shown in Fig. 14.17.

The points p_1 and p_2 on the transfer characteristic where $dv_{OUT}/dv_{IN} = -1$ are of special interest. The quantities V_{IL} and V_{IH} are defined at these operating points. In the enhancement-load inverter, the transition from cutoff to the constant-current region of the transfer characteristic is a sharp one that coincides with point of -1 slope, hence the location of p_1 is given by

$$V_{IL} = V_{TR1} \tag{14.34}$$

We also note from the transfer characteristic that

$$V_{OH} = V_{DD} - V_{TR2} \tag{14.35}$$

This same value of v_{OUT} also happens to apply to p_1 for this transfer characteristic.

Figure 14.17
Complete transfer
characteristic of the
enhancement-load
NMOS inverter of
Fig. 14.16.

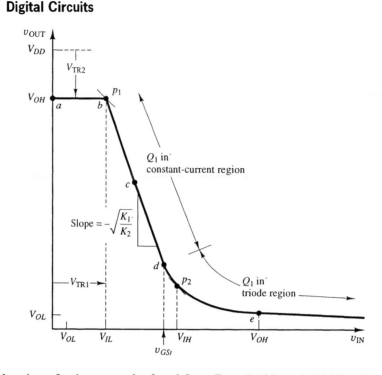

The location of point p_2 can be found from Eqs. (14.32) and (14.33). These equations together define the triode-region portion of the transfer characteristic. Equating i_{D1} with i_{D2} and taking the derivative with respect to v_{IN} yields an expression relating dv_{OUT}/dv_{IN} to v_{OUT} and v_{IN}:

$$2K_1[\dot{v}_{OUT}(v_{IN} - V_{TR}) + v_{OUT} - v_{OUT}\dot{v}_{OUT}] = -2K_2(V_{DD} - v_{OUT} - V_{TR})\dot{v}_{OUT} \quad \textbf{(14.36)}$$

where $V_{TR1} = V_{TR2} = V_{TR}$, $\dot{v}_{OUT}$ symbolizes the derivative dv_{OUT}/dv_{IN}, and v_{OUT} at p_2 is equal to $V(0)_{max}$. In principle, Eq. (14.36) can be solved simultaneously with Eqs. (14.32) and (14.33) to find a value for V_{IH}, but the algebra involved in the solution is tedious. Alternatively, the location of point p_2, along with the value of V_{OL}, can be estimated graphically, as shown in the next example.

EXAMPLE 14.3　An enhancement-load NMOS inverter with aspect ratio $K_1/K_2 = 8$ is fabricated by setting device dimensions to $W_1 = 2\,\mu\text{m}$, $L_1 = 2\,\mu\text{m}$, $W_2 = 2\,\mu\text{m}$, and $L_2 = 16\,\mu\text{m}$, so that $W_1/L_1 = 1$ and $W_2/L_2 = 1/8$.

 (a) If $V_{TR1} = V_{TR2} = 1\,\text{V}$ and $V_{DD} = 5\,\text{V}$, find the critical points p_1 and p_2 on the inverter transfer characteristic and compute the noise margins NM_H and NM_L.

 (b) Repeat if the aspect ratio is changed to 2 by making $L_2 = 4\,\mu\text{m}$, so that $W_2/L_2 = 1/2$.

Solution

• Find the transfer characteristic of the inverter using the graphical technique

In Fig. 14.18, the load curve of V_{DD} and Q_2 that corresponds to an aspect ratio of 8 is plotted against the $v–i$ characteristics of Q_1. The i_D-axis has been normalized to the value $K_1 = 1\,\mu\text{A/V}^2$. The resulting transfer characteristic, obtained graphically, is shown in Fig. 14.19.

• Determine the values of V_{IL} and V_{OH}

The values of V_{IL} and V_{OH} for this inverter, corresponding to point p_1 on the transfer characteristic of Fig. 14.19, can be found directly from Eqs. (14.33) and (14.34), and are equal to

$$V_{IL} = V_{TR1} = 1\,\text{V} \quad \textbf{(14.37)}$$

and

$$V_{OH} = V_{DD} - V_{TR2} = 5\,\text{V} - 1\,\text{V} = 4\,\text{V} \quad \textbf{(14.38)}$$

Figure 14.18
Load curve of V_{DD} and Q_2 for an aspect ratio of $K_1/K_2 = 8$. The value of V_{DD} is 5 V, and $V_{TR1} = V_{TR2} = 1$ V. The current scale has been normalized to $K_1 = 1 \mu A/V^2$.

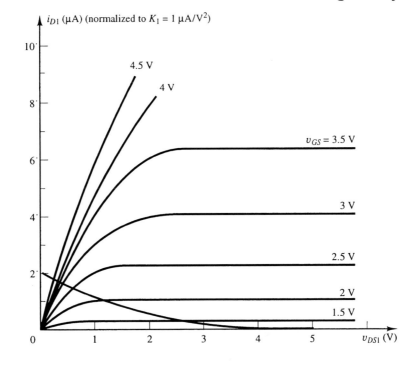

• **Graphically determine the values of V_{IH} and V_{OL}**

The location of point p_2 can be estimated from the graph, where

$$V_{IH} \approx 2.2 \text{ V} \tag{14.39}$$

Similarly, the value of v_{OUT} at $v_{IN} = V_{OH} = 4$ V can also be estimated from the graph:

$$V_{OL} \approx 0.25 \text{ V} \tag{14.40}$$

Figure 14.19
Transfer characteristic for an aspect ratio of $K_1/K_2 = 8$. The values of V_{IH} and V_{OL} are obtained graphically.

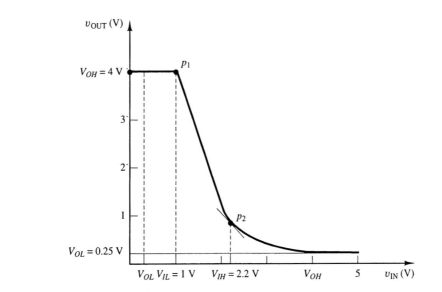

• **Find the noise margins NM_H and NM_L using the values found for V_{OH}, V_{IH}, V_{IL}, and V_{OL}**

For this inverter, the high noise margin is equal to

$$NM_H = V_{OH} - V_{IH} = 4\,V - 2.2\,V = 1.8\,V \qquad (14.41)$$

This value is reasonably large and is adequate for a working logic system operating with a V_{OH} of 4 V. The low noise margin

$$NM_L = V_{IL} - V_{OL} = 1\,V - 0.25\,V = 0.75\,V \qquad (14.42)$$

is smaller, but acceptable.

• **Compute the degradation in the noise margins when the aspect ratio is changed from 8 to 2**

Changing the aspect ratio from 8 to 2 modifies the load curve of V_{DD} and Q_2, as shown in Fig. 14.20. Note that the intercept of each load curve with the i_D-axis can be found by substituting the value $v_{OUT} = 0$ into Eq. (14.33), resulting in

$$i_{D2}\bigg|_{v_{OUT}=0} = K_2(V_{DD} - V_{TR2})^2 \qquad (14.43)$$

Because V_{DD} and V_{TR2} are fixed, the location of this intercept in Fig. 14.20 changes inversely as the ratio K_1/K_2.

Figure 14.20
Modified load curve for $K_1/K_2 = 2$. Dashed curve: original load curve for $K_1/K_2 = 8$. The vertical scale has been normalized to $K_1 = 1\mu A/V^2$.

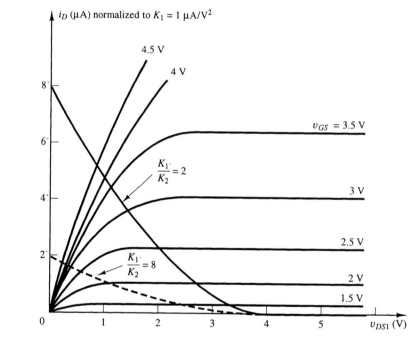

The modified load curve for the new aspect ratio of 2 causes a degradation in inverter performance, as shown in Fig. 14.21. The slope of the transfer characteristic is reduced, and the values of V_{IH} and V_{OL} are shifted to 2.8 V and 0.9 V, respectively. For these new values of V_{IH} and V_{OL}, the noise margins become

$$NM_H = V_{OH} - V_{IH} = 4\,V - 2.8\,V = 1.2\,V \qquad (14.44)$$

and $\qquad\qquad NM_L = V_{IL} - V_{OL} = 1\,V - 0.9\,V = 0.1\,V \qquad (14.45)$

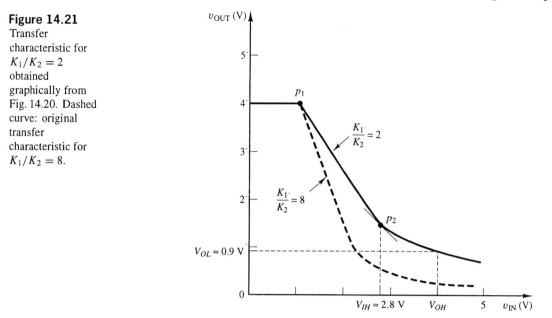

Figure 14.21
Transfer characteristic for $K_1/K_2 = 2$ obtained graphically from Fig. 14.20. Dashed curve: original transfer characteristic for $K_1/K_2 = 8$.

The resulting low noise margin obtained in this case approaches zero and is unacceptable in a practical logic system. The high noise margin has been reduced from 1.8 V to 1.2 V. The reason for the degradation in performance can be understood by examining the graph of Fig. 14.20. The smaller aspect ratio raises the intercept at which the $V_{DD}-Q_2$ load curve crosses the i_D-axis. Larger values of v_{GS1} are required to force v_{DS1} to its low-output state.

EXERCISE 14.15 For the NMOS inverter of Example 14.3, with $K_1/K_2 = 8$, solve Eqs. (14.32), (14.33), and (14.36) to find values for V_{IH} and $V(0)_{max}$.   **Answer:** 2.2 V, 0.8 V

14.16 For the NMOS inverter of Example 14.3 with $K_1/K_2 = 8$, show that the values $v_{IN} = V_{IH} = 2.2$ V and $v_{OUT} = V(0)_{max} = 0.8$ V yield a slope of $dv_{OUT}/dv_{IN} = -1$ at point p_2 by substituting these values into Eq. (14.36) and solving for v_{OUT}.

14.3.2 NMOS Inverter with Depletion Load

As discussed in Chapter 6, the NMOS inverter with depletion-mode pull-up load of Fig. 6.45 (see page 358) has a transfer characteristic with a very steep transition region and sharp corners, as shown in Fig. 6.47. The value of v_{IC} for the curve, derived in Section 6.4.5, is given by

$$v_{IC} = \sqrt{K_2/K_1}|V_{TR2}| + V_{TR1} \tag{14.46}$$

This transfer characteristic has a much steeper slope than that of the equivalent enhancement-load inverter of Fig. 14.16. The steeper slope arises because the transition region is traversed with both Q_1 and Q_2 operating in the constant-current region—a situation similar to that encountered in the CMOS inverter of Section 14.2.1. The steep slope causes the depletion-load inverter to have higher noise margins and causes its transfer characteristic to more closely approach that of an ideal logic inverter.

Unfortunately, the body effect, introduced in Section 6.4.6, causes the threshold voltage of the depletion-mode device to increase according to the equation

$$V_{TR} = V_{TRo} + \gamma \left[(v_{SB} + 2\phi_F)^{1/2} - (2\phi_F)^{1/2} \right] \tag{14.47}$$

where v_{SB} is the source-to-substrate voltage ($v_{SB} = v_{OUT}$ for the inverter of Fig. 6.48), ϕ_F is the Fermi potential, and V_{TRo} the MOSFET threshold voltage when $v_{SB} = 0$. This modification to V_{TR} causes the inverter transfer characteristic to depart from the excellent curve of Fig. 6.47. The actual transfer characteristic, computed for the case $K_R = 8$ and $\gamma = 0.5 \, \mathrm{V}^{1/2}$, is shown by the solid curve in Fig. 6.50. As this plot shows, the near-ideal nature of the depletion-load transfer characteristic is significantly compromised by the body effect. Nevertheless, the transfer characteristic of the depletion-load inverter is still superior to that of an equivalent enhancement-load inverter. In addition, a depletion-load inverter designed to have the *same* slope as its enhancement-load counterpart will occupy less chip area because a smaller aspect ratio $K_R = K_1/K_2$ will be required to achieve the same slope. This feature makes the depletion-load inverter a popular choice in many VLSI circuits.

14.3.3 Dynamic Behavior of NMOS Inverter

The dynamic behavior of an NMOS gate is strongly influenced by both internal device and external load capacitances. The latter are contributed by the interconnect paths that connect one gate to another. Together these capacitances determine the rise time, fall time, and propagation delay of an NMOS gate.

An estimate of the general dynamic performance of NMOS circuits can be made by examining the case of a depletion-load NMOS inverter driving a single capacitive load, as shown in Fig. 14.22. Such an analysis illustrates a method applicable to enhancement-load gates as well. The load capacitance C_L represents the relevant internal MOSFET capacitances that appear between v_{OUT} and ground. It also represents the interconnection capacitances seen by the inverter output terminal. For the purpose of illustration, we again assume the input v_{IN} to be an ideal digital signal with zero rise and fall times. This assumption simplifies the calculation of the rise time, fall time, and propagation delay of v_{OUT}. The circuit is a nonlinear one in which the operating regions of Q_1 and Q_2 change with the values of v_{IN} and v_{OUT}. Consequently, the dynamic behavior for the high-to-low transition differs from that of the low-to-high transition.

Figure 14.22
Depletion-load NMOS inverter driving a single load capacitance.

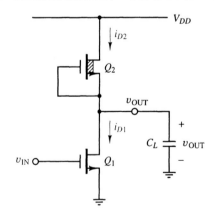

We first discuss the events that follow a low-to-high-output transition. Suppose the inverter output to be initially low with the capacitor charged to the voltage $V(0) = V_{OL}$. When v_{IN} is switched from $V(1)$ to $V(0)$, Q_1 will go into cutoff and the capacitor will begin to be charged through Q_2. Because Q_2 is not a linear resistive load, this charging will not be exponential in time. Rather, the charging of the capacitor will be governed by the MOSFET v–i characteristics of Q_2 and the v–i equation for the capacitor. With $v_{GS2} = 0$, the v–i equation of the depletion-mode MOSFET Q_2 is given by either

$$i_{D2} = K_2(-V_{TR2})^2 \quad (Q_2 \text{ in constant-current region}) \tag{14.48}$$

or $\quad i_{D2} = K_2[2(-V_{TR2})^2(V_{DD} - v_{OUT}) - (V_{DD} - v_{OUT})^2] \quad (Q_2 \text{ in triode region})$

The v–i equation of the capacitor is given by

$$\frac{dv_{OUT}}{dt} = \frac{i_{D2}}{C_L} \tag{14.49}$$

Because v_{DS2} is equal to $(V_{DD} - v_{OUT})$, the constant-current region characteristic will apply when $(V_{DD} - v_{OUT}) \geq -V_{TR2}$. The triode region characteristic will apply when $(V_{DD} - v_{OUT}) < -V_{TR2}$.

The nonlinear set of equations (14.48) through (14.49) can be solved by numerical iteration or can be estimated by hand calculation. Both methods are illustrated in the following example.

EXAMPLE 14.4

Suppose that the circuit of Fig. 14.22 has an aspect ratio K_1/K_2 of 4 and a load capacitance C_L of 0.1 pF. For a low-to-high-output transition, find the rise time of v_{OUT} and the propagation delay between v_{IN} and v_{OUT}. The approximate device parameters are $K_1 = 8\ \mu A/V^2$, $K_2 = 2\ \mu A/V^2$, $V_{TR1} = 1\ V$, and $V_{TR2} = -2\ V$. Assume the high and low logic levels to be given by $V(0) = V_{OL} = 0.3\ V$ and $V(1) = V_{OH} = V_{DD} = 5\ V$, so that t_r is defined in terms of the 10% value: $V(0) + 0.1(5\ V - 0.3\ V) \approx 0.8\ V$, and the 90% value: $V(0) + 0.9(5\ V - 0.3\ V) \approx 4.5\ V$. In order to specifically focus on the properties of the gate, assume that v_{IN} makes an instantaneous high-to-low transition with zero fall time.

Solution

• Estimate the rise time t_r and propagation delay t_{PLH} by hand calculation

Approximate values for t_r and t_{PLH} can be found by estimating the average capacitor charging current. When v_{IN} first makes its high-to-low transition, the capacitor voltage v_{OUT} cannot change instantaneously from its initial value of $V(0)$. Immediately after the transition of v_{IN}, the voltage across Q_2 will remain at $V_{DD} - V(0)$, and Q_2 will operate in the constant-current region. The capacitor current at this point in time will be equal to

$$i_{D2} = K_2(-V_{TR2})^2 = (2\mu A/V^2)(-2\ V)^2 = 8\ \mu A \tag{14.50}$$

As the capacitor charges up, v_{DS2} will be reduced, and Q_2 will eventually operate in the triode region. When the capacitor charging has been completed, so that $v_{OUT} = V(1)$, the current i_{D2} will fall to zero. Over a large portion of the charging time, however, Q_2 will operate in its constant-current region with i_{D2} given by Eq. (14.50). If i_{D2} is approximated as being constant for the entire charging time, Eq. (14.49) can be integrated over time to yield

$$\int_{V(0)}^{v_{OUT}(t)} dv_{OUT} = \int_0^t \frac{\langle i_{D2} \rangle}{C_L} dt \tag{14.51}$$

where $\langle i_{D2} \rangle$ is the average value of $i_{D2}(t)$. Integrating this equation over the indicated limits and using the constant-current region current (14.50) to approximate $\langle i_{D2} \rangle$ results in

$$\begin{aligned} v_{OUT}(t) &= V(0) + \frac{\langle i_{D2} \rangle}{C_L} t \\ &= 0.3\ V + \frac{8\ \mu A}{0.1\ pF} t = 0.3\ V + (0.08\ V/ns)t \end{aligned} \tag{14.52}$$

where t is in nanoseconds. For a transition from the 10% value of 0.8 V to the 90% value of 4.5 V, the estimated rise time is given by

$$t_r = \frac{4.5\,\text{V} - 0.8\,\text{V}}{0.08\,\text{V/ns}} \approx 46\,\text{ns} \tag{14.53}$$

The propagation delay time t_{PLH} can be estimated in a similar manner. Because the jump in v_{IN} is instantaneous, t_{PLH} becomes equivalent to the time required for v_{OUT} to rise to 2.65 V, which is halfway between $V(0) = 0.3$ V and $V(1) = 5$ V. Applying the equation analogous to Eq. (14.53) yields

$$t_{PLH} = \frac{2.65\,\text{V} - 0.3\,\text{V}}{0.08\,\text{V/ns}} \approx 29\,\text{ns} \tag{14.54}$$

• Compute the rise time and propagation delay by numerical integration

A numerical solution can be obtained by following the flowchart program of Fig. 14.23. A numerical calculation based on this flowchart produces the rising portion of the plot of $v_{OUT}(t)$ shown in Fig. 14.24. The values $t_r = 54$ ns and $t_{PLH} = 29$ ns can be noted graphically. These results should be compared to the values $t_r = 46$ ns and $t_{PLH} = 29$ ns obtained by hand calculation.

Figure 14.23
Flowchart for computing the charging of a capacitive load C_L during a low-to-high output transition.

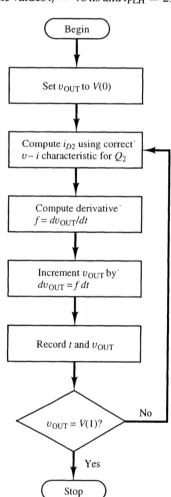

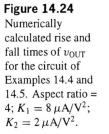

Figure 14.24
Numerically
calculated rise and
fall times of v_{OUT}
for the circuit of
Examples 14.4 and
14.5. Aspect ratio =
4; $K_1 = 8\,\mu A/V^2$;
$K_2 = 2\,\mu A/V^2$.

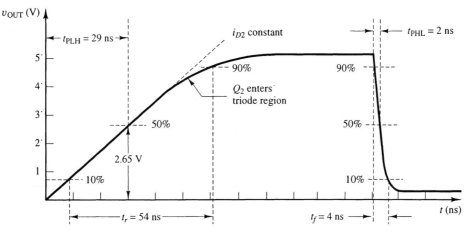

The values of t_{PLH} match because Q_2 does not enter the triode region until after the 50% point at which t_{PLH} is evaluated. Q_2 will enter the triode region only after v_{OUT} rises above 3 V, causing v_{DS2} to fall below $v_{GS2} - V_{TR2} = -V_{TR2} = 2\,V$. The current charging the capacitor is thus indeed constant at $8\,\mu A$ while C_L is charged up to the 50% point. The numerical and hand-calculated values of t_r, however, are not identical. This result is expected because the charging current falls below the assumed constant value of $8\,\mu A$ as Q_2 enters the triode region. The numerically computed charging curve departs from the $i_{D2} = $ constant curve just prior to the 90% point, as shown in Fig. 14.24.

EXERCISE 14.17   For the low-to-high output transition described in the preceding example, trace the operating point of Q_2 as a function of time along the V_{DD}–Q_2 load curve. Show that the approximation of constant i_{D2} over the charging interval is a reasonable one.

We next discuss the dynamic behavior of the inverter during a high-to-low output transition. Suppose that the capacitor voltage is equal to $V(1)$ when v_{IN} switches from low to high. After the input step function, Q_1 will be turned on and will begin to discharge C_L toward the voltage $V(0)$. At any given instant during this discharge period, C_L will appear to Q_1 as an instantaneous voltage source of value $v_{OUT}(t)$. The operating point of Q_1 will be determined by this voltage. The discharge of C_L will be governed by the capacitor equation

$$\frac{dv_{OUT}}{dt} = \frac{i_{D2} - i_{D1}}{C_L} \qquad (14.55)$$

where the values of i_{D2} and i_{D1} are determined by the MOSFET v–i equations. For Q_2, the relevant expressions are given by Eqs. (14.47) and (14.48). When $v_{IN} = V(1)$, similar expressions apply for Q_1 with $v_{GS1} = V(1)$:

$$i_{D1} = K_1[V(1) - V_{TR1}]^2 \qquad (Q_1 \text{ in constant-current region}) \qquad (14.56)$$

or $\qquad i_{D1} = K_1\left\{2[V(1) - V_{TR1}]v_{OUT} - v_{OUT}^2\right\} \qquad (Q_1 \text{ in triode region}) \qquad (14.57)$

Equation (14.56), valid in the constant-current region, applies for $v_{OUT} \geq V(1) - V_{TR1}$. The triode-region expression (14.57) applies for $v_{OUT} < V(1) - V_{TR1}$. Note that with $K_1 > K_2$ and $v_{GS1} > v_{GS2}$, i_{D1} will be greater than i_{D2} for all values of v_{OUT}, so that the capacitor current will be negative and the capacitor will discharge.

Equations (14.56)–(14.57) and (14.48) can again be estimated by hand calculation or solved by numerical iteration. Both methods are illustrated in the following example.

EXAMPLE 14.5

For a high-to-low output transition in the circuit of Fig. 14.22, find the fall time of v_{OUT} and the propagation delay between v_{IN} and v_{OUT}. The inverter has the same parameters and aspect ratio given in Example 14.4. Again assume the jump in v_{IN} to be instantaneous.

Solution

• **Estimate the fall time t_r and propagation delay t_{PHL} by hand calculation**

The values of t_f and t_{PHL} can be estimated by finding the average capacitor discharge current. At the beginning of the transition, v_{OUT} will be initially equal to $V(1)$; therefore, Q_1 will operate in its constant-current region with $v_{GS1} = V(1)$ and $v_{DS1} > v_{GS1} - V_{TR1}$. With the value of v_{OUT} initially at $V(1) = 5$ V, the voltage across Q_2 will be zero, so that $i_{D2} = 0$. After some discharge has taken place, however, Q_2 will enter its constant-current region. The average capacitor current i_C can be estimated by taking the difference between the values of i_{D1} and i_{D2}, with each computed using the appropriate constant-current region equation:

$$
\begin{aligned}
i_C &= i_{D2} - i_{D1} \\
&= K_2(-V_{TR2})^2 - K_1[V(1) - V_{TR1}]^2 \\
&= (2\,\mu\text{A/V}^2)(-2\,\text{V})^2 - (8\,\mu\text{A/V}^2)(5\,\text{V} - 1\,\text{V})^2 = -120\,\mu\text{A}
\end{aligned}
\tag{14.58}
$$

Integrating Eq. (14.55) with respect to time then results in

$$
\begin{aligned}
v_{OUT}(t) &= V(1) + \frac{\langle i_C \rangle t}{C_L} = 5\,\text{V} - \frac{120\,\mu\text{A}}{0.1\,\text{pF}}t \\
&= 5\,\text{V} - (1.2\,\text{V/ns})t
\end{aligned}
\tag{14.59}
$$

where t is again in nanoseconds, and the current $i_{D2} - i_{D1}$ has been used to approximate the average value of i_C. The resulting values of t_f and t_{PHL} become

$$
t_f \approx \frac{4.5\,\text{V} - 0.8\,\text{V}}{1.2\,\text{V/ns}} \approx 3\,\text{ns}
\tag{14.60}
$$

and

$$
t_{PHL} = \frac{5\,\text{V} - 2.65\,\text{V}}{1.2\,\text{V/ns}} \approx 2.0\,\text{ns}
\tag{14.61}
$$

• **Compute the fall time and propagation delay by numerical integration**

A numerical solution for $v_{OUT}(t)$ can be found by following a flowchart similar to that of Fig. 14.23. An extra step must be added for the computation of i_{D1}. Performing the computation with $K_1 = 8\,\mu\text{A/V}^2$ and $K_2 = 2\,\mu\text{A/V}^2$ produces the falling portion of the plot of Fig. 14.24. The values $t_f \approx 4$ ns and $t_{PHL} \approx 2$ ns are obtained graphically. These values are again reasonably close to those obtained by hand calculation. Note that the form of the rising waveform differs significantly from that of the falling waveform. For this nonlinear circuit, the charging and discharging of the capacitor are not symmetrical functions in time.

EXERCISE 14.18

For the high-to-low output transition described in the preceding example, trace the operating point of Q_1 as a function of time along the $v_{GS1} = V(1)$ v–i characteristic.

14.19 Write a computer program to obtain the plots of Fig. 14.24. Follow the flowchart of Fig. 14.23.

14.20 Apply the analysis of Examples 14.4 and 14.5 to the case of the enhancement-load inverter. Perform both the numerical calculation and the hand estimate.

14.3.4 NMOS Logic Gates

The basic NMOS logic inverter is easily modified to produce multi-input NOR and NAND logic gates. The NOR function can be implemented by connecting parallel input transistors to a common pull-up load, as in Fig. 14.25(a). This figure depicts a two-input NOR gate; more inputs can be created by adding more parallel input transistors.

Figure 14.25
Two-input NMOS NOR gate:
(a) circuit diagram;
(b) geometrical device layout for an aspect ratio of 3;
(c) logic symbol.

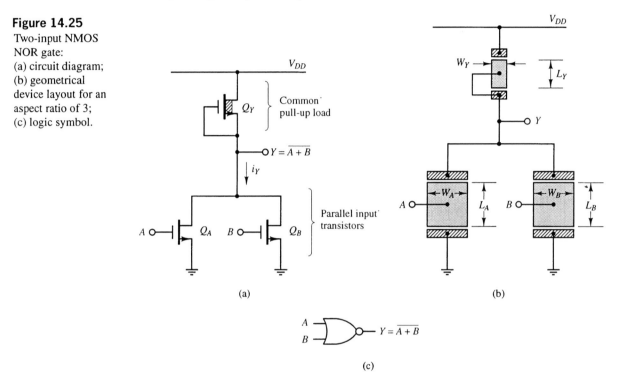

(a)

(b)

(c)

When either input to the circuit is high, current will be pulled down through Q_Y, causing the output voltage to be low. Only when *both* inputs are simultaneously low will the current i_Y become zero. When i_Y is zero, the output voltage will attain its high state.

The input transistors in a NOR gate are geometrically identical. Their width-to-length ratios are chosen so that each device will be independently capable of causing the output to fall to V_{OL} in the low state, where V_{OL} is defined in terms of a single-input inverter. The width and length of Q_A and Q_B are thus individually set to values appropriate for a single-input logic inverter. The geometrical layout of a two-input NOR gate based on an inverter with an aspect ratio of 3, for example, is shown in Fig. 14.25(b). In this case, $W_A = W_B = 3W_Y$ and $L_A = L_B = L_Y$.

The NAND logic function is implemented by connecting input transistors in series, as in Fig. 14.26(a). This figure depicts a two-input gate; more inputs can be created by adding more input transistors in *series*. The output of the circuit will be low only if both inputs are high so that both Q_A and Q_B conduct. If either A or B is low, the conduction path for i_Y will be broken and i_Y will become zero.

Like the NOR configuration, the input transistors in the NAND configuration are geometrically identical. Unlike the case of the NOR gate, however, the width-to-length ratios of the input devices must be larger than those of an inverter designed to have the same aspect ratio. For the two-input gate of Fig. 14.26(a), for example, the effective channel length $L_A + L_B$ of both devices acting in series is double that of either device acting alone. If W_1 and L_1 are the required dimensions of the input device in a single-transistor inverter, Q_A and Q_B must be fabricated with

Figure 14.26
Two-input NMOS
NAND gate:
(a) circuit diagram;
(b) geometrical
device layout for an
aspect ratio of 3;
(c) equivalent
width-to-length
ratio of the series
combination of Q_A
and Q_B; (d) logic
symbol.

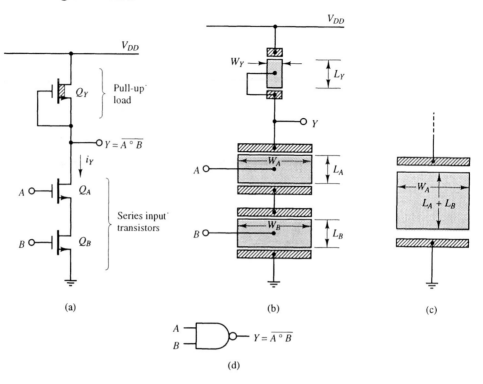

(a)

(b)

(c)

(d)

a width of $2W_1$ and a length of L_1. As illustrated in Figs. 14.26(b) and 14.26(c), the effective width-to-length ratio of both Q_A and Q_B acting together will then be equal to

$$\left(\frac{W}{L}\right)_{A+B} = \frac{W_A}{L_A + L_B} = \frac{2W_1}{2L_1} = \frac{W_1}{L_1} \tag{14.62}$$

This selection of width-to-length ratios for Q_A and Q_B guarantees that the output voltage will fall to V_{OL} when inputs A and B are high. For the more general case of a NAND gate with N inputs, the width of each input device must be made N times larger than the width of the equivalent transistor in the single-output inverter. This requirement causes NMOS NAND gates to occupy a larger surface area than a comparable NOR gate having the same number of inputs. For this reason, multiple-input NAND gates are used sparingly in densely packed digital NMOS circuits.

EXERCISE 14.21   Draw the circuit and geometrical layout diagrams for three-input and four-input NMOS NOR gates. Choose width-to-length ratios based on an inverter with an aspect ratio of 8. The smallest dimension in the gate should be $2\,\mu$m.

14.22   Design a two-input NMOS NAND gate with an aspect ratio of 8 and a smallest scale dimension of $2\,\mu$m.

14.23   Draw the circuit diagram of a three-input NMOS NAND gate. Draw the geometrical layout of the gate based on an inverter with an aspect ratio of 3. Specify the value of W and L for each device in microns if the smallest dimension is $2\,\mu$m.

14.24   Compare the surface area of an N-input NAND gate with that of a single-input inverter designed for the same aspect ratio.

14.4 TTL LOGIC FAMILY

The TTL (transistor–transistor logic) family is made from *npn* BJTs and resistors. TTL was one of the first digital logic families to be invented and is still popular in many digital applications. Many MSI-level TTL logic gates are available as off-the-shelf integrated circuits suitable for rapid prototype design. At the heart of the TTL gate lies the basic BJT inverter with resistive pull-up load introduced in Chapter 6. Other components are added to the inverter to transform it into a complete TTL logic gate. These additional components significantly improve the switching speed of the inverter and add other important features. To appreciate the need for the complexity of the complete TTL gate, we first discuss the switching speed limitations of the basic BJT inverter with resistive pull-up load.

14.4.1 Dynamic Behavior of BJT Inverter

The switching-speed limitations of BJT circuits can be understood by considering the simple BJT inverter of Fig. 14.27. At first we shall ignore the effect of the load capacitance C_L and concentrate on stored charge effects inside the BJT itself. The latter contribute fundamental limitations to device switching speed.

Figure 14.27
BJT inverter with resistive pull-up load connected to a load capacitance.

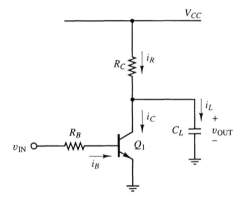

Suppose that the circuit of Fig. 14.27, with C_L disconnected, is driven by an ideal, instantaneous step input that rises from $V(0)$ to $V(1)$ with zero rise time. As the inverter output switches from its logic high- to its logic-low-output state, the BJT will be forced to operate over all three regions of operation. It begins in cutoff, with $i_B = 0$, traverses the active region of operation, then reaches saturation, with $v_{CE} = V_{sat}$. The transitions between these regions occur over time intervals related to the charge that must be stored in the BJT before it can conduct current. Specifically, the ability of Q_1 to conduct requires that a gradient of excess minority-carrier electrons $n'_p(x)$ exist within the *p*-type base region, as depicted in Fig. 14.28. In Chapter 9, this electron gradient was shown to be responsible for the nonlinear diffusion capacitance C_d of the base–emitter junction, which is modeled by the bias-dependent, incremental capacitance C_π in the linear small-signal model. When v_{IN} switches from low to high, the buildup of the excess electron gradient in the base (i.e., the charging of C_d), requires a finite amount of time. It can be shown[1] that the evolution of the stored excess charge Q_F over time is governed by the equations

$$i_C = Q_F/\tau_F \tag{14.63}$$

and
$$i_B = \frac{Q_F}{\beta_F \tau_F} + \frac{dQ_F}{dt} \tag{14.64}$$

---

[1] See, for example, D. A. Hodges and H. G. Jackson, *Analysis and Design of Digital Integrated Circuits*, 2nd ed. New York: McGraw-Hill, 1988, p. 196.

where τ_F is a material- and geometry-dependent parameter called the *forward transit time*, and i_B, determined from circuit considerations, is equal to $(v_{IN} - v_{BE})/R_B$ in this case. Solution of the differential equation (14.64) subject to the initial condition $Q_F = 0$ at $t = 0$ results in

$$Q_F(t) = \beta_F \tau_F i_B (1 - e^{-t/\beta_F \tau_F}) \tag{14.65}$$

The collector current as a function of time, determined from (14.63), then becomes

$$i_C(t) = \beta_F i_B (1 - e^{-t/\beta_F \tau_F}) \tag{14.66}$$

As this equation shows, the BJT can fully reach the active region, where $i_C = \beta_F i_B$, only after several time constants $\beta_F \tau_F$.

Figure 14.28
Evolution of the electron concentration gradient in the base region of Q_1 following the application of a voltage step function to the base–emitter junction.

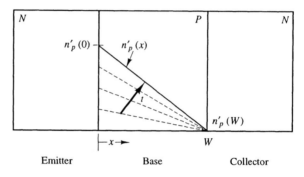

Equation (14.66) can be solved for t in terms of $i_C(t)$ to obtain an equation useful for computing the inverter's rise time:

$$t = \beta_F \tau_F \ln \frac{\beta_F i_B}{\beta_F i_B - i_C(t)} \tag{14.67}$$

If the steady-state collector current $i_C(t = \infty)$ is designated I_C, where $I_C = \beta_F i_B$, then Eq. (14.67) can be solved for times t_{10} and t_{90} at which $i_C(t)$ is equal to 10 and 90%, respectively, of I_C. Specifically,

$$t_{10} = \beta_F \tau_F \ln \frac{\beta_F i_B}{\beta_F i_B - 0.1 I_C} = \beta_F \tau_F \ln \frac{I_C}{0.9 I_C} \tag{14.68}$$

and

$$t_{90} = \beta_F \tau_F \ln \frac{\beta_F i_B}{\beta_F i_B - 0.9 I_C} = \beta_F \tau_F \ln \frac{I_C}{0.1 I_C} \tag{14.69}$$

The rise time t_r can be computed from the time difference $t_{90} - t_{10}$:

$$t_r = t_{90} - t_{10} = \beta_F \tau_F \ln \frac{0.9 I_C}{0.1 I_C} = 2.2 \beta_F \tau_F \tag{14.70}$$

As this result shows, the rise time of the inverter is on the order of twice the product of β_F and the forward transit time τ_F. With τ_F on the order of tens of nanoseconds for the typical BJT, the value of t_r can approach the microsecond level if β_F is large. This delay contributes to the generally poor dynamic performance of resistive-load BJT logic circuits.

During a low-to-high output transition, the charge storage phenomenon also contributes to the inverter's poor dynamic performance. When v_{OUT} is low, Q_1 operates in saturation. In the saturated state, the base–emitter and the collector–emitter junctions of Q_1 are both forward-biased. The profile of excess minority carriers in the base region under saturation conditions appears as in Fig. 14.29. As shown by the shaded region, additional stored electrons appear in the base region as the BJT is driven deeper into saturation by i_B. This additional stored charge significantly increases the turn-off time of the BJT. When v_{IN} switches from high to low, the electron gradient *and* the additional stored electrons must be removed or allowed to recombine with holes before Q_1 can return to cutoff. Since recombination is a relatively slow process, the stored electrons will disappear slowly unless allowed to flow *out* of the base through R_B. This reverse base current must be "sinked" by the circuit that provides the v_{IN} voltage.

Figure 14.29
Concentration gradient after the BJT has been driven into saturation.

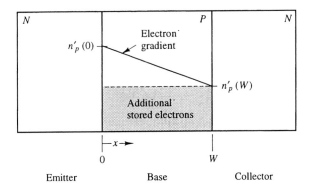

In the moments just after an input step from high to low, the excess electrons stored in the base region will flow backward out of the base as reverse base current. Unless R_B is very small, this discharge process will be slow. If the value R_B is made too small, however, large currents will be required to drive the gate input when v_{IN} is high. The need to deplete the excess electrons stored during saturation is one of the main factors responsible for the slow switching time of the resistive-load BJT logic inverter of Fig. 14.27.

Discussion. For the BJT inverter of Fig. 14.27, precise hand computation of the output voltage as a function of time during a high-to-low or a low-to-high transition is a difficult task. The charging and discharging of the base–emitter junction diffusion capacitance C_d (i.e., the storage and removal of excess electrons from the base region) is a nonlinear process that is voltage-dependent. Precise calculations are best performed by numerical iteration using a computer-aided circuit-simulation program such as SPICE.

The rise and fall times of the BJT inverter of Fig. 14.27 are also affected by the load capacitance C_L. The latter represents the capacitance of the interconnections between gates and also the device capacitance presented by the inputs of other gates. During a low-to-high output transition, the circuit must charge C_L to the voltage $V(1)$. Conversely, during a high-to-low output transition, the circuit must discharge C_L to the voltage $V(0)$. These charge and discharge intervals contribute, respectively, to the rise and fall times of the inverter output. In the discussion that follows, we ignore the contribution of charge–store phenomenon to the rise and fall times, and focus exclusively on the role of C_L. Suppose that the inverter output is in the low output state with $v_{IN} = V(1)$ and Q_1 in saturation. Applying an input step from $V(1)$ to $V(0)$ will force Q_1 into cutoff. The output will eventually attain the value $V(1)$, but before v_{OUT} can reach this final value, C_L must be charged from $V(0)$ to $V(1)$ via the series load resistance R_C. With Q_1 in cutoff, the combination of V_{CC}, R_C, and C_L forms a simple RC circuit with output voltage given by

$$v_{OUT} = V(0) + [V(1) - V(0)](1 - e^{-t/R_C C_L}) \qquad (14.71)$$

The rise time of this waveform is governed by the exponential time constant $R_C C_L$. If R_C is made small, the charging time constant $R_C C_L$ can be made short as well. If R_C is made too small, however, excessive current will flow through the inverter when Q_1 is saturated in the low-output state. At the same time, an R_C that is too small will limit the steep slope of the inverter's transfer characteristic. The time required to resistively charge C_L through R_C constitutes another major factor responsible for the slowness of the BJT inverter with resistive pull-up load.

Next suppose the inverter output to be in the high state with $v_{IN} = V(0)$ and Q_1 in cutoff. Application of an input step from $V(0)$ to $V(1)$ will eventually saturate Q_1 and force v_{OUT} to the value $V(0) = V_{sat}$. The voltage across the load capacitor cannot change instantaneously, however. The initial capacitor voltage of $V(1)$ will cause Q_1 to enter the active region. As it begins to discharge C_L, the BJT will remain in its active region with constant i_C until the voltage across C_L is reduced to V_{sat}. After the capacitor has discharged to V_{sat}, the BJT will enter saturation with $v_{CE} = V_{sat}$, $i_C = (V_{CC} - V_{sat})/R_C$, and $v_{BE} = V_f$. Over the discharge interval, the capacitor voltage will be governed by the equation

$$\frac{dv_{OUT}}{dt} = \frac{i_L}{C_L} = \frac{i_R - i_C}{C_L} \tag{14.72}$$

where the collector current i_C is given by $i_C = \beta_F i_B$ and the current through R_C by

$$i_R = \frac{V_{CC} - v_{OUT}}{R_C} \tag{14.73}$$

Because β_F is large, i_C will be much greater than i_R, and i_L will be negative. This large, negative i_L rapidly discharges the capacitor. The time required for discharging the capacitor with a constant current is much shorter than that required for exponential charging through a passive resistor. Consequently, the effect on the fall time caused by C_L is minor compared to the effect of C_L on the rise time. The need to discharge C_L does not contribute significantly to the slowness of the inverter. ∎

EXERCISE 14.25   For the BJT logic inverter with capacitor load shown in Fig. 14.27, plot the voltage across C_L during high-to-low and low-to-high output transitions. Neglect the contribution of the stored charge in the base region. Assume that $R_B = 500\,\Omega$, $R_C = 500\,\Omega$, $\beta_F = 100$, and $C_L = 10\,\text{pF}$.

14.26   Sketch the rising waveform of the circuit described in Exercise 14.25 if $\tau_F = 20\,\text{ns}$. Do not ignore the contribution of the stored charge in the base region.

14.4.2   Basic Structure of TTL

The poor dynamic performance of the BJT inverter with resistive load can be improved significantly by adding other components to the basic gate topology. These modifications transform the simple BJT inverter into the fundamental building block of the TTL logic family. The TTL configuration preserves the function of the resistive inverter while reducing the delays caused by stored charge effects and load capacitance.

TTL Input Stage

The limitations inherent to stored charge effects can be significantly moderated by the TTL input driver stage of Fig. 14.30. A basic inverter is formed by Q_2 and pull-up load R_2. Base current is provided to Q_2 by Q_1; the latter has its emitter driven by v_{IN}. The function of Q_1 and the purpose of R_3 will be explained shortly.

Figure 14.30
TTL input driver stage reduces gate rise and fall times and generates the complementary current signals i_{B3} and i_{B4}.

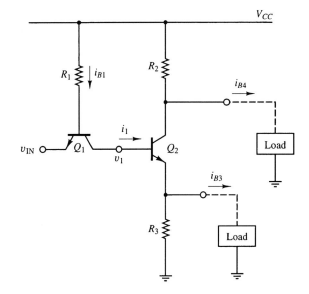

In order to understand the mechanism by which this configuration speeds up gate switching time, we must first understand its operation. In summary, the input voltage v_{IN} drives Q_2 to the extremes of cutoff and saturation via the action of Q_1. When v_{IN} is low, the base–emitter junction of Q_1 becomes forward-biased. Ordinarily, this forward bias would cause collector current to flow through Q_1. In this case, however, the collector current of Q_1 must flow backward through the base–emitter junction of Q_2, which it cannot do. Consequently, Q_1 is forced into saturation with i_1 equal to the reverse saturation current of the base–emitter junction of Q_2. Note that the voltage v_1 at the base of Q_2 attains the value $v_{IN} + V_{sat}$ under these conditions, with V_{sat} contributed by Q_1.

With i_1 forced toward zero, Q_2 is forced into cutoff. The current i_{B3} becomes zero and the current i_{B4} flows via R_2 to the load. The specific value of i_{B4} is determined by the v–i characteristic of the load connected to the i_{B4} terminal. The choice of the labels i_{B3} and i_{B4} will become evident shortly.

When v_{IN} is high, the base–emitter junction of Q_1 becomes reverse-biased and the base–collector junction of Q_1 becomes *forward-biased*. This combination of junction voltages forces Q_1 into the *reverse-active* mode, in which the roles of collector and emitter are exchanged. In the reverse-active mode, the collector and emitter currents are related to the base current via the reverse beta parameter β_R. For the terminal current directions defined in Fig. 14.31, these relationships become

$$i_{IN} = \beta_R i_B \qquad (14.74)$$

and
$$i_1 = (\beta_R + 1)i_B \qquad (14.75)$$

In principle, the *npn* BJT can function with either *n*-type region playing the role of the collector. In practice, the collector–base junction always has a larger cross-sectional area than the base–emitter junction; hence the device has inherent asymmetry. Because of the geometrical asymmetry between the collector and emitter regions in a planar BJT (large collector, small emitter), the parameter β_R is much smaller than the parameter β_F. Depending on transistor size and geometry, β_R may have a value anywhere between 0.01 and 5.[2]

---

[2] The reverse-active mode is discussed in more detail in Appendix A using the Ebers–Moll transistor model.

Figure 14.31
Current
relationships for a
BJT in the
reverse-active
mode.

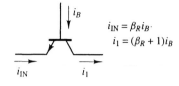

Despite the small value of β_R, the current supplied to Q_2 by the reverse-active Q_1 is sufficient to drive Q_2 into saturation. When v_{IN} in Fig. 14.30 is high, the base current through Q_1 becomes

$$i_{B1} = \frac{V_{CC} - v_1 - V_f}{R_1} \tag{14.76}$$

where V_f is contributed by the forward-biased base–*collector* junction of Q_1. The resulting current flowing out of Q_1 becomes

$$i_1 = (\beta_R + 1)\frac{V_{CC} - v_1 - V_f}{R_1} \tag{14.77}$$

With Q_2 saturated by i_1, the current i_{B4} will be forced toward zero. The current i_1 will flow through the base–emitter junction of Q_2 and become the current i_{B3}. Thus, when v_{IN} is high, i_{B3} will be turned on and i_{B4} will be turned off. The specific value of v_1, and hence the value of i_{B3}, will depend on the load connected to the i_{B3} terminal. In any case, the currents i_{B3} and i_{B4} are used to drive the TTL output stage, discussed in the next section, to its logic-high and logic-low output states.

The action of Q_1 is instrumental in improving the dynamic performance of the TTL inverter during a low-to-high output transition. When the base–emitter junction of Q_1 is forward-biased by a low v_{IN}, the collector current of Q_1 quickly removes the stored electrons from the base region of Q_2. The value of i_{C1} during this discharge interval is much greater than the current that would flow were the base of Q_2 to be connected to $V(0)$ directly through a resistor. It is possible to compute the time required to remove the stored base charge, but the nonlinear calculations are difficult. Such a computation is readily carried out using a circuit simulation package such as SPICE (see Example 14.8 at the end of the chapter). The results show the rise time of a gate with the TTL input configuration of Fig. 14.30 to be much shorter than that of an unmodified BJT inverter.

TTL Output Stage

The dynamic performance of the TTL gate is further improved by the addition of the output driver stage of Fig. 14.32. The arrangement shown, in which two *npn* BJTs are placed on top of each other, is sometimes called the *totem-pole* output configuration. For its inputs, the circuit requires the two currents i_{B3} and i_{B4} produced by the input driver stage of Fig. 14.30. As discussed previously, these currents have the special relationship that one is always off while the other is on. The purpose of R_4, which appears in series with the collector of the voltage follower Q_4, will be explained shortly.

When the load in Fig. 14.32 contains capacitance C_L, the totem-pole topology significantly reduces the time required to charge C_L to the logic value $V(1)$. When i_{B3} is off and i_{B4} on, as in Fig. 14.33(a), the lower device Q_3 is forced into cutoff and the upper device Q_4 is turned on by i_{B4}. The constant current through Q_4 rapidly charges C_L to the value $V(1)$, thereby greatly reducing the output rise time. Because of its role in pulling up the output voltage to a high value, Q_4 is sometimes called the *active pull-up* device of the totem-pole output stage. When i_{B3} is on and i_{B4} off, as in Fig. 14.33(b), the upper device Q_4 is forced into cutoff. The lower device Q_3 is turned on by i_{B3} and rapidly discharges C_L to the value $V(0) = V_{\text{sat}}$.

Figure 14.32
TTL totem-pole output stage. The input current signals i_{B3} and i_{B4} provided by the input stage of Fig. 14.30 are complementary.

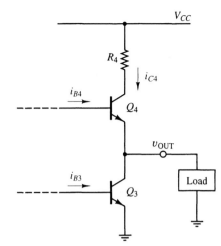

Note that the use of Q_4 as a follower buffer between i_{B4} and v_{OUT} increases the fan-out capability of the TTL gate when the output voltage is high. Current flow to other gates is provided through the base–emitter junction of Q_4, rather than through R_C, thereby minimizing the drop in v_{OUT} caused by load currents.

Figure 14.33
Operation of the totem-pole output stage for different values of i_{B3}:
(a) i_{B3} off, i_{B4} on;
(b) i_{B3} on, i_{B4} off.

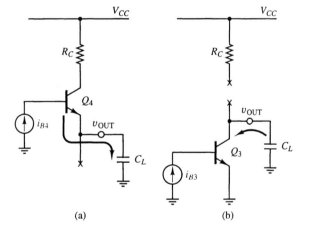

14.4.3 Complete TTL Logic Inverter

The preceding section provides a qualitative description of the component parts of the TTL logic inverter. In Fig. 14.34, the TTL input driver circuit of Fig. 14.30 is combined with the totem-pole output stage of Fig. 14.32 to produce a complete TTL logic inverter. A detailed examination of the terminal voltages and the current flow within the gate is provided in this section. Proper operation of the circuit requires the addition of the diode D_1 to the output stage. The purpose of this diode will soon become evident. Our first objective is to find the node voltages and current flow when the input is low, with $v_{IN} = V(0) \approx V_{sat}$. For the purpose of illustration, we assume the BJTs to have parameters $\beta_F = 100$, $\beta_R = 0.1$, $V_f = 0.7\,\text{V}$, and $V_{sat} \approx 0.2\,\text{V}$. These parameters yield voltage and current values that are typical of a TTL gate.

Figure 14.34

Complete TTL
logic inverter with
v_{IN} low.

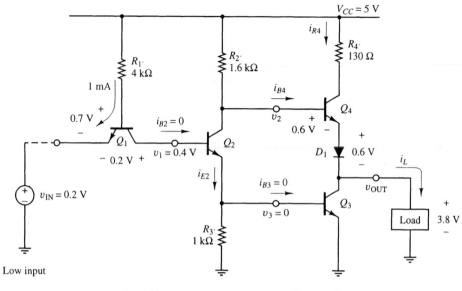

Low input

Input driver stage   Totem-pole output stage

When v_{IN} is low, the base–emitter junction of Q_1 becomes forward-biased, so that the base current of Q_1 becomes

$$i_{B1} = \frac{V_{CC} - [V(0) + V_f]}{R_1} = \frac{5\,\text{V} - (0.2\,\text{V} + 0.7\,\text{V})}{4\,\text{k}\Omega} \approx 1\,\text{mA} \tag{14.78}$$

After the stored charge has been removed from the base of Q_2, the collector current of Q_1 will become zero and Q_1 will go into saturation with $v_{CE1} = V_{\text{sat}}$. The node voltage v_1 therefore becomes

$$v_1 = v_{\text{IN}} + v_{CE1} = 0.2\,\text{V} + 0.2\,\text{V} = 0.4\,\text{V} \tag{14.79}$$

This voltage is too small to forward bias the base–emitter junction of Q_2; this condition is consistent with the result $i_{B2} = -i_{C1} = 0$.

With Q_2 in cutoff, the emitter current of Q_2 becomes zero, with the result that

$$v_3 = (i_{E2} - i_{B3})R_3 = 0 \tag{14.80}$$

This result implies that i_{B3} will be zero as well, so that Q_3 will be in cutoff.

With Q_2 in cutoff, the base current i_{B4} into Q_4 will be equal to the current flowing through R_2. If the output terminal is unloaded, the only current flowing through Q_4 with Q_3 in cutoff will be any leakage current that flows through the v_{OUT} terminal. Under such conditions, the voltage drop across R_2 will be minimal, and v_{OUT} will be equal to V_{CC} minus the two V_f voltage drops contributed by Q_4 and D_1. With the current through Q_4 small, these drops will be less than 0.7 V. If a value of 0.6 V is assumed, then v_{OUT} becomes

$$V(1) = V_{CC} - 2V_f = 5\,\text{V} - 1.2\,\text{V} = 3.8\,\text{V} \tag{14.81}$$

If the v_{OUT} terminal is loaded, so that i_L is nonzero, v_{OUT} will be given instead by

$$v_{\text{OUT}} = V_{CC} - i_{B4}R_2 - 2V_f \equiv V_{CC} - \frac{i_L}{\beta_F + 1}R_2 - 2V_f \tag{14.82}$$

Since β_F is large, the load current i_L can become reasonably large before the output voltage drops significantly below the value given by Eq. (14.81).

If i_L becomes too large, the drop across R_4 will cause Q_4 to go into saturation when v_{CE4} falls to the value V_{sat}. The value of i_L corresponding to the entry of Q_4 into saturation can be found by taking KVL in the output loop with $v_{CE4} = V_{sat}$ and $i_{C4} = \beta_F i_{B4}$:

$$
\begin{aligned}
v_{CE4} &= (V_{CC} - i_{C4}R_4) - \overset{\text{contribution of } D_1}{(v_{OUT} + V_f)} \\
&= \left(V_{CC} - \frac{\beta_F}{\beta_F + 1} i_L R_4 \right) - \left(V_{CC} - \frac{i_L}{\beta_F + 1} R_2 - V_f \right)
\end{aligned}
\tag{14.83}
$$

In writing this expression, Eq. (14.82) has been substituted for v_{OUT}. Solving Eq. (14.82) for i_L with $v_{CE4} = V_{sat}$ yields the maximum value that i_L can have before Q_4 goes into saturation:

$$
i_L = \frac{V_f - V_{sat}}{[\beta_F/(\beta_F + 1)]R_4 - [R_2/(\beta_F + 1)]} \approx \frac{V_f - V_{sat}}{R_4}
\tag{14.84}
$$

Note that with Q_4 approaching saturation, V_f will approach the value 0.75 V. For the assumed values $V_{sat} \approx 0.2\,\text{V}$ and $V_f \approx 0.72\,\text{V}$, the maximum i_L given by Eq. (14.84) becomes (0.72 V − 0.2 V)/(130 Ω) = 4 mA. From this calculation, we note the purpose of R_4, which is to limit the load current when Q_4 saturates.

For load currents above this value, v_{OUT} will be further reduced according to the equation

$$
v_{OUT} = V_{CC} - i_L R_4 - V_{sat} - V_f
\tag{14.85}
$$

The reduction of v_{OUT} described by Eq. (14.85) will be rapid for small increases in i_L. This feature protects the TTL gate from output short circuits by limiting the short-circuit load current.

EXERCISE 14.27   For the TTL gate of Fig. 14.34, plot v_{OUT} as a function of i_L when v_{OUT} is high.

Discussion.   The resistor R_4 has the role of limiting the output current during an unexpected short circuit. It also serves another important current limiting function in the TTL gate. When the inverter output undergoes a low-to-high output transition, Q_2 goes into cutoff and the pull-up device Q_4 is turned on quickly by the current i_{B4}. At the same time, Q_3 must come out of saturation before it can stop conducting. As Q_3 comes out of saturation, the stored charge in the base of Q_3 will flow to ground relatively slowly through R_3. This process often results in a momentary condition in which Q_3 is still conducting when Q_4 is turned on. The resistor R_4 limits the current through the series connection of Q_3 and Q_4 when both devices are in a conducting state.

Even with the current limiting provided by R_4, the current that flows during this brief transition period may be substantial. If Q_3 and Q_4 are both in saturation, for example, the current through R_4 will be approximately equal to

$$
i_{R4} = \frac{V_{CC} - V_{sat4} - V_f - V_{sat3}}{R_4} \approx 30\,\text{mA}
\tag{14.86}
$$

This current must be supplied by V_{CC}. It will flow in the form of a short (10–100-ns) pulse of current called a "spike". If V_{CC} is not supplied by a perfect voltage source, this current spike may lead to momentary dips in the actual V_{CC} delivered to the TTL gate. Such a momentary dip in supply voltage is often referred to as a "glitch".

Voltage glitches in a digital circuit can be transferred to other logic gates in the system and cause erroneous digital signals. The scenario by which such voltage spikes might take place is illustrated in Fig. 14.35. The resistive and inductive impedances in series with V_{CC} represent the distributed resistance and inductance of the power-supply connection paths between gates. A short pulse of current through an inductance is capable of causing a sizable unwanted voltage drop.

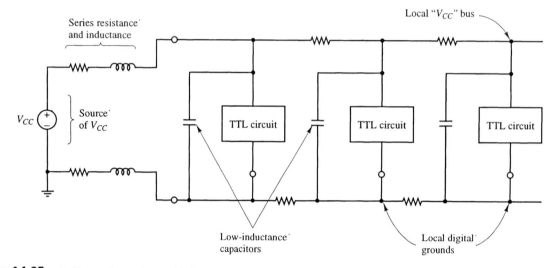

Figure 14.35 Addition of capacitors with low parasitic series inductance across power-supply connections of each TTL circuit. These capacitors act as momentary voltage sources to locally supply "spike" currents and help to reduce global transmission of voltage transients across the power-supply bus.

As a precaution against the effects of transient-current spikes, several procedures are common in the design of TTL systems. Power-supply connections to the various integrated circuits in the system are made as short as possible to minimize series inductance. Similarly, these connections are made with wires of large cross-sectional area to minimize series resistance. In addition, low-inductance capacitors are often connected between the power-supply bus and ground at the site of each logic gate. These capacitors act as momentary "voltage sources" that temporarily provide voltage through a low impedance for the duration of the current spike. ∎

We next find the node voltages and currents when the input is high, that is, when $v_{IN} = V(1) = 3.8\,\text{V}$. This choice of input voltage assumes v_{IN} to be provided by the output of some other TTL gate, with $V(1)$ given by Eq. (14.81). We again assume the BJTs to have parameters $\beta_F = 100$, $\beta_R = 0.1$, $V_f = 0.7\,\text{V}$, and $V_{\text{sat}} \approx 0.2\,\text{V}$.

When v_{IN} is high, as in Fig. 14.36, the base–emitter junction of Q_1 becomes reverse-biased and Q_1 enters the reverse-active mode. The base current flowing through R_1 exits via the forward-biased base–collector junction of Q_1. If Q_2 and Q_3 conduct, v_1 will acquire the voltage $2V_f$ (two base–emitter drops), which is sufficiently below V_{CC} to allow the base–collector junction of Q_1 to become forward-biased. Under these conditions, the base current into Q_1 becomes

$$i_{B1} = \frac{V_{CC} - 3V_f}{R_1} = \frac{5\,\text{V} - 2.1\,\text{V}}{4\,\text{k}\Omega} \approx 0.73\,\text{mA} \tag{14.87}$$

where a value of $V_f = 0.7\,\text{V}$ has also been assumed for the forward-biased base–collector junction.

Figure 14.36
TTL logic inverter
with v_{IN} high. The
load current flows
into Q_3.

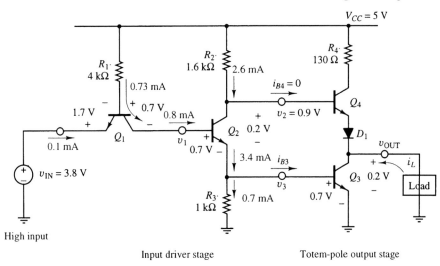

High input

Input driver stage   Totem-pole output stage

The resulting base current into Q_2 will be equal to $(\beta_R + 1)i_{B1}$. Given the low value of $\beta_R = 0.1$, the value of i_{B2} will be approximately equal to i_{B1}:

$$i_{B2} = (\beta_R + 1)i_{B1} = (0.1 + 1)\left(\frac{5\,\text{V} - 2.1\,\text{V}}{4\,\text{k}\Omega}\right) \approx 0.8\,\text{mA} \qquad (14.88)$$

This base current is sufficient to saturate the driver transistor Q_2. It will be proven shortly that Q_4 is forced into cutoff when Q_2 saturates. With the base current Q_4 zero, the emitter current flowing through the saturated Q_2 will be equal to the sum of i_{B2} and the current through R_2:

$$
\begin{aligned}
i_{E2} &= i_{B2} + i_{R2} \\
&= i_{B2} + \frac{V_{CC} - (V_f + V_{\text{sat}})}{R_2} = 0.8\,\text{mA} + \frac{5\,\text{V} - 0.7\,\text{V} - 0.2\,\text{V}}{1.6\,\text{k}\Omega} \approx 3.4\,\text{mA} \qquad (14.89)
\end{aligned}
$$

where the V_f drop is provided by the base–emitter junction of Q_3. The voltage across R_3 under these conditions, as determined by the forward-biased base–emitter junction of Q_3, is equal to V_f; hence the current through R_3 is equal to 0.7 mA. The remainder of the emitter current i_{E2}, equal to $3.4\,\text{mA} - 0.7\,\text{mA} = 2.7\,\text{mA}$, flows into the base of Q_3. This current is large enough to saturate Q_3 and drive the output to the logic-low state.

Note that the voltage properties of the v_{OUT} terminal under logic-low conditions are determined by the load current flowing into the collector of Q_3. This current is called the *sink* current of the logic gate. As i_L is increased, Q_3 will be forced less deeply into saturation, causing the saturation voltage of Q_3 to increase. A TTL gate has a maximum specified sink current that must not be exceeded if the voltage of the logic-low output state is to stay below $V(0)_{\text{max}}$. This maximum sink current in part determines the maximum fan-out capability of the TTL gate.

Given the distribution of voltages in the TTL gate with v_{IN} high, the purpose of the series diode D_1 becomes evident. With Q_2 saturated, the voltage v_2 appearing at the base of Q_4, as determined by the base–emitter voltage of Q_3 and the collector–emitter voltage of Q_2, becomes $V_f + V_{\text{sat}} = 0.9\,\text{V}$. With v_{OUT} in the logic-low state, the voltage across Q_3 becomes equal to V_{sat}. Without D_1 in place, the voltage across the base–emitter junction of Q_4 would thus be equal to V_f, placing Q_4 just on the edge of conduction. The diode in series with the emitter of Q_4 absorbs some of this forward-biasing voltage and ensures that Q_4 remains in cutoff when v_{OUT} is low.

EXERCISE 14.28

For the circuit of Fig. 14.36, calculate i_{C2} if Q_2 is assumed to operate in the active region with v_{IN} high and $i_{B2} = 0.8\,\text{mA}$. Show that the voltage drop across R_2 instead causes the saturation of Q_2.

14.4.4 Transfer Characteristic of the TTL Logic Inverter

The input–output transfer characteristic of the TTL inverter of Figs. 14.34 and 14.36 can be determined by examining circuit behavior for inputs between the extremes of $V(0)$ and $V(1)$. The resulting transfer characteristic is represented by the piecewise linear plot of Fig. 14.37. This graph has four distinct sectors that are related to the operating regions of the devices within the gate.

Figure 14.37

Transfer characteristic of the TTL inverter. For clarity, the horizontal scale has been expanded by a factor of 2.

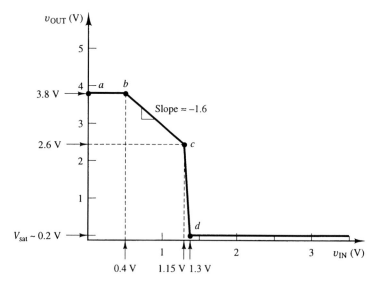

Over segment a–b, the input voltage is low enough to forward bias the base–emitter junction of Q_1. In the steady state, Q_1 is driven to saturation, forcing Q_2 and Q_3 into cutoff and Q_4 into saturation. The resulting output voltage $V(1)$ lies at approximately 3.8 V, as derived in Eq. (14.81).

As v_{IN} increases beyond the value at point b, the base–emitter junction of Q_1 remains forward-biased and Q_1 remains saturated, but the voltage v_1 at the base of Q_2 becomes large enough to forward bias the base–emitter junction of Q_2. With Q_1 still saturated, the voltage v_1 is given by

$$v_1 = v_{IN} + V_{sat} \tag{14.90}$$

The value of v_{IN} at the transition point b can be determined by finding the value of v_1 that is sufficient to forward bias the base–emitter junction of Q_2. Specifically, v_{IN} at point b must result in a v_1 equal to V_f. Solving Eq. (14.90) for the condition $v_1 = V_f$ results in

$$v_{IN}(b) = V_f - V_{sat} \approx 0.6\,\text{V} - 0.2\,\text{V} = 0.4\,\text{V} \tag{14.91}$$

where the value $V_f \approx 0.6\,\text{V}$, valid when Q_2 just begins to conduct, has been assumed. For v_{IN} larger than the value at point b, Q_1 supplies base current to Q_2 while Q_1 remains in the saturation mode. Note that this action requires both the base–emitter junction *and* the base–collector junction of Q_1 to be forward-biased.

Between points b and c, incremental increases in v_{IN} are linearly amplified by Q_2. The amplifier configuration, as determined by R_2 and R_3, is essentially that of an inverting amplifier in the feedback-bias configuration. The gain of the amplifier, as derived in Chapter 7, is given approximately by

$$\frac{\Delta v_2}{\Delta v_1} \approx \frac{-R_2}{R_3} \tag{14.92}$$

The voltage decrement Δv_2 is passed on to the output by the follower action of Q_4, causing v_{OUT} to fall by $(-R_2/R_3)\, \Delta v_1$ as well. The slope dv_{OUT}/dv_{IN} of the transfer characteristic over segment b–c is therefore equal to $-R_2/R_3 = -1.6$. Note that the voltage developed across R_3 over this region of the transfer characteristic is insufficient to forward bias the base–emitter junction of Q_3, which remains in cutoff.

The onset of conduction in Q_3 occurs when its base–emitter junction becomes forward-biased by the drop across R_3. This point is marked on the graph by point c. The value of v_{OUT} at point c can be found from the condition $v_{BE3} = V_f$. This relation can be expressed in terms of i_{E2} as

$$v_{BE3} = i_{E2} R_3 = V_f \tag{14.93}$$

or

$$i_{E2} = \frac{V_f}{R_3} = \frac{0.6\,\text{V}}{1\,\text{k}\Omega} = 0.6\,\text{mA} \tag{14.94}$$

where the value $V_f \approx 0.6\,\text{V}$, valid when Q_3 just begins to conduct, has again been assumed. For a collector current of $i_{C2} \approx i_{E2} = 0.6\,\text{mA}$, the output voltage at point c becomes

$$\begin{aligned} v_{OUT}(c) &= V_{CC} - i_{C2}R_2 - 2V_f \\ &= 5\,\text{V} - (0.6\,\text{mA})(1.6\,\text{k}\Omega) - 2(0.7\,\text{V}) \approx 2.6\,\text{V} \end{aligned} \tag{14.95}$$

In this case, a value of $V_f \approx 0.7\,\text{V}$ is chosen for Q_4 and D_1, because conduction is well established in both devices at point c. The corresponding value of v_{IN} at point c can be found by considering the slope implicit in Eq. (14.92). Graphical triangulation over the segment b–c implies that

$$\frac{\Delta v_{OUT}}{\Delta v_{IN}} = \frac{v_{OUT}(c) - v_{OUT}(b)}{v_{IN}(c) - v_{IN}(b)} = -1.6 \tag{14.96}$$

$$\overset{\displaystyle \uparrow}{\underset{\displaystyle \text{slope}}{\rule{0pt}{0pt}}}$$

or

$$\begin{aligned} v_{IN}(c) &= \frac{v_{OUT}(c) - v_{OUT}(b)}{-1.6} + v_{IN}(b) \\ &= \frac{2.6\,\text{V} - 3.8\,\text{V}}{-1.6} + 0.4\,\text{V} \approx 1.15\,\text{V} \end{aligned} \tag{14.97}$$

This value of $v_{IN}(c)$ can also be approximately derived by taking KVL around the loop containing v_{IN}, the saturation voltage of Q_1, and the base–emitter junctions of Q_2 and Q_3:

$$\begin{aligned} v_{IN}(c) &= -V_{\text{sat}} + v_{BE2} + v_{BE3} \\ &= -0.2\,\text{V} + 0.7\,\text{V} + 0.6\,\text{V} = 1.1\,\text{V} \end{aligned} \tag{14.98}$$

The results (14.97) and (14.98) are more or less equivalent.

As v_{IN} is increased beyond point c, the output falls rapidly as Q_3 begins to conduct. The output continues to fall until Q_2 and Q_3 saturate at point d, where v_{OUT} reaches the value $V_{\text{sat}} \approx 0.2\,\text{V}$. The value of $v_{IN}(d)$ can be estimated by noting that the base–emitter junctions of both Q_2 and Q_3 are well forward-biased at point d with $V_f \approx 0.75\,\text{V}$, and Q_1 is just leaving saturation. Consequently,

$$v_{\text{IN}}(d) = v_{BE3} + v_{BE2} - V_{\text{sat}}$$
$$= 0.75\,\text{V} + 0.75\,\text{V} - 0.2\,\text{V} = 1.3\,\text{V} \tag{14.99}$$

For values of v_{IN} above point d, Q_1 enters the reverse-active mode and continues to drive the TTL gate in its low-output state with $v_{\text{OUT}} = V_{\text{sat}}$.

From the transfer characteristic of Fig. 14.37, the values of V_{OH}, V_{OL}, V_{IL}, and V_{IH} are readily identified:

$$\begin{aligned} V_{OH} &\approx v_{\text{OUT}}(a) = 3.8\,\text{V} \\ V_{OL} &= V_{\text{sat}} = 0.2\,\text{V} \\ V_{IL} &= v_{\text{IN}}(b) = 0.4\,\text{V} \\ V_{IH} &= v_{\text{IN}}(d) = 1.3\,\text{V} \end{aligned} \tag{14.100}$$

The noise margins that correspond to these voltage parameters become

$$\text{NM}_L = V_{IL} - V_{OL} = 0.2\,\text{V} \tag{14.101}$$

and

$$\text{NM}_H = V_{OH} - V_{IH} = 2.5\,\text{V} \tag{14.102}$$

The value of NM_L is a realistic one for standard, commercially available TTL logic gates. The computed value of NM_H, on the other hand, represents the ideal case. Its value was calculated with no external load connected to the gate output. When the output is loaded by other gates, the transfer characteristic of Fig. 14.37 exhibits a much lower value of V_{OH}. This loading effect tends to decrease the actual value of NM_H. Standard TTL gates typically have a specified maximum fan-out of 10. Under maximum loaded output conditions, the high noise margin NM_H acquires approximately the same value as the low noise margin value given by Eq. (14.101).

14.4.5 TTL Logic Gates

The basic TTL inverter of Fig. 14.34 can be easily modified to produce a multi-input NAND gate. The key feature in such a modification is the multiemitter BJT depicted in Fig. 14.38. Each emitter in this device can independently create a forward-biased base–emitter junction and cause the transistor to enter its active or saturation region. Two or more emitters can act in parallel to form a combined emitter of larger overall surface area.

Figure 14.38
Multiemitter *npn* BJT used for the TTL input device.

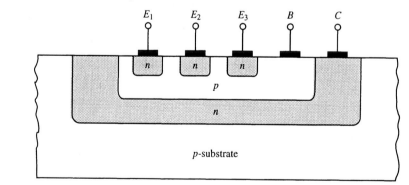

The circuit of Fig. 14.39 illustrates the use of a multiemitter BJT as the input device in a three-input TTL NAND gate. If any one input is low, Q_1 will conduct with a forward-biased base–emitter junction and Q_2 will be forced into cutoff. The resulting output will be high. Only if *all* inputs are high will Q_1 be forced into reverse-active mode, thereby saturating Q_2 and setting the output low.

Figure 14.39
Three-input TTL
NAND gate with a
multiemitter input
transistor:
(a) actual circuit;
(b) logical
equivalent.

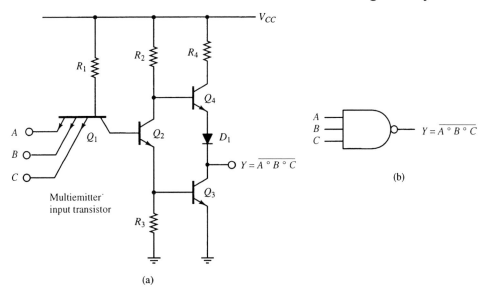

(a)

(b)

The TTL input configuration of Q_1 is not conducive to the direct implementation of the NOR function. NOR logic capability is more easily introduced in the driver stage formed by Q_2 where multiple transistors can be connected in parallel. The circuit of Fig. 14.40 shows one possible implementation of a two-input NOR gate. If both inputs A and B are low, then Q_{2A} and Q_{2B} will both be forced into cutoff and the totem-pole output will be high with $i_{B3} = 0$. If either input is high, then Q_{2A} or Q_{2B} will saturate, forcing i_{B4} to zero and allowing i_{B3} to saturate Q_3. The resulting output will be low.

Figure 14.40
A two-input TTL
NOR gate is
implemented by
combining driver
transistors in
parallel: (a) actual
circuit; (b) logical
equivalent.

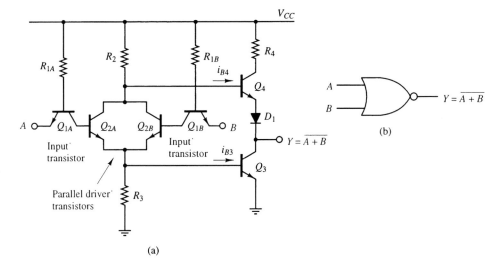

(a)

(b)

Discussion. In designing a digital system from commercially available logic gates, one must often use a gate with more than the required number of inputs. In theory, an unused TTL input left floating should be interpreted as a high input, because no current will flow through the associated base–emitter junction. In reality, a floating input is capable of picking up transient induced voltages by the mechanism of capacitive coupling. Such spurious coupling, fed by noise or other voltage signals within the circuit, is capable of momentarily forcing a floating input to a low logic value, causing unwanted logic signals and erroneous digital data. For this reason, unused inputs in a TTL system must be connected to the

V_{CC} bus. Inputs terminated in this way will be unambiguously interpreted as high outputs. Alternatively, depending on the needs of the designer, unused inputs can be connected to ground to be interpreted as low logic signals. ∎

EXERCISE 14.29 Use the NAND and NOR configurations of Figs. 14.39 and 14.40 to synthesize a single TTL circuit that implements the logic function

$$Y = \overline{A \cdot B + C \cdot D + E \cdot F} \tag{14.103}$$

Note that DeMorgan's law of Boolean algebra can be used to express Y as

$$Y = (\overline{A} + \overline{B}) \cdot (\overline{C} + \overline{D}) \cdot (\overline{E} + \overline{F}) \tag{14.104}$$

14.4.6 Tristate Output

It is often desirable to connect more than one gate output to the same digital signal path. Such a connection cannot be made directly in TTL. A low voltage at the output of any one gate will sink the current from other outputs that are high, causing all output states to be low with current flow limited only by the resistances R_4. One possible method for connecting more than one gate output to the same signal path involves the use of the tristate configuration of Fig. 14.41.

Figure 14.41
Tristate configuration in a two-input TTL NAND gate.

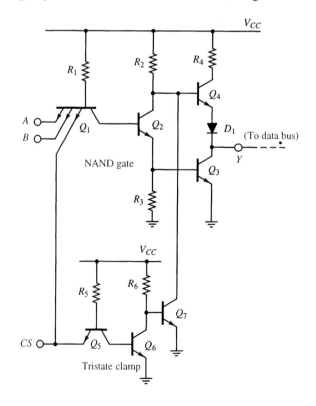

The gate is selected for connection to the data bus via the chip-select (CS) input. When the CS input is high, Q_5 enters the reverse-active mode and saturates Q_6. Consequently, Q_7 enters cutoff and the collector of Q_7 (i.e., the output of the tristate clamp) becomes an open circuit, allowing the NAND gate to function normally. When CS is low, Q_7 conducts and clamps off the input to Q_4. At the same time, the low CS signal fed to Q_1 forces Q_2 and Q_3 into cutoff. With both Q_3 and Q_4 in cutoff, the output Y of the gate appears as an open circuit. This "third state" (neither high nor low) can assume the status of the data bus as set by the output stage of another gate. In operation, the CS signal is made high when the gate output Y is called upon to drive the data bus. When CS is low, the gate is effectively disconnected from the data bus, allowing the bus to be driven by another gate.

14.4.7 Improved Versions of TTL

The standard TTL family performs adequately in many applications, but in modern digital systems, newer versions of TTL with even better properties are the norm. In this section, several modifications that enhance the overall performance of the TTL family are examined.

Schottky TTL

As discussed in Section 14.4.2, the basic TTL configuration reduces the time delay caused by stored base charge in a saturated BJT logic inverter. The dynamic performance of the gate can be improved even more by connecting a Schottky diode across the base–collector junction of all transistors that saturate. This connection is depicted in Fig. 14.42(a). The Schottky diode, introduced in Chapter 3, is formed by a metal-to-semiconductor junction. The resulting diode has a turn-on voltage of about 0.3 V and negligible stored charge under forward-biased conditions. Schottky diodes are small and do not significantly increase the area of an integrated circuit. Because the Schottky diode becomes an integral part of the transistor, the Schottky-clamped BJT is often represented by the composite symbol of Fig. 14.42(b).

Figure 14.42

Addition of Schottky diode across the base–collector junction of a BJT: (a) diode connection; (b) symbolic representation of Schottky-clamped BJT.

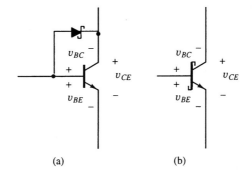

(a)   (b)

During cutoff and active operation, the Schottky diode in Fig. 14.42(a) is reversed-biased and does not affect transistor behavior. When the BJT is driven toward saturation, the collector voltage falls below the voltage of the base. In a regular *npn* BJT under saturation conditions, the base–collector junction becomes forward-biased. The Schottky clamp prevents saturation by shunting current away from the base–collector junction. When the collector voltage falls below that of the base, the Schottky diode becomes forward-biased and clamps the base–collector voltage to about $v_{BC} \approx 0.3$ V. The BJTs used to make Schottky-clamped transistors are small in size and thus have a V_f on the order of 0.75 V. For a v_{BE} in this range, the resulting v_{CE} becomes

$$v_{CE} = v_{BE} - v_{BC} = 0.75\,\text{V} - 0.3\,\text{V} \approx 0.45\,\text{V} \qquad \textbf{(14.105)}$$

This voltage is sufficient to cause the BJT to enter saturation, but only by a small amount. Further increases in v_{BE} serve only to increase the current through the Schottky diode but do not drive the BJT deeper into saturation. The action of the Schottky clamp prevents excess stored charge from building up in the base region of the BJT. Without the need to remove this charge during switching operations, the switching time becomes much faster. Under saturation conditions, a Schottky-clamped BJT is functionally equivalent to a regular saturated transistor.

The Schottky diode can be added to the BJT during device fabrication by extending the metallic contact electrode of the base over to the collector region. This construction is shown in Fig. 14.43. The contact electrode must be made of a suitable metal, usually aluminum or platinum.

Figure 14.43
A metal-semiconductor Schottky diode is formed by extending the metallic base electrode to the collector region of an *npn* BJT.

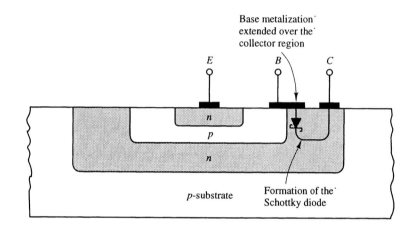

Low-Power Schottky TTL

Although the use of Schottky-clamped transistors greatly improves the speed of TTL, the resulting circuits draw more current from the power supply. The action of the Schottky clamps causes the power consumption of Schottky TTL to be significantly higher than that of standard TTL. One way of alleviating this situation is to increase the value of all resistors in the TTL circuit. In low-power Schottky (LS) TTL, key resistances are made about 5 to 10 times larger. This modification results in a proportional decrease in power consumption. The use of large-value resistors inside the TTL gate increases gate switching times; this drawback is compensated for by adding other design features to the TTL gate.

These additional design features are included in the representative three-input low-power Schottky NAND gate of Fig. 14.44. The modifications to the standard TTL gate are shown in color. Because Q_2 is a Schottky-clamped device, the need to remove stored charge rapidly from its base no longer exists; hence the input transistor of standard TTL is no longer required. In the circuit of Fig. 14.44, the multiemitter input transistor has been replaced by Schottky diodes D_A, D_B, and D_C. With all inputs high, these diodes become reversed-biased and the current i_{R1} is allowed to flow via R_1 into the base of Q_2. If any one input is low, its input diode will become forward-biased and shunt the current i_{R1} away from Q_2. The input diodes D_A to D_C have smaller junction areas than the base–emitter junctions of a multiemitter BJT and thus have smaller junction capacitances. Smaller input capacitances, when functioning as loads to other gates, lead to shorter rise and fall times.

The shunting diodes D_1 to D_3 minimize spurious voltage signals that may appear at the gate inputs. Unwanted signals can be caused by the "ringing" of digital signals along the interconnecting paths between gates. Such behavior is exhibited when the interconnecting paths of high-speed digital circuits behave more like transmission lines than simple wires. The term "ringing" refers to the resonant behavior of voltage signals that interact with the distributed in-

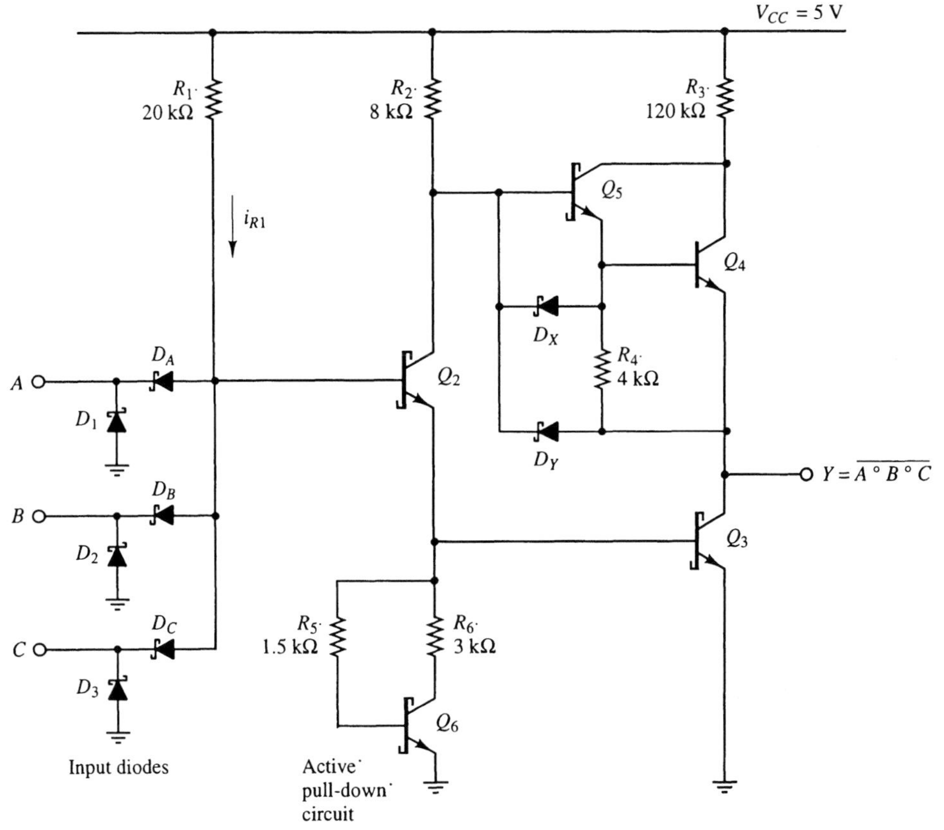

Figure 14.44
Three-input
low-power
Schottky TTL
NAND gate. The
modifications made
to standard TTL are
shown in color.

ductance and capacitance of improperly terminated transmission lines. Such behavior resembles the damped sinusoidal response of an *RLC* circuit.

The low-power Schottky gate is further improved by the addition of Q_6 to the circuit. This device replaces the simple 1-kΩ resistor found in the standard TTL gate. One function of this resistor in the standard gate is the removal of the excess stored charge in the base of Q_3 during a low-to-high output transition. The combination of R_5, R_6, and Q_6 shown in Fig. 14.44 is capable of removing this stored charge much more quickly. The function of Q_6, which acts as an *active pull-down* device, is explored in Problem 14.104.

One other modification acts to improve the dynamic performance of low-power Schottky TTL. The function of the diode between the emitter of Q_4 and collector of Q_3 is replaced by the base–emitter junction of Q_5. When the gate output is low and Q_2 is saturated, v_{BE5} absorbs some of the voltage drop between the collector of Q_2 and the output terminal. As a result, the base–emitter junction of Q_4 is prevented from becoming forward-biased. The combination of Q_5 and Q_4 also increases the current-sourcing capability of the gate. Diodes D_X and D_Y aid in speeding up the turn-off of Q_4 and the turn-on of Q_3 during a high-to-low output transition. Note that Q_4 need not be a Schottky-clamped BJT because it is never forced into saturation.

14.5 EMITTER-COUPLED LOGIC FAMILY

The speed of digital BJT circuits based on the TTL configuration is limited by the need to remove excess stored charge from the base regions of saturated devices. As discussed in Section 14.4.7, the problem can be minimized by adding Schottky diodes to the circuit; however, the need to charge and discharge transistor capacitances still represents a major limitation to gate switching speeds. In the emitter-coupled logic (ECL) family, the stored-charge phenomenon is minimized by operating all BJTs out of saturation. Logical operations are instead implemented by using the differential-amplifier topology of Chapter 8. Additional speed improvements are realized by restricting signal swings to small values. In this way, base charge stored during active-region operation remains small. The ECL logic family is characterized by very fast switching times and short propagation delays. This high speed comes at the expense of increased power consumption, which must be expended to keep the differential pair biased in the active region. The power-supply current to an ECL gate is roughly constant.

14.5.1 Basic ECL Logic Inverter

The topology of a simplified and ideal ECL logic inverter is shown in Fig. 14.45. The input signal is applied to the base of Q_1, which is connected in a differential pair configuration with Q_2. The devices split the bias current I_o. The base of Q_2 is fixed at the reference voltage V_R, which is set to a negative value. The output of the amplifier is tapped in single-ended fashion at v_{C1} and fed to the follower Q_3. The latter provides an output buffer and increases the fan-out capabilities of the gate.

Figure 14.45
Basic structure of an ideal ECL logic inverter with a single output taken from v_{C1}.

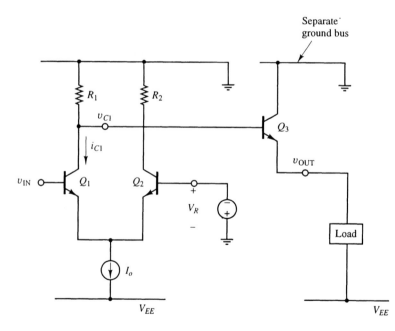

Note that the upper power-supply bus of the circuit is connected to ground (i.e., V_{CC} is equal to zero). For convenience, emitter-coupled logic circuits are designed to operate from a single V_{EE} power-supply voltage. In theory, the ECL family also could be designed to function from bipolar V_{CC} and V_{EE} power supplies. For reasons to be explained later, the output follower Q_3 is connected to its own separate, isolated ground bus.

When an input voltage is applied to the base of Q_1, the amplifier responds to the differential-mode component $v_{IDM} = v_{IN} - V_R$. The logic levels $V(1)$ and $V(0)$ of v_{IN} are assigned to the voltage values that cause Q_1 and Q_2 to arrive at their cutoff limits. Specifically, the ECL gate is designed so that applying an input voltage of value

$$v_{IN} = V(0) = V_R - \Delta V \qquad (14.106)$$

creates a v_{IDM} of value $-\Delta V$ that causes Q_1 to go into cutoff. Similarly, applying an input voltage of value

$$v_{IN} = V(1) = V_R + \Delta V \tag{14.107}$$

creates a v_{IDM} of value $+\Delta V$ that causes Q_2 to go into cutoff. Neither $V(0)$ nor $V(1)$ is large enough to saturate either device in the differential pair.

The minimum value of ΔV required to drive Q_1 or Q_2 to the edge of cutoff is determined by I_o and by the values of R_1 and R_2. In a well-designed ECL gate, these parameters are chosen so that the output levels $V(0)$ and $V(1)$ measured at v_{OUT} result in a ΔV larger than this minimum value. This condition permits the successful operation of a digital system in which gate outputs feed the inputs to other gates. In the ideal ECL gate, the current I_o and the resistor R_1 are chosen so that v_{C1} is biased at $(V_R + V_f)$ when $v_{IN} = V_R$ (i.e., when $v_{IDM} = 0$). The resulting value of v_{OUT}, obtained by subtracting the base–emitter drop of Q_3 from v_{C1}, is equal to V_R.

Note that the voltage at the collector of Q_2 can be buffered by its own follower to reproduce a second gate output. Given the differential symmetry of the circuit, the output derived from v_{C2} becomes the logical complement of v_{OUT}. The availability of complementary logic outputs greatly simplifies the design and implementation of ECL digital systems.

14.5.2 Detailed Analysis of ECL Logic Inverter

In this section, the characteristics of the standard ECL logic inverter of Fig. 14.46 are examined. The component values shown in this circuit are those of a specific, commercially available ECL logic gate that is precision engineered for operation in a multielement digital system.

This section provides only an approximate analysis of circuit operation. More detailed analyses can be performed using computer-based design aids, such as the SPICE program described in Appendix C.

Note that the BJTs in a standard ECL gate are fabricated with small dimensions to minimize internal capacitances. As a result, current densities in the device are higher than those in comparable BJT circuits. For this reason, the value $V_f \approx 0.75\,\text{V}$ is normally assumed for all forward-biased base–emitter and pn junctions in ECL circuits.

Reference Voltage V_R

In the circuit of Fig. 14.46, the reference voltage $V_R = -1.32\,\text{V}$ is generated by Q_R and its surrounding components. The current through the bias leg of R_3, R_5, D_1, and D_2 can be calculated approximately by neglecting the base current into Q_R:

$$I_5 \approx \frac{-V_{EE} - 2V_f}{R_3 + R_5} = \frac{5.2\,\text{V} - 2(0.75\,\text{V})}{907\,\Omega + 4.98\,\text{k}\Omega} \approx 0.63\,\text{mA} \tag{14.108}$$

This bias current establishes a voltage V_{BB} at the base of Q_R, which is down-shifted by the base–emitter voltage of Q_R and applied to Q_2 as V_R. This value of V_R is given by

$$V_R = V_{BB} - V_f = -I_5 R_3 - V_f = -(0.63\,\text{mA})(907\,\Omega) - 0.75\,\text{V} = -1.32\,\text{V} \tag{14.109}$$

This value of V_R is standard for the reference voltage in ECL systems. The circuit surrounding Q_R is designed to compensate for changes in temperature. This feature of the circuit is explored in Problems 14.107 and 14.108.

Differential-Pair Bias

The bias current I_o to the differential pair is established by R_E. An approximate expression for I_o can be obtained by taking KVL around the base loop of Q_2:

$$I_o = \frac{V_R - V_f - V_{EE}}{R_E} = \frac{-1.32\,\text{V} - 0.75\,\text{V} - (-5.2\,\text{V})}{779\,\Omega} \approx 4\,\text{mA} \tag{14.110}$$

Note that the current source formed by the R_E network is not ideal; rather, it has a finite parallel Norton resistance equal to R_E.

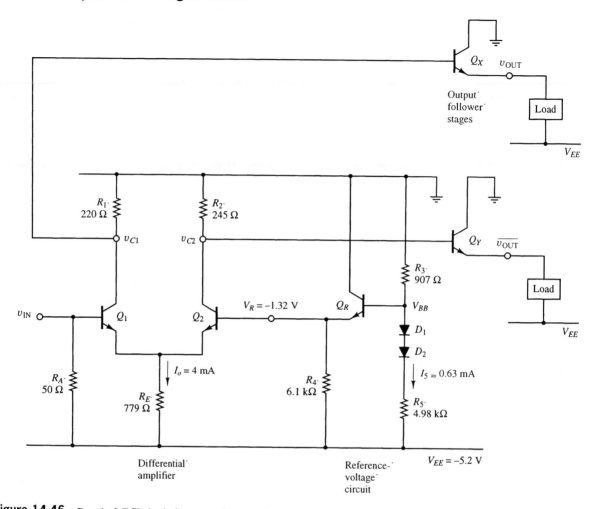

Figure 14.46 Practical ECL logic inverter with complementary outputs v_{OUT} and $\overline{v_{OUT}}$.

Operation with Input Low

When v_{IN} is low ($v_{IN} = V_R - \Delta V$), Q_1 is forced into cutoff and the voltage v_{C1} is changed from its bias value of $-I_o R_1/2$ to a voltage of zero (no current drawn through R_1). This change in v_{C1} is also transmitted to v_{OUT}, where the gate output v_{OUT} is equal to v_{C1} minus the V_f drop of the base–emitter junction of Q_X. The ΔV by which v_{C1} rises when it reaches cutoff is equal to

$$\Delta V = \frac{I_o R_1}{2} = (2\,\text{mA})(220\,\Omega) = 0.44\,\text{V} \qquad \textbf{(14.111)}$$

This voltage increment defines the ΔV required for ECL gate operation.

Operation with Input High

When v_{IN} is high ($v_{IN} = V_R + \Delta V$), the current through Q_2 is forced to zero and Q_2 is driven into cutoff. With Q_2 in cutoff, all of I_o flows through Q_1, so that the current through Q_2 increases by the net amount $I_o/2$. This increase causes v_{C1} to fall by an increment $I_o R_1/2$, which is again equal in magnitude to $\Delta V = 0.44\,\text{V}$.

Logic High and Logic Low Voltage Values

With $\Delta V = 0.44$ V and with $V_R = -1.32$ V representing the mean input, or "bias" level, of v_{IN}, the $V(1)$ logic level given by Eq. (14.107) becomes

$$V(1) = -1.32\,\text{V} + 0.44\,\text{V} = -0.88\,\text{V} \qquad \textbf{(14.112)}$$

Similarly, the $V(0)$ logic level given by Eq. (14.106) becomes

$$V(0) = -1.32\,\text{V} - 0.44\,\text{V} = -1.76\,\text{V} \qquad \textbf{(14.113)}$$

Discussion. In the ECL circuit of Fig. 14.46, the collectors of Q_X and Q_Y are connected to their own separate ground bus, which leads directly to the grounded side of V_{EE}. This separate ground connection is made intentionally to isolate the output follower stages from the rest of the circuit. A ground bus often contributes series resistance or distributed inductance between its various connection points. When Q_X and Q_Y drive low-impedance loads, large currents are drawn from their ground bus. Significant voltage drops can occur across the impedances of this ground bus, causing a change in the voltage of the bus, which ideally should be held at a voltage of zero. This problem is particularly noticeable in high-speed circuits, where fast current rise times cause inductive effects to dominate. The voltage levels in the differential amplifier of the ECL gate must be maintained at precise levels for proper circuit operation. To help meet this requirement, the output stages are provided with a separate ground bus through which large load currents can flow. The separate ground buses of the differential amplifier and driver stages are connected to a common ground at the source of V_{EE}, as illustrated in Fig. 14.47.

Figure 14.47
Connection of ground buses in an ECL circuit.

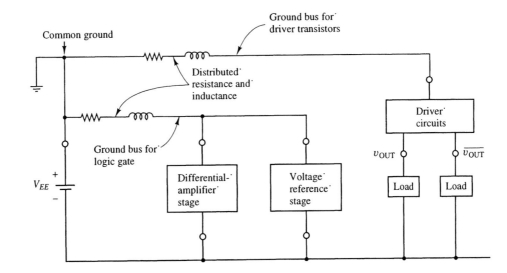

The low-output-impedance followers Q_X and Q_Y that drive v_{OUT} and $\overline{v_{\text{OUT}}}$ serve a second important function. Emitter-coupled logic circuits are often so fast that the signal paths between the various logic gates must be treated as transmission lines rather than as simple wire connections. This scenario is depicted in Fig. 14.48. In carefully engineered ECL systems, interconnection paths are designed to have a transmission-line impedance of $50\,\Omega$. The output devices Q_X and Q_Y are capable of driving such low-impedance loads.

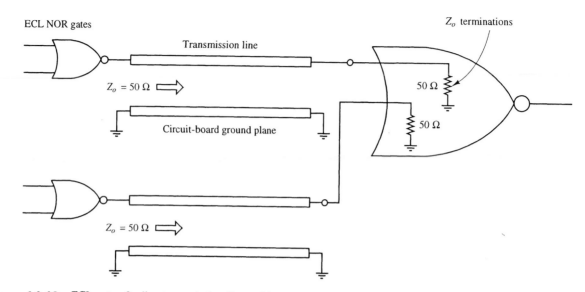

Figure 14.48 ECL gates feeding transmission lines of impedance $Z_o = 50\,\Omega$.

Note that the input terminal of the ECL inverter of Fig. 14.46 is shunted by a 50-Ω resistor R_A. This resistor serves two functions. First, it provides a matching termination when a 50-Ω transmission line feeds the v_{IN} input. Second, R_A connects the base of Q_1 to V_{EE}. If v_{IN} is not connected to another gate, the input will be pulled down to a logic low state, thus preventing indeterminate logic levels and noise due to capacitive coupling from affecting the gate output. In most ECL systems, unused gate input terminals may be left unconnected with no adverse effect on circuit operation. ∎

14.5.3 ECL Inverter Transfer Characteristics

The ECL logic inverter of Fig. 14.46 has two transfer characteristics: one for the v_{OUT} output and one for the $\overline{v_{OUT}}$ output. These transfer characteristics, shown in Fig. 14.49, are more or less the mirror images of each other and resemble the large-signal differential-amplifier characteristics derived in Chapter 8. As can be seen in the figure, the transfer characteristic measured at v_{OUT} exhibits an asymmetrical dip for large v_{IN} above the value $v_{IN} = V_{IH}$. This feature arises because the left half of the differential amplifier in Fig. 14.46 is driven by v_{IN} while the right half is driven by the fixed reference voltage V_R. This asymmetry creates a common-mode input component that is responsible for the difference in the v_{OUT} and $\overline{v_{OUT}}$ transfer characteristics. Note that resistors R_1 and R_2 in Fig. 14.46 have slightly different values. The value of R_1 is lower than that of R_2 to compensate for this common-mode input component.

Transfer Characteristic at v_{OUT}

The v_{OUT} transfer characteristic shown in Fig. 14.49 can be verified by examining the large-signal characteristics of the differential amplifier. These large-signal characteristics were derived in Section 8.5.1. For a differential-mode input v_{IDM}, the current i_{C1}, given by Eq. (8.206) becomes

$$i_{C1} \approx i_{E1} = \frac{I_o}{1 + e^{-v_{IDM}/\eta V_T}} \tag{14.114}$$

In ECL circuits, the parameter V_{IH} is defined at the point where i_{C1} reaches 99% of the available bias current I_o. This definition is more convenient than the definition of $dv_{OUT}/dv_{IN} = -1$

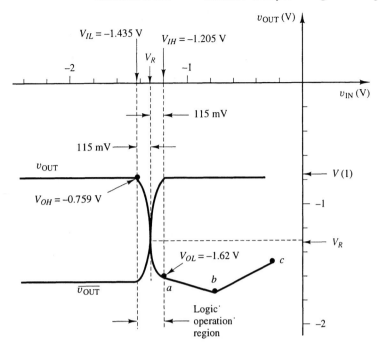

Figure 14.49
Transfer characteristics of an ECL logic inverter measured at the v_{OUT} and $\overline{v_{OUT}}$ outputs.

and can be shown to produce almost identical results. Solving Eq. (14.114) at room temperature for the case $i_{C1} = 0.99I_o$ (positive v_{IDM}) results in

$$v_{IDM}^+\big|_{v_{IN}=V_{IH}} = -\eta V_T \ln\left(\frac{I_o}{i_{C1}} - 1\right) = -(1)(0.025\,\text{V})\ln\left(\frac{1}{0.99} - 1\right) = 0.115\,\text{V} \quad \textbf{(14.115)}$$

Similarly, V_{IL} is defined at the point where i_{C1} reaches 1% of the available bias current I_o. Solving Eq. (14.114) for the case $i_{C1} = 0.01I_o$ results in

$$v_{IDM}^-\big|_{v_{IN}=V_{IL}} = -\eta V_T \ln\left(\frac{1}{0.01} - 1\right) = -0.115\,\text{V} \quad \textbf{(14.116)}$$

The input values (14.115) and (14.116) define the voltages V_{IH} and V_{IL}, that is,

$$V_{IH} = V_R + v_{IDM}^+ = -1.32\,\text{V} + 0.115\,\text{V} = -1.205\,\text{V} \quad \textbf{(14.117)}$$

and

$$V_{IL} = V_R + v_{IDM}^- = -1.32\,\text{V} - 0.115\,\text{V} = -1.435\,\text{V} \quad \textbf{(14.118)}$$

The values of V_{OL} and V_{OH} can be found by computing v_{C1} at $i_{C1} = 0.99I_o$ and at $i_{C1} = 0.01I_o$. Because they involve the same magnitude of v_{IDM}, these output levels will self-consistently reproduce each other. When $i_{C1} = 0.99I_o$, v_{OUT} becomes

$$V_{OL} = v_{OUT}\big|_{v_{IN}=V_{IH}} = -0.99I_o R_1 - V_f$$
$$= -0.99(4\,\text{mA})(220\,\Omega) - 0.75\,\text{V} = -1.62\,\text{V} \quad \textbf{(14.119)}$$

Similarly,

$$V_{OH} = v_{OUT}\big|_{v_{IN}=V_{IL}} = -0.01I_o R_1 - V_f$$
$$= -0.01(4\,\text{mA})(220\,\Omega) - 0.75\,\text{V} = -0.759\,\text{V} \quad \textbf{(14.120)}$$

The resulting noise margins of the gate as measured at the v_{OUT} output become

$$NM_H = V_{OH} - V_{IH} = -0.759\,V - (-1.205\,V) = 0.446\,V \qquad (14.121)$$

and

$$NM_L = V_{IL} - V_{OL} = -1.435\,V - (-1.62\,V) = 0.185\,V \qquad (14.122)$$

The dip and subsequent rise in the v_{OUT} transfer characteristic occurs after Q_2 reaches cutoff. For further increases in v_{IN} beyond this point, Q_1 appears as a linear amplifier with R_1 as its collector resistor and R_E as its emitter resistor. The incremental output of the circuit is given approximately by the linear amplification factor $dv_{OUT}/dv_{IN} = -R_1/R_E$. The voltage v_{C1} becomes a linearly amplified version of v_{IN} until Q_1 saturates. This behavior accounts for the portion of the transfer characteristic between points a and b in Fig. 14.49. As v_{IN} is increased still further, Q_1 goes into saturation and its voltage v_{CE1} becomes constant. The voltage v_{C1} rises with v_{IN}, thus accounting for the portion of the v_{OUT} transfer characteristic between points b and c. Note that in working ECL systems, a gate is never operated beyond point a, so that Q_1 is never actually driven into saturation.

Transfer Characteristics at $\overline{v_{OUT}}$

The transfer characteristic measured at the $\overline{v_{OUT}}$ terminal is the logical inverse of the v_{OUT} characteristic. Unlike the v_{OUT} curve, however, the $\overline{v_{OUT}}$ curve displays no asymmetrical dip for large v_{IN}. When Q_2 is driven into cutoff (v_{IN} high), the voltage v_{C2} remains fixed at zero, and $\overline{v_{OUT}}$ at $-V_f$, for further increases in v_{IN}. Similarly, when Q_1 is driven into cutoff (v_{IN} low), further decreases in v_{IN} do not lead to changes in current through either Q_1 or Q_2. Thus, the voltage v_{C2} remains fixed at the value $-I_o/R_2 = -0.98\,V$ for low v_{IN}, and $\overline{v_{OUT}}$ remains fixed at $v_{C2} - V_f = -1.73\,V$.

14.5.4 ECL Logic Gates

The ECL inverter configuration is easily modified to produce NOR and OR logic gates. The NOR function is implemented by connecting input transistors in parallel, as shown in Fig. 14.50. This figure depicts a two-input ECL gate. A gate with more inputs can be created by adding more transistors in parallel. When either input is high, the differential pair swings to the extreme in which Q_2 is cut off and most of the bias current I_o flows through R_1. Under these conditions, v_{OUT} will attain its logic low state. If *both* inputs are high, the current I_o will be shared by the input devices, and the output will still be low. Conversely, if both inputs are low, the current through R_1 will be zero and v_{OUT} will be driven to its logic high state. The output v_{OUT} thus functions as the logical NOR of the inputs A and B.

Since the $\overline{v_{OUT}}$ output is the logical inverse of the v_{OUT} output, it functions as the logical OR of inputs A and B. The ECL configuration produces NOR and OR outputs simultaneously without the need for additional gate circuitry. This feature is advantageous in the design of high-speed digital systems. The availability of complementary outputs from each gate reduces the total number of gates required to implement a large digital system. With fewer gates in the system, the overall speed of operation is increased.

14.6 BICMOS LOGIC CIRCUITS

The bipolar-CMOS, or BiCMOS, configuration, in which BJTs are combined with n-channel and p-channel MOSFETs on the same IC, was introduced in the context of analog differential amplifiers in Chapter 8. In this section, we examine the BiCMOS digital logic family. BiCMOS combines the high speed of bipolar circuits with the infinite input resistance and small static

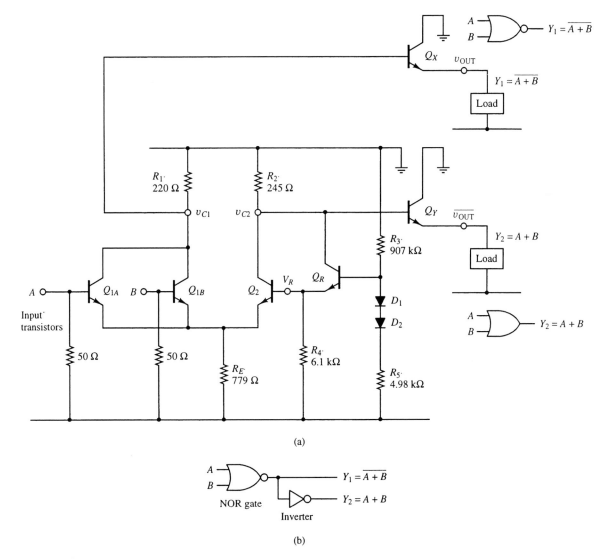

(a)

(b)

Figure 14.50 Two-input ECL NOR gate. The output at v_{OUT} functions as the NOR output, and the output at $\overline{v_{OUT}}$ as the OR output. (a) circuit diagram; (b) alternative logic representation.

power dissipation of CMOS circuits, thus drawing upon the advantages of each. As illustrated in Fig. 14.51, where various logic families are compared for gate delay versus power dissipation, BiCMOS circuits fill an important performance gap. CMOS circuits are relatively slow, but consume minimal power. ECL circuits are extremely fast, but require large amounts of power. TTL and NMOS circuits have speeds somewhat faster than CMOS but significantly larger power dissipation. The BiCMOS family falls somewhere between CMOS and ECL with respect to performance and has several added advantages over either of these logic families. BiCMOS circuits can directly drive most other logic families and are much less sensitive to both capacitive and resistive loads, thus providing very flexible input–output capabilities. Process technologies now routinely yield BiCMOS circuits with submicron device size, leading to larger packing densities. Compared to CMOS, NMOS, and ECL, BiCMOS circuits are relatively insensitive to fabrication process variations. Their principal disadvantage is the increased cost of fabrication. As BiCMOS technology has matured, many of the fabrication steps needed to make BJTs and

MOSFETs have merged, but not all process steps are shared. Hence the overall fabrication process for a BiCMOS circuit involves considerably more steps than a comparable CMOS or bipolar circuit. Despite its greater fabrication cost, many digital designers feel that BiCMOS will ultimately become the dominant digital technology. Regardless of its future, BiCMOS is now an important logic family that is worthy of detailed study.

Figure 14.51
Plot of gate delay versus power dissipation for various logic families.

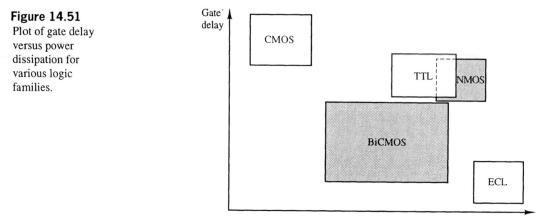

14.6.1 Standard BiCMOS Inverter

The most common BiCMOS inverter topology, sometimes called the BiCMOS "totem-pole" gate, is illustrated in Fig. 14.52. This configuration yields relative low static power dissipation and is easily integrated onto an IC. For clarity, MOSFETs have been labeled with the letter M and BJTs with the letter Q. The inverter of Fig. 14.52 is an extension of earlier BiCMOS designs in which the shunting MOSFETs M_1 and M_2 were absent or replaced by resistors. The function of the CMOS input stage consisting of M_P, M_N, M_1, and M_2 is very similar to that of a standard CMOS logic inverter. The input signal drives M_P and M_N simultaneously, so that M_P is in cutoff and M_N conducts when v_{IN} is high, and M_N is in cutoff and M_P conducts when v_{IN} is low. The BJTs of the totem-pole output stage provide large output-current capability. When the inverter drives another BiCMOS gate, Q_1 provides the current necessary to rapidly charge the load to $V(1)$. Similarly, Q_2 provides the current to quickly discharge the load capacitance to $V(0)$.

Figure 14.52
Basic BiCMOS logic inverter with "totem-pole" bipolar output stage.

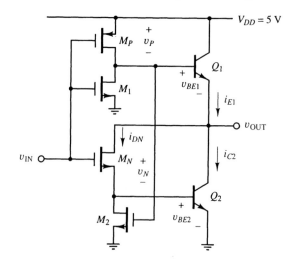

The speed limitations caused by stored base charge in Q_1 and Q_2 are reduced by MOSFETs M_1 and M_2. M_1 turns on when v_{IN} is high and quickly removes the stored charge from the base of Q_1 as v_{OUT} goes low. In this role, M_1 performs the same function as the input driver transistor of a TTL gate. Similarly, MOSFET M_2, which is driven by M_P, turns on when v_{IN} is low and removes the stored charge from the base of Q_2 as v_{OUT} goes high.

14.6.2 DC Transfer Characteristic of the BiCMOS Inverter

The transfer characteristic of the BiCMOS inverter is complicated and difficult to derive in analytical form, but the principles of its operation are easily understood. The general form of the transfer characteristic is shown in Fig. 14.53. This plot has several regions of operation, labeled 1 through 6, which we now describe qualitatively. For the sake of simplicity, we assume that $V_{TRN} = |V_{TRP}| = V_{TR}$ and $K_N = |K_P|$. We also assume the circuit to drive only other BiCMOS gates, so that the load is purely capacitive.

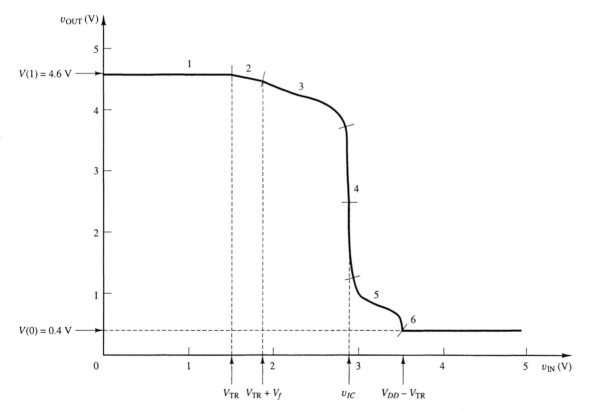

Figure 14.53 Transfer characteristic of BiCMOS inverter of Fig. 14.52 with $|K| = 0.5\,\text{mA/V}^2$, $|V_{TR}| = 1.5\,\text{V}$, $\lambda = 0.005\,\text{V}^{-1}$, and $\beta_F = 100$.

Operation for $v_{IN} < V_{TR}$ (Region 1)

When v_{IN} is low and below V_{TR}, M_N remains in cutoff, and M_P is driven hard into conduction in the triode region with its drain-source voltage, represented by v_P in Fig. 14.52, close to zero. In this respect, M_P functions in the same manner as the p-channel transistor of a CMOS inverter. The current through M_P flows via the base–emitter junction of Q_1, through M_N, and through M_2 to ground. The latter MOSFET is driven into conduction by its connection to V_{DD} via M_P. With M_N in cutoff, it would seem that the current i_{DN} should be zero. In actuality, M_N operates in a mode called the *subthreshold* regime in which an extremely small current, exponentially related to v_{GS}, flows through the device. The subthreshold current that flows in cutoff is typically in the tens of picoamperes range and is so small that it usually can be ignored. In this case, however, subthreshold current is sufficient to create a forward drop across the base–emitter junction of Q_1. The resulting value of v_{BE1} can be estimated by assuming a typical subthreshold current of 10 pA for M_N and the parameter $I_{EO} = 10^{-15}$ mA for Q_1, so that

$$v_{BE1} = \eta V_T \ln \frac{i_{E1}}{I_{EO}} = (1)(0.025\,\text{V}) \ln \frac{10\,\text{pA}}{10^{-15}\,\text{mA}} \approx 0.4\,\text{V} \tag{14.123}$$

where $i_{E1} = i_{DN}$. Note that Q_2 remains in cutoff with insignificant collector current i_{C2} when v_{IN} is below V_{TR}.

With $v_{BE1} = 0.4$ V and $v_P \approx 0$, the inverter output becomes

$$v_{\text{OUT}} = V_{DD} - v_P - v_{BE1} = 5\,\text{V} - 0 - 0.4\,\text{V} = 4.6\,\text{V} \tag{14.124}$$

The voltage value 4.6 V constitutes $V(1)$ for the BiCMOS inverter. Over region 1 of the transfer characteristic, the principal current through the circuit consists of the subthreshold current that flows through M_N via M_P. Because this current is very small, the static power drawn from V_{DD} is also small.

Operation for $V_{\text{TR}} < v_{\text{IN}} < V_{\text{TR}} + V_f$ (Region 2)

When v_{IN} exceeds V_{TR}, MOSFET M_N is turned on and operates in the constant-current region with small i_D. Until v_{IN} reaches $V_{\text{TR}} + V_f$, however, v_{IN} will be too small to strongly forward bias Q_2, which remains in cutoff. The increase in current though M_N, however, causes v_{BE1} to increase above the value (14.123) produced by subthreshold current flow. As a result, the output voltage begins to drop slightly.

Operation for $V_{\text{TR}} + V_f < v_{\text{IN}} < v_{IC}$ (Region 3)

When v_{IN} exceeds the value $V_{\text{TR}} + V_f$, Q_2 is driven into conduction with its base current provided by M_N. The current required by Q_2 is pulled down through Q_1, which has its base current provided by the still-conducting M_P. The increased current flow through M_P causes v_P to increase, further reducing v_{OUT}, which is equal to $V_{DD} - v_P - v_{BE1}$. Although M_N is brought into the constant-current region by the increasing v_{IN}, M_P continues to operate in the triode region until the output voltage is sufficiently reduced. Hence, although the output voltage decreases over this portion of the transfer characteristic, the point of maximum slope is not reached until M_P is brought into the constant-current region.

Operation for $v_{\text{IN}} \approx v_{IC}$ (Region 4)

As v_{IN} further increases, approaching the "crossover" voltage v_{IC}, M_P is drawn into the constant-current region while M_N also remains in the constant-current region. Over this portion of the transfer characteristic, the value of v_{OUT} decreases rapidly as the operating points of M_P and M_N slide across their intersecting constant-current region v–i characteristics. The two devices, connected in series via the relatively constant voltage drop of the base–emitter junction of Q_1, function in the same manner as the devices in a conventional CMOS inverter. Over region 4, the constant-current-region operating points of M_N and M_P share the same i_D, hence v_N across M_N decreases rapidly as v_P increases, causing M_N to approach the triode region. This scenario can be understood by reexamining Fig. 14.9 from Section 14.2.1. The slope of the transfer characteristic over region 4 is determined by the r_o values of M_P and M_N. As v_{IN} passes through v_{IC}, M_N moves from the constant-current region toward the triode region, and M_P is moved farther into the constant-current region. Note that Q_1 continues to act as a voltage follower with output equal to $V_{DD} - v_P - v_{BE1}$, where $v_{BE1} \approx V_f$ is a relatively constant voltage.

Operation for $v_{IC} < v_{IN} < V_{DD} - V_{TR}$ (Region 5)

As v_{IN} is increased still further, the gate-to-source voltage of M_P is decreased in magnitude, so that its current begins to fall. Note that v_P remains sizable even under reduced current flow. The decreased current through M_P results in a smaller current through M_N. As M_N is driven deeper into the triode region, its drain–source voltage is reduced, causing v_{OUT} to be reduced as well. When v_{IN} reaches $V_{DD} - V_{TR}$, M_P is driven into cutoff with current flow limited to the subthreshold regime.

Operation for $V_{DD} - V_{TR} < v_{IN} < V_{DD}$ (Region 6)

Over this region, M_P operates in the subthreshold regime, providing a very small current that flows via the base–emitter junction of Q_1, through M_N, and through the base–emitter junction of Q_2. The voltage drop across M_N, which remains in the triode region, approaches zero, so that v_{OUT} becomes approximately equal to v_{BE2}. With only the subthreshold current of M_P flowing through M_N, v_{BE2} attains the value given by Eq. (14.123). The logic-low value of v_{OUT} therefore is given by $V(0) = 0.4\,\text{V}$. Note that the static current drawn by the gate is again very small when $v_{OUT} = V(0)$.

The Roles of M_1 and M_2

MOSFETs M_1 and M_2 do not significantly affect the dc transfer characteristic of the BiCMOS inverter but are important to its dynamic performance. When v_{IN} is low, M_1 operates in cutoff while M_2 is driven into conduction via V_{DD} and M_P. This action effectively grounds the base of Q_2. When v_{IN} is high, M_1 is driven into conduction, thereby grounding the base of Q_1, while M_P and M_2 are forced into cutoff, thereby releasing the base of Q_2.

Now consider the BiCMOS inverter during an output transition. When v_{IN} undergoes a low-to-high transition, forcing v_{OUT} to switch from $V(1)$ to $V(0)$, Q_1 must make a transition from a conducting to a nonconducting state. As v_{IN} becomes high, it turns on M_1, which rapidly removes the base charge stored in Q_1, thereby accelerating the transition of v_{OUT} to $V(0)$.

When v_{IN} undergoes a high-to-low transition, forcing v_{OUT} to switch from $V(0)$ to $V(1)$, Q_2 must become nonconducting. As v_{IN} becomes low, it turns off M_1 and turns on M_P, thereby activating M_2. The latter quickly removes the charge stored in the base of Q_2 and accelerates the transition of v_{OUT} to $V(1)$.

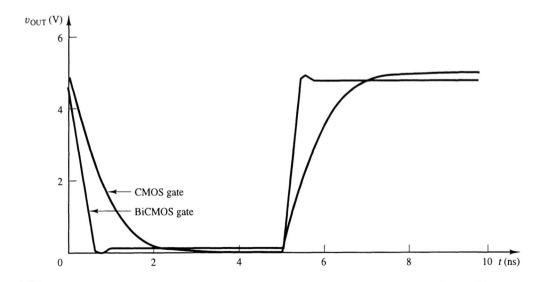

Figure 14.54 Comparison of dynamic performance of BiCMOS and CMOS inverters with 2-pF capacitive load.

As the preceding description shows, the *discharge transistors* M_1 and M_2 enable the BiC-MOS inverter to attain its fast switching speeds. As an illustration of the relative speed of the BiCMOS inverter, consider the SPICE-generated plot of Fig. 14.54, which compares the output of the gate of Fig. 14.52 to that of a similar CMOS gate without the BJT totem-pole output stage or discharge transistors. In both cases, the gate drives a 2-pF load capacitance. As this plot shows, the BiCMOS gate is considerably faster.

14.6.3 BiCMOS Logic Gates

The NAND and NOR functions are easily implemented in BiCMOS using the same techniques employed in the CMOS family. If a NOR function is desired, logic inputs are used to drive n-channel MOSFETs in parallel and p-channel MOSFETs in series. This concept is illustrated in Fig. 14.55, where a two-input NOR gate is shown. When both A and B are low, both M_{PA} and M_{PB} are turned on, and both M_{NA} and M_{NB} are turned off. The output is thus driven to its $V(1)$ logic state. At the same time, M_2 is driven by V_{DD} and the series combination of M_{PA} and M_{PB} and provides the discharge path for the base of Q_2.

Figure 14.55
Two-input
BiCMOS NOR
gate.

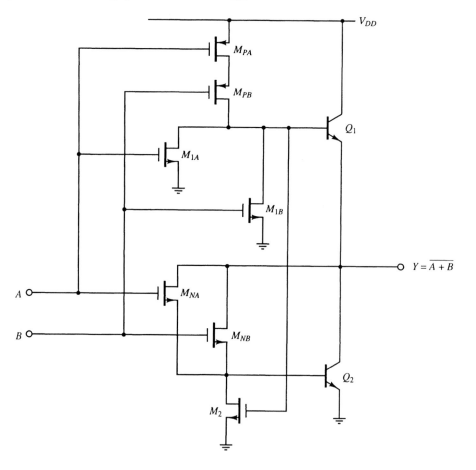

When either A or B is high, the path via M_{PA} and M_{PB} is broken, and either M_{NA} or M_{NB} drives v_{OUT} to its $V(0)$ logic state. At the same time, M_{1A} or M_{1B} provides a means to discharge the base of Q_1. This description shows the gate to implement the logical NOR function in which the output is forced low when either input is high.

If a NAND function is desired, the reverse procedure is followed. Logic inputs are used to drive n-channel MOSFETs in series and p-channel MOSFETs in parallel. This arrangement is illustrated in Fig. 14.56, where a two-input NAND gate is shown. When both A and B are high, the path via M_{NA} and M_{NB} is completed, and both M_{PA} and M_{PB} are driven into cutoff. This action forces v_{OUT} to its $V(0)$ logic state. At the same time, the series combination of M_{1A} and M_{1B} provides a means to discharge the base of Q_1.

Figure 14.56
Two-input
BiCMOS NAND
gate.

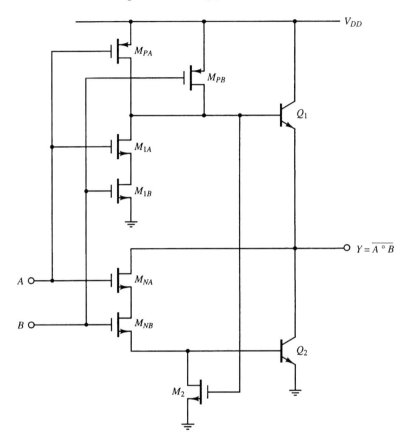

When either A or B is low, M_{PA} or M_{PB} is turned on, the series connection of M_{NA} and M_{NB} is broken, and the output is driven to its $V(1)$ logic state. At the same time, M_2 is driven via M_{PA} or M_{PB} and provides the discharge path for the base of Q_2. This description shows the gate to implement the logical NAND function, in which the output is forced low only when both inputs are high.

Note the n-channel MOSFET M_2, which discharges the base of Q_2 during a high-to-low output transition, need not be replicated in the NAND and NOR gates. This device is automatically driven to the correct state by the M_P transistors in both gate configurations.

EXERCISE 14.30 Draw the circuit diagram of a three-input BiCMOS NOR gate using the topology of Fig. 14.55.

14.31 Draw the circuit diagram of a three-input BiCMOS NAND gate using the topology of Fig. 14.56.

14.32 Implement the Boolean expression $Y = (A + B) \cdot C$ using an appropriate BiCMOS circuit.

SUMMARY

- A digital system can perform logic operations in Boolean algebra.

- A digital system manipulates digital signals using AND, OR, NAND, NOR, NOT, and XOR gates. A digital signal can have the value **1** or **0**.

- A digital circuit falls into one of several logic families, including NMOS, PMOS, CMOS, TTL, ECL, and BiCMOS.

- A digital circuit can be fabricated with an MSI, LSI, VLSI, or ULSI size scale.

- The basic digital building block of any logic family is the digital inverter.

- The three-terminal devices in a digital inverter operate at the limits of their swing ranges most of the time. These limits are associated with the digital values **0** and **1**.

- The analog region of device operation in a digital inverter is traversed only during a transition between high and low logic values.

- A digital inverter is characterized by its noise margin, fan-in, fan-out, propagation delay, power dissipation, and delay–power product.

- A CMOS inverter is made from complementary n-channel and p-channel MOSFETs.

- Series transmission gates are easily made in the CMOS logic family.

- An NMOS inverter is made from an n-channel enhancement-mode MOSFET input device and an enhancement-mode or depletion-mode pull-up load.

- The nonzero source-to-substrate voltage of an NMOS pull-up load affects the transfer characteristics of an NMOS inverter.

- The npn BJT inverter forms the basis of the TTL logic family.

- In a TTL inverter, the input transistor operates in the cutoff, saturation, active, and reverse-active modes, depending on the status of the inverter input and output.

- The TTL logic family can be improved by the use of Schottky devices and low-power design principles.

- The emitter-coupled logic family is based on the differential-amplifier configuration.

- Emitter-coupled logic is characterized by very high switching speeds.

- The devices in an ECL inverter are never driven into saturation. Active region operation is responsible for the superior speed of the ECL family.

- The ECL logic family consumes more static power than other logic families because its devices are always biased in the active region.

- The BiCMOS family achieves the performance advantages of CMOS and ECL circuits. BiCMOS logic circuits have fast switching speed and low power dissipation.

- NAND and NOR logic gates based on the basic digital inverter can be made in the NMOS, CMOS, TTL, ECL, and BiCMOS logic families.

◆ SPICE EXAMPLES

EXAMPLE 14.6

Simulate the CMOS inverter with capacitive load shown in Fig. 14.10 using SPICE and the parameter and elements values specified in Example 14.2. Specifically, let $C_L = 2\,pF$, and assume the MOSFETs to be symmetrical with $|K| = 0.2\,mA/V^2$ and $|V_{TR}| = 1\,V$. Show that the propagation delay times t_{PHL} and t_{PLH} are both equal to 1.6 ns, as computed in Example 14.2.

Solution

A suitable SPICE input file, in which node 1 is V_{DD}, node 2 is the inverter input, node 3 the output, and node 0 ground, is shown below. The input voltage consists of a piecewise linear (PWL) voltage source that simulates a 0 to 5-V, 20-ns pulse with 50% rise and fall times of 0.05 ns. These rise and fall times are negligible compared to the expected values of t_{PHL} and t_{PLH}. Note that SPICE does permit the specification of a pulsed waveform with zero rise or fall time.

The various MOSFET parameters are set in the two .MODEL statements. The choice of the channel-length modulation parameter $\lambda = 0.1\,V^{-1}$ is arbitrary. Note that the SPICE parameter KP is a positive number, even though K for a p-channel MOSFET is negative, and that KP is set to twice the desired K. The substrate of the p-channel device is connected to V_{DD}; the substrate of the n-channel device is connected to ground.

Input File

```
CMOS INVERTER with CAPACITOR LOAD
      VDD   1 0 5V

*Set vIN to a 0- to 5-V, 20-ns pulse with 0.1-ns rise time:
      vIN 2 0 PWL(0 0  0.1n 5V  20n 5V  20.1n 0)

*Specify the connections of the p-channel and n-channel
MOSFETs:
      MP 3 2 1 1 PMOS
      MN 3 2 0 0 NMOS

*Add the load capacitance CL:
      CL 3 0 2pF

*Set the parameters of the MOSFETs:
      .MODEL PMOS PMOS(VTO=-1 KP=0.4e-3 LAMBDA=0.01)
      .MODEL NMOS NMOS(VTO=1  KP=0.4e-3 LAMBDA=0.01)

*Perform a transient analysis from t = 0 to t = 40 ns using a 0.1-ns time step;
*record values every 1 ns:
      .TRAN 1n 40n 0 0.1n
      .PROBE V(2) V(3)
      .END
```

Results. A SPICE-produced plot of $v_{OUT}(t)$, superimposed on $v_{IN}(t)$, is shown in Fig. 14.57. As predicted by the analysis of Example 14.2, the 50% fall and rise times t_{PHL} and t_{PLH} are both equal to about 1.6 ns. ■

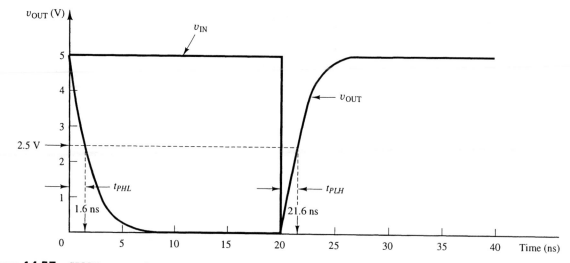

Figure 14.57 SPICE-generated output waveforms for the circuit of Fig. 14.10.

EXAMPLE 14.7 Simulate the standard TTL inverter of Fig. 14.36 using SPICE. Assume the circuit to drive a 40-kΩ load resistor. This value of resistance will draw a current on the same order as that drawn by the input of another TTL gate. Show that the circuit yields a transfer characteristic similar to that of Fig. 14.37.

Solution

The circuit of Fig. 14.36 is redrawn in Fig. 14.58 with all nodes suitably numbered for SPICE. A load resistor $R_L = 40 \text{ k}\Omega$ has also been included. An input file that performs the requested analysis follows.

Input File

```
TRANSFER CHARACTERISTIC OF TTL GATE

*Set the voltage sources in the circuit:
        VCC 1 0 5V
        vIN 5 0 dc 0V

*Enter all resistor values:
        R1 1 2 4k
        R2 1 3 1.6k
        R4 1 4 130
        R3 7 0 1k
        RL 9 0 40k

*Enter all transistors and the diode D₁;
*assume a βF of 100 and a βR of 0.1:
        Q1 6 2 5 BJT
        Q2 3 6 7 BJT
        Q3 9 7 0 BJT
        Q4 4 3 8 BJT
        .MODEL   BJT   NPN(BF=100 BR=0.1)
        D1 8 9 DIODE
        .MODEL DIODE D

*Obtain the dc transfer characteristic by varying vIN over the range 0 to 5 V:
.DC vIN 0 5 0.001
        .PROBE V(9)        ; (or use .PLOT DC V(9) in generic SPICE)
        .END
```

Figure 14.58
TTL logic inverter
with nodes suitably
numbered for
SPICE.

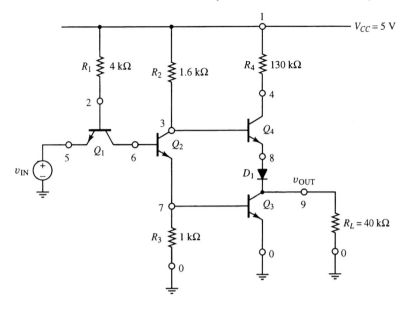

Results. The transfer characteristic produced by SPICE is shown in Fig. 14.59. This plot exhibits the four principal piecewise-linear regions described in Section 14.4.4. For the chosen set of transistor and diode parameters, points b, c, and d cited in Fig. 14.37 occur at 0.6, 1.5, and 1.6 V, respectively, and $V(0)$ is equal to about 0.1 V. These values differ somewhat from those obtained by hand calculation—a consequence of the arbitrarily assumed values of V_f and V_{sat}. ∎

Figure 14.59
TTL inverter
transfer
characteristic
generated by
SPICE.

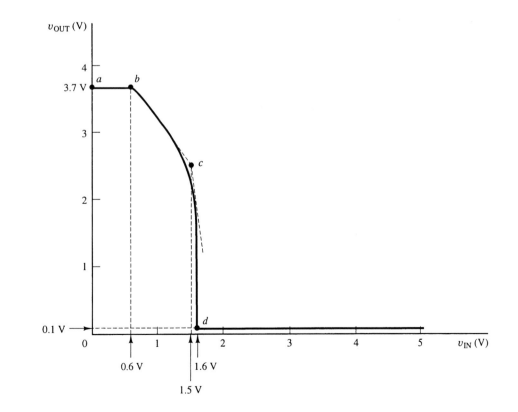

EXAMPLE 14.8

Consider the connection of input transistor Q_1 in the TTL gate of Fig. 14.34. Use SPICE to assess the effectiveness of this arrangement in removing charge stored from the base of Q_2 during a high-to-low input transition. Assume all BJTs to have parameters $C_\mu = 0.5\,\text{pF}$, $C_\pi = 2\,\text{pF}$, $\tau_F = 20\,\text{ns}$, $\beta_F = 100$, and $\beta_R = 0.1$.

Solution

The standard TTL gate can be analyzed using the input file of Example 14.7 with a few statements changed. The input voltage can be set to a 5 to 0-V step function using a PWL voltage source, and a transient analysis must be requested. The .MODEL statement for the BJT also must be modified so that the dynamic behavior caused by C_μ, C_π, and τ_F can be included. The modifications to the input file of Example 14.7 are indicated below.

Figure 14.60
Modified TTL inverter in which v_{IN} drives the base of Q_2 directly via series resistor R_1.

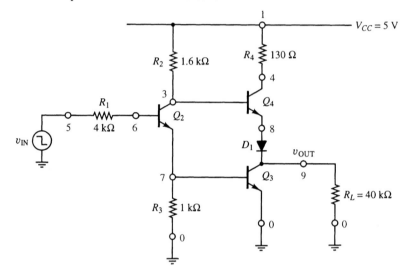

For comparison, the file can be further modified to simulate the revised circuit of Fig. 14.60. In this latter case, v_{IN} drives Q_2 directly via the current-limiting resistor R_1, thus eliminating the charge-storage removal function of Q_1.

Modified Input File from Example 14.7

```
DYNAMIC RESPONSE OF TTL GATE with INPUT TRANSISTOR:
        ⎡ Same input statements as in Example 14.7 ⎤
        ⎣          except for the following:       ⎦

*Change vIN to a 5 to 0-V step function of negligible fall time:
     vIN 5 0 PWL(0 5 1p 0)

*Add Cμ, Cπ, and τF to the .MODEL statement for the BJT:
     .MODEL BJT NPN(CJC = 0.5pF CJE = 2pF TF=20n BF=100 BR=0.1)

*Specify a transient analysis:
     .TRAN 1n 100n 0
     .END
```

Results. A plot of v_{OUT} versus time for the circuits of Figs. 14.58 and 14.60 is shown in Fig. 14.61. We first note a delay of about 30 ns with Q_1 present (and about 50 ns with Q_1 absent) before any rise in v_{OUT} occurs at all. This delay is caused by the time required to remove the additional stored electrons depicted in Fig. 14.29, as associated with the forward transit time parameter τ_F. These stored electrons are present in the base of Q_2, which is saturated at $t = 0$. After this initial delay, which is much smaller when Q_1 is present, the gate exhibits a rise time that is significantly reduced by the action of Q_1. ∎

Figure 14.61
TTL inverter output versus time in response to 5 to 0-5 V step input. The action of Q_1 substantially reduces the propagation delay and the rise times.

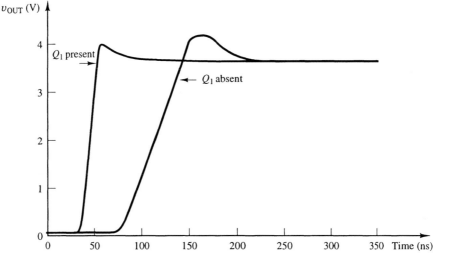

EXAMPLE 14.9   In the discussion at the end of Section 14.5.2, the importance of input matching in high-speed ECL circuits was addressed. In this example, we illustrate the problem using SPICE simulation. Create a subcircuit definition for the ECL inverter of Fig. 14.46, then connect two such gates as in Fig. 14.62. The output of the first inverter feeds the input of the second via a 50-Ω, 100-ns delay transmission line. The entire circuit sits atop the V_{EE} voltage bus. Show the effect of omitting and including the input matching resistor R_A in Fig. 14.46. When R_A is included, create a matched load for the transmission line by setting R_A to the same value as the transmission line impedance. Assume the value $\beta_F = 200$ for the transistors in the ECL gate.

Figure 14.62
Two ECL inverters connected via a 50-Ω transmission line.

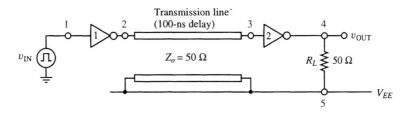

Solution

The subcircuit definition for the ECL gate, with nodes defined as in Fig. 14.63, is contained within the input file listing that follows. Only the inverting output buffer Q_X, but not Q_Y, is included because the noninverting output is not of interest in this example. The transmission line connection is made using a two-port lossless transmission line element T1, which is supported in both SPICE and PSPICE. The input is excited by a voltage pulse that begins at $V(0) = -1.76$ V, rises to $V(1) = -0.88$ V for 10 ns, and then returns to $V(0)$. These values chosen for $V(1)$ and $V(0)$ are those derived in Eqs. (14.112) and (14.113). The simulation is run twice, once with R_A absent in the subcircuit definition, and once with R_A present.

Input File

```
TWO ECL INVERTERS WITH TRANSMISSION LINE

*Specify the voltage sources of system:
        VEE 5 0 -5.2
        vIN 1 0 PWL(0 -1.76 1n -0.88  10n -0.88  11n -1.76)
```

Figure 14.63
Subcircuit
definition for an
ECL inverter with
nodes suitably
numbered for
SPICE.

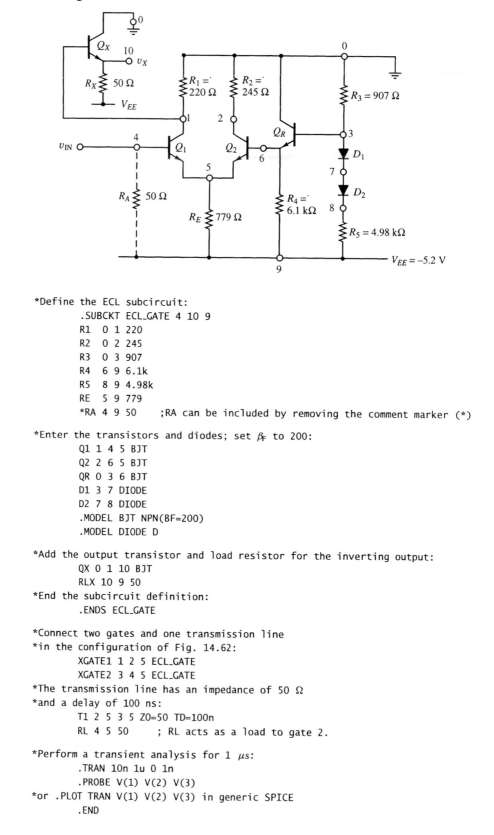

```
*Define the ECL subcircuit:
      .SUBCKT ECL_GATE 4 10 9
      R1   0 1 220
      R2   0 2 245
      R3   0 3 907
      R4   6 9 6.1k
      R5   8 9 4.98k
      RE   5 9 779
      *RA 4 9 50      ;RA can be included by removing the comment marker (*)

*Enter the transistors and diodes; set βF to 200:
      Q1 1 4 5 BJT
      Q2 2 6 5 BJT
      QR 0 3 6 BJT
      D1 3 7 DIODE
      D2 7 8 DIODE
      .MODEL BJT NPN(BF=200)
      .MODEL DIODE D

*Add the output transistor and load resistor for the inverting output:
      QX 0 1 10 BJT
      RLX 10 9 50
*End the subcircuit definition:
      .ENDS ECL_GATE

*Connect two gates and one transmission line
*in the configuration of Fig. 14.62:
      XGATE1 1 2 5 ECL_GATE
      XGATE2 3 4 5 ECL_GATE
*The transmission line has an impedance of 50 Ω
*and a delay of 100 ns:
      T1 2 5 3 5 Z0=50 TD=100n
      RL 4 5 50     ; RL acts as a load to gate 2.

*Perform a transient analysis for 1 μs:
      .TRAN 10n 1u 0 1n
      .PROBE V(1) V(2) V(3)
*or .PLOT TRAN V(1) V(2) V(3) in generic SPICE
      .END
```

Figure 14.64
Output of the circuit of Fig. 14.63 with the matching resistor R_A absent. The signal launched by gate 1 onto the transmission line is reflected back and forth many times, causing a v_{OUT} pulse each time it arrives at the input to gate 2.

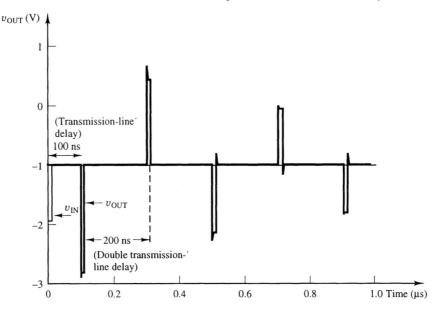

Results. The results of this simulation with R_A absent are shown in Fig. 14.64. As this plot shows, the signal launched at the left-hand end of the transmission line by gate 1 reflects off the right-hand end of the line, travels back to gate 1, and is reflected back to gate 2. The reflection process is repeated every 200 ns (twice the delay length of the transmission line) until the pulse is absorbed by the nonmatching resistances at the ends of the line. Each time the original pulse or one of the subsequent reflections reaches the right-hand end of the line, an additional output pulse is generated by gate 2. For the element values specified in this simulation, each subsequent pulse arriving at the input to gate 2 is of alternating polarity because the driving impedance of gate 1 is less than the line impedance $Z_o = 50\ \Omega$. Hence inverting reflections occur at the left-hand end of the line. The pulse will require many reflections to decay to zero—many more than are shown in the plot of Fig. 14.64.

Figure 14.65
Output of the circuit of Fig. 14.63 with $R_A = 50\ \Omega$. The output consists of a single delayed pulse.

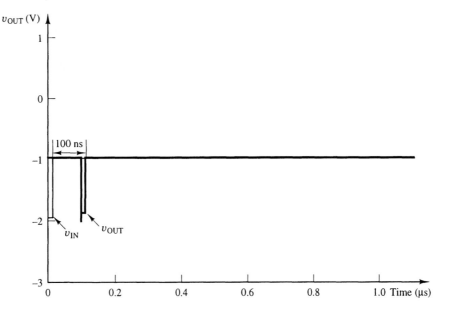

The plot of Fig. 14.65 shows the output when the matching resistor $R_A = 50\,\Omega$ is included at the input to gate 2. The initial pulse arriving on the transmission line from gate 1 is absorbed in R_A with no further reflections. Consequently, v_{OUT} displays just one pulse, delayed 100 ns by the transmission line, in response to v_{IN}. ■

◆

PROBLEMS

○ More difficult	◐ Even more difficult	● Most difficult

14.1 Fundamental Concepts of Digital Circuits

14.1.3 Definition of Logic Levels

14.1 A particular logic inverter operates from a 5-V supply. The circuit has an input threshold voltage of 2 V and an output of 0.5 V when $v_{IN} = 5$ V. If $v_{OUT} = 5$ V when $v_{IN} = 0$, what is the minimum required average slope of the transfer characteristic in the transition region?

14.2 A particular logic inverter operating from a 5-V supply is constructed with an average transition-region slope of -2. If $v_{OUT} = 5$ V when $v_{IN} = 0$, and $v_{OUT} = 1$ V when $v_{IN} = 5$ V, what is the maximum input threshold voltage the inverter can have if it is to be used in a logic circuit?

14.3 A simple BJT inverter has the form shown in Fig. 14.27. If $V_{DD} = 5$ V, $V_f = 0.7$ V, and $V_{sat} = 0.2$ V, what is the maximum ratio R_B/R_C that will enable the circuit to function as a logic inverter if $\beta_F = 100$?

14.4 A BJT inverter with resistive pull-up load has parameters $\beta_F = 50$, $V_f = 0.6$ V, $V_{sat} \approx 0.3$ V, $R_C = 1\,k\Omega$, and $R_B = 4.7\,k\Omega$. If $V_{DD} = 5$ V, what are the values of V_{OH} and V_{OL} for this inverter?

14.5 ○ Consider the BJT inverter with parameters listed in Problem 14.4. If $R_B = 100\,k\Omega$, find the values of V_{OH} and V_{OL}. Comment on the suitability of this circuit as a logic inverter.

14.6 A primitive logic system is made from NMOS inverters with resistive pull-up loads. Consider one such inverter made from a MOSFET with parameters $K = 4\,mA/V^2$ and $V_{TR} = 2$ V. If the pull-up load has value $5\,k\Omega$, and if $V_{DD} = 5$ V, find the values of V_{OH} and V_{OL}.

14.1.4 Noise Margins

14.7 A particular logic family operates from a 5-V power supply. The gates in the family are designed so that V_{IL} is equal to 10% of V_{CC} and V_{IH} is equal to 90% of V_{CC}. If the minimum high and low noise margins are to be

0.4 V, what must be the minimum and maximum values, respectively, of V_{OH} and V_{OL}?

14.8 A logic family is powered by a supply voltage of value V_{DD}. The circuits are designed so that $V_{OH} = 0.9V_{DD}$, $V_{IH} = 0.8V_{DD}$, $V_{OL} = 0.1V_{DD}$, and $V_{IL} = 0.2V_{DD}$.

(a) Express the noise margins as a percentage of V_{DD}.

(b) Express the width of the undefined transition region as a percentage of V_{DD}.

(c) How large must V_{DD} be if the noise margins are to have a minimum value of 0.5 V?

14.9 A digital inverter has perfectly flat high and low regions and a straight-line transition-region slope of -4. What are the maximum possible values of the noise margins for this inverter if it operates from a 5-V supply?

14.10 Compute the approximate noise margins of the resistive BJT inverter of Fig. 14.27 if $V_{CC} = 5$ V, $R_B = 10\,k\Omega$, $R_C = 1\,k\Omega$, and $\beta_F = 100$. Assume that the gate drives other similar logic gates.

14.11 Consider the resistive BJT inverter of Fig. 14.27 with $V_{CC} = 5$ V, $R_B = 10\,k\Omega$, and $R_C = 2\,k\Omega$.

(a) Find the noise margins if $\beta_F = 50$.

(b) Redesign the inverter so that the noise margins are improved by at least a factor of 2 over those of part (a).

14.1.5 Fan-Out and Fan-In

14.12 A logic circuit made from AND gates must implement the function $Y = A \cdot B \cdot C \cdot D$. What is the minimum fan-out requirement for the logic family used?

14.13 ○ A logic circuit made from AND gates, OR gates, and inverters must implement the function $Y = (A \cdot \overline{B} + C) + (B \cdot \overline{A} + C) + (C \cdot \overline{A} + B)$. What is the minimum fan-out requirement for the logic family used?

14.14 ○ A logic circuit made from AND gates, OR gates, and inverters must implement the function $Y = (A \cdot \overline{B} \cdot \overline{C} \cdot \overline{D}) + (B \cdot \overline{A} \cdot \overline{C} \cdot \overline{D}) + (C \cdot \overline{A} \cdot \overline{B} \cdot \overline{D}) + (D \cdot \overline{A} \cdot \overline{B} \cdot \overline{C})$. What is the minimum fan-out requirement for the logic family used?

14.15 ◑ The resistive BJT inverter of Fig. 14.27, made with $R_B = 10 \, \text{k}\Omega$, $R_C = 1 \, \text{k}\Omega$, and $V_{CC} = 5 \, \text{V}$, is designed to operate with $V(1) = 5 \, \text{V}$ and $V(0) = V_{\text{sat}} \approx 0.1 \, \text{V}$.

(a) For an input of $V(1)$, how large must β_F be if the output is to reach the $V(0)$ value with the output unloaded?

(b) The same gate is to be designed so that it can feed similar gates with a fan-out of 5. For the value of β_F found in part (a), select a new appropriate value for R_C such that $V(1)$ will be at least 4.5 V under maximum fan-out load conditions.

(c) If the value of $V(0)$ under maximum fan-out conditions is specified to be 0.3 V, what is the fan-out of the gate with the R_C chosen in part (b)?

14.16 ◑ Consider the switched resistor logic circuit of **Fig. P14.16**. When v_{IN} is low, Q_1 is in cutoff, and v_{OUT} is determined solely by the resistors in the circuit. When v_{IN} is high, Q_1 is saturated and behaves as a constant-voltage source of value $V_{\text{sat}} \approx 0.2 \, \text{V}$. If $V_f = 0.7 \, \text{V}$, $V_{IL} = 0.5 \, \text{V}$, and $V_{IH} = 3 \, \text{V}$, find the maximum fan-out of the inverter.

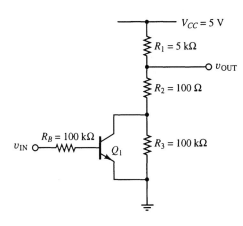

Fig. P14.16

14.1.6 Propagation Delay

14.17 A logic inverter has propagation delays of $t_{PHL} = 25 \, \text{ns}$ and $t_{PLH} = 40 \, \text{ns}$. Four such gates are connected in a series cascade. What is the total overall propagation delay of the cascade?

14.18 A combinatorial logic circuit with six cascaded stages is to operate at 10 MHz (the circuit must process logic signals with pulse widths on the order of 50 ns). The circuit configuration is such that the output is used as an input at various points within the same circuit. What should be the maximum propagation delay times of the gates used to create the circuit?

14.19 A digital ring oscillator is created by cascading five logic inverters in series, with the output of each stage driving the input of its successor. The output of the fifth inverter is fed back to the input of the first. If $t_{PLH} = t_{PHL} = 40 \, \text{ns}$, what will be the frequency of the resulting square wave?

14.20 A digital ring oscillator is to be constructed from logic inverters that have propagation delay times of $t_{PLH} = 28 \, \text{ns}$ and $t_{PHL} = 42 \, \text{ns}$. The output of each stage drives the input of its successor, and the output of the last inverter is fed back to the input of the first. If a square wave with a frequency of about 1 MHz is desired, how many gates must be connected in the ring?

14.21 In the most general case, the propagation delay times t_{PHL} and t_{PLH} describe the time delays that occur when the input and output signals of a logic inverter both have nonzero rise and fall times. As shown in Fig. 14.5(b), t_{PHL} is defined as the time interval between the 50% levels of a rising input signal and a falling output signal, where the subscript refers to the transition direction of the output. Similarly, t_{PLH} is defined as the interval between the 50% levels of a falling input and a rising output.

Suppose that a logic gate has transition times of $t_{PHL} = 15 \, \text{ns}$ and $t_{PLH} = 20 \, \text{ns}$. Two such gates are connected in sequence. The input to the first is driven by an ideal digital square wave having zero rise and fall times. If the rising and falling transitions of the gates can be modeled as linear ramps, calculate the time required for the output to reach 90% of its final value during a rising transition and the time required for the output to reach 10% of its final value during a falling transition.

14.22 Suppose that a logic inverter is constructed using a voltage source $V_{CC} = 5 \, \text{V}$ and a pull-up resistor $R_C = 1 \, \text{k}\Omega$. In this case, let the inverter transistor be replaced by a physical switch that is closed to produce a logic low and open to produce a logic high. When the switch is closed, it exhibits a physical resistance of $10 \, \Omega$. The circuit drives a load consisting of a 10-kΩ resistor and 100-pF capacitor in parallel. Compute the rise and fall times of the inverter.

14.23 The resistive-load MOSFET inverter of **Fig. P14.23** is driven by a square wave. Suppose that the only capacitance of significance is the gate-to-source capacitance C_{gs}.

(a) Show that the rise time t_r and fall time t_f have the same value for this inverter.

(b) Find expressions for v_{OUT} valid for rising and falling output transitions.

(c) Evaluate t_r and t_f in terms of the various circuit parameters.

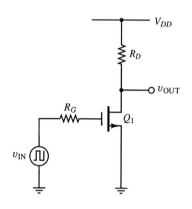

Fig. P14.23

14.1.7 Power Dissipation

14.24 A logic circuit operating from a 5-V supply requires power supply currents of $I(1) = 1\,\text{mA}$ at logic-output high [$V(1) = 4.8\,\text{V}$] and $I(0) = 0.4\,\text{mA}$ at logic-output low [$V(0) = 0.4\,\text{V}$]. What is the static power dissipation in the circuit for a 50% output duty cycle? For an 80% high duty cycle?

14.25 A logic circuit requires power supply currents of $I(1) = 100\,\mu\text{A}$ at logic-output high and $I(0) = 2\,\mu\text{A}$ at logic-output low regardless of power-supply voltage. If the output duty cycle for the circuit is 50%, what is the largest supply voltage that can be used if the static power dissipation is to be kept below 350 μW?

14.26 A particular logic circuit with rise and fall times of 10 ns is switched at 1 MHz. At this frequency, the circuit exhibits a static power dissipation of 400 μW and a dynamic power dissipation of 150 μW. If the circuit is instead switched at 5 MHz, what will be the total approximate power dissipation?

14.27 A combined capacitive and resistive load is charged to a voltage V_{Th} through a series resistance R_{Th} by closing a switch at $t = 0$, as shown in **Fig. P14.27**.

(a) What are the values of load voltage at $t = 0$ and $t = \infty$?

(b) Derive an expression for the load voltage as a function of time for $t > 0$.

(c) What is the energy stored in the capacitor at $t = \infty$?

(d) Derive expressions for the total energy dissipated in R_{Th} and R_L over the time interval $0 < t < \infty$. Compare the sum of these energies with the stored capacitor energy of part (c).

(e) For what value of R_{Th} will the energy dissipated in the resistors be a minimum? A maximum?

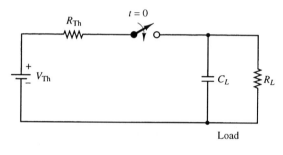

Fig. P14.27

14.1.8 Delay–Power Product

14.28 A gate draws 4 mA when the output is low and 2 mA when the output is high. The propagation delays for rising and falling output voltages are equal to 10 ns and 15 ns, respectively. Estimate the value of the delay–power product of the gate with a 5-V supply.

14.29 A gate has propagation delay times of $t_{PLH} = 10\,\text{ns}$ and $t_{PHL} = 5\,\text{ns}$. The gate draws 10 mA from a +5-V supply when the output is low and draws 2 mA when the output is high.

(a) Find the average static power dissipation if the gate spends half of its time in the logic high state.

(b) Estimate the value of delay–power product if the gate drives a 10-pF load capacitance and is switched at a 500-kHz rate.

14.30 A gate with a 75% low, 25% high output-duty cycle operates from a 5-V supply. At $V(1) = 4.8\,\text{V}$, the gate draws 0.1 mA from its supply. At $V(0) = 0.4\,\text{V}$, the gate draws 0.02 mA. If $t_{PLH} = 28\,\text{ns}$ and $t_{PHL} = 37\,\text{ns}$, find the delay–power product if the gate is switched at 10 MHz and drives a 50-pF load capacitance.

14.31 A gate operating from a 5-V supply requires supply currents of $I(1) = 1\,\text{mA}$ and $I(0) = 0.4\,\text{mA}$. Assume a 50% high/low duty cycle. If $V(1) = 4.8\,\text{V}$,

$V(0) = 0.2$ V, $t_{PLH} = t_{PHL} = 10$ ns, and the gate is switched at 10 MHz, what is the largest capacitive load that can be driven if the delay–power product is to be kept under 60 pJ?

14.32 Show that the dynamic power dissipation of a logic inverter that feeds a capacitive load C_L and is switched by a periodic square wave can be expressed by the quantity $V_{CC}(V_{OH} - V_{OL})C_L/T$, where V_{CC} is the supply voltage, and T is the period of the square wave. Evaluate the dynamic power dissipation for the case $V_{CC} = 5$ V, $V_{OH} = 4.8$ V, $V_{OL} = 0.2$ V, $C_L = 40$ pF, and $T = 0.5$ μs.

14.2 CMOS Logic Family

14.2.1 CMOS Inverter Transfer Characteristic

14.33 The CMOS logic inverter of Fig. 14.7 is made with $K_N = |K_P| = 2.2$ mA/V^2 and $V_{TRN} = |V_{TRP}| = 1.8$ V. If $V_{DD} = 5$ V, find the values of NM_H and NM_L.

14.34 The CMOS logic inverter of Fig. 14.7 is made with $K_N = |K_P| = 50$ μA/V^2 and $V_{TRN} = |V_{TRP}| = 2.3$ V. If $V_{DD} = 5$ V, find the values of NM_H and NM_L and the values of v_{OUT} at the points p_1 and p_2 where the slope is -1 on the inverter transfer characteristic.

14.35 The CMOS logic inverter of Fig. 14.7 is constructed with $K_N = 0.5$ mA/V^2, $K_P = -0.2$ mA/V^2, and $V_{TRN} = |V_{TRP}| = 1.2$ V. If $V_{DD} = 5$ V, find NM_H and NM_L.

14.36 A CMOS logic inverter of the type shown in Fig. 14.7 is constructed with $K_N = 2$ mA/V^2, $K_P = -1$ mA/V^2, $V_{TRN} = 1$ V, and $V_{TRP} = -1.5$ V. If $V_{DD} = 5$ V, find NM_H and NM_L.

14.37 ◨ Specify the parameters of the devices in a CMOS logic inverter such that its noise margins are at least 2 V. The inverter operates from a $V_{DD} = 5$ V supply.

14.38 ◨ The MOSFETs produced by a particular CMOS process have threshold voltages of $V_{TRN} = 1.5$ V and $V_{TRP} = -1.8$ V. If $V_{DD} = 5$ V, choose W and L values for the devices in a CMOS inverter such that the noise margins are at least 2 V.

14.2.2 Dynamic Behavior of CMOS Inverter

14.39 The CMOS inverter of Fig. 14.10 drives a 4-pF load capacitance. If $V_{DD} = 5$ V, $V_{TRN} = |V_{TRP}| = 2.2$ V, and $K_N = |K_P| = 0.7$ mA/V^2, find the propagation delay times t_{PHL} and t_{PLH}.

14.40 The CMOS inverter of Fig. 14.10 drives a 5-pF load capacitance. If $V_{DD} = 5$ V, $V_{TRN} = 1.2$ V, $V_{TRP} = -2$ V, and $K_N = |K_P| = 0.4$ mA/V^2, find the propagation delay times t_{PHL} and t_{PLH}.

14.41 ◨ ○ A CMOS inverter must drive a 10-pF load capacitance with a maximum t_{PHL} of 4 ns. Specify suitable values for the transistor parameters.

14.2.3 CMOS Logic Gates

14.42 ◨ Design a CMOS logic gate that can implement each of the following logic functions:

(a) $Y = \overline{A + B + C}$

(b) $Y = \overline{A + (B \cdot C)}$

(c) $Y = \overline{(A \cdot B) + (C \cdot D)}$

(d) $Y = \overline{(A + B) \cdot (C + D)}$

(e) $Y = \overline{A + (B \cdot C \cdot D)}$

14.43 ◨ For each of the logic functions listed in Problem 14.42, design and draw the geometrical layout of a gate that will accept the indicated inputs and produce the specified output. Estimate the total surface area occupied by the gate.

14.44 ◨ Design a CMOS logic circuit that can be used to control traffic lights at a busy intersection. Consider the road map of **Fig. P14.44**. Traffic flow should be allowed over route A only if no car stops at position B or C.

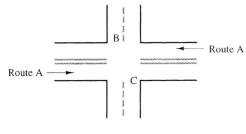

Fig. P14.44

14.45 ◨ ○ Design a CMOS logic circuit that can be used to control traffic lights the intersection shown in **Fig. P14.44**. Traffic flow should be allowed over route A until three cars appear at position B, but only if no car is stopped at position C. If more that three cars become stopped along route A, it should be open to traffic flow regardless of the status of positions B and C.

14.46 ◨ ○ Design a CMOS logic circuit that can serve as a three-digit password decoder for an alarm system. Each of the digits entered into the alarm should be represented in binary form. Choose the first three digits of your telephone number as the decoded password.

14.47 🔄 ● Design a CMOS logic circuit that can tally the voting of a five-person city council. The output should indicate the majority of "aye" or "nay" votes with a logic-high or logic-low output, respectively.

14.48 Consider the two-input CMOS NAND gate of Fig. 14.13. Suppose that $V_{DD} = 5\,\text{V}$, with $|K| = 1\,\text{mA/V}^2$ and $|V_{TR}| = 2\,\text{V}$ for all devices. If $C_L = 2.5\,\text{pF}$, estimate the rise and fall times of output Y. Assume step inputs with zero rise and fall times at A and B.

14.49 🔄 A particular CMOS fabrication process yields a K parameter of $0.1\,\text{mA/V}^2$ for all n-channel devices with $W = L$ and all p-channel devices with $W = 2L$. The threshold voltages all have a magnitude of $1.2\,\text{V}$. Design a logic inverter that can drive load capacitances of up to $400\,\text{pF}$ with rise and fall times of no more than $100\,\text{ns}$. Assume a supply voltage of $5\,\text{V}$, and ignore internal MOSFET capacitances.

14.50 🔄 ○ A CMOS fabrication process yields K parameters of $0.2\,\text{mA/V}^2$ for all n-channel devices with $W = L$ and for all p-channel devices with $W = 2L$. The process also yields threshold voltages of $V_{TRN} = 2\,\text{V}$ and $V_{TRP} = -1\,\text{V}$. Design a logic inverter that can drive load capacitances of up to $250\,\text{pF}$ with rise and fall times of no more than $100\,\text{ns}$. Assume a supply voltage of $5\,\text{V}$, and ignore internal MOSFET capacitances.

14.2.4 CMOS Transmission Gate

14.51 The B output of the CMOS transmission gate of Fig. 14.15 drives a 10-pF load capacitance in parallel with a 10-kΩ resistor. If $K_N = 0.5\,\text{mA/V}^2$, $K_P = -1.2\,\text{mA/V}^2$, $V_{TRN} = 1\,\text{V}$, and $V_{TRP} = -2\,\text{V}$, estimate the rise and fall times of output B if input A is fed by a constant voltage $V(1) = 4.8\,\text{V}$. Assume that the clock signals C and $\overline{C}$ rise and fall instantaneously.

14.52 ○ The B output of the CMOS transmission gate of Fig. 14.15 drives a 10-pF load capacitance in parallel with the drain-to-source terminals of an n-channel MOSFET whose gate is driven by the clock signal C. If $K_N = 1\,\text{mA/V}^2$, $K_P = -0.2\,\text{mA/V}^2$, $V_{TRN} = 1.5\,\text{V}$, and $V_{TRP} = -1.2\,\text{V}$, estimate the rise and fall times of output B if A is fed by a constant voltage $V(1) = 4\,\text{V}$. Assume that the clock signals C and $\overline{C}$ rise and fall instantaneously.

14.53 🔄 Design a circuit based on the CMOS transmission gate of Fig. 14.15 that can implement the logic function $Y = (A{\cdot}C) + (B{\cdot}\overline{C})$.

14.54 🔄 Design a circuit based on the CMOS transmission gate of Fig. 14.15 that can implement the logic function $Y = (A{\cdot}C{\cdot}D) + (B{\cdot}C{\cdot}\overline{D})$.

14.3 NMOS Logic Family

14.3.1 NMOS Inverter with Enhancement Load

14.55 Plot the transfer characteristic of an enhancement-load NMOS inverter for aspect ratios of 1, 2, 4, and 9. Find the noise margins when $V_{TR} = 2\,\text{V}$ and $V_{DD} = 5\,\text{V}$.

14.56 ○ An enhancement-load NMOS inverter is to have an aspect ratio of 4 and relative dimensions $W_1/L_1 = 1/4$, where Q_1 is the transistor to which v_{IN} is connected. The inverter is to be fabricated on an NMOS integrated circuit in which the smallest scale dimension is $1\,\mu\text{m}$.

(a) Draw the geometrical layout of the inverter. Label all dimensions.

(b) If $V_{TR} = 2\,\text{V}$ for each device and $V_{DD} = 5\,\text{V}$, evaluate the noise margins of the inverter.

(c) Estimate the static power dissipation of the inverter if it is operated with a 50% low output duty cycle.

(d) Estimate the static power dissipation of the inverter if it is operated with a 25% low output duty cycle.

(e) Repeat for an aspect ratio of 2 and relative dimensions $W_1/L_1 = 1$ and $W_2/L_2 = 1/2$.

14.57 The enhancement-load NMOS inverter of Fig. 14.16 is constructed with $K_1 = 125\,\mu\text{A/V}^2$, $K_2 = 25\,\mu\text{A/V}^2$, and $V_{TR} = 1.5\,\text{V}$.

(a) If $V_{DD} = 5\,\text{V}$, plot the inverter transfer characteristic. Ignore the body effect.

(b) Find values for V_{OH}, V_{OL}, V_{IH}, and V_{IL}, and evaluate the noise margins.

14.58 Consider the enhancement-load NMOS inverter of Fig. 14.16 with $K_1 = 0.1\,\text{mA/V}^2$, $K_2 = 0.5\,\text{mA/V}^2$, and $V_{TR} = 1.2\,\text{V}$.

(a) If $V_{DD} = 5\,\text{V}$, plot the inverter transfer characteristic. Ignore the body effect.

(b) Find values for V_{OH}, V_{OL}, V_{IH}, and V_{IL}, and evaluate the noise margins.

14.59 The NMOS inverter of Fig. 14.16 operates from a supply voltage of value V_{DD}. The MOSFETs have threshold voltages of $V_{TR} = 0.3V_{DD}$. Assuming the body effect to be negligible, find the aspect ratio required to yield $V_{OL} = 0.1V_{DD}$.

14.60 The enhancement-load NMOS inverter of **Fig. P14.60** uses a second bias voltage V_1 to alter the inverter transfer characteristic. Suppose that $K = 1\,\text{mA/V}^2$ and $V_{\text{TR}} = 2\,\text{V}$ for both transistors.

(a) If $V_{DD} = 5\,\text{V}$ and $V_1 = 6\,\text{V}$, plot the inverter transfer characteristic. Ignore the body effect.

(b) Plot the transfer characteristic for $V_1 = 7$, 8, and $9\,\text{V}$.

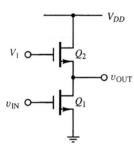

Fig. P14.60

14.61 ○ The enhancement-load NMOS inverter of Fig. 14.16 is constructed with $V_{DD} = 5\,\text{V}$ using devices having parameters $V_{\text{TR}o} = 1\,\text{V}$, $K_o = 20\,\mu\text{A/V}^2$, $\phi_F = 0.3\,\text{V}$, and $\gamma = 0.5\,\text{V}^{1/2}$, where K_o is the conductance parameter of a device for which $W = L$.

(a) Find V_{OH}, V_{OL}, NM_H, and NM_L while ignoring the body effect.

(b) Find new values for V_{OH} and NM_H if the body effect is considered.

14.62 The NMOS circuit of **Fig. P14.60** can be treated as a two-input logic gate. What Boolean logic function is realized by this gate? Assume inputs V_1 and v_{IN} to be driven to either V_{DD} or to zero.

14.3.2 NMOS Inverter with Depletion Load

14.63 For a particular NMOS logic family, the optimal aspect ratio of a single-input depletion load inverter is $K_R = 4$.

(a) If the smallest device fabrication dimension is $2\,\mu\text{m}$, specify the length and width of the driven device and the pull-up load.

(b) Estimate the surface area occupied by the inverter.

14.64 Plot the transfer characteristic of a depletion-load NMOS inverter for aspect ratios of 1, 2, 4, and 8 when $V_{\text{TR-DM}} = -1\,\text{V}$, $V_{\text{TR-EM}} = 2\,\text{V}$, and $V_{DD} = 5\,\text{V}$.

14.65 ○ Consider the depletion-load inverter of Fig. 14.22 with $V_{DD} = 5\,\text{V}$, $K_1 = K_2 = 1\,\text{mA/V}^2$, $V_{\text{TR1}} = 2\,\text{V}$, and $V_{\text{TR2}} = -1\,\text{V}$. Suppose that the gate of Q_2 is connected to its drain instead of its source.

(a) Prove that Q_2 will operate in the triode region.

(b) Find the resulting v–i characteristics of Q_2 as seen from its drain–source terminals.

(c) Find the load curve superimposed by Q_2 and V_{DD} on Q_1.

(d) Plot the inverter's v_{IN}–v_{OUT} transfer characteristic under these conditions.

14.66 A depletion-load NMOS inverter is to have an aspect ratio of 4 and relative dimensions $W_1/L_1 = 1/4$. The inverter is to be fabricated on an NMOS integrated circuit in which the smallest scale dimension is $1\,\mu\text{m}$.

(a) If $V_{\text{TR-EM}} = 1\,\text{V}$ and $V_{\text{TR-DM}} = -2\,\text{V}$, estimate the static power dissipation of the inverter for a 50% low output duty cycle.

(b) Estimate the static power dissipation of the inverter if it is operated with a 25% low output duty cycle.

14.67 A depletion-load inverter has device threshold voltages of $V_{\text{TR-EM}} = 1\,\text{V}$ and $V_{\text{TR-DM}} = -2\,\text{V}$. Plot the high-to-low crossover voltage of the transfer characteristic as a function of the inverter aspect ratio K_R for $1 < K_R < 14$.

14.68 Consider the depletion-load inverter of Fig. 14.22. If the body effect is ignored, show that V_{OL} can be approximately expressed by $(V_{\text{TR1}})^2[2K_R(V_{DD}-V_{\text{TR2}})]^{-1}$, where K_R is the aspect ratio of the inverter.

14.3.3 Dynamic Behavior of NMOS Inverter

14.69 The enhancement-load NMOS inverter of Fig. 14.16 drives a capacitive load. The circuit is constructed from devices with parameters $V_{\text{TR}} = 1\,\text{V}$ and $K_o = 20\,\mu\text{A/V}^2$, where K_o is the conductance parameter of a device for which $W = L$. If $V_{DD} = 5\,\text{V}$ and $v_{\text{IN}} = V_{OH}$, plot the current available to discharge the load capacitance versus v_{OUT}.

14.70 Consider the enhancement-load NMOS inverter of Fig. 14.16 with $V_{OH} = 4\,\text{V}$, $V_{OL} = 0.3\,\text{V}$, $K_1 = 1.2\,\text{mA/V}^2$, $K_2 = 0.2\,\text{mA/V}^2$, and $V_{\text{TR}} = 2\,\text{V}$. Compute t_{PLH} and t_{PHL} for the inverter if it drives a 0.5-pF on-chip load capacitance.

14.71 ◑ An NMOS inverter with enhancement load is fabricated on silicon with an aspect ratio of 8. The smallest device dimension is $1\,\mu\text{m}$. The area of the gate-oxide layers, for which $\epsilon_{\text{ox}}/t_{\text{ox}} = 4 \times 10^{-4}\,\text{pF}/\mu\text{m}^2$, is as small as possible. For silicon, $\mu_e = 1350\,\text{cm}^2/\text{V-s}$.

(a) What is the capacitance per unit area between the gate region and the substrate for each device?

(b) What are the values of K for each device?

(c) Assume a threshold voltage of $V_{TR} = 1\,V$. Estimate the rise and fall times of the gate when it drives a 1-pF capacitive load.

(d) If depletion-mode devices on the same IC have a threshold voltage of $-2\,V$, repeat part (c) for a depletion-load inverter.

14.72 The two-input depletion load NMOS NAND gate of Fig. 14.26(a) is fabricated with equal values of K for all devices. After fabrication, the enhancement-mode devices on the integrated circuit have a measured threshold voltage of $0.5\,V$, and the depletion-mode device has a threshold voltage of $-1\,V$.

(a) Draw the geometrical layout of the gate.

(b) What is the aspect ratio of the equivalent single-input inverter? The upper input device is driven by a constant voltage equal to $V(1)$. The lower input device is driven by a square wave that switches periodically from $V(0)$ to $V(1)$. In the calculations to follow, neglect the body effect in Q_A and Q_Y.

(c) Estimate the static power consumption of the gate.

(d) Plot v_{OUT} versus time if the gate drives a 0.2-pF load capacitance.

(e) Estimate the dynamic power consumption with the capacitive load of part (d) if $T = 10\,\mu s$.

(f) Repeat parts (c) through (e) for the case where the lower input device is driven by a constant voltage of value $V(1)$ and the upper input device is driven by a square wave.

14.3.4 NMOS Logic Gates

14.73 Find the logic function implemented by the NMOS circuit of **Fig. P14.73**.

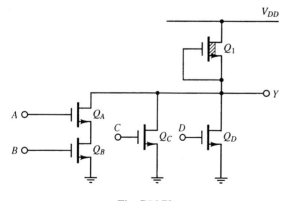

Fig. P14.73

14.74 Design NMOS gates that can implement each of the logic functions listed in Problem 14.42.

14.75 Design an NMOS logic circuit that can be used to sound an alarm if the key is removed from the ignition of an automobile while the headlights are still on or the parking brake has not been set.

14.76 Design an NMOS logic circuit that can be used to turn on a garden watering system if the temperature rises above 30°C, the sun is not shining on the plants, and the time is not before noon. Assume that sensors are available that can produce logic 1 outputs for temperature above 30°C, sun out, and time after noon.

14.77 ○ Design an NMOS logic circuit that can be used to sound an alarm in a four-passenger automobile if the ignition is energized but the driver has not put on a seat belt. The alarm should also sound if a passenger is located in any of the remaining three seats and has not put on a seat belt.

14.78 ● Design an NMOS logic circuit that can be used by a scientific investigator to assess the probability of various events. The circuit should accept five binary input signals and provide an output with high or low status corresponding to majority of the inputs.

14.79 A two-input silicon enhancement-load NMOS NAND gate is to be fabricated based on an inverter aspect ratio of 8. The smallest device dimension is $1\,\mu m$ and the thickness of the oxide over the gate region is about $500\,\text{Å}$.

(a) Draw a circuit diagram and geometrical layout of the NAND gate. Specify the length and width parameters of each device in microns.

(b) What is the capacitance per unit area between the gate region and the substrate?

(c) What are the relative values of K for each device?

14.80 A two-input enhancement-load NMOS NOR gate is designed for an inverter aspect ratio of 4. When one input is high, the output attains the value $V(0) = 0.2\,V$. Draw the load curve of the pull-up load and V_{DD} over the v–i characteristics of the input transistor(s) with one input high and with both inputs high. Estimate the value of $V(0)$ with both inputs high.

14.81 The optimal device parameters of a particular depletion-load NMOS inverter are such that the driven device has dimensions $W_1 = 4\,\mu m$ and $L_1 = 2\,\mu m$, and the pull-up load has dimensions $W_2 = 2\,\mu m$ and $L_2 = 4\,\mu m$.

(a) What is the aspect ratio of the inverter?

(b) A two-input NAND gate is fabricated using the inverter as a basis. Specify the optimum length and width of each device in the gate.

(c) Repeat for a two-input NOR gate.

14.82 A three-input enhancement load NMOS NOR gate is to be fabricated in silicon based on an inverter aspect ratio of 8. The smallest device dimension is $2\,\mu m$ and the area of the oxide over the gate regions is as small as possible. For this device, $\epsilon_{ox}/t_{ox} = 4 \times 10^{-4}\,\mathrm{pF}/\mu m^2$ and $\mu_e = 1350\,\mathrm{cm^2/V\text{-}s}$.

(a) Draw a circuit diagram and geometrical layout of the NOR gate. Specify the length and width parameters of each device in microns.

(b) What is the capacitance per unit area between the gate region and the substrate?

(c) What are the values of K for each device?

14.83 ⚡ Each of the logic functions specified in Problem 14.42 is to be implemented as an NMOS gate.

(a) Specify the W/L ratios for each device so that an inverter aspect ratio of 8 is preserved.

(b) If the smallest device dimension is to be $2\,\mu m$, draw the geometrical layout of the gate and estimate its total surface area.

14.84 A three-input enhancement-load NMOS NOR gate is designed for an inverter aspect ratio of 4. When one input is high, the output attains the value $V(0) = 0.3\,\mathrm{V}$.

(a) Draw the load curve of the load device and the v–i characteristic of the driver stage with one input high.

(b) Draw the load curve and v–i characteristic with two and three inputs high.

(c) Graphically estimate the value of $V(0)$ when two and three inputs are high.

14.85 In the NMOS NOR gate of Fig. 14.25, the output with both inputs high will be lower than the value obtained with only one input high. Compare the value of $V(0)$ at the output terminal for these two cases. Ignore the body effect. The silicon devices all have a gate-oxide thickness such that $\epsilon_{ox}/t_{ox} \approx 15\,\mathrm{nF/cm^2}$. The K parameter for each device can be computed from the relation $K = W\mu_e\epsilon_{ox}/2Lt_{ox}$, where $\mu_e = 1350\,\mathrm{cm^2/V\text{-}s}$. The device dimensions are listed in the following table:

	$W\,(\mu m)$	$L\,(\mu m)$	V_{TR}
Q_A	2	4	2
Q_B	2	4	2
Q_Y	4	2	-1

14.4 TTL Logic Family

14.86 ◯ Consider the simple BJT inverter of Fig. 14.27 with $V_{CC} = 5\,\mathrm{V}$, $R_B = 10\,\mathrm{k\Omega}$, $R_C = 1\,\mathrm{k\Omega}$, and $\beta_F = 50$. If the inverter drives a 0.5-pF load capacitance, estimate the rise and fall times of the output voltage. Plot v_{OUT} versus time when the input is driven by a perfect square wave. Estimate the inverter propagation delay times.

14.87 ⚡ Consider the simple BJT inverter of Fig. 14.27. Choose values for all parameters such that the inverter can drive a 10-pF load capacitance with a time constant of no longer than 5 ns.

14.88 What transit time is required if the base current in a BJT inverter with $\beta_F = 200$ is to increase with a rise time of no more than 50 ns?

14.89 A step input is applied to the BJT inverter of Fig. 14.27 with $R_B = R_C = 1\,\mathrm{k\Omega}$ and $V_{CC} = 5\,\mathrm{V}$. Estimate the fall time for a high-to-low output transition if no load capacitance is connected and the BJT has parameters $\beta_F = 100$ and $\tau_F = 10\,\mathrm{ns}$. Revise your estimate if the inverter output drives a 10-pF load capacitance.

14.90 What is the fan-out of a TTL gate based on the circuit of Fig. 14.34 if $\beta_F = 50$?

14.91 Consider the TTL inverter of Fig. 14.36. Estimate the minimum value of β_F that will ensure proper operation if $\beta_R = 0.2$.

14.92 Compute the output voltage of the TTL gate of Fig. 14.34 if the load current is equal to 2 mA.

14.93 Consider the TTL gate with low input of Fig. 14.34. If $\beta_F = 250$ and $V_{sat} = 0.2\,\mathrm{V}$, find the value of i_L at which Q_4 first goes into saturation.

14.94 Consider the TTL inverter of Fig. 14.34 with the value of resistor R_3 doubled. Plot the resulting voltage transfer characteristic. Compare to the plot of Fig. 14.37.

14.95 ◯ The TTL gate of Fig. 14.34 is constructed with all resistor values doubled. For v_{IN} low, calculate the values of all transistor currents and node voltages if $\beta_F = 100$, $\beta_R = 0.2$, and $V_f = 0.7\,\mathrm{V}$. Assume the load to consist of a 500 Ω resistor.

14.96 ○ The TTL gate of Fig. 14.36 is constructed with all resistor values doubled. For v_{IN} high, calculate the values of all transistor currents and node voltages if $\beta_F = 100$, $\beta_R = 0.2$, and $V_f = 0.7\,\text{V}$. Assume the load to consist of a 500-Ω resistor.

14.97 ⊿ Design TTL gates that can implement each of the logic functions specified in Problem 14.42.

14.98 ⊿ Design a TTL gate circuit that can be used in a railroad system to sound an alarm if a train approaches an intersection with a road while a car approaches from either direction, and either the gate is not down, the warning lights are not flashing, or the warning bells have not been turned on.

14.99 ⊿ ◐ Design and specify the layout of a two-input exclusive-OR (XOR) TTL gate. The XOR function is high only if one, but not both, of its inputs is high.

14.100 The output of a three-input TTL gate based on the TTL inverter of Fig. 14.34 is accidentally shorted to ground. Calculate the current through the short circuit for the case of three, two, one, and zero inputs high.

14.101 (a) Draw the voltage transfer characteristic of the BJT inverter of Fig. 14.27 with $V_{CC} = 5\,\text{V}$, $R_B = 10\,\text{k}\Omega$, and $R_C = 1\,\text{k}\Omega$. Consider the cases where $\beta_F = 50$, 100, and 200.

(b) Draw the new transfer characteristic for the specified values of β_F if a Schottky diode is connected from the base to the collector of the BJT.

14.102 Compare the low-power Schottky TTL gate of Fig. 14.44 to the standard TTL gate of Fig. 14.34. Estimate the static power consumption of each gate and compare values.

14.103 ○ Examine the low-power Schottky TTL gate of Fig. 14.44.

(a) If all the inputs are high, determine the current flowing through R_1, R_2, and R_3 for $\beta_F = 100$. Show that the output voltage is in the low logic state and calculate the approximate value of $V(0)$.

(b) Repeat the calculation for the case where one of the inputs is driven low by a voltage equal to the $V(0)$ value calculated in part (a). Show that the output voltage is in the high logic state.

14.104 ○ Examine the subcircuit formed by Q_6, R_5, and R_6 in the low-power Schottky TTL gate of Fig. 14.44. For the case $\beta_F = 50$, derive an expression for the v–i characteristic of the circuit as measured between the upper R_5–R_6 terminal and ground.

14.105 A simplified three-input Schottky-clamped TTL gate is shown in **Fig. P14.105**. Find the Boolean expression that relates the output Y to inputs A, B, and C.

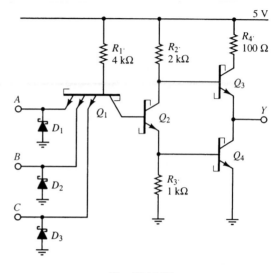

Fig. P14.105

14.106 ◑ For the Schottky-clamped TTL gate of **Fig. P14.105**, assume that all the BJTs have $\beta_F = 50$.

(a) Under open-circuit load conditions, what is the value of $V(0)$ at the Y terminal?

(b) Determine the maximum fan-out possible if Q_4 is to remain in saturation when the output is low.

(c) Find the average static power dissipated in the gate. Assume a 50% duty cycle for the high and low output states.

14.5 Emitter-Coupled Logic Family

14.107 In this problem, the temperature variation of the reference voltage in the ECL logic inverter of Fig. 14.46 is examined. The voltage drop across each pn junction in the circuit is assumed to change at the rate of $\Delta V = a\,\Delta T$, where $a = -2\,\text{mV/°C}$ and ΔT is the variation in degrees Centigrade from room temperature. The junction voltage variations can be considered as small signals that contribute incrementally to the reference voltage V_R, which is measured at the top of R_4, as shown in **Fig. P14.107**.

(a) Draw the small-signal model of the circuit of Fig. P14.107.

(b) Find an expression for the variation in V_R as a function of temperature. It may be helpful to use the superposition theorem.

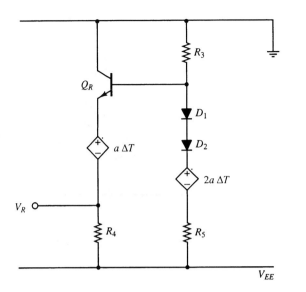

Fig. P14.107

14.108 ◗ In this problem, the temperature variations in the output levels of the ECL logic inverter of Fig. 14.46 are examined. Specifically, the change in the midpoint between V_{OH} and V_{OL}, as measured at the v_{OUT} output, is derived as a function of temperature. As in Problem 14.107, the voltage drop across each pn junction in the inverter is assumed to change at the rate of $\Delta V = a\,\Delta T$, where $a = -2\,\text{mV/°C}$ and ΔT is the variation in degrees Centigrade from room temperature.

(a) When the input is low, so that Q_1 is in cutoff and $\overline{v_{OUT}} = V(0)$, the left-hand side of the ECL inverter appears as in **Fig. P14.108(a)**. Find an expression for the variation $\Delta \overline{v_{OUT}}$ as a function of temperature.

(b) When the input is high, so that Q_2 is in cutoff and $\overline{v_{OUT}} = V(1)$, the left-hand side of the ECL inverter appears as in **Fig. P14.108(b)**. Find the variation $\Delta \overline{v_{OUT}}$ as a function of temperature.

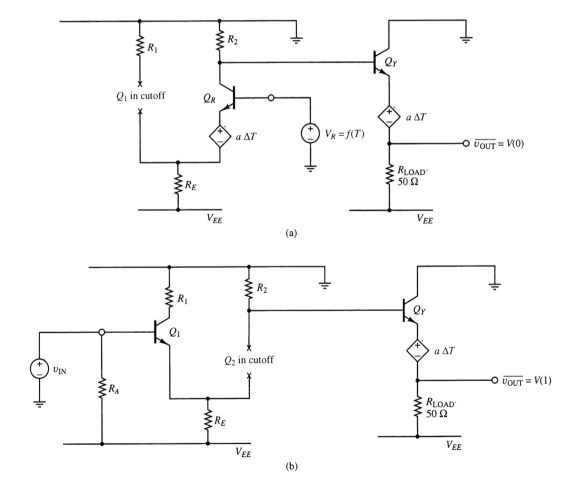

Figure P14.108 (a) v_{IN} **low; (b)** v_{IN} **high**

14.109 The transition points p_1 and p_2 of the ECL logic inverter are normally defined as the points where 99% of the available bias current flows through one of the differential pair devices. Compute the values of V_{IL} and V_{IH} if the transition points are defined as 90% of the bias current.

14.110 ⊠ ○ Design ECL gates that can implement each of the logic functions specified in Problem 14.42.

14.6 BiCMOS Logic Circuits

14.111 Consider the BiCMOS inverter of Fig. 14.52. Compute the value of $V(1)$ at the output terminal if the n-channel MOSFETs have subthreshold currents of 25 pA and the BJTs have scale currents $I_{EO} = 2 \times 10^{-14}$ mA.

14.112 Consider the BiCMOS inverter of Fig. 14.52. Compute the value of $V(0)$ at the output terminal if the n-channel MOSFETs have subthreshold currents of 12 pA

and the BJTs have scale currents $I_{EO} = 5 \times 10^{-16}$ mA.

14.113 Estimate the time required for the BiCMOS gate of Fig. 14.52 to charge a 5-pF load capacitance if $K_N = |K_P| = 1$ mA/V^2, $V_{\text{TR}N} = |V_{\text{TR}P}| = 1$ V, and $\beta_F = 100$.

14.114 Compute the value of $V(1)$ at the output of the BiCMOS NOR gate of Fig. 14.55 if the n-channel MOSFETs have subthreshold currents of 10 pA and the BJTs have scale currents $I_{EO} = 2 \times 10^{-15}$ mA.

14.115 Compute the value of $V(0)$ at the output of the BiCMOS NOR gate of Fig. 14.55 with one input high and with both inputs high. Assume the n-channel MOSFETs to have subthreshold currents of 12 pA and the BJTs to have scale currents $I_{EO} = 10^{-15}$ mA.

14.116 ⊠ ○ Design BiCMOS gates that can implement each of the logic functions specified in Problem 14.42.

◆ SPICE PROBLEMS

14.117 Use SPICE to determine the propagation delay times of a capacitive-load CMOS inverter if $C_L = 5$ pF and $V_{DD} = 5$ V. Use the device parameters specified in Example 14.2.

14.118 Use SPICE to compute the dynamic power dissipation of a CMOS inverter if its input consists of a square wave with 50% duty cycle and it drives a 10-pF load. Use the device parameters specified in Example 14.2.

14.119 Consider a digital CMOS inverter with symmetrical p-channel and n-channel devices. If $|K| = 0.6$ mA/V^2 and $|V_{\text{TR}}| = 1.2$ V, estimate the values of V_{OH}, V_{IH}, V_{OL}, and V_{IL} using SPICE analysis.

14.120 Use SPICE to assess the effect of source-to-substrate voltage on the CMOS NAND gate of Fig. 14.13. Consider the transfer characteristic of the Y output versus input A with input B held at $V(1)$. Obtain two plots, one with the source-to-substrate voltage of Q_2 neglected and one with the source-to-substrate voltage included.

14.121 Use SPICE to find the rise and fall times of the enhancement load and depletion load NMOS inverters described in Problem 14.71.

14.122 Use SPICE to find the rise and fall times of an enhancement-load NMOS inverter with an aspect ratio of 5 and a 1-pF capacitive load.

14.123 A two-input NAND gate is implemented using NMOS technology. The gate is fabricated based on an

inverter aspect ratio of 8. Use SPICE to plot the transfer characteristics of the gate output versus each gate input. The plots should be made with the unused input held at the logic-high voltage $V(1)$. Consider the body effect in your simulation. Estimate the values of any required device parameters.

14.124 The calculations of Problem 14.85 concerning a two-input depletion-load NMOS NAND gate were performed while neglecting the MOSFET source-to-substrate voltage.

(a) Use SPICE to verify the calculations of Problem 14.85 by connecting the substrate of each MOSFET to its source.

(b) Run the simulation again with all substrates connected to a common ground. Compare to the results of part (a).

14.125 Consider a depletion-load NMOS inverter with an aspect ratio of 8. The fabrication process guarantees that $V_{\text{TR-EM}}$ and $|V_{\text{TR-DM}}|$ will differ by 2 V. The exact value of $V_{\text{TR-EM}}$ cannot be predicted exactly but will lie within the range 0.2 to 5 V. Use SPICE to plot the inverter transfer characteristic for a variety of threshold voltages within this range. Consider the body effect.

14.126 Use SPICE to obtain the plots requested in Problem 14.84 for a three-input enhancement-load NMOS NOR gate.

14.127 Consider the TTL logic inverter of Fig. 14.34. Use SPICE to obtain the results of Section 14.4.3, in which the voltage and current levels with v_{IN} low were obtained.

14.128 Consider the TTL logic inverter of Fig. 14.36. Use SPICE to obtain the results of Section 14.4.3, in which the voltage and current levels with v_{IN} high were obtained.

14.129 Consider the TTL logic inverter of Fig. 14.34. Use SPICE to determine the inverter transfer characteristic if the power supply is reduced to 3 V.

14.130 Use SPICE to assess the dynamic performance of the TTL inverter of Fig. 14.34 if it drives a 10-pF load capacitance. Assume device parameters $\beta_F = 150$, $\beta_R = 0.02$, $\tau_F = 1$ ns, $V_A = 50$ V, $C_\pi = 1$ pF, and $C_\mu = 0.5$ pF.

14.131 Plot the dc transfer characteristic of the ECL inverter of Fig. 14.46 using SPICE.

14.132 Use SPICE to find the temperature variations described in Problems 14.107 and 14.108. Use the component values shown in the ECL inverter diagram of Fig. 14.46.

14.133 Obtain the BiCMOS inverter transfer characteristic of Fig. 14.53 by simulating the circuit of Fig. 14.52 in SPICE. Use the parameter values $K_N = |K_P| = 0.2$ mA/V^2, $V_{TRN} = |V_{TRP}| = 1.5$ V, $\lambda = 0.05$ V^{-1}, $\beta_F = 100$, and $V_A = 50$ V.

Chapter **15**

Fundamentals of Digital Systems

T he digital circuits and gates described in Chapter 14 form the basis from which working digital systems are designed. Such systems may vary in complexity from the simplest NAND and NOR combinational logic circuits to the intricate inner workings of a modern microprocessor or computer-interface chip. In this chapter, we expand upon the topics covered in Chapter 14 by studying some of the more complex circuit configurations used in large-scale digital systems.

15.1 SEQUENTIAL LOGIC CIRCUITS

The digital NAND and NOR gates embodied in the simple circuit topologies of Chapter 14 are examples of *combinational* logic circuits in which the output is determined solely by the instantaneous logic values of its inputs. Other types of gates, called *sequential* logic circuits, also can be made using combinational gates as the fundamental building blocks. The output of a sequential logic circuit depends not only on its inputs, but also on the previous history of its inputs. This feature gives the sequential logic circuit digital *memory*. Sequential circuits are often driven by timing clocks or other time-varying digital signals.

Figure 15.1
Set–reset (SR)
flip-flop:
(a) implementation
from two
cross-coupled NOR
gates; (b) logic
symbol; (c) binary
truth table.

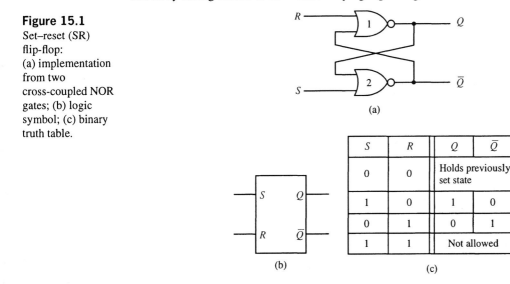

S	R	Q	$\bar{Q}$
0	0	Holds previously set state	
1	0	1	0
0	1	0	1
1	1	Not allowed	

(b) (c)

15.1.1 Set–Reset Flip-Flop

The simplest form of sequential logic circuit is called the *set–reset* (SR) flip-flop. An SR flip-flop can be made in any of the logic families introduced in Chapter 14, including CMOS, BiCMOS, NMOS, TTL, and ECL. To make an SR flip-flop, two inverters, each configured as a two-input NOR gate, are connected so that the output of each serves as one input to the other, as illustrated in Fig. 15.1(a). The logic symbol for the resulting SR flip-flop is shown in Fig. 15.1(b). Examples of NMOS and CMOS versions of the SR flip-flop are illustrated in Fig. 15.2.

Figure 15.2
Circuit diagram of cross coupled NOR-gate SR flip-flop:
(a) NMOS version;
(b) CMOS version.

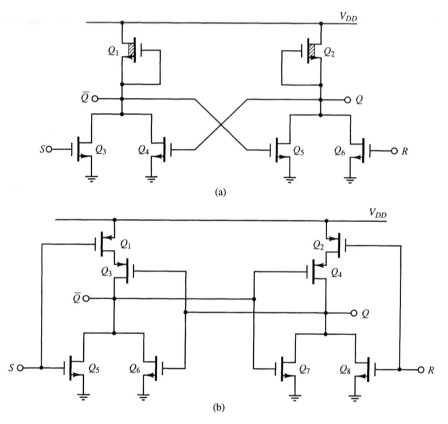

The cross-coupling, or *feedback*, between the input and output terminals gives the flip-flop its memory capabilities. The operation of the flip-flop can be understood by assuming an initial condition in Fig. 15.1(a) in which the S and R inputs are both low. Under these conditions, suppose the outputs of the flip-flop to have the initial values $Q = 1$ and $\overline{Q} = 0$. In this state, the high value of Q fed to gate 2 forces $\overline{Q}$ low; similarly, the low value of $\overline{Q}$ forces Q high via gate 1. The outputs are thus self-sustaining and correspond to one stable state of the circuit. A second stable state also exists in which $Q = 0$ and $\overline{Q} = 1$. In this second state, the low value of Q forces $\overline{Q}$ high via gate 2, and a high value of $\overline{Q}$ self-consistently forces Q low via gate 1.

With both S and R inputs low, the output state of the flip-flop is, in principle, randomly determined when the circuit is first energized. (In practice, inevitable imbalances in the circuit favor one output state over the other during power-up.) Thereafter, outputs Q and $\overline{Q}$ can be readily changed from one state to another by applying digital signals to the gate inputs labeled S and R. If the flip-flop resides in the $\overline{Q} = 1$ state, a high signal applied to the S input of NOR gate 2 will force $\overline{Q}$ low, thereby forcing Q high via gate 1. Upon removal of the S input, the outputs Q and $\overline{Q}$ will retain their respective high and low values. The application of a momentary

$S = 1$ input is called *setting* the flip-flop. In a similar manner, applying a high signal to the R input of NOR gate 1 will force Q low, thereby forcing $\overline{Q}$ high via gate 2. Upon removal of the R input, Q and $\overline{Q}$ will retain their respective low and high values. The application of a momentary $R = 1$ input is called *resetting* the flip-flop. Note that simultaneous $S = 1$ and $R = 1$ inputs cause the output state to be undetermined. This input combination is not allowed. A truth table for the SR flip-flop, which summarizes the operations just described, is shown in Fig. 15.1(c).

EXERCISE 15.1 Draw the circuit diagram of a TTL SR flip-flop. Use the TTL NOR gate of Fig. 14.40 as the basic building block.

15.2 Draw the circuit diagram of an ECL SR flip-flop. Use the ECL NOR gate of Fig. 14.50 as the basic building block.

15.3 Draw the circuit diagram of a BiCMOS SR flip-flop. Use the BiCMOS NOR gate of Fig. 14.55 as the basic building block.

15.1.2 Clocked SR Flip-Flop

A variation of the SR flip-flop, called the *clocked* SR flip-flop, has the logic symbol shown in Fig. 15.3(a). A clocked SR flip-flop can be formed by adding two AND gates to a basic NOR-gate SR flip-flop, as shown in Fig. 15.3(b). The clocked SR flip-flop has an additional input, called the *clock* input, which allows Q and $\overline{Q}$ to assume the output states set by S and R *only* when the clock signal C is high. When the clock is low, the inputs S' and R' are forced to zero by gates 1 and 2. The data outputs Q and $\overline{Q}$ thus remain unchanged regardless of inputs S and R when C is low. The operation of the clocked SR flip-flop is summarized by the truth tables of Fig. 15.3(c).

Figure 15.3
Clocked SR
flip-flop: (a) logic
symbol; (b) one
possible logic-level
implementation;
(c) truth table for
gates 1 and 2.

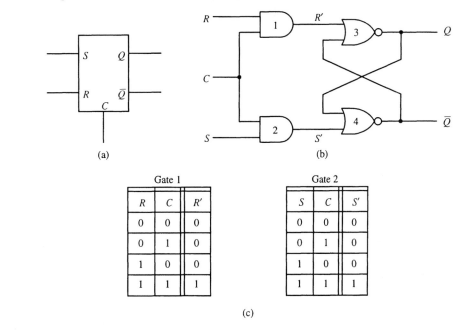

(a)   (b)

Gate 1		
R	C	R'
0	0	0
0	1	0
1	0	0
1	1	1

Gate 2		
S	C	S'
0	0	0
0	1	0
1	0	0
1	1	1

(c)

The C input of the clocked SR flip-flop is used to precisely time the appearance of new data at the Q and $\overline{Q}$ outputs, even when changes at the S and R inputs appear at imprecisely synchronized times. An example of the Q output for specific S and R input data streams is shown in Fig. 15.4.

Figure 15.4

Q output versus time for the S, R, and clock (C) inputs shown.

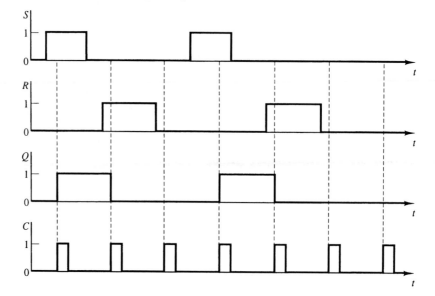

In many applications, it is advantageous to have a flip-flop that is made from identical gates. An alternative version of the clocked SR flip-flop, made from four NAND gates, is shown in Fig. 15.5. It can be shown (see Problem 15.13) that this version is functionally equivalent to the flip-flop of Fig. 15.3(b).

Figure 15.5

Four-NAND-gate version of the clocked SR flip-flop that is functionally equivalent to the circuit of Fig. 15.3(b).

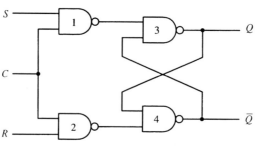

EXERCISE 15.4 Draw the output of a clocked SR flip-flop if the S input is held high and the R input low at all times. Assume the clock input to be a periodic square wave.

15.5 Draw the output of a clocked SR flip-flop if the S input is held low and the R input high at all times. Assume the clock input to be a periodic square wave.

15.1.3 JK Flip-Flop

The output of an SR flip-flop becomes ambiguous when both S and R inputs are high. This ambiguity can be eliminated by connecting two additional AND gates to the input of a clocked SR flip-flop, as shown in Fig. 15.6. Adding these connections turns the SR flip-flop into a JK flip-flop. Like the basic SR flip-flop, the JK flip-flop has two stable output states. The *set* state, with $Q = 1$ and $\overline{Q} = 0$, is reached when S is made high. Conversely, the *reset* state, with $Q = 0$ and $\overline{Q} = 1$, is reached when R is made high. In this case, the S and R signals are equal to the outputs of AND gates 1 and 2, and are thus given by

$$S = J \cdot \overline{Q} \qquad (15.1)$$

and

$$R = K \cdot Q \qquad (15.2)$$

Figure 15.6
A JK flip-flop is formed by adding two AND gates to a clocked SR flip-flop. The Q and $\overline{Q}$ outputs are cross-connected as the second inputs to these AND gates.

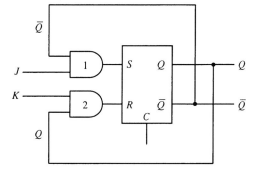

When *both* J and K are low, the outputs of AND gates 1 and 2 will be low regardless of status of the Q and $\overline{Q}$ signals. For this condition, S and R will both equal **0**, and the state of the SR flip-flop will be preserved. Suppose that the SR flip-flop initially resides in its reset state with $Q = \mathbf{0}$ and $\overline{Q} = \mathbf{1}$. If the J input is forced high with $K = \mathbf{0}$, the output of AND gate 1 will become high, while the output of AND gate 2 remains low. The SR flip-flop will thus be set to the state $Q = \mathbf{1}$ and $\overline{Q} = \mathbf{0}$. Once $\overline{Q}$ is changed to **0**, the output S of AND gate 1 will return to its value of **0** regardless of the value of J, hence the SR flip-flop will retain its newly set state. Now consider the opposite scenario in which the SR flip-flop initially resides in its set state with $Q = \mathbf{1}$ and $\overline{Q} = \mathbf{0}$. If the K input is forced high with $J = \mathbf{0}$, the output of AND gate 2 will become high, thereby resetting the SR flip-flop to the state $Q = \mathbf{0}$ and $\overline{Q} = \mathbf{1}$. Once Q is reset to **0**, the R output of AND gate 2 will return to **0**, again allowing the SR flip-flop to retain is newly reset state.

A fourth scenario exists in which the J and K inputs are simultaneously forced high. Under these conditions, outputs Q and $\overline{Q}$ change to the opposite, or complement, of their previous values. Suppose, for example, that the flip-flop outputs are initially set to $Q = \mathbf{1}$ and $\overline{Q} = \mathbf{0}$. Applying simultaneous high signals to J and K will, according to Eqs. (15.1) and (15.2), apply inputs $S = \mathbf{0}$ and $R = \mathbf{1}$ to the SR portion of the flip-flop. These inputs will *toggle* the flip-flop, producing the output state $Q = \mathbf{0}$ and $\overline{Q} = \mathbf{1}$. Similarly, if the flip-flop is initially in the state $Q = \mathbf{0}$ and $\overline{Q} = \mathbf{1}$, applying simultaneously high J and K signals will produce the inputs $S = \mathbf{1}$ and $R = \mathbf{0}$, thereby toggling the outputs to the values $Q = \mathbf{1}$ and $\overline{Q} = \mathbf{0}$.

The internal design of the JK flip-flop must overcome a problem called the *race-around* condition that occurs when J and K are made high simultaneously. As discussed above, AND-gating the simultaneous values $J = \mathbf{1}$ and $K = \mathbf{1}$ with Q and $\overline{Q}$ creates the S and R inputs needed to toggle the values of Q and $\overline{Q}$. The new outputs will be established at Q and $\overline{Q}$ after signals have propagated though all the various elements of the flip-flop; hence, when K are forced high, the toggled Q and $\overline{Q}$ signals will appear after a time delay on the order of several gate propagation delay times τ_p, where τ_p is defined as $(t_{PLH} + t_{PHL})/2$. If the J and K inputs retain their high values too long, the arrival of the newly set Q and $\overline{Q}$ values will force the flip-flop to toggle itself again, returning Q and $\overline{Q}$ to their original values. This undesirable anomaly will continue until at least one of the J or K inputs is returned to zero. During the time that J and K are both high, the race-around condition will produce square-wave oscillations with a period on the order of several τ_p.

In practical JK flip-flops, the race-around problem is overcome by a connection called the *master–slave* flip-flop. The latter is produced by cascading a clocked SR flip-flop with a second clocked SR flip-flop, as shown in Fig. 15.7. The first SR flip-flop is called the *master*, and the second is called the *slave*. When the clock input C is high, AND gates 1 and 2 are enabled. Because the signal $\overline{C}$ is an inverted version of the clock input C, AND gates 3 and 4 are disabled while the clock is high. When the clock falls to zero, AND gates 1 and 2 are disabled and gates 3 and 4 enabled. Only when C is low are the newly set Q and $\overline{Q}$ outputs of the master applied to the slave flip-flop. The Q and $\overline{Q}$ outputs of the slave are then set and fed back to the disabled AND gates 1 and 2, where they are ready to interact with input signals J and K upon the next rise of the clock.

Figure 15.7
Master–slave JK flip-flop.

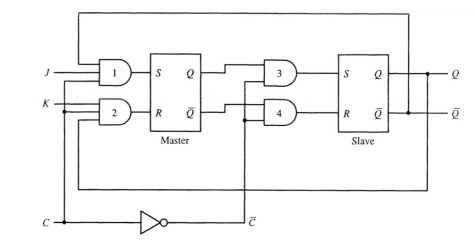

15.1.4 Type D and Type T Flip-Flops

If the J and K inputs of a JK flip-flop are joined by an inverter, as in Fig. 15.8(a), the resulting circuit is called a *delay*, or *type D*, flip-flop. The type D flip-flop serves the function of setting Q to the value of D whenever the clock input C goes through a complete cycle. This action introduces a sequenced delay, determined by the clock period, between the input and output data signals. As in the JK flip-flop, the output $\overline{Q}$ assumes the complement of Q.

If the J and K inputs are connected together, as in Fig. 15.8(b), the JK flip-flop becomes a *toggle*, or *type T*, flip-flop. If the input T is low, outputs Q and $\overline{Q}$ retain their values regardless of the clock signal. If T is high, the inputs J and K are set high simultaneously. Under these conditions, the output of the flip-flop changes state, or toggles, when the clock signal goes through a complete cycle.

Figure 15.8
Variations of the JK flip-flop: (a) type D (delay) flip-flop; (b) type T (toggle) flip-flop.

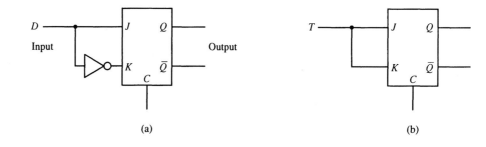

(a)   (b)

15.1.5 Preset and Clear Inputs

Most clocked flip-flops include two other inputs that serve as overrides to the clocked data functions. These inputs, called *preset* and *clear*, are depicted in Fig. 15.9 for several types of flip-flop. If the PR (preset) input is forced high, the flip-flop is set to the state $Q = 1$; $\overline{Q} = 0$ regardless of all other inputs. Similarly, if the CL (clear) input is forced high, the flip-flop is reset to the state $Q = 0$; $\overline{Q} = 1$. When both PR and CL are low, the flip-flop functions normally.

Figure 15.9
Preset (PR) and
clear (CL) inputs.

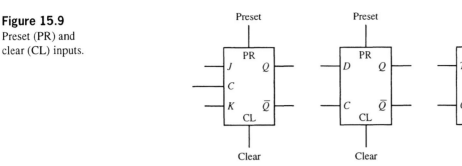

Preset–clear logic capability adds a dimension of versatility to the flip-flop and greatly facilitates the design of logic systems. The implementation of the preset and clear inputs in specific flip-flop circuits is investigated in Problem 15.17.

EXERCISE 15.6   The S input signal of Fig. 15.4 is applied to a type D flip-flop. Draw the resulting Q and $\overline{Q}$ outputs as a function of time.

15.7   Repeat Exercise 15.6 for a type T flip-flop.

15.8   The R input signal of Fig. 15.4 is applied to a type D flip-flop. Draw the resulting Q and $\overline{Q}$ outputs as a function of time.

15.9   Repeat Exercise 15.8 for a type T flip-flop.

15.2 MULTIVIBRATOR CIRCUITS

A multivibrator is a circuit that has two independent states—one with output high and one with output low. The two states of a multivibrator circuit are produced by feeding back the output to the input, thus causing the output to be affected by itself. The flip-flops of the previous section and the Schmitt-trigger op-amp circuit of Chapter 2 are examples of *bistable* multivibrator circuits in which the output state holds itself in place and can only be changed by the application of a counteracting input signal. In this section, we examine other multivibrator circuits in which at least one output state is unstable and controlled by the charging of an *RC* network. Multivibrator circuits of this type produce timing signals, digital pulses, and square waves of fixed period.

15.2.1 Monostable Multivibrator

A *monostable* multivibrator, sometimes called a "one-shot" circuit, produces a single pulse of fixed duration after it receives an input trigger pulse. As its name implies, the monostable multivibrator has only one stable, self-sustaining output state. The other output state is entered momentarily while the pulse is being produced, and only after an input signal has been received. The time duration of the output pulse is determined solely by the characteristics of the multivibrator and is not affected by the duration of the input pulse. After producing its output pulse, the one-shot resets itself and prepares to receive another input trigger.

A simple monostable one-shot is formed by interconnecting two NOR gates with a simple *RC* circuit, as illustrated in Fig. 15.10. In principle, NOR gates from any of the logic families discussed in Chapter 14 can be used, but the inherent symmetry and sharp transition region of CMOS gates makes them ideal for the job. In addition, the input terminals of most CMOS gates are protected by diode limiting circuits that play an important role in the operation of the monostable circuit. We assume the CMOS gates to have sharp inverter transfer characteristics with logic levels of $V_{OH} = V_{DD}$, $V_{OL} = 0$, and a logic transition voltage (the input voltage at which v_{OUT} jumps between V_{OH} and V_{OL}), of value v_{IC}.

Figure 15.10
Digital "one-shot" monostable multivibrator made from two NOR gates and a single RC circuit. The latter is solely responsible for setting the one-shot's pulse duration.

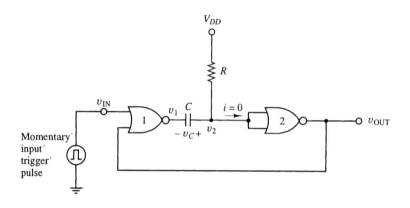

Before the arrival of an input pulse, the circuit lies at rest, with $v_{IN} = 0$ and no voltage fluctuations in time. Under these conditions, the timing capacitor C behaves as an open circuit. The input leads to gate 2 are connected via R to V_{DD}, forcing the output of gate 2 low. With both v_{IN} and v_{OUT} low, the two inputs to gate 1 are also low, hence its output is driven to logic high. In this state, both capacitor terminals lie at V_{DD}, so that $v_C = 0$.

When the input line is made high, the NOR function of gate 1 is initiated, causing its output to be driven low ($v_1 \approx 0$). This action momentarily "grounds" the left-hand side of the capacitor in Fig. 15.10. Because v_C cannot change instantaneously in this circuit, the voltage v_2, which is equal to $v_1 + v_C$, is also momentarily pulled to ground by the uncharged capacitor, thereby causing the output of gate 2 to be driven high. After this initial switching operation, the capacitor begins to charge toward V_{DD} through the resistor R. Until it charges significantly, the low value of v_2 will be sufficient to keep the output of gate 2 high, thus sustaining the low output state of gate 1. The status initiated by v_{IN} thus will remain in effect even when v_{IN} returns to zero.

The status of the circuit *just* after the high input signal forces v_1 low is depicted in Fig. 15.11. The capacitor voltage v_C is initially zero and begins to charge exponentially, following the equation

$$v_C = V_{DD}(1 - e^{-t/RC}) \tag{15.3}$$

Note that the current into gate 2 is zero, so that all of the current flowing through R flows through C. When v_2 rises to the logic threshold level v_{IC} of gate 2, the output of gate 2 returns to zero, signifying the end of the pulse produced by the circuit. If the input signal v_{IN} has returned to zero by the time gate 2 switches low, the output v_1 of NOR gate 1 will again become high. At this point in the sequence, the capacitor voltage will have been charged to the value v_{IC}, so that v_2 will be equal to

$$v_2 = v_1 + v_C = V_{DD} + v_{IC} \tag{15.4}$$

Note that this v_2 exceeds the power-supply voltage V_{DD}. With the left-hand side of the capacitor held at V_{DD} by v_1, the capacitor will discharge through the loop consisting of R, the output terminal of gate 1, and V_{DD}. Once the capacitor has fully discharged, the circuit will reach its initial rest condition with $v_1 = V_{DD}$, $v_C = 0$, and $v_{OUT} = 0$.

Figure 15.11
Status of the one-shot circuit immediately after the application of a $v_{IN} = 1$ input signal at $t = t_1$. The capacitor voltage charges toward V_{DD} and reaches the switching level v_{IC} of gate 2 at time t_3. The current i_2 into the CMOS NOR gate is zero.

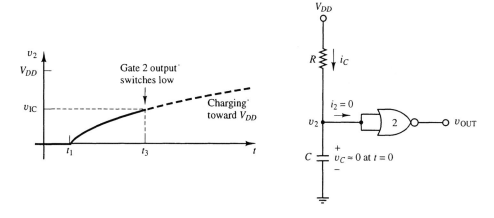

The duration T of the one-shot's output pulse can be precisely determined by computing the time interval required for v_C to charge from zero to the value v_{IC}. Substituting $v_C = v_{IC}$ and $t = T$ into Eq. (15.3) results in

$$v_{IC} = V_{DD}(1 - e^{-T/RC}) \tag{15.5}$$

Solving this latter equation for the pulse period T yields

$$T = -RC \ln(1 - v_{IC}/V_{DD}) \tag{15.6}$$

If gate 2 has a symmetrical transfer characteristic with $v_{IC} = V_{DD}/2$, Eq. (15.6) becomes

$$T = -RC \ln(1 - 0.5) = RC \ln 2 = 0.69RC \tag{15.7}$$

The time required to discharge the capacitor and return the circuit to its rest condition can also be computed. The status of the circuit just after gate 2 switches to zero is depicted in Fig. 15.12. With the resistor and left-hand side of the capacitor both held at V_{DD}, the capacitor will discharge according to the equation

$$v_C = v_{IC}e^{-t/RC} \tag{15.8}$$

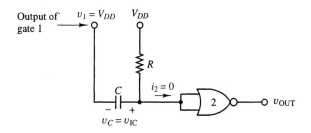

Figure 15.12 Status of the one-shot circuit immediately after gate 2 switches back to its low output state, thereby causing v_1 to enter its logic high ($v_1 = V_{DD}$) output state. The capacitor discharges to zero through R.

Plots of v_C, v_1, v_2, and v_{OUT} versus time are shown in Fig. 15.13. In these plots, the input pulse rises at the arbitrary time t_1 and falls sometime later at t_2; the output pulse rises at t_1 and falls at t_3, where $T = t_3 - t_1 = 0.69RC$.

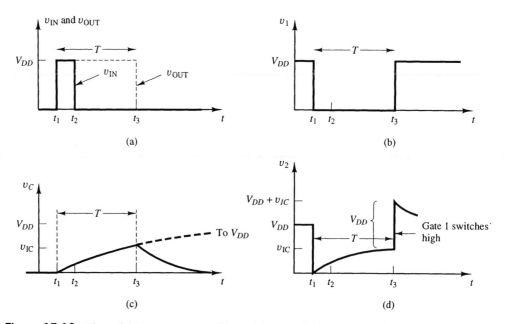

Figure 15.13 Plots of (a) v_{IN} and v_{OUT}; (b) v_1; (c) v_C; and (d) v_2 versus time. The input trigger signal goes high at an arbitrary time $t = t_1$ and falls to zero sometime later at t_2. The output returns to zero at t_3, where $t_3 - t_1 = T$. The values of R and C must be chosen so that $t_3 > t_2$. The jump in v_2 at $t = t_3$ occurs when v_1 switches high, superimposing an additional voltage V_{DD} in series with v_C.

EXERCISE 15.10   Show that connecting the resistor R in Fig. 15.10 to ground, rather than to V_{DD}, results in a one-shot circuit for which logic high is the stable output state.

15.11   Compute the output-pulse duration of the one-shot circuit of Fig. 15.10 if CMOS gate 2 is asymmetric with a logic threshold voltage equal to $v_{IC} = 2V_{DD}/3$. Repeat for $v_{IC} = V_{DD}/4$.
Answer: $1.1RC$; $0.29RC$

15.12   Compute the output-pulse duration of the one-shot circuit of Fig. 15.10 if CMOS gate 1 is asymmetric with a logic threshold voltage equal to $v_{IC} = 2V_{DD}/3$.   **Answer:** $0.69RC$

Clamping Diode

As indicated in the plot of Fig. 15.13(d), the output v_1 of gate 1 jumps from zero to V_{DD} at t_3. Because v_1 and v_C combine in series to produce v_2, the latter voltage will jump to the value $v_{IC} + V_{DD}$ at t_3. This voltage level, which exceeds V_{DD}, may damage gate 2, depending on how the CMOS gate is fabricated. To help alleviate the problem, a clamping diode D_1 is often connected in parallel with R, as illustrated in Fig. 15.14. During most of the one-shot cycle, D_1 is reverse-biased by V_{DD} and remains in the off condition. During the portion of the sequence when v_2 attempts to exceed V_{DD}, D_1 becomes forward-biased and holds v_2 to the value $V_{DD} + V_f$. Although this voltage level is higher than V_{DD}, it will not cause damage to gate 2. With D_1 in place, the capacitor will discharge primarily through the forward-biased diode, rather than through the resistor R. Note that D_1 is often included inside gate 2 as part of the CMOS gate's input-protection circuitry.

Figure 15.14
Clamping diode
added in parallel
with resistor R
prevents v_2 from
rising significantly
above V_{DD}.

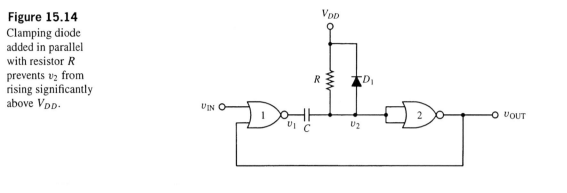

EXERCISE 15.13 Suppose that a clamping diode with forward resistance $r_d \approx 5\,\Omega$ is connected as in Fig. 15.14. If $C = 0.1\,\mu F$ and $R = 10\,k\Omega$, determine the pulse duration T and the time constant associated with the capacitor discharge.   **Answer:** $0.69\,ms$; $0.5\,\mu s$

15.2.2 Astable Multivibrator

The monostable multivibrator of Section 15.2.1 produces a single digital pulse of fixed time duration. If the charging resistor R is connected to the output of gate 2, rather than to V_{DD}, as in Fig. 15.15, the circuit becomes an *astable* multivibrator capable of producing sustained square-wave oscillations. The square-wave period is determined solely by the time constant RC. Note that the circuit has no input signal; the input terminal at gate 1 has been connected to ground.

Figure 15.15
Astable
multivibrator.
Charging resistor R
is connected to the
output of gate 2,
rather than to V_{DD}.

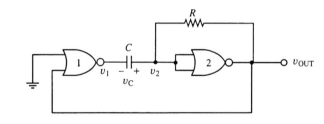

The operation of the multivibrator can be analyzed by supposing it to begin in a state with the output of gate 2 high ($v_{OUT} = V_{DD}$), output of the gate 1 low ($v_1 = 0$), and the capacitor voltage precharged to the negative value $v_C = v_{IC} - V_{DD}$, where v_{IC}, which is less than V_{DD}, is the logic transition level of gate 2. Under these conditions, the status of the capacitor appears as in Fig. 15.16(a). With v_1 low, the input v_2 to gate 2, given by $v_1 + v_C = 0 + v_{IC} - V_{DD}$, will be initially negative and will indeed force the output of gate 2 high. The capacitor will charge toward V_{DD}, causing v_C to increase. When v_2 reaches v_{IC}, the output of gate 2 will be forced low, in turn causing the output v_1 of gate 1 to be forced high. Just prior to this switching operation, the

Figure 15.16
Configuration of
the astable
multivibrator when
(a) v_{OUT} has just
turned high and v_1
low; (b) v_{OUT} has
just turned low and
v_1 high.

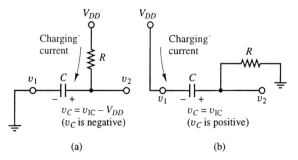

capacitor will have been charged to the value $v_C = v_{IC}$. After the switching operation, the circuit will appear as in Fig. 15.16(b) with the capacitor precharged to the value $v_C = v_{IC}$. With v_1 now equal to V_{DD} and with v_{OUT} equal to zero, v_C will begin to charge in the opposite direction toward $-V_{DD}$, thereby causing v_2 to fall. When v_2 falls below the logic transition level v_{IC}, the output of gate 2 will switch high, forcing v_1 low. With v_2 equal to $V_{DD} + v_C$, the value $v_2 = v_{IC}$ will be reached in this second case when $v_C = v_{IC} - V_{DD}$. After the switching operation, the circuit will again appear as in Fig. 15.16(a), with v_C precharged to $v_{IC} - V_{DD}$, which is the circuit condition originally assumed. The cycle will thus repeat itself, continuing indefinitely. The circuit will produce a square-wave output at v_{OUT} and its logical complement at v_1. Plots of the four principal voltages in the circuit are shown in Fig. 15.17.

Figure 15.17
Waveforms associated with the astable multivibrator of Fig. 15.16:
(a) output of gate 2;
(b) output of gate 1;
(c) capacitor voltage; (d) net input to gate 2.

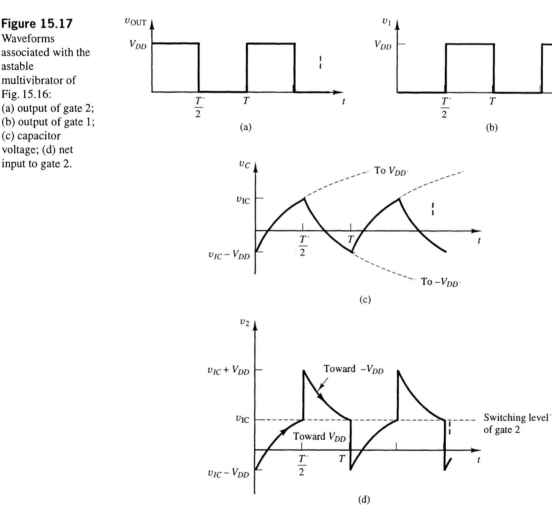

The period of the square wave produced by the circuit can be computed by determining the time required for v_2 to charge from $v_{IC} - V_{DD}$ to v_{IC} in the circuit of Fig. 15.16(a). Over the time interval when this circuit configuration is valid, v_2 will equal v_C and will be described by the equation

$$v_2 = (v_{IC} - V_{DD})e^{-t/RC} + V_{DD}(1 - e^{-t/RC})$$
$$\equiv v_{IC}e^{-t/RC} + V_{DD}(1 - 2e^{-t/RC}) \tag{15.9}$$

where $t = 0$ is defined as the point where the output of gate 2 just switches high. As a check, we note that Eq. (15.9) yields the values $v_2 = v_{IC} - V_{DD}$ at $t = 0$ and $v_2 = V_{DD}$ for $t = \infty$. (The capacitor voltage v_C never reaches V_{DD}, but merely charges *toward* V_{DD} until the circuit switches state and begins to charge the capacitor toward $-V_{DD}$). As indicated in the plot of Fig. 15.17(c), the capacitor voltage will reach v_{IC} after a half period $T/2$. Substituting the values $v_2 = v_C = v_{IC}$ and $t = T/2$ into Eq. (15.9) and solving for T results in

$$T = 2RC \ln \frac{2V_{DD} - v_{IC}}{V_{DD} - v_{IC}} \qquad (15.10)$$

For a symmetrical CMOS gate with $v_{IC} = V_{DD}/2$, Eq. (15.10) reduces to

$$T = 2RC \ln \frac{2V_{DD} - V_{DD}/2}{V_{DD} - V_{DD}/2} = 2RC \ln \frac{3/2}{1/2} = 2RC \ln 3 \approx 2.2RC \qquad (15.11)$$

Note that this period is the same one produced by the Schmitt-trigger oscillator of Chapter 13 when $R_1 = R_2$.

The astable multivibrator circuit of Fig. 15.15 produces both negative voltages and voltages that exceed V_{DD} at the input of gate 2. These voltages may damage the gate, depending on which CMOS process is used to fabricate it. As in the monostable circuit of Fig. 15.14, clamping diodes are often used to limit v_2 to values close to 0 and V_{DD}.

EXERCISE 15.14   Show that the capacitor voltage in the astable multivibrator circuit of Fig. 15.15 will have a dc component if $v_{IC} \neq V_{DD}/2$.

15.15   Find an expression for the capacitor voltage over the interval $T/2 < t < T$ in Fig. 15.17(c). Show that the result (15.11) can also be derived using this latter equation for v_C.

15.16   Show that the operation of the astable multivibrator of Fig. 15.15 will be the same if both inputs of gate 1 are connected to v_{OUT}.

15.17   Show that the operation of the astable multivibrator of Fig. 15.15 will be the same if only one input to gate 2 is connected to the capacitor, with the remaining input connected to ground.

15.2.3 The 555 IC Timer

The 555 IC timer chip is a multivibrator circuit available in premade, integrated-circuit form. Originally introduced by Signetics Corporation, a version of the 555 is now produced in bipolar or CMOS form by just about every major IC manufacturer. This versatile IC can be connected as an astable or monostable multivibrator and can perform many additional digital timing and switching functions.

The basic architecture of the 555 is shown in Fig. 15.18, where the components inside the shaded box are located on the IC. In order to perform its various functions, the 555 requires a number of external components, shown outside the box in Fig. 15.18 for a monostable application. At the heart of the timer lies a simple set–reset (SR) flip-flop. Only the $\overline{Q}$ output is shown in Fig. 15.18. The S and R inputs are fed by open-loop analog comparators A_1 and A_2, which are similar to the op-amps of Chapter 2. The v_- input of A_1 and the v_+ input of A_2 are connected to a voltage divider formed by the three resistors labeled R_1. The divider produces the voltages $V_1 = 2V_{CC}/3$ and $V_2 = V_{CC}/3$. The $\overline{Q}$ output of the flip-flop feeds v_{OUT} via an inverting buffer, so that v_{OUT} is the inverse of $\overline{Q}$. The $\overline{Q}$ output also feeds the base of transistor Q_1 via a current-limiting resistor R_2.

Figure 15.18
Architecture of the
555 timer. The
components R_A
and C result in
monostable, or
"one-shot"
operation.

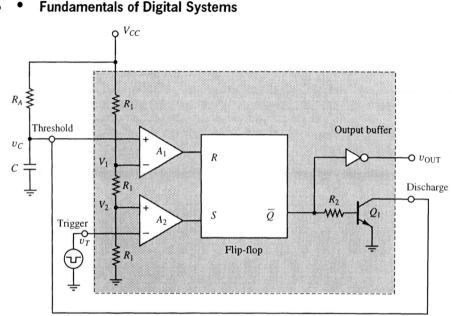

Monostable Mode

The circuit of Fig. 15.18 is connected as a monostable, or "one-shot", multivibrator. It produces a single output pulse in response to a single input trigger signal. For this particular circuit, the trigger input v_T is held high until the pulse is desired. Allowing v_T to fall initiates the timing sequence. Suppose that the circuit begins in a rest condition with v_T *high* at V_{CC}, the flip-flop in the *reset* state ($\overline{Q}$ high), transistor Q_1 saturated by $\overline{Q}$, and the capacitor held at the voltage $v_C = V_{sat}$ by Q_1. Under these conditions, the voltage $V_1 = 2V_{CC}/3$ will exceed $v_C = V_{sat}$, forcing the output of A_1 low. Similarly, the trigger input $v_T = V_{CC}$ will exceed $V_2 = V_{CC}/3$, forcing the output of A_2 low. With both S and R signals low, the flip-flop will remain in its reset state until the circuit conditions are changed.

Now suppose that the trigger input v_T is momentarily forced *low*. This action will cause V_2 to exceed v_T, thereby forcing the output of A_2 high and forcing the flip-flop into its set state. With $\overline{Q}$ low, transistor Q_1 will go into cutoff, in turn releasing C to become charged by the combination of V_{CC} and R_A. The charging of C will continue, with the flip-flop remaining in its set state, until v_C becomes more positive than V_1. When this latter threshold is reached, the output of A_1 will be forced high, thereby resetting the flip-flop and turning on Q_1 via the $\overline{Q}$ output. The conducting transistor will quickly discharge the capacitor to the voltage V_{sat}, returning the circuit to its original condition.

Plots of the various waveforms associated with the monostable capacitor charging cycle are shown in Fig. 15.19. Note that v_{OUT} goes high while the capacitor charges and $\overline{Q}$ is low. The timing of the cycle—that is, the duration of the output pulse—will depend only on the values of R_A and C, and not on the duration of the inverted trigger pulse v_T. The only requirement is that the v_T pulse be low for a duration shorter than the output pulse.

A value for the pulse duration T can be found by straightforward calculation. Suppose that the trigger input v_T is forced low at $t = 0$. The capacitor will then follow a charging trajectory given by

$$v_C = V_{sat}e^{-t/R_A C} + V_{CC}(1 - e^{-t/R_A C}) \tag{15.12}$$

as it attempts to charge toward V_{CC}. The time at which v_C reaches $V_1 = 2V_{CC}/3$, signifying the resetting of the flip-flop and the end of the output pulse, can be found by substituting $v_C = 2V_{CC}/3$

Figure 15.19
Waveforms
associated with the
monostable pulse
produced by the
circuit of
Fig. 15.18.

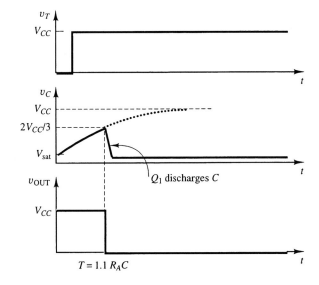

into Eq. (15.12) and solving for T. Performing this operation results in

$$\frac{2}{3}V_{CC} = V_{\text{sat}}e^{-T/R_A C} + V_{CC}(1 - e^{-T/R_A C}) \tag{15.13}$$

or

$$T = R_A C \ln \frac{V_{CC} - V_{\text{sat}}}{V_{CC}/3} \tag{15.14}$$

In the limit $V_{\text{sat}} \ll V_{CC}$, Eq. (15.14) reduces to $T = R_A C \ln 3 \equiv 1.1 R_A C$.

Astable Mode

If the 555 timer is connected in the *astable* configuration of Fig. 15.20, it will produce a square wave indefinitely. In this configuration, the v_+ input of A_1 and the v_- input of A_2 are both connected to the capacitor. Suppose that the circuit begins with $\overline{Q}$ high, Q_1 in saturation, and the capacitor voltage falling toward V_{sat}. When v_C falls just below $V_2 = V_{CC}/3$, the output S of A_2 will be forced high and the output R of A_1 will remain low. The switching of S will force the flip-flop into its set state with $\overline{Q}$ low, and transistor Q_1 will be forced into cutoff. With Q_1 in cutoff, the capacitor will be released and will begin to charge toward V_{CC} via the series combination of R_A and R_B. When v_C surpasses V_2, A_2 will be forced low, leaving S and R both low and preserving the set state of the flip-flop.

As the capacitor continues to charge, v_C will eventually reach the value $V_1 = 2V_{CC}/3$, at which point the output R of A_1 will be forced high and the flip-flop will be reset, forcing $\overline{Q}$ high. This action turns on Q_1, which will discharge the capacitor via R_B toward the saturation voltage V_{sat}. Note that, unlike the monostable circuit of Fig. 15.18, Q_1 is connected to C via R_B, so that the discharging of C occurs with approximate time constant $R_B C$, rather than instantaneously. After the capacitor has been discharged for some time, v_C will fall just below the value $V_2 = V_{CC}/3$, thereby forcing the output of A_2 high and beginning the cycle all over again. The process will continue indefinitely, producing a square-wave output at the v_{OUT} terminal.

The waveforms and timing intervals associated with the astable cycle are shown in Fig. 15.21. The period of the square-wave output can be determined by considering the capacitor charge and discharge cycles. Suppose that $t = 0$ is assigned to the time at which v_C just falls to $V_2 = V_{CC}/3$, causing the S input to switch high. During the subsequent capacitor charging interval, v_C will follow a trajectory toward V_{CC}, given by

$$v_C = \frac{V_{CC}}{3}e^{-t/(R_A+R_B)C} + V_{CC}(1 - e^{-t/(R_A+R_B)C}) \tag{15.15}$$

Figure 15.20
Connections to the
555 timer for
astable operation.
The output consists
of a continuous
square wave.

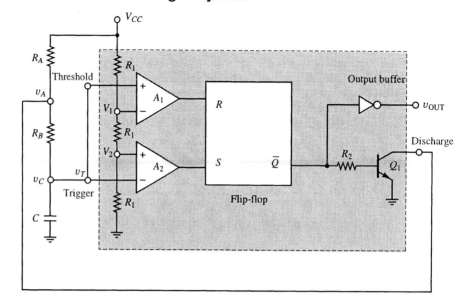

The time t_1 at which v_C reaches $V_1 = 2V_{CC}/3$ can be found by substituting $v_C = V_1$ into Eq. (15.15), resulting in

$$2V_{CC}/3 = \frac{V_{CC}}{3}e^{-t_1/(R_A+R_B)C} + V_{CC}(1 - e^{-t_1/(R_A+R_B)C}) \qquad (15.16)$$

This latter equation can be simplified to the form

$$\frac{2}{3}V_{CC}e^{-t_1/(R_A+R_B)C} = \frac{V_{CC}}{3} \qquad (15.17)$$

and solved for t_1 to yield

$$t_1 = (R_A + R_B)C \ln 2 = 0.69(R_A + R_B)C \qquad (15.18)$$

Figure 15.21
Waveforms
associated with
astable operation of
the circuit of
Fig. 15.20.

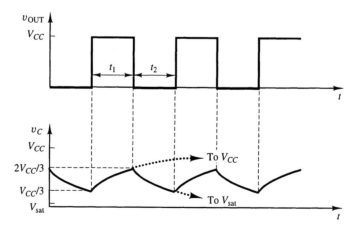

When v_C reaches V_1 at time t_1, the flip-flop will be reset and Q_1 will be forced into saturation. The capacitor will then be discharged by Q_1 via R_B and the capacitor voltage will follow a new trajectory toward V_{sat}, given approximately by

$$v_C = \frac{2}{3}V_{CC}e^{-t/R_BC} + V_{\text{sat}}(1 - e^{-t/R_BC}) \qquad (15.19)$$

where $t = 0$ in this equation has been newly assigned to the old time t_1. Note that only R_B appears in the equation in this case. The equation is approximate, because it assumes the transistor voltage to remain at a constant V_{sat} for the entire discharge interval. Solving Eq. (15.19) for the time t_2 required for v_C to fall from $2V_{CC}/3$ to $V_2 = V_{CC}/3$ results in

$$\frac{V_{CC}}{3} = \frac{2}{3} V_{CC} e^{-t_2/R_B C} + V_{sat}(1 - e^{-t_2/R_B C}) \tag{15.20}$$

or
$$t_2 = R_B C \ln \frac{2V_{CC}/3 - V_{sat}}{V_{CC}/3 - V_{sat}} \tag{15.21}$$

In the limit $V_{sat} \ll V_{CC}$, Eq. (15.21) reduces to

$$t_2 \approx R_B C \ln 2 = 0.69 R_B C \tag{15.22}$$

Under these conditions, the overall period of the square wave becomes

$$T = t_1 + t_2 = 0.69(R_A + R_B)C + 0.69 R_B C = 0.69(R_A + 2R_B)C \tag{15.23}$$

EXERCISE 15.18 The 555 timer of Fig. 15.18 is connected as a monostable multivibrator with $R_A = 10\,k\Omega$ and $C = 0.01\,\mu F$. What is the duration of the pulse that occurs when the trigger input falls low?
Answer: 0.11 ms

15.19 A pulse duration of approximately 10 ms is desired from the 555 timer circuit of Fig. 15.18. Choose values for R_A and C so that this objective can be met.
Answer: (one possible choice) $R_A = 27\,k\Omega$, $C = 0.33\,\mu F$

15.20 The 555 timer of Fig. 15.20 is connected as an astable multivibrator with $R_A = R_B = 10\,k\Omega$ and $C = 0.27\,\mu F$. What are the period and frequency of the resulting square wave?
Answer: 5.6 ms; 0.18 kHz

15.21 A square-wave frequency of approximately 10 kHz is desired from the 555 timer circuit of Fig. 15.20. Choose values for R_A, R_B, and C so that this objective can be met.
Answer: (one possible choice) $R_A = 820\,\Omega$, $R_B = 470\,\Omega$, $C = 0.082\,\mu F$

15.3 DIGITAL MEMORY

Memory circuits play an important role in much of the digital world. The widespread use of computers has been made possible by the development of inexpensive and reliable VLSI memory chips utilizing NMOS, CMOS, BJT, and BiCMOS technologies. The very first computer memories consisted of minute magnetic toroids, called *core memory*, that were written to and accessed by tiny wires thread through the centers of the toroids. Core memory required large, bulky circuit boards stored in large cabinets. Semiconductor electronic memory, the modern offspring of core memory, is made entirely from transistors on very-large-scale (VLSI) integrated circuits. Compared to its archaic ancestor, semiconductor memory is very compact and suitable for installation on printed circuit boards. It can be accessed at very high speeds and can store data in extremely high densities (measured in the number of digital bits per chip). All modern computer and microprocessor systems contain semiconductor memory in one form or another.

Electronic memory differs in function from the storage media found in other parts of a computer system. Information from magnetic and optical storage devices such as hard disks, floppy disks, CD-ROM, and digital tape must be accessed sequentially, starting at the beginning of a data file or track. In contrast, data stored in an electronic memory cell can be accessed at random and on demand using *direct addressing*. Direct addressing eliminates the need to process a large stream of irrelevant data in order to find the desired data word. In this section, we examine several circuit configurations and addressing schemes relevant to the most common types of VLSI memory.

Electronic memory cells are most often arranged in ordered arrays in which each cell is assigned an individual digital address. In a *read-only memory*, or ROM, array, data bits are stored once during chip fabrication using a process called *masking*. The data stored in a ROM array can be rapidly accessed by the computer or microprocessor at any time. Information stored in ROM is retained even when power to the memory circuit is removed. ROM circuits are typically used to provide the computer with resident programs and key operating functions needed to "boot" the computer's operating system.

Random-access memory, or RAM, has the same characteristics as ROM but allows the computer to both read and write information to each memory cell. The information stored in a RAM array is easily accessed but is lost when power to the memory circuit is removed. Random-access memory is the most abundant type found in modern computers and is used extensively during program execution.

Erasable-programable read-only memory, or EPROM, provides a medium in which the stored information remains after power is removed. Unlike ROM, however, the information in an EPROM array is stored electrically and can be erased by applying strong ultraviolet light to the surface of the integrated circuit. A UV-transparent quartz window is provided on the top of the IC package by the manufacturer for this purpose. EPROM ICs are used extensively in the prototype-development stages of microprocessor and computer systems. As a general rule, EPROM circuits are more expensive per unit than mass-produced ROM circuits of similar density, but the ability to quickly change their stored information makes EPROM ICs invaluable to the design process.

Erasing and reprogramming an EPROM IC takes several minutes and requires the application of external light. The reprogramming operation is not suitable as a routine computer task. In contrast, the information stored in an electrically erasable-programable read-only memory array, or EEPROM, can be erased electrically as part of normal system operation. Like ROM and EPROM circuits, the information stored on an EEPROM circuit is preserved when power is removed. Unlike an EPROM, however, the information stored in an EEPROM array can be changed relatively rapidly using electrical signals generated by the host computer or microprocessor. EEPROM circuits are typically found in user-programmable devices such as computer terminals, laser printers, video recorders, microwave oven control panels, and compact-disc players. The principal disadvantage of EEPROM circuits is their cost, which can be several times that of ROM or EPROM arrays of similar size. EEPROM arrays also cannot be manufactured with the same large packing density found in conventional ROM and RAM circuits.

15.3.1 Read-Only Memory

Read-only memory, or ROM, circuits can be made using NMOS, CMOS, BJT, or BiCMOS technology. A typical ROM circuit made in NMOS is illustrated in Fig. 15.22. Data bits are stored in a rectangular array in which each horizontal row of cells represents a digital word. An array of 4-bit words is shown in the figure. In principle, ROM arrays of any row length can be made, although 8, 16, 32, and 64-bit row lengths are typical. A given bit represents a logic **0** if a transistor is present in its cell; a cell with no transistor represents a logic **1**. If a transistor is present in the cell, its gate is connected to the cell's row line (the horizontal wires in the figure) and its drain to the data, or column line (the vertical wires in the figure). Each data line is connected to V_{DD} via an NMOS pull-up load. If a transistor is present in a given cell, it will form an inverter with the data line's pull-up load, with output taken at one of the corresponding data terminals D_0 through D_3. When the transistor's gate lead is energized to V_{DD} by its row line, its gate-to-source voltage will rise above V_{TR} and the MOSFET will conduct. The activated transistor will then cause the voltage on the data line to go low. The transistors with gates connected to all nonactivated row lines will remain in cutoff and will not affect the voltage on the data lines. Only one row line is activated at any given time, hence the data bits stored in a given row are made to appear on data output terminals D_0 through D_3 when the row line is forced high. A given row line is addressed by a combinational circuit that accepts an address in the form of a binary number on lines A_0 through A_3 and forces high the solitary row line that corresponds to the address number. Those data lines connected by row transistors are then forced low. For the array depicted in Fig. 15.22, for example, the first row holds the binary word **0100**, and the second through fourth rows the words **1001**, **1110**, and **1011**, respectively. The presence or absence of a transistor in a given ROM cell is determined by a masking operation that occurs when the integrated circuit is fabricated.

Figure 15.22
NMOS read-only memory (ROM) array.

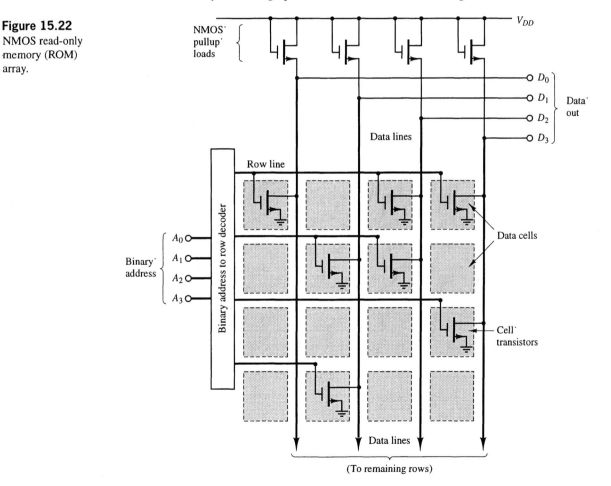

ROM Fabrication

In order to avoid repeating costly refabrication processes each time a new chip is programmed, all ROM arrays of a given type are made using the same basic layout configuration. Specifically, strips of drain and source area implants are brought to each cell location, including cells that eventually will hold no transistor. The drain strip implants become the data lines, as indicated in

Fig. 15.23, and the source strip implants are grounded. For each row in the array, a thin oxide layer that will hold future gates is deposited across the drain and source strips, thus forming possible sites for transistor placement. A thick oxide layer covered with polysilicon (a conductor) is also included beneath each row to act as a row line. The oxide beneath the row line is so thick that the threshold voltage of any MOS devices formed by the intersection of the polysilicon and the drain and source strips will have threshold voltages much higher than V_{DD}; these devices will always remain in cutoff.

Figure 15.23

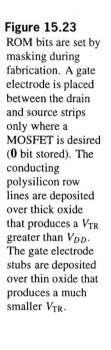

ROM bits are set by masking during fabrication. A gate electrode is placed between the drain and source strips only where a MOSFET is desired (**0** bit stored). The conducting polysilicon row lines are deposited over thick oxide that produces a V_{TR} greater than V_{DD}. The gate electrode stubs are deposited over thin oxide that produces a much smaller V_{TR}.

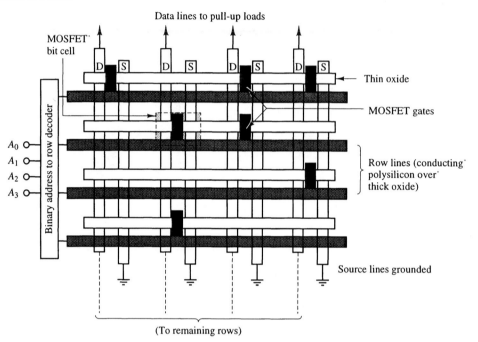

Working MOSFETs are placed in the **0**-bit cell sites of the array during fabrication. When the conducting polysilicon layer is deposited over the row-line oxide, a user-specified mask is used to simultaneously deposit polysilicon over the thin oxide in the gate regions of those cells where a working MOSFET is desired. The gates so deposited in Fig. 15.23 are shown as short stubs extending vertically from the row lines. Because the oxide beneath the stubs is thin, these gates form transistors with $V_{TR} < V_{DD}$ between the drain and source strips. In locations where a transistor is not desired, the gate region between the drain and source strips is left bare. In this way, the array is filled with transistors in all desired **0**-bit locations. This fabrication technique requires the user to pay the cost of creating only the gate programming mask. The same masks are used for all other stages of the fabrication process regardless of the ROM contents.

15.3.2 Static Random-Access Memory

Static random-access memory, or SRAM, is constructed by placing a complete flip-flop in each cell location of a memory array. The added complexity of a flip-flop over a single-transistor memory cell greatly increases the amount of chip area required. As a result, static RAM chips are much less compact and hold fewer data bits per unit area than do ROM arrays of comparable storage size. A static RAM array also requires more time to read and write data from its cells. The use of *dynamic* RAM, to be discussed in Section 15.3.3, drastically reduces the transistor count, but has the disadvantage that its contents must be refreshed periodically during use. Hence its support circuitry is more complicated.

In order to minimize the number of transistors in a static RAM array, the flip-flop cell is kept as simple as possible. Two traditional six-transistor flip-flop cells—one NMOS and the other CMOS—are shown in Fig. 15.24. In each case, the cross-connection of inverters formed by transistors Q_1 to Q_4 resembles the SR flip-flop of Section 15.1.1. The MOSFETs labeled Q_A, sometimes called *access transistors*, are used to access the data cell during read and write operations. The status of the flip-flop is set not by energizing separate set and reset lines, but rather by forcing the data outputs D and $\overline{D}$ to opposite values using external driving circuitry. The D and $\overline{D}$ outputs of a memory cell are similar to the Q and $\overline{Q}$ outputs of an SR flip-flop, but also serve as the forced inputs. If the pass transistors Q_A are turned on via the row select line, and the data lines are forced into a state with D high and $\overline{D}$ low, MOSFET Q_1 in Fig. 15.24 will be turned on and Q_2 off. When the forcing signals are removed, Q_1 will continue to hold $\overline{D}$ low, which will, in turn, keep Q_2 off and the D output high. The forced state is thus self-sustaining and stable. A similar stable state exists with D low and $\overline{D}$ high. For either case, the stored data will be held by the flip-flop until it is either changed by new forcing signals or until the power to the circuit is removed.

Figure 15.24
Two versions of a flip-flop data cell: (a) NMOS with depletion-mode load; (b) symmetrical CMOS.

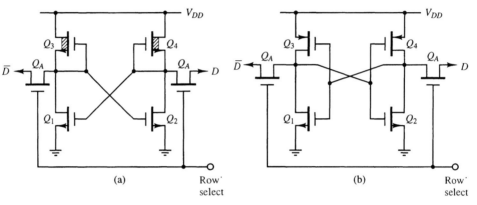

The alternative CMOS memory cell of Fig. 15.25 uses a single data bus connected to the flip-flop via transmission gates 1 and 2. Gate 1 is made conducting during a write operation via the complementary signals W and $\overline{W}$; gate 2 is made conducting during a read operation by the signals R and $\overline{R}$. During write operations, transmission gate 3 is used to disconnect the two inverters of the flip-flop, thereby breaking the feedback loop and simplifying the writing of new data into the cell.

Figure 15.25
CMOS flip-flop with loop interrupter and read/write transmission gates.

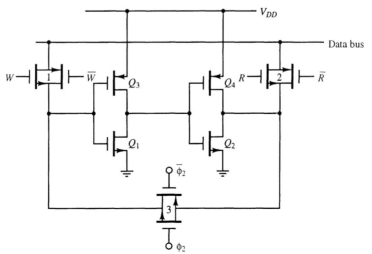

A complete RAM array is made by adding precharge and sense circuitry to a matrix of data cells. Precharge circuitry is used during both read and write operations, and sensing circuitry, included in the form of a flip-flop-connected sense amplifier, is used during read operations. One common RAM configuration is illustrated in CMOS form in Fig. 15.26, where only one of the many flip-flop data cells is shown. To save chip space, one sense amplifier is used to service an entire column of data cells. Each cell is accessed when both its row-select and column-select lines are forced high using the address decoding scheme discussed in Section 15.3.1. This action causes the access transistors Q_A to conduct, thereby connecting the flip-flop cell to the D and $\overline{D}$ lines. Note that the Q_A transistors do not have an arrow symbol indicating the source terminal. The drain and source regions of most MOSFETS on a digital IC are physically identical, so that the device has geometric symmetry and is able to conduct freely in both directions.

Figure 15.26
One possible architecture for a static CMOS RAM.

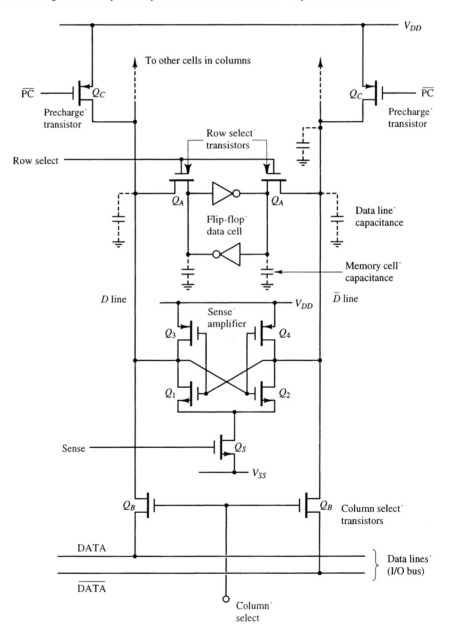

When it is desired to write a **1** to the cell, write control circuitry on the RAM chip first forces the I/O bus into a state with DATA = **1** and $\overline{\text{DATA}}$ = **0**. The cell is then accessed by forcing its row-select and column-select lines high, thus connecting the cell, via the D and $\overline{D}$ data lines, to the I/O bus. This action forces the flip-flop into a state with the D side high and the $\overline{D}$ side low. This logic **1** state will remain after the write operation has been completed and the cell is disconnected from the D lines. Writing a **0** into the cell requires a similar sequence in which the I/O lines are first forced into a state with DATA low and $\overline{\text{DATA}}$ high.

The cell is read by accessing it via the Q_A access transistors while the column's SENSE line is energized. With SENSE at logic high, MOSFET Q_S is turned on and the sense amplifier is activated. Suppose that the data cell has been storing a logic **1**. When the access transistors Q_A are turned on, the capacitance of the D line will begin to charge toward V_{DD}, and the capacitance of the $\overline{D}$ line will begin to discharge toward zero. This action causes a voltage differential to develop between the D and $\overline{D}$ lines. The voltage difference is amplified by the sense amplifier, which supplies the additional current needed to quickly establish the values V_{DD} and zero on the DATA and $\overline{\text{DATA}}$ I/O lines. The latter are then ready for the subsequent reading of data. A similar process occurs when the cell has been storing a logic **0**. In this latter case, the voltage difference develops in the opposite direction, and the I/O bus is charged to the state DATA = **0**; $\overline{\text{DATA}}$ = **1**.

The most time-consuming part of the read operation is the charging and discharging of the capacitance of the data lines. To help speed up this process, the data lines are first precharged to the value $V_{DD}/2$ by the *precharge* transistors Q_C. This action moves the voltages of the data lines "halfway" to their ultimate destinations of V_{DD} or zero, thus greatly reducing the charge/discharge times. With D and $\overline{D}$ both precharged to $V_{DD}/2$, the sense amplifier is immediately driven to its crossover point, allowing it to more quickly respond to the voltage differential that develops between the D and $\overline{D}$ lines as the data bit is read from the memory cell.

15.3.3 Dynamic Random Access Memory

The static memory described in the previous section has the disadvantage that it requires several transistors for each data cell. This feature greatly reduces the amount of data that can be stored per unit area. The problem can be overcome by using the dynamic random-access memory (DRAM) configuration of Fig. 15.27. Because of its high packing density and relatively low power consumption, the DRAM array has become the de facto memory element of choice in many computer applications, including desktop personal computers and workstations. The principal disadvantage of DRAM is the more complex circuitry required to read, write, and refresh its data.

The circuit in Fig. 15.27 shows the key elements of a single DRAM memory cell. The bit storage capacitor C_B is charged to V_{DD} when the cell stores a **1**; this same capacitor is left uncharged when the cell stores a **0**. The capacitor C_B is fabricated on the integrated circuit and is very small—usually on the order of 0.05 pF. Typically, C_B is formed using some type of three-dimensional structure that extends vertically into the semiconductor substrate. Popular construction methods include the trench cell and stack cell configurations.

In order to access the cell, its row and column select lines must both be driven high. Suppose that a **1** has been previously stored in the cell, so that C_B is charged to V_{DD}. During a read operation, the cell transistor Q_1 is turned on, thereby connecting C_B to the load capacitance C_L. The latter represents the distributed parasitic capacitance of the data line and is not a fabricated on-chip capacitor. Nevertheless, C_L is much larger than C_B, so that most of the charge on the latter is transferred to the former, causing the voltage on the sense line to rise by the value

$$\Delta v = V_{DD}\frac{C_B}{C_B + C_L} \tag{15.24}$$

Figure 15.27
Dynamic RAM
data cell. A **1** bit is
stored by charging
the bit capacitor
C_B. A **0** bit is
stored by leaving
C_B uncharged.

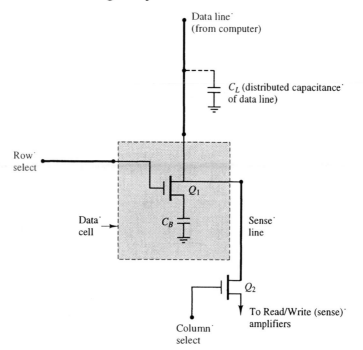

This change in voltage is fed to a sense amplifier via the column-access transistor Q_2. Upon receiving the Δv voltage signal, the sense amplifier produces a high output signal equal to V_{DD}. The latter is then applied to the column's data line where its content is read by the computer. At the same time, the high data line is used to recharge C_B to V_{DD}, thus refreshing the **1** bit stored in the cell.

If a **0** bit has been previously stored in the cell, the cell capacitor C_B will be uncharged, and no change in voltage Δv will occur when the cell is accessed. When $\Delta v = 0$, the sense amplifier sets the column's data line to zero, and the capacitor in the cell remains uncharged. Note that the contents of the cells in a given row are automatically refreshed during a read operation using this method, because the data lines of those cells storing a **1** will be forced high during read, while the data lines of those cells storing a **0** will be forced to zero.

In order to write a **1** to a particular cell, the cell is first accessed via its row- and column-select lines. A voltage of V_{DD} is then sent in over the sense line via Q_1 and Q_2, thereby charging the cell capacitor. If a **0** is to be written to the cell, the sense line is forced to zero, and any charge previously stored on the capacitor is discharged via Q_1 and Q_2.

One key problem associated with the DRAM cell configuration of Fig. 15.27 is that any voltage stored on unread cell capacitors will slowly discharge with time over leakage paths that will inevitably be present on the chip surface and substrate. In order to preserve high data bits, all the cell capacitors in a DRAM array must be refreshed periodically by reading their data one row at a time. During this refresh operation, the memory array is not available for random-access read/write operations. The typical DRAM array is refreshed about once every 1 to 2 ms.

EXERCISE 15.22   Estimate the maximum resistive leakage path that can be tolerated by a 0.05-pF cell storage capacitor if its cell data bit is refreshed every 1 ms and its voltage can fall to no more than 50% of V_{DD}.

15.3.4 EPROM and EEPROM Memory Elements

A compromise between the permanence of ROM and the erasability of RAM systems is realized by the use of erasable-programmable read-only memory elements, or EPROM. An EPROM memory array has many of the same features as the ROM array of Section 15.3.1, and has the general appearance of Fig. 15.22. In an EPROM array, however, row lines and data lines are connected using a *floating-gate* MOSFET, depicted in ideal form in Fig. 15.28. The floating-gate device has two gates. The inner, or floating, gate is completely surrounded by oxide on all sides. This construction isolates the inner electrode and virtually eliminates any conducting leakage paths that would otherwise connect it to the outside world. In addition, the floating gate contains a section where it lies in close proximity to the substrate, separated only by a very thin oxide layer. The outer gate is contacted via an external gate terminal and functions in a manner similar to the gate of an ordinary MOSFET. Applying a voltage between the outer gate and the source produces an electric field that couples capacitively through the inner gate to the substrate surface, causing the MOSFET to follow the usual drain-to-source v–i characteristics.

Figure 15.28
Cross-sectional diagram of the floating-gate MOSFET used in EPROM circuits.

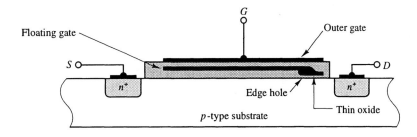

In an EPROM array, a floating-gate transistor is present in each cell location. If the floating gate is left uncharged, the MOSFET will conduct when its gate is energized via its row line, causing a logic **0** to appear on the cell's data line. The floating-gate MOSFET can be made nonconducting, however, by forcing negative charge onto its floating-gate electrode. This task is accomplished during a write operation by grounding the source and energizing the drain to a high (16 to 20 V) voltage and the gate to a somewhat higher (25 to 28 V) voltage. As shown in Fig. 15.29, this arrangement produces a conducting channel between the drain and source. Electrons flowing beneath the thin oxide region under the floating gate acquire a large velocity and also experience the field extending from the floating gate to the substrate. These conditions cause a few energetic electrons to migrate through the oxide layer, where they are attracted to and flow onto the floating gate. The process is self-limiting, in that any negative charge accumulated on the floating gate reduces the field extending to the substrate surface. When the programming voltages are removed, the charge collected by the floating gate persists. Because this charge is negative, it acts to reduce the effect of any v_{GS} applied to the outer gate, thereby increasing the effective threshold voltage of the device. The device is designed such that the threshold voltage exceeds 5 V when the floating gate holds charge. Under these conditions, the device will not conduct when its gate is energized by its row line, and a logic **1** will appear on its data line.

The charge stored on the floating gate of an EPROM device will remain for an extremely long time. Extrapolated estimates suggest that an EPROM cell can hold its data bit anywhere from 10 to 100 years. Conversely, the charge on a floating-gate MOSFET can be removed by shining ultraviolet light of the correct wavelength on the floating gate. This procedure transfers additional energy to the electrons, allowing them to overcome the energy barrier of the oxide and relax back to the substrate. The typical EPROM chip contains a quartz window and an exposed circuit

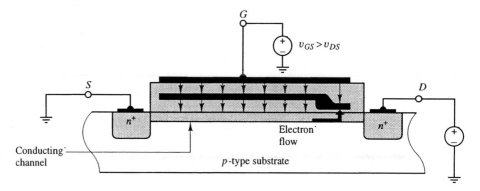

Figure 15.29 EPROM circuit during a write operation. The programming voltages v_{GS} (25 to 28 V) and v_{DS} (16 to 20 V) jointly cause electrons to flow to the floating gate, thereby increasing the equivalent threshold voltage of the MOSFET.

pattern that allows all devices to be erased simultaneously. The procedure usually takes several minutes to an hour. Once erased, the devices in the EPROM array are ready to be reprogrammed and to accept new data. The write/erase process, though time-consuming, can be repeated many times. An EPROM memory array is thus ideal for development and prototype work, where an occasional change in stored data is required.

For applications where erasure and reprogramming must be performed on a regular basis, the electrically erasable-programmable read-only memory, or EEPROM, adds additional flexibility. Like an EPROM array, an EEPROM array also contains a floating-gate MOSFET in each cell location. In this case, the floating gate is charged by applying a large voltage to the gate with the drain and source grounded. A process known as *tunneling* causes electrons to flow through the thin oxide region, thereby charging the floating gate, raising the device's threshold voltage, and causing it to store a logic **1** in its data cell. The floating gate is discharged by applying a large positive voltage to the drain with the gate grounded. This action reverses the process, causing electrons to flow from the charged floating gate to the substrate, thereby restoring the floating gate to its uncharged condition. The charge/discharge process takes a second or so rather than minutes or hours, making EEPROM memory extremely useful in many applications. The physical geometry of the floating-gate EEPROM transistor differs somewhat from that of a MOSFET designed for an EPROM cell. The former occupies more surface area, and is thus not available in as large a memory density as the latter. The use of EEPROM in a digital system also requires that the programming voltages and the circuitry needed to apply them also be present. In the typical EPROM application, programming and erasure are performed by removing the IC from the digital circuit, hence the programming voltages need not be included as part of the circuit.

15.4 ANALOG-TO-DIGITAL INTERFACING

A complete treatment of digital circuits must also include a discussion of digital-to-analog (D/A) and analog-to-digital (A/D) conversion. Although physical measurement usually involves analog variables, most data collection, information transmission, and signal analysis are performed digitally. Analog-to-digital and digital-to-analog circuits provide the all-important interfaces between the analog and the digital worlds.

15.4.1 Digital-to-Analog Conversion

A digital-to-analog (D/A) converter produces a single analog output voltage from a multibit digital input. The decoding can be performed using a variety of D/A algorithms. One common algorithm produces an analog output proportional to a fixed reference voltage, as determined by the equation

$$v_{\text{OUT}} = \frac{n}{2^N - 1} V_{\text{REF}} \tag{15.25}$$

In this equation, N is the number of bits in the digital input word, and n is the decimal integer represented by the input bits that are set to **1**. The voltage V_{REF} can be either positive or negative. Other D/A converters, which produce output voltages over a bipolar voltage range, employ the two's complement decoding scheme, to be described in Section 15.4.3.

EXAMPLE 15.1   The input to a 10-bit D/A converter with a reference voltage of 5 V is (**00 1001 0001**). Find the resulting analog output.

Solution

The decimal integer n represented by the specified digital input is $128 + 16 + 1 = 145$. In this case, $N = 10$ and 2^{10} is equal to 1024; hence the output given by Eq. (15.25) becomes

$$v_{\text{OUT}} = \frac{145}{1023}(5\text{ V}) = 0.709\text{ V} \tag{15.26}$$

EXERCISE 15.23   An 8-bit D/A converter has an input of (**0010 1111**) and a reference voltage of 5 V. According to Eq. (15.25), what is the output of the converter?   **Answer:** 0.922 V

15.24   What is the smallest increment of analog output that can be produced by a 12-bit D/A converter with a reference voltage of 5 V if the algorithm of Eq. (15.25) is used?   **Answer:** 1.22 mV

15.25   What is the largest analog output that can be produced by an 8-bit D/A converter with a reference voltage of 5 V if the algorithm of Eq. (15.25) is used?   **Answer:** 5.00 V

Summing Amplifier D/A Converter

One circuit that can perform D/A conversion using the algorithm of Eq. (15.25) utilizes the op-amp summation amplifier of Chapter 2. In a summing converter, the digital input bits to be decoded control the parallel input nodes of a summation amplifier, as depicted in Fig. 15.30. The input terminals of this circuit are connected to either ground or to V_{REF} by a set of two-position switches labeled D_0 through D_7. The status of each switch is determined by the value of its corresponding digital input bit. If a given input bit is equal to **1**, the switch represented by the bit is connected to V_{REF}; if the bit is equal to **0**, the switch is connected to ground. In practice, switches D_0 through D_7 are made from MOSFET devices driven by the digital input signals. An 8-bit converter is shown in Fig. 15.30; a converter with any number of bits could be made by changing the number of input channels.

Figure 15.30
Eight-bit D/A
converter made
from summation
amplifier. This
converter
essentially
implements the
decoding algorithm
given by
Eq. (15.25) but
creates a negative
output due to the
inversion of the
summation
amplifier.

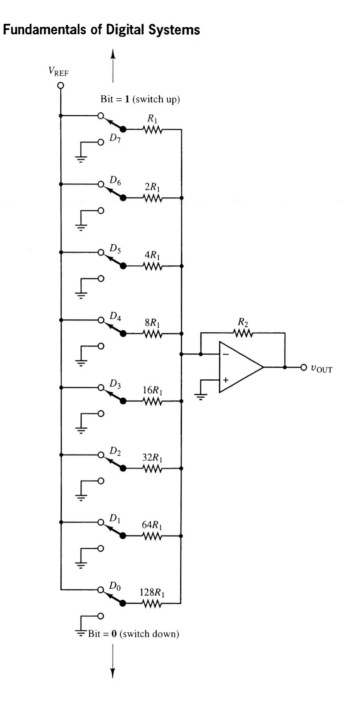

The D/A decoding performed by this circuit is realized by selecting proper values for the resistors. Specifically, beginning with the most significant bit, the input resistor of each successive channel must be made two times larger than its predecessor as the weighting of the corresponding binary bit is decreased. The arrangement is illustrated in Fig. 15.30, where, for example, the input resistor to D_7 has value R_1, the input resistor to D_6 is $2R_1$, the resistor to D_5 is $4R_1$, and so on. This ordering of resistor values allows the bit with the highest weighting (the D_7 bit) to amplify V_{REF} by the largest gain, and the bit with the smallest weighting (the D_0 bit) to amplify V_{REF} by the smallest gain. An 8-bit converter like the one shown in Fig. 15.30 has input resistors that vary between the values R_1 and $128R_1$. Given this geometrical progression of resistor values, the output of the circuit can be expressed by the equation

$$v_{\text{OUT}} = -\frac{R_2}{R_1}\left(D_7 + \frac{D_6}{2} + \frac{D_5}{4} + \cdots + \frac{D_1}{64} + \frac{D_0}{128}\right)V_{\text{REF}} \tag{15.27}$$

The variables D_0 through D_7 in this equation represent integers with values of **1** or **0**, as determined by the status of the eight digital input bits.

For the general case of an N-bit converter, Eq. (15.27) becomes

$$v_{\text{OUT}} = -\frac{R_2}{R_1}\sum_{n=0}^{N-1}\frac{D_n}{2^K}V_{\text{REF}} \tag{15.28}$$

where $K = 2^{N-1-n}$.

The principal disadvantage of the summing D/A converter shown in Fig. 15.30 is its sensitivity to the input resistor values. These resistors must be fabricated to close tolerances over a wide range of values. If the converter is fabricated on an integrated circuit, adequate control of resistor values may not be possible.

EXERCISE 15.26 The input to an 8-bit summing D/A converter of the type shown in Fig. 15.30, in which $R_2 = 20\,\text{k}\Omega$ and $R_1 = 10\,\text{k}\Omega$, is **(0011 0011)**. Find the analog output if the reference voltage is 10 V.
Answer: $-7.97\,\text{V}$

15.27 An 8-bit summing D/A converter of the type shown in Fig. 15.30 with $R_2 = 100\,\text{k}\Omega$ has a reference voltage of 2 V. If the output is to be limited to a maximum negative value of $-10\,\text{V}$, find the smallest allowed value of R_1. **Answer:** $39.8\,\text{k}\Omega$

15.28 Repeat Exercise 15.27, but find the largest allowed value of R_2 if $R_1 = 10\,\text{k}\Omega$.
Answer: $25.1\,\text{k}\Omega$

15.29 Repeat Exercise 15.27, but let the reference voltage equal 10 V. **Answer:** $199\,\text{k}\Omega$

Resistive Ladder Converter

The D/A converter shown in Fig. 15.31 is called a *resistive ladder converter*. Its central component is a ladder network in which horizontally drawn resistors have a value of R and vertically drawn resistors have a value of $2R$. This network is sometimes called an R–$2R$ network. The output buffer in Fig. 15.31 (usually an op-amp voltage follower) minimizes the current drawn from the ladder network, thereby reducing the loading at the v_3 node (node 3). For simplicity, a 4-bit converter is shown in Fig. 15.31. The number of bits could be increased to any number by extending the length of the ladder network.

Figure 15.31
Four-bit resistive ladder D/A converter.

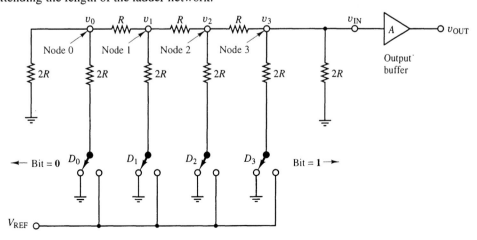

The digitally controlled switches in the circuit connect the lower resistor terminals to either V_{REF} or ground. Each of the switches is controlled by one bit of the digital input word, with D_0 representing the least significant bit. In practice, the entire converter is fabricated on a single integrated circuit, and the switches are constructed from digitally driven MOSFETs.

An examination of the circuit reveals one of its more interesting characteristics. With all switches connected to ground, each of the numbered nodes 0 through 3 is connected directly to ground via a resistor of value $2R$. At the same time, the equivalent resistance to ground seen to either the right or to the left of each numbered node is equal to $2R$. For example, the resistance seen to the right of node 3 consists of a single resistor $2R$. The resistance seen to the right of node 2 (including the resistance $2R$ between node 3 and ground) becomes the sum of R plus the parallel combination of $2R$ and $2R$. Altogether, this combination is equal to $2R$. The resistance seen to the right of node 1 then becomes R plus the parallel combination of $2R$ and $2R$, which again is equal to $2R$. This same reasoning can be applied to node 0 and can also be applied when computing the resistance seen to the left of any node.

If the nth switch is connected to V_{REF}, with all other switches connected to ground, the voltage v_n appearing at the node directly above it can be determined by applying the voltage-divider relation, that is,

$$v_n = V_{REF} \frac{2R \| 2R}{2R + 2R \| 2R} = \frac{V_{REF}}{3} \qquad (15.29)$$

The factor $2R \| 2R$ in Eq. (15.29) represents the parallel combination of resistances seen to the right and to the left of the nth node; the single factor of $2R$ in the denominator represents the resistance between the nth switch and its associated node.

The fraction of v_n that appears as v_{IN} at the input to the buffer depends on the node's position along the ladder. The node voltage v_3, for example, is applied directly to the v_{IN} terminal. The voltage v_2, however, is attenuated by a voltage divider formed by R and $2R \| 2R$, so that $v_{IN} = v_2/2$. A similar consideration shows that v_1 is attenuated at node 2 by a voltage divider formed from the resistance R and the combination $2R \| (R + 2R \| 2R)$. The latter combination is equivalent to a resistance of value R, so that

$$v_2 = v_1 \frac{R}{R + R} = \frac{v_1}{2} \qquad (15.30)$$

The fraction of v_1 that appears at v_{IN} thus becomes

$$v_{IN} = \frac{v_2}{2} = \frac{v_1}{4} \qquad (15.31)$$

Applying this attenuation algorithm to all the nodes in the ladder, together with the superposition principle and Eq. (15.29), yields the total output voltage of the circuit as a function of all the switch values:

$$v_{OUT} = \frac{V_{REF}}{3} \left(D_3 + \frac{D_2}{2} + \frac{D_1}{4} + \frac{D_0}{8} \right) \qquad (15.32)$$

Equation (15.32) assumes that v_{OUT} is equal to v_{IN}, that is, that the output buffer has unity gain. Each of the variables D_0 through D_3 in this expression is equal to either **0** or **1**. As Eq. (15.32) suggests, the ladder network applies appropriate weighting to each digital bit in determining the analog output voltage.

For a ladder converter with N input bits, Eq. (15.32) can be generalized to the expression

$$v_{OUT} = \frac{V_{REF}}{3} \left(D_{N-1} + \frac{D_{N-2}}{2} + \cdots + \frac{D_1}{2^{N-2}} + \frac{D_0}{2^{N-1}} \right) \qquad (15.33)$$

In contrast to the summing converter, which is sensitive to the absolute values of its resistors, the ladder converter is sensitive only to the ratio of the resistors R and $2R$. In the environment of an integrated circuit, resistors with precisely fixed ratios of 2:1 are easily fabricated.

15.4.2 Sample-and-Hold Circuit

The sample-and-hold (S/H) circuit is an important component in analog-to-digital (A/D) conversion systems. When A/D conversion is performed, an analog signal must be sampled at a momentary instant in time, then held until conversion to digital form can be completed. This operation is readily performed by the circuit depicted in Fig. 15.32. The input and output signals to this circuit are both analog voltages.

Figure 15.32
Basic elements of a sample-and-hold (S/H) circuit.

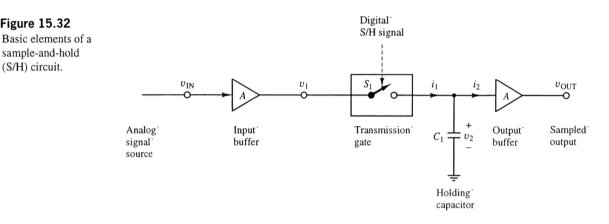

The circuit of Fig. 15.32 includes an input buffer stage, input transmission gate, holding capacitor, and output buffer stage. The input buffer replicates v_{IN} as v_1 while minimizing the current drawn from the analog signal source. The input buffer also provides the current required to charge the holding capacitor C_1. The output buffer replicates the capacitor voltage v_2 as v_{OUT} when S_1 is open. During this interval, the output buffer also minimizes the discharge of the capacitor. With the exception of the holding capacitor, which must be connected externally, the S/H circuit is usually contained on a single IC chip.

The status of switch S_1 is controlled by a digital signal. When S_1 is closed, the capacitor is charged and tracks the analog voltage v_1, which is a replica of v_{IN}. This function constitutes the *sample* mode of operation. When S_1 is opened, the capacitor voltage v_2 retains the value of v_1 that appears at the moment of switch opening. This function is called the *hold* mode of operation. Because the output buffer draws minimal current from the holding capacitor, the capacitor can hold v_2 for a long time. In an ideal sample-and-hold circuit, i_2 is zero, and the voltage v_2 is held indefinitely. In a practical sample-and-hold circuit, i_2 is made as small as possible.

Important parameters of a sample-and-hold circuit include its *settling time*, *output droop*, and *capacitor charging current*. The settling time is a measure of the time required to charge the holding capacitor via the input buffer and input transmission gate. The latter components contribute incremental series resistance between v_{IN} and C_1, thus increasing the capacitor charging time. The digital signal that closes S_1 must have a duration at least as long as the sample-and-hold's settling time. The output droop, expressed in volts per unit time, expresses the decay of the capacitor voltage when the circuit is in the *hold* mode (i.e., when S_1 is open). The droop parameter increases with i_2 and also depends on the capacitance C_1. Note that i_2 depends not only on the internal design of the output buffer, but also on leakage currents that flow inside the capacitor and between the connections to the capacitor and ground via the surfaces of the IC and capacitor packaging materials. The maximum current available to charge C_1, which is determined by the characteristics of the input buffer, limits the maximum permissible value of C_1.

15.4.3 Analog-to-Digital Conversion

An analog signal is converted to digital form by an *analog-to-digital (A/D) converter*. The typical A/D converter compares its analog input to a fixed reference voltage and then provides a binary output equal to

$$B = \text{int}\left[\frac{v_{IN}}{V_{REF}}(2^N - 1)\right] \tag{15.34}$$

where N is an integer equal to the number of bits in the binary output word. In this equation, the *int* operation truncates the expression in brackets to the nearest integral value. The encoding operation expressed by Eq. (15.34) is called *binary-weighted* encoding. A full-scale binary output (all N bits set to **1**) occurs when v_{IN} becomes equal to V_{REF}. Only positive values of v_{IN} are allowed. The *resolution* of a binary-weighted converter, that is, the voltage increment represented by the least significant bit of the digital output, is equal to $V_{REF}/(2^N - 1)$.

Some binary-weighted A/D converters can accommodate input voltages of negative as well as positive polarity. Converters of this type provide an additional *polarity* bit, which indicates the sign of v_{IN}. Alternatively, a bipolar A/D converter may utilize the two's complement binary system, in which an input of $+V_{REF}$ is represented by a digital output with the most significant bit (MSB) set to **0** and all other bits set to **1**, and an input of $-V_{REF}$ is represented by a digital output with the MSB set to **1** and all other bits set to **0**. In the two's complement system, an analog input of zero is represented by a digital output with all bits set to **0**. The two's complement number scheme facilitates the interfacing of the A/D converter to computers, microprocessors, and other digital processing systems.

A second type of A/D converter, called a *bin-encoded* converter, compares its analog input to a reference voltage, then sets high the first m least significant bits of its N-bit binary output. The value of m is determined by the algorithm

$$m\frac{V_{REF}}{N+1} < v_{IN} < (m+1)\frac{V_{REF}}{N+1} \tag{15.35}$$

The output bits of a bin-encoded converter are set in sequential order, beginning with the least significant bit. Hence the values (**0000 0001**), (**0000 0011**), and (**0000 0111**) are valid outputs of an 8-bit converter, but the values (**0000 0010**) or (**1000 1111**) are not. The resolution of an N-bit bin-encoded converter is equal to $V_{REF}/(N + 1)$. A binary-weighted converter has much finer resolution than a bin-encoded converter, but the latter can perform its analog-to-digital conversion on the order of N times faster. The reasons for this increased speed are explored later in this section.

EXAMPLE 15.2   An 8-bit binary-weighted A/D converter has a reference voltage of 5 V.

(a) Find the analog input corresponding to the binary outputs (**1111 1110**) and (**0001 0000**).

(b) Find the binary output if $v_{IN} = 1.1$ V.

(c) Find the resolution of the converter.

(d) Find the additional voltage that must be added to a 1-V analog input if the digital output is to be incremented by 1 bit.

Solution

The binary number (**1111 1110**) corresponds to the decimal integer 254. Applying this number to Eq. (15.34) with $N = 8$ and $V_{REF} = 5$ V yields the analog value

$$v_{IN} = \frac{B V_{REF}}{2^N - 1} = \frac{254}{255}(5\,\text{V}) = (0.996)(5\,\text{V}) = 4.98\,\text{V} \tag{15.36}$$

Similarly, the binary number (**0001 0000**) corresponds to the decimal integer 16; hence

$$v_{IN} = \frac{16}{255}(5 \text{ V}) = 0.314 \text{ V} \tag{15.37}$$

If $v_{IN} = 1.1$ V, then

$$B = \text{int}\left[\frac{v_{IN}}{V_{REF}}(2^N - 1)\right] = \text{int}\left[\frac{1.1 \text{ V}}{5 \text{ V}}(255)\right] = \text{int}(56.1) = 56 \tag{15.38}$$

Converting this number to binary yields the digital value (**0011 1000**).

As Eq. (15.34) suggests, the resolution of this A/D converter, defined as the voltage increment represented by the least significant digital bit, is equal to

$$\frac{1}{2^N - 1}V_{REF} = \frac{1}{255}(5 \text{ V}) \approx 0.02 \text{ V} \tag{15.39}$$

Note, however, that the next binary output following (**0011 1000**) is (**0011 1001**). This output corresponds to an analog voltage of 1.118 V, implying that v_{IN} must only be increased by 0.018 V to increment the output of the A/D converter by one bit. The discrepancy between this additional 0.018-V increment and the 0.02-V resolution of the converter arises because the binary number derived from Eq. (15.38) represents the truncated integral value of 56.1. The smallest analog input corresponding to a binary output of (**0011 1000**) is actually 1.098 V.

EXAMPLE 15.3 An 8-bit bin-encoded A/D converter has a reference voltage of 5 V.

(a) Find the range of analog inputs that correspond to the binary output (**0001 1111**).

(b) Find the binary output if $v_{IN} = 1$ V.

(c) Find the resolution of the converter.

Solution

The analog input range of this bin encoder, for which $N = 8$ and $V_{REF} = 5$ V, can be found for the specified output from Eq. (15.35). Specifically, the digital output (**0001 1111**), for which $m = 5$, corresponds to an input somewhere between the limits

$$\frac{5}{9}(5 \text{ V}) < v_{IN} < \frac{6}{9}(5 \text{ V}) \tag{15.40}$$

or

$$2.78 \text{ V} < v_{IN} < 3.33 \text{ V} \tag{15.41}$$

Each output bit of the converter corresponds to a voltage increment of $V_{REF}/(N + 1) = 0.56$ V, which also corresponds to the voltage difference between the right- and left-hand sides of the inequality (15.41). If $v_{IN} = 1$ V, application of the algorithm (15.35) yields the value $m = 1$, so that the binary output becomes (**0000 0001**).

EXERCISE 15.30 Repeat Example 15.2 if the reference voltage is 12 V.

15.31 Repeat Example 15.2 if the A/D converter has 12 bits of resolution.

15.32 Repeat Example 15.3 if the reference voltage is 12 V.

15.33 Repeat Example 15.3 if the A/D converter has 12 bits of resolution.

15.34 Find the analog voltages corresponding to the five successive binary-weighted outputs beginning with (**0001 0000**). The A/D converter has an 8-bit output and $V_{REF} = 5$ V.

15.35 Find the analog voltages corresponding to the five successive bin-encoded outputs beginning with (**0000 1111**). The A/D converter has a 12-bit output and $V_{REF} = 5$ V.

Counting A/D Converter

A binary-weighted A/D converter can be implemented in several ways. One method utilizes a digital-to-analog (D/A) converter, sequential binary counter, and analog comparator, as illustrated in Fig. 15.33. The analog input v_{IN} to this circuit typically consists of the output of a sample-and-hold circuit. The D/A converter typically used in this system is similar to the resistive ladder converter discussed in Section 15.4.1. The A/D converter depicted in Fig. 15.33 is capable of encoding only positive input voltages over the range $0 \leq v_{IN} \leq V_{REF}$.

Figure 15.33
Analog-to-digital converter made from a binary counter, analog comparator, and D/A converter.

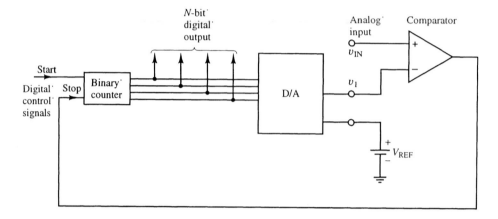

When the binary counter receives a digital *start* signal, it resets each of its output bits to zero, then begins to count sequentially in binary. The output from the counter is fed to the D/A converter, which produces an analog voltage v_1 proportional to V_{REF} using the binary-weighting algorithm. When v_1 surpasses the analog input v_{IN}, the comparator changes its output state, sending a *stop* signal to the binary counter. The counter output at that instant corresponds to a binary-weighted representation of the analog input signal v_{IN}. A new start signal sent to the system resets the counter and begins the process anew.

Dual-Slope A/D Converter

A second A/D conversion method that is very accurate is called the *dual-slope*, or *ratiometric*, conversion method. The basic block diagram of a dual-slope A/D system is shown in Fig. 15.34. In practice, S_1 and S_2 are implemented using MOSFET transmission gates. Before the conversion cycle begins, switch S_2 is closed and the output of the integrator set to zero. The start of a conversion cycle begins with the simultaneous opening of S_2 and the connection of S_1 to the analog input v_{IN}. The integrator is allowed to function for a fixed time interval τ_1 over which v_{IN} must be held constant. This latter requirement is usually met by taking v_{IN} from the output of a sample-and-hold circuit which samples the analog data at the beginning of the conversion cycle. After the integration interval τ_1, the output of the integrator becomes

$$v_1 = -\frac{\tau_1}{RC} v_{IN} \tag{15.42}$$

At time τ_1, S_1 is switched from v_{IN} to a fixed reference voltage V_{REF}, where V_{REF} has polarity *opposite* that of v_{IN}. At the same instant, a reset signal is sent to the sequential binary counter, resetting its digital output to zero. This binary counter is run by a clock of fixed frequency.

The polarity of V_{REF} is opposite that of v_{IN}, causing the integrator output v_1 to head back toward zero while S_1 is connected to V_{REF}. When v_1 reaches zero, the comparator changes state, sending a *stop* signal to the binary counter via a digital pulse generator. The output of the integrator over this second interval is given by

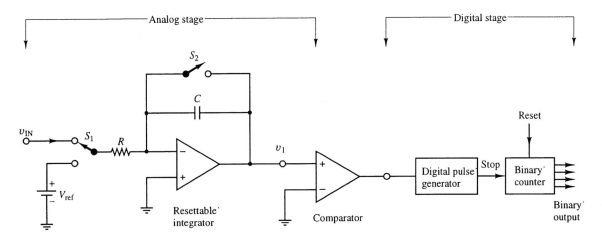

Figure 15.34 Dual-slope A/D conversion system.

$$v_1 = -\underbrace{\frac{\tau_1}{RC}v_{\text{IN}}}_{\text{initial value at } t = \tau_1} - \underbrace{\frac{t - \tau_1}{RC}V_{\text{REF}}}_{\text{added integration for } t > \tau_1} = -\frac{1}{RC}[\tau_1 v_{\text{IN}} + (t - \tau_1)V_{\text{REF}}] \qquad (15.43)$$

Because V_{REF} and v_{IN} have opposite polarity, the integrator output v_1 will return to zero and the counter will stop counting after an additional integration interval τ_2, given by

$$\tau_2 = -\frac{v_{\text{IN}}}{V_{\text{REF}}}\tau_1 \qquad (15.44)$$

Equation (15.44) can be obtained by solving Eq. (15.43) for $v_1 = 0$, with $\tau_2 = t(v_1=0) - \tau_1$. Note that the value of τ_2 is not affected by R and C; the latter components affect only the final magnitude of v_1 at time τ_1.

The key to the dual-slope conversion technique lies in the selection of τ_1, which is set to the time required for the counter to reach its full-scale binary output (i.e., the output with all bits set to **1**). In practice, a combinational logic circuit, fed by the output of the binary counter, is used to sense the arrival of this condition. This combinational circuit generates the counter reset pulse and the logic signals required to control S_1 and S_2. Since the counter output is reset to zero at τ_1, the binary output at τ_2 will be proportional to v_{IN}. Specifically, the ratio of the binary output at τ_2 to the full-scale binary output will be equal to the ratio $v_{\text{IN}}/V_{\text{REF}}$. This result occurs because the binary counter is sequenced by the same fixed-frequency clock over time intervals τ_1 and τ_2.

The dual-slope conversion method is commonly employed in digital-readout analog voltmeters. Such a voltmeter must include calibrated input amplifiers and a digital-display driving circuit, and also a polarity-recognition circuit that applies a V_{REF} of proper polarity to the integrating amplifier during the τ_1-to-τ_2 interval.

Bin-Encoded A/D Converter

A circuit capable of bin-encoded conversion is shown in Fig. 15.35. In a bin-encoded conversion system, the analog input is simultaneously fed to a number of parallel analog comparators. For simplicity, a 6-bit system ($N = 6$) is shown here. The number of conversion bits could be increased to any number by adding more comparator circuits in parallel. The reference voltage to the mth comparator, taken from the mth tap of a resistive voltage divider in which all resistors have the same value, is equal to $mV_{\text{REF}}/(N + 1)$. The output of the mth comparator remains low when v_{IN} lies below the analog reference voltage provided by the mth tap. Conversely, the output of a given comparator goes high when v_{IN} exceeds its individual analog reference voltage. The digital output is derived from the parallel combination of comparator outputs. An examination of the system will show the digital output to be a bin-encoded version of the input, as given by the algorithm (15.35).

This system is sometimes called a *flash*, or *strobe*, A/D converter. Unlike the counting and dual-slope converters, a flash converter performs its conversion operation in a single instant with almost no time delay. The price paid for this conversion speed is a much lower resolution, which is limited to $V_{REF}/(N + 1)$ instead of the value $V_{REF}/(2^N - 1)$ obtained from a binary-weighted conversion method.

Figure 15.35
Six-bit parallel comparator A/D bin converter. The seven resistors in the voltage-divider network all have the same value.

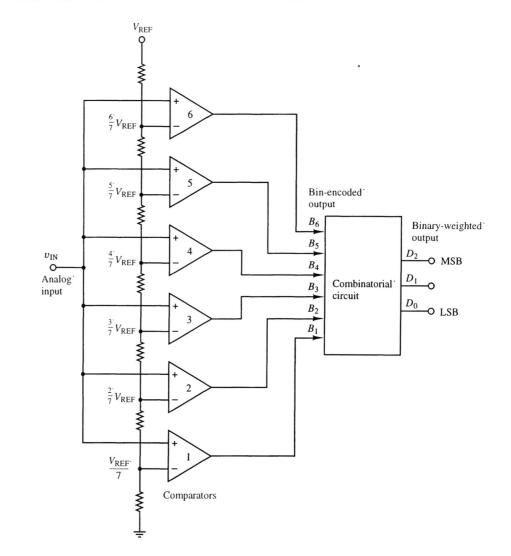

Most digital processing requires that binary numbers be expressed in binary-weighted form. Consequently, a flash converter typically includes a combinational output circuit that converts the bin-encoded output to a binary-weighted number. This combinational circuit has fewer output bits than it has input bits, since the binary-weighted output can be expressed using a smaller number of digits.

EXERCISE 15.36   A 12-bit encoded converter feeds a combinational circuit that converts its output to binary-weighted form. How many output bits must this combinational circuit have?   **Answer:** 4

15.37   Repeat Exercise 15.36 if the bin-encoded circuit has 16 bits.   **Answer:** 5

SUMMARY

- A combinational logic circuit is sensitive only to the instantaneous values of its input voltages.

- A sequential logic circuit is sensitive to the instantaneous values of its input voltages and to the values of its previous input and output voltages.

- Flip-flops function as the basic memory elements of a digital system.

- Flip-flop types include the SR, clocked SR, JK, type D, and type T flip-flops.

- A basic SR flip-flop is formed from two cross-connected logic gates.

- A clocked SR flip-flop is made by adding two AND gates to a NOR gate SR flip-flop.

- A JK flip-flop is made by coupling the outputs of a clocked SR flip-flop to its input via a pair of AND gates.

- The master–slave configuration eliminates the race around condition in a JK flip-flop.

- A type D (delay) flip-flop is made from a JK flip-flop by connecting the J input to the K input with a NOT gate.

- A type T (toggle) flip-flop is made from a JK flip-flop by connecting together the J and K inputs.

- A monostable multivibrator produces a single pulse after a trigger signal is received.

- An astable multivibrator produces a continuous, periodic square-wave type signal.

- Digital memory is available in ROM, RAM, EPROM, and EEPROM varieties.

- ROM memories are nonvolatile; they do not lose their data when power is removed. RAM memories are volatile; they lose their data when power is removed.

- Static RAM employs a flip-flop in each data cell; dynamic memory employs a single transistor in each cell but must be refreshed periodically.

- Analog-to-digital interfacing provides the link between analog and digital worlds.

- A digital-to-analog converter produces an analog output from a digital input.

- D/A circuit types include the summing amplifier and resistive ladder configurations.

- A sample-and-hold circuit holds the sampled value of an analog signal until analog-to-digital conversion can be completed.

- An analog-to-digital converter produces a digital output from an analog input.

- A/D conversion can be performed using a binary-weighted or a bin-encoded algorithm.

- A/D circuit types include the counting, dual-slope, and flash converter configurations.

DESIGN AND ANALYSIS PROBLEMS

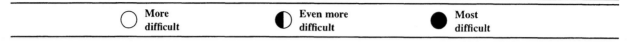

15.1 Sequential Logic Circuits

15.1.1 Set–Reset Flip-Flop

15.1 A 1-kHz square wave is first connected to the S input of the SR flip-flop of Fig. 15.1 during the time interval when the square wave is low. Sketch the resulting Q output.

15.2 A 10-kHz square wave and its complement are connected to the S and R inputs, respectively, of the SR flip-flop of Fig. 15.1. Sketch the resulting Q and $\overline{Q}$ outputs.

15.3 A 1-kHz square wave is connected to the S input of the SR flip-flop of Fig. 15.1. A 5-kHz square wave, whose leading edge coincides with that of the 1-kHz input, is connected to the R input. Sketch the resulting Q and $\overline{Q}$ outputs.

15.4 The SR flip-flop of Fig. 15.1 is made from two NOR gates. Show that an SR flip-flop function can also be made by cross-connecting two NAND gates. Verify its operation by constructing a binary truth table.

15.5 Show that the CMOS SR flip-flop of Fig. 15.2(b) will also work if MOSFETs Q_1 and Q_2 are omitted. This configuration saves valuable chip space on a complicated integrated circuit.

15.6 ○ Consider the CMOS SR flip-flop of Fig. 15.2(b). Suppose that the devices have a threshold voltage of $|V_{TR}| = 1$ V, with K parameters such that $|K_1| = |K_2| = |K_3| = |K_4| = 2K_6 = 2K_7$. If $V_{DD} = 5$ V, find values for K_5 and K_8 such that the flip-flop changes state when the S or R input is raised to $V_{DD}/2$.

15.7 A number of SR flip-flops of the type shown in Fig. 15.2(a) are connected such that one of the outputs of each flip-flop feeds one of the inputs of the next flip-flop in the cascade. One output from the last flip-flop is connected back to an input of the first. The total number of flip-flops is N. Discuss the status of the cascade for each of the following cases:

(a) Each Q output feeds an S input.

(b) Each Q output feeds an R input.

(c) Each Q output alternately feeds an S input and an R input.

(d) Each Q output feeds an R input, each $\overline{Q}$ output feeds an S input, and the total number of flip-flops is odd.

(e) Each Q output feeds an R input, each $\overline{Q}$ output feeds an S input, and the total number of flip-flops is even.

15.8 Consider the NMOS SR flip-flop of Fig. 15.2(a). The flip-flop resides in the reset state with $Q = \mathbf{0}$. Suppose that the pull-up loads have parameters $K = 0.5\,\text{mA/V}^2$ and $V_{TR} = -1$ V, and that $K = 4\,\text{mA/V}^2$ and $V_{TR} = 2$ V for all other MOSFETs. If the dominant capacitance in each device consists of $C_{gs} = 2$ pF, determine the approximate rise time of the Q output if the S input is excited by a step function having zero rise time.

15.1.2 Clocked SR Flip-Flop

15.9 Draw the Q output of a clocked SR flip-flop versus time if the R input is held high and the S input low at all times. Assume the clock to consist of a 1-kHz square wave.

15.10 Draw the Q output of a clocked SR flip-flop versus time if the R input is held low at all times, the clock is a 1-kHz square wave, and the S input is a 10-kHz square wave synchronized to the clock signal (the leading edge of the clock signal is coincident with the leading edge of the S input).

15.11 ○ Draw the Q output of a clocked SR flip-flop versus time if the clock is a 1-kHz square wave, the S input is a 2-kHz square wave, and the R input in a 3-kHz square wave. The three signals are synchronized to begin with a low-to-high transition at $t = 0$.

15.12 A clocked SR flip-flop can be formed by cross-coupling two NOR gates, as in Fig. 15.3. Show that cross-coupling two NAND gates will produce a flip-flop in which the signal S' becomes the input of the gate that has Q as its output.

15.13 By constructing a binary truth table of gates 1 through 4, show that the clocked SR flip-flop of Fig. 15.5, which is made from four NAND gates, is functionally equivalent to the gate of Fig. 15.3(b), which is made from two AND gates and two cross-coupled NOR gates.

15.14 ◨ Design a 4-bit serial-to-parallel converter using the clocked SR flip-flop of Fig. 15.3.

15.15 ◨ ○ Using the clocked SR flip-flop of Fig. 15.3 as a basis, design a pattern-recognition circuit that produces a logic **1** when it receives the coded digital word **11001** in serial form.

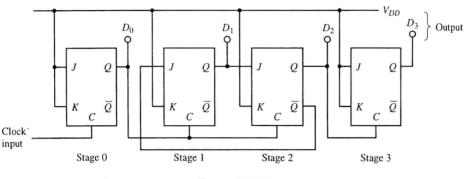

Figure P15.18

15.16 ⓢ ◗ Using the clocked SR flip-flop of Fig. 15.3 as a basis, design a circuit that will allow the ignition of an automobile to be energized only after the binary sequence **101010** has been entered via a mechanical switch, where each digit is entered only after a beep has been sounded by the circuit. Such a system could be used as an in-vehicle driver-sobriety tester.

15.1.3 JK Flip-Flop

15.17 Using Fig. 15.5 as a model for the master and slave sections, draw the logic diagram of a JK flip-flop that includes PR (preset) and CL (clear) inputs. These additional inputs can be added to the master flip-flop by adding input lines to NAND gates 3 and 4 in Fig. 15.5.

15.18 Consider the circuit of **Fig. P15.18**. Assume that the system begins with all Q outputs set to **0**. Construct the truth table for Q_0 through Q_3 for each subsequent clock pulse. What function does this circuit perform?

15.19 ⓢ Using a JK flip-flop and other gates, design a circuit that can be used to provide a tally of two simultaneously flipped coins. The output should consist of four digital outputs that can send pulses to four separate digital counters, one for each of the four possible outcomes.

15.1.4 Type D and Type T Flip-Flops

15.20 The circuit of **Fig. P15.20** is called a *ripple counter*. The binary output of the T flip-flops indicate how many pulses have arrived at the input since the cascade was reset with all Q outputs low. Plot the state of outputs D_0 through D_3 if the input is a square wave that shifts between zero and V_{DD} at a 1-kHz rate. What practical limitation exists in the operation of this circuit?

15.21 ⓢ ○ Using the type T flip-flop of Fig. 15.8(b) and the ripple counter of **Fig. P15.20** as a basis, design an "up–down" ripple counter that will either increment or decrement the parallel output, depending on the status of a control bit D_X.

15.22 ⓢ ○ Use four type T flip-flops to design a decade ripple counter. The circuit should receive a digital pulse train and then reset itself after 10 pulses have been received.

15.23 The circuit of **Fig. P15.23** is called a *serial-carry, synchronous* counter. The T inputs to all flip-flops but the first in the cascade consist of AND combinations of previous Q outputs. Analyze the operation of the circuit, construct a binary truth table of D_0 through D_4, and show that the circuit counts the number of pulses applied to the clock input.

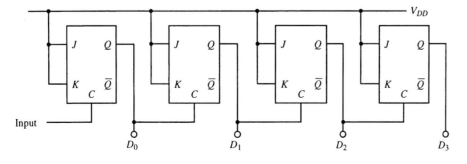

Figure P15.20

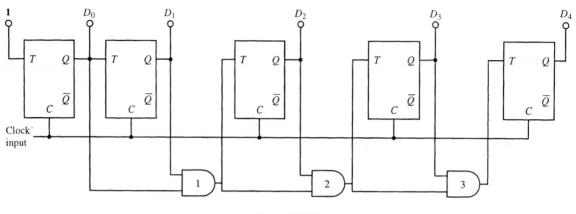

Figure P15.23

15.24 Use the type D flip-flop of Fig. 15.8(a) to design a 4-bit by 4-word memory array that allows the user to read and write data upon application of appropriate input signals.

15.1.5 Preset and Clear Inputs

15.25 Using JK flip-flops with preset and clear inputs, design a 6-digit counter that resets itself when the binary number **100000** is reached.

15.26 Using an SR flip-flop with preset and clear inputs as a basis, design a 5-bit serial/parallel converter. The register should be capable of receiving serial input and producing parallel output. It should also be capable of receiving parallel input and producing serial output. In addition, the register should be reset to **00000** by the application of a single reset pulse.

15.27 Design a bidirectional shift register that can receive a 4-bit parallel data word and shift it either right or left, depending on the status of a "direction control" line.

15.2 Multivibrator Circuits

15.2.1 Monostable Multivibrator

15.28 Compute the output-pulse duration of the monostable multivibrator of Fig. 15.10 if $R = 12\,k\Omega$, $C = 0.01\,\mu F$, $V_{DD} = 5\,V$, and the gates are CMOS gates with a crossover voltage of $v_{IC} = 3\,V$.

15.29 Compute the output-pulse duration of the monostable multivibrator of Fig. 15.10 if $R = 39\,k\Omega$, $C = 0.0022\,\mu F$, $V_{DD} = 5\,V$, the first gate has a crossover voltage of $v_{IC} = 2\,V$, and the second gate has a crossover voltage of $v_{IC} = 3.5\,V$.

15.30 Consider the monostable multivibrator of Fig. 15.10 and the timing diagram of Fig. 15.13. Determine the output for the case where $t_2 > t_3$.

15.31 Consider the monostable multivibrator of Fig. 15.10 and the timing diagram of Fig. 15.13. Determine the output for the case where a second input pulse occurs prior to time t_3.

15.32 The monostable multivibrator of Fig. 15.10 is constructed from TTL gates for which the crossover voltage v_{IC} is equal to 2.8 V at room temperature. Temperature variations may cause v_{IC} to change by $\pm 0.2\,V$. What is the range of the resulting output-pulse duration?

15.33 The monostable multivibrator of Fig. 15.10 is constructed from CMOS gates for which the crossover voltage v_{IC} is equal to 2.5 V at room temperature. Temperature variations may cause v_{IC} to change by $\pm 0.3\,V$. What is the range of the resulting output-pulse duration?

15.34 In this problem, the duration of the monostable pulse produced by the circuit of Fig. 15.10 is derived while taking into account the nonzero output resistance of each gate. Suppose that the CMOS gates have a small-signal output resistance of value r_{out} in both high- and low-output states and a crossover voltage v_{IC}. Following the analysis leading to Eqs. (15.3) to (15.7), derive an expression for the pulse duration T.

15.35 Design a circuit that can produce an adjustable pulse of duration 1 to 100 ms to be used in a heart defibulator machine. The pulse will be used to set the duration of the high-voltage dc supply applied to the chest electrodes.

15.36 Design a circuit that can produce three pulses in sequence, each of duration twice its predecessor.

15.37 ⊠ Design a circuit that can produce the pulses indicated in **Fig. P15.37**.

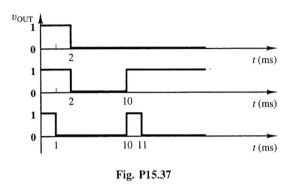

Fig. P15.37

15.38 ⊠ Design a circuit that can provide the "line" trigger for an oscilloscope. The circuit should accept a 12-V rms, 60-Hz ac input signal and produce a 0.1-ms pulse every 1/60 second.

15.2.2 Astable Multivibrator

15.39 Consider the astable multivibrator of Fig. 15.15. Show that the output of the circuit will be an asymmetrical square wave if the crossover voltage v_{IC} of the gates is not equal to $V_{DD}/2$. Find the duration of the high and low portions of the output if $v_{IC} = V_{DD}/3$.

15.40 An astable multivibrator is made from TTL gates that have the transfer characteristic of Fig. 14.37. Estimate the period of the multivibrator output.

15.41 Suppose that the astable multivibrator circuit of Fig. 15.15 is made from CMOS gates, for which $V(1) = 5$ V and $V(0) = 0$ V, using component values $R = 8.2$ kΩ and $C = 0.047\,\mu F$. What are the maximum currents that gate 1 must source and sink?

15.42 For the astable multivibrator circuit of Fig. 15.15 plot the output period T as a function of the supply voltage V_{DD} for 3 V $< V_{DD} <$ 7 V if the crossover voltage v_{IC} remains constant at 2.5 V. Show that T remains constant if v_{IC} is equal to a fixed fraction f of the supply voltage V_{DD}.

15.43 The circuit of **Fig. P15.43** is made from CMOS gates with symmetrical transfer characteristics and voltage levels $V(1) = V_{DD}$, $V(0) = 0$, and $v_{IC} = V_{DD}/2$. Show that the circuit produces an asymmetrical square wave if $R_1 \neq R_2$. Find an expression for the duration of the high and low portions of the output cycle.

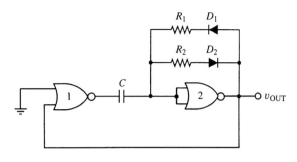

Fig. P15.43

15.44 ⊠ Use the astable multivibrator circuit of Fig. 15.15 as a basis to design an adjustable square-wave clock to be used inside a function generator. The clock period should vary by a factor of 10 over each of three ranges, beginning at 100 Hz, 1 kHz, and 10 kHz.

15.45 ⊠ Design a circuit that can produce a sequence of impulse like functions. Specifically, a pulse of duration $0.1\,\mu s$ should be produced every 1 ms.

15.46 ⊠ ◐ Using a combination of astable multivibrator circuits, monostables, and other logic gates, design a circuit that can provide the warning siren for an alarm system. Upon receiving an input pulse, the circuit should continuously emit a 5-kHz tone for 0.5s, and then be silent for the next 0.5s. The warning siren should cease only when the system has received a separate clear signal from the operator.

15.2.3 The 555 IC Timer

15.47 The 555 timer of Fig. 15.18 is connected as a monostable multivibrator with $R_A = 22$ kΩ and $C = 0.47\,\mu F$. What is the duration of the pulse that occurs when the trigger input falls low?

15.48 A pulse duration of approximately $150\,\mu s$ is desired from the 555 timer circuit of Fig. 15.18. Choose appropriate values for R_A and C.

15.49 A 555 timer is connected in monostable mode, as in Fig. 15.18. Suppose that $C = 1\,\mu F$ and R_A is a 1- to 10-kΩ variable resistor. What is the range of the pulse duration obtained at v_{OUT}? Assume that $V_{sat} \approx 0$ for Q_1.

15.50 The 555 timer of Fig. 15.20 is connected in astable mode with $R_A = 15$ kΩ, $R_B = 10$ kΩ, and $C = 0.5\,\mu F$. What are the period and frequency of the resulting square wave?

15.51 ⊠ Using a 555 timer in astable mode, design a circuit that can produce 10- to 50-kHz square waves.

15.52 Consider the astable 555 timer circuit of Fig. 15.20 with $R_A = 22$ kΩ, $R_B = 120$ kΩ, $C = 0.05\,\mu F$, and

$V_{CC} = 6$ V. Suppose that the top of R_A is connected to a variable voltage supply V_1 instead of to V_{CC}. Determine the frequency of the output for the cases $V_1 = 4, 5$, and 6 V. This circuit forms the basis of a crude *voltage-controlled oscillator*.

15.53 A 555 timer is connected in the astable mode using the configuration of Fig. 15.20 with $R_A = 10$ kΩ, $R_B = 68$ kΩ, $C = 10$ nF, and $V_{CC} = 5$ V. Suppose that the upper terminal of R_A is connected to a 10-V supply. Find the period of the resulting output.

15.54 ⚡ A square-wave frequency of approximately 500 kHz is desired from the 555 timer circuit of Fig. 15.20. Choose values for R_A, R_B, and C so that this objective can be met.

15.55 ⚡ Design a 555 timer circuit whose output is a 10-kHz square wave with a 40% high, 60% low duty cycle.

15.56 ⚡ Using a 555 timer and any other necessary components, design a circuit that can produce a 1-ms pulse after each zero crossing of a 100-Hz signal.

15.57 Consider the 555 timer circuit inside the box in Fig. 15.18 (without the external components). Suppose that the trigger and threshold inputs are driven by a voltage source v_{IN}. Plot the resulting v_{IN}–v_{OUT} transfer characteristic assuming $V_{sat} \approx 0$ for Q_1.

15.58 ⚡ ◯ Using 555 timers, design a circuit that can produce the ϕ_1 and ϕ_2 signals used by a switched-capacitor network for which $T_c = 100$ kHz.

15.3 Digital Memory

15.59 Draw the layout of a 5×5 ROM circuit storing the words **11001**, **10011**, **11000**, **01000**, and **11011**.

15.60 How many address lines are required for the address-to-row decoder of a 256-word $\times$ 8-bit ROM array?

15.61 Design a simple NMOS, depletion-load logic circuit that can function as the four-row address-to-row decoder used in the ROM circuit of Fig. 15.22. Assume that the address bus consists of complementary A and $\overline{A}$ lines.

15.62 Consider the NMOS ROM circuit of Fig. 15.22. Suppose that the pull-up loads have W/L ratios of 0.25. If the array contains 12 rows, what must be the W/L ratios of each transistor in the cell if an inverter aspect ratio of 8:1 is to be maintained? Can you think of a simple modification that would help to minimize chip area?

15.63 Consider the dynamic RAM cell of Fig. 15.27 with $C = 10$ fF. Suppose that the system provides a refresh cycle every 1 ms, and the logic operates with $V_{OH} = 5$ V and $V_{IH} = 4.2$ V. How large a leakage current from the capacitor can be tolerated on the memory chip?

15.64 The dynamic RAM cell of Fig. 15.27 is located on an integrated circuit with $C = 20$ fF. The leakage currents over the surface of the IC chip can be as high as 1 pA when the capacitor is charged to 5 V. What is the maximum allowed refresh interval if the stored capacitor voltage can fall no lower than 3.7 V?

15.4 Analog-to-Digital Interfacing

15.4.1 Digital-to-Analog Conversion

15.65 A 12-bit D/A converter operating from a 10-V reference has a digital input of **100000 100000**. What is the analog output?

15.66 A D/A converter operates from a 5-V reference. Determine the digital input required to produce an analog output of 4 V if the circuit performs a 4-bit conversion. Repeat for 8-, 10-, and 12-bit conversions.

15.67 What are the largest and smallest analog outputs that can be produced by a 14-bit D/A converter operating from a 10-V reference? What is the resolution per bit?

15.68 Consider the 8-bit summing D/A converter of Fig. 15.30. Determine the value of v_{OUT} if $V_{REF} = 12$ V, $R_1 = 10$ kΩ, $R_2 = 22$ kΩ, switches D_3 and D_5 are in the up position, and all other switches are in the down position.

15.69 ⚡ Choose resistor values for the summing D/A converter of Fig. 15.30 and suitably modify the circuit so that the maximum possible excursion of v_{OUT} is obtained. The saturation limits of the op-amp are ± 10 V, and $V_{REF} = 5$ V.

15.70 Determine the general output algorithm for the summing D/A converter of Fig. 15.30 if the down switch position connects the input resistors to a voltage $-V_{REF}$, rather than to ground.

15.71 ◯ Consider a 4-bit version of the summing D/A converter of Fig. 15.30. To what percentage tolerance must the resistor values be held if the output is to vary by no more than one least significant bit (LSB) for a given set of switch settings?

15.72 ⚡ Modify the summing D/A converter of Fig. 15.30 so that it becomes an analog amplifier with a digitally selectable gain. What is the ratio of the largest possible gain to the smallest possible gain? What are the largest and smallest values of input resistance seen by the analog signal in your circuit?

15.73 ⚡ Design a 3-bit D/A converter based on the differential-amplifier configuration. Its output should be equal to a current whose magnitude is determined by the various binary inputs.

15.74 ◯ Determine the output of a 3-bit resistive ladder converter similar to that of Fig. 15.31. In this case, let all resistors labeled $2R$ instead be equal to $4R$.

15.75 ⚡ Design a D/A converter that produces an analog output equal to the binary addition of two 4-bit input words.

15.76 ⚡ Design a D/A converter that produces an analog output equal to the binary subtraction of two 3-bit input words.

15.4.2 Sample-and-Hold Circuit

15.77 A sample-and-hold circuit draws a current of 100 nA from its holding capacitor during the hold operation. If the output of the circuit is to droop by no more than 10 mV, find the minimum holding capacitance required. Perform this calculation for sampling rates of 1 and 10 ms.

15.78 A sample-and-hold circuit draws current from its 10 pF holding capacitor during the hold operation. If the output of the circuit is to droop by no more than 5 mV, find the minimum necessary sampling rate. Perform this calculation for discharge currents of 10, 1, and 0.1 nA.

15.79 A sample-and-hold circuit has a holding capacitor of 0.01 μF. If the circuit draws 100 nA from its holding capacitor, find the maximum time interval between sample operations if the held signal is to droop by no more than 1 mV. Repeat for 0.1 mV.

15.80 A sample-and-hold circuit can provide a maximum of 2 mA of holding-capacitor charging current. If a 0.001-μF capacitor is used, what is the minimum required duration of the digital sample signal? This parameter is called the *acquisition time* of the sample-and-hold circuit. The maximum analog voltage to be encountered is 5 V.

15.81 The acquisition time of a sample-and-hold circuit (the time required to charge its holding capacitor) is to be no more than 50 μs. If a 0.001-μF capacitor is used, what must be the minimum capacitor-charging current? The maximum analog voltage to be encountered is 10 V.

15.4.3 Analog-to-Digital Conversion

15.82 What is the minimum resolution of a 14-bit, binary-weight-encoded A/D converter operating with a 10-V reference voltage?

15.83 An analog voltage of value 4.2 V is applied to an 8-bit, binary-weight-encoded A/D converter for which $V_{REF} = 5$ V. What is the resulting binary output?

15.84 A 3.3-V analog voltage is applied to a 4-bit, binary-weight-encoded A/D converter for which $V_{REF} = 12$ V. What is the resulting binary output?

15.85 Determine the binary output of a 12-bit, binary-weight-encoded A/D converter with a 10-V reference voltage if its analog input is equal to 8.7 V.

15.86 An analog voltage of value 5.1 V is applied to an 8-bit, bin-encoded A/D converter for which $V_{REF} = 7$ V. What is the resulting binary output?

15.87 A 2.2-V analog voltage is applied to a 16-bit, bin-encoded A/D converter for which $V_{REF} = 12$ V. What is the resulting binary output?

15.88 A 10-bit, binary-weighted A/D converter with a reference voltage of 10 V has an output of **11110 00000**. Find the corresponding analog input. What voltage increment must be added to the analog input if the binary output is to be incremented by **1**?

15.89 An 8-bit, binary-weighted A/D converter has a reference voltage of 10 V. If its output is **1010 1010**, what is the range of possible values for the analog input?

15.90 ⚡ Consider the binary counter used in the 4-bit A/D converter of Fig. 15.33. Design an appropriate counter using a combination of flip-flops and an astable multivibrator.

15.91 Consider the dual-slope A/D converter of Fig. 15.34. Draw the output of the integrator versus time if $RC = 100$ ms, $\tau_1 = 10$ ms, $v_{IN} = 2$ V, and $V_{REF} = -10$ V.

15.92 Consider the dual-slope A/D converter of Fig. 15.34. Find the digital output of the 8-bit binary counter if $V_{REF} = -2$ V and $v_{IN} = 1.5$ V.

15.93 Consider the dual-slope A/D converter of Fig. 15.34. If the binary counter is a 12-bit circuit, find its output if $V_{REF} = -5$ V and $v_{IN} = 2.3$ V.

Chapter 16

Electronic Design

The principal job of most engineers is the practice of design. Design involves the creation of a device, component, circuit, or system that meets some set of desired specifications. In practicing design, an engineer makes choices based on a thorough understanding of engineering fundamentals, but also must consider the constraints of feasibility, cost, manufacturability, and human impact when making design choices. Good design skills require experience, an understanding of the context of the problem, knowledge of what has worked in the past, and a considerable amount of intuition. Although intuition can be learned from textbooks, it is more readily acquired through practice, repetition, and hard work. Developing intuition also requires "seasoning"—the process by which a novice engineer gradually learns the "tricks of the trade" from other, more experienced engineers. Such information is often conveyed by an oral tradition of practices and procedures that everyone in the company "knows" but no one has written down. Only by practicing real design in a real engineering environment can an engineer gain knowledge of these practices.

The experience of failure is also a very important part of the design process. When the first attempt at a design fails, the engineer gains valuable insight into what changes and alterations may be needed to make the design successful. Acquiring proficiency in design requires practice in testing real designs on the lab bench, observing failures, iterating through design choices, and observing the results of design decisions. There is no better way to learn design than through actual experience.

Because of the importance of design in engineering education, this chapter, written in the context of electronics, is devoted to addressing some of the key elements of the design process. The chapter includes a selection of open-ended problems devised to help you, the student of electronics, become familiar with the essential elements of design. Each problem outlines a set of technical specifications to be met using the concepts and techniques developed in this book. In each case, multiple solutions exist, requiring you to weigh the relative advantages and disadvantages of each design alternative. The systems described can be designed on paper, but you will learn a great deal more if you attempt to built and test your design in the laboratory. Each of the projects outlined in this chapter can be built and tested using commonly found electronic components and instruments.

16.1 AN OVERVIEW OF THE DESIGN PROCESS

The process of designing an electronic circuit or system involves several important steps. The first step always should be a definition of the overall design objectives. Although this task may seem trivial to the student eager to build and test hardware, it is one of the most important. Only by first considering the "big picture" can the engineer determine all the factors relevant to the design effort. Good design involves more than making technical design choices. Key questions must be answered: Who will use the system being designed? What are the needs of the end user? Which features are critical, and which are desirable but not crucial to the success of the project? Can it be easily manufactured? How much will it cost?

To answer these questions, the designer must be familiar with the end user and the environment in which the circuit or system will be used. "Human factors" such as physical appearance, ease of use, size and weight constraints, and affordability must be considered along with such issues as method of construction, type of circuit technology, and method of manufacture. These latter considerations should share equal weight with technical design decisions.

The second step in the design process involves the selection of a design strategy. At this stage, the electronic designer might decide, for example, whether a discrete or integrated-circuit approach, an analog or digital circuit is best, whether a BJT or MOSFET circuit is warranted, what type of power supply will be required, or whether an inverter, follower circuit, or differential amplifier will best meet user specifications. If the system is complicated, it should be broken up into simpler, smaller modules that can be interconnected to form the complete product. Modules should be designed so that they can be individually completed and tested before the entire system is put together. Organizing the task in this way simplifies synthesis, testing, and evaluation, and helps to subdivide the problem into tasks easily performed by one person. The modular approach is critical in a team design effort.

The design strategy should also consider the results of research on previous related work. In many cases, solutions to similar problems may already have been tried or may exist in commercial form. The wise designer will gladly make use of products that simplify the design process. There is no shame in using a ready-made circuit to accomplish an electronic task, and a designer should always look for shortcuts to the final design objectives. Imagine how needlessly complex the task, for example, of building a simple feedback circuit from scratch without using a commercially available monolithic IC operational amplifier.

After the design strategy has been solidified, it is time to make a "first cut" at each module in the system. A detailed layout of each circuit is formulated on paper and tentative values for its elements are specified. At this stage, key quantities—voltages, currents, resistor values, time durations, and transistor parameters, for example—are estimated and evaluated. This step in the design process may involve rough approximations and gross estimates. Its primary purpose is to determine whether or not the design approach has a chance of working, and it should result in a circuit layout and tentative choice of components for each module in the system.

The next step in the design process is a crucial one. The paper design is evaluated, built, tested, and evaluated again. The design may be revised at any time in the process. It is in this stage that the principal work of the engineer occurs. A good engineer will review a design many times, often proceeding through numerous iterations until the best circuit configuration and element values are found. The process of evaluation can take many forms. Prototype circuits can be built in temporary "breadboard" form or fabricated using the wirewrap method. Portions or subsections of the design can be individually tested. Computer-aided design tools such as SPICE can be used to simulate circuit operation. Circuit simulations save time and expense by helping to identify fundamental design flaws *before* the circuit is actually built. Despite the usefulness of computer-aided design tools, however, there is no substitute for constructing an actual prototype to evaluate and test the design.

In evaluating a circuit, its performance should be assessed from many points of view. Suppose, for example, that we wish to design an amplifier to meet a specific gain requirement. A thorough evaluation of the circuit would also assess the change in gain with temperature, the power dissipated in each component in the circuit and the effect of power dissipation on temperature, the total current required by the circuit from its power supply, the mean time required before battery replacement (if battery-operated), and the actual commercial availability of the components used in the circuit.

In the last step of the electronic design process, the thoroughly tested and "debugged" circuit or system is built in final form. Individual modules, if they exist, are interconnected to form a complete system. At this stage, the designer must decide what type of wiring technique to employ and which packaging method to use. If the circuit is intended for actual delivery to a customer, the designer may wish to etch a permanent and reliable printed-circuit board and build an attractive case or housing for the circuit. One-of-a-kind prototypes can use one of the wiring methods described in the next section. In any case, the finished circuit should be thoroughly tested and subjected to a "burn-in" period during which the circuit is left energized and functioning. This procedure will help to identify any latent component defects that might cause the circuit to fail in the field. Only after a comprehensive test period is the circuit ready to be put into actual service.

16.2 THE TOOLS OF ELECTRONIC DESIGN

In this section, we review several of the techniques, methods, and procedures that are commonly used in electronic design. This section has two objectives. One is to acquaint the reader with the methods and procedures practiced in the actual workplace. The other is to describe design tools and methods that are easily obtained in even the simplest educational laboratory setting.

Laboratory Workbench

Efficient and sound design requires proper laboratory tools. A basic electronics lab suitable for design work should be equipped with a dual-voltage power supply, multimeter, function generator, and basic analog oscilloscope. Other useful lab instruments include a frequency counter, digital storage scope, logic analyzer, capacitance meter, and computer. A good set of hand tools, including long-nose and diagonal pliers, slotted and Phillips-head screwdrivers of various sizes, wire cutters and strippers, and a soldering iron, is also very useful. Of prime importance is an uncluttered workplace with plenty of space to locate instruments, circuit prototypes, notes, and parts. The importance of an uncluttered workplace to the design process and to the attitude of the designer cannot be overemphasized.

The Parts Inventory

When a design is ready for mass production, a list is created of all parts, no matter how small or minor, that are used in the circuit and its package. This list is used by many people, including those responsible for ordering or making the parts, determining overall cost, and assembling the finished product. During the initial and prototype stages of development, however, it is difficult to work from a fixed parts list. Even minor changes in the design may require components to be changed or altered. Waiting for new parts to be ordered on an as-needed basis can significantly delay the design process. A well-equipped laboratory will include a running inventory of commonly used parts. Such parts typically include a set of fixed resistors extending over several decades, a good selection of ceramic and electrolytic capacitors, some variable resistors, general-purpose *npn* and *pnp* signal transistors, power transistors, LM741 operational amplifiers, CMOS or TTL logic gates, 555 timers, light-emitting diodes, and mechanical switches. The small quantities of these parts required for prototype development can be purchased from one of several electronics retailers or consumer-oriented mail-order vendors. Maintaining a parts inventory may involve added expense, but the latter is offset by a considerable savings in time.

Wiring Methods

Virtually every commercial, mass-produced circuit is wired together using an etched, copper-clad *printed-circuit board*. As illustrated in Fig. 16.1, the wiring paths on a "PC" board (abbreviation not to be confused with "personal computer") consist of thin strips of copper, called *traces*, bonded to an insulated board made from phenolic or epoxy glass of approximately 1 mm thickness. In conventional PC board design, component leads pass through holes drilled in the board and are soldered to the copper wiring paths. Technology exists to make PC boards with several separated layers of copper, so that wiring paths may cross without making contact. More modern boards make use of *surface mount* technology in which the "leads" of components consist of short, stubby connection pads that solder directly to the copper PC board traces.

Figure 16.1
Printed-circuit board showing copper wire traces on one side and components on the other. Component leads protrude through holes in the board and are soldered to the copper traces.

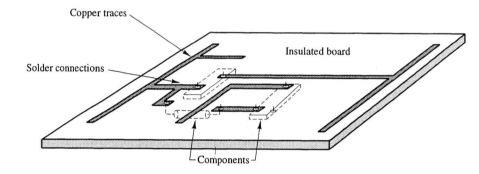

Printed-circuit board methods are ideal for mass-producing circuits. Many commercial software packages exist to aid the designer in laying out the board for subsequent etching by a commercial PC board vendor. These software programs generate files that can be fed directly to a board fabrication machine. The per-unit cost of a mass-produced, professionally made PC board is low, but the setup charge is high and sophisticated fabrication equipment is required. Hence PC board methods are generally used only for circuits that will be produced in large quantities. One alternative for prototyping involves a system in which PC boards are produced by hand in "one-at-a-time" fashion. The paths of copper traces are first marked with a resist pen or marking tape on a copper-clad board. The board is then etched by hand in an appropriate solution that dissolves unmarked copper. After etching, the traces are tinned (coated with a thin layer of solder) and holes are drilled for component leads. Relatively inexpensive kits for hand etching boards are available from most electronics parts vendors. Hand-etched boards provide the physical robustness of mass-produced PC boards, but are of significantly poorer quality, require traces of much larger line width, and require much more time to make. Correcting errors or making changes in the circuit layout is also difficult.

A form of PC board, called copper trace board, or sometimes "experimenter's prototype" board, is also commercially available. This product is similar to a PC board and contains preetched, predrilled strip traces to which component leads may be soldered. Connections across traces are made by soldering wire jumpers into place. Changes to the circuit can be made, but a given connection can be resoldered only a few times before the trace begins to separate from the board.

One-of-a-kind prototypes may also be made using wirewrap technology. As depicted in Fig. 16.2, components are inserted into special IC sockets that have long pins protruding through holes in a prepunched insulated board. Components may also be soldered to the ends of special wirewrap pins. Connections between pins are made by tightly winding special wire (usually #30 gage) around the pins with a wirewrap tool. The pins used for wirewrap have a square cross section that digs into the wrapping wire and makes a good metallurgical bond. Wirewrapped circuits are not as robust nor nearly as compact as PC-board wired circuits, but still provide good, nearly permanent connections. Their principal advantage is the relative ease with which connections can be altered to accommodate design changes and alterations.

Figure 16.2
Basic elements of a wirewrap system. Pins with square cross-section protrude through perforated board. Connections are made by tightly wrapping special wire around the posts.

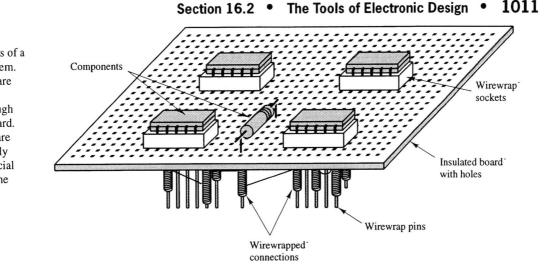

For rapid prototype development in which many design changes will be made, component breadboards are ideal. The breadboard of Fig. 16.3 is typical of most versions. Holes along each short row in the board reside over conducting metal spring clips into which component leads can be inserted. All leads inserted into the same row become electrically connected. Additional connections can be made using jumper wires. Most breadboards also contain long outer rows of interconnected holes that are used as power-supply buses. The holes on the typical breadboard are spaced at 0.10 inch, which is the same spacing as the pins of a dual-inline integrated-circuit package. The leads of discrete components such as resistors, capacitors, and transistors can be cut to size and bent at right angles, so that the components are easily inserted.

Figure 16.3
A neatly laid out component breadboard. Each row of holes resides over a conducting spring clip that electrically connects component leads forced into the holes.

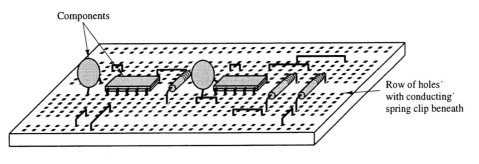

Avoiding the "Bird's Nest" Pitfall

Good wiring practice requires that a breadboard circuit be compact, neat, and orderly, with all leads cut as short as possible. Component bodies should physically rest on or just above the board surface, and wires should be easy to trace and touch with the probe of an oscilloscope or multimeter. The breadboard depicted in Fig. 16.3 is example of a well laid out circuit. The "bird's nest" approach illustrated in Fig. 16.4 should be avoided at all costs (photo taken in the lab at Boston University). When a circuit consists of a disorderly tangle of wires leading haphazardly in every direction, component leads may short together, wiring errors are likely, and circuit testing becomes extremely difficult. One easily becomes lost in such a circuit. Long leads also create unnecessary stray capacitances and inductances that may adversely affect circuit behavior. A sloppy circuit affects the attitude of the designer, who is likely to take the design process less seriously if work on the circuit is difficult. The wise designer produces circuits that are neat, compact, tidy, and easily accessible.

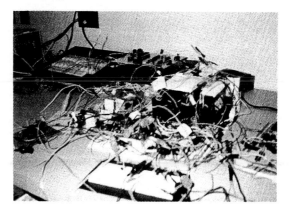

Documentation

When designing a circuit or system, the engineer must keep careful records of all tests performed and design elements completed. It is a good strategy to write down everything, even if an item seems unimportant at the time. Documentation should be written in such a way that another engineer who is only slightly familiar with the project can repeat all work done by simply reading the notebook. Careful documentation will aid in the task of writing product literature and technical manuals if the design is destined for commercial sale. Above all, good documentation will provide the engineer with an overview of the design history and the key questions that were addressed during the design process.

SUMMARY OF DESIGN PRINCIPLES

- Design is an acquired skill best learned by practicing on real problems.
- Good design involves a combination of experience, knowledge, and intuition.
- Engineers become "seasoned" by working with other, more experienced engineers and learning accepted engineering practice.
- The design process requires that the engineer do the following:
    - Assess the goals of the problem from an overall point of view.
    - Choose a design strategy.
    - Divide the problem into smaller, more manageable pieces.
    - Attempt a "first-cut" design by choosing values for all components and elements in the system.
    - Evaluate the design and revise; evaluate the design and revise.
    - Test and retest the design until the best possible solution is obtained.
    - Build the product in finished form.
    - Keep good documentation at all times.

16.3 OPEN-ENDED DESIGN PROBLEMS

This section contains a series of open-ended problems chosen to help the student of electronics practice design skills. Each problem has more than one, and often many, possible solutions. None has a single, or "correct" answer. Each circuit or system described can be designed to specifications using the concepts and techniques developed in this book and will require you to confront the many design issues raised in the previous two sections and throughout the text.

16.3.1 High-Current Adjustable DC Power Supply

As discussed in Chapter 4, a power-supply circuit transforms an ac input voltage taken from the local power line into a dc voltage suitable for powering other electronic circuits. A well-designed power supply has minimal ripple and maintains a constant output voltage regardless of load current. In this project, you are to design, build, and test a bipolar power supply that produces both a positive and a negative output voltage. These voltages are to be simultaneously adjustable, via a single variable resistor control, from zero to full output value. Your supply should accept its input from a transformer with a secondary voltage between 12 V rms and 24 V rms. (You should adjust your design parameters to accommodate whatever transformer is available.) Your design must meet the following specifications:

- Output voltage range: 0 to 10 V and 0 to −10 V.
- Maximum output current from each voltage polarity: 300 mA.
- Ripple voltage: Less than 50 mV at full load current.

The positive and negative voltage outputs may have a common ground connection, but such a connection is not a requirement.

16.3.2 Dynamic Microphone Amplifier

Dynamic microphones are used in many sound processing applications, including public address systems, radio and television broadcasting, music recording, and audio signal detection. A dynamic microphone consists of a diaphragm that supports a coil of very fine wire in the field of a permanent magnet. When sound waves strike the diaphragm, the coil vibrates in the magnetic field and produces a small voltage that replicates the sound striking the diaphragm. A typical dynamic microphone produces a 10-mV peak signal in response to the human voice. Its coil has a series resistance of a few tens or hundreds of ohms.

Design a circuit that accepts the signal from a dynamic microphone and delivers an amplified version to an 8-Ω loudspeaker at a maximum power level of 0.25 W. The gain of the amplifier should be constant to within ±1.5 dB, with no obvious clipping or distortion, over a minimum frequency range of 50 Hz to 10 kHz.

If you choose to build the circuit and a dynamic microphone is not available, you can simulate one using a small 8-Ω loudspeaker and a 100-Ω series resistor. This combination will approximate the properties of a real dynamic microphone, but will have poorer frequency response.

16.3.3 Rotating Shaft Controller

Using one or more operational amplifiers, design a control circuit that adjusts the stationary angle of a rotatable shaft in response to an input signal voltage. The principal sensing element in the system should be a 100-kΩ potentiometer whose control arm is connected to the rotating shaft. The shaft should be coupled via a gear box or other mechanical reduction system to a motor that turns in the clockwise direction for one voltage polarity and in the counterclockwise direction for the opposite voltage polarity. One of the key design parameters in this problem is the gear ratio connecting the motor to the shaft. This parameter can be specified only after considering the torque capabilities and speed range of whatever motor is chosen.

If a gear box is not available, the required mechanical reduction can be implemented by coupling the shaft of the motor to the shaft of the potentiometer via a rubber band and an empty thread spool, as shown in Fig. 16.5. Design your system to meet the following specifications:

- Angle response: 0 to 270° for an input control voltage of 0 to 5 V. (Note that the typical potentiometer can be turned only over an angle of about 270°.)
- Signal input resistance: > 10 kΩ.
- Response speed: < 10 seconds over an angle of 270°.

Figure 16.5
Rotating shaft
controller.

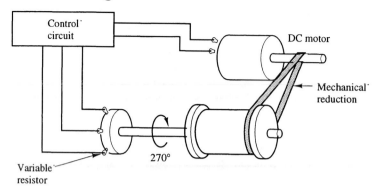

16.3.4 Motor Shaft Speed Indicator (Analog Tachometer)

Low-power dc electric motors are used in countless applications to provide controlled mechanical motion. Examples of systems in which small motors are found include compact disc (CD) players, hard and floppy disk controllers, CD-ROM drives, digital and audio tape drives, video players (VCRs), video cameras, laser printers, photocopiers, model electric trains, model cars, and medical equipment. Over a certain range, the speed of a dc motor is roughly proportional to applied voltage. Its actual speed for a given voltage, however, will depend upon mechanical loading. If the motor speed is to be controlled precisely, independent of load, it must be incorporated into a negative-feedback loop.

In this project, you will design a tachometer circuit capable of providing an analog signal proportional to instantaneous motor speed. Such a signal is required if the motor is to be included in a negative-feedback loop. Your system should include a sensor coupled to the motor shaft plus a circuit that accepts the sensor signal and produces the required output voltage. Small dc motors designed for 3- to 12-V operation at speeds of 1000 to 10,000 rpm are readily available from electronic supply stores or hobby shops. For test purposes, the speed of the motor can be varied by simply connecting it to a variable voltage source. A stroboscopic light, if available, can be used to calibrate the tachometer circuit.

Your system will require some means of sensing the actual rotational speed of the motor shaft. One possible approach might involve the use of a commercially available optical sensor coupled to a homemade shaft encoder. Such a system is illustrated in Fig. 16.6. One arm of the optical sensor contains a light-emitting diode (LED), and the other contains a sensitive phototransistor. If the optical path between the LED and phototransistor is not interrupted, the phototransistor will receive a light signal and become conducting. Conversely, if the optical path is interrupted by a solid object, the phototransistor will receive no light and will become nonconducting. The shaft encoder has the job of periodically interrupting the optical path of the sensor. It consists of a disk of paper, metal, or plastic attached to the motor shaft and perforated with holes around its perimeter. As the motor shaft turns, the holes in the disk interrupt the optical path of the LED–phototransistor sensor. A suitably designed processing circuit can transform the alternate conducting and nonconducting status of the phototransistor into the required analog signal. Another approach to sensing shaft speed might involve the use of a Hall-effect device and a small permanent magnet mounted on the motor shaft. When the magnet is positioned in close proximity to the Hall-effect device, the device becomes conducting.

To meet the requirements of the feedback control system, your tachometer must conform to the following specifications:

- Analog output: Linearly proportional to motor shaft speed; 10 V at full motor speed.
- Output Impedance: $< 50\,\Omega$.

Figure 16.6
Optical sensor and
shaft encoder.

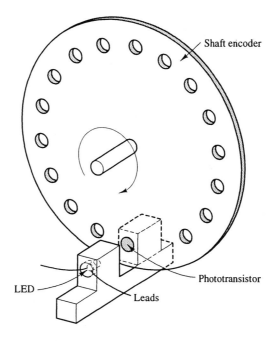

Note that response time is seldom an issue in designing an electronic tachometer. The inertial response of the motor to changing conditions is usually so slow that the response of the electronic circuit can be considered instantaneous.

> *Optional:* Use your tachometer circuit create a feedback control loop that precisely controls motor speed. A block diagram of one possible system is illustrated in Fig. 16.7. The motor speed should be directly proportional to a 0- to 5-V analog reference signal.

Figure 16.7
Possible block
diagram of a motor
tachometer in an
analog feedback
control system.

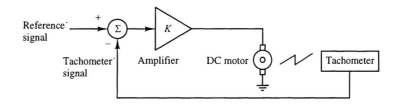

16.3.5 Electronic Odometer for a Bicycle

Use the Hall-effect sensor or shaft encoder described in Section 16.3.4 to design a battery-operated speedometer and/or odometer for a bicycle. (A speedometer measures instantaneous velocity, and an odometer measures total distance traveled.) The output of your device should be digital, rather than analog. Specifically, the bicycle speed or distance readings should be displayed on a seven-segment numerical display containing an appropriate number of digits. A single momentary pushbutton switch should be used to toggle the display between speedometer and odometer readings. A second momentary switch should be used to reset the odometer reading to zero. The unit should be housed in a case suitable for mounting on bicycle handlebars.

16.3.6 Voice-Actuated Light Switch

Design a circuit capable of turning on a light or other electrical appliance when a loud sound, such as a clap or finger snap, is first received by a microphone. The light should turn off when the sound is received a second time. The circuit should energize a mechanical relay that requires a coil voltage between 5 and 24 V. The relay contacts would ordinarily be used to connect the light bulb or appliance to the ac power line. To avoid electrocution hazards, however, you should develop and test your circuit using a flashlight bulb and an appropriate battery or low-voltage power supply. If you wish, the mechanical relay may be replaced by one or more silicon controlled rectifiers (SCRs) that can be used to switch the light bulb or appliance on and off. A small loudspeaker can be used in lieu of a microphone if necessary, as explained in Section 16.3.2.

16.3.7 Amplitude Modulator

Amplitude-modulation (AM) circuits are used in a variety of control and communications applications. An AM modulator can be thought of as an amplifier whose gain is controlled by an externally applied voltage. Such circuits are typically used in radio stations to encode the high-frequency rf carrier with an audio signal in preparation for radio transmission. Other applications include the modulation of video signals used to drive the scanning beam of a television screen and the encoding of the signals produced by infrared remote-control devices. The objective of this project is the design of a basic amplitude modulator capable of operating at audio frequencies. The finished circuit should function as an amplifier with voltage-controlled gain and should meet the following specifications:

- Range of gain: 0 to 10 for a control voltage of 0 to +5 V.

- Signal input resistance: $> 1\,\text{M}\Omega$.

- Control-voltage input resistance: $> 1\,\text{k}\Omega$.

- Amplifier output resistance: $< 50\,\Omega$.

- Amplifier bandwidth: At least 10 Hz to 20 kHz.

- Midband response: "Flat" to within ± 3 dB into a 50-Ω load at 1 V p–p output.

- Output swing range: ± 10 V minimum.

16.3.8 Audio-Frequency Analog Wattmeter

A wattmeter is an instrument that measures the time-average power delivered by a power source to a circuit, system, or device. In this project, you are to design an ac wattmeter capable of measuring the time-average audio-frequency ac power delivered by an audio amplifier to its loudspeaker load. For test purposes, you may use your meter to measure the power delivered by a signal generator or power amplifier to a comparable resistive load.

Your circuit should include two pairs of terminals, one for sensing the voltage across the loudspeaker or load, and the other for sensing the current through the load. One possible connection arrangement is illustrated in Fig. 16.8. The *apparent* power delivered to the load is defined as the product $v_{\text{LOAD}}(t)i_{\text{LOAD}}(t)$. Your wattmeter must measure the *real* power delivered to the load and hence must respond to the vector dot product $\mathbf{V}_{\text{load}} \cdot \mathbf{I}_{\text{load}}$, where $\mathbf{V}_{\text{load}}$ and $\mathbf{I}_{\text{load}}$ are the phasors representing v_{LOAD} and i_{LOAD}.

Figure 16.8
Analog wattmeter
connections.

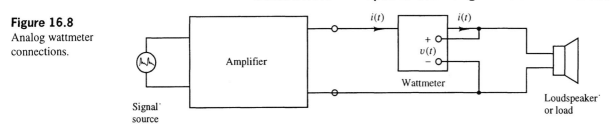

Your wattmeter should conform to the following specifications:

- Power range: 0 to 5 W.

- Frequency range: 20 Hz to 20 kHz.

- Maximum voltage across voltage-sensing terminals: ± 20 V.

- Maximum current through current-sensing terminals: ± 500 mA.

16.3.9 Ambient Temperature Monitor

Design a circuit capable of providing an analog voltage signal that is proportional to temperature. Your sensor could consist of a forward-biased diode, for which the forward voltage V_f is temperature-sensitive. Alternatively, you could use a BJT operating in the constant-current region. In the latter case, the temperature sensitivity of β_F would make the device current change with temperature. In either case, the temperature dependency will be nonlinear, hence some means of interpreting the reading and converting it to a signal linearly proportional to temperature must be devised. Other possibilities for a temperature sensor include a commercially available thermistor or thermocouple. The former device is essentially a temperature-sensitive resistor. The latter consists of a metallurgical bond of two dissimilar metals that produces a thermally driven open-circuit voltage in proportion to absolute temperature.

> *Optional:* Add a final stage to your system that converts the analog output signal into a digital signal and displays it on a three-digit LCD or LED display. This portion of the project will require you to design an appropriate A/D converter, decoding circuit, and display driver.

16.3.10 Hand-Proximity Electronic Musical Instrument

The hand-proximity detector shown in Fig. 16.9 can be used as the basis for a solid-state version of the Theremin, an electronic musical instrument that is played by subtle movement of the hands.[1] The circuit of Fig. 16.9 consists of a noninverting amplifier with a phase-shifting positive-feedback network. The capacitor C_X in the circuit diagram represents the stray capacitance between the sensing plate and ground. If the circuit is built using an LM741 or equivalent op-amp, it will be unstable and will oscillate when the noninverting gain is reduced via R_2 below some critical level. In practice, the position of the wiper arm of R_2 is adjusted until oscillation at a suitable output magnitude occurs. The frequency of oscillation will be related to the values of C_X, C_1, and R_1. The capacitance C_X is altered by approaching (but not touching) the sensing plate with the palm of the hand.

---

[1] The Theremin was invented by Leon Theremin in 1910.

Figure 16.9

Hand-proximity detector made from an LM741 operational amplifier.

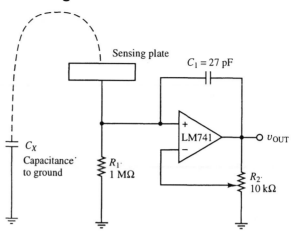

If the component values shown in Fig. 16.9 are used with a sensing plate of approximately 10 cm × 10 cm in size, the resulting frequency of oscillation will cause the output signal to be slew-rate limited. This slew-rate limitation will prevent v_{OUT} from reaching the saturation limits of the op-amp between oscillation cycles and will create a triangular output waveform. If the frequency of oscillation is changed over this slew-rate limited range, the magnitude of the output will change with frequency. The output of the hand-proximity detector thus becomes a triangular voltage waveform whose peak magnitude is a function of the distance between the operator's hand and the sensing plate. In practice, the sensing plate can be made from a 10-cm square of aluminum foil or from a 10-cm diameter loop of wire connected to the v_+ terminal of the op-amp.

One possible outline of a musical instrument based on the hand-proximity circuit of Fig. 16.9 is shown in Fig. 16.10. The hand-proximity detector is used twice, once for controlling the pitch of the musical tone, and once for controlling the volume. The output of each hand-proximity circuit is fed to an appropriate peak detector and level-adjusting circuit that produces a dc signal related to the operator's hand proximity. One of these output signals is fed to a voltage-controlled oscillator (VCO), and the other is used as the input to an automatic gain control (AGC) circuit, or amplitude modulator. The VCO and AGC circuits are to be designed as part of the project. The gain- and pitch-controlled output can be fed to an audio amplifier driving a loudspeaker or headphones.

Figure 16.10

Theremin musical instrument based on the detector of Fig. 16.9.

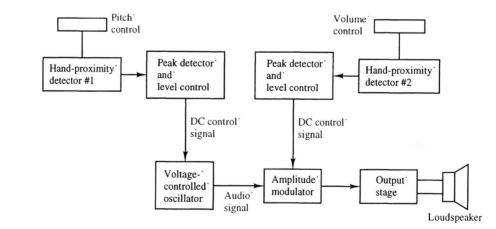

It is possible to build the musical instrument described using only op-amps, resistors, capacitors, diodes, and transistors. Digital components can also be used, however, to design some of the circuit stages. Be sure to use an LM741 or equivalent in the hand-proximity detector. Other op-amps may not oscillate with the indicated component values.

16.3.11 Hand-Operated Light Switch

The basic theory and operation of the hand-proximity circuit of Fig. 16.9 is described in Section 16.3.10. Use this circuit to design a light switch that is activated by hand proximity. The circuit should energize a mechanical relay requiring a coil voltage of about 12 V. The relay contacts can be used to connect a light bulb or appliance to the ac power line. To avoid electrocution hazards, you should test your circuit using a flashlight bulb and an appropriate battery or low-voltage power supply. If you wish, the mechanical relay may be replaced by one or more silicon controlled rectifiers (SCRs) that can be used to switch the light bulb on and off.

16.3.12 Heart Rate Monitor

In this project, you are to build a circuit capable of measuring the heart rate of an adult human. The system will require a sensor that responds to the human heartbeat. One possible sensor arrangement might consist of an infrared light-emitting diode and photodetector placed on opposite sides of the index finger. Because fingertip tissue is partially transparent to infrared light but blood is not, the sensor will receive a varying signal when the heart pulse reaches the fingertip. This sensing method is the one used in many commercial heart-rate monitors. An alternative sensor could consist of a microphone placed on the chest cavity, essentially functioning as an electronic stethoscope. The signal from the microphone could be amplified and conditioned to provide the required pulse signal. Regardless of which sensing method you use, the sensor signal is likely to be noisy and erratic. Your circuit must process it until it becomes a cleanly defined digital pulse that can drive a digital counter or other suitable rate-monitoring circuit.

One approach to determining the pulse rate would be to feed the heartbeat pulses to a digital counter that is reset each minute (or smaller fraction of a minute) by a second timing circuit. A second approach might be to use the heartbeat pulse to charge a capacitor via a series resistor. If the RC time constant is suitably chosen, the voltage to which the capacitor decays between pulses will be indicative of the time between pulses. Alternatively, an RC circuit can be used as a low-pass filter that provides the dc component of the heart-rate pulse train. The more frequent the pulses, the larger their dc component.

16.3.13 DC Nanoammeter

Many signal-processing applications in biology, physics, and chemistry require the measurement of dc currents in the picoampere range. This task is not a trivial one, because the minimum current detectable using commonly available laboratory instruments is only about 1 μA. In this project, the measurement task will focus on currents in the nanoampere range, although the design principles are identical to those required for the picoampere range.

Design a circuit that accepts a dc current in the 0- to 10-nA range as its input. The circuit should then produce an output voltage in the 0- to 10-V range, for a current-to-voltage conversion ratio of 1 V/nA. The input resistance to the circuit should not exceed 1 MΩ, so that the maximum voltage drop across the circuit's input terminals will not be greater than 10 mV.

You can test your circuit using the nanoampere current source shown in Fig. 16.11. For the suggested component values shown, and with $V_1 = 9\,V$, I_o will be approximately equal to 9 nA. The value of I_o can be altered by changing R_2. The voltage source V_1 should be a self-contained battery, rather than a benchtop power supply. The entire current-source circuit should be enclosed in a grounded, conducting box to avoid problems with noise and line-frequency (60 or 50 Hz) interference. To help reduce the latter, the current-source circuit should be connected to your measuring circuit via two shielded, coaxial cables. The two center conductors carry the reference current I_o; the shields of these cables should be connected to the grounded box.

Figure 16.11
Nanoampere
current source.

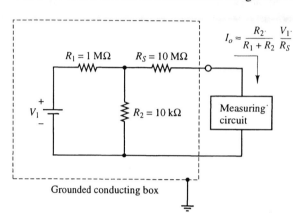

$$I_o \approx \frac{R_2}{R_1 + R_2} \frac{V_1}{R_S}$$

Grounded conducting box

16.3.14 Low-Battery-Level Indicator

In this project, you will develop a circuit capable of indicating a "low-battery" condition for a battery-operated appliance. Essentially, you must design a circuit that can detect a voltage drop on its own power supply. Your circuit should be powered by a standard "transistor radio" type battery whose voltage, when new, is about 9 V. When the battery voltage drops below 7 V, your circuit should activate a light-emitting diode or other indicating device. Your circuit should continue to operate even when the battery voltage falls as low as 3 V. No other power supplies are available for powering your circuit—it should be designed to operate from the same battery that it tests.

16.3.15 Function Generator

Design a three-waveform function generator that operates in the 1-Hz to 100-kHz frequency range. The frequency of the output signal should be adjustable over a factor of 10 via a single variable-resistor control. Different frequency decades should be selected by changing the values of one or more resistors and/or capacitors. In your initial prototype design, the values of these elements may be changed by substituting components. In the actual instrument, these elements should be selected via a master rotary switch or digitally activated switching circuit.

Your function generator should have three parallel outputs: square wave, triangle wave, and sinusoid. You might use the shaping circuit of Fig. P4.151 to generate the sinusoid. Other methods of producing a sinusoid include the Wien-bridge and phase-shift oscillators of Chapter 13. Ideally, the amplitude of all three outputs should be simultaneously adjustable over the range 0- to 10-V peak via a single variable resistor. The output resistance at each output terminal should be at most 1 kΩ. Note that this latter requirement is less stringent than that encountered in practice. Most function generators are designed to have an output resistance of precisely 50 Ω.

16.3.16   Nickel–Cadmium (NiCad) Battery Charger

The goal of this project is to design a circuit capable of recharging nickel–cadmium (NiCad) batteries in the AAA to D size range. Charging NiCads properly is a bit tricky because NiCads can be damaged by overcharging. Charging should be terminated abruptly when the battery is fully charged with a voltage across each cell of precisely 1.2 V. A charger with two modes of operation—one for rapid charging at high current levels of about 100 mA per cell and one for "trickle" charging at current levels of about 10 mA per cell—is also desirable. The latter mode will charge batteries more slowly but will extend their overall lifetime. Because NiCads are best charged only after being fully discharged, a well-designed battery charger will also first fully discharge the battery before charging it to full capacity.

Your battery charger should be designed to meet the following specifications:

- Number of cells charges simultaneously: one to four.
- Battery size: AAA through D; special connections for battery packs of odd size.
- Charging modes: Rapid rate at 100 mA; "trickle" rate at 10 mA.

Optional Features

- Full-charge detector with shutdown option.
- Full discharge and recharge sequence.

16.3.17   AC Magnetic Field Meter

Design a battery-powered, hand-held instrument capable of measuring ac magnetic fields in the range 1 to 100 milligauss (mG) at frequencies of 50 to 60 Hz. The instrument will be used to measure the magnetic fields produced by household wiring and electrical appliances. Such fields are very small and are difficult to measure accurately. For comparison, the earth's dc magnetic field is on the order of 500 milligauss, and the magnetic field deep inside a typical electric motor is on the order of 10 kilogauss.

Your primary sensor may consist of a flat coil of wire of appropriate diameter and number of turns. Alternatively, you might consider using a semiconductor Hall-effect sensor. Note that dc fields, such as those produced by the earth or any nearby permanent magnets, are not of interest, hence any signal produced by them in your instrument should be filtered out.

16.3.18   Model Train Controller

The speed of a model train is controlled by changing the voltage applied to its electric motor. A model engine is so lightweight that it arrives at its newly set throttle speed almost instantaneously. In contrast, the speed of a real, full-sized train increases slowly when the engineer applies the throttle because the train's large mass must be accelerated up to speed. Similarly, the large inertia of a real train causes it to come to a stop very slowly when the throttle is reduced, even when the brakes are applied. A model train can be made more realistic if the inertial effects of a real train are modeled by the control system.

Your goal in this project is to design and build a model electric train controller that simulates the inertia and braking effects of a real train. This goal can be most easily met if the small motor inside the model train is powered a periodic voltage signal of constant magnitude and varying duty cycle. The motor will respond with a speed that is proportional to the average applied voltage.

Design your train controller to simulate a full-sized train with the following properties:

- Total train weight: 2000 metric tons.
- Maximum speed: 50 km/h.
- Maximum distance required to come to a stop from full speed: 2.5 km.
- Maximum distance to accelerate to full speed: 5 km.

16.3.19 Low-Resistance Ohmmeter

A common laboratory ohmmeter measures resistance in the range $1\,\Omega$ to $20\,M\Omega$. The goal of this project is the design of an ohmmeter capable of measuring resistance in the range $1\,m\Omega$ to $1\,\Omega$. Making such a measurement is difficult because the resistance of the wires and contacts that make up the typical electronic circuit also lie in the milliohm range; hence they are not negligible. In order to function properly, your low-resistance ohmmeter will have to incorporate some sort of balancing or nulling technique so that the resistance of the circuit's internal wiring can be easily taken into account.

16.3.20 Wireless Microphone

Design a wireless microphone that will enable you to transmit over an unused frequency in the AM radio broadcast band (550 kHz to 1.6 MHz). If you connect the radio output to a power amplifier and loudspeaker, your transmitter can also be used as a wireless public address system. The audio signal from the microphone element should modulate the amplitude of an rf carrier signal. The latter must be generated by a sinusoidal oscillator circuit. A square-wave oscillator should not be used to produce the rf carrier because its higher harmonic content will interfere with other radio bands. You will also need an antenna to radiate the modulated rf signal. A single length of wire approximately 1 m long can serve as a crude antenna in many cases.

Your wireless microphone should draw a relatively small amount of power so that it does not drain its battery too rapidly. The load on a standard 9-V rectangular battery, for example, should not exceed about 50 mA if the battery is to last for more than a few hours. In addition, the power consumed by the stage driving the antenna should be no more than about 100 mW so that your transmitter can be officially designated as a low-power rf source. In this way, you should not need any special licenses or permits to operate your transmitter.

16.3.21 Transistor Curve Tracer

A curve tracer is an instrument that displays the v–i characteristics of two- and three-terminal devices. The component parts of the typical curve-tracer are summarized by the block diagram of Fig. 16.12. At the core of the instrument is an oscilloscope operating in the x–y mode. A set of driving circuits probes the device under test and sends appropriate signals to the oscilloscope inputs.

Figure 16.12

Block diagram description of transistor curve tracer.

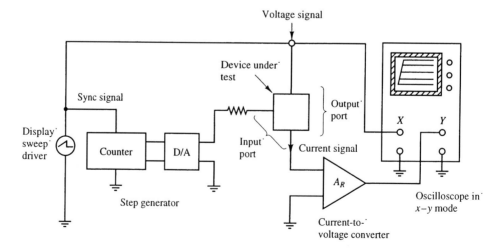

The display-sweep driver applies a periodic voltage signal to the output port of the measured device (e.g., to the collector–emitter port of a BJT), monitors the resulting current, and converts the latter to a second voltage signal proportional to the device current. The voltage applied to the device is monitored by the x-axis of the oscilloscope, and the voltage signal representing the device current is monitored by the y-axis. At any given moment, the instantaneous response of the oscilloscope will thus be a single dot at the v–i coordinates of the device's operating point. If the sweeping frequency of the display-sweep driver is suitably high, the scanned v–i characteristic will persist on the oscilloscope screen. Note that the sweeping voltage applied to the device need only be unipolar (e.g., zero to positive voltage or zero to negative voltage), because the v–i characteristics of the device need be displayed in a single quadrant only.

The input-port driver forms a second important part of the curve tracer. This circuit has the job of establishing fixed operating conditions at the input port of the measured device while the output port is being scanned by the display sweep circuit. For example, the input port driver will typically hold i_B of a BJT or v_{GS} of a MOSFET constant while the i_C–v_{CE} or i_D–v_{DS} curve is being displayed.

The input-port driver must be designed so that the operating conditions at the input port may be altered by the user, thus enabling different operating curves of the device to be displayed. A well designed input-port driver will incorporate a step generator that fixes the input port at some desired value for one sweep of the display driver, then steps the input variable by some fixed increment for the next sweep of the display driver. This feature enables the curve tracer to display the device's entire family of v–i characteristics on the same screen. Designing a step generator is nontrivial, because the initiation of the step change in i_B or v_{GS} must coincide with either the minimum or maximum extreme of each swept v–i characteristic. If this synchronization is not provided, the family of v–i characteristics will be randomly segmented and "choppy" on the oscilloscope screen. In the block diagram of Fig. 16.12, step generation is provided by a digital counter that drives a D/A converter. Synchronization in ensured by allowing the counter to sequence only when the display sweep driver reaches either a minimum or maximum extreme of its sweeping voltage.

Your goal in this project is the design of a set of driving circuits that can transform a common benchtop oscilloscope into a transistor curve tracer. Your system should be designed to meet the following specifications:

- Output port sweeping voltage: 0 to 20 V maximum.
- Output port current-measuring capability: 0.05 to 50 mA.
- Input port range: 0 to 1 mA (for BJTs) or −5 to 5 V (for MOSFETs and JFETs).

If you choose to incorporate a step generator into the input-port driving circuit, its step magnitude should be selectable over a range from 0.01 to 0.1 mA or 0.1 to 1 V.

16.4 ANALOG INTEGRATED-CIRCUIT DESIGN PROBLEMS

The analog integrated circuit problems in this section follow the topics of Chapter 12. They require that the material in Chapter 12 be covered beyond the general survey level. These problems can serve as capstone assignments to readers studying analog IC design methods in detail. Note that the problems are intended for paper design only. Actual circuit construction would require access to an IC fabrication facility.

16.4.1 BJT Operational Amplifier #1

The basic, but unfinished, circuit diagram of an integrated-circuit BJT operational amplifier is shown in Fig. 16.13. Your job is to complete the design by choosing values for all resistors such that the amplifier meets or exceeds the following open-loop specifications:

- Differential-mode input resistance: $> 20\,\mathrm{k\Omega}$.
- Common-mode input resistance: $> 100\,\mathrm{M\Omega}$.
- Output resistance: $< 20\,\Omega$.
- Differential-mode gain: > 1600.
- DC level of output voltage with both inputs at ground: $0 \pm 0.2\,\mathrm{V}$.
- Allowed common-mode swing of input voltage: $\pm 2\,\mathrm{V}$.
- Output voltage swing: $\pm 1.5\,\mathrm{V}$.
- High-frequency -3-dB endpoint of differential-mode response: $> 900\,\mathrm{kHz}$.

You should assume that the transistors have been fabricated with the following set of parameters: $\beta_o = 50$ to 150, $C_\mu = 0.3\,\mathrm{pF}$, $V_A = 50\,\mathrm{V}$, $f_T = 400\,\mathrm{MHz}$, $r_x = 10\,\Omega$, $V_{BE} \approx 0.7\,\mathrm{V}$ when forward biased, and $\eta = 1$.

The operation of the circuit may be explained as follows: Transistors Q_1 and Q_2 form a differential input stage biased by the modified Widlar, or Wilson, current source of Q_7 and Q_8. The outputs of Q_1 and Q_2 are fed to a second differential stage formed by Q_3 and Q_4. This second stage is biased by R_{11}. The output of Q_4 is tapped in single-ended fashion and fed to the double voltage-follower stage formed by Q_5 and Q_6. Level shifting is provided by R_5. Transistors Q_5 and Q_6 are biased by the Q_9–Q_{10} current mirror. As part of the design process, you should determine which input is v_+ and which is v_-.

Figure 16.13
BJT operational amplifier #1.

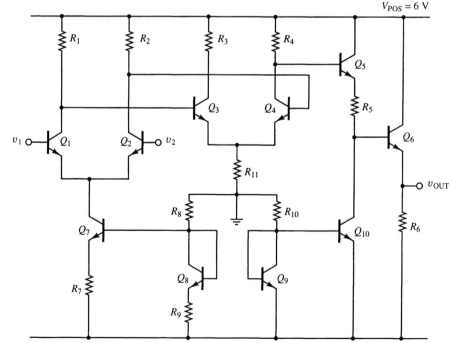

16.4.2 BJT Operational Amplifier #2

The basic, but unfinished, circuit diagram of an integrated circuit BJT operational amplifier is shown in Fig. 16.14. Your job is to choose values for all resistors in the circuit such that the amplifier meets or exceeds the following open-loop specifications:

- Differential-mode input resistance: $\geq 20\,k\Omega$.

- Common-mode input resistance: $\geq 100\,M\Omega$.

- Output resistance: $\leq 100\,\Omega$.

- Differential-mode gain: $\geq 10^4$.

- DC level of output voltage with both inputs at ground: 0 ± 0.5 V.

- Output voltage swing: ± 5 V.

- High-frequency -3-dB endpoint to differential-mode response: $\geq 400\,kHz$.

- CMRR: $\geq 40\,dB$.

Assume the transistors in the op-amp to have the following parameters: $\beta_o = 50$ to 150, $C_\mu = 0.3\,pF$, $V_A = 50\,V$, $f_T = 400\,MHz$, $r_x = 10\,\Omega$, $V_{BE} \approx 0.7\,V$ when forward biased, and $\eta = 1$.

Figure 16.14
BJT operational amplifier #2.

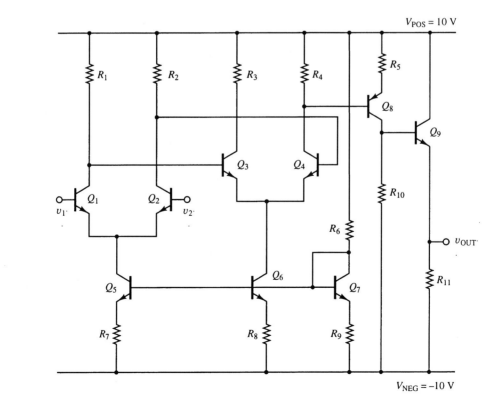

16.4.3 BiFET Operational Amplifier

Design an operational amplifier cascade using the BiFET input stage of Fig. 16.15. Add an appropriate bias network, middle-gain stage, output stage, and compensation capacitor to produce an op-amp that meets the following set of specifications:

- Minimum open-loop differential-mode gain: 10^4.
- Unity-gain frequency: $\geq 500\,\text{kHz}$.
- Input resistance: $\geq 10\,\text{M}\Omega$.
- Output resistance: $\leq 100\,\Omega$.

Assess the common-mode rejection ratio and stability performance of your op-amp. Design for power-supply voltages of $V_{\text{POS}} = 15\,\text{V}$ and $V_{\text{NEG}} = -15\,\text{V}$.

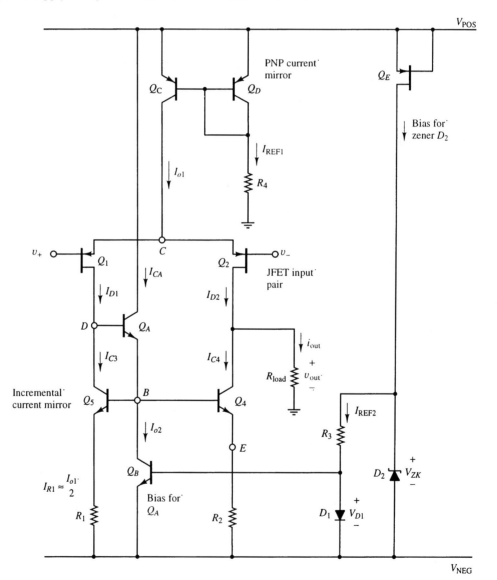

Figure 16.15 BiFET (bipolar-FET) op-amp stage incorporating p-channel JFETs and npn BJTs.

16.4.4 CMOS Operational Amplifier

The circuit of Fig. 16.16 is a CMOS operational amplifier made entirely from *n*-channel and *p*-channel MOSFETs. The circuit contains no resistors. Choose *W* and *L* values for each device in the circuit such that the following specifications are met:

- Differential-mode gain: $\geq 20\,\mathrm{dB}$.
- Maximum drivable load capacitance: $1000\,\mathrm{pF}$.
- DC value of output voltage with both inputs grounded: -5 to $+5$ V.
- Maximum bias current through any device: $1\,\mathrm{mA}$.
- All MOSFETs biased in the constant-current region.

Assume that each MOSFET is fabricated simultaneously on the same IC with identical parameters ϵ_{ox} and t_{ox}, a common threshold voltage magnitude somewhere in the range $1\,\mathrm{V} < |V_{TR}| < 2\,\mathrm{V}$, a conductance of $|K| = (W/L)(0.05\,\mathrm{mA/V^2})$, and an Early voltage of $V_A = 100\,\mathrm{V}$. The circuit should be designed to operate from ± 15-V power supplies. The smallest device dimension should be $2\,\mu\mathrm{m}$. (The op-amp is to be made using a 2-$\mu\mathrm{m}$ line-width fabrication process.) No device dimension should be larger than $500\,\mu\mathrm{m}$. Evaluate the common-mode gain and CMRR of your circuit.

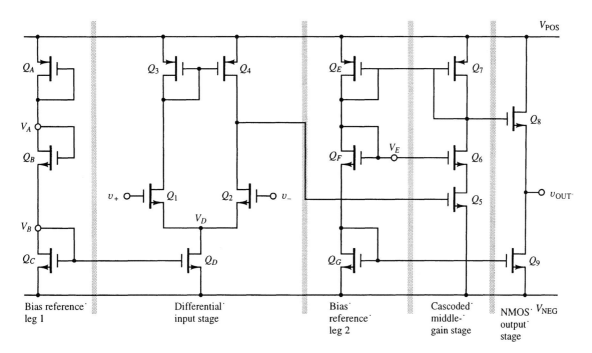

Figure 16.16 CMOS operational-amplifier cascade.

Physics of Semiconductor Devices

I n this appendix, the basic principles of semiconductor physics are summarized. An understanding of these principles is necessary to explain the operation of most electronic devices. A detailed treatment of semiconductor physics could rightfully occupy an entire book. The goal of this appendix is simply to introduce the reader to the basic concepts involved. The reader interested in more detail can consult one of the many books sited in the references of Appendix E.

A.1 ELECTRONIC MATERIALS

Most modern electronic devices are made from a combination of semiconductors, metals, and insulators. Semiconductors important to electronics include silicon, gallium arsenide, indium phosphide, and germanium. The conductivity of a semiconductor increases with temperature and can be altered by the addition of *dopants*. In contrast, the conductivity of a metal decreases with temperature and cannot be altered by the addition of dopants. These differing properties make electronic devices possible and can be explained by examining the properties of crystalline solids on an atomic level.

Figure A.1
Two-dimensional crystal lattice. Positive ion cores are joined to nearest neighbors by covalent bonds, each requiring two electrons.

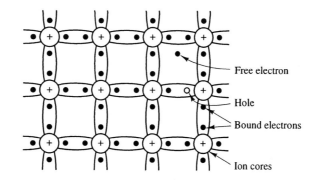

Free electron

Hole

Bound electrons

Ion cores

Metals, semiconductors, and many insulators are formed by ordered, crystalline arrays of atoms in which nearest neighbors are joined by covalent bonds, as depicted in two-dimensional form in Fig. A.1. The structure of an actual crystalline solid, called the crystal *lattice*, is three-dimensional. The ordered structure of the crystal causes the electrons in the outermost shell of any atom to be influenced by all the atoms in the crystal. This effect, called *long-range ordering*, transforms the discrete single-atom electron energy levels characteristic of the element into energy bands available to all the electrons in the crystal. The highest occupied energy band in a crystalline solid is called the *conduction band* and the next highest the *valence band*. Similarly, the jumps in energy between the discrete electron levels of the single atom transform into energy *gaps*, or *forbidden bands*, in the crystalline form of the element. The energy gap just below the conduction band is called the *band gap*. A simplified schematic diagram representing the allowed and forbidden energy bands in a typical crystalline solid is shown in Fig. A.2.

Figure A.2
Energy-band diagram for a typical crystalline solid showing the valence band, the conduction band, and the band gap of energy E_G.

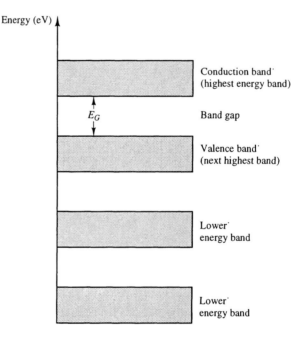

The band structure of a crystal can also be described in terms of chemical bonds. The valence band contains those electrons bound up in covalent bonds. The conduction band contains those electrons that are not bound and are free to move about the crystal. The band gap represents the energy required to break the covalent bonds of the valence electrons.

In a single atom, the Pauli exclusion principle of quantum mechanics limits the population of a given electron level to two electrons (spin up and spin down). A band in a crystalline solid is made up of one energy level from each atom in the crystal, so the Pauli principle limits the population of a given band to twice as many electrons as there are atoms in the crystal. This limitation strongly affects the ability of the crystal to conduct current. When a voltage is applied to the crystal, the resulting electric field tends to move any free electrons. Since motion adds an increment of energy, a given electron can move only by jumping up to a higher unoccupied energy level. The energy added to an electron by such an external voltage is typically very small compared to the crystal's band gap. Consequently, electrons can participate in conduction only if nearby unoccupied energy states are available within the same energy band. Such a situation can occur only in a partially filled energy band.

A metal is made from atoms whose outermost orbitals are half full. As a result, the conduction band of a metal crystal, which contains one electron from each atom, is only half full. When an external voltage is applied to the metal, the electrons in the conduction band can jump easily to states in the unoccupied upper half of the band and contribute to conduction processes. This situation is depicted in Fig. A.3, where the occupied portions of the valence and conduction bands are represented by shaded regions.

Figure A.3
Energy-band diagram for a typical metal. The valence band is completely full and the conduction band is half full.

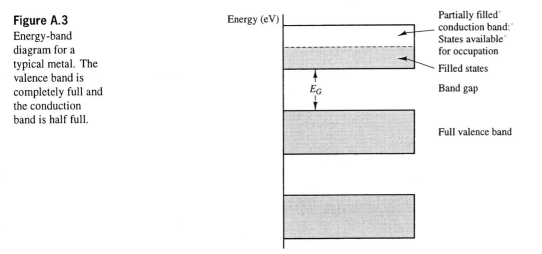

In contrast to a metal, a semiconductor is made from atoms whose outer orbitals are completely full. In effect, the outer shell of a single semiconductor atom resembles the electron configuration of a noble gas. As a result, the highest occupied energy band in a semiconductor crystal is completely filled with electrons. This completely filled energy band constitutes the valence band of the semiconductor. The next highest energy band, which constitutes the conduction band, is essentially empty. This situation is depicted in Fig. A.4. Energy cannot be easily added to valence-band electrons for conduction processes because the nearby energy states within the valence band are already filled. Adding extra energy to valence-band electrons requires that they be excited all the way up through the band gap to the conduction band. This energy is much higher than the energy available from any practical voltage.

Figure A.4
Energy-band diagram for a typical semiconductor near zero temperature. The valence band is completely full and the conduction band is empty.

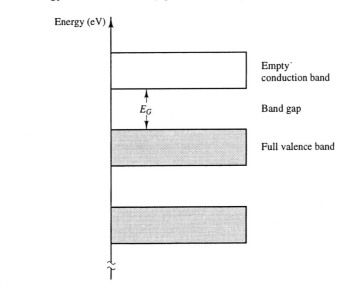

The previous description suggests that a semiconductor, with its outermost electrons locked in the valence band, should be incapable of conducting electricity. Fortunately, this picture of semiconductors is valid only near absolute-zero temperature (zero Kelvin). At higher temperatures, a small number of electrons in the valence band acquire enough random thermal vibrational energy to exceed the energy span of the band gap. These thermally excited electrons break out of their covalent bonds and jump spontaneously from the valence band into the conduction band, as depicted in Fig. A.5. In the nearly empty conduction band of a semiconductor, a multitude of closely spaced unoccupied energy states are available. The electrons that are thermally excited into the conduction band can easily acquire an additional increment of energy from an externally applied voltage and can thus contribute to the conduction of current in the crystal. This thermal-excitation mechanism is responsible for the conductivity of a pure semiconductor.

Figure A.5

Energy-band diagram for a typical semiconductor at higher temperatures. Thermally excited electrons jump from the valence band into the conduction band.

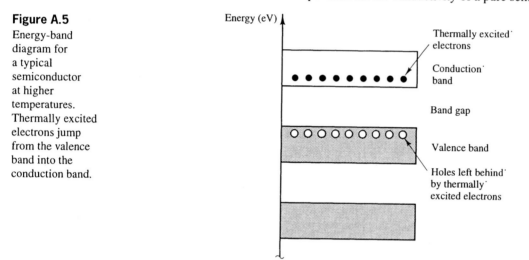

A.2   QUALITATIVE DESCRIPTION OF HOLES

When an electron is thermally excited into the conduction band, it leaves behind a singular vacancy in the valence band, as depicted in Fig. A.5. This vacancy, which has a net positive charge due to the missing electron, is called a *hole*. A hole can be thought of as the vacancy left behind when an electron breaks out of its covalent bond. If the electrons in the conduction band of a semiconductor are caused only by thermal agitation, an equal number of holes will exist in the valence band, because every thermally excited free electron leaves behind a hole. A semiconductor with equal numbers of thermally generated holes and electrons is called *intrinsic*.

In a metal, the number of electrons in the conduction band is on the order of the number of atoms in the crystal, that is, on the order of Avogadro's number ($6 \times 10^{23} \, \text{cm}^{-3}$). In contrast, the number of thermally generated hole–electron pairs in silicon at room temperature (about 300 K) is on the order of $10^{10} \, \text{cm}^{-3}$, or about one per 6×10^{13} crystal atoms. By analogy, this number is roughly equivalent to one "thermally generated" pair of people per entire population of 10,000 Earths—obviously, a small number. Nevertheless, this small number of thermally generated hole–electron pairs is sufficient to cause a semiconductor to conduct at finite temperatures. Moreover, because the number of thermally generated carrier pairs increases with temperature, the conductivity of a semiconductor increases with temperature as well. In contrast, the conductivity of a metal decreases with temperature as increased crystal lattice vibrations impede the flow of electrons.

As discussed in Section 3.3.1, a hole in a semiconductor represents more than the simple absence of an electron. Like the electron, the hole can contribute to the conduction of current. When voltage is applied across a semiconductor, a bound electron adjacent to a given hole is forced into the vacancy of the hole, causing the hole to migrate in the direction of the electric field associated with the applied voltage. Because the net charge of the hole vacancy is positive, the hold can be thought of as a discrete positive particle capable of supporting conduction. A detailed quantum mechanical analysis shows that the hole has a finite effective mass and has a positive charge equal in magnitude to the charge of an electron.

A.3 IMPURITIES

A metal has a fixed electron concentration and no holes. In contrast, the hole and electron concentrations in a semiconductor can be altered by the addition of small amounts of dopants to the crystal composition. This feature of semiconductors makes electronic devices possible. A semiconductor with added impurities is called *extrinsic*. In this section, the mechanisms by which impurities control the hole and electron concentrations in an extrinsic semiconductor are examined.

A.3.1 Acceptor Atoms

By adding impurities called *acceptors* to the semiconductor, the holes created by thermal generation can be augmented by additional holes. Acceptor atoms have one less electron in their outer valence shell than do the semiconductor atoms of the host crystal. Typical acceptor atoms added to silicon and germanium, for example, include boron, gallium, and indium. When an acceptor atom is added to the crystal, it provides an unoccupied electron state in the band gap just above the valence band. With a minimal amount of thermal energy, an electron from the valence band can easily jump to fill this vacant state. When it does so, the electron becomes trapped in the covalent bond of the acceptor atom, creating a fixed, immobile, negative ion at the site of the acceptor atom. In turn, the electron from the valence band leaves behind a hole that is free to migrate around the crystal. This hole is created in the valence band without the creation of a corresponding free electron in the conduction band.

A.3.2 Donor Atoms

By adding impurities called *donors* to the semiconductor, the electrons created by thermal generation can be augmented by additional electrons. Donor atoms have one more electron in their outer valence shell than do the semiconductor atoms of the host crystal. Typical donor atoms added to silicon and germanium, for example, include antimony, phosphorus, and arsenic.

When a donor atom is added to the crystal, the energy level of its extra electron lies in the band gap of the host crystal, just below the conduction band. From there, the extra electron is easily thermally excited into the conduction band, where it becomes a newly liberated electron that is free to participate in conduction processes within the nearly empty conduction band of the semiconductor. The excited electron leaves behind a fixed positive ion core at the site of the donor atom, but no corresponding hole in the valence band.

A.4 CARRIER DENSITIES WITHIN A SEMICONDUCTOR

When donor and acceptor atoms are added to a semiconductor, the net equilibrium concentrations of holes and electrons are not given by a simple algebraic addition of the dopant concentrations. In any semiconductor at finite temperature, electron–hole pairs are constantly generated by thermal agitation. On a statistical basis, holes also recombine with free electrons, thereby destroying hole–electron pairs. At a given temperature, these generation and recombination mechanisms precisely balance so that the equilibrium hole and electron concentrations satisfy the relationship

$$n_o p_o = n_i^2 \tag{A.1}$$

where n_o and p_o are the equilibrium electron and hole concentrations, respectively, and n_i represents the semiconductor's intrinsic carrier concentration. The value of n_i depends on temperature and is given by

$$n_i^2 = A_o T^3 e^{-E_{Go}/kT} \tag{A.2}$$

where T is in Kelvin. The constant A_o and the zero-temperature band gap E_{Go} are properties of the specific semiconductor. Equation (A.1) is called the *mass-action law* and is a basic principle of physics. In an intrinsic semiconductor (i.e., one with no donors or acceptors), n_o and p_o are equal, and Eq. (A.1) yields

$$n_o = p_o = n_i \tag{A.3}$$

In an extrinsic, or doped, semiconductor, n_o and p_o are not equal. Nevertheless, it can be shown that the product $n_o p_o$ is still equal to n_i^2.

A second constraint on the values of n_o and p_o comes from the general requirement that charge neutrality be maintained everywhere in the crystal. This requirement may be argued on fundamental physical grounds. Were charge neutrality to be violated somewhere within the crystal, an internal electric field would develop from the unbalanced charges. Electrons or holes, as needed, would quickly flow to the unbalanced site and equalize the charge imbalance. Charge neutrality requires that the positive-charge density at any point in the crystal be equal to the negative-charge density. The total positive-charge density is given by the combined local densities of the holes and the positively charged donor ion cores. Similarly, the total negative-charge density is given by the combined local densities of the electrons and the negatively charged acceptor ion cores. The charge neutrality condition thus becomes

$$p_o + N_D = N_A + n_o \tag{A.4}$$

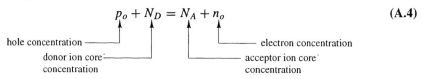

hole concentration ——————— electron concentration
donor ion core ——————— acceptor ion core
concentration ——————— concentration

The charge-neutrality condition can be combined with the mass action law to determine the hole and electron concentrations in a doped semiconductor. This calculation is illustrated in the following example.

EXAMPLE A.1

Intrinsic silicon at room temperature has a carrier concentration of $n_o = p_o = n_i = 1.5 \times 10^{10}\,\mathrm{cm}^{-3}$. A sample of silicon is uniformly doped with donor phosphorus atoms to a density of $N_D = 10^{11}\,\mathrm{cm}^{-3}$. No acceptor atoms are added, so that $N_A = 0$. Find the equilibrium electron and hole concentrations n_o and p_o in the resulting extrinsic material.

Solution

The product $n_o p_o$ must equal n_i^2, hence

$$p_o = \frac{n_i^2}{n_o} \tag{A.5}$$

This relation can be substituted into the charge neutrality condition (A.4), with $N_A = 0$, to yield

$$\frac{n_i^2}{n_o} + N_D = n_o \tag{A.6}$$

or

$$n_o^2 - n_o N_D - n_i^2 = 0 \tag{A.7}$$

This quadratic equation has solutions

$$n_o = \frac{N_D}{2} \pm \frac{(N_D^2 + 4n_i^2)^{1/2}}{2} \tag{A.8}$$

Taking the minus sign in Eq. (A.8) yields a negative n_o, which represents a mathematical, but not a physical, solution. The equilibrium concentration of electrons n_o is found by taking the plus sign. Substituting appropriate values for N_D and n_i results in

$$n_o = \frac{10^{11}}{2} + \frac{[10^{22} + 4(1.5 \times 10^{10})^2]^{1/2}}{2} = 1.02 \times 10^{11} \text{ cm}^{-3} \tag{A.9}$$

Substituting this result into Eq. (A.5) yields the value of p_o:

$$p_o = \frac{n_i^2}{n_o} = \frac{(1.5 \times 10^{10})^2}{1.02 \times 10^{11}} = 2.2 \times 10^9 \text{ cm}^{-3} \tag{A.10}$$

The semiconductor of Example A.1 is heavily doped with donors only so that $N_D \gg n_i$. Consequently, the electron concentration n_o is approximately equal to N_D. If the material is instead heavily doped only with acceptors, so that $N_A \gg n_i$, the hole concentration p_o will be approximately equal to N_A. In either case, an increase in temperature will cause the value of n_i to rise. Because the number of dopant atoms in the crystal is fixed and the number of thermally generated holes and electrons an increasing function of temperature, the semiconductor will approach an intrinsic state at elevated temperatures. This phenomenon negates the effect of the dopant atoms.

A logarithmic plot of n_i versus temperature, as given by Eq. (A.2), is shown for silicon in Fig. A.6. Note that n_i rises steeply with temperature. This dependency is moderated on the graph because the vertical axis is logarithmic. For comparison, the values of n_o and p_o for the extrinsic silicon of Example A.1 are also plotted in Fig. A.6. As computed in Example A.1, the carrier concentrations at 300 K are equal to $n_o = 1.02 \times 10^{11} \text{ cm}^{-3}$ and $p_o = 2.2 \times 10^9 \text{ cm}^{-3}$. Above about 340 K, n_i exceeds N_D and the material begins to appear intrinsic with $n_o \approx p_o \approx n_i$.

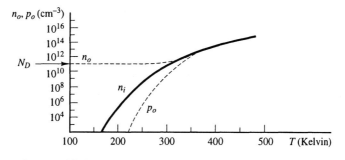

Figure A.6 Plot of the equilibrium concentrations n_o and p_o versus temperature. Solid line: intrinsic silicon ($A_o = 1.74 \times 10^{33} \text{ cm}^{-6} K^{-3}$; $E_{G_o} = 1.15 \text{ eV}$). Dashed lines: extrinsic silicon with $N_D = 10^{11} \text{ cm}^{-3}$ and $N_A = 0$. Above about 330 K, n_i exceeds N_D and the material approaches the intrinsic state.

A.5 CURRENT FLOW IN A SEMICONDUCTOR

In this section, the various mechanisms by which holes and free electrons contribute to current flow are discussed. Specifically, the current-flow mechanisms of drift and diffusion are examined in detail. Familiarity with these physical processes is critical to an understanding of semiconductor devices such as diodes and transistors. The concepts presented here are applied to specific devices in Sections A.8 and A.9.

A.5.1 Drift-Current Density

Holes and electrons can contribute to current conduction by a mechanism called *drift*. The flow of drift current is the fundamental physical process described by Ohm's law. The drift-current mechanism can be understood by considering the electrons in the volume of material shown in Fig. A.7. This material might represent a metal or a semiconductor; in the latter case, holes would also be present. For clarity, only the motion of electrons is considered here. The motion of holes occurs in a similar manner.

Figure A.7

Current flows when an electric field is applied across the crystal. This mechanism is called *drift*.

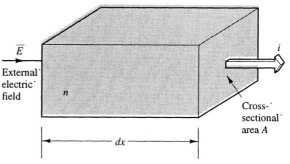

If an external voltage is applied across the material, an internal electric field will be created, causing the free electrons to accelerate.[1] After traveling a short distance, a given electron will collide with a stationary ion core, losing much of its accumulated velocity. After the collision, the electron will be free to accelerate again before its next collision. On a microscopic scale, a given electron will experience many such collisions. If its motion is averaged over many collisions, it will appear to move at an average *drift* velocity v_e proportional to the applied electric field:

$$v_e = -\mu_e E \qquad \text{(A.11)}$$

The constant of proportionality μ_e is called the *electron mobility*. Because an electron has negative charge, its drift velocity v_e is opposite the direction of E.

The current associated with drifting electrons in Fig. A.7 is equal to the total number of charges passing per unit time through the area A. This current can be expressed as follows:

$$i = \frac{\text{total number of charges within the volume } A dx}{\text{total time required to move the charges through the distance } dx} \qquad \text{(A.12)}$$

This ratio can also be expressed by the equation

$$i = \frac{(\text{charge per unit volume})(\text{volume } A dx)}{\text{length of box/velocity}}$$

$$= \frac{(-qn)(A\, dx)}{dx/v_e} = -qnv_e A = qn\mu_e EA \qquad \text{(A.13)}$$

---

[1] Although the material is conducting, it must not be considered a "perfect" conductor as is often done in electromagnetics. As such, it can support a field inside.

where n is the concentration of free electrons. The charge per unit volume $-qn$ is preceded by a negative sign because an electron has a negative charge. The unit magnitude electronic charge q is equal to 1.6×10^{-19} coulomb.

Current density J is defined as the total current flow per unit cross-sectional area. The drift current density due to the electric field in Fig. A.7 thus becomes

$$J_{\text{drift}} = \frac{i}{A} = -qnv_e = qn\mu_e E = \sigma E \qquad \text{(A.14)}$$

The quantity $\sigma = qn\mu_e$ is called the electron *conductivity* of the material. Equation (A.14) is actually a fundamental form of Ohm's law.

Both electrons and holes are present in a semiconductor; hence the expression for drift-current density has two components. In general, the hole mobility μ_h will differ from the electron mobility μ_e because holes and electrons have different effective masses. Consequently, total drift-current density must be expressed by separate hole and electron terms:

$$J_{\text{drift}} = J_{e\text{-drift}} + J_{h\text{-drift}} = \underbrace{qn\mu_e E}_{\text{electron component}} + \underbrace{qp\mu_h E}_{\text{hole component}} \qquad \text{(A.15)}$$

Note that holes and electrons are oppositely charged carriers, and thus move in opposite directions under the influence of the same electric field. Because electrons have negative charge, however, their negative motion against the field results in positive current flow in the direction of the field. The expressions for hole and electron drift-current densities thus have the same algebraic sign.

A.5.2 Diffusion-Current Density

The current-density components represented by Eq. (A.15) describe holes and electrons that move at constant velocity under the influence of the electric field created by an applied voltage. Another mechanism called *diffusion* can also cause current to flow in a semiconductor. If the concentration of a carrier species varies between different regions of the crystal, particles will tend to migrate from the region of high concentration to the region of low concentration. This carrier migration, depicted in Fig. A.8, is a purely statistical process driven by the random thermal motion of the carriers. It is unrelated to the charge on the carriers or to the presence of an electric field. Diffusion does not depend on the absolute value of carrier concentration, but only on its spatial derivative, or *gradient*.

Figure A.8

Region of high carrier concentration on the left causes a migration of carriers to the right. This mechanism is called *diffusion*.

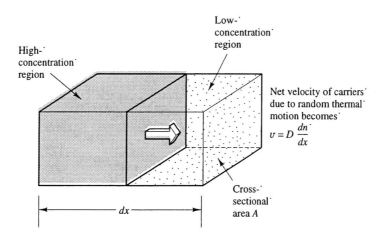

High-concentration region

Low-concentration region

Net velocity of carriers due to random thermal motion becomes

$v = D \dfrac{dn}{dx}$

Cross-sectional area A

dx

In a metal, diffusion is not a relevant process because there is no mechanism by which a density gradient can originate. Since only negative charge carriers are present in a metal, any carrier gradient that might form would upset charge neutrality. The resulting local electric field would cause a drift current of electrons that would instantly eliminate the gradient before any diffusion processes could take place. In contrast, a semiconductor has both positive and negative charge carriers; hence it is possible for a density gradient of holes and electrons to exist while charge neutrality is maintained.

In a semiconductor, the diffusion current density components can be expressed in one-dimensional form by the equation

$$J_{\text{diff}} = J_{e\text{-diff}} + J_{h\text{-diff}} = \underbrace{q D_e \frac{dn}{dx}}_{\text{electron component}} - \underbrace{q D_h \frac{dp}{dx}}_{\text{hole component}} \tag{A.16}$$

The second term in Eq. (A.16) has a negative sign because a negative slope of holes drives a positive current.

Total Current Flow

The total current density of either carrier in a semiconductor is equal to the sum of its own drift and diffusion components, as expressed by the electron and hole *transport equations*:

$$J_e = q n \mu_e E + q D_e \frac{dn}{dx} \qquad \text{(electrons)} \tag{A.17}$$

$$J_h = q p \mu_h E - q D_h \frac{dp}{dx} \qquad \text{(holes)} \tag{A.18}$$

$$\underbrace{}_{\text{drift component}} \underbrace{\phantom{- q D_h \frac{dp}{dx}}}_{\text{diffusion component}}$$

The total overall current density (i.e., that measurable by external instruments) is equal to $J_e + J_h$.

The diffusion and mobility constants D and μ for a given carrier species are related by a basic equation of quantum statistics called the *Einstein relation*:

$$\frac{D_e}{\mu_e} = \frac{D_h}{\mu_h} = \frac{kT}{q} = V_T \tag{A.19}$$

In this equation, k is Boltzmann's constant, q the electronic charge, and T the temperature in Kelvin. The parameter $V_T = kT/q$, called the *thermal voltage*, has a value of about 25 mV at room temperature, and is the same V_T used in the diode and transistor v–i equations throughout the book.

A.5.3 Recombination of Excess Carrier Concentrations

One concept important to the operation of electronic devices concerns the role of equilibrium versus nonequilibrium carrier concentrations. At a given temperature, the hole and electron concentrations at a given location within the material each can be described as the sum of equilibrium and excess components:

$$p = p_o + p' \tag{A.20}$$

$$n = n_o + n' \tag{A.21}$$

As discussed in Section A.4, the equilibrium concentrations n_o and p_o are determined by the equations

$$n_o p_o = n_i^2 \tag{A.22}$$

and
$$n_o + N_A = p_o + N_D \tag{A.23}$$

The quantities p' and n' express the presence of any *excess* hole or electron concentrations that may be introduced by physical processes to be discussed in Section A.6.

When local excess carrier concentrations n' or p' do appear, the semiconductor eventually reverts to its equilibrium state because the excess carriers recombine with each other over time. This physical process, called *recombination*, can be approximately described by the equations

$$\frac{dp'}{dt} = -\frac{p'}{\tau_h} \tag{A.24}$$

and
$$\frac{dn'}{dt} = -\frac{n'}{\tau_e} \tag{A.25}$$

The recombination time constants τ_h and τ_e describe the recombination per unit time per unit volume for holes and electrons, respectively. Each recombination event involves the mutual annihilation of a hole and an electron. Equations (A.24) and (A.25) are valid only when p' and n' are much smaller than p_o and n_o, a condition called *low-level injection*.

Note that Eqs. (A.24) and (A.25) are valid regardless of the algebraic signs of p' and n'. Negative values of p' and n' describe local decreases in carrier concentration below the equilibrium values n_o and p_o.

A.5.4 Current-Density Gradient

Another term must be added to Eqs. (A.24) and (A.25) to account for local changes in carrier concentration due to a net divergence of carriers. If one of the physical processes to be described in Section A.6 causes more holes to flow out of a given incremental volume than into it, for example, the excess hole concentration p' will decrease over time. This divergence of carriers can be accounted for by adding a second term to Eq. (A.24) to describe the spatial derivative of current density:

$$\frac{dp'}{dt} = -\frac{p'}{\tau_h} - \frac{1}{q}\frac{dJ_h}{dx} \tag{A.26}$$

The same argument applies to the electrons, leading to a modified version of Eq. (A.25):

$$\frac{dn'}{dt} = -\frac{n'}{\tau_e} + \frac{1}{q}\frac{dJ_e}{dx} \tag{A.27}$$

Note that the derivative of J_e appears with a positive sign in Eq. (A.27). Because electrons are negative carriers, an algebraically positive net current flow out of the volume (positive dJ_e/dx) describes an increase in n'.

Equations (A.26) and (A.27) are called the *conservation equations*. They describe the respective effects of recombination and diffusion on the excess hole and electron concentrations within a semiconductor. According to these equations, holes and electrons can be neither created nor destroyed except by recombination events occurring on a time scale of order τ. The conservation equations imply that in steady state, when the time derivative is zero, the spatial current derivatives will be exactly balanced by recombination.

A.5.5 Summary of Properties

Metal	Semiconductor
Crystalline	Crystalline
Good conductivity; decreases with temperature as lattice vibrations create more opportunities for electron collisions	Fair conductivity; increases with temperature as more hole–electron carrier pairs are thermally generated
Single charge carrier (electrons)	Bipolar charge carriers (electrons and holes)
Large carrier density (order of Avogardro's number	Moderate carrier density varies with temperature and number of dopants
n fixed	n and p not fixed; can be adjusted by addition of dopants
Drift current important; no diffusion current	Drift and diffusion current important

Table A.1. Comparison of Metals and Semiconductors

The principal differences between metals and semiconductors, as discussed in Section A.1, are summarized in Table A.1. Numerical values of key parameters for silicon, germanium, and gallium arsenide are provided in Table A.2.

Property	Symbol	Units	Si	Ge	GaAs
Intrinsic carrier concentration at 300 K	n_i	cm^{-3}	1.5×10^{10}	2.5×10^{13}	9.0×10^6
Mobility					
Electron	μ_e	cm^2/ V-s	1350	3900	8500
Hole	μ_h	cm^2/ V-s	480	1900	400
Diffusion constant					
Electron	D_e	cm^2/s	34	98	212
Hole	D_h	cm^2/s	12	48	10
Energy gap					
300 K	E_G	eV	1.11	0.67	1.43
0 K	E_{Go}	eV	1.153	0.744	1.53
Relative dielectric permittivity	ϵ/ϵ_o		11.7	15.8	13.1
Oxide permittivity	ϵ_{ox}/ϵ_o		3.9	—	—
Atomic or molecular density	N_o	cm^{-3}	5.00×10^{22}	4.42×10^{22}	2.21×10^{22}
Atomic or molecular weight	MW	g/mole	28.09	72.59	144.64
Density		g/cm^3	2.328	5.323	5.316
Melting point		°C	1415	936	1238
Crystal structure			Diamond	Diamond	Zinc blende

* Values given at 300 K unless otherwise noted.

Table A.2. Properties of Silicon, Germanium, and Gallium Arsenide*

A.6 DIFFUSION GRADIENT WITHIN A SEMICONDUCTOR

The principles of drift, diffusion, and carrier recombination outlined in Sections A.1 to A.5 are central to the physical operation of most semiconductor devices. In the following sections, these principles are used to derive the v–i equations of the diode, the BJT, and the MOSFET. We begin with a description of diffusion and its role in producing current flow. Diffusion of holes and electrons plays an important role in many electronic devices, including the *pn* junction diode and the bipolar junction transistor (BJT). In the discussion to follow, the equations that describe the flow of diffusion current in a carrier gradient are explored. The hypothetical situation to be described does not represent a specific electronic device, but it illustrates several of the important features about diffusion that are critical in deriving the v–i equations of the diode and the BJT.

Figure A.9
Holes are continuously injected into an *n*-type bar at $x = 0$.

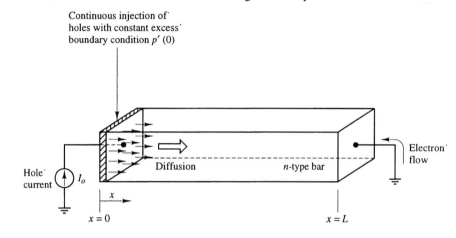

Figure A.9 shows a piece of *n*-type semiconductor of cross-sectional area A and length L, where L is much longer than the bar's lateral dimensions. In an *n*-type semiconductor, the free electrons are called *majority* carriers and the holes *minority* carriers. Suppose that holes are continuously injected into the left-hand edge of the bar by some unspecified process represented by the current source I_o. The injection is performed so as to maintain a constant excess concentration $p'(0)$ at $x = 0$. Under such conditions, the total hole concentration at $x = 0$ becomes $p(0) = p_o + p'(0)$, where p_o is the hole concentration prior to injection. If the concentration of injected holes is small compared to p_o, the injection mechanism will satisfy the low-level injection condition discussed in Section A.5.3. The excess holes will diffuse down the length of the bar, recombining along the way, until the hole concentration reverts to the equilibrium value p_o at some distance from $x = 0$. The electrons necessary for recombination flow from the ground connection at $x = L$, satisfying KCL for the bar as a whole. The situation depicted in Fig. A.9 is somewhat analogous to the pan of viscous liquid shown in Fig. A.10. If extra liquid, representing excess holes, is continuously poured into one end of the pan, the excess will diffuse toward the other end of the pan as the liquid seeks its own equilibrium level. If the amount of excess "injected" liquid is small, the equilibrium level will remain essentially unchanged, but a continuous diffusion of liquid from left to right will be established.

A mathematical expression for the steady-state distribution of diffusing holes in Fig. A.9 can be calculated from the conservation equation

$$\frac{dp'}{dt} = -\frac{p'}{\tau_h} - \frac{1}{q}\frac{dJ_h}{dx} \tag{A.28}$$

and the transport equation

$$J_h = qp\mu_h E - qD_h\frac{dp}{dx} \tag{A.29}$$

Figure A.10
Viscous liquid
analogous to the
excess hole
injection of
Fig. A.9.

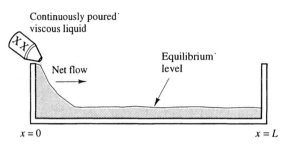

In this n-type bar, holes are the minority carriers; hence $p \ll n$. The hole drift-current term $qp\mu_h E$, which depends on the value of p, will be small compared to other current components. The hole diffusion current $qD_h dp/dx$ need not be small, however, since it depends on the derivative of p. To a first approximation, therefore, the first term on the right in Eq. (A.29) can be neglected in comparison to the second.

Given this simplification, a steady-state solution for the excess hole distribution $p'(x)$ can be found by setting the time derivative in Eq. (A.28) to zero. Equations (A.28) and (A.29) can then be combined by taking the derivative of the second equation with respect to x and substituting it into the first. Taking the x-derivative of Eq. (A.29) results in

$$\frac{dJ_h}{dx} \approx -qD_h \frac{d^2 p'}{dx^2} \tag{A.30}$$

Note that dp/dx is equivalent to dp'/dx, because $p = p_o + p'$, and p_o is a constant. Setting the time derivative in Eq. (A.28) to zero results in

$$\frac{p'}{\tau_h} = -\frac{1}{q}\frac{dJ_h}{dx} \tag{A.31}$$

Combining Eqs. (A.30) and (A.31) yields

$$\frac{d^2 p'}{dx^2} = \frac{p'}{L_h^2} \tag{A.32}$$

where the parameter $L_h = \sqrt{D_h \tau_h}$ is called the hole *diffusion length*.

The hole concentration at $x = 0$, as determined by the injection mechanism, is fixed at $p'(0)$. With this boundary condition, the differential equation (A.32) has the solution

$$p'(x) = p'(0)e^{-x/L_h} \tag{A.33}$$

The mathematically correct but nonphysical positive exponential solution has been omitted in Eq. (A.33). Equation (A.33) describes an excess hole concentration that decays exponentially with distance as the carriers diffuse away from the end of the bar at $x = 0$. The distribution $p'(x)$ is exponential because the recombination of diffusing holes with electrons is proportional to the local density of holes. If the bar length L is long compared to the diffusion length L_h, then $p'(x)$ will essentially decay to zero at $x = L$. Under such conditions, the holes revert to their equilibrium concentration p_o before reaching the end of the bar. A plot of the $p'(x)$ expressed by Eq. (A.33) for the condition $L \gg L_h$ is shown in Fig. A.11.

Figure A.11
Distribution
$p(x) = p_o + p(x)$
for an n-type
semiconductor
subject to hole
injection. The
boundary condition
$p'(0)$ at $x = 0$ is a
constant. Because
$p_o \gg p'(0)$, the
vertical scale is
broken.

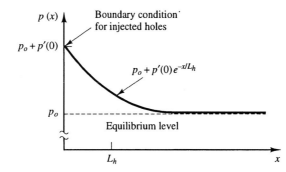

The hole current associated with this $p'(x)$ distribution can be found from the diffusion term in the transport equation:

$$J_{h\text{-diff}} = -qD_h\frac{dp}{dx} = \frac{qD_h}{L_h}p'(0)e^{-x/L_h} \tag{A.34}$$

This current expression indicates a flow of holes to the right in Fig. A.9.

The solutions obtained in the preceding discussion were obtained for an n-type material in which the holes are the minority carriers. The entire discussion could also apply to the case in which electrons are injected as minority carriers into a p-type semiconductor. In such a case, an analogous calculation leads to expressions for the electron distribution

$$n'(x) = n'(0)e^{-x/L_e} \tag{A.35}$$

and the electron current density

$$J_{e\text{-diff}} = qD_e\frac{dn}{dx} = -\frac{qD_e}{L_e}n'(0)e^{-x/L_e} \tag{A.36}$$

where $L_e = \sqrt{D_e\tau_e}$. In p-type material, the electron drift current can be assumed negligible compared to the electron diffusion current. Note that $J_{h\text{-diff}}$ and $J_{e\text{-diff}}$ have opposite signs, because the particle motion occurs in the same direction, but holes are positive and electrons are negative.

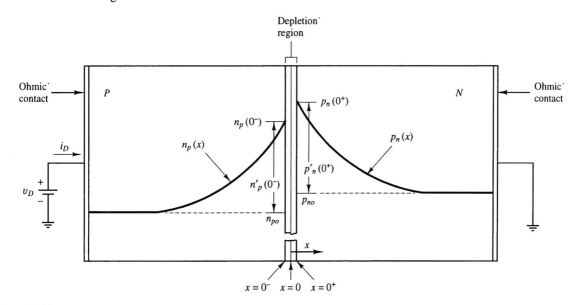

Figure A.12 Basic pn junction showing hole and electron distributions under forward-biased conditions. This one-dimensional structure models the central portion of the pn junction of Fig. A.13.

A.7 DERIVATION OF THE v–i CHARACTERISTIC OF THE PN JUNCTION DIODE

In this section, the solutions of the preceding section are used to derive the v–i characteristic of the *pn* junction diode. A qualitative discussion of the diode was given in Sections 3.3.2 and 3.3.3. The reader is urged to review these sections before proceeding with the present discussion. The representative geometry of a simple *pn* junction is depicted in Fig. A.12. The geometry shown in Fig. A.12 is a one-dimensional representation of the more realistic junction structure shown in Fig. A.13. The diagram of Fig. A.12 represents the core region (shaded in Fig. A.13), where most of the junction current flows.

Figure A.13
Portion of an actual *pn* junction modeled by the structure of Fig. A.12.

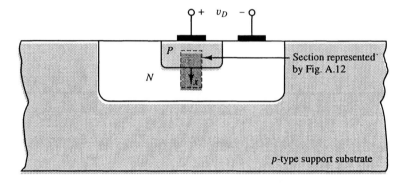

A.7.1 Boltzmann Relations

The first step in the derivation of the diode v–i equation is the identification of the injection mechanism that operates across the diode's depletion region. As we shall soon see, this injection mechanism is related to the applied diode voltage. With no external voltage applied, the drift and diffusion currents of the holes and electrons must balance so that the net diode current is zero. Under such conditions, a built-in electric field develops in the depletion region, as discussed in Chapter 3. The magnitude of this field can be found from either the hole or the electron transport equation by setting the current density to zero. The transport equation for holes with $J_h = 0$, for example, becomes

$$J_h = qp\mu_h E - qD_h \frac{dp}{dx} = 0 \tag{A.37}$$

This equation can be solved for E, yielding

$$E = \frac{D_h}{\mu_h} \frac{1}{p} \frac{dp}{dx} \equiv \frac{kT}{q} \frac{1}{p} \frac{dp}{dx} \tag{A.38}$$

where the Einstein relation $kT/q = D_h/\mu_h$ has been used in writing the right-hand side of the equation.

The electric field E, which extends across the depletion region, creates a drift current of holes that exactly balances the corresponding hole diffusion current. A similar drift current balances the diffusion of electrons across the depletion region. In a *pn* junction with no externally applied voltage, the drift and diffusion components of the hole and electron currents must sum to zero separately. This requirement is called the principle of *detailed balance*.

By definition, an electric field is equal to the negative gradient of potential. For the junction of Fig. A.12, the potential varies only with x, so that

$$E = -\frac{dV(x)}{dx} \tag{A.39}$$

Applying this definition of E to Eq. (A.38) results in

$$-\frac{kT}{q}\frac{1}{p}\frac{dp}{dx} = \frac{dV(x)}{dx} \tag{A.40}$$

Multiplying both sides of Eq. (A.40) by dx results in

$$-\frac{kT}{q}\frac{dp}{p} = dV(x) \tag{A.41}$$

Equation (A.41) can be integrated from the p side of the depletion region to the n side (i.e., from $x = 0^-$ to $x = 0^+$):

$$\int_{p(0^-)}^{p(0^+)} -\frac{kT}{q}\frac{dp}{p} = \int_{V(0^-)}^{V(0^+)} dV(x) \tag{A.42}$$

where the depletion region is assumed to have negligible thickness. In this case, $V(0^-)$ and $V(0^+)$ represent the values of $V(x)$ on the p and n sides of the depletion region, respectively. Performing the indicated integration yields

$$\frac{kT}{q}\ln\frac{p(0^-)}{p(0^+)} = V(0^+) - V(0^-) \triangleq \Psi_o \tag{A.43}$$

The voltage Ψ_o, defined as $V(0^+) - V(0^-)$, is called the *built-in voltage*. It is equal to the potential difference across the depletion region, as defined from the n side to the p side. The built-in field associated with Ψ_o extends from the positive donor ions on the n side of the depletion region to the negative acceptor ions on the p side of depletion region (see Figs. 3.16 and 3.17).

With no external voltage applied, the boundary conditions $p(0^-)$ and $p(0^+)$ become just the equilibrium hole concentrations on the p and n sides of the junction, respectively. Under these conditions, Eq. (A.43) can be expressed in the form

$$\Psi_o = \frac{kT}{q}\ln\frac{p_{po}}{p_{no}} \tag{A.44}$$

In this latter equation, a second subscript, referring to either the p side or the n side of the junction, has been added to the symbol for p_o. Specifically, p_{po} refers to the equilibrium hole concentration on the p side of the junction (where holes are the majority carrier), and p_{no} refers to the equilibrium hole concentration on the n side of the junction (where holes are the minority carrier). Equation (A.44), called the *Boltzmann relation*, can be inverted to express the carrier concentrations at the edges of the depletion region in terms of the built-in voltage:

$$p_{no} = p_{po}e^{-\Psi_o/V_T} \tag{A.45}$$

The quantity $V_T = kT/q$ again defines the *thermal voltage*.

Note that the built-in voltage Ψ_o, as given by Eq. (A.44), depends only on the carrier concentrations p_{po} and p_{no} at the edges of the depletion region. It does not depend on the general distribution $p(x)$. The Boltzmann relation is generally applicable to hole concentrations measured at any two points in the diode; in this case, it has been specifically applied to the two sides of the depletion region.

A parallel analysis yields the corresponding Boltzmann relation for electrons. Specifically, for the case $J_e = 0$, the transport equation for electrons becomes

$$J_e = qn\mu_e E + qD_e\frac{dn}{dx} = 0 \tag{A.46}$$

or

$$E = -\frac{D_e}{\mu_e}\frac{1}{n}\frac{dn}{dx} = -\frac{kT}{q}\frac{1}{n}\frac{dn}{dx} \tag{A.47}$$

This electric field is the same one described by Eq. (A.38). Applying the relation $E = -dV/dx$ to Eq. (A.47) yields

$$\frac{kT}{q}\frac{dn}{n} = dV(x) \tag{A.48}$$

Equation (A.48) can be integrated from $x = 0^+$ to $x = 0^-$ (again from the n side of the junction to the p side), to yield

$$\Psi_o \triangleq V(0^+) - V(0^-) = \frac{kT}{q}\ln\frac{n_{no}}{n_{po}} \tag{A.49}$$

This second Boltzmann relation also can be expressed in the form

$$n_{po} = n_{no}e^{-\Psi_o/V_T} \tag{A.50}$$

The beauty of the Boltzmann relations (A.45) and (A.50) is that they also apply to the diode even when an external voltage is applied. The Boltzmann relations thus can be used to relate the carrier concentrations at either edge of the depletion region to an externally applied voltage. This concept is explored in the next section.

A.7.2 Carrier Injection Mechanism

The depletion region of a pn junction is void of carriers, hence it behaves like a minimally conductive region compared to the rest of the semiconductor material in the diode. Consequently, when an external voltage v_D is applied to the diode, it appears mainly across the depletion region, where it becomes superimposed on the built-in voltage. When the applied voltage is of such a polarity that the field it produces counteracts the built-in field, the diode becomes forward biased. The reduction in the net field upsets the zero-current balance of the junction and allows carriers to diffuse across the depletion region. The flow of holes under this mechanism, which constitutes an injection of holes into the n side of the diode, can be quantified by adding an external voltage term to the Boltzmann relation for holes. The substitution of $(\Psi_o - v_D)$ for Ψ_o in Eq. (A.45) yields the total hole concentration p_n at the $x = 0^+$ side of the depletion region:

$$p_n(0^+) = p_p(0^-)e^{(-\Psi_o + v_D)/V_T} \tag{A.51}$$

A similar consideration involving the Boltzmann relation for electrons yields the total electron concentration n_p measured at the $x = 0^-$ side of the depletion region:

$$n_p(0^-) = n_n(0^+)e^{(-\Psi_o + v_D)/V_T} \tag{A.52}$$

The holes are the majority carriers on the p side of the diode. If the low-level injection condition is met, $p_p(0^-)$ will be approximately equal to p_{po}. Equation (A.51) therefore may be expressed as

$$p_n(0^+) = p_{po}e^{-\Psi_o/V_T}e^{v_D/V_T} \tag{A.53}$$

The Boltzmann relation (A.35) can be used to further transform Eq. (A.53) into

$$p_n(0^+) = p_{no}e^{v_D/V_T} \tag{A.54}$$

Similarly, Eq. (A.52) for electrons becomes

$$n_p(0^-) = n_{no}e^{-\Psi_o/V_T}e^{v_D/V_T} \tag{A.55}$$

which can be transformed using the Boltzmann relation (A.50) into

$$n_p(0^-) = n_{po}e^{v_D/V_T} \tag{A.56}$$

Equations (A.54) and (A.56) relate the injected carrier concentration boundary conditions $p_n(0^+)$ and $n_p(0^-)$ to the applied diode voltage v_D.

A.7.3 Diode Current Components

In a forward-biased diode, holes and electrons are injected in opposite directions across the depletion region by the externally applied voltage. This injection results in exponential carrier gradients on either side of the diode, subject to the carrier boundary conditions $n(0^-)$ and $p(0^+)$ given by Eqs. (A.56) and (A.54). As these equations show, the boundary conditions are dependent on the applied diode voltage v_D. Equations (A.54) and (A.56) express the values of the *total* minority concentrations $p_n(0^+)$ and $n_p(0^-)$ at the edges of the depletion region. These quantities can be transformed into boundary conditions for the *excess* concentrations by subtracting the equilibrium concentrations p_{no} and n_{po}:

$$p'_n(0^+) = p_n(0^+) - p_{no} = p_{no}(e^{v_D/V_T} - 1) \tag{A.57}$$

and
$$n'_p(0^-) = n_p(0^-) - n_{po} = n_{po}(e^{v_D/V_T} - 1) \tag{A.58}$$

Equations (A.57) and (A.58) can be combined with the expressions (A.33) and (A.35) derived in Section A.6 to find the diode current flow under forward-biased conditions. Specifically, the distribution of excess injected holes on the n-side of the diode can be expressed as a function of position by

$$p'_n(x) = p'_n(0^+)e^{-x/L_h} \tag{A.59}$$

where $p'_n(0^+)$ is given by Eq. (A.57), and $L_h = \sqrt{D_h\tau_h}$. As was done in Section A.6, the diffusion current density associated with this distribution can be found by substituting it into the transport equation for holes:

$$J_{h\text{-diff}} = -qD_h\frac{dp_n(x)}{dx} = \frac{qD_h}{L_h}p_{no}(e^{v_D/V_T} - 1)e^{-x/L_h} \tag{A.60}$$

Equation (A.60) is approximately equal to the total hole current on the n side of the diode, because the drift-current component of the minority holes is negligible in comparison.

The electron current density injected from the n side of the diode to the p side can be found in a similar manner. The excess electron distribution becomes

$$n'_p(x) = n'_p(0^-)e^{x/L_e} \tag{A.61}$$

where $n'_p(0^-)$ is given by Eq. (A.58), and $L_e = \sqrt{D_e\tau_e}$. The exponent in Eq. (A.61) has a positive coefficient because the distribution decays in the negative x-direction defined in Fig. A.12. Applying Eq. (A.61) to the transport equation for electrons yields the diffusion component of the electron current density on the p side of the diode:

$$J_{e\text{-diff}} = qD_e\frac{dn_p}{dx}(x) = \frac{qD_e}{L_e}n_{po}(e^{v_D/V_T} - 1)e^{x/L_e} \tag{A.62}$$

Equations (A.60) and (A.62) describe the minority-carrier diffusion-current densities on either side of the diode. These expressions are plotted in Figs. A.14(a) and A.14(b). The majority-carrier current densities on each side, also plotted in Fig. A.14, are composed of two components. The first component consists of the majority carriers that flow from the outer ends of the diode to recombine with the injected minority carriers flowing from the depletion region. The second component consists of the majority carriers destined to be injected across the depletion region. At any point along the length of the diode, which has a constant cross-sectional area, KCL must be satisfied; hence the sum of all the current-density components must be independent of x.

Figure A.14
Current components in a forward-biased *pn* junction diode: (a) current flow associated with hole injection; (b) current flow associated with electron injection.

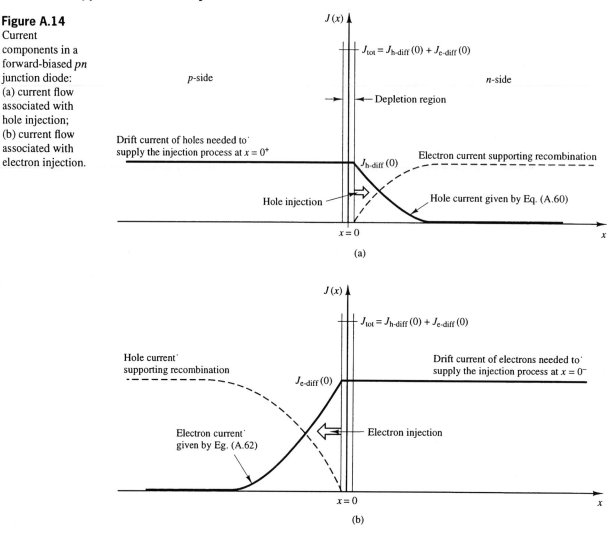

The total current density in the diode is equal to the sum of the hole and electron currents at any position x. It is easiest to evaluate the total current at the depletion region, where the majority-carrier components associated with recombination are equal to zero. Only those carriers associated with injection are nonzero at $x = 0$. The total current density flowing in the diode will thus be equal to the sum of $J_{h\text{-diff}}$ and $J_{e\text{-diff}}$ evaluated at $x = 0$. Performing the summation of Eqs. (A.60) and (A.62) with $x = 0$ results in

$$J_{\text{tot}} = \underbrace{\left(\frac{qD_e}{L_e}n_{po}\right.}_{\text{electron component}} + \underbrace{\left.\frac{qD_h}{L_h}p_{no}\right)}_{\text{hole component}} (e^{v_D/V_T} - 1) \tag{A.63}$$

Multiplying this current-density equation by the cross-sectional area A to obtain the total current through the diode yields the familiar diode equation:

$$i_D = I_s(e^{v_D/V_T} - 1) \tag{A.64}$$

where

$$I_s = qA\left(\frac{D_e}{L_e}n_{po} + \frac{D_h}{L_h}p_{no}\right) \tag{A.65}$$

Note that $p_{no} = n_i^2/n_{no} \approx n_i^2/N_D$ on the n side of the diode. Similarly, $n_{po} = n_i^2/p_{po} \approx n_i^2/N_A$ on the p side. The parameter I_s thus can be expressed in the alternative form

$$I_s = q A n_i^2 \left(\frac{D_e}{N_A L_e} + \frac{D_h}{N_D L_h} \right) \tag{A.66}$$

As Eq. (A.66) shows, I_s depends on the donor and acceptor concentrations N_D and N_A, where N_D applies to the n side of the diode and N_A to the p side. The parameter I_s is also a strong function of temperature due to the temperature dependency of n_i^2. Even though the thermal voltage V_T appears in the exponent of the diode equation (A.64), the exponential rise of n_i with temperature dominates the temperature dependence of the diode.

7.4 Correction for Depletion-Region Recombination

In the derivation of Section A.7.3, the depletion region was assumed to have negligible width, and recombination was neglected within the depletion region. In reality, the depletion region has finite width; hence some recombination does occur within it. A detailed analysis that includes depletion-region combination results in a minor modification to the diode equation that can be accommodated by adding an empirical constant η to the exponential term:

$$i_D = I_s(e^{v_D/\eta V_T} - 1) \tag{A.67}$$

The parameter η, called the *emission coefficient*, has a value between 1 and 2 for silicon diodes and is approximately equal to 1 for germanium and gallium arsenide diodes. The value of η also changes slightly with the width of the depletion region, and hence is a mild function of the diode voltage. In some of the literature, the emission coefficient is represented by the letter n.

7.5 The *pn* Junction at Extreme Operating Points

The physical model developed in Sections A.7.1 to A.7.3 must be expanded to accommodate diode operation at extremes of voltage and current. The v–i equation for the diode derived in Section A.7.3 uses the diffusion-limited model for hole and electron flow in the *pn* junction under forward-biased conditions. The effect of ohmic resistance in the semiconductor materials and diode contacts is ignored. This omission is valid if the diode is operated with low forward-bias current. When the diode current is small, the voltage drop across its internal ohmic resistance is minimal compared to the principal voltage drop across the depletion region. At larger values of current, the voltage drop across the diode's internal ohmic resistance becomes comparable to, or greater than, the depletion-region voltage drop. This ohmic resistance is a real, physical resistance and should not be confused with r_d, the equivalent incremental resistance that represents the slope of the diode's v–i characteristic in the piecewise linear model. At very high currents, the diode's ohmic resistance limits the exponential rise of current with voltage. In essence, a diode at very high currents begins to behave more like the series combination of a diode and a small-valued resistor.

In the reverse-biased region of operation, other processes limit the validity of the simple diode equation. As the reverse-bias diode voltage is increased, resistive leakage paths around the physical extremities of the diode act in parallel to pass current through the diode terminals. At even moderately large voltages, this reverse leakage current can exceed the reverse saturation current $-I_s$ by several orders of magnitude. In the reverse-biased mode, a diode often behaves more like the parallel combination of a diode and a large-valued resistor.

If the reverse bias voltage is very large, the diode may succumb to reverse breakdown via the avalanche or zener effects discussed in Section 3.3.5. The avalanche and zener mechanisms thus act to further limit the general validity of the diode equation (A.67).

A.8 THE BIPOLAR JUNCTION TRANSISTOR

The physical operation of the *npn* BJT was described qualitatively in Section 5.3.1. In this section, the BJT *v–i* characteristic is derived quantitatively. The reader is urged to review the discussion of Section 5.3.1 before proceeding with the analysis presented here.

The *v–i* characteristic of the *npn* BJT is easily found by adapting the equations governing current flow in the *pn* junction, as derived in Section A.7.3. Of particular importance is the distribution of electrons injected from the *n* side to the *p* side of a forward-biased diode. If the *x*-direction is defined as positive from the *n* side to the *p* side, this distribution can be expressed by

$$n'_p(x) = n_p(x) - n_{po} = n_{po}(e^{v_D/\eta V_T} - 1)e^{-x/L_e} \tag{A.68}$$

The diffusion current density that flows in response to this electron distribution is given by

$$J_{e\text{-diff}} = qD_e\frac{dn}{dx} = -\frac{qD_e}{L_e}n_{po}(e^{v_D/\eta V_T} - 1)e^{-x/L_e} \tag{A.69}$$

In Fig. A.15(a), a *pn* junction obeying this equation is forward-biased by a voltage source V_1 and a current-limiting resistor. A second piece of *n*-type material is added to this forward-biased junction, thereby forming an *npn* "sandwich". The junction formed by the second *n*-type region and the central *p*-type region is reverse-biased by a separate voltage source V_2 and current-limiting resistor. In this hypothetical structure, the length W of the central *p*-type region is much longer than the electron diffusion length L_e.

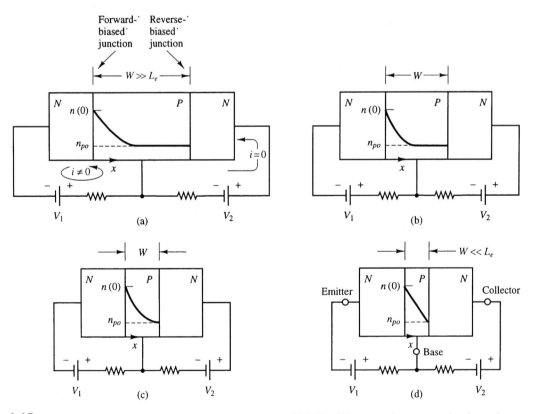

Figure A.15 The *npn* structure of (a) becomes an *npn* BJT as width W of the central *p*-type region is made smaller than the electron diffusion length L_e. (a) *npn* structure is not a transistor because $W \gg L_e$; (b) and (c) reducing W; (d) *npn* structure becomes a transistor when $W \ll L_e$. Electron collection takes place at the reverse-biased collector-base junction.

Because W is much longer than L_e, the *npn* sandwich of Fig. A.15(a) behaves like two diodes wired "back to back". The current flowing through the forward-biased junction simply circulates within its own current loop, and no current flows through the reverse-biased junction. The two junctions do not "talk" to each other in any way. The structure of Fig. A.15(a) resembles an *npn* BJT, but it is not a transistor.

An actual BJT can be formed by decreasing the width of the central *p*-type region. When W is made small, the structure becomes a transistor, and this *p*-type region becomes the base. The *n*-type region on the left of the forward-biased junction functions as the emitter, and the *n*-type region on the right of the reverse-biased junction becomes the collector.

Figures A.15(b) through A.15(d) depict the gradual reduction of the central base region. As W becomes shorter than the electron diffusion length L_e, as in Figs. A.15(c) and A.15(d), a point is reached where some of the diffusing electrons injected from the emitter reach the reverse-biased collector–base junction before recombining. These electrons find themselves on the *p* side of the reverse-biased collector–base depletion region. Because the electric field inside this depletion region points to the left, from the *n* side to the *p* side, these electrons are pulled across the depletion region into the collector, where they become majority carriers that contribute to collector current.

The collector acquires its name because it collects any electrons that enter its reverse-biased collector–base depletion region from the base. Similarly, the emitter acquires its name because it injects, or emits, electrons into the base via the forward-biased base–emitter junction. A BJT operating under these conditions is said to be in the *forward-active* mode. If the base region is narrow, many of the injected electrons will reach the reverse-biased collector–base junction before recombining. These collected electrons become current in the external collector circuit.

As Fig. A.16 shows, the emitter current in an *npn* BJT operating under forward-active conditions is composed of three components. The first component is formed by those electrons injected into the base and destined to diffuse all the way to the collector–base junction. A second emitter-current component is associated with the electrons that do not make it to the collector, but instead recombine in the base and contribute to base current. The third emitter-current component is caused by the injection of holes from the *p*-type base into the *n*-type emitter. These injected holes are unaffected by the presence of the reverse-biased collector–base junction. The last two current components described above constitute the total external base current.

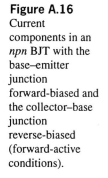

Figure A.16
Current components in an *npn* BJT with the base–emitter junction forward-biased and the collector–base junction reverse-biased (forward-active conditions).

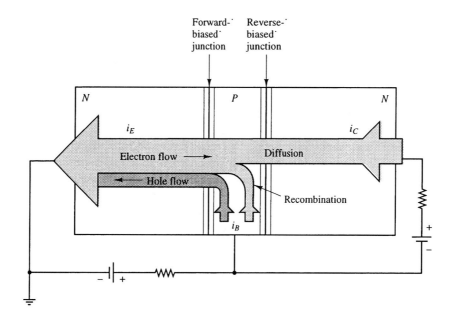

A.8.1 BJT $v-i$ Characteristic

In this section, the general concepts discussed in Section A.8 are applied to a quantitative derivation of the $v-i$ characteristic of the *npn* BJT. The total emitter current, which includes both the electron and hole components injected across the base–emitter junction, is given by

$$i_E = I_{EO}(e^{v_{BE}/\eta V_T} - 1) \tag{A.70}$$

where I_{EO} is the saturation, or scale, current of the base–emitter junction. Note that current is defined as positive *out* of the emitter by this equation.

The electron diffusion current captured by the collector–base junction can be expressed as a fraction α_F of this total emitter current:

$$i_C = \alpha_F i_E = \alpha_F I_{EO}(e^{v_{BE}/\eta V_T} - 1) \tag{A.71}$$

Note that electrons following *out* of the collector constitute a positive current *into* the collector. Equation (A.71) neglects the reverse saturation current of the reverse-biased collector–base junction.

The base current is equal to the remaining fraction of the total emitter current not described by Eq. (A.71). This current can be described by the equation

$$i_B = (1 - \alpha_F)i_E \tag{A.72}$$

The base current given by Eq. (A.72) includes the hole current injected into the emitter from the base and the hole-current component required to support electron recombination in the base. If the base width W is small, the recombination current component will also be small.

Given Eq. (A.72), the relation between the collector and base currents becomes

$$\frac{i_C}{i_B} = \frac{\alpha_F i_E}{(1 - \alpha_F)i_E} = \frac{\alpha_F}{1 - \alpha_F} \triangleq \beta_F \tag{A.73}$$

where β_F constitutes the large-signal current gain of the transistor.

The emitter saturation current I_{EO} is a function of doping densities, junction area, and the width of the base region, as shown in Section A.7.3. An expression for I_{EO} can be obtained by substituting W for L_e into Eq. (A.66), yielding

$$I_{EO} = qAn_i^2 \left(\underbrace{\frac{D_e}{N_A W}}_{\text{electron component}} + \underbrace{\frac{D_h}{N_D L_h}}_{\text{hole component}} \right) \tag{A.74}$$

The base width W replaces L_e in Eq. (A.74) because the magnitude of the electron diffusion current injected by the emitter depends on the base width W, rather than on the diffusion length L_e. This substitution is based on a linear approximation for $n_p'(x)$ in the base region. For an *npn* transistor, N_D in Eq. (A.74) is evaluated on the emitter side and N_A on the base side. In a well-designed BJT, the component of base current associated with holes injected into the emitter is made small, and the value of β_F made large, by lightly doping the base region with acceptors (small N_A) and heavily doping the emitter region with donors (large N_D). This asymmetrical doping causes the second term in Eq. (A.74) to become small.

The device depicted in Fig. A.15(d) is an idealized, one-dimensional BJT structure that models the central portion of an actual transistor, as shown in Fig. A.17. Despite the simplicity of the model, the derivation of this section is reasonably accurate.

Figure A.17
Core portion of the BJT that is represented by the one-dimensional structure of Fig. A.15(d).

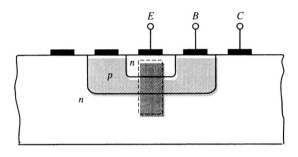

A.8.2 Narrow Base Region of the BJT

In the analysis of Section A.8.1, the diffusion and recombination of electrons in the base region of an *npn* BJT were described qualitatively. In this section, an explicit expression for the electron distribution is derived by considering the effect of the base width W. The derivation leads to a more accurate expression for β_F.

When the base–emitter junction of the BJT is forward-biased, the diffusion equation for electrons in the base becomes

$$\frac{d^2 n'_p}{dx^2} = \frac{n'_p}{L_e^2} \tag{A.75}$$

where $L_e = \sqrt{D_e \tau_e}$. The electrons are subject to the usual excess-carrier boundary condition at the edge of the emitter–base junction at $x = 0^+$:

$$n'_p(0^+) = n_{po}(e^{v_{BE}/\eta V_T} - 1) \tag{A.76}$$

The boundary condition for electrons on the other side of the base region at $x = W$, where the reverse-biased base–collector junction captures them, becomes

$$n'_p(W) \approx 0 \tag{A.77}$$

When W is much larger than L_e, as in the diode, the boundary condition (A.77) is automatically satisfied by the negative exponential solution alone. When W is much less than L_e, both the positive and negative exponential solutions to the diffusion equation (A.75) must be utilized to satisfy the boundary conditions (A.76) and (A.77). With both exponential solutions included, the general expression for the electron distribution in the base region becomes

$$n'_p(x) = Ae^{-x/L_e} + Be^{x/L_e} \tag{A.78}$$

where A and B are constants to be determined. Subjecting Eq. (A.78) to the boundary conditions (A.76) and (A.77) results in the equations

$$A + B = n'_p(0^+) \quad (\text{at } x = 0^+) \tag{A.79}$$

and
$$Ae^{-W/L_e} + Be^{W/L_e} = 0 \quad (\text{at } x = W) \tag{A.80}$$

Solving Eqs. (A.79) and (A.80) for the constants A and B yields

$$A = \frac{n'_p(0^+)}{1 - e^{-2W/L_e}} \tag{A.81}$$

and

$$B = \frac{n'_p(0^+)}{1 - e^{+2W/L_e}} \tag{A.82}$$

where $n'_p(0^+)$ is given by Eq. (A.76). Given these values of A and B, the general solution (A.78) becomes

$$n'_p(x) = n'_p(0^+)\left[\frac{e^{-x/L_e}}{1 - e^{-2W/L_e}} + \frac{e^{x/L_e}}{1 - e^{2W/L_e}}\right] \tag{A.83}$$

This equation can be simplified by multiplying the numerator and denominator of the first term in brackets by $\exp(W/L_e)$ and those of the second term in brackets by $\exp(-W/L_e)$, yielding

$$
\begin{aligned}
n'_p(x) &= n'_p(0^+)\left[\frac{e^{(W-x)/L_e}}{e^{W/L_e} - e^{-W/L_e}} + \frac{e^{(x-W)/L_e}}{e^{-W/L_e} - e^{W/L_e}}\right] \\
&= n'_p(0^+)\left[\frac{e^{(W-x)/L_e} - e^{-(W-x)/L_e}}{e^{W/L_e} - e^{-W/L_e}}\right] \\
&= n'_p(0^+)\frac{\sinh[(W-x)/L_e]}{\sinh(W/L_e)}
\end{aligned} \tag{A.84}
$$

To the extent that the boundary conditions (A.76) and (A.77) are accurate, Eq. (A.84) is an exact solution for the electron distribution $n'_p(x)$ when the base region has width W. A plot of $n'_p(x)$ for the case $W \ll L_e$ is shown by the solid line in Fig. A.18. The dashed line in this figure represents the linear approximation to $n'_p(x)$ used in Section A.8.1.

Figure A.18
Excess electron distribution in the base region of width W. Solid line: actual distribution given by Eq. (A.84). Dashed line: linear approximation given by Eq. (A.86).

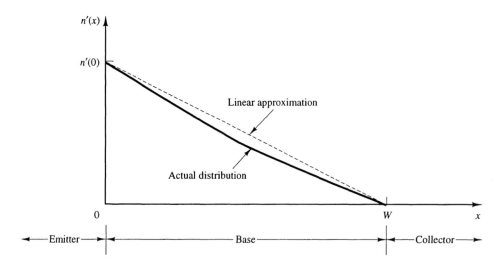

The portion of emitter current collected by the collector can be found by evaluating the electron diffusion current in the base at $x = W$. This current represents the electrons that have diffused across the base and are about to enter the collector region to become collector current. In this case, i_C must be computed from the negative of $J_{e\text{-diff}}$. The latter current density is defined as positive in the positive x-direction in Fig. A.15, but i_C is defined as positive into the collector.

Substituting $n'_p(x)$ from Eq. (A.84) into the diffusion term of the electron transport equation thus results in

$$
\begin{aligned}
i_C = -AJ_{e\text{-diff}}\Big|_{x=W} &= -qAD_e\frac{dn'_p(x)}{dx}\Big|_{x=W} \\
&= qA\frac{D_e}{L_e}n'_p(0^+)\frac{\cosh[(W-x)/L_e]}{\sinh(W/L_e)}\Big|_{x=W} \\
&= \frac{qAD_e n'_p(0^+)}{L_e \sinh(W/L_e)}
\end{aligned}
\tag{A.85}
$$

where $\cosh(0) = 1$.

If the base width is extremely narrow, so that $W \lll L_e$, the $n'_p(x)$ given by Eq. (A.84) approaches the expression

$$
n'_p(x) \approx n'_p(0^+)\frac{W-x}{W}
\tag{A.86}
$$

where the approximation $\sinh(x) \approx x$ has been used for $x \ll 1$. This equation represents a linear approximation to Eq. (A.84) and is shown by the dashed line in Fig. A.18. In such a case, the electron diffusion current collected by the collector is given by the simpler expression

$$
i_C = -AJ_{e\text{-diff}}\Big|_{x=W} = -qAD_e\frac{dn'_p(x)}{dx} = \frac{qAD_e n'_p(0^+)}{W}
\tag{A.87}
$$

Equation (A.85) or (A.87) can be combined with the two components of the base current to find an expression for β_F. One component of i_B is equated with the holes that are injected from the base into the emitter. This hole current is identical to the hole component of the forward-biased *pn* junction of Section A.7.3, given by

$$
i_B\Big|_{\substack{\text{injected} \\ \text{holes}}} = \frac{qAD_h}{L_h}p'_n(0^-) = \frac{qAD_h}{L_h}p_{no}(e^{v_{BE}/\eta V_T} - 1)
\tag{A.88}
$$

These injected holes contribute to the total emitter current i_E but not to the collector current i_C.

The second component of base current arises from recombination in the base. The recombination associated with the excess electron distribution $n'_p(x)$ in the base can be expressed by

$$
\frac{dn'_p}{dt} = -\frac{n'_p}{\tau_e}
\tag{A.89}
$$

Electrons that have recombined are continuously replenished by electrons newly injected at the base–emitter junction. The flow of holes into the base required to recombine with these electrons is equal to the total charge that recombines in the base per unit time. This quantity can be expressed by

$$
i_B\Big|_{\text{recombination}} = \frac{\text{total excess charge } Q \text{ in the base region}}{\text{recombination time } \tau_e}
\tag{A.90}
$$

The total excess charge in the base can be found by integrating $n'_p(x)$ over the volume of the base region from $x = 0^+$ to $x = W$. Performing this integration on the simplified $n'_p(x)$ given by Eq. (A.86) results in

$$
Q = A\int_{x=0^+}^{x=W} \underbrace{q n'_p(0^+)\frac{W-x}{W}}dx = q A n'_p(0^+)\frac{W}{2}
\tag{A.91}
$$

integral over — total volume of base   charge per unit volume

If the recombination time in the base region is equal to τ_e, the recombination component of base current, called the *base defect*, can be found by combining Eqs. (A.90) and (A.91):

$$i_B\Big|_{\text{recombination}} = \frac{Q}{\tau_e} = \frac{qAn'_p(0^+)}{2\tau_e}W \tag{A.92}$$

In the general case, a third component of base current, contributed by the saturation current of the reverse-biased base–collector junction, must also be considered in evaluating β_F. In a well-designed BJT, however, this current is usually small and can be neglected in a first-order computation of β_F.

The expressions (A.88) and (A.92) for the components of i_B can be combined with Eq. (A.87) for i_C to yield an approximate expression for β_F:

$$\beta_F \triangleq \frac{i_C}{i_B} \approx \frac{qAD_e n'_p(0^+)/W}{qAD_h p'_n(0^-)/L_h + qAn'_p(0^+)W/2\tau_e} \tag{A.93}$$

Factoring out common terms and applying Eqs. (A.57) and (A.58) results in

$$\begin{aligned}
\beta_F &\approx \frac{D_e n_{po}/W}{D_h\, p_{no}/L_h + n_{po}W/2\tau_e} \\
&= \frac{D_e/N_A W}{D_h/N_D L_h + W/2N_A\tau_e} = \frac{1}{D_h N_A W/D_e N_D L_h + W^2/2D_e\tau_e}
\end{aligned} \tag{A.94}$$

where $n_i^2 = n_o p_o$, N_D applies to the emitter region, and N_A to the base region. Note that β_F depends on W, which decreases with increasing voltage across the base–collector junction. This dependance causes β_F to increase slightly with v_{CE}—a phenomenon manifest as the Early voltage parameter in the BJT's large-signal v–i characteristics.

A.8.3 The Ebers–Moll Transistor Model

Although the simple physical transistor model of Section A.8.1 is a good one, it is not complete. The model is valid for forward-active operation, but not for cutoff and saturation. It also omits a mode of operation called *reverse active* in which the roles of the collector and emitter are reversed. Similarly, the model of Section A.8.1 overlooks secondary components of transistor current that can become sizable under these other modes of operation or at elevated operating temperatures. The reverse-bias saturation current across the collector–base junction, for example, becomes significant as the junction temperature increases, and the injected electron current from collector to base comes important in the reverse-active mode. These additional characteristics of the BJT are all accommodated by the *Ebers–Moll* transistor model.[2]

In the Ebers–Moll model, the *total* collector current for an *npn* BJT, where i_C is defined as positive into the collector, is given by

$$i_C = \alpha_F I_{EO}(e^{v_{BE}/\eta V_T} - 1) - I_{CO}(e^{v_{BC}/\eta V_T} - 1) \tag{A.95}$$

Here α_F is the fraction of injected emitter electron current collector by the collector, as discussed in Section A.8.1, and I_{CO} is the saturation, or *scale*, current of the collector–base junction. The electron collection described by the first term in Eq. (A.95) can occur even when the collector–base junction is forward-biased, because the electrons injected from the emitter will still diffuse to the base–collector depletion region and will be captured by its electric field. The second term in Eq. (A.95) describes the current component contributed by the collector–base junction, whether forward- or reverse-biased.

---

[2] J. J. Ebers, and J. L. Moll, "Large Signal Behavior of Junction Transistors," *Proceedings of the IRE*, Vol. 42, No. 12, 1761–1772, December 1954.

A similar consideration yields the equation for the total emitter current of an *npn* BJT, where i_E is defined as positive out of the emitter:

$$i_E = I_{EO}(e^{v_{BE}/\eta V_T} - 1) - \alpha_R I_{CO}(e^{v_{BC}/\eta V_T} - 1) \tag{A.96}$$

The constant α_R describes the fraction of electron current injected from the collector that arrives at the emitter. These electrons, when injected by a forward-biased base–collector junction, are collected by the depletion region field of the base–emitter junction.

Consideration of Eqs. (A.95) and (A.96) and the fact that $i_B = i_E - i_C$ yields an expression for the base current of an *npn* BJT:

$$i_B = (1 - \alpha_F)I_{EO}(e^{v_{BE}/\eta V_T} - 1) + (1 - \alpha_R)I_{CO}(e^{v_{BC}/\eta V_T} - 1) \tag{A.97}$$

These expressions for i_B, i_C, and i_E motivate the equivalent circuit of Fig. A.19, which constitutes a complete model for the *npn* BJT. A similar model valid for *pnp* transistors is shown in Fig. A.20. For a *pnp* BJT, the correct current expressions are given by

$$i_C = -\alpha_F I_{EO}(e^{-v_{BE}/\eta V_T} - 1) + I_{CO}(e^{-v_{BC}/\eta V_T} - 1) \tag{A.98}$$

$$i_E = -I_{EO}(e^{-v_{BE}/\eta V_T} - 1) + \alpha_R I_{CO}(e^{-v_{BC}/\eta V_T} - 1) \tag{A.99}$$

and

$$i_B = -(1 - \alpha_F)I_{EO}(e^{-v_{BE}/\eta V_T} - 1) - (1 - \alpha_R)I_{CO}(e^{-v_{BC}/\eta V_T} - 1) \tag{A.100}$$

In the circuit models of Figs. A.19 and A.20, i_F represents the principal current flow through the BJT under forward-active operation; i_R represents the principal current flow under reverse-active operation. The coefficients α_F and α_R correspond to the beta parameters:

$$\beta_F \triangleq \frac{i_C}{i_B}\bigg|_{\substack{\text{forward} \\ \text{active}}} = \frac{\alpha_F}{1 - \alpha_F} \tag{A.101}$$

and

$$\beta_R \triangleq -\frac{i_E}{i_B}\bigg|_{\substack{\text{reverse} \\ \text{active}}} = \frac{\alpha_R}{1 - \alpha_R} \tag{A.102}$$

In the second equation, i_E is again defined as positive out of the device.

Figure A.19
The Ebers–Moll circuit model for an *npn* BJT.

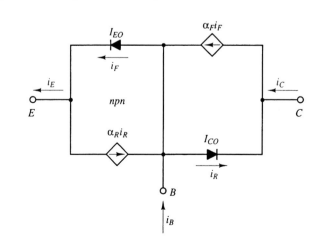

Figure A.20
The Ebers–Moll
circuit model for a
pnp BJT.

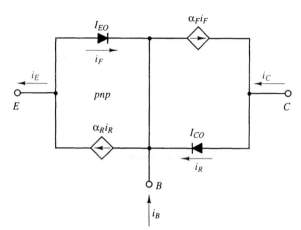

A.9 THE METAL-OXIDE-SEMICONDUCTOR FIELD-EFFECT TRANSISTOR

The physical operation of the metal-oxide-semiconductor field-effect transistor (MOSFET) was introduced qualitatively in Chapter 5. As discussed in Section 5.2.1, the v–i characteristic of the MOSFET's output port depends on the electric field in the gate-to-substrate oxide layer. This field is created by the gate-to-source voltage v_{GS}. In this section, the output port v–i characteristic of the n-channel MOSFET is derived quantitatively from fundamental physical principles. As a prelude to this section, the reader is encouraged to review the physical description given in Section 5.2.1.

A.9.1 Derivation of n-Channel Enhancement-Mode MOSFET v–i Characteristic for Small v_{DS}

When v_{DS} is less than the quantity $(v_{GS} - V_{TR})$ and v_{GS} is greater than V_{TR}, the MOSFET operates in the triode region. Because v_{GS} exceeds V_{TR}, the electric field in the gate oxide layer terminates at the oxide–substrate interface on negative surface charge composed of both exposed acceptor ion cores and inversion-layer electrons. In the limit of small v_{DS}, the density of this surface charge will not vary significantly with position along the channel and can be expressed by

$$Q_s = -q(\underbrace{N_{As}}_{} + \underbrace{n_s}_{}) \qquad \text{(A.103)}$$

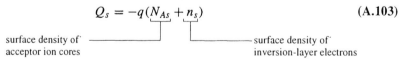

surface density of
acceptor ion cores

surface density of
inversion-layer electrons

In this equation, n_s is the surface density of the inversion-layer electrons and N_{As} the surface density of the exposed acceptor ion cores. Both of these quantities are measured in carriers per square centimeter. The unit of electronic charge magnitude q is equal to 1.6×10^{-19} coulomb.

Gauss's law of electromagnetics can be applied in the device geometry of Fig. A.21 to express the oxide field E_o in terms of the substrate surface charge density Q_s. With E_o defined as positive into the substrate surface, this relationship becomes

$$E_o = -\frac{Q_s}{\epsilon_{ox}} = \frac{q}{\epsilon_{ox}}(N_{As} + n_s) \qquad \text{(A.104)}$$

The constant ϵ_{ox} is the permittivity of the oxide layer between the gate and the substrate.

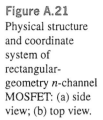

Figure A.21
Physical structure
and coordinate
system of
rectangular-
geometry n-channel
MOSFET: (a) side
view; (b) top view.

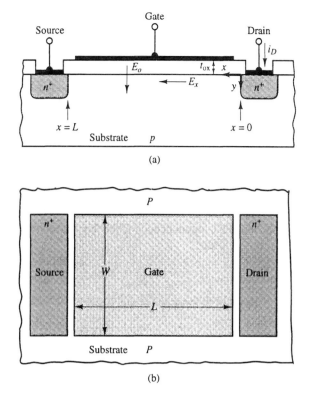

As discussed in Section 5.2.1, conduction in the channel can occur via electrons but not by holes. Only the electrons in the substrate are capable of bridging the reverse-biased depletion region between the drain and the substrate. In the channel region, therefore, the expression for the drain-to-source surface current density J_s contains the surface density of electrons only:

$$J_s = q n_s \mu_e E_x \qquad \text{(A.105)}$$

where J_s has the units of current per unit depth into the page in Fig. A.21(a). In this equation, μ_e is the mobility of the electrons, and E_x is the electric field parallel to the surface of the channel. The horizontally directed E_x arises as a result of v_{DS} and should not be confused with the downward-directed E_o caused by v_{GS}. The p-type substrate of an n-channel MOSFET is lightly doped, hence the field E_x is not confined to just the drain-to-substrate depletion region, as it is in a pn junction diode. Rather, E_x is distributed along the entire length of the substrate.

The drain current i_D is equal to the surface current J_s multiplied by the lateral width W of the device:

$$i_D = W J_s \qquad \text{(A.106)}$$

Equation (A.106) can be combined with Eq. (A.105), and Eq. (A.104) can be used to obtain an expression for $q n_s$. The net result becomes

$$i_D = W q n_s \mu_e E_x = W (\epsilon_{ox} E_o - q N_{As}) \mu_e E_x \qquad \text{(A.107)}$$

For small v_{DS}, E_x will be approximately uniform in the x-direction and can be expressed by

$$E_x \approx \frac{v_{DS}}{L} \qquad \text{(A.108)}$$

Similarly, if v_{GS} is much larger than v_{DS}, the field in the oxide layer will be constant and can be approximated by

$$E_o \approx \frac{v_{GS}}{t_{\text{ox}}} \qquad \text{(A.109)}$$

where t_{ox} is the thickness of the oxide layer. Implicit in Eq. (A.109) is the assumption that the substrate is at the same potential everywhere along its length. This assumption is valid only if $v_{DS} \ll v_{GS}$.

Combining Eqs. (A.107) through (A.109) yields an expression for i_D as a function of v_{GS} in the regime $v_{DS} \ll v_{GS}$:

$$i_D = W\mu_e \left(\epsilon_{\text{ox}} \frac{v_{GS}}{t_{\text{ox}}} - qN_{As} \right) \frac{v_{DS}}{L} \qquad \text{(A.110)}$$

Equation (A.110) can be rearranged to arrive at an expression for the v–i characteristic of the n-channel MOSFET for small v_{DS}. Over this portion of triode-region operation, the v–i characteristic is almost linear:

$$\frac{i_D}{v_{DS}} \triangleq \frac{1}{R_{\text{chan}}} = \mu_e \frac{\epsilon_{\text{ox}}}{t_{\text{ox}}} \frac{W}{L} \left(v_{GS} - qN_{As} \frac{t_{\text{ox}}}{\epsilon_{\text{ox}}} \right) \qquad \text{(A.111)}$$

The second term in parentheses in this equation defines the device threshold voltage V_{TR}:

$$V_{\text{TR}} = qN_{As} \frac{t_{\text{ox}}}{\epsilon_{\text{ox}}} \qquad \text{(A.112)}$$

Equation (A.112) represents the net negative surface charge contributed by the p-type acceptor ion cores in the substrate region when all the holes have been pushed out by the oxide-layer field. Any further increase in v_{GS} requires that inversion-layer electrons be pulled into the channel to help terminate the increased oxide-layer field.

If v_{GS} lies above the threshold voltage, the equivalent channel resistance for $v_{DS} \ll v_{GS}$ can be expressed by

$$\frac{1}{R_{\text{chan}}} = \left[\frac{\mu_e}{2} \frac{\epsilon_{\text{ox}}}{t_{\text{ox}}} \frac{W}{L} \right] 2(v_{GS} - V_{\text{TR}}) \qquad \text{(A.113)}$$

The expression in brackets in Eq. (A.113) defines the parameter K used in the MOSFET v–i equations. Its value is proportional to the geometric width-to-length ratio W/L of the device.

A.9.2 Derivation of n-Channel Enhancement-Mode MOSFET v–i Characteristic for Large v_{DS}

When v_{DS} is comparable to or greater than v_{GS}, the way in which v_{DS} is distributed along the channel becomes important. For the coordinate system defined in Fig. A.21, this distribution can be described by a position-dependent potential $V(x)$ measured relative to the source. The value of $V(x)$ at $x = 0$ is equal to v_{DS}; its value at $x = L$ is equal to zero. When v_{DS} is comparable to v_{GS}, the distribution of $V(x)$ between these two boundary limits will not be linear. In this regime of operation, the strength of the oxide-layer field becomes proportional to the difference between the gate voltage and the local channel potential $V(x)$. Near the source end of the channel, where $V(L) = 0$, the field is large. Conversely, near the drain end of the channel, where $V(0) = v_{DS}$, the field is smaller. This situation is illustrated for the specific case $v_{DS} = (v_{GS} - V_{\text{TR}})$ in Fig. A.22. Note that if v_{DS} becomes larger than v_{GS}, the oxide-layer field actually reverses direction near the drain end of the channel. The oxide-layer field at any point x can be approximated by the voltage difference between the gate electrode and the local value of $V(x)$:

$$E_o(x) = \frac{v_{GS} - V(x)}{t_{\text{ox}}} \qquad \text{(A.114)}$$

where t_{ox} is the oxide-layer thickness.

Figure A.22

The n-channel MOSFET with $v_{DS} = (v_{GS} - V_{TR})$. At any position along the channel, the oxide-layer field is proportional to the difference between v_{GS} and $V(x)$.

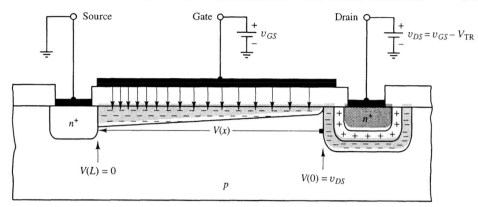

This equation expresses E_o as a function of channel position x; it can be substituted into the right-hand side of Eq. (A.107) to yield

$$i_D = W\mu_e\epsilon_{ox}\left[\frac{v_{GS} - V(x)}{t_{ox}} - \frac{qN_{As}}{\epsilon_{ox}}\right]E_x \tag{A.115}$$

or

$$i_D = W\mu_e\epsilon_{ox}\left[\frac{v_{GS} - V(x) - V_{TR}}{t_{ox}}\right]E_x \tag{A.116}$$

where V_{TR} is defined by Eq. (A.112).

For the case of v_{DS} comparable to v_{GS}, the distribution of $V(x)$ must be considered in computing E_x, which also becomes a function of x. Specifically, E_x is related to $V(x)$ by the definition

$$E_x(x) \triangleq -\frac{dV(x)}{dx} \tag{A.117}$$

According to KCL, the i_D given by Eq. (A.116) cannot be a function of x. This fact can be used to find i_D in terms of v_{GS} and v_{DS}. Specifically, with E_x expressed by Eq. (A.117), Eq. (A.116) becomes

$$i_D = -\frac{W\mu_e\epsilon_{ox}}{t_{ox}}[(v_{GS} - V_{TR}) - V(x)]\frac{dV(x)}{dx} \tag{A.118}$$

Equation (A.118) can be integrated along the length of the channel from $x = 0$ to $x = L$:

$$\int_{x=0}^{x=L} i_D\, dx = -\frac{W\mu_e\epsilon_{ox}}{t_{ox}}\int_{V(0)=v_{DS}}^{V(L)=0}[(v_{GS} - V_{TR}) - V(x)]\, dV(x) \tag{A.119}$$

Performing the integration with i_D independent of x results in

$$i_D L = \frac{W\mu_e\epsilon_{ox}}{t_{ox}}\left[(v_{GS} - V_{TR})v_{DS} - \frac{v_{DS}^2}{2}\right] \tag{A.120}$$

or

$$i_D = 2K\left[(v_{GS} - V_{TR})v_{DS} - \frac{v_{DS}^2}{2}\right] \tag{A.121}$$

where $K = W\mu_e\epsilon_{ox}/2t_{ox}L$, as before. Equation (A.121) is the i_D–v_{DS} relationship valid for the case where v_{DS} is not negligible compared to v_{GS} but is less than $(v_{GS} - V_{TR})$. It describes the MOSFET's triode region and reduces to Eq. (A.113) in the limit $v_{DS} \ll (v_{GS} - V_{TR})$.

When v_{DS} becomes larger than $(v_{GS} - V_{TR})$, a region develops near the drain where the oxide-layer field falls below the threshold value required for inversion layer formation, as depicted in Fig. A.23. The larger v_{DS} causes an electron-free region to develop between $x = 0$ and $x = \delta$. Over this region of the channel, the voltage difference $v_{GS} - V(x)$ falls below V_{TR}. The inversion layer disappears over this region because a voltage drop of at least V_{TR} must exist between gate and substrate if an inversion layer is to form. This zone functions as an extension of the drain-to-substrate depletion region, as depicted in Fig. A.23.

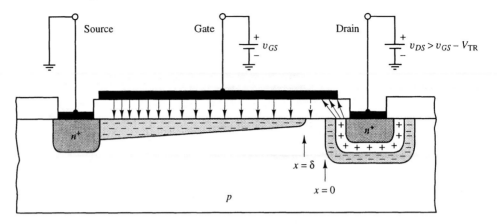

Figure A.23 An n-channel MOSFET with $v_{DS} > (v_{GS} - V_{TR})$. The electron inversion layer stops at $x = \delta$, just to the left of the drain-to-substrate depletion region. The gate-oxide electric field reverses direction near the drain end of the channel.

The depletion region surrounding the drain functions as a collector of any electrons that are swept into it from the inversion-layer channel. The rightward flow of these electrons, which is driven by E_x, is governed solely by the voltage distribution $V(x)$ in the inversion layer to the left of $x = \delta$. This $V(x)$ will be approximately equal to the $V(x)$ that exists when v_{DS} is exactly equal to $(v_{GS} - V_{TR})$. For v_{DS} greater than $(v_{GS} - V_{TR})$, the excess in v_{DS} is dropped across the drain-to-substrate depletion region between $x = 0$ and $x = \delta$. The current i_D thus can be found by substituting $(L - \delta)$ for L in Eq. (A.120) and by substituting the value $V(\delta)$ for v_{DS}. Note that $V(\delta)$ is equal to $(v_{GS} - V_{TR})$, which is the value of $V(x)$ at which the inversion layer disappears. Under these conditions, the drain current becomes

$$i_D = \frac{W \mu_e \epsilon_{ox}}{(L - \delta) t_{ox}} \frac{(v_{GS} - V_{TR})^2}{2} \tag{A.122}$$

The value of i_D in this equation, which describes the constant-current region of operation, is independent of v_{DS}.

In most device geometries, δ is much smaller than L; hence Eq. (A.122) can be written approximately as

$$i_D \approx K(v_{GS} - V_{TR})^2 \tag{A.123}$$

where K is defined, as before, by

$$K = \frac{1}{2} \frac{W}{L} \frac{\mu_e \epsilon_{ox}}{t_{ox}} \tag{A.124}$$

Equation (A.122) describes the MOSFET's constant-current region as given in Chapter 5. As discussed in Section 5.2.3, an n-channel depletion-mode MOSFET is made by implanting a thin layer of donor ions just beneath the substrate surface. This implanted layer has the effect of internally biasing the MOSFET with a v_{GS} that exceeds V_{TR}.

The preceding discussion provides an idealized picture of MOSFET operation. In a real MOSFET, the depletion width δ increases with v_{DS}, causing the denominator of Eq. (A.122) to decrease slightly as v_{DS} increases. Consequently, the MOSFET's $v–i$ characteristic slopes upward slightly with increasing v_{DS} in the constant-current region. This effect is described by the Early voltage parameter of Section 5.4.1.

PROBLEMS

A.1 Describe at least three properties of semiconductors that differ from those of metals. Why are semiconductors so sensitive to temperature?

A.2 The resistivity of a semiconductor is given by $\rho = 1/\sigma = 1/q(n\mu_e + p\mu_h)$, where n and p are the equilibrium concentrations of holes and electrons.

(a) Find the resistivity of intrinsic silicon at 300 K.

(b) If a donor-type impurity is added to the extent of 1 atom per 10^8 silicon atoms, find the resistivity.

A.3 A sample of silicon is doped such that the drift current components of holes and electrons are of equal magnitude. What are the values of n_o and p_o in this case?

A.4 A piece of silicon at room temperature is doped with $2 \times 10^{16}/\mathrm{cm}^3$ concentration of boron atoms (an acceptor) and $5 \times 10^{15}/\mathrm{cm}^3$ of phosphorus atoms (a donor).

(a) Find the hole and electron concentrations in this material.

(b) Is the silicon p- or n-type?

A.5 A piece of silicon is doped with a donor impurity until its resistivity is $\rho = 2.5\ \Omega$-cm.

(a) Calculate the density of holes and electrons in the sample.

(b) To what temperature (in Kelvin) must the silicon be heated before n_i becomes larger than the electron concentration of part (a)?

(c) What conductivity must the silicon have at 300 K to assure that less that 0.1% of the current will be carried by holes?

A.6 A certain doped semiconductor at room temperature has the following properties: $n_o = 9 \times 10^{14}/\mathrm{cm}^3$, $p_o = 4 \times 10^{14}/\mathrm{cm}^3$, $\mu_e = 800\ \mathrm{cm}^2/\mathrm{V\text{-}s}$, $\mu_h = 400\ \mathrm{cm}^2/\mathrm{V\text{-}s}$, and $(D_h\tau_h)^{1/2} = 10^{-4}\ \mathrm{cm}$.

(a) What is the intrinsic carrier concentration n_i for this material?

(b) Is the semiconductor p- or n-type?

(c) If an electric field is applied, what fraction of the resulting drift current flow will be due to electrons?

A.7 A semiconductor is doped with donor and acceptor concentrations N_D and N_A, respectively.

(a) If $N = N_D - N_A$, prove that

$$n_o = \frac{|N|\sqrt{1 + 4n_i/N^2}}{2} + \frac{N}{2}$$

and

$$p_o = \frac{|N|\sqrt{1 + 4n_i/N^2}}{2} - \frac{N}{2}$$

(b) In practice, a semiconductor is usually doped such that $|N| >>> n_i$. For such a case, show that the above equations reduce to $n_o \approx N$; $p_o \approx n_i^2/N$ for $N > 0$, and $p_o \approx N$; $n_o \approx n_i^2/N$ for $N < 0$.

A.8 Copper has a free-electron concentration of approximately $10^{22}/\mathrm{cm}^3$, a conductivity of about $10^6 (\Omega\text{-m})^{-1}$ (siemen/meter), and no free holes.

(a) Calculate the mobility of electrons in copper.

(b) The ends of a 0.1-cm radius copper wire 10 km long are connected to a 5-V battery. Calculate the approximate values of the electric field and current in the wire.

(c) Determine the drift velocity of an average electron in the wire and the length of time required for that electron to travel from one end of the wire to the other. How do you explain the fact that, from a circuit point of view, the current reaches a steady-state value in less than $1\ \mu s$ when the battery is connected to the wire?

A.9 (a) An n-type slab of semiconductor ($n_i = 10^{10}/\mathrm{cm}^3$, $D_e = 40\ \mathrm{cm}^2/s$, $D_h = 12\ \mathrm{cm}^2/s$) is doped uniformly with donors to a concentration of $N_D = 10^{16}/\mathrm{cm}^3$. Calculate the equilibrium hole and electron concentrations.

(b) Light shines on the edge of the slab at $x = 0$, generating excess hole–electron pairs at a concentration of $10^{12}/\mathrm{cm}^3$. These holes and electrons diffuse to the right, recombining as they go. Find expressions for $n'(x)$ and $p'(x)$ for $x > 0$.

(c) Suppose the diffusion lengths are $L_e = 10\ \mu m$ and $L_h = 5\ \mu m$. For the case L much longer than L_e and L_h, find the distance x_1 at which the hole concentration falls to $1/e$ of its value at $x = 0$.

(d) Find an expression for the hole current density, ignoring the effect of any electric fields.

(e) Calculate the hole current density J_h at $x = 0$ and at $x = L_h$.

A.10 A p-type bar of silicon has length $2L$. Under low-level injection conditions, the bar is subject to the minority-carrier boundary conditions $n'(L) = K$ and $n'(-L) = -K$.

(a) Find expressions for $n'(x)$ and $J_e(x)$.

(b) Find $n'(x)$ in the limit $L \ll L_e$.

A.11 A 1-cm cube of p-type silicon ($\rho = 0.1\,\Omega$-cm) acquires, via some external process, a linear electron distribution in the x-direction, such that $n = 10^{14}/\mathrm{cm}^3$ at one side and $n = 10^5/\mathrm{cm}^3$ at the opposite side.

Wires are attached to the sides of the cube via ohmic contacts, and a 0.1-mV voltage source is applied. Find the values of the electron, hole, and total currents that flow in the external circuit.

A.12 The derivation presented in Section A.7 for the v–i equation of the pn junction relies on the low-level injection condition. When the injection process is low-level, the conductivity of the region does not change significantly due to the injected carriers. Consider a pn junction with $N_n = |N_p| = 10^{18}$, where $N = N_D - N_A$. How large a forward voltage can be applied to the diode before low-level injection conditions are violated? Specifically, at what value of v_D will the holes injected into the n-side reach about 1% of the concentration of the majority-carrier electrons?

A.13 Excess holes are injected into a semiinfinite n-type semiconductor bar at $x = 0$ such that the steady-state excess-hole concentration at $x = 0$ is $p'(0)$. Assume that low-level injection conditions apply ($p' \ll n_o$).

(a) Using the continuity and transport equations, find an expression for the excess-hole distribution $p'(x)$ as a function of $x > 0$.

(b) Find an expression for the resulting hole diffusion current.

(c) Suppose that the bar, instead of extending to "infinity" for $x > 0$, has finite length L. How long must L be if the answers of parts (a) and (b) are to be valid?

A.14 For the pn junctions of **Figs. PA.14(a)** and PA. 14(b), indicate whether the following statements are true or false:

(a) $|N_n| > |N_p|$, where $N = N_A - N_D$.

(b) A depletion region exists at $x = 0$.

(c) The diode is forward-biased.

(d) $n_o p_o = n_i^2$ for all x.

(e) Current flows through the diode.

(f) A net charge is contained within the diode.

A.15 A silicon pn junction diode at room temperature is doped such that $N_A = 10^{16}$ on the p-side and $N_D = 10^{15}$ on the n-side. The diode is forward-biased with a voltage $v_D = 0.5\,\mathrm{V}$. The diode has a cross-sectional area of $1\,\mathrm{mm} \times 1\,\mathrm{mm}$, an emission coefficient $\eta = 1$, and recombination time constants $\tau_e = \tau_h = 1\,\mu\mathrm{s}$.

(a) Find the diode current i_D.

(b) Find the minority concentration n_p at the edge of the depletion region on the p-side.

(c) Find the minority concentration p_n at the edge of the depletion region on the n-side.

A.16 A pn junction diode has net doping concentrations of $10^{18}/\mathrm{cm}^3$ on the n-side and $N_A = 10^{15}/\mathrm{cm}^3$ on the p-side. The width of each side is $W = 10\,\mu\mathrm{m}$, with $L_e = 50\,\mu\mathrm{m}$ and $L_h = 2\,\mu\mathrm{m}$. Suppose that the diode is forward-biased with $v_D = 0.5\,\mathrm{V}$.

(a) Make a rough plot of the hole and electron densities as a function of distance x from the depletion region (positive $x = n$-side).

(b) Calculate the electron density at $x = 0^-$ and the hole density at $x = 0^+$, where $x = 0$ is the center of the depletion region.

(c) Consider only the electron diffusion current on the p-side ($x < 0$). What is the ratio of $J_{e\text{-diff}}$ at $x = 0^-$ to its value at $x = -W/2$?

A.17 A germanium diode of cross-sectional area $A = 10^{-2}\,\mathrm{cm}^2$ is to be made with $N_A = N_D$. Specify the value of N_A such that $I_s = 10^{-12}A$. Assume recombination time constants of $\tau_e = \tau_h = 1\,\mu\mathrm{s}$.

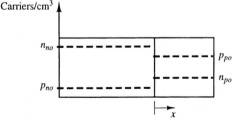

(a)

(b)

Figure PA.14

A.18 Choose the net doping levels $|N_D - N_A|$ on both sides of a long silicon *pn* junction diode such that the conductivity of the *n*-type side is $0.1(\Omega\text{-cm})^{-1}$ and 99% of the diode current is carried by holes. Specify the value of $|N_D - N_A|$ on each side of the diode and the diode parameter I_s. Calculate the hole, electron, and total diode currents when $v_D = 0.5$ V.

A.19 In the *npn* transistor of **Fig. PA.19**, $|N_B| = 10^{17}/\text{cm}^3$, $|N_E| = 10^{20}/\text{cm}^3$, $W_B = 0.5\,\mu\text{m}$, and $W_E = 2\,\mu\text{m}$, where $N = N_D - N_A$. The transistor has a cross-sectional area of $10^{-6}\,\text{cm}^2$ and is operated in the forward-active (constant-current) region with $V_{BE} = 0.73$ V. Calculate the collector current at 300 K.

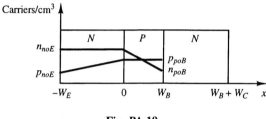

Fig. PA.19

A.20 In the *npn* transistor of **Fig. PA.19**, $I_C = 0.6$ mA when $V_{BE} = 0.72$ V and $V_{CB} = 1$ V. The average base width is $0.3\,\mu\text{m}$. When V_{CB} is increased to 10 V, the collector–base depletion layer width increases by $0.02\,\mu\text{m}$ on both sides of the junction.

(a) Find I_C at $V_{CB} = 10$ V.

(b) Find the incremental output resistance r_o for this transistor.

(c) V_{CB} is returned to 1 V and V_{BE} increased to 0.76 V. Find the new I_C.

(d) If the acceptor concentration is increased in the base, what will happen to β_F?

A.21 Consider an *npn* transistor made from a semiconductor with $n_i = 10^{10}/\text{cm}^3$, $D_e = 2D_h = 50\,\text{cm}^2/s$, and $L_e = 2L_h = 100\,\mu\text{m}$. The doping levels are such that $|N| = |N_D - N_A| = 10^{17}/\text{cm}^3$ in the emitter, $|N| = 8 \times 10^{14}/\text{cm}^3$ in the base, and $|N| = 10^{13}$ in the collector. The emitter, base, and collector regions have widths $W_E = 3\,\mu\text{m}$, $W_B = 1.5\,\mu\text{m}$, and $W_C = 2\,\mu\text{m}$, respectively.

(a) Assume that recombination is negligible. Sketch the minority-carrier concentrations everywhere in the transistor for $v_{BE} > 0$ and $v_{CB} = 0$. These voltages will place the transistor in the forward-active region of operation.

(b) The current flowing across the base–emitter junction can be expressed as $i_E = I_{EO}[\exp(v_{BE}/\eta V_T) - 1]$. Derive an expression for I_{EO}.

(c) With no recombination in the base, all electrons injected across the base–emitter junction will flow by diffusion to the collector–base junction. Using this fact, derive expressions for the collector and base currents i_C and i_B.

(d) Determine numerical values for α_F and β_F for this transistor.

A.22 A *pnp* transistor structure is made with $W_E = W_B = W_C = 2\,\mu\text{m}$. For the semiconductor used, $L_e = L_h = 20\,\mu\text{m}$, and the dopant levels are such that $|N_E| = 10^{17}/\text{cm}^3$, $N_B = 2 \times 10^{16}/\text{cm}^3$, and $|N_C| = 10^{16}/\text{cm}^3$ in the emitter (E), base (B), and collector (C) regions, respectively, where $N = N_D - N_A$. The voltage V_{BC} is set to zero and V_{BE} is set to a value that yields $n_p(0^-) = 10^{14}/\text{cm}^3$ on the emitter side of the base–emitter junction. Plot the minority-carrier concentrations in each of the three regions of the transistor if $n_i^2 = 10^{10}/\text{cm}^3$.

A.23 When the v_{DS} of a MOSFET is low, the device will behave as a resistor whose resistance is inversely proportional to v_{GS}. Suppose that $V_{TR} = 1$ V, $t_{ox} = 500$ Å, $\epsilon_{ox} = 3.5 \times 10^{-13}$ F/cm, and $\mu_e = 1500\,\text{cm}^2/\text{v-s}$. What value of W/L will yield an NMOS transistor whose channel resistance changes from $1\,\text{k}\Omega$ to $200\,\Omega$ for $3\,\text{V} < v_{GS} < 11\,\text{V}$?

A.24 Draw the cross-sectional view of an *n*-channel MOSFET with $V_{TR} = 4$ V for each of the terminal conditions that follow. Your sketch should show the gate, source, and drain regions, the inversion layer, and the electric field in the oxide layer.

(a) $v_{GS} = 9$ V; $v_{DS} = 7$ V.

(b) $v_{GS} = 8$ V; $v_{DS} = 1$ V

(c) $v_{GS} = 6$ V; $v_{DS} = 2$ V.

A.25 At constant v_{GS}, the point between the source and drain at which the conducting channel in a MOSFET is completely constricted moves as v_{DS} changes. In spite of this movement, the channel voltage V_{chan} (measured relative to the source terminal) at the point of constriction remains constant. What is the value of this voltage?

A.26 The threshold voltage of an *n*-channel MOSFET increases as the surface concentration of acceptors on the *p*-type substrate increases. For a particular silicon MOSFET, $qN_{As} = 10^{-9}$ C/cm^2 and $V_{TR} = 0.5$ V. If additional acceptor atoms are added to the substrate in the concentration of one per 10^3, what will be the new V_{TR}? For silicon, the volume concentration of atoms is about $10^{21}/\text{cm}^3$.

A.27 Suppose that you wish to make a depletion-mode p-channel MOSFET by adding dopant atoms at the silicon-oxide interface. What kind of dopant would you use, and why?

A.28 The deviation from constant-current behavior in the "constant-current" region of the MOSFET arises from the change in effective channel length with changes in v_{DS}. Is the slope of the curve constant or is it a function of v_{DS}? (Hint: Consider the change in diode depletion-region width with increasing reverse-bias voltage.)

A.29 A planar silicon n-channel MOSFET is fabricated with a gate oxide of 500 Å thickness. Specify the length and width of the gate if the device is to have a K parameter of $2\,\text{mA/V}^2$.

A.30 An n-channel MOSFET can be thought of as a parallel-plate capacitor formed by the gate electrode, silicon-dioxide dielectric, and silicon substrate. What is the effective capacitance if $v_{DS} \ll (v_{GS} - V_{TR})$?

Semiconductor Device and Integrated-Circuit Fabrication

I n this appendix, the steps required to manufacture semiconductor devices and integrated circuits are summarized. The simplified treatment presented here is far from comprehensive, but it exposes the reader to the basics of semiconductor fabrication. No "standard" manufacturing process really exists because each manufacturer may incorporate separate proprietary features. The basic processes described here, however, are typical of those used in most fabrication facilities.

B.1 AN OVERVIEW OF THE FABRICATION PROCESS

A typical IC chip is about 1 mm thick and a few millimeters on each side. As of the writing of this text, most integrated circuits are still made from silicon, although other materials, including gallium arsenide and other III–V compounds, are likely to play important roles in the future. The actual electronic circuit penetrates only about $10\,\mu$m into the crystal surface. The remaining 99% of the chip depth acts merely as a supporting substrate. A single IC chip is fabricated simultaneously with many identical circuits on a 1-mm-thick single-crystal wafer of large diameter. The wafer is subjected to many processing steps that ultimately produce the finished integrated circuits. After processing, individual circuits are cut from the wafer in a step called *dicing*.

The use of doping plays an important role in IC manufacturing. Heavily doped regions are created on the wafer to serve as conducting paths and as sites for external connections, while moderately doped regions form resistors. *PN* junctions are produced by doping adjacent regions of the chip with donors and acceptors. Some *pn* junctions are used as diodes or parts of BJTs and MOSFETs, while others are used as isolation barriers to suppress conduction between adjacent regions of the IC. The latter junctions are reverse biased during normal circuit operation. Conductors are often added to the IC to form electrical contacts, circuit interconnections, and capacitor electrodes.

Dielectric materials such as silicon dioxide or silicon nitride are grown on the wafer surface to form capacitors and MOSFET gates. Thicker layers of these dielectrics may also be used to isolate underlying conductive layers or to create masking layers for selective modification of the underlying wafer during the fabrication process. Such masking layers are typically formed and removed several times during a fabrication sequence.

Extreme cleanliness is vital to the fabrication process. Impurities can cause crystalline defects in the wafer or significant changes in wafer conductivity. Microscopic dust particles can interfere with the high-resolution patterning of devices and interconnections on the wafer surface. The environment in which the work is done is carefully controlled to minimize airborne particles, and the purity of the chemical reagents used for all fabrication steps is higher than that encountered in any other kind of manufacturing process. The typical wafer is cleaned many times during the IC fabrication sequence.

B.2 EPITAXIAL GROWTH

Integrated circuits are often fabricated within an upper layer of extremely pure semiconductor that is chemically deposited on the wafer surface using a process called *epitaxial growth*. The use of an epitaxially grown layer relieves some of the stringent requirements on the purity of the underlying wafer. A layer is considered epitaxial if it possesses the same single-crystal structure as the substrate; this condition is necessary for optimum circuit performance.

Using a process called *molecular-beam epitaxy*, it is possible to produce an epitaxial layer of one semiconductor over a substrate made from a different semiconductor. Molecular-beam epitaxy also permits the creation of tailor-made "superlattices" in which complex crystal structures are deposited one atomic layer at a time.

B.3 OXIDATION

The oxidation of silicon and other semiconductors is crucial to the fabrication process. Oxide films are used to make devices and are also used as temporary masking layers at intermediate fabrication steps. Thorough study of the oxidation process, most notably that of silicon, has given the IC manufacturer the ability to precisely control the thickness of the oxide layers by choosing the correct combination of type of reaction, oxidation temperature, and reaction time.

Oxidation and other high-temperature processes are carried out in cylindrical silica tubes placed horizontally in a furnace. A batch of wafers is mounted on a silica "sled", or *boat*, while reactant gases are introduced. By processing a large number of wafers simultaneously, each of which contains many identical circuits, a high through-put of circuits with minimal variation in characteristics can be obtained.

B.4 WAFER DOPING

As noted throughout this book, doping plays an important role in the function of all semiconductor devices. Selected areas of a processed wafer can be made *p*-type by replacing some of the atoms in the crystal lattice with acceptor atoms from group III of the periodic table. For silicon, boron is the acceptor most often used. Similarly, a wafer can be made *n*-type by doping it with donor atoms from group V of the periodic table. Of these elements, phosphorus and arsenic are most often used with silicon; antimony is used in special cases.

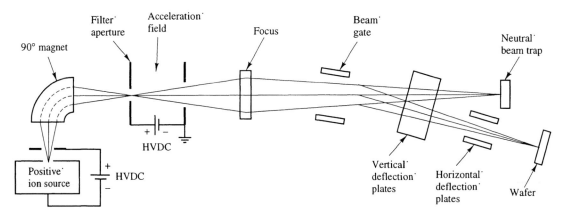

Figure B.1 Schematic diagram of an ion implanter. The entire system operates in a high-vacuum chamber.

The process by which a dopant atom replaces a host crystal atom is not a chemical reaction. Rather, the dopant need only be deposited on the wafer surface, then allowed to diffuse into the crystal under the influence of high temperature. Diffusion temperatures in the range 1000 to 1200°C are most often employed. Dopant diffusion is performed in the same type of tube furnace used for oxidation, as described in Section B.3. The most common dopant sources are gaseous compounds such as BCl_3 (boron doping) and $POCl_3$ (phosphorus doping).

When the silica wafer boat is placed in the furnace tube, the surface of each wafer quickly becomes saturated with the dopant, with the concentration reaching a level on the order of $10^{21}/cm^3$. (Recall that 100% corresponds to a concentration of about $10^{22}/cm^3$ for a typical semiconductor.) After completion of this predeposition step, the diffusion step is performed in one of two ways. The wafers may remain in the furnace tube at the same temperature while diffusion takes place. In this case, the surface remains saturated because it is constantly exposed to the dopant gas. In an alternative method, the wafers are removed from the predeposition furnace and placed into another high-temperature tube, where an inert atmosphere of nitrogen or argon is maintained. The diffusion thus takes place under the condition of constant total amount of dopant instead of constant surface concentration.

The technique of *ion implantation* provides an alternative to thermal diffusion as a means of doping selected regions of the wafer. Ion implantation allows both the concentration and the penetration depth of specific dopants to be accurately controlled. The implantation process, which is performed in a high-vacuum system, is depicted schematically in Fig. B.1. A gaseous compound of the dopant is bombarded with high-energy electrons to create a positive ion source. These ions are accelerated by a high-voltage supply and deflected 90° by the field of an electromagnet through a filter aperture. After leaving the aperture, the ions are further accelerated by an electric field to an energy of up to several hundred kilovolts. The ions are then focused into a narrow beam and passed via a beam gate to sets of vertical and horizontal deflection plates. A synchronized set of voltages applied to the plates, similar to those found in a TV picture tube, cause the beam to scan the wafer surface. The wafer is usually covered with a masking layer that permits implantation only over selected areas. Commercial ion implanters are provided with sophisticated wafer-handling mechanisms that permit the processing of several hundred wafers per hour. Because of its precision and cleanliness, ion implantation can also be used to replace the furnace predeposition step that precedes high-temperature thermal doping. Some fabrication processes use ion implantation exclusively.

Ion implantation cannot introduce dopants far below the wafer surface because the beam penetration depth is typically only several tenths of a micrometer. Even with special megavolt-range implanters, it is not possible to form junctions more than several micrometers deep. If greater depth is needed, a high-temperature postimplantation diffusion step must be used to drive the dopant into the wafer. The shallow junctions found in modern VLSI circuits are well within the range of ion implantation, however.

When an accelerated ion enters the wafer, its energy is lost in collisions with the host crystal atoms. As the ion collides, it creates defects and disordered imperfections in the lattice structure of the wafer. A short anneal at low temperature allows the ion-implanted wafer to recrystallize into a near-perfect crystal suitable for electronic devices. No significant diffusion of dopants occurs during this annealing step.

B.5   FILM DEPOSITION

Thin films of various materials are deposited on wafer surfaces to act as temporary masking layers during ion implantation, thermal doping, and other steps in the fabrication process. Thin films can also serve as ohmic contacts, interconnections between contacts, and insulators that separate multilevel interconnections. A specific deposition method is associated with each type of film. *High-vacuum evaporation*, for example, is used to create films of metals such as aluminum and gold. In this process, the metal is heated until it evaporates. The metal vapor then condenses on the cooler wafer surface, forming a thin-film layer. The source of heat may be a resistively heated filament, a graphite crucible heated by induction from an rf (radio-frequency) coil, an electron beam, or a laser beam focused on a target made from the metal to be deposited. *Sputtering* is also used to create films and is useful for practically all materials, including insulators. Sputtering is performed in a vacuum chamber that contains a target made from the material to be deposited. The target is negatively charged to several hundred volts with respect to the wafer. An inert gas such as argon is introduced and ionized. The positive ions bombard the target, dislodging atoms of the target material. These dislodged atoms migrate toward and adhere to the wafer surface. Because the energy needed to vaporize the target comes from an electric field, the target need not be heated to the high temperature needed for evaporation. Sputtering therefore can be used for low-vapor-pressure refractory metals. It can also be used for insulators such as silicon dioxide.

Chemical vapor deposition (CVD) creates films from the chemical reaction of gases at the wafer surface. The process of epitaxial growth described in Section B.2, for example, is often performed by CVD. Chemical vapor deposition reactions have been developed for all types of films used in IC fabrication.

B.6   WAFER ETCHING

For many years, *wet chemical etching* was the only method for dissolving and removing materials from the wafer surface. Although more modern processes have replaced most etching steps, one common reaction still in use involves the dissolving of silicon dioxide by dilute hydrofluoric acid (HF). The use of this reaction allows areas of the wafer surface to be selectively doped. The process begins by growing an oxide layer over the entire wafer surface, then covering the oxide surface with a masking layer. The masking layer is patterned by lithography and then developed to expose the oxide only over areas of the underlying wafer that are to be doped. The HF solution reacts more rapidly with the exposed SiO_2 than with the masking-layer material, allowing the HF to etch away the oxide layer over the lithographically defined areas. The etching process does not attack unoxidized silicon, so that the etching of the oxide stops at the silicon surface. This process thus leaves the silicon surface exposed over lithographically selected areas; other areas remain covered by the masking layer and oxide. If the wafer is subsequently subjected to dopants, only the exposed areas of the silicon will become doped.

In a similar etching process, aluminum films can be removed over selected areas by a mixture of nitric and phosphoric acids. This process is sometimes used to pattern the metallic conduction lines over interconnected regions of an IC.

Some films are difficult to remove with liquid reagents. The removal of silicon nitride, for example, requires boiling in phosphoric acid. Selective removal with this reagent requires a highly resistant material for the masking layer. The deposition and patterning of these special masking layers seriously complicates the fabrication process.

Another serious limitation of liquid etchants is their isotropic action, which causes etching at comparable rates in both horizontal and vertical directions. The resultant undercutting of the masking layer and widening of etched openings places a minimum limit on device size.

Plasma etching is a useful substitute for liquid etching. In this gaseous process, wafers are placed in a vacuum chamber and reactants are introduced at low pressure in the presence of an rf field. The reactant molecules are dissociated and partially ionized by the field, creating extremely reactive molecular fragments. In one version of the process, called *reactive ion etching*, the ions are accelerated into the wafer surface. This process produces vertical etching, which proceeds at a much faster rate than does horizontal etching, yielding steep-walled openings with the same area as the mask aperture. The anisotropic etching possible with plasma processes permits the fabrication of smaller device sizes on the IC chip.

B.7 LITHOGRAPHIC PROCESSING

The masking layers mentioned in previous sections are made using a photolithographic process. A thin film of a light-sensitive organic material, called a *resist*, is deposited on the wafer surface. In the simplest version of the process, the resist is exposed to ultraviolet light through a photographic plate, or *mask*. If the resist is *positive*, the exposed regions of the film are subsequently dissolved by a developing solution. If the resist is *negative*, the exposed regions are left intact by the developing solution and all other regions are dissolved. In either case, the resist still attached to the wafer after developing serves as a mask for etching the exposed regions of the underlying material or for growing an oxide layer over exposed areas of the wafer. An SiO_2 film patterned in this way, for example, can subsequently be used as a dopant diffusion mask. The resist layer itself cannot be used as a diffusion mask because it cannot withstand the high temperatures required for diffusion.

The first requirement of the lithographic process is high resolution. State-of-the-art manufacturing processes employ feature sizes as small as $0.5\,\mu$m or less. Equally important is the need for overlay accuracy. The fabrication of a finished device or chip can require as many as 16 photolithographic steps. In many of these steps, the mask pattern must be aligned precisely with features that already exist on the wafer. The alignment tolerance is often no more than a few tenths of a micrometer. Each mask contains special alignment marks, usually located at the corners of the chip, that are transferred to the wafer along with the rest of the pattern. Successive masks are aligned to these marks with an extremely sensitive set of manipulators called a *mask aligner*. Mask alignment is performed while observing the position of the mask under a microscope. All photoresist processing is done in a specially designed cleanroom that excludes contaminant dust particles that might fall on critical mask or wafer areas.

Improvements in optical technology have extended the definition capability of ultraviolet lithography to the 1-μm range. For work of still higher resolution, focused electron beams are used instead of ultraviolet light because of the much shorter wavelength of electrons. The improved resolution is obtained at the cost of processing time. An electron-beam-generated pattern requires that the narrow electron beam be scanned over the resist-coated wafer. No masks are needed, however, because the electron beam can be switched on and off under digital control.

B.8 A MOS FABRICATION SEQUENCE

In this section, the sequence of process steps required to fabricate a MOS circuit is described from start to finish. The circuit to be fabricated, part of a larger IC, is shown in Fig. B.2. It consists of an NMOS inverter Q_1 with depletion-mode load Q_2. Although the illustrations will show only these devices, the same process steps are used to create many other similar components on the chip simultaneously.

Figure B.2
Circuit diagram of an NMOS inverter with depletion-mode load.

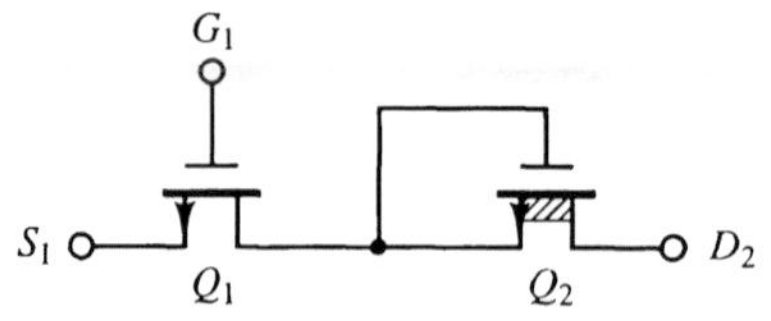

Each of the devices to be illustrated has a length L of 1.5 μm and a width W of 4.5 μm. The other dimensions are chosen to be consistent with the mask-overlay accuracy requirements that are typical of MOS fabrication processes. When a mask step is invoked, it implies a complete lithographic sequence of positive resist application, prebaking, exposure, development, postbaking, and inspection. Also implied are the many cleaning steps used throughout the process.

The wafers used to make this MOS circuit consist of boron-doped p-type silicon with an initial resistivity of about 10 Ω-cm. The major steps are numbered and referenced to Figs. B.3(a) through B.3(h).

Step 1: Oxidize the wafer to form a 10-nm-thick buffer oxide layer.

Step 2: Deposit 0.1 μm of silicon nitride by CVD; cover the wafer with positive photoresist.

Step 3: [***Fig. B.3(a)***] ***Mask 1.*** Define the working area of the wafer where the circuit will reside. Remove the nitride and oxide layers by plasma etching everywhere *except* where transistors will appear. The working area shown in the figure is a 4.5-μm × 11.5-μm rectangle. Positive resist is used, hence the working area is defined by the opaque part of the lithographic mask. This first mask can be placed anywhere on the wafer.

Step 4: Strip the remaining resist by plasma etching in oxygen.

Step 5: Ion implant boron at a dose of 5×10^{12}/cm^2 and an energy of 50 keV. The implantation beam is masked from the working area by the nitride layer defined in step 3.

Step 6: [***Fig. B.3(b)***]. Oxidize the wafer in water vapor at 950°C for 5 h to grow a 0.60-μm field oxide on portions of the wafer not covered by the nitride layer. The oxide penetrates slightly under the edges of the nitride layer.

Step 7: Remove the remaining nitride by plasma etching. Part of the field oxide is also removed because the etching selectivity is not perfect.

Step 8: Grow a 25-nm-thick oxide layer within the active area by exposing to dry oxygen at 900°C for 1 h. Follow with an anneal in argon for 30 min. This thin oxide layer will be used to form the MOSFET gates.

Step 9: Implant 5×10^{11}/cm^2 boron ions into the active area (masked by the field oxide) to adjust the enhancement-mode threshold voltage to 0.7 V. In this step, the entire active area is implanted, hence no photoresist mask is needed.

Step 10: [***Fig. B.3(c)***] ***Mask 2.*** Adjust the threshold voltage of the depletion-mode device. The depletion-mode device will occupy the right-hand side of the working area. A resist layer placed over the enhancement-mode device area masks the implantation of 10^{12}/cm^2 arsenic ions. This implant compensates for the boron that was implanted in the preceding step, and adjusts the depletion-mode threshold voltage to −0.7 V. Mask alignment is not critical for this step.

Step 11: Strip the remaining resist and deposit polycrystalline silicon by CVD.

Step 12: **[Fig. B.3(d)] Mask 3.** Plasma etch to remove the polysilicon deposited in step 11 except under the opaque areas of the mask. This step forms the gates and defines the channel regions of the devices. The mask leaves sufficient polysilicon outside the working area to form electrical contacts with the metal layer to follow in step 17. The mask alignment in this step must ensure overlap of both sides of the active area by the polysilicon. This step also determines the channel length for each device in the circuit. For the devices shown, the gate length is $1.5\,\mu$m.

Step 13: **[also Fig. B.3(d)].** Implant $10^{16}/$cm^2 arsenic ions at $100\,$keV to form the n^+ sources and drains. The polysilicon patterned in step 12 serves as the mask for this step by protecting the two channels beneath the gates from the arsenic. The polysilicon becomes heavily doped at the same time. Note that the MOS devices are geometrically symmetric; hence a given n^+ layer can serve as either a drain or a source contact.

Step 14: Heat the wafer in nitrogen at $950°$C for 30 min and then in oxygen at $950°$C for 15 min. This combined treatment anneals the implants, diffuses the junctions to a depth of about $0.3\,\mu$m, and grows an oxide layer about $0.1\,\mu$m thick over the drain, source, and gate.

Step 15: Deposit silicon dioxide by CVD to a thickness of $0.35\,\mu$m.

Step 16: **[Fig. B.3(e)] Mask 4.** Define regions for electrical contact to the drain, source, and gate. Open windows to sources, drains, and gates in the 0.1-μm oxide layer by plasma etching. Alignment in this step is critical. Because the source and drain contacts must not overlap the gate, the spacing between them is $0.5\,\mu$m to allow for some misalignment of the mask.

Step 17: Deposit the first-level aluminum layer by sputtering. This metal layer makes contact with the drain, source, and gate regions through the contact windows formed in step 16.

Step 18: **[Fig. B.3(f)] Mask 5.** Define the detail of the first metal layer. Pattern the aluminum by plasma etching. This step forms the leads to the drain, source, and gate, to the interconnection lines that lead to other points in the circuit. As required of the circuit of Fig. B.2, the source of Q_2 is connected to the gate of Q_2 just outside the active device area, on top of the thick isolation oxide layer.

Step 19: Anneal in hydrogen at $470°$C for 15 min to form ohmic contacts where the aluminum meets the n^+ silicon regions.

Step 20: Deposit an SiO$_2$ interlevel dielectric by low-temperature CVD, possibly with plasma enhancement.

Step 21: **[Fig. B.3(g)] Mask 6.** Define the locations of *contact vias* by plasma etching. A contact via is a window in the interlevel dielectric layer that provides a site where the levels of metal can be interconnected. In the circuit shown, vias are provided at S_1 and D_2.

Step 22: Deposit the second-level aluminum layer by sputtering.

Step 23: **[Fig. B.3(h)] Mask 7.** Define the detail of the second metal layer. Pattern the second level of aluminum by plasma etching. This level usually contains the wider metallization lines that are used for power-supply buses.

Step 24: Deposit silicon dioxide or silicon nitride by low-temperature CVD, possibly with plasma enhancement. This layer serves as a protective cover for the entire chip. The only openings in it will lie above the bonding pads at the chip edges, where wires will be attached.

Step 25: **Mask 8.** Passivation. Expose the bonding pads by plasma etching. This step is not shown in Fig. B.3.

An electron micrograph of an NMOS circuit based on the inverter of Figs. B.2 and B.3 is shown in Fig. B.4. The portion of the circuit shown contains about 40 transistors. The horizontal and vertical lines between devices are metal interconnect paths separated from the underlying wafer surface by an oxide layer.

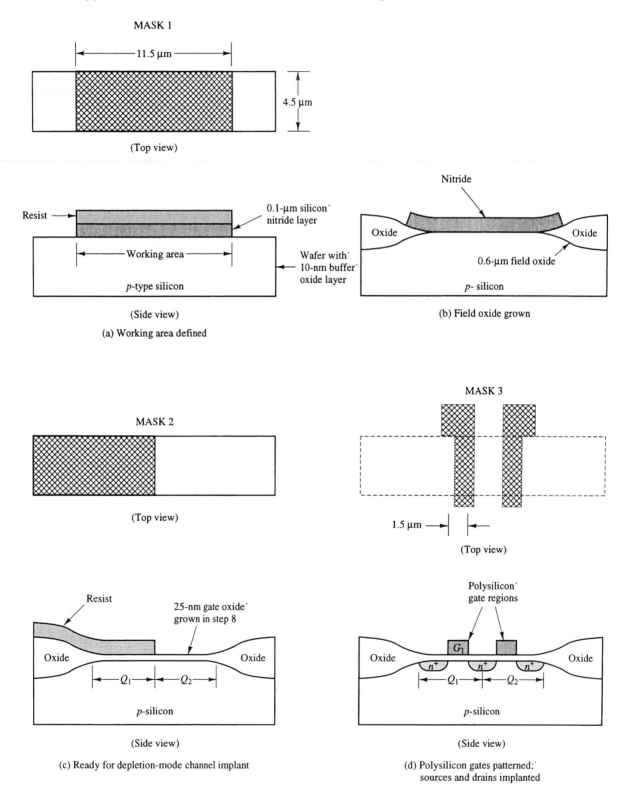

Figure B.3 Fabrication sequence for an enhancement-depletion MOS process. Vertical dimensions are not to scale.

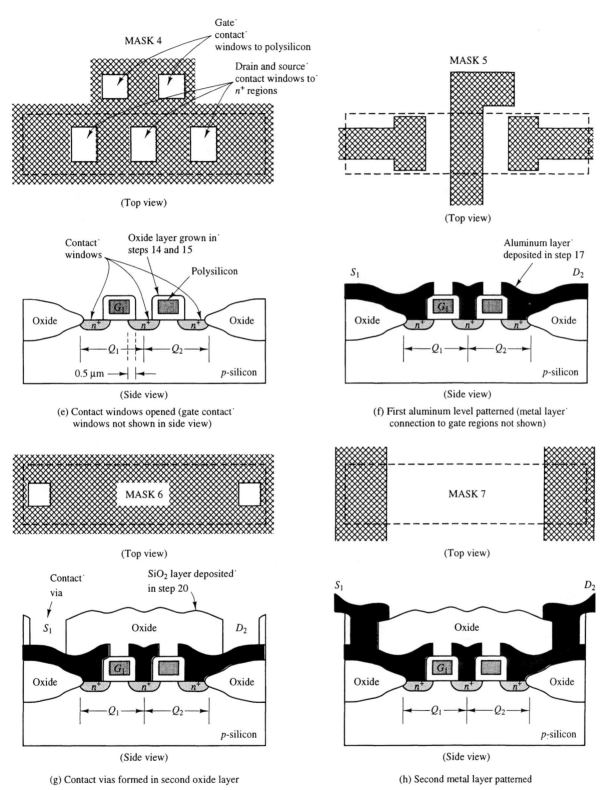

(e) Contact windows opened (gate contact windows not shown in side view)

(f) First aluminum level patterned (metal layer connection to gate regions not shown)

(g) Contact vias formed in second oxide layer

(h) Second metal layer patterned

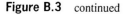

Figure B.3   continued

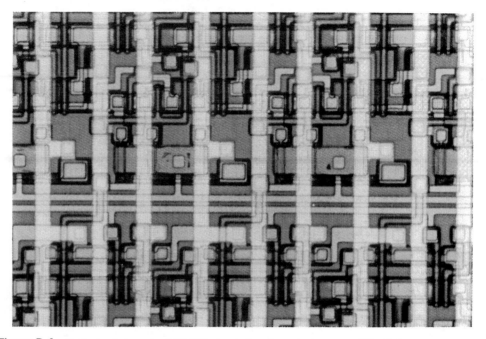

Figure B.4 Optical micrograph of NMOS circuit based on the inverter of Fig. B.2. Courtesy of GTE Laboratories.

B.9 A BJT FABRICATION SEQUENCE

The following illustration of an *npn* bipolar fabrication sequence employs positive resist and uses oxide layers for isolation between devices. The principal difference between the bipolar process described here and the NMOS process of the preceding section is the orientation of the fabricated devices. MOS devices are build horizontally on the chip with the critical length and width parameters determined by mask dimensions. Conversely, bipolar devices have a vertical structure because the emitter, base, and collector regions are produced by diffusion into the water surface. The critical base-width dimension is determined by the relative depths of diffused or implanted junctions. The MOS device is symmetrical; either doped region can serve as the source or the drain without a change in device characteristics. The bipolar transistor is not symmetrical because the emitter is doped much more heavily than the collector and has a much smaller area.

The illustrations of Fig. B.5 show only one transistor but are meant to represent the processes used to fabricate an entire bipolar IC. The starting material is lightly doped *p*-type silicon ($10^{15}/cm^3$ acceptor atoms).

Step 1:   Grow an oxide layer to be used in the masking step to follow.

Step 2:   [*Fig. B.5(a)*] *Mask 1.*   Create a heavily doped n^+ region by ion implantation of antimony. This region will serve as a *buried layer* that defines the transistor area and reduces the series resistance in the collector. Slowly diffusing antimony is used so that its profile will not be disturbed by the high temperature required in the next step. The mask is opaque everywhere *outside* the region enclosed by the dashed line.

Step 3: Strip the resist and drive the implanted layer into the substrate by heating in oxygen. A discontinuity forms in the wafer surface because the heavily doped region oxidizes more quickly. This step also serves as a convenient alignment pattern for the masks that follow.

Step 4: Strip the oxide and grow a 1- to 2-μm-thick epitaxial n-type layer with $N_D = 10^{17}/\text{cm}^3$.

Step 5: Oxidize to form a buffer oxide layer about 10 nm thick; deposit 0.1 μm of silicon nitride by CVD. This latter deposited layer will become the selective oxidation mask.

Step 6: ***[Figs. B.5(b) and B.5(c)] Mask 2.*** Define the working portions of the transistor. Remove the nitride and oxide layers, plus about half the thickness of the epitaxial silicon layer, everywhere outside the mask windows. The windows define the areas where the emitter, base, and collector contacts will emerge from the surface of the wafer.

Step 7: Implant boron to form a p^+ layer over the regions etched in step 6. This p^+ layer forms *channel stops*, or isolation zones, between adjacent transistors. The locations of the p^+ channel stops are shown in Fig. B.5(d).

Step 8: Strip the resist and oxidize. The remaining epitaxial layer in the exposed regions is converted to oxide, creating isolated islands where the layer was not etched away in step 6. The working regions of the transistor will occupy these isolated islands.

Step 9: Etch the nitride and buffer oxide layers and grow a thin oxide layer on top of the exposed islands.

Step 10: ***[Fig. B.5(d)] Mask 3.*** Define the base region. Protect all but the region where the p-type base implantation will be performed. As before, the mask is opaque everywhere outside the region enclosed by the dashed line. Implant boron through the existing layer to turn the upper half of the n-type epitaxial layer into p-type material. The remaining lower half of the n-type epitaxial layer will serve as the active collector region. The dose is chosen so that after diffusion, the boron concentration will be about $2 \times 10^{17}/\text{cm}^3$ in the region between emitter and collector, but greater than $10^{18}/\text{cm}^3$ at the surface where the contact will be formed. At this point in the process, resistors are made in other areas of the chip using p-type material that has received the same implantation as the transistor bases.

Step 11: ***[Fig. B.5(e)] Mask 4.*** Define the contact regions. Remove the oxide layer grown in step 9 from the areas where the emitter, base, and collector contacts will be made.

Step 12: ***[Fig. B.5(f)] Mask 5.*** Define the emitter and collector regions. Expose the emitter and collector; implant the arsenic at high dose and low energy to form n^+ layers just beneath the surface of the epitaxial layer. The collector is doped with this n^+ concentration to prepare the surface for ohmic contact formation.

Step 13: Drive the emitter to the desired depth in an inert atmosphere. This step is a critical one. If the drive is too deep, the emitter will extend all the way to the n-type collector region beneath the base. If the drive is too shallow, the base width will be reduced and the transistor will have a small beta.

Step 14: Deposit aluminum over the entire chip by sputtering.

Step 15: ***[Fig. B.5(g)] Mask 6.*** Define the details of the aluminum layer. Plasma etch the aluminum to form contact regions and interconnection lines. This figure shows the finished *npn* transistor.

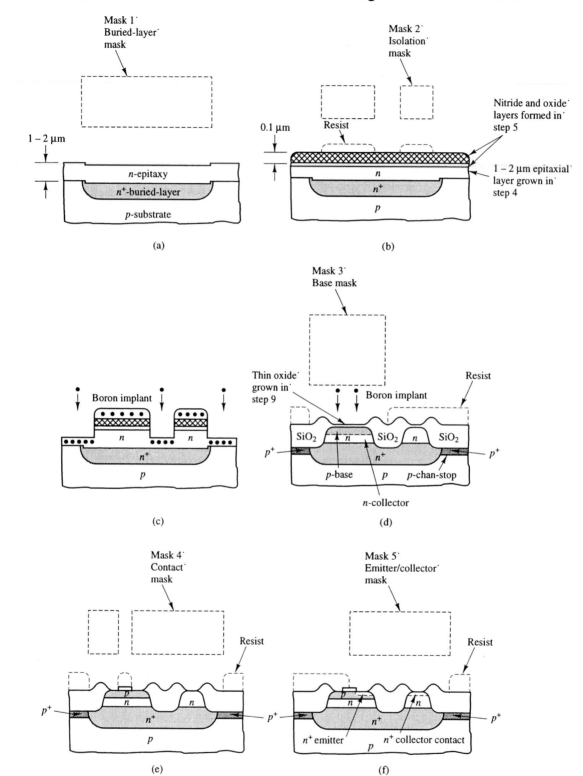

Figure B.5 Fabrication sequence for an oxide-isolated bipolar process. (From S.M. Sze, Ed., *VLSI Technology*, New York: McGraw-Hill Inc., 1983. Reprinted by permission.)

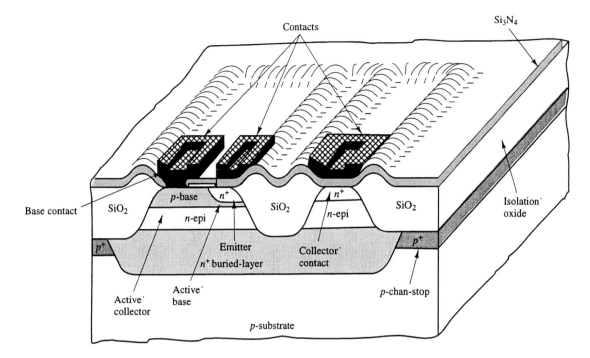

Three-dimensional view of oxide-isolated bipolar transistor

(g)

Figure B.5   continued

Computer-Aided Circuit Design Using SPICE and PSpice®

A t the time of the writing of this book, several computer software packages exist that enhance the electronic design process. Of these software packages, SPICE is by far the most widely used and universally accepted. The SPICE program, short for *Simulation Program* with *Integrated Circuit Emphasis*, was first developed at the University of California at Berkeley in the mid–1970s. The software has since become public domain, and versions of SPICE reside on computers and workstations at universities, companies, and institutions worldwide. Several upgrades have been added to the original program, and a popular version called PSPICE[1] has been written for desktop personal computers. Other electronic simulation software may appear in the future, but at the present time, SPICE is the generally accepted industry standard.

C.1 USE OF SPICE

SPICE should be used to fine tune the design process, not to replace it. Although approximate "first-cut" circuit designs can often be made by hand, an exact analysis of circuit behavior is sometimes required. A complicated IC design, for example, must be perfected before it is actually fabricated, since refabrication to accommodate even minor design changes is costly. In such situations, SPICE can provide valuable assistance in testing a tentative design before it is actually fabricated.

What follows is a simplified description of SPICE designed to help the reader get started using the program. The reader is encouraged to consult one or more of the references listed in Section C.6 to obtain a full explanation of the program's many features.

---

[1] PSPICE is a registered trademark of MicroSim Corporation.

C.2   CAPABILITIES OF SPICE AND PSPICE

SPICE can model independent voltage and current sources, resistors, capacitors, inductors, transformers, transmission lines, nonlinear magnetics, dependent sources, BJTs, MOSFETs, JFETs, MESFETs, and diodes. A means also exists for explicitly describing to SPICE the v–i characteristics of any arbitrary nonlinear element. SPICE permits the indirect modeling of operational amplifiers via the use of dependent voltage sources.

SPICE can analyze the transient, small-signal, or steady-state frequency response of a circuit. It can also provide the dc transfer characteristic of any output variable versus any input variable. SPICE simulations can be performed at different operating temperatures, and thermal component noise can be included, if desired. Output is provided in various forms, including tabular listings, two-dimensional plots and graphs, and Bode plots. The program PSPICE includes a graphical output interface called Probe and a library of devices with predetermined characteristics called Parts.

Most versions of SPICE are not interactive. The input file is edited, the SPICE simulation is performed, and the output is stored in an output data file for later inspection by the user. This mode of computer interaction is called the *batch* mode. Some of the more recent versions of SPICE allow *real-time* interaction and graphical entry of data. The software package MICRO-CAP[2] uses the SPICE computational algorithms but presents a comprehensive graphical user interface for entering data and viewing output.

C.3   CIRCUIT DESCRIPTION

In batch-mode versions of SPICE, the circuit is described by a set of program statements stored in a data file labeled FILENAME.CIR, where FILENAME is a name chosen by the user. This data file is retrieved by SPICE during program execution. Each line of the input file must contain one program statement only. A single program statement can be continued onto additional lines by entering a "+" (plus) sign in the first column of the additional lines. The typical input file consists of a *title line* (mandatory), a set of *element statements* that describes the circuit, and a set of *control statements* that instructs SPICE during program execution. Control statements always begin with a period. Except for the title line, which must always be first, element and control lines may appear in any order. The entire input file must be terminated by an .END control statement. Any blank lines that follow the .END statement are read as input data and may generate an error message. Input lines in which the first character is an asterisk (*) are interpreted as comment lines, not as input data. Similarly, all characters following a semicolon (;) on a given line are treated as a comment. The wise programmer sprinkles comments liberally throughout the input file.

The first step in preparing a suitable input file is the numbering of all the nodes in the circuit. All nodes must be numbered, even if not of interest to the user. One node must be designated as the ground node and assigned the number 0. The assignment of other node numbers is arbitrary. Each node in the circuit must have a dc path to ground. An isolated branch of two capacitors in series, for example, is not permitted in SPICE and will generate an error message.

---

[2]   MICRO-CAP is a trademark of Spectrum Software, Inc.

C.3.1 Resistors, Capacitors, and Inductors

Each resistor, capacitor, and inductor in the circuit is described to SPICE by an element statement in the form

$$\text{NAME} \quad \text{NODE A} \quad \text{NODE B} \quad \text{VALUE} \tag{C.1}$$

where each entry is separated by one or more spaces. The first letter of NAME identifies the type of component, with R = resistor, C = capacitor, and L = inductor. The remaining characters of NAME are arbitrary but must uniquely identify the circuit element. The node designations A and B specify how the component is connected to the rest of the circuit. The value of the component is specified by the VALUE entry, which must be a number written in integer, decimal, or scientific format. The suffixes listed in Table C.1 are recognized by SPICE and can be used in specifying VALUE. The value entries 1000, 0.001MEG, and 1.0E3, for example, all signify a component value of 1000.

Abbreviation	Meaning	Implied Multiplier
F	femto	10^{-15}
P	pico	10^{-12}
N	nano	10^{-9}
U	micro	10^{-6}
M	milli	10^{-3}
K	kilo	10^{3}
MEG	mega	10^{6}
G	giga	10^{9}
T	tera	10^{12}

Table C.1. Multiplying Suffixes Recognized by SPICE

An initial voltage condition can be specified for a capacitor by adding the entry IC = *voltage value* after the VALUE parameter. Similarly, an initial current condition can be specified for an inductor by adding IC = *current value*. When an initial condition is specified, the positive node of the element must be listed first. Thus the statement

$$\text{CS} \quad 3 \quad 4 \quad 1.0\text{U} \quad \text{IC} = 10 \tag{C.2}$$

specifies a 1.0-μF capacitor CS precharged to 10 V with the positive terminal connected to node 3 and the negative terminal to node 4.

C.3.2 Independent Sources

The element statement for an independent voltage or current source is written in the form

$$\text{NAME} \quad \text{POS-NODE} \quad \text{NEG- NODE} \quad \text{TYPE} \quad \text{SPECIFICATION} \tag{C.3}$$

The first letter of NAME identifies the source as an independent voltage source (V) or current source (I). The remaining characters of NAME are arbitrary. The positive and negative terminals of the source are specified by POS-NODE and NEG-NODE, respectively. The current through an independent source in SPICE is defined as positive *into* the positive terminal. Thus, when a source supplies power to the circuit, its voltage–current product is negative. The current supplied by a positive current source flows *out* of the terminal designated NEG-NODE.

Figure C.1
Explanation of
SPECIFICATION
parameters for SIN
(sinusoidal) source.

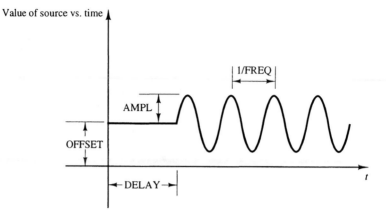

The TYPE entry in (C.3) specifies the general nature of the source, with details included in the SPECIFICATION entry. Five commonly used source types and their SPECIFICATION formats are shown in Table C.2. As the table shows, the information included in the SPECIFICATION entry depends on the type of source. The SPECIFICATION entry for DC and AC sources follows the TYPE entry with spaces and no parentheses. The data for SIN, PULSE, and PWL sources are included in parentheses with no space after the TYPE entry; entries within the parentheses are separated by spaces. The SIN, PULSE, and PWL sources are used exclusively for transient analyses that yield circuit response as a function of time. The significance of the SPECIFICATION parameters for SIN, PULSE, and PWL source types is summarized in Figs. C.1, C.2, and C.3.

Type	Specification Parameters	Description	Type of Analysis in Which Used
DC	Voltage or current value	Fixed dc source	Operating-point analysis; dc transfer characteristic
AC	MAG PHASE	Fixed ac source	Sinusoidal steady-state frequency response
SIN	(OFFSET AMPL FREQ DELAY)	Transient sinusoidal source	Transient analyses involving ac sources
PULSE	(V1 V2 DELAY TRISE TFALL DUR PERIOD)	Periodic pulse train; if PERIOD is omitted, a single pulse is generated	Transient analyses
PWL	(TI VAL1 T2 VAL2 T3 VAL3 ...)	Piecewise linear source; varies linearly from VAL_n to VAL_{n+1} over the time interval t_n to t_{n+1}	Transient analyses

Table C.2. Independent Sources Recognized by SPICE

Figure C.2
Explanation of
SPECIFICATION
parameters for a
PULSE source.

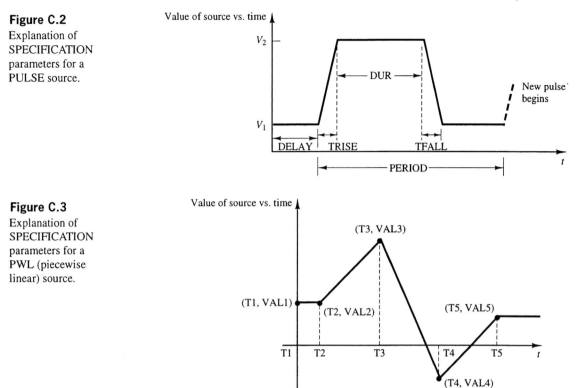

Figure C.3
Explanation of
SPECIFICATION
parameters for a
PWL (piecewise
linear) source.

C.3.3 Dependent Sources

The element statement for a linear dependent source has the form

$$\text{NAME} \quad \text{POS-NODE} \quad \text{NEG- NODE} \quad \textit{control} \quad \text{COEFFICIENT} \qquad \textbf{(C.4)}$$

The first letter of NAME specifies the type of dependent source according to the designations shown in Table C.3. The NAME parameter is followed by the POS-NODE and NEG-NODE entries. The *control* information identifies the voltage or current on which the source depends.

Source Type	First Letter of NAME	POS-NODE	NEG-NODE	Units of the Coefficient Specified by VALUE
Voltage-controlled voltage source	E	Positive terminal	Negative terminal	V/V
Current-controlled current source	F	Current in	Current out	A/A
Voltage-controlled current source	G	Current in	Current out	A/V
Current-controlled voltage source	H	Positive terminal	Negative terminal	V/A

Table C.3. Dependent-Source Specifications for SPICE

Element Statement	Meaning
EPHI 1 2 5 6 0.01	Dependent voltage source connected between nodes 1 (positive terminal) and 2 (negative terminal). Output voltage is equal to 0.01 times the voltage between nodes 5 and 6.
FX 8 7 VIN 50	Dependent current source connected between node 8 (current in) and node 7 (current out). Output current is equal to 50 times the current flowing into the positive terminal of the voltage source labeled VIN.

Table C.4. Examples of Dependent-Source Statements

For a source that depends on a voltage, *control* consists of two additional node designations. The output of the dependent source is set by SPICE to the voltage between these additional nodes (positive node listed first) multiplied by COEFFICIENT. For a source that depends on a current, *control* consists of the name of a voltage source located somewhere in the circuit. The output of the dependent source is set to the current flowing through the named source multiplied by COEFFICIENT. If a voltage source does not exist in the branch of the circuit through which the controlling current flows, an independent voltage source of zero value must be inserted into the branch so that the controlling current can be monitored.[3] Examples of correctly specified element statements for dependent sources are given in Table C.4.

C.3.4 The .MODEL Statement

The SPICE program models its semiconductors, as well as other elements in the circuit, using default values for all pertinent coefficients and parameters. To provide maximum flexibility, SPICE allows these parameters to be changed for any device in the circuit. The diode model

DEVICE-TYPE	Device Modeled
D	Diode
NPN	*npn* bipolar transistor
PNP	*pnp* bipolar transistor
NMOS	*n*-Channel MOSFET
PMOS	*p*-Channel MOSFET
NJF	*n*-Channel JFET
PJF	*p*-Channel JFET
GASFET	*n*-Channel GaAs MESFET
RES	Resistor
CAP	Capacitor
IND	Inductor

Table C.5. Semiconductor Device Designations for the DEVICE-TYPE Entry in .MODEL Statements

---

[3] Some newer versions of SPICE permit the direct specification of a branch current without reference to a voltage source.

Device Modeled	Parameter Label	Parameter Description	Default Value
Diode	IS	Saturation current	10^{-14} A
	RS	Series ohmic resistance	0
	N	Emission coefficient (η)	1
	CJO	Junction capacitance at $v_D = 0$	0
	VJ	Built-in voltage (Ψ)	1 V
	BV	Reverse-breakdown voltage	∞
	IBV	Current at $v_D = $ BV	10^{-10} A
Bipolar junction transistor	BF	Forward beta (β_F)	100
	NF	Forward emission coefficient(η)	1
	VAF	Forward Early voltage	∞
	BR	Reverse beta (β_R)	1
	NR	Reverse emission coefficient (η_R)	1
	VAR	Reverse Early voltage	∞
	CJE	Base–emitter depletion capacitance	0
	CJC	Base–collector depletion capacitance	0
MOSFET	VTO	Threshold voltage	0
	KP	Conductance parameter (K)	0.02 mA/V^2
	CBD	Drain-to-substrate capacitance	0
	CBS	Source-to-substrate capacitance	0
	TOX	Oxide-layer thickness (t_{ox})	$10^{-7} m$
	CGDO	Gate-to-drain capacitance per unit channel width W	0
	CGSO	Gate-to-source capacitance per unit channel width W	0
	CGBO	Gate-to-substrate capacitance per unit channel length L	0
JFET	VTO	Threshold (pinch-off) voltage	-2 V
	BETA	Conductance parameter (K)	0.1 mA/V^2
	LAMBDA	Channel-length modulation parameter	0
	RD	Ohmic drain resistance	0
	RS	Ohmic source resistance	0
	CGS	Gate-to-source capacitance	0
	CGD	Gate-to-drain capacitance	0
GaAs MESFET	VTO	Threshold (pinch-off) voltage	-2.5 V
	BETA	Transconductance parameter (K)	0.1
	IS	Gate *pn* saturation current	1e $-$ 14
	(LAMBDA, RD, RS, CGS, CGD are the same as in the JFET)		
Resistor	R	Resistance multiplier	1
Capacitor	C	Capacitance multiplier	1
Inductor	L	Inductance multiplier	1

Table C.6. Partial Listing of Assignable Parameters for .MODEL Statements

within SPICE, for example, utilizes 14 parameters, all of which can be adjusted by the user. Similarly, the BJT model uses 40 parameters, the MOSFET model up to 48 parameters, and the JFET model 14 parameters. To avoid the necessity of adjusting numerous model parameter values for similar devices, SPICE makes use of a construct called the .MODEL statement. Each .MODEL statement in the input file describes a uniquely defined device that can be referenced by any number of element statements. A .MODEL statement must be written in the following format:

$$\text{.MODEL MODELNAME DEVICE-TYPE}(parameter = value, parameter = value, \dots)$$
$$(\text{C.5})$$

At the very least, the MODELNAME and DEVICE-TYPE entries must be present. The label MODELNAME identifies the .MODEL statement and must be included in any element statement that references the device described. The entry DEVICE-TYPE can be one of approximately 30 device types, including the designations listed in Table C.5. The parameter value declarations that follow DEVICE-TYPE are optional. Only those parameter values to be reset need be listed; unlisted parameters retain their default values. A partial listing of some of the more important device parameters and their default values are included in Table C.6. In the case of resistors, capacitors, and inductors, the element value is set to the number specified in the element statement multiplied by the multiplier parameter (R, L, or C) specified in the .MODEL statement. Note that these linear elements may, and usually are, specified in SPICE without the use of a .MODEL statement; the specification of a semiconductor device in SPICE *requires* the use of a .MODEL statement. A complete listing of all the assignable device parameters can be found in the references of Section C.6.

C.3.5 Semiconductors

The element statement for a semiconductor is written in the form

$$\text{NAME NODE A NODE B NODE C } \dots \text{ MODELNAME AREA} \qquad (\text{C.6})$$

The first letter of NAME specifies the type of device according to the designations shown in Table C.7. The number of nodes listed in the element statement depends on the type of semiconductor. A diode, for example, has two terminals; hence its element statement contains two node connections. In all cases, the terminals of the semiconductor must be listed in the order indicated in Table C.7. SPICE allows an optional substrate node connection in the BJT and MOSFET element statements. If this fourth node is omitted, the substrate is assumed to be connected to ground (node 0).

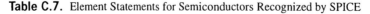

DEVICE	First Letter of NAME	NODE A	NODE B	NODE C	NODE D
Diode	D	p-Side	n-Side		
BJT	Q	Collector	Base	Emitter	(Substrate)
MOSFET	M	Drain	Gate	Source	(Substrate)
JFET	J	Drain	Gate	Source	
MESFET	B*	Drain	Gate	Source	

* Z in SPICE3

Table C.7. Element Statements for Semiconductors Recognized by SPICE

The MODELNAME in the element statement (C.6) must correspond to the label of an appropriate .MODEL statement. The class of device designated by NAME must match the class of device described by the .MODEL statement. The AREA parameter indicates how many of the devices described by the .MODEL statement are connected in parallel to make up the named device. If AREA is omitted, it is set to a default value of 1. The use of the AREA parameter allows devices with differing geometries, but otherwise identical parameters, to be described by the same .MODEL statement. Note that the AREA parameter for a MOSFET is broken up into separate length and width specifications and is written in the form

$$W = value \qquad L = value \tag{C.7}$$

where the channel width W and the channel length L are in meters. If W and L are omitted from the element statement, they are assigned default values of 10^{-4} m.

Note that SPICE computes the current through a MOSFET using the transconductance parameter $K = \frac{1}{2}KP(W/L)$, where $KP = \mu\varepsilon_{ox}/t_{ox}$ is the parameter that appears in the .MODEL statement (or otherwise has a default value of 0.02 mA/V^2). The value of KP must therefore be set to *twice* the K value desired of a MOSFET with $W = L$.

C.3.6 Modeling Op-Amps in SPICE

A single op-amp can be modeled in SPICE by detailing every component of its internal circuitry. This approach becomes cumbersome in a circuit with several op-amps, and is often unnecessary. A simpler approach involves the use of a voltage-controlled voltage source to model the op-amp, as shown in Fig. C.4. This model includes the op-amp's input resistance r_{in} and output resistance r_{out}. For an op-amp with an open-loop gain of 10^6 and the node numbering shown in the figure, an appropriate element statement for the dependent source EOUT would be

$$EOUT \quad 3 \quad 0 \quad 1 \quad 2 \quad 1MEG \tag{C.8}$$

where the voltage V(1,2) is equivalent to $(v_+ - v_-)$.

Figure C.4
Op-amp model for SPICE.

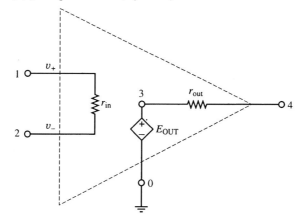

If the op-amp's external feedback network provides a dc path to ground for the v_+ and v_- nodes, it is possible to set r_{in} to an open circuit. Under most conditions, it is also possible to set r_{out} to a short circuit. These modifications allow the model of Fig. C.4 to approach the ideal op-amp approximation.

The linear op-amp model of Fig. C.4 does not account for nonideal op-amp behavior. Non-ideal behavior, including output saturation, input offset voltage, input bias current, finite slew rate, and finite frequency response, can be accommodated by adding additional components to the op-amp model. These modifications are explored in end-of-chapter computer-aided design problems.

C.3.7 Arbitrary Nonlinear Devices (Polynomial Sources)

A nonlinear device whose v–i characteristics can be expressed (or approximated) by a polynomial of the form

$$i = a_0 + a_1 v + a_2 v^2 + a_3 v^3 + \ldots + a_n v^n \tag{C.9}$$

or

$$v = b_0 + b_1 i + b_2 i^2 + b_3 i^3 + \ldots + b_n i^n \tag{C.10}$$

can be modeled in SPICE using the *polynomial source* representation. Polynomial sources form a general set of devices in SPICE, of which the linear dependent sources of Section C.3.3 are a subset.

The element statement for a nonlinear voltage-controlled current source is written in the form

$$\text{Gxxxx} \quad \text{POS-NODE1} \quad \text{NEG-NODE1} \quad \text{POLY(1)} \quad \text{POS-NODE2, NEG-NODE2} \atop \text{A0} \quad \text{A1} \quad \text{A2} \ldots \text{AN} \tag{C.11}$$

where the source name Gxxxx must begin with the letter G. POS-NODE1 and NEG-NODE1 identify the source connections; POLY(1) signifies that the nonlinear source is a function of one controlling voltage only. POS-NODE2 and NEG-NODE2 (separated by a comma) identify the controlling voltage. The coefficients A0 ... AN correspond to the polynomial coefficients $a_0 \ldots a_n$ in Eq. (C.9). If POS-NODE1 = POS-NODE2 and NEG-NODE1 = NEG-NODE2, the current through the source depends on its own terminal voltage, and the source becomes a passive nonlinear device. Note that SPICE also permits the specification of a nonlinear source that is a function of n variables. Such a source must include the parameter POLY(n) in place of POLY(1). Details concerning the use of *n-dimensional* sources may be found in the references cited in Section C.6.

The element statement for a nonlinear current-controlled voltage source is written in the form

$$\text{Hxxxx} \quad \text{POS-NODE1} \quad \text{NEG-NODE1} \quad \text{POLY(1)} \quad \text{VX} \quad \text{B0} \quad \text{B1} \quad \text{B2} \ldots \text{BN} \tag{C.12}$$

This element statement is similar in form to (C.11). In this case, VX identifies the voltage source through which the controlling current flows. The coefficients B0 ... BN correspond to the polynomial coefficients $b_0 \ldots b_n$ in Eq. (C.10).

C.3.8 Subcircuit Definitions

SPICE permits a collection of elements used repeatedly in a circuit to be described by a subcircuit definition. A set of elements defined in subcircuit format can be called repeatedly by the main program listing, thus saving numerous program lines. A subcircuit definition is written in the form

$$\begin{array}{l} \text{.SUBCKT} \quad \text{SUBNAME} \quad \text{NODE X} \quad \text{NODE Y} \ldots \text{NODE N} \\ \textit{element statements that} \\ \textit{define the subcircuit module} \\ \text{.ENDS} \quad \text{SUBNAME} \end{array} \tag{C.13}$$

More than one subcircuit definition can be included in the SPICE input file. Each one must be terminated by a separate .ENDS statement and have a unique SUBNAME. Subcircuit definitions are typically included after the main program listing, but before the final .END statement. Note that the .END statement is always the last statement in the SPICE input file. A subcircuit definition many not contain control lines.

Any nodes contained in the element statements of the subcircuit module are local to the subcircuit definition and do not correspond to node numbers in the main circuit. The one exception to this rule is node 0 (ground), which is assumed to be common to all subcircuits and the main circuit. The nodes listed in the .SUBCKT statement specify which nodes of the subcircuit connect to the "outside world".

A subcircuit is used by the main circuit description via an element statement of the form

$$\text{XNAME} \quad \text{NODE A} \quad \text{NODE B} \quad \dots \quad \text{NODE N} \quad \text{SUBNAME} \qquad \textbf{(C.14)}$$

where XNAME must begin with the letter X. The SUBNAME entry of the element statement (C.14) must match the SUBNAME label of a .SUBCKT statement. The nodes listed in the element statement (C.14) indicate the numbered nodes of the main circuit to which the subcircuit is connected. The connections are made to the nodes of the subcircuit in the order specified in the .SUBCKT statement, that is, NODE X of the subcircuit is connected to NODE A of the main circuit, NODE Y to NODE B, and so forth.

C.4 TYPES OF ANALYSES

The four principal analyses performed by SPICE are invoked by separate control statements. Other types of analyses are possible, as discussed in this section and in the references of Section C.6.

C.4.1 .DC (Large-Signal Transfer Characteristic)

The .DC control statement instructs SPICE to compute the circuit's large-signal dc transfer characteristic. This statement is written in the form

$$\text{.DC} \quad \text{SOURCE} \quad \text{START} \quad \text{STOP} \quad \text{INCR} \qquad \textbf{(C.15)}$$

where SOURCE is a designated independent voltage or current source that serves as the input to the circuit. The transfer characteristic of one or more output variables is computed by incrementing SOURCE from START to STOP in steps of INCR. the statement

$$\text{.DC} \quad \text{VIN} \quad -5 \quad 5 \quad 0.1 \qquad \textbf{(C.16)}$$

for example, instructs SPICE to compute the dc transfer characteristic by incrementing VIN from -5 V to $+5$ V in equal steps of 0.1 V. The output variable(s) of the circuit are specified using one of the output statements described in Section C.5. Note that if a dc voltage source is to be incremented using the .DC control function, the VALUE entry in the source's element statement can be omitted or set to zero if no other circuit analyses are to be performed.

C.4.2 .AC (Sinusoidal Steady-State Frequency Response)

If an .AC control statement is included in the input file, SPICE first computes the dc operating point of all devices, then substitutes small-signal linear models for all nonlinear devices. The sinusoidal steady-state response of the circuit is then computed by simultaneously changing the frequency of all independent sources that have been designated as AC in their element statements. The results can be printed in tabular form or plotted in Bode plot form using one of the output statements described in Section C.5. The .AC control statement is written in the form

$$\text{.AC} \quad \text{SCALE-TYPE} \quad \text{NPOINTS} \quad \text{FSTART} \quad \text{FSTOP} \tag{C.17}$$

The parameter SCALE-TYPE must be set to LIN, DEC, or OCT, indicating that the frequency scale is to be expressed in linear increments, logarithmic increments, or octaves of frequency. For a LIN scale, NPOINTS specifies the number of frequencies between FSTART and FSTOP at which the response is computed. If a DEC or OCT scale is chosen, NPOINTS defines the number of points per decade or octave. The statement

$$\text{.AC} \quad \text{DEC} \quad 50 \quad 1\text{K} \quad 1\text{MEG} \tag{C.18}$$

for example, causes SPICE to perform a sinusoidal steady-state analysis over three decades of frequency from 1 kHz to 1 MHz, computing the response at 50 points within each decade. If analysis at a single frequency is desired, NPOINTS should be set to 1 and FSTART and FSTOP should be set to the same value.

C.4.3 .TRAN (Transient Analysis)

The .TRAN control statement instructs SPICE to compute the output of the circuit as a function of time in response to sources designated as SIN, PULSE, or PWL.[4] The .TRAN statement is written in the form

$$\text{.TRAN} \quad \text{TSTEP} \quad \text{TSTOP} \quad \text{TSTART} \quad \text{TINCR} \quad \text{UIC} \tag{C.19}$$

The last three entries are optional. The analysis begins at $t = 0$ and ends at $t = $ TSTOP, where times are specified in seconds. The output is printed or plotted after every time increment TSTEP; TSTART specifies the time at which output begins. If TSTART is nonzero, the analysis begins at $t = 0$ but the output does not begin until TSTART. If TSTART is omitted, its value defaults to zero. Computations are performed using a time increment no larger than TINCR. If TINCR is omitted, its value is set to (TSTOP − TSTART)/50.

The transient analysis is performed by first computing the dc operating point of all elements in the circuit while considering capacitors to be open circuits and inductors to be short circuits. If the entry UIC (*User Initial Conditions*) is added, the initial conditions specified in capacitor and inductor element statements will be used instead. An .IC control line can also be used to set an initial condition for any node in the circuit (see Section C.4.6).

C.4.4 .TF (Small-Signal Transfer Function)

If the .TF control statement is included in the input file, SPICE will first compute the dc operating point of all elements in the circuit, then perform a small-signal analysis using small-signal linear models for all nonlinear devices in the circuit. The .TF statement is written in the form

$$\text{.TF} \quad \text{OUTPUT-VARIABLE} \quad \text{INPUT-VARIABLE} \tag{C.20}$$

---

[4] Or other transient sources supported by SPICE but not described here.

where INPUT-VARIABLE must refer to a voltage or current source in the circuit. This statement instructs SPICE to find the small-signal gain as expressed by the ratio d(OUTPUT-VARIABLE)/d(INPUT-VARIABLE). The input resistance at the terminals of INPUT-VARIABLE and the output resistance at the terminals of OUTPUT-VARIABLE are also printed. Thus, the statement

$$.TF \quad V(1,2) \quad VIN \tag{C.21}$$

instructs SPICE to find the small-signal voltage gain with the voltage between nodes 1 and 2 as the output variable and the voltage source VIN as the input. Similarly, the statement

$$.TF \quad I(V1) \quad VS \tag{C.22}$$

instructs SPICE to find the small-signal transconductance using the current through the source V1 as the output variable and voltage source VS as the input variable.

C.4.5 .TEMP (Temperature Specification)

The statement
$$.TEMP \quad VAL1 \quad VAL2 \quad VAL3 \ldots \tag{C.23}$$

instructs SPICE to perform the analysis at each of the temperatures VAL_n. If the .TEMP statement is omitted from the input file, the simulation is performed at room temperature (27°C). The temperatures VAL_n are specified in °C (Celsius).

C.4.6 .IC (Initial Conditions)

The initial condition command line .IC may be used to preset the voltage of any node, or the voltage across any two nodes, to a known, initial value. The specified voltage is used for all bias calculations and is then set free for any subsequent circuit simulation initiated by a .TRAN or .DC command. The .IC command is particularly useful for setting initial conditions on capacitor voltages in transient analyses. The .IC command line has the form

$$.IC \quad V(N) = VALUE \quad V(J,K) = VALUE \tag{C.24}$$

where N, J, and K are node numbers, and VALUE is a voltage.

C.4.7 .STEP (Parameter Step)

The .STEP command has the syntax

$$.STEP \quad (TYPE) \quad STEPVAR \quad START \quad STOP \quad INCR \tag{C.25}$$

This command instructs SPICE to perform the simulation while stepping the parameter STEPVAR between the values START and STOP in increments of INCR. If another command line such as .AC, .DC, or .TRAN is included in the input file, SPICE will perform these analyses for each value of STEPVAR. The optional (TYPE) entry may be set to one of the designations LIN (linear increments), DEC (INCR points per decade), OCT (INCR points per octave), or LIST. If TYPE is set to LIST, a sequence of STEPVAR values (separated by spaces) must follow LIST in either ascending or descending order; the START, STOP, and INCR entries are omitted in this case. If TYPE is omitted, its default is LIN.

The variable STEPVAR can be a voltage or current source, the temperature TEMP, any global parameter designated in a .PARAM statement, or any parameter from a .MODEL statement. In this last case, the STEPVAR entry consists of the two elements DEVICETYPE and MODELNAME(PARAM) separated by spaces, where DEVICETYPE is one of the entries specified in Table C.5 (or one of the device-type entries permitted by SPICE but not listed in the table). MODELNAME must correspond to the name of a .MODEL statement in the input file, and PARAM must be one of the element parameters of the modeled device (e.g., one of the entries in Table C.6). The following are all examples of valid .STEP commands:

$$.STEP \quad VS \quad 1V \quad 10V \quad 0.5V \quad\quad\quad\quad (C.26)$$

$$.STEP \quad DEC \quad Io \quad 1nA \quad 1mA \quad 10 \quad\quad\quad\quad (C.27)$$

$$.STEP \quad TEMP \quad LIST \quad 10 \quad 50 \quad 85 \quad 150 \quad\quad\quad\quad (C.28)$$

$$.STEP \quad NMOS \quad device1(VTO) \quad 0.5 \quad 2 \quad 0.25 \quad\quad\quad\quad (C.29)$$

Note that .STEP and .DC commands may not attempt to change the same variable. Such a situation will generate an error message in SPICE.

C.4.8 Other Analyses

Several other control statements recognized by SPICE are discussed in detail in the various references of Section C.6. These statements include the commands .OP (print bias point information), .DISTO (distortion analysis), .SENS (sensitivity analysis), .OPTIONS (set options), .NOISE (noise analysis), .FOUR (Fourier transform analysis), .MC (Monte Carlo statistical analysis), .NODESET (initial guess at bias point as a convergence aid for complex circuits), and .PARAM (set values of key parameters).

C.5 GENERATING OUTPUT

Output can be generated in older forms of SPICE using the traditional .PRINT and .PLOT control statements. When used, these statements define the output variables of the circuit. There is no limit to the number of .PRINT or .PLOT lines that can be included in a single input file, but their use should be limited to one or two to avoid excessively large output files.

The .PRINT command causes SPICE to print data in tabular form to the file FILENAME.OUT, where FILENAME is the same file name as the input file. The .PRINT statement has the form

$$.PRINT \quad ANALYSIS\text{-}TYPE \quad VAR1 \quad VAR2 \quad VAR3 \ldots \quad\quad (C.30)$$

The entry ANALYSIS-TYPE must be set to one of the designations DC, AC, or TRAN described in Section C.4. A control statement corresponding to the specified ANALYSIS-TYPE must also exist somewhere in the input file. Note that the ANALYSIS-TYPE entry in the .PRINT statement does not have a period as its first character. The entries VAR_n specify the voltages and currents to be printed. The instructions used for generating output are contained in the control statement corresponding to ANALYSIS-TYPE. For example, the combined statements

$$.DC \quad VIN \quad -5 \quad 5 \quad 0.1 \quad\quad\quad\quad (C.31)$$

and $\quad\quad\quad\quad .PRINT \quad DC \quad V(5) \quad V(3,4) \quad I(VIN) \quad\quad\quad\quad (C.32)$

will cause SPICE to compute three dc transfer characteristics using VIN as the input variable. The output variables will be the voltage from node 5 to ground, the voltage between nodes 3 and 4, and the current flowing through the VIN source. The transfer characteristics will be computed while incrementing VIN from -5 V to $+5$ V in steps of 0.1 V.

When the .PRINT command is used with the .AC control statement, output voltages can be designated using the notation VM (magnitude), VP (phase), VR (real component), VI (imaginary component), and VDB (value expressed in decibels). Similar designations can be used for output current variables. Thus, the statement

$$\text{.PRINT} \quad \text{AC} \quad \text{VM(1)} \quad \text{VP(1)} \quad \text{IM(VIN)} \quad \text{IP(VIN)} \qquad \textbf{(C.33)}$$

will cause SPICE to print the magnitude and phase of the voltage from node 1 to ground and the magnitude and phase of the current through the VIN source as a function of frequency.

The .PLOT command causes SPICE to plot data in graphical form in the output file using ASCII characters. Its format is essentially identical to that of the .PRINT command, except that the .PLOT command also allows vertical scale limits to be specified. Thus the .PLOT command has the format

$$\text{.PLOT} \quad \text{ANALYSIS-TYPE} \quad \text{VAR1 (LIM-HI LIM- LO)} \quad \text{VAR2 (LIM-HI LIM-LO)} \dots$$
$$\textbf{(C.34)}$$

The limits LIM-HI and LIM-LO are optional and are added in parentheses as needed. A given set of plot limits applies to all variables that follow until a new set of plot limits is provided. Output data that exceed the limits LIM-HI and LIM-LO are computed but not plotted. If plot limits are not specified, the plot limits are set to the highest and lowest values computed for the output variable. In most cases, SPICE will generate a separate vertical scale for each output variable. Note that the .PLOT command also causes SPICE to .PRINT the data for the first output variable listed.

The horizontal axis generated by the .PLOT command depends on the type of analysis specified by ANALYSIS-TYPE. A DC analysis will generate a horizontal axis expressed in units of voltage or current. An AC analysis will generate a horizontal axis expressed in units of frequency, and a TRAN analysis will generate a horizontal axis expressed in units of time.

Output can be generated on a personal computer or workstation running PSPICE by running the Probe postprocessor utility. This powerful graphic interface is invoked by including the command line .PROBE in the input file. When included without parameters, .PROBE stores the values of all voltages and currents in a data file during the simulation. When the simulation is completed, the Probe program is loaded, then prompts the user for information about what data to display in graphical form. The size of the data file created by Probe can be large if the circuit is complex. Considerable file space (and program running time) can be saved if the .PROBE command line is followed by a listing of only those variables that are of interest to the user. The command line

$$\text{.PROBE} \quad \text{V(1)} \quad \text{V(4)} \quad \text{I(VCC)} \qquad \textbf{(C.35)}$$

for example, will cause Probe to save only the voltages of nodes 1 and 4 and the current through the source VCC during program execution.

The graphical interface screen displayed by Probe is self-explanatory. The user can utilize its many features by simply invoking the various command prompts on the screen. The functions possible in Probe include graphing voltages and currents with respect to other voltages, currents, or time, scaling data in linear or logarithmic form, adding comments, text, and line labels to graphs, analyzing plots using cursors, determining maximum and minimum values, and generating hard-copy output. Most of the SPICE plots included in this book were generated from PSPICE using the Probe utility.

C.6 REFERENCES

A complete explanation of the many features of SPICE can be found in each of the following references. These books also discuss the origins of the semiconductor device models that are resident within the SPICE program.

Banzhaf, W., *Computer-Aided Circuit Analysis Using PSpice*, 2nd Ed. Englewood Cliffs, NJ: Regents/ Prentice Hall, 1992.

Keown, J., *PSpice and Circuit Analysis*. New York: Macmillan, 1993.

Rashid, M., *SPICE for Circuits and Electronics Using PSpice*. Englewood Cliffs, NJ: Prentice Hall, 1990.

Tuinenga, P., *A Guide to Circuit Simulation and Analysis Using PSpice*, 2nd Ed. Englewood Cliffs, NJ: Prentice Hall, 1992.

Vladimirescu, A., *The SPICE Book*. New York: John Wiley, 1994.

Resistor Color Code and Standard Values

Resistors can be described in many ways, including their nominal resistance, tolerance, wattage (heat dissipation capability), and construction (carbon film, metal film, wire-wound). For most common carbon and metal-film resistors, the nominal resistance value (the intended value for the resistance) and tolerance (the maximum percentage by which the actual resistance value will deviate from the intended value) are indicated by a series of colored rings painted on the resistor body, as shown in Fig. D.1. The Resistor Color Code (RCC) of the Electronic Industries Association defines a convention for the color-coding system, shown in Table D.1. Each band color signifies a numerical digit. The band closest to the end of the resistor body is designated as the first band. For a resistor with 5% and higher tolerance, the first and second bands indicate the two significant digits used to specify the resistance value. The third band designates the power-of-10 multiplier that follows the two significant digits. The colors brown, black, red, for example, indicate a nominal resistance value of $1000\,\Omega$ (digits 1 and 0 followed by a multiplier of 10^2).

Figure D.1
Resistor color band coding system.

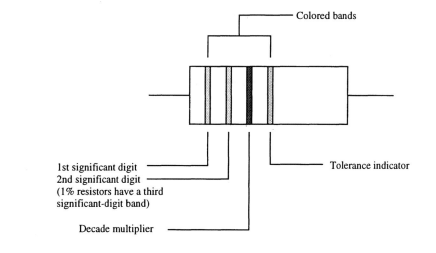

Colored bands

1st significant digit
2nd significant digit
(1% resistors have a third significant-digit band)

Decade multiplier

Tolerance indicator

The fourth band on the resistor body indicates the percent tolerance of the resistor. Tolerance values of 10%, 5%, and 1% are common. The value of a smaller-tolerance resistor is usually printed in text on the resistor body, rather than using the color-code system. Resistors with 1% tolerance require three significant-digit bands to indicate a resistance value. For these resistors, the tolerance indicator (black, in this case) follows as the fifth band.

Color	As a Digit (1st and 2nd Bands)	As a Multiplier (3rd Band)	As a Tolerance (4th or 5th Band)
Black	0	1	1%
Brown	1	10	2%
Red	2	10^2	
Orange	3	10^3	
Yellow	4	10^4	
Green	5	10^5	
Blue	6	10^6	
Violet	7	10^7	
Gray	8	10^8	
White	9	10^9	
Gold		10^{-1}	5%
Silver		10^{-2}	10%
Empty			20%

Table D.1. Resistor Color-Code Table

Commercial resistors are available in a set of standard nominal values that bridge the complete resistance spectrum. A list of standard significant digits for various tolerance values is shown in Table D.2. Resistors corresponding to the numbers in the table are generally available over the range $10\,\Omega$ to $22\,M\Omega$ for common carbon and metal-film resistors. Note that the numbers in Table D.2 are also used to designate the values of various types of capacitors.

The standard values are chosen so that the possible resistance spreads for the given tolerance limit overlap. The actual value of 5%-tolerance resistor, for example, will lie anywhere between 95% and 105% of its stated value. A 5% tolerance, 1-kΩ resistor will thus have an actual value lying somewhere between $950\,\Omega$ and $1.05\,k\Omega$. The next lowest standard 5% resistor of $910\,\Omega$ will have a value between $860\,\Omega$ and $955\,\Omega$, which overlaps the low end of the 1-kΩ resistor range. Similarly, the next highest standard resistor of $1.1\,k\Omega$ will have a value between $1.045\,k\Omega$ and $1.16\,k\Omega$. This latter range overlaps the high end of the 1-kΩ resistor range.

One area where the novice engineer often makes errors is in the specification of resistor values in circuit design. Suppose, for example, that a particular design calculation calls for a resistor of value $1.033\,k\Omega$. The novice engineer might be tempted to connect a 1-kΩ and a 33-Ω resistor in series in an attempt to match the required resistor value "exactly." There would be little value to this approach, however, because the chances of two 5%-tolerance resistors in series actually equaling $1.033\,k\Omega$ is small. If the design did truly call for a resistor value of precisely $1.033\,k\Omega$, a more expensive (and much more difficult to find) resistor of at least 0.1% tolerance would be required. If a design calculation yields a resistance value of $1.033\,k\Omega$, enough error margin should be allowed in the design such that an inexpensive 5%- or 10%-tolerance resistor in the 1-kΩ range can be used.

5% Tolerance	10% Tolerance	20% Tolerance
10	10	10
11		
12	12	
13		
15	15	15
16		
18	18	
20		
22	22	22
24		
27	27	
30		
33	33	33
36		
39	39	
43		
47	47	47
51		
56	56	
62		
68	68	68
75		
82	82	
91		

Table D.2. Standard Resistor Values

Suggestions for Further Reading

OTHER BOOKS ABOUT ELECTRONICS

Fonstad, C. G., *Microelectronic Devices and Circuits*. New York: McGraw-Hill, 1994.

Gaussi, M. S., *Electronic Devices and Circuits: Discrete and Integrated.* New York: Holt, Rinehart and Winston, 1985.

Gray, P. E., and C. L. Searle, *Electronic Principles*. New York: John Wiley, 1969.

Horowitz, P., and W. Hill, *The Art of Electronics*, 2nd ed. Cambridge, England: Cambridge University Press, 1989.

Millman, J., and A. Graybel, *Microelectronics*, 2nd ed. New York: McGraw-Hill, 1987.

Mitchell, F. H., Jr., and F. H. Mitchell, Sr., *Introduction to Electronics Design*, 2nd ed. Englewood Cliffs, NJ: Prentice Hall, 1992.

Rips, E. M., *Discrete and Integrated Electronics*. Englewood Cliffs, NJ: Prentice Hall, 1986.

Savant, C. J., Jr., M. S. Roden, and G. L. Carpenter, *Electronic Design: Circuits and Systems*, 2nd ed. Redwood City, CA: Benjamin-Cummings, 1991.

Schilling, D. L., C. Belove, T. Apelewicz, and R. Saccardi, *Electronic Circuits: Discrete and Integrated*, 3rd ed. New York: McGraw-Hill, 1989.

Sedra, A. S., and K. C. Smith, *Microelectronic Circuits*, 3rd ed. Philadelphia: Saunders, 1991.

Senturia, S. D., and B. D. Wedlock, *Electronic Circuits and Applications*. New York: Kreiger, 1993.

LINEAR CIRCUIT THEORY

Balabanian, N., *Electric Circuits*. New York: McGraw-Hill, 1994.

Boctor, S. A., *Electric Circuit Analysis*, 2nd ed. Englewood Cliffs, NJ: Prentice Hall, 1992.

Bode, H. W., *Network Analysis and Feedback Amplifier Design*. Princeton, NJ: D. Van Nostrand, 1945.

Budak, A., *Circuit Theory Fundamentals and Applications*, 2nd ed. Englewood Cliffs, NJ: Prentice Hall, 1987.

Hayt, W. H., Jr., and J. E. Kemmerly, *Engineering Circuit Analysis*, 5th ed. New York: McGraw-Hill, 1993.

Huelsman, L. P., *Basic Circuit Theory*, 3rd ed. Englewood Cliffs, NJ: Prentice Hall, 1991.

Johnson, D. E., J. R. Johnson, and J. L. Hilburn, *Electric Circuit Analysis*, 2nd ed. Englewood Cliffs, NJ: Prentice Hall, 1992.

Nilsson, J. W., *Electric Circuits*, 4th ed. Reading MA: Addison-Wesley, 1993.

Thomas, R. E., and A. J. Rosa, *The Analysis and Design of Linear Circuits*. Englewood Cliffs, NJ: Prentice Hall, 1994.

Van Valkenburg, M. E., *Network Analysis*, 3rd ed. Englewood Cliffs, NJ: Prentice Hall, 1974.

Vlach, J., *Basic Network Theory with Computer Applications*. New York: Van Nostrand Reinhold, 1992.

OPERATIONAL AMPLIFIER CIRCUITS

Bell, D. A., *Operational Amplifiers: Applications, Troubleshooting, and Design*. Englewood Cliffs, NJ: Prentice Hall, 1990.

Clayton, G. B., *Operational Amplifiers*, 3rd ed. Oxford: Newnes, 1992.

Irvine, R. G., *Operational Amplifier Characteristics and Applications*, 2nd ed. Englewood Cliffs, NJ: Prentice Hall, 1987.

Roberge, J. K., *Operational Amplifiers: Theory and Practice*. New York: John Wiley, 1975.

Rutkowski, G. B., *Integrated Circuit Operational Amplifiers*, 2nd ed. Englewood Cliffs, NJ: Prentice Hall, 1984.

Rutkowski, G. B., *Operational Amplifiers: Integrated and Hybrid Circuits*. New York: John Wiley, 1993.

ANALOG INTEGRATED CIRCUITS

Allen, P. E., and D. R. Holberg, *CMOS Analog Circuit Design*. New York: Saunders, 1987.

Chirlian, P. M., *Analysis and Design of Integrated Electronic Circuits*, 2nd ed. New York: Harper & Row, 1987.

Geiger, R. L., P. E. Allen, and N. R. Strader, *VLSI Design Techniques for Analog and Digital Circuits*. New York: McGraw-Hill, 1990.

Gray, P. R., B. A. Wooley, and R. W. Broderson (editors), *Analog MOS Integrated Circuits*, 2nd ed. New York: IEEE Press, 1989.

Gray, P. R., and R. G. Meyer, *Analysis and Design of Analog Integrated Circuits*, 3rd ed. New York: John Wiley, 1993.

Grebene, A. B., *Bipolar and MOS Analog Integrated Circuit Design*. New York: John Wiley, 1984.

IEEE Journal of Solid-State Circuits. The December issue of each year features analog integrated circuits.

Widlar, R. J., "Design Techniques for Monolithic Operational Amplifiers", *IEEE Journal of Solid-State Circuits*, Vol SC–4, No. 4, August 1969, pp. 184–191.

Wilson, G. R., "A Monolithic Junction FET–NPN Operational Amplifier," *IEEE Journal of Solid-State Circuits*, Vol. SC–2, No. 4, December 1968, pp. 341–348.

ACTIVE FILTERS

Allen, P. E., and E. Sanchez-Sinencio, *Switched Capacitor Circuits*. New York: Van Nostrand Reinhold, 1984.

Brodersen, R. W., P. R. Gray, and D. A. Hodges, "MOS Switched Capacitor Filters," *Proceedings of the IEEE*, Vol. 67, No. 1, January 1979, pp. 61–75.

Huelsman, L. P., *Active and Passive Analog Filter Design: An Introduction*. New York: McGraw-Hill, 1993.

Meiksin, Z. H., *Complete Guide to Active Filter Design, Op-Amps, and Passive Components*. Englewood Cliffs, NJ: Prentice Hall, 1990.

Schaumann, R., M. Ghausi, and K. R. Laker, *Design of Analog Filters: Passive, Active RC, and Switched Capacitor*. Englewood Cliffs, NJ: Prentice Hall, 1990.

Van Valkenburg, M. E., *Analog Filter Design*. New York: Holt, Rinehart and Winston, 1982.

POWER ELECTRONICS AND DEVICES

Fisher, M. J., *Power Electronics*. Boston: PWS-Kent, 1991.

Kassakian, J. G., M. F. Schlecht, and G. C. Verghese, *Principles of Power Electronics*. Reading, MA: Addison-Wesley, 1991.

DIGITAL CIRCUITS AND DEVICES

Hodges, D. A., and H. G. Jackson, *Analysis and Design of Digital Integrated Circuits*. New York: McGraw-Hill, 1988.

IEEE Journal of Solid-State Circuits. The October issue of each year features digital circuits.

Mano, M. M., *Digital Design*, 2nd ed. Englewood Cliffs, NJ: Prentice Hall, 1991.

Mead, C., and L. Conway, *Introduction to VLSI Systems*. Reading, MA: Addison-Wesley, 1980.

Uyemura, J. P., *Circuit Design for CMOS VLSI*. Boston: Kluwer Academic, 1992.

Uyemura, J. P., *Fundamentals of MOS Digital Integrated Circuits*. Reading, MA: Addison-Wesley, 1988.

BICMOS CIRCUITS

Alvarez, A. R., (editor), *BiCMOS Technology and Applications*. Boston: Kluwer Academic, 1993.

Buchanan, J. E., *BiCMOS/CMOS Systems Design*. New York: McGraw-Hill, 1991.

Embabi, S., A. Bellaouar, and M. Elmasry, *Digital BiCMOS Integrated Circuit Design*. Boston: Kluwer Academic, 1993.

SEMICONDUCTOR DEVICES

Early, J. M., "Effects of Space-Charge Layer Widening in Junction Transistors," *Proceedings of the IRE*, Vol. 40, No. 11, November 1952, pp. 1401–1406.

Streetman, B. G., *Solid State Electronic Devices*, 3rd ed. Englewood Cliffs, NJ: Prentice Hall, 1990.

Sze, S. M., *Semiconductor Devices, Physics, and Technology*. New York: John Wiley, 1985.

Sze, S. M., *High-Speed Semiconductor Devices*. New York: John Wiley, 1990.

Answers to Selected Problems

CHAPTER 1

1.2 $V_1 - v_2$ **1.3** $2.67\,\text{V}$ **1.8** $-gR_1v_1/(1 + gR_2)$ **1.14** (a) $8.5\,\text{V}$ (b) $V_3 = 6\,\text{V}$; $R_A = 5\,\text{k}\Omega$, $V_4 = 2.5\,\text{V}$; $R_B = 5\,\text{k}\Omega$ (c) $0.85\,\text{mA}$ **1.21** slope $= -1/R_1$ **1.23** horizontal line at $i_X = 5\,\text{mA}$ **1.27** $-2.73\,\text{mA}$; $7.27\,\text{mA}$ **1.31** one possible solution: $R_2 = R_1 = 1\,\text{k}\Omega$, $R_3 = 2\,\text{k}\Omega$ **1.35** $-0.23\,\text{V}$ **1.42** $\beta > 10$ **1.44** $r(v_1 - v_2)/R_1$ **1.49** $4.76\,\text{V}$; $8.92\,\text{k}\Omega$ **1.53** $V_3 = V_4$; $R_A + R_B$ **1.56** $0.11\,\text{V rms}$ **1.60** (a) $15\,\text{V}$; $500\,\Omega$ (b) $R_M > 38\,\text{k}\Omega$ **1.63** $4\,\text{V}$; $3.3\,\text{k}\Omega$ **1.65** $v_{\text{Th}} = -gv_1R_1^2/(R_S + R_1)$; $R_{\text{Th}} = R_1$ **1.69** $i_N = 1.2\,\text{mA}$; $R_N = 5\,\text{k}\Omega$ **1.74** $i_N = 0.167\,\text{mA rms}$, $100\,\text{Hz}$; $R_N = 15\,\text{k}\Omega$ **1.76** $1.26\,\text{V}$; $2.84\,\text{V}$; $5.89\,\text{V}$ **1.81** $I_1(R_2 + R_3)/(R_1 + R_2 + R_3)$ **1.83** $v_{\text{OUT}} = 5[1 - \exp(-8.12t)]\,\text{V}$ **1.86** $V_{R1} = 5e^{-8.12t}\,\text{V}$ **1.90** $V_o\{\exp(-t/\tau) + [1 - \exp(-t/\tau)][R_2/(R_1 + R_2)]\}$ where $\tau = (R_1 \| R_2)C$ **1.94** $V_{\text{out}}/V_{\text{in}} = j\omega R_2 C_1/[1 + j\omega C_1(R_1 + R_2)]$ **1.99** $C_1/(C_1 + C_2) = R_2/(R_1 + R_2)$; $v_{\text{OUT}}/v_{\text{IN}} = R_2/(R_1 + R_2)$

CHAPTER 2

2.9 18.2 **2.12** one possible solution: $R_2 = 270\,\text{k}\Omega$; $R_1 = 10\,\text{k}\Omega$ **2.16** $1.55\,\text{mA}$ **2.20** (a) $6v_{\text{IN}} + V_R$ (b) $6v_{\text{IN}} - 5V_R$ **2.23** (a) $0\,\text{mA}$ (b) $0\,\text{mA}$ (c) v_{IN} **2.27** $1 + 3.3\{1 - \exp[-t/(33\,\text{ms})]\}\,\text{V}$ **2.30** (a) $6\,\text{V}$; $-6\,\text{V}$ (b) maximum $|I_{\text{POS}} - I_{\text{NEG}}| = 25\,\text{mA}$ **2.34** $-R_2/R_1$ **2.37** $v_{\text{OUT}} = v_{\text{IN}}$ **2.43** (a) $v_{\text{OUT}} = (-R_2/R_1) \exp(-t/R_1C_1)v_S$ (b) $\mathbf{V}_{\text{out}}/\mathbf{V}_{\text{in}} = -j\omega R_2 C_1/(1 + j\omega R_1 C_1)$ **2.45** $v_1 : R_1 + R_2$; $v_2 : R_1$ **2.49** R **2.54** (a) v_-/r_{in} (b) $v_-(1 + A_o)/R_2$ (c) $v_{\text{IN}}/[1 + (R_1/r_{\text{in}}) + (1 + A_o)R_1/R_2)]$ **2.57** $15\,\mu\text{V}$; $0.3\,\text{mA}$ **2.60** $r_{\text{in}}(1 + A_o) + r_{\text{out}}$ **2.65** $2R_1$; $v_{\text{IN}}/2R_1$ **2.70** R_2/R_1 **2.76** R_n **2.78** one possible solution: summation amplifier with $R_1 = 36\,\text{k}\Omega$, $R_2 = 10\,\text{k}\Omega$, $R_F = 180\,\text{k}\Omega$ **2.83** $v_{\text{OUT}} = 2.8(v_4 + v_5) - 2(v_1 + v_2 + v_3)$ **2.87** $-(R_2/R_B)(R_A + R_B)i_{\text{IN}}$ **2.92** linear ramp in $1\,\text{ms}$ from $5\,\text{V}$ to $15\,\text{V}$; constant $15\,\text{V}$ thereafter **2.102** $v_{\text{OUT}} = -25v_{\text{IN}}$ **2.110** (a) symmetric ± 15-V square wave (b) $+15\,\text{V}$ for $v_{\text{IN}} > 4\,\text{V}$, $-15\,\text{V}$ otherwise (c) $+15\,\text{V}$ for $v_{\text{IN}} > 8\,\text{V}$, $-15\,\text{V}$ otherwise **2.118** (a) $8\,\text{V}$; $-4\,\text{V}$ **2.120** transitions at $v_{\text{IN}} = -V_{\text{POS}}$ and $-V_{\text{NEG}}$; modified: $-V_{\text{POS}}/2$ and $-V_{\text{NEG}}/2$ **2.126** $0.254\,\text{V}$ **2.130** $13\,\text{mV}$ **2.135** $2.005\,\mu\text{A}$; $1.995\,\mu\text{A}$ **2.140** $R_F \| R_1 \| \cdots \| R_N$ **2.145** (a) $35\,\text{mV}$ for symmetric input; $11.4\,\text{kHz}$ **2.148** (a) $-A_o/(1 + j\omega/\omega_p)$ (b) $V_{\text{in}}(1 + A_o + j\omega/\omega_p)/[R(1 + j\omega/\omega_p)]$ (c) $R/(1 + A_o + j\omega/\omega_p) + j\omega R/\omega_p(1 + A_o + j\omega/\omega_p)$ **2.150** $4.9\,\text{kHz}$

CHAPTER 3

3.1 3.41 V **3.5** 21.1 V **3.10** 2.29 V **3.12** 23.3 kΩ **3.20** (a) −50 V; 0 mA
3.26 1.2 V; 4 mA **3.31** 2.4 V; 4.9 mA **3.34** $n = p$; $n = p = 0$ at 0 K
3.42 $p_n = 4.85 \times 10^{14}$; $n_p = 4.85 \times 10^{12}$ **3.50** 0.69 V; 0.75 V; 0.81 V **3.57** (a) 0.2 (b) 0.67 V
3.62 (a) −6.3 V (b) −5.7 V (c) −3 V **3.67** 60 mA **3.71** (a) no (b) 6.42 mA
3.75 one possible solution: $V_{ZK} = 6$ V; $R_1 = 56\,\Omega$ **3.80** 27.3 V **3.83** 0.3 V; 9.4 mA
3.89 20 pF; 63 fF **3.93** 1.6 V $< v_{Th} < 11$ V **3.99** 20 V; 2 mA **3.118** $(5 + 8 \sin \omega t)$
V clipped at 6 V and −0.7 V **3.122** 0.7657 V; 0.4469 mA **3.125** 5.394 V; 4.606 mA
3.131 At 1 V: 330 Ω and −0.5 mA in parallel; at 10 V: 330 Ω and −0.167 V in series
3.136 4.52 V

CHAPTER 4

4.8 (a) horizontal at $v_{OUT} = 0$ for $v_{IN} < V_f$; slope of +1 for $v_{IN} > V_f$ **4.11** horizontal at
$v_{OUT} = 0$ for $- V_f < v_{IN} < V_f$; slope of +1 for $v_{IN} > V_f$ and $v_{IN} < -V_f$ **4.17** slope of +1
for $v_{IN} > 2.3$ V; horizontal at $v_{OUT} = 2.3$ V for $v_{IN} \le 2.3$ V **4.29** slope of v_{OUT} vs v_{IN} equals
1/2 for $v_{IN} < 4$ V; slope equals 1 thereafter **4.32** For $v_{IN} = 10$ V : $i_{D1} = 4.3$ mA; $i_{D2} = 0$; for
$v_{IN} = -10$ V: $i_{D1} = 0$; $i_{D2} = 9.3$ mA **4.45** 5.89 V; −1.83 V **4.51** one possible design:
10-kΩ resistor in series with two zeners pointing in opposite directions; v_{OUT} across both zeners
4.57 one possible design: diode in series with two equal resistors; v_{OUT} across one resistor
4.65 (a) 0 (c) 0.3 V **4.68** 84 Ω **4.72** (a) c (b) 0 Ω; 36 Ω **4.76** (a) 9.3 V (b) 0.41 V
4.81 (a) 7.64 V (b) 0.73 V **4.85** 694 μF → 820 μF **4.87** (a) one possible choice: $V_1 = 4.7$ V; $V_2 = 0$
4.95 one possible design: $V_{ZK} = 9$ V, $r_z = 10\,\Omega$, $|V_2| = 17$ V, $R_1 = 125\,\Omega$, C = 520 μF for 80-mV ripple
4.101 (a) 170 Ω; (b) 64 mA **4.105** 9.5 V, 9.3 Ω **4.108** V_m must be positive; 20 kHz **4.118** 25 V
4.121 rectifying "greater-than" circuit **4.126** one possible design: diode in series with −0.7 V voltage
source and capacitor; v_{OUT} across C

CHAPTER 5

5.1 7.5 V; 2 mA **5.5** For $v_1 = 6$ V: 9.6 V; 4 mA **5.9** $v_{PK} = 180$ V; $v_{GK} = -1.3$ V; $i_P = 1.2$ mA
5.15 (a) 0.25 mA/V^2; 2 V **5.20** 2.14 V **5.25** 3.13 mA/V^2 **5.30** (a) triode (b) CCR
5.35 1.2 mA/V^2; −2.58 V **5.39** (a) 0.19 mA/V^2 (b) 1.15 mA **5.44** 3.25 kΩ
5.50 10.12 V **5.53** 50 V **5.58** 4 mA; 6 V **5.61** (a) 4 mA (b) 0 V **5.68** 3 V
5.72 (a) −1.2 mA/V^2 (b) −2.5 V **5.77** (b) 1.17 V **5.84** $v_{DS} \ge 2$ V; $i_D = 6$ mA **5.86** 4 mA; 6 V
5.91 $i_B = 19.6\,\mu$A; $i_C = 0.98$ mA **5.95** 0.5% **5.96** (a) 200 (b) 100 μA (c) $> V_{sat}$
5.101 125 μA; 25 mA; 1 V **5.106** ≈ 5 V **5.111** 65 μA; 3.25 mA; 3.5 V **5.114** −20 μA
5.116 −3.9 V $< v_B < 9.3$ V **5.121** 120 V **5.125** 122 V **5.126** (a) 833 kΩ (b) 0.12 mW/cm^2
5.129 $R_L = 556\,\Omega$; $V_{Th} = 5$ V **5.136** polarity indicator; $R_1 + R_2 = 540\,\Omega$
5.140 $v_1 = 3.48$ V; $V_2 \ge 6.32$ V **5.142** 356 Ω → 360 Ω **5.146** (a) 0.8626 mA, 0.687 V;
0.9025 mA, 0.488 V; D_2 at 400 K (b) 593 μW; 440 μW (c) 29 Ω; 37 Ω **5.153** $I_s \exp(P_{max}/\eta V_T i_D)$
5.157 51 V; 14.6 V

CHAPTER 6

6.2 20 μA; 2 mA; 3 V **6.8** i_B ramps from 0 to 0.22 mA for 0.7 V $< v_{\text{IN}} < 5$ V
6.12 (a) 3.3 V (b) 3.3 mA (d) 13.4 V (e) 3.3 V (f) 6.9 V **6.16** (a) 0.7 V (b) 10.5 V (c) 5.6 V (d) 4.9 V
6.22 (a) 8.48 V for $V_{\text{sat}} = 0.2$ V and $V_f = 0.7$ V (b) 0.7 V (c) $10.1 + 2.5 \sin \omega t$ V **6.26** -8
6.32 (a) 5 V (b) 2 V (c) 3.4 V (d) 2.95 V **6.37** 1 mA; 1 mA; 10.5 V **6.46** 5 V; 2.5 V; approx 0.25 V
6.51 $v_{IC} = (|K_P|/K_N)^{1/2}(V_{DD} - |V_{\text{TRP}}|) + V_{\text{TRN}}$ **6.56** 110 ns **6.60** 0.67 **6.63** 10.4 V
6.68 3.4 mA **6.74** (a) 2.97 Ω (b) 0.7 W; 6.37 W; 3.03 W; 6.9% (c) 40 Ω (d) 0.7 W; 0.144 W; 0.42 W;
0.29 W; 45% **6.77** $v_{\text{OUT}} = 0$ for $0 < t < 5$ ms; $v_{\text{OUT}} = -10$-V peak triangle for $50 < t < 10$ ms
6.81 (b) BJT: 9.28 V; MOSFET: 8 V **6.86** $K_1/K_2 = 4$ **6.88** $v_{\text{OUT}} = v_{\text{IN}} + (K_2/K_1)^{1/2}V_{\text{TR2}} - V_{\text{TR1}}$
6.94 -0.5 V; 9 V **6.100** one possible solution: $V_{CC} = 10$ V; $V_{EE} = -10$ V;
$R_L = 5$ kΩ between collector of Q_1 and V_{CC}; 15 V zener between V_{CC} and base of Q_1;
$R_1 = 50$ kΩ between base of Q_1 and V_{EE}. **6.105** (a) $i_{C2} = (v_{\text{IN}} - V_f) \beta_{F2}/(\beta_{F2} + 1)R_E$
6.109 $-R_D$ **6.113** $v_{\text{OUT}} = V_{DD} - (K_3/K_1)^{1/2}v_{\text{IN}} + (K_3/K_1)^{1/2}V_{\text{TR3}} - V_{\text{TR1}}$
6.115 (b) 7.3 V **6.124** 0.9 V; 4.4 V **6.131** 0.7 V; V_{sat} **6.132** -14.25
6.136 11 **6.143** one possible solution: $V_{CC} = 15$ V; $R_C = 15$ kΩ; $R_B = 43$ kΩ; $\beta_F \geq 10$
6.146 (a) $V_{\text{ILT}} = (2.5 \text{ V})(R_S/\beta_F R_C) + V_f$ at $V_{CC} = 5$ V (b) $R_S = 38$ kΩ; $R_C = 1$ kΩ **6.150** approx 2.8 V
6.155 0.53 mW

CHAPTER 7

7.2 signal: $1.0 \sin \omega t$ V; bias: 2.2 V; $2 \sin^2 \omega t$ mW **7.7** 0.72 W; 0.24 A
7.10 $0.91 \sin \omega t + 0.45$ V; power gain $= 41.4$ **7.16** 186 kΩ **7.21** -1.67 mA; -4.48 V; 5.52 V
7.27 one possible solution: $R_C = 4$ kΩ; $R_E = 1$ kΩ; $R_1 = 110$ kΩ; $R_2 = 18$ kΩ
7.31 6.9 kΩ; 0.83 mA **7.36** (a) 40 kΩ; 4.8 V (b) 42 kΩ; 1.8 mA; 6.1 V **7.43** -7.15 mA; 0.7 V; -7.85 V
7.45 (a) $I_C = \beta_F(V_{CC} - V_f)/[R_B + (\beta_F + 1)R_C] \rightarrow (V_{CC} - V_f)/R_C$ (b) 4.63 mA; 5.33 V (c) 3.7 mA; 4.4 V
7.55 17 V **7.60** 0.47 mA; 0.69 mA **7.64** 0.93 mA; 6.42 V **7.69** 1 mA; 4 V
7.74 3.05 mA; 8.9 V; 0.25 V **7.79** 3.13 mA; 3.3 V; 8.7 V **7.88** (a) 2.5 V; 2.5 kΩ (b) ± 1.2 mA peak
7.93 -0.83 **7.99** -0.17 **7.104** $R_E \| [r_\pi/(\beta_o + 1)]$ **7.116** 2.6 mA/V
7.120 (a) 6.68 V (b) -2.83; ∞; 79 Ω **7.124** input port: open circuit; output port: dependent voltage source
of value $\mu_p v_{SO}$ in series with 2-Ω resistor, where $\mu_p = 1$ V/V. **7.126** (a) 7.34 V (b) -1.41
7.133 $r_{\text{th}} \approx r_{o2}[1 + \beta_o R_2/(R_2 + r_{\pi 2})]$ **7.137** $R_E \| [(r_\pi + R_1 \| R_2)/(\beta_o + 1)]$
7.142 $r_{\text{in}} = \infty$; $r_{\text{out}} = r_{o1} \| r_{o2} \| (1/g_{m2})$; $a_v \approx -g_{m1}[r_{o1} \| r_{o2} \| (1/g_{m2})]$

CHAPTER 8

8.2 (a) gain ratio $= [r_{\pi 1} + (\beta_o + 1)R_2]/r_{\pi 1}$ (b) $[r_{\pi 1} + (\beta_o + 1)R_2]/2r_{\pi 1}$ **8.5** (a) $1/g_{m2}$
8.12 $v_{dm} = 5$ V; $v_{cm} = 7.5$ V **8.18** (a) $-(R_2/R_1)v_1 + [R_4(R_1 + R_2)/R_1(R_3 + R_4)]v_2$
(c) $A_{dm} = -R_2/R_1$; $A_{cm} = 0$ **8.24** -11.5 **8.29** -0.045; -0.055; 0.01
8.35 (a) $-20 \sin \omega t$ mV; $25 \sin \omega t$ mV (b) -200; -0.05 (c) $3.999 \sin \omega t$ V **8.41** 1500
8.45 100 **8.48** $I_C = 0.713$ mA **8.53** 7.45 mA **8.57** (a) 3.9 mA (b) 3.9 mA; 6.8 V; -0.7 V
8.61 one possible solution: $R_1 = R_2 = 15$ kΩ; $R_3 = R_4 = 2$ kΩ; $R_C = 50$ kΩ; $V_{ZK} = 7.1$ V
8.66 (a) $I_{C1} = I_{C2} \approx 0.25$ mA; $I_{C3} = I_{C4} \approx 0.5$ mA; $I_{C5} \approx 1.43$ mA (b) 7260 **8.70** 0.32 mA; 356 kΩ
8.76 $I_{\text{ref}} \approx -(V_{EE} + V_f)/R_A$; $I_{E2} \approx I_{\text{ref}} \exp[I_{\text{ref}} R_A/\eta V_T]$ **8.79** 2 mA **8.83** (a) ± 3.46
8.87 $I_{D1} \approx 0.715$ mA; $I_{D2} \approx 1.43$ mA; $I_{D3} \approx 2.15$ mA; $I_{D4} \approx 2.86$ mA **8.93** (a) 3 mA (b) 1.84 (c) -0.032
8.100 one possible solution: $R_{D1} = 5$ kΩ; $R_{D2} = 0$; $I_{DSS} = 10$ mA; $V_P = -5$ V **8.103** 68.8
8.108 $-g_{m1}(r_{o1} \| r_{o2})/(1 + 2g_{m1}r_{oD})$ **8.111** (a) $-(I_o/2K)^{1/2}/4\eta V_T$ (b) $(2KI_o)^{-1/2}$ (c) $2r_{\pi 1}$
8.115 (a) 5 V (b) ≈ 5.5 V; -9.3 V

CHAPTER 9

9.1 0.37 pF **9.5** 420 MHz **9.13** −6 dB **9.17** (a) R_{SCOPE} (b) 9 MΩ (d) 1.56 pF
9.23 $500(j\omega/\omega_1)/(1 + j\omega/\omega_1)(1 + j\omega/\omega_2)$; $\omega_1 = 62.8$ rad/s; $\omega_2 = 1.26 \times 10^5$ rad/s;
9.26 1670 Hz (1760 Hz if true midband gain is computed) **9.31** 3.33 krad/s
9.35 275 rad/s; 3.33 krad/s **9.39** (e) 16.6 rad/s; 238×10^6 rad/s **9.45** (c) true; $f_s = 0.028$ Hz
9.48 gain: 1.41; $f_L = 612$ Hz; $f_H = 32$ MHz (692 Hz and 30 MHz with superposition
of poles) **9.52** midband gain $\approx$ 31 dB; high-frequency poles at 520 kHz; 33 MHz
9.56 $C_S = 74\,\mu$F; $C_C = 5.1\,\mu$F; $C_E = 330\,\mu$F **9.60** 575 kHz **9.63** 17 kHz
9.70 (a) 20 kΩ (c) poles at 2, 6100, 1.25×10^5 rad/s; zero at 200 rad/s; midband gain = 39.5 dB
9.74 (a) 58 Ω (b) 274 Hz (c) poles at 5.6, 16, 271, 523×10^3, 33×10^6 Hz; zero at 4.8 Hz;
midband gain = 31.4 dB **9.78** −195; 515 kHz **9.82** 17 MHz **9.83** $f_Z = 1.6$ MHz
9.88 (b) 0.5 mA for $\eta = 1$ (c) one possible solution: $R_E = 22.6$ kΩ; $R_C = 15.4$ kΩ; $R_1 = 10$ kΩ; $R_2 = 10$ kΩ
9.93 (a) $5\left[1 - \exp(-10t)\right]$ V (b) $3.33\left[1 - \exp(-15t)\right]$ V (c) $5 - 1.67\left[1 - \exp(-15t)\right]$ V
9.97 $4.81\exp\left[-(1.9 \times 10^6)t\right]$ V **9.103** (a) 2.6 kΩ (b) $\tau_r = 2.6\,\mu$s; $\tau_{droop} = 4.2$ ms; dc gain = −100
9.112 (a) one possible solution: $R_A = 270$ kΩ; $R_B = 560$ kΩ (b) $\tau_r = 40$ ns; $\tau_{droop} = 6$ s

CHAPTER 10

10.1 9900 **10.5** $\beta = R_1/(R_1 + R_2)$ **10.10** +0.23; −0.28 **10.15** (b) $A_{fb1} = 9.998$ for
$v_{OUT} < 7$ V; $A_{fb2} = 9.995$ for 7 V $< v_{OUT} < 15$ V **10.20** BW = 5×10^4 Hz; GB = 5×10^6 Hz
10.24 4 MHz **10.30** 1000; 0.009 **10.36** 0.0099; 10.1; 99 Ω **10.40** 49.5; 0.2 Ω; 303 kΩ
10.47 20 GΩ; 1 mΩ **10.53** $i_{out}/v_s = g_m/(1 + g_m R_F + R_F/r_o)$; $R_{out} = r_o + R_F + g_m r_o R_F$
10.60 10^3 V/mA; 5 Ω; 0.56 mΩ **10.66** −9.7 V/mA; 75 Ω; 125 Ω **10.69** 0.01 μW; $2.5 \times 10^{-9}\,\mu$W
10.76 (a) 1.1×10^5 rad/s (b) 0.024 (c) 17.5 kHz **10.81** (b) 1.6 μF **10.86** 5.3 kHz

CHAPTER 11

11.3 4 mV p–p **11.7** 17 kΩ; 1.5 kΩ; 3×10^5 **11.10** $v_{out}/v_g = 90.9$ **11.16** (a) −1.2 V
(b) 1.05 mA (c) $R_{in} = 115$ kΩ; $R_{out} \approx 150$ Ω; $a_v \approx -10$ **11.21** (a) $R_{in} = \infty$; $R_{out} = (1/g_{m4})\|r_{o3}$;
$a_v = g_{m1}g_{m3}\left[r_{o1}\|(1/g_{m2})\right]\left[r_{o3}\|(1/g_{m4})\right]$ (b) $I_{D1} = I_{D2} = 4$ mA; $I_{D3} = I_{D4} = 8$ mA; $v_{OUT} = 1$ V (c) 2
11.25 $I_{C1} = I_{C2} \approx 0.4$ mA; $I_{E2} = 0.41$ mA; $V_{C1} = 6.8$ V; $V_{E2} = 6.1$ V **11.38** $|A_{dm}| = 1000$; $A_{cm} = 0$
11.42 94 kHz **11.46** $v_{OUT} = 0$ for $|v_{IN}| < V_{TR}$; slope = 1 for $|v_{IN}| > V_{TR}$
11.54 (a) one possible solution: $R_1 = R_2 = 36$ kΩ (b) $R_1 = R_2 = 960$ kΩ **11.62** 197°C **11.64** 7.5°C
11.66 10.5 W **11.70** (a) 35 W (b) 45 W

CHAPTER 12

12.2 0.733 mA **12.6** $i_{out}/v_{idm} = -\beta_o/[2(r_{\pi1} + r_{\pi3})]\{1 + [r_{\pi5} + (\beta_o + 1)R_1]/[r_{\pi6} + (\beta_o + 1)R_2]\}$
12.9 $-V_f$ **12.12** one possible solution: $V_{ZK} = 3$ V; $R_3 = 23$ kΩ; $|I_{DSS}| >$
100 μA; $V_P > 3$ V **12.17** 12.9 μA $+ I_{B9}$ **12.19** (c) $(V_{CC} - V_{EE})/2$
12.23 (a) $(\beta_{o8} + 1)i_{in}R_6/[R_6 + r_{\pi9} + (\beta_{o9} + 1)R_7]$ (d) $-[\beta_{o9} + (\beta_{o9} + 1)R_7/r_{o9}]$
$i_{b9}r_{ox}/\{(1 + r_{ox}/r_{o9})[R_7/(R_7 + r_{o9})]\}$ where $i_{b9} \approx (\beta_{o8} + 1)R_6i_{in}/[R_6 + r_{\pi9} + (\beta_{o9} + 1)R_7]$
12.28 (a) ±22.2 mA (c) 12 μA **12.31** approx 3 Hz **12.35** 166 MHz **12.39** 0.37 V/μs

12.46 $V_A = \{(V_{DD} - V_{TR})[(K_A/K_B)^{1/2} + (K_A/K_C)^{1/2}] + V_{SS} + 2V_{TR}\}/[1 + (K_A/K_B)^{1/2} + (K_A/K_C)^{1/2}];$
$V_B = \{V_{DD}(K_A/K_C)^{1/2} + (V_{SS} + V_{TR})[1 + (K_A/K_B)^{1/2}] - 2V_{TR}(K_A/K_C)^{1/2}\}/[1 + (K_A/K_B)^{1/2} + (K_A/K_C)^{1/2}]$
12.49 $A_{\text{dm-se}} = \pm(K_1/4K_3)^{1/2};$ $A_{\text{cm-se}} = -g_{m1}/[g_{m3}(1 + 2g_{m1}r_{oD})]$
12.55 $(g_{m2}g_{m4}r_{o3}r_{o4})/[(1 + g_{m2}r_{o3})(1 + g_{m4}r_{o4})]$

CHAPTER 13

13.1 one possible solution: $R_1 = R_2 = 10\,\text{k}\Omega$; C = 16 nF **13.3** 1590 Hz; 14 dB; −20 dB/dec
13.7 (a) $(R_1 + R_2)[1 + j\omega(R_1\|R_2)C]/R_1(1 + j\omega R_2 C)$ (c) −2
13.11 (a) $-j\omega R_2 C_1/[(1 + j\omega R_1 C_1)(1 + j\omega R_2 C_2)]$ **13.14** 1 **13.17** 18
13.21 (a) −20 dB (b) $a_0 = 200$; $b_2 = 1$; $b_1 = 210$; $b_0 = 2000$ **13.25** $0.2(s + 5)(s - 5)/(s^2 + s + 1)$
13.30 (a) 500 rad/s; 0.87 (b) $(5 \times 10^5)/[s^2 + 576s + (2.49 \times 10^5)]$ **13.32** (a) 6285 rad/s (b) 6285 rad/s
13.34 one possible solution: $C_1 = 5\,\text{nF}$; $C_2 = 10\,\text{nF}$; $R_1 = R_2 = 22.5\,\text{k}\Omega$
13.43 (a) 5384 rad/s (b) $Q = 0.46$ **13.47** $f_o = 0.1125/RC$ **13.52** (a) 3.1 (b) 2350 Hz; 3250 Hz
13.55 (a) $2\omega_o$ (b) set R_2 to $4R_1/3$; C_2 to $3C_1$ **13.59** one possible solution: $R_1 = 1\,\text{k}\Omega$;
$R_2 = 2\,\text{M}\Omega$; $C_1 = C_2 = 356\,\text{pF}$; $Q = 22.4$ **13.61** $(1 - j\omega RC)/(1 + j\omega RC)$ **13.63** −69.9 dB
13.67 one possible solution: a single Sallen-Key low-pass section with $R_1 = R_2 = 22.5\,\text{k}\Omega$;
$C_1 = 1\,\mu\text{F}$; $C_2 = 0.5\,\mu\text{F}$ **13.72** n = 5; $−508 \pm j369$; $−194 \pm j597$; −628; 1158 Hz
13.75 7 **13.81** poles at $(−0.382 \pm j0.924)\omega_o$; $(−0.926 \pm j0.378)\omega_o$ lie close to
required Butterworth values **13.87** $R_1 = R_2 = 1.9\,\text{k}\Omega$ **13.89** 1720 rad/s; 0.39
13.94 one possible solution: $T_c = 5\,\mu\text{s}$; C = 500 pF **13.97** 0.23 μs **13.98** 0.63 μW each
13.103 $(s/R_F C)/[s^2 + s(3/R_F C) + (1/R_F C)^2]$; $\omega_o = 1/R_F C$; $Q = 3$ **13.106** 8.49
13.109 1.73R/L **13.112** one possible solution: $C_2 = 100\,\text{nF}$; $C_1 = 5\,\text{nF}$; L = 0.53 mH
13.117 $f_s = 3.5588\,\text{MHz}$; $f_p = 3.5677\,\text{MHz}$ **13.119** 1.88 mH **13.121** 8.7 ms
13.125 (a) 16.7 kHz

CHAPTER 14

14.1 −1.5 **14.4** 5 V; 0.3 V **14.7** 4.9 V; 0.1 V **14.13** 3 (consider inputs)
14.17 130 ns **14.20** 14 **14.23** (c) $t_r = t_f = 2.3R_G C_{gs}$ **14.25** 6.86 V
14.30 378 pJ **14.33** 2.33 V **14.34** 2.45 V; p_1: 4.95 V; p_2 : 0.05 V **14.39** 5.1 ns
14.49 $W_N/L_N > 1.43$; $W_P/L_P > 2.86$ **14.56** (b) $NM_H = 0.5\,\text{V}$; $NM_L = 0.8\,\text{V}$ (c) 60 μW (d) 30 μW
14.61 (a) 4.0 V; 0.16 V; 2.1 V; 0.84 V (b) 3.4 V; 1.5 V **14.63** (a) $L_1 = W_2 = 2\,\mu\text{m}$;
$L_2 = W_1 = 4\,\mu\text{m}$ (b) 16 μm² **14.71** (a) $4 \times 10^{-4}\,\text{pF}/\mu\text{m}^2$ (b) $K_1 = 76.4\,\mu\text{A/V}^2$;
$K_2 = 9.5\,\mu\text{A/V}^2$ (c) $t_r \approx 38\,\text{ns}$; $t_f \approx 2.4\,\text{ns}$ **14.72** (b) 1/2 (c) 5 μW (e) 0.41 μW (f) same
14.73 $Y = \overline{A \cdot B} + C + D$ **14.79** one possible solution: (a) $W_A = W_B = 4\,\mu\text{m}$; $L_A = L_B = 1\,\mu\text{m}$;
$L_2 = 4\,\mu\text{m}$; $W_2 = 1\,\mu\text{m}$ (b) $6.9 \times 10^{-4}\,\text{pf}/\mu\text{m}^2$ (c) $K_A = K_B = 186\,\mu\text{A/V}^2$; $K_2 = 11.6\,\mu\text{A/V}^2$
14.83 (a) one possible solution: $W_2 = 2\,\mu\text{m}$; $L_2 = 8\,\mu\text{m}$; $W_A = W_B = W_C = 4\,\mu\text{m}$;
$L_A = L_B = L_C = 2\,\mu\text{m}$ (b) 52 μm² **14.85** one high: 0.76 V; two high: 0.35 V
14.86 $t_r \approx 53\,\text{ps}$; $t_f \approx 90\,\text{ps}$ **14.90** 57 **14.91** 3 **14.100** 34.7 mA; 34.7 mA; 0
14.102 11 mW; 2.1 mW for 50% duty cycle **14.105** $Y = \overline{A \cdot B \cdot C}$ **14.106** (a) 0.5 V (b) 91 (c) 8.6 mW
14.107 (b) $\Delta V_R \approx (−0.14\,\text{mV/°C})\Delta T$ **14.109** −1.375 V; −1.265 V **14.111** 4.65 V
14.112 0.423 V

CHAPTER 15

15.7 a, b, c, e are stable for all N; d is unstable   **15.8** ~ 20 ns   **15.28** $110 \, \mu$s
15.32 $0.91RC$ to $0.73RC$ for $V_{DD} = 5$ V   **15.33** $(R + r_{\text{out}})C \ln [V_{DD}R/(R + r_{\text{out}})(V_{DD} - v_{IC})]$
15.39 $0.92RC$; $1.39RC$ **15.41** 0.61 mA   **15.47** 11.4 ms   **15.50** 12 ms
15.52 6 V: 9.1 ms; 5 V: 12 ms; 4 V: won't work   **15.60**   eight   **15.65** 10 nF; 100 nF
15.69 0.2 mA   **15.70** 5.078 V   **15.72** 0.61 mV   **15.82** 0.61 mV   **15.87**   first three $LSB = 1$
15.92   **1011 1111**

APPENDIX A

A.2   (a) 2.28×10^5 Ω-cm (b) 9.3 Ω-cm   **A.6**   (a) 6×10^{14}/cm^3 (b) n-type (c) 0.82 **A.9**   (a) $n_o = 10^{16}$/cm^3;
$p_o = 10^4$/cm^3 (b) $n'(x) = n_o \exp(-x/L_e)$; $p'(x) = p_o \exp(-x/L_h)$ where $n_o = p_o = 10^{12}$/cm^3
(c) $5 \, \mu$m (d) $(q D_h/L_h)p'(0) \exp(-x/L_h)$ (e) 3.84×10^{-3} A/cm^2; 1.41×10^{-3} A/cm^2
A.14   For Fig. PA.14(a), a, b, d are true; for Fig. PA.14(b), a, b, c, e are true.   **A.17** 6.1×10^{15}/cm^3
A.20   (a) 0.64 mA (b) 2 kΩ (c) 2.97 mA   **A.26** 8.5 V   **A.29** $W/L = 10.7$

Index